IET MATERIALS, CIRCUITS AND DEVICES SERIES 79

CMOS-Based Sensors and Actuators for Life Science Applications

Other volumes in this series:

Volume 2	**Analogue IC Design: The current-mode approach** C. Toumazou, F.J. Lidgey and D.G. Haigh (Editors)
Volume 3	**Analogue-Digital ASICs: Circuit techniques, design tools and applications** R.S. Soin, F. Maloberti and J. France (Editors)
Volume 4	**Algorithmic and Knowledge-based CAD for VLSI** G.E. Taylor and G. Russell (Editors)
Volume 5	**Switched Currents: An analogue technique for digital technology** C. Toumazou, J.B.C. Hughes and N.C. Battersby (Editors)
Volume 6	**High-frequency Circuit Engineering** F. Nibler et al.
Volume 8	**Low-power High-frequency Microelectronics: A unified approach** G. Machado (Editor)
Volume 9	**VLSI Testing: Digital and mixed analogue/digital techniques** S.L. Hurst
Volume 10	**Distributed Feedback Semiconductor Lasers** J.E. Carroll, J.E.A. Whiteaway and R.G.S. Plumb
Volume 11	**Selected Topics in Advanced Solid State and Fibre Optic Sensors** S.M. Vaezi-Nejad (Editor)
Volume 12	**Strained Silicon Heterostructures: Materials and devices** C.K. Maiti, N.B. Chakrabarti and S.K. Ray
Volume 13	**RFIC and MMIC Design and Technology** I.D. Robertson and S. Lucyzyn (Editors)
Volume 14	**Design of High Frequency Integrated Analogue Filters** Y. Sun (Editor)
Volume 15	**Foundations of Digital Signal Processing: Theory, algorithms and hardware design** P. Gaydecki
Volume 16	**Wireless Communications Circuits and Systems** Y. Sun (Editor)
Volume 17	**The Switching Function: Analysis of power electronic circuits** C. Marouchos
Volume 18	**System on Chip: Next generation electronics** B. Al-Hashimi (Editor)
Volume 19	**Test and Diagnosis of Analogue, Mixed-signal and RF Integrated Circuits: The system on chip approach** Y. Sun (Editor)
Volume 20	**Low Power and Low Voltage Circuit Design with the FGMOS Transistor** E. Rodriguez-Villegas
Volume 21	**Technology Computer Aided Design for Si, SiGe and GaAs Integrated Circuits** C.K. Maiti and G.A. Armstrong
Volume 22	**Nanotechnologies** M. Wautelet *et al.*
Volume 23	**Understandable Electric Circuits** M. Wang
Volume 24	**Fundamentals of Electromagnetic Levitation: Engineering sustainability through efficiency** A.J. Sangster
Volume 25	**Optical MEMS for Chemical Analysis and Biomedicine** H. Jiang (Editor)
Volume 26	**High Speed Data Converters** A.M.A. Ali
Volume 27	**Nano-Scaled Semiconductor Devices** E.A. Gutiérrez-D (Editor)
Volume 28	**Security and Privacy for Big Data, Cloud Computing and Applications** L. Wang, W. Ren, K.R. Choo and F. Xhafa (Editors)
Volume 29	**Nano-CMOS and Post-CMOS Electronics: Devices and modelling** Saraju P. Mohanty and Ashok Srivastava
Volume 30	**Nano-CMOS and Post-CMOS Electronics: Circuits and design** Saraju P. Mohanty and Ashok Srivastava
Volume 32	**Oscillator Circuits: Frontiers in design, analysis and applications** Y. Nishio (Editor)
Volume 33	**High Frequency MOSFET Gate Drivers** Z. Zhang and Y. Liu
Volume 34	**RF and Microwave Module Level Design and Integration** M. Almalkawi
Volume 35	**Design of Terahertz CMOS Integrated Circuits for High-Speed Wireless Communication** M. Fujishima and S. Amakawa
Volume 38	**System Design with Memristor Technologies** L. Guckert and E.E. Swartzlander Jr.
Volume 39	**Functionality-Enhanced Devices: An alternative to Moore's law** P-E. Gaillardon (Editor)
Volume 40	**Digitally Enhanced Mixed Signal Systems** C. Jabbour, P. Desgreys and D. Dallett (Editors)

Volume 43	**Negative Group Delay Devices: From concepts to applications** B. Ravelo (Editor)
Volume 45	**Characterisation and Control of Defects in Semiconductors** F. Tuomisto (Editor)
Volume 47	Understandable Electric Circuits: Key concepts, 2nd Edition **M. Wang**
Volume 48	**Gyrators, Simulated Inductors and Related Immittances: Realizations and applications** R. Senani, D.R. Bhaskar, V.K. Singh, A.K. Singh
Volume 49	**Advanced Technologies for Next Generation Integrated Circuits** A. Srivastava and S. Mohanty (Editors)
Volume 51	**Modelling Methodologies in Analogue Integrated Circuit Design** G. Dundar and M.B. Yelten (Editors)
Volume 53	**VLSI Architectures for Future Video Coding** M. Martina (Editor)
Volume 54	**Advances in High-Power Fiber and Diode Laser Engineering** Ivan Divliansky (Editor)
Volume 55	**Hardware Architectures for Deep Learning** M. Daneshtalab and M. Modarressi
Volume 57	**Cross-Layer Reliability of Computing Systems** Giorgio Di Natale, Alberto Bosio, Ramon Canal, Stefano Di Carlo and Dimitris Gizopoulos (Editors)
Volume 58	**Magnetorheological Materials and their Applications** S. Choi and W. Li (Editors)
Volume 59	**Analysis and Design of CMOS Clocking Circuits for Low Phase Noise** W. Bae and DK. Jeong
Volume 60	**IP Core Protection and Hardware-Assisted Security for Consumer Electronics** A. Sengupta and S. Mohanty
Volume 63	**Emerging CMOS Capacitive Sensors for Biomedical Applications: A multidisciplinary approach** Ebrahim Ghafar-Zadeh and Saghi Forouhi
Volume 64	**Phase-Locked Frequency Generation and Clocking: Architectures and circuits for modem wireless and wireline systems** W. Rhee (Editor)
Volume 65	**MEMS Resonator Filters** Rajendra M Patrikar (Editor)
Volume 66	**Frontiers in Hardware Security and Trust: Theory, design and practice** C.H. Chang and Y. Cao (Editors)
Volume 67	**Frontiers in Securing IP Cores: Forensic detective control and obfuscation techniques** A Sengupta
Volume 68	**High Quality Liquid Crystal Displays and Smart Devices: Vol. 1 and Vol. 2** S.Ishihara, S. Kobayashi and Y. Ukai (Editors)
Volume 69	**Fibre Bragg Gratings in Harsh and Space Environments: Principles and applications** B. Aïssa, E.I. Haddad, R.V. Kruzelecky, W.R. Jamroz
Volume 70	**Self-Healing Materials: From fundamental concepts to advanced space and electronics applications, 2nd Edition** B. Aïssa, E.I. Haddad, R.V. Kruzelecky, W.R. Jamroz
Volume 71	**Radio Frequency and Microwave Power Amplifiers: Vol. 1 and Vol. 2** A. Grebennikov (Editor)
Volume 72	**Tensorial Analysis of Networks (TAN) Modelling for PCB Signal Integrity and EMC Analysis** Blaise Ravelo and Zhifei Xu (Editors)
Volume 73	**VLSI and Post-CMOS Electronics Volume 1: VLSI and post-CMOS electronics and Volume 2: Materials, devices and interconnects** R. Dhiman and R. Chandel (Editors)
Volume 75	**Understandable Electronic Devices: Key concepts and circuit design** M. Wang
Volume 76	**Secured Hardware Accelerators for DSP and Image Processing Applications** Anirban Sengupta
Volume 77	**Integrated Optics Volume 1: Modeling, material platforms and fabrication techniques and Volume 2: Characterization, devices, and applications** G. Righini and M. Ferrari (Editors)

CMOS-Based Sensors and Actuators for Life Science Applications

Ebrahim Ghafar-Zadeh, Saghi Forouhi and Tayebeh Azadmousavi

The Institution of Engineering and Technology

Published by The Institution of Engineering and Technology, London, United Kingdom

The Institution of Engineering and Technology is registered as a Charity in England & Wales (no. 211014) and Scotland (no. SC038698).

First published 2023

The Institution of Engineering and Technology
Futures Place
Kings Way, Stevenage
Herts, SG1 2UA, United Kingdom

www.theiet.org

British Library Cataloguing in Publication Data
A catalogue record for this product is available from the British Library

ISBN 978-1-83953-673-1 (hardback)
ISBN 978-1-83953-674-8 (PDF)

Typeset in India by MPS Limited

Cover Image: Yuichiro Chino/Moment via Getty Images

Contents

About authors

Saghi Forouhi[1] is a postdoctoral researcher at York University (Department of Electrical Engineering and Computer Science (EECS), Biologically Inspired Sensors and Actuators (BioSA) Laboratory), Canada. She received her B.Sc. and M.Sc. degrees in electrical engineering from Guilan University, Iran, in 2010 and 2012, respectively. Then, she completed her Ph.D. at Isfahan University of Technology (IUT) in 2019 in an active collaboration between IUT and York University. Her research interests lie in the area of biologically inspired micro-systems, CMOS sensors, circuits, and systems.

Tayebeh Azadmousavi[1,2] earned both her B.Sc. and M.Sc. degrees in electronic engineering from Urmia University, located in Urmia, Iran, in 2011 and 2013 respectively. She later achieved her Ph.D. degree in electronic engineering from Sahand University of Technology in Sahand New Town, Tabriz, Iran, in 2019. During her academic journey, Dr. Azadmousavi's exceptional accomplishments stood out. She was notably recognized as a distinguished Ph.D. researcher by the Faculty of Electrical Engineering at Sahand University of Technology in 2019. Additionally, she secured the distinction of being the top Ph.D. graduate from Iran's National Elites Foundation in the same year.

Currently serving as an Assistant Professor at the University of Bonab in Bonab, Iran. Dr. Azadmousavi has also taken on the role of a Research Visiting Professor at the Biologically Inspired Sensors and Actuators Laboratory (BioSA), situated within the esteemed Lassonde School of Engineering at York University in Toronto, ON, Canada. Her research pursuits encompass a diverse array of topics, including low voltage/low power analog and digital integrated circuits, analog integrated circuits operating in current mode and voltage mode, wireless RFIC design, as well as the innovative field of biosensors.

Ebrahim Ghafar-Zadeh[1] received his B.Sc. and M.Sc. in Electrical Engineering from the KNT University of Technology (Tehran, Iran) and the University of Tehran (Tehran, Iran), respectively. He then continued his studies in the Polytechnique of Montreal (Montreal, Canada), where he received his Ph.D. degree in Electrical Engineering in 2008. His graduate studies focused on complementary metal-oxide semiconductor (CMOS)-based sensors for lab-on-chip applications. In recognition of his research achievements, he received several fellowship awards including a Postdoctoral Fellowship (PDF) from the Natural Sciences and Engineering Research Council of Canada (NSERC). Then he continued his PDF research studies in Electrical Engineering at McGill University (Montreal, Canada) and in Bioengineering, at the University of California, Berkeley. As Assistant Professor, in 2013, Ebrahim joined the Department of Electrical Engineering and Computer Science (EECS) in the Lassonde School of Engineering at York University where currently he is an Associate Professor, Member of Graduate Programs of the Departments of EECS and Biology, and Director of Biologically Inspired Sensors and Actuators (BioSA) research laboratory. His research is aimed at exploring novel integrated sensors and actuators for life science applications. He is a Senior Member of the IEEE and a licensed Professional Engineer in the province of Ontario.

Affiliations:
[1]Biologically Inspired Sensors and Actuators (BioSA), Department of Electrical Engineering and Computer Science (EECS), Lassonde School of Engineering, York University, Toronto, Ontario, Canada
[2]Department of Electrical Engineering, University of Bonab, Bonab, Iran

Email: s.forouhi@cse.yorku.ca, tayebehazadmousavi@gmail.com, egz@yorku.ca

Table of terminology and definitions

Context	Definition
Absorption photometry	The change of photon absorption through chemical reactions in response to the analyte concentration.
Accuracy	The accuracy of thermal sensors is how close the sensor output is to the physical temperature being measured.
Action potential (AP)	A brief, rapid change in the electrical potential across the cell membrane of a neuron or muscle cell. It is a crucial process in the transmission of electrical signals in the nervous system and enables communication between neurons and the contraction of muscles.
Adenosine triphosphate (ATP)	Energy-carrying molecules found in the cells of all living things.
Adenovirus	A type of virus with a double-stranded DNA genome that can infect both humans and animals. It can cause a wide range of illnesses, including respiratory infections, conjunctivitis, and gastrointestinal issues, and has been widely studied for its use as a viral vector in gene therapy and vaccine development.
Affibody	A type of synthetic protein engineered from a small, stable, and robust scaffold that can specifically bind to target molecules, such as proteins or peptides. These affibodies are used in various applications, such as molecular imaging, diagnostic assays, and targeted therapies, due to their high binding specificity and affinity for their respective targets.
Affimer	A small, synthetic protein engineered to bind to specific target molecules with high affinity and selectivity. Similar to antibodies, affimers are used in various applications, such as diagnostics, research tools, and therapeutic development, due to their ability to recognize and bind to specific proteins or other biomolecules.
AlamarBlue	A cell viability assay reagent used to assess the metabolic activity and viability of cells. It measures the reduction of a non-toxic, cell-permeable dye by living cells, producing a color change that can be quantified to determine cell viability and proliferation in various biological and biomedical research applications.

(Continues)

(*Continued*)

Context	Definition
Alcohol	A broad class of organic compounds that contain a hydroxyl group (–OH) bonded to a carbon atom. Alcohols can range from simple structures like methanol and ethanol to more complex molecules. They are characterized by their ability to undergo various chemical reactions, including oxidation and esterification, and are widely used in industries such as pharmaceuticals, solvents, and fuels. Additionally, some alcohols, like ethanol, have significant applications in beverages and alcoholic drinks.
Aldehyde	A functional group (–CHO) consisting of a carbon atom attached to a hydrogen atom and an oxygen atom. It is commonly found in organic compounds and is known for its characteristic reactivity, such as in oxidation-reduction reactions and its role in the synthesis of various chemical compounds.
Alkanethiol	Organic compounds consisting of a hydrocarbon chain with a thiol (–SH) functional group at one end. They are commonly used in self-assembled monolayers (SAMs) on metal surfaces, providing a stable and well-ordered surface for various applications in nanotechnology, surface chemistry, and sensor development.
Alkyl benzene	A group of organic compounds derived from benzene by replacing one or more of its hydrogen atoms with alkyl groups. Alkyl benzenes are commonly used as chemical intermediates, solvents, and raw materials in the production of detergents and other industrial chemicals.
Amino acid moiety	The core structural unit of an amino acid molecule, consisting of an amino group (–NH2), a carboxyl group (–COOH), and a variable side chain (R group). This moiety is the common feature among all 20 standard amino acids and plays a crucial role in peptide and protein synthesis, as well as in their diverse functions in biological systems.
Amino group	A functional group (–NH_2) consisting of a nitrogen atom bonded to two hydrogen atoms. It is a crucial component of amino acids, the building blocks of proteins, and is involved in various chemical reactions and biological processes, including protein synthesis, enzyme catalysis, and neurotransmitter function.
Amplicon	Short segments of DNA that are generated through a process called polymerase chain reaction (PCR) amplification. PCR amplifies a specific target DNA sequence by using DNA primers and a DNA polymerase enzyme. Amplicons serve as replicated copies of the original DNA sequence and are widely used in genetic analysis, diagnostics, and research applications, providing valuable insights into genetic variations, gene expression, and other molecular processes.

(Continues)

(*Continued*)

Context	**Definition**
Amplitude-locked loop (ALL)	A control system that synchronizes the output amplitude of an oscillator with the amplitude of a reference signal. It is commonly used in communication systems and signal processing to maintain a stable and constant output amplitude for various applications, such as modulators and demodulators in wireless communication.
Anisotropic Magnetoresistance	A phenomenon where the electrical resistance of a material changes based on the angle between the direction of electric current and an applied magnetic field. This effect is utilized in various electronic devices, such as magnetic field sensors, to measure the direction and strength of magnetic fields.
Anisotropical etching	A type of etching process that selectively removes material along specific crystallographic planes, resulting in anisotropic (direction-dependent) material removal. It is commonly used in microfabrication and semiconductor manufacturing to create precise and well-defined features with high aspect ratios in certain crystallographic directions.
Antidig-labeled magnetic nanoparticle (MNP)	A magnetic nanoparticle that has been functionalized with an antibody or an antibody fragment specific to digoxigenin (dig). This complex is used in various immunoassays and biosensing applications to capture and detect digoxigenin-labeled biomolecules, providing a sensitive and specific method for detection and analysis.
Apoptosis	A programmed cell death process that occurs naturally in multicellular organisms. It plays a crucial role in maintaining tissue homeostasis, eliminating damaged or unwanted cells, and shaping proper development by activating a series of cellular events that lead to controlled cell death without causing inflammation.
Aptamer	A short single-stranded DNA or RNA molecule that can bind to specific target molecules with high affinity and selectivity. These synthetic or naturally occurring oligonucleotides are used in various applications, such as biosensors, targeted drug delivery, and diagnostic assays, due to their ability to recognize and bind to specific molecules, including proteins, small molecules, and pathogens.
AQUA stain	A fluorescent DNA dye used in cell biology and cytometry to measure DNA content and cell cycle analysis. It is cell-permeable and binds to DNA, allowing researchers to quantitatively analyze DNA content and assess cell cycle distribution in a sample.
Artifact (in neural recording)	An unwanted signal or interference that is recorded along with the desired neural activity. Artifacts can arise from various sources, such as electrical noise, muscle activity, or movement artifacts, and can distort or obscure the accurate

(Continues)

(*Continued*)

Context	Definition
	representation of neural signals, requiring careful signal processing and artifact removal techniques for accurate analysis.
Astrocyte	A type of glial cell found in the central nervous system (CNS). They play essential roles in supporting and maintaining the health and function of neurons, regulating neurotransmitter levels, providing metabolic support, and contributing to the blood–brain barrier, among other functions.
Avalanche photodiode	A p–n junction diode reverse-biased above its breakdown voltage, which uses impact ionization (avalanche effect) to amplify the photogenerated current.
Axon	A long, slender extension of a neuron that carries nerve impulses away from the cell body towards other neurons, muscles, or glands. It plays a crucial role in transmitting electrical signals between different parts of the nervous system, enabling communication and coordination within the body.
Back-end-of-line (BEOL)	The final stages of semiconductor device manufacturing, where the individual devices on the wafer are interconnected and packaged. It involves the creation of metal interconnects, dielectric layers, and vias to link the active components of the chip, enabling the proper functioning of the integrated circuit.
Bacteriophage (phage)	Viruses that specifically infect and replicate within bacteria. They can inject their genetic material into bacterial cells, leading to the production of new phages and ultimately causing the destruction of the host bacterium, which makes them potentially useful in targeted antibacterial therapies.
Barkhausen criteria	Conditions that must be satisfied for a system to exhibit sustained oscillations. In electronics and control systems, the criteria state that the loop gain should be equal to or greater than one, and the phase shift around the loop should be a multiple of 360 degrees for the system to oscillate at a specific frequency. This concept is essential in understanding the conditions required for the stable operation of oscillators and other feedback systems.
Betaine (also known as trimethylglycine)	A naturally occurring compound found in various foods, especially in plants, and also synthesized by the human body. It acts as a methyl donor and osmoprotectant, playing important roles in metabolism and cellular protection against osmotic stress.
Bioluminescence	Light produced by an organism using a chemical reaction.
Biotinylated	A molecule or biomolecule that has been covalently linked to biotin, a vitamin B7 derivative. Biotinylated molecules are commonly used in various biological and molecular biology assays as they can be specifically detected and

(*Continues*)

(*Continued*)

Context	Definition
	isolated using streptavidin or avidin, which exhibit strong binding affinity for biotin.
Blood Oxygen Level Dependent (BOLD)	The presence of deoxygenated hemoglobin (Hb) results in a drop in T2 and T2* in magnetic resonance imaging (MRI), known as the blood oxygenation level-dependent (BOLD) effect.
Blood Oxygen Level Dependent (BOLD)	The presence of deoxygenated hemoglobin (Hb) results in a drop in T2 and T2* in magnetic resonance imaging (MRI).
Bovine capillary endothelial (BCE)	A type of cultured cells derived from the endothelial lining of small blood vessels in cows (bovines). They are commonly used in scientific research to study vascular biology, angiogenesis, and various physiological processes related to blood vessels and endothelial cells.
Bovine serum albumin (BSA)	A protein derived from cow's blood, and it is structurally similar to human serum albumin. BSA is widely used in biochemical and molecular biology applications as a blocking agent to prevent non-specific binding in assays, a stabilizer for enzymes and other proteins, and a standard in protein quantification assays.
Brownian motion	Brownian motion is the random motion of particles suspended in a fluid (a liquid or a gas) resulting from their collision with the fast-moving atoms or molecules in the gas or liquid.
Buried oxide (BOX)	Buried oxide layer is an excellent electrical insulating layer and an effective etch-stop but can be used as a sacrificial layer as well.
Carbon fiber	A lightweight and strong composite material composed of thin, strong fibers made mostly of carbon atoms. It is widely used in various industries, including aerospace, automotive, and sporting goods, to create high-performance and durable products due to its exceptional strength-to-weight ratio and resistance to corrosion.
Carbon nanotube (CNT)	Cylindrical nanostructures made of carbon atoms arranged in a tubular fashion. They have exceptional mechanical, electrical, and thermal properties, making them promising materials for various applications, including electronics, composites, and nanotechnology.
Cardiomyocyte	A specialized muscle cell that makes up the majority of the heart tissue. These cells contract rhythmically to pump blood throughout the body and are essential for the heart's function as a vital organ in the circulatory system.
Carr–Purcell–Meiboom–Gill (CPMG) sequence	Sequence of pulses used for spin-echo experiments in nuclear magnetic resonance (NMR) in which the initial pulse is 90° followed by a series of 180° pulses.
Chemiluminescence	A chemical reaction that produces light emission without the need for an external light source. It occurs when certain chemical reactions release energy in the form of light, resulting in the generation of a luminescent glow,

(Continues)

(*Continued*)

Context	Definition
	commonly used in analytical chemistry, immunoassays, and bioluminescent imaging.
Chromophore	A chemical group or moiety within a molecule that is responsible for its color or ability to absorb and emit light at specific wavelengths. It is commonly found in dyes, pigments, and fluorescent molecules, where its unique electronic structure allows it to interact with light and give rise to distinctive colors or fluorescence.
Chronocoulometric	A technique used in electrochemistry to study the behavior of electrode reactions by measuring the current as a function of time during a controlled potential or current application. It provides valuable information about the kinetics and mechanism of redox processes and is commonly used in the characterization of electrochemical systems and the study of electrode reactions.
Collagenase	An enzyme that breaks down collagen, the primary structural protein in connective tissues such as skin, tendons, and bones. It is widely used in research and medicine to dissociate tissues, isolate cells, and facilitate cell culture processes by digesting the collagen-rich extracellular matrix and releasing cells for further analysis or transplantation.
Collimator	An optical device used to make a beam of light or other electromagnetic radiation parallel and well-defined. It is commonly used in various applications, including in telescopes, cameras, and spectrometers, to ensure that light or radiation is properly focused and directed.
Computer-aided design (CAD) tools	Software applications used by engineers, architects, and designers to create, modify, and analyze 2D or 3D models of products, buildings, or mechanical components. These tools enable efficient and precise design, visualization, and simulation, helping to streamline the design process and improve overall productivity.
Correlated double sampling (CDS)	A technique used in electronic sensor systems to reduce noise and remove unwanted signal offsets. It involves taking two consecutive measurements of a signal, with the second measurement being taken while the sensor's reset noise is subtracted from the first measurement, resulting in a cleaner and more accurate signal reading.
Covalent interaction	A chemical bond formed when two atoms share electrons to achieve a stable electron configuration. This type of bond is strong and stable, resulting in the formation of molecules or compounds with shared electron pairs between the atoms involved.
Cross-linker solution	A chemical agent used to form covalent bonds between molecules or functional groups. It is commonly used in various biological and chemical applications to stabilize protein–protein interactions, immobilize biomolecules on

(Continues)

(*Continued*)

Context	Definition
	surfaces, or create three-dimensional networks in polymers or hydrogels, among other uses.
Cyanine-3 (Cy3)	An orange-fluorescent label
Dark current	Minimum amount of reverse current presented in the absence of light illumination.
Debye length	A parameter used in plasma physics and electrochemistry to measure the extent of charge screening or the distance over which charged particles influence each other's electric fields. It is a characteristic length scale that determines the behavior of charged particles in a plasma or electrolyte solution, and it is inversely proportional to the square root of the electron density and temperature.
Deep reactive ion etching (DRIE)	An advanced etching technique used in microfabrication to create deep and high-aspect-ratio features in silicon or other materials. It involves using a combination of chemical and physical processes, such as alternating etch and passivation steps, to achieve precise and highly anisotropic etching, commonly used in the manufacturing of MEMS devices and microstructures.
Deionized (DI) water	Deionized water is water that has had all of the ions removed.
Demyelination	The loss or destruction of the myelin sheath that surrounds nerve fibers in the central nervous system (CNS) or peripheral nervous system (PNS). It disrupts the normal conduction of nerve impulses, leading to impaired communication between neurons and various neurological symptoms, such as weakness, sensory disturbances, and coordination difficulties. Demyelination is a characteristic feature of several neurological disorders, including multiple sclerosis.
Dendrimer	Highly branched, nanoscale macromolecules with a well-defined structure. They consist of multiple repeating units arranged in a radially symmetric manner, making them valuable in various applications, such as drug delivery, nanotechnology, and as templates for nanomaterial synthesis due to their tunable size, shape, and surface functionality.
Dendrite	A branch-like extension of a neuron that receives incoming signals from other neurons or sensory receptors. It plays a crucial role in transmitting these signals towards the cell body, where they are integrated and processed before further transmission along the neuron's axon.
Deoxynucleotide triphosphate (dNTP)	The building blocks of DNA. They are the monomers that are used by DNA polymerase enzymes during DNA replication to add nucleotides to the growing DNA strand, forming the complementary strand to the original DNA

(*Continues*)

(*Continued*)

Context	Definition
	template. dNTPs consist of deoxyribose sugar, a phosphate group, and one of the four nitrogenous bases (adenine, guanine, cytosine, or thymine) that are used to encode genetic information in DNA.
Deoxyribonucleic Acid (DNA)	A molecule that contains the genetic instructions used in the growth, development, functioning, and reproduction of all known living organisms and many viruses. It consists of a double helix structure made up of nucleotide building blocks, each containing a sugar (deoxyribose), a phosphate group, and one of four nitrogenous bases (adenine, thymine, cytosine, or guanine).
Depletion region	A region in a semiconductor device, usually at the juncture of P-type and N-type materials, in which there is neither an excess of electrons nor of holes.
Design rule check (DRC)	A crucial step in semiconductor manufacturing and integrated circuit (IC) design. It involves automated software-based checks that verify if the layout of the circuit adheres to the specific manufacturing rules and design guidelines to ensure proper functionality, manufacturability, and reliability of the final semiconductor product.
Digoxigenin (dig)-modified	A molecule, typically a nucleotide or an antibody, that has been chemically modified with digoxigenin, a steroid-like compound. Digoxigenin-modified molecules are widely used in molecular biology techniques, such as *in situ* hybridization and immunodetection, where they serve as tags to label and detect specific target sequences or proteins in biological samples.
DNA hybridization	A process in which two single-stranded DNA molecules with complementary base sequences come together and form a stable double-stranded DNA molecule. It is a fundamental technique used in molecular biology and genetics for various applications, including detecting specific DNA sequences, studying genetic relationships, and identifying gene expression levels.
DNA polymerase	An enzyme that plays a crucial role in DNA replication and repair. It catalyzes the synthesis of a new DNA strand by adding deoxynucleotide triphosphates (dNTPs) to the growing DNA chain, using a DNA template as a guide to ensure accurate copying of genetic information.
DNA sequencing	A laboratory technique used to determine the precise order of nucleotides in a DNA molecule. It enables the identification of genetic information, mutations, and variations in the DNA sequence, playing a crucial role in various fields, including genetics, genomics, and biomedical research.
Dopamine (DA)	A hormone and a neurotransmitter with the molecular formula $C_8H_{11}NO_2$ that constitutes about 80% of the catecholamine content in the brain.

(*Continues*)

(*Continued*)

Context	Definition
Doping	The intentional introduction of impurities into a semiconductor material to modify its electrical properties. By adding controlled amounts of specific atoms, the semiconductor's conductivity can be altered, making it either more conductive (n-type doping) or less conductive (p-type doping), which is a fundamental process in semiconductor device manufacturing.
Dry etching	A semiconductor manufacturing process that uses plasma or reactive gases to remove material from a substrate, allowing precise and controlled pattern transfer for creating microstructures and integrated circuits. Unlike wet etching, it does not involve a liquid chemical solution.
Dynabead (also known as Dynal bead)	Superparamagnetic beads coated with specific ligands, such as antibodies or streptavidin, that allow them to bind to specific target molecules. They are widely used in various biomedical applications, including cell separation, immunoassays, and nucleic acid purification, due to their ease of use, flexibility, and gentle magnetic handling.
Electrochemiluminescence	A luminescent phenomenon that occurs when electrochemical reactions generate light emission. It is commonly used in analytical chemistry and immunoassays, where specific molecules or biomolecules are detected based on their ability to produce light during electrochemical reactions, providing sensitive and quantitative measurement capabilities.
Electrodeposition (also known as electroplating)	An electrochemical process used to coat a conductive surface with a thin layer of metal or other materials. It involves passing an electric current through an electrolyte solution containing metal ions, causing the metal ions to be reduced and deposited onto the electrode's surface, resulting in a uniform and adherent coating.
Electroencephalogram	A non-invasive test that measures and records the electrical activity of the brain. It involves placing electrodes on the scalp to detect and analyze brainwave patterns, providing valuable information about brain function and neurological conditions.
Electrogenic cell	A type of cell that can generate an electrical potential or voltage across its membrane, typically due to the movement of ions. This electrical potential is essential for various physiological processes and cellular functions, such as nerve impulses in neurons and muscle contractions.
Electroless nickel immersion gold (ENIG) plating	A surface finishing process used in electronics manufacturing. It involves depositing a thin layer of nickel on a substrate through a chemical reaction, followed by an immersion in gold to improve the surface conductivity, solderability, and corrosion resistance of the substrate, commonly used in printed circuit boards (PCBs) and semiconductor packaging.

(*Continues*)

(*Continued*)

Context	Definition
Electrolyte	A substance, typically an ionic solution or molten salt, that conducts electricity due to the presence of freely moving ions. It is essential in various electrochemical processes, such as batteries, fuel cells, and electrolysis, where ions enable the flow of electric current between electrodes.
Electrophysiology	A branch of physiology that studies the electrical activity of living cells, tissues, and organs, typically focusing on the nervous system, heart, and muscles. It involves measuring and analyzing electrical signals generated by these systems to understand their functions, behavior, and potential abnormalities.
Electrostatic interaction	The attraction or repulsion between charged particles or molecules due to the presence of electrostatic forces. It plays a crucial role in various biological and chemical processes, such as protein-protein interactions, DNA binding, and the stabilization of ionic compounds.
Emulsion polymerization	A process in which water-insoluble monomers are dispersed as droplets in an aqueous medium containing surfactants. Through the use of initiators, these monomer droplets undergo polymerization to form polymer particles suspended in the water.
Endothelial cell	A type of specialized cells that line the inner surface of blood vessels and lymphatic vessels. They play a crucial role in regulating blood flow, controlling the exchange of nutrients and waste products between blood and tissues, and maintaining the integrity of the vascular system.
Enterococcus faecalis	*Enterococcus faecalis* is a species of enterococci bacteria that normally lives in your gastrointestinal tract and can also be found in the oral cavity and vaginal tract.
Enzyme-Linked Immunosorbent Assay (ELISA)	A widely used laboratory technique for detecting and quantifying specific proteins or other target molecules in a sample. It involves the use of antibodies and enzymes to generate a measurable signal, providing sensitive and specific results in various research, clinical, and diagnostic applications.
Epilepsy	A neurological disorder characterized by recurrent and unpredictable seizures, which are caused by abnormal electrical activity in the brain. Seizures can manifest in various ways, ranging from brief periods of altered consciousness or muscle twitching to more severe convulsions, and may be triggered by different factors such as genetic predisposition, brain injury, or certain medical conditions.
Epithelial cell	A type of cell that line the surfaces of tissues and organs in the body. They form protective barriers and play essential roles in absorption, secretion, and other physiological functions, serving as a vital interface between the internal and external environments.

(*Continues*)

(*Continued*)

Context	Definition
Epitope	A specific region or site on an antigen (usually a protein or peptide) that is recognized and bound by an antibody or a T cell receptor. Epitopes are essential for immune responses, as they trigger the activation of immune cells and the production of antibodies, leading to the elimination of foreign substances or infected cells.
Ester	An organic compound derived from an acid (usually a carboxylic acid) and an alcohol, resulting in the formation of the characteristic ester functional group (–COO–). Esters have a wide range of applications, including as fragrances, flavorings, solvents, and plasticizers, and they contribute to the pleasant aromas and tastes found in various fruits and flowers.
Evanescent wave	An oscillating electric and/or magnetic field that does not propagate as an electromagnetic wave, but whose energy is spatially concentrated near the source.
Falcon tube (also known as a conical tube or centrifuge tube)	A plastic laboratory tube with a conical bottom and a screw cap. It is commonly used in biological and chemical research for holding, storing, and centrifuging samples, such as cell cultures, reagents, and solutions.
Fiber-optic faceplate (FOF)	A device used to convert an incoming image or light distribution into a uniform output on the other side. It consists of a bundle of optical fibers arranged in a specific pattern to transmit light and is commonly employed in imaging and display systems, medical devices, and scientific instruments for efficient light transfer and image transmission.
Fibroblast cell	A type of connective tissue cell that plays a critical role in tissue repair and wound healing. They are responsible for producing and organizing the extracellular matrix, which provides structural support and integrity to tissues in the body.
Finite-difference time-domain (FDTD)	Method for solving Maxwell's equations in complex geometries
Flip-chip bonding	A semiconductor packaging technique where an integrated circuit (IC) chip is directly mounted face-down onto a substrate, such as a printed circuit board (PCB). The electrical connections between the chip and the substrate are established using solder bumps or other conductive materials, allowing for a more compact and efficient package with improved electrical performance.
Flow cytometry	A powerful technique used to analyze and quantify the physical and chemical characteristics of individual cells or particles in a fluid suspension. It involves passing cells or particles through a laser-based detection system, allowing researchers to gather information about cell size, granularity, and various surface or intracellular markers, enabling a wide range of applications in immunology, cancer

(*Continues*)

(*Continued*)

Context	Definition
	research, and cell biology.
Fluorescein	A synthetic organic compound and a fluorophore commonly used as a fluorescent dye. It emits bright green fluorescence when excited by light, making it valuable in various applications, including fluorescence microscopy, flow cytometry, and fluorescent labeling of biomolecules for detection and visualization.
Fluorescence	The emission of light by a substance that has absorbed light or other electromagnetic radiation
Fluorophore	A molecule or compound that can absorb light energy at one wavelength and emit it as light at a longer wavelength, typically in the visible or ultraviolet spectrum. Widely used in fluorescence microscopy and various biological assays, fluorophores enable the visualization and tracking of specific molecules within cells and tissues.
Functionalization	The process of modifying the sensor's surface with specific molecules or biological elements (e.g., antibodies, enzymes, or DNA probes). This modification allows the biosensor to selectively and sensitively detect and interact with the target analyte, enabling accurate and precise detection of various biomolecules or pathogens in complex samples.
Galinstan	A brand name for a eutectic alloy composed of gallium, indium, and tin. It is a non-toxic liquid metal that remains in a liquid state over a wide temperature range, making it useful for various applications in electronics, thermometers, and medical devices.
Ganglion cell	A type of neuron found in the retina of the eye. It receives visual information from other retinal cells and transmits the processed signals to the brain via the optic nerve, playing a crucial role in relaying visual information from the retina to the visual processing centers in the brain.
Giant Magnetoimpedance	A phenomenon observed in certain magnetic materials where the electrical impedance of the material changes significantly in response to an applied magnetic field. This effect is utilized in GMI sensors for detecting and measuring magnetic fields, finding applications in fields like non-destructive testing and medical diagnostics.
Glial cell (also known as neuroglia)	Non-neuronal cells that provide crucial support and protection to neurons in the central nervous system. They play essential roles in maintaining the structural integrity of the nervous system, providing metabolic support to neurons, and regulating the microenvironment for optimal neuronal function.
Glucose	A simple sugar and an essential source of energy for living organisms. It is commonly found in various foods and serves as a primary fuel for cellular processes in the body.

(Continues)

(*Continued*)

Context	Definition
GlutaMAX	A cell culture supplement that contains L-alanyl-L-glutamine, a stable dipeptide of glutamine and alanine. It is used to provide a more stable and consistent source of glutamine in cell culture media, promoting better cell growth and viability compared to traditional L-glutamine supplements.
Glutaraldehyde	A chemical compound with two aldehyde groups, and it is commonly used as a cross-linking agent and fixative in various biomedical and industrial applications. In the medical field, it is often utilized for tissue fixation and sterilization, while in industry, it is employed for cross-linking proteins and as a disinfectant and preservative.
Glycol	A class of organic compounds that contain two hydroxyl groups (–OH) on adjacent carbon atoms. These compounds are widely used as coolants, antifreeze agents, and solvents due to their ability to lower the freezing point and raise the boiling point of water-based solutions, making them valuable in various industrial and automotive applications. Ethylene glycol and propylene glycol are two common examples of glycols used in antifreeze formulations and other products.
Homeostasis	The body's ability to maintain a stable and constant internal environment despite external changes. It involves a complex series of physiological processes that regulate various factors, such as temperature, pH, blood glucose levels, and fluid balance, to keep the body's internal conditions within a narrow range, ensuring optimal function and survival.
Homogeneity	The state or quality of being uniform, consistent, and the same throughout. In various contexts, it can refer to the uniformity of a mixture, the distribution of properties within a sample, or the similarity of elements or components in a system.
Human serum albumin (HSA)	A protein found in the blood plasma, and it is the most abundant protein in human blood. It plays essential roles in maintaining osmotic pressure, transporting various substances, and regulating pH and buffering capacity in the bloodstream. HSA is also used in medical settings as a plasma expander and a carrier for drugs and therapeutic agents due to its biocompatibility and long half-life in circulation.
Hydrophilicity	The property of a material or surface to have a strong affinity for water, allowing it to readily absorb or interact with water molecules. Hydrophilic surfaces are characterized by their ability to attract and spread water, making them useful in various applications, such as wetting agents, filtration, and biomedical devices.

(*Continues*)

(*Continued*)

Context	Definition
Hydrophobic interaction	The tendency of nonpolar molecules or groups to associate with each other in an aqueous environment due to the exclusion of water. This phenomenon plays a significant role in the folding of proteins, the formation of lipid bilayers, and the assembly of hydrophobic molecules in biological systems.
Immunoglobulin (also known as antibodies)	Y-shaped proteins produced by the immune system in response to foreign substances (antigens). They play a crucial role in the immune response by recognizing and binding to specific antigens, marking them for destruction by other immune cells, and providing immunity against pathogens and foreign invaders.
In Vitro	Experiments or processes that are conducted outside of a living organism, typically in a controlled laboratory environment. These experiments are often performed using isolated cells, tissues, or biological molecules in artificial conditions, such as test tubes, Petri dishes, or culture plates. *In vitro* studies allow researchers to investigate biological processes and interactions under controlled and reproducible settings.
In Vivo	Experiments or observations that are conducted within a living organism, whether it's an animal, plant, or human. *In vivo*, studies involve studying biological processes, responses to treatments, or the effects of interventions in their natural context. This type of research provides insights into how biological systems function within a whole organism and can help researchers better understand complex interactions and potential outcomes in a more realistic setting.
Incubator	A controlled environment chamber used in laboratories and medical facilities to maintain optimal conditions for the growth and cultivation of cells, tissues, or microorganisms. It provides precise control over temperature, humidity, and sometimes CO_2 levels, creating a stable and suitable environment for various biological experiments and applications.
Ischemia	A condition characterized by an inadequate blood supply to an organ or tissue, leading to reduced oxygen and nutrient delivery. It can result from various factors, such as blood vessel blockage, atherosclerosis, or decreased blood flow, and is a common underlying cause of tissue damage and organ dysfunction, particularly in the heart and brain.
Isotropic conductive adhesive (ICA)	A type of adhesive material that contains conductive particles, allowing electrical conduction in all directions. It is commonly used in electronics for bonding components such as flip chips or surface-mount devices, providing both

(*Continues*)

(*Continued*)

Context	Definition
	mechanical attachment and electrical connectivity between the components and the substrate.
Kapton film	A polyimide film known for its high-temperature resistance and excellent electrical insulation properties. It is commonly used in various applications, such as in electronics, aerospace, and industrial settings, where a thin, flexible, and durable insulating material is required to withstand extreme temperatures and harsh environments.
Ketone	A type of organic compound that contains a carbonyl group (C=O) bonded to two carbon atoms. Ketones are commonly found in various natural substances and have significant industrial applications as solvents, flavoring agents, and in the synthesis of pharmaceuticals and other chemicals.
Laminar flow	A type of fluid flow in which the fluid moves in parallel layers with minimal mixing between them. It occurs at lower flow rates and is characterized by smooth, predictable flow patterns, making it useful in various applications, such as in the design of pipelines, airfoil shapes, and certain laboratory processes.
Langmuir–Blodgett film	A thin film made by depositing a monolayer of molecules or nanoparticles onto a solid substrate. The deposition process involves transferring the molecules or nanoparticles from the air-water interface onto the substrate in a controlled and organized manner, resulting in a uniform and well-ordered thin film with precise control over its thickness and molecular arrangement. These films find applications in nanotechnology, surface engineering, and various optical and electronic devices.
Laser etching	A process that uses a focused laser beam to remove material from a surface, creating a permanent and precise marking or pattern. It is widely used in various industries for engraving, marking, and etching on materials like metal, plastics, glass, and ceramics.
Layer-By-Layer (LBL) coating technique	A versatile method used to create thin films or coatings by sequentially depositing alternating layers of different materials onto a substrate. Each layer is typically formed through self-assembly, electrostatic interactions, or other chemical bonding, allowing precise control over film thickness, composition, and properties, making it valuable in various applications, including drug delivery, sensors, and surface engineering.
Leukocyte cell line	A group of white blood cells that are derived from a common progenitor in the bone marrow. These cells are crucial components of the immune system and play a key role in defending the body against infections and foreign invaders.
Lift-off	A microfabrication technique used to pattern thin films by depositing them over a sacrificial material and then

(*Continues*)

(*Continued*)

Context	Definition
	dissolving the sacrificial layer, leaving behind the desired pattern on the substrate. It is widely employed in semiconductor and nanotechnology industries for creating precise and intricate structures in various applications such as integrated circuits, MEMS devices, and sensors.
Ligand	A molecule or ion that binds to a specific site on a larger biomolecule, such as a receptor or enzyme, forming a stable complex. Ligand-receptor interactions play a critical role in cellular signaling, enzymatic activity, and many biological processes, influencing the behavior and function of the target biomolecule.
Light shield	A physical barrier or structure designed to block, redirect, or reduce the transmission of light. It's commonly used in photography, astronomy, and architectural design to prevent unwanted light from interfering with sensitive equipment or affecting the desired lighting conditions.
Linker	A molecule or chemical structure that connects the sensing element (e.g., antibodies, aptamers, or enzymes) to the transducer or the surface of the sensor. The linker is designed to facilitate efficient and specific binding between the sensing element and the target molecule, ensuring accurate detection and measurement in the biosensing process.
Local field potential (LFP)	A measure of the low-frequency electrical activity generated by a group of neurons in a specific brain region. It represents the summed electrical potentials of nearby neurons and is often used in neuroscience research to study the collective activity and communication patterns of neural populations.
Loop-mediated isothermal amplification (LAMP)	A molecular biology technique used to amplify DNA rapidly and efficiently under isothermal conditions, typically at a single temperature. It is a powerful method for the detection of specific DNA sequences and has various applications in research, clinical diagnostics, and field-based testing due to its simplicity, sensitivity, and speed.
Low-temperature co-fired ceramic (LTCC)	A versatile multilayer substrate technology used in the electronics industry for packaging and interconnecting electronic components. It involves forming ceramic layers and conductive traces, which are stacked and fired at relatively low temperatures, enabling the integration of complex electronic circuits in a compact and reliable package.
Luciferin	Organic compound whose oxidation in the presence of the enzyme luciferase produces light.
Magnetorelaxometry (MRX)	A scientific method where magnetic nanoparticles are exposed to a magnetic field and their subsequent relaxation behavior upon field removal is studied.

(Continues)

(*Continued*)

Context	Definition
Membrane	In thermal sensors, membrane made of dielectric layers of the CMOS process that act as a thermally isolated stage.
Microglia	Specialized immune cells in the central nervous system (CNS) that act as the primary immune defense in the brain. They function to protect the brain from infections, remove damaged neurons and debris, and play a role in regulating neural connections during brain development and in response to injury or disease.
Molecularly imprinted polymer (MIP)	A synthetic polymer with selective binding sites that are designed to recognize and bind to specific target molecules. These polymers are created through a process called molecular imprinting, where the target molecule is used as a template during polymerization, resulting in the formation of complementary binding sites that can selectively capture and detect the target molecule in various applications such as sensors, separations, and drug delivery.
Monoclonal	A type of antibody produced by a single clone of identical immune cells. Monoclonal antibodies are highly specific and uniform, making them valuable tools in various research, diagnostics, and therapeutic applications, including targeted cancer treatments and immunotherapy.
Multi-walled carbon nanotubes (MWCNT)	Cylindrical nanostructures made of multiple layers of carbon atoms arranged in concentric tubes. They possess unique mechanical, electrical, and thermal properties, making them valuable in a wide range of applications, including nanotechnology, electronics, materials science, and biomedical research.
Myocyte (also known as a muscle cell)	A specialized cell found in muscle tissue responsible for contraction and movement. These cells contain contractile proteins that allow them to generate force and enable movement in the body.
Nano-imprint lithography (NIL)	A high-resolution manufacturing technique used to create extremely small and intricate patterns on the surface of sensor materials. It involves pressing a template or mold with nanoscale features onto a substrate coated with a material that can be solidified, such as a polymer. When the template is removed, it leaves behind a precise pattern that can be used to functionalize the sensor surface with biomolecules or other sensing elements. This method enables the fabrication of nanostructured surfaces for biosensors, enhancing their sensitivity and performance by providing a large surface area for interactions with target molecules.
Nanorod (NR)	A nanostructure with a rod-like shape, typically with one dimension in the nanoscale range. These nanoscale rods can be made from various materials and are widely used in nanotechnology and nanomaterials research for their unique properties and applications, including as catalysts, sensors, and building blocks for more complex nanostructures.

(*Continues*)

(*Continued*)

Context	Definition
Near-infrared (near-IR)	A region of the electromagnetic spectrum with wavelengths ranging from approximately 700 to 2500 nm. Near-IR light is used in various applications, including spectroscopy, imaging, and communication, due to its ability to penetrate through biological tissues and many materials while providing valuable information about molecular structures and characteristics.
Neurotransmitter	Chemical messengers that transmit signals between nerve cells (neurons) in the nervous system, enabling communication and coordination of various physiological and cognitive functions in the body.
Neutrophil	A type of white blood cell that plays a crucial role in the body's immune response against bacterial infections. They are the most abundant type of white blood cell and are highly mobile, capable of migrating to infected tissues and engulfing and destroying invading bacteria through a process called phagocytosis.
N-terminal pro-brain natriuretic peptide (NT-proBNP)	A biomarker used to assess heart function and diagnose heart failure. It is released into the bloodstream in response to increased stretching of heart muscle cells, and elevated levels of NT-proBNP are indicative of heart stress or dysfunction.
Ohm's law	A fundamental principle in electrical engineering and physics that states the relationship between voltage (V), current (I), and resistance (R) in an electrical circuit: $V = I \times R$. This law demonstrates that the voltage across a conductor is directly proportional to the current flowing through it and inversely proportional to its resistance.
Oligodendrocyte	A type of glial cell in the central nervous system (CNS) responsible for producing myelin, a fatty substance that wraps around axons to insulate and facilitate the conduction of nerve impulses. They play a critical role in promoting efficient nerve signal transmission and supporting the overall function and integrity of the nervous system.
Oligodendroglia	A type of glial cell found in the central nervous system (CNS). Their main function is to produce myelin, a fatty substance that insulates and speeds up the conduction of nerve impulses along axons, supporting efficient communication between neurons in the CNS.
Oligonucleotide	Short chains of nucleotides, the building blocks of DNA and RNA, typically composed of around 10 to 50 nucleotide units. They play a crucial role in various biological processes, including genetic research, diagnostics, and therapeutic applications, such as antisense therapy and gene editing.
Optical density (OD)	A measure of the absorbance or attenuation of light as it passes through a sample. It is the ratio of the intensity of light incident on the filter to the intensity of the light passing through it.

(Continues)

(*Continued*)

Context	Definition
Optical spectrometer	A scientific instrument used to analyze the wavelengths of light emitted, absorbed, or scattered by a sample. It separates the different wavelengths of light and measures their intensities, providing valuable information about the sample's composition, molecular structure, and other optical properties.
Organosulfur	Organic compounds that contain one or more sulfur atoms in their molecular structure. These compounds play diverse roles in various biological, pharmaceutical, and chemical processes, and they are often responsible for the distinctive odors and flavors of certain foods, such as garlic and onions.
Organotypic slice culture	A laboratory technique used to maintain and study living tissue slices from organs or tissues in an artificial environment. This method allows researchers to investigate the physiology and interactions of cells and tissues outside of the body, enabling a closer representation of the *in vivo* conditions and providing valuable insights into various biological processes and disease models.
Osmium-polyvinylpyridine gel	A fixative used in electron microscopy to preserve biological samples for imaging. It contains osmium tetroxide (O_sO_4) and polyvinylpyridine, which work together to fix the cellular structures and enhance contrast for electron microscopy analysis.
Oxygen plasma	A type of plasma generated by energizing oxygen gas with an electric field. It is commonly used in various applications, such as cleaning and surface modification of materials, semiconductor processing, and plasma etching in microelectronics, due to its strong reactivity and ability to remove organic contaminants and functionalize surfaces.
Paraffinophilic Mycobacterium Avium Complex	A group of environmental mycobacteria that have a strong affinity for paraffin wax. They are known for their ability to colonize and grow in the paraffin wax used to embed tissue samples in histopathology, leading to potential contamination of laboratory specimens and challenging diagnostic processes.
Parainfluenza virus	A group of single-stranded RNA viruses that can cause respiratory infections in humans, particularly in young children and infants. These viruses are a common cause of croup, bronchiolitis, and pneumonia, and they are distinct from influenza viruses, which cause seasonal flu.
Patch-clamp	A laboratory technique used to measure the electrical activity of individual cells, particularly neurons or ion channels, by placing a glass microelectrode (patch pipette) on the cell membrane. It enables the study of ion channel currents and membrane potential changes, providing valuable insights into cellular physiology and neuronal communication.

(*Continues*)

(*Continued*)

Context	Definition
Pathogen	A microorganism, such as a virus, bacteria, fungus, or parasite, that can cause disease in its host. These infectious agents can lead to a wide range of illnesses, from mild to severe, by invading and damaging the host's cells or disrupting normal physiological functions.
Peptide	A short chain of amino acids linked together by peptide bonds. Peptides can vary in length, and they play essential roles in biological processes, including as signaling molecules, enzyme inhibitors, and components of proteins.
Peptide bond	Strong chemical links between amino acids, building blocks of proteins. Formed through a condensation reaction, they connect an amino group of one amino acid to a carboxyl group of another, releasing water. These bonds create the linear sequence of amino acids, which defines a protein's structure and function. The planar nature of peptide bonds limits rotation, influencing the protein's overall shape and interactions.
Peptide nucleic acid (PNA)	An artificial nucleic acid analog that contains a peptide backbone instead of the deoxyribose-phosphate backbone found in DNA or RNA. PNA can hybridize with complementary DNA or RNA sequences and exhibits high binding affinity and stability, making it a valuable tool in molecular biology, diagnostics, and potential therapeutic applications.
Petri dish	A shallow, flat, cylindrical container made of glass or plastic, often with a lid, used for culturing and observing living microorganisms or cells in a laboratory setting. It provides a controlled environment for studying cell growth, bacterial colonies, and other biological processes.
Petroleum jelly	A semi-solid mixture of hydrocarbons derived from petroleum. It is commonly used as a moisturizer, lubricant, and protective barrier for skin, lips, and other surfaces, due to its non-comedogenic and occlusive properties.
pH (also known as potential of hydrogen)	A measure of the acidity or alkalinity of a solution. It quantifies the concentration of hydrogen ions (H^+) present in the solution. The pH scale ranges from 0 to 14, where a pH of 7 is considered neutral. Solutions with a pH less than 7 are acidic, indicating a higher concentration of hydrogen ions, while solutions with a pH greater than 7 are alkaline (basic), indicating a lower concentration of hydrogen ions. The pH of a solution plays a crucial role in various chemical and biological processes, influencing the behavior and properties of substances dissolved in the solution.
Pharmacokinetics	The study of how the body processes a drug, including its absorption, distribution, metabolism, and elimination. It helps to understand how drugs are absorbed, distributed in the body, metabolized, and eventually excreted, providing valuable information for drug dosing, efficacy, and safety.

(Continues)

(*Continued*)

Context	Definition
Pharmacological intervention	The use of drugs or medications to treat, manage, or prevent medical conditions or diseases. It involves the administration of specific chemical substances that interact with the body's physiological processes to achieve a desired therapeutic effect, such as relieving symptoms, curing infections, or regulating biological functions.
Phase-locked loop (PLL)	An electronic control system that compares the phase of an input signal with that of a feedback signal and generates an output signal with a frequency and phase locked to the input signal. The PLL aims to maintain a stable and precise relationship between the input and output signals, making it useful in various applications such as clock generation, frequency synthesis, and synchronization.
Photodiode	A photodiode is a light-sensitive semiconductor diode that converts light energy into electrical energy (voltage or current).
Photolithography	A process used in semiconductor manufacturing and microfabrication, where light is used to transfer patterns onto a substrate coated with a photosensitive material, enabling precise and controlled fabrication of intricate structures for electronic circuits and other microdevices.
Photomask	A transparent plate or glass substrate containing a pattern of opaque and transparent areas. It is a crucial tool in photolithography used to project the pattern onto a semiconductor wafer or other substrates during the manufacturing of integrated circuits and microdevices.
Photomultiplier tube (PMT)	A highly sensitive detector that converts photons to electricity and consists of a photocathode followed by an electron multiplier.
Photosensitive	A property of a material that reacts to light or electromagnetic radiation. Photosensitive materials can undergo chemical changes or transformations upon exposure to light, making them useful in various applications, such as photography, photolithography, and photoresists in semiconductor manufacturing.
Physical vapor deposition (PVD)	A family of thin-film deposition techniques where materials are vaporized and then condensed onto a substrate to form a thin film. It involves processes such as sputtering and evaporation, commonly used in electronics, optics, and coatings to produce films with controlled thickness, composition, and properties.
Piezoresistive	The property of certain materials that change their electrical resistance when subjected to mechanical stress or strain. This effect is commonly utilized in piezoresistive sensors, such as pressure sensors and strain gauges, where the change in resistance is measured and correlated with the applied force or deformation, enabling the sensing and measurement of physical quantities.

(*Continues*)

(*Continued*)

Context	Definition
PIN photodiode	A p–n junction diode with an intrinsic (i.e., undoped) region in between the n- and p-doped region.
Plasma-enhanced atomic layer deposition (PEALD)	A thin-film deposition technique used in nanotechnology and semiconductor manufacturing. It involves using plasma to enhance the reaction between precursor gases, enabling precise and conformal growth of thin films with atomic-level control over thickness and composition.
Plasma-enhanced chemical vapor deposition (PECVD)	A thin-film deposition technique that uses a plasma to enhance the chemical reactions between gaseous precursors, leading to the formation of thin films on a substrate. It is commonly used in semiconductor manufacturing and other industries to deposit various materials with improved control over film properties, uniformity, and adhesion.
Plasmon	Oscillation of free electron density with respect to the fixed positive ions in a metal.
Plasmonic filter	A nanoscale device that exploits the interaction between light and free electrons in metal nanostructures to selectively manipulate and control the transmission of specific wavelengths of light. It's used in various applications like optical communication and sensing due to its ability to enhance light–matter interactions at the nanoscale.
Pneumatic valve	A type of valve that is controlled by compressed air or gas. It uses the pressure of the air or gas to open or close the valve, allowing or blocking the flow of fluids or gases in a system, making it commonly used in various industrial and automation applications.
Polycarbonate	A durable and transparent thermoplastic material known for its high impact resistance and excellent thermal properties. It is widely used in various applications, including eyewear lenses, safety helmets, electronic components, and automotive parts.
Polyclonal	A mixture of antibodies produced by multiple clones of different immune cells in response to a specific antigen. Polyclonal antibodies recognize multiple epitopes on the target antigen, providing a broader range of binding capabilities compared to monoclonal antibodies and are commonly used in research, diagnostics, and immunoassays.
Polyelectrolyte multilayer (PEM)	Thin films or coatings formed by alternating layers of oppositely charged polyelectrolytes. They are widely used in various applications, including drug delivery systems, sensors, and surface modification, due to their tunable properties and the ability to control the film's thickness and surface charge.
Polymer	Large molecules composed of repeating subunits called monomers. They are versatile materials with a wide range of applications, including plastics, fibers, adhesives, and coatings, due to their unique physical and chemical properties.

(Continues)

(*Continued*)

Context	Definition
Polymerase chain reaction (PCR)	A powerful molecular biology technique used to amplify and generate multiple copies of a specific DNA sequence. It involves a cyclic process of denaturation, annealing, and extension, and has various applications in research, diagnostics, and forensics, among others.
Polymerase chain reaction (PCR) primers	Short, single-stranded DNA sequences that are designed to anneal to specific regions of the target DNA during the PCR. They serve as starting points for DNA amplification, allowing the DNA polymerase enzyme to replicate the target sequence, leading to the exponential amplification of the desired DNA fragment for further analysis or experimentation.
Polymorphism	The occurrence of different crystal structures or forms of a substance while having the same chemical composition. It is a common phenomenon in materials science and chemistry, where the arrangement of atoms or molecules in a solid can vary, resulting in different physical properties and behaviors of the material.
Positive photoresist	A type of photosensitive material used in photolithography processes. When exposed to ultraviolet light, the positive photoresist becomes more soluble in a developer, allowing the exposed regions to be removed, leaving behind the desired pattern on the substrate.
Presynaptic bouton (also known as an axon terminal)	A specialized structure at the end of an axon. It contains synaptic vesicles filled with neurotransmitters and is responsible for releasing these neurotransmitters into the synapse to communicate with the postsynaptic neuron or target cell.
Probe	A specific biomolecule or receptor that selectively binds to a target analyte, such as a protein, enzyme, DNA, or other biological molecule of interest. The binding event generates a signal that can be detected and quantified, enabling the biosensor to measure the presence or concentration of the target analyte in a sample.
Proteolytic enzyme (also known as proteases)	Enzymes that break down proteins by catalyzing the hydrolysis of peptide bonds. They play essential roles in various biological processes, including protein digestion, cellular regulation, and tissue remodeling, and are widely used in research, medicine, and industrial applications.
Purkinje cell	Large, specialized neurons found in the cerebellum of the brain. They play a crucial role in motor coordination and balance by receiving inputs from various parts of the brain and spinal cord and modulating the activity of other neurons to fine-tune motor movements.
Pyrex glass	A type of borosilicate glass known for its high thermal resistance and low coefficient of thermal expansion. It is commonly used in laboratory glassware, cookware, and industrial applications where resistance to thermal shock and chemical durability are essential requirements.

(*Continues*)

(*Continued*)

Context	Definition
Quantum dot (Qdot)	Semiconductor nanocrystals that exhibit unique optical and electronic properties due to quantum confinement effects. They can emit bright, tunable fluorescence, making them valuable in applications such as biological imaging, medical diagnostics, and display technologies.
Quantum efficiency (QE)	The probability of an incident photon generating an electron-hole pair contributing to the detection signal, which is affected by material properties and device geometry.
Quencher molecules	Substances used in various scientific and technological applications to dampen or suppress the fluorescence or luminescence of a fluorophore. They work by absorbing and dissipating the energy emitted by the fluorophore, thereby reducing its signal. Quencher molecules are commonly used in fluorescence-based assays, such as molecular biology experiments and fluorescence resonance energy transfer studies, to investigate molecular interactions, probe conformational changes, and measure distances between molecules.
Reactive etching process	A type of dry etching process that utilizes chemically reactive gases, such as fluorine-based or chlorine-based compounds, to remove material from a substrate. The reactive nature of the gases allows for highly selective and controlled material removal in semiconductor manufacturing and microfabrication processes.
Reactive ion etching (RIE)	A dry etching process used in microfabrication and semiconductor manufacturing to remove material from a substrate. It involves bombarding the substrate with reactive ions in a plasma, which chemically reacts with the material to be etched, resulting in precise and anisotropic material removal, commonly used for patterning in integrated circuits and MEMS devices.
Reactive physical vapor deposition (RPVD)	A thin-film deposition technique that combines physical vapor deposition (PVD) with a reactive gas to form compound films. It involves the deposition of a metal target material in a reactive gas environment, leading to the formation of compound films with controlled stoichiometry and enhanced properties, commonly used in semiconductor and optical coating applications.
Receptor	The biological recognition element (BRE) that interacts with the target analyte of interest. This recognition event leads to a measurable signal, such as an electrical or optical response, which is then used to detect and quantify the presence of the target molecule or substance in the sample being analyzed. The receptor in a biosensor can be an antibody, enzyme, nucleic acid, aptamer, or other biomolecular entity with high selectivity and affinity for the target analyte.

(*Continues*)

(*Continued*)

Context	Definition
Reduction-oxidation (Redox)	A chemical reaction that involves the transfer of electrons between reactants. The process includes one substance losing electrons (oxidation) and another substance gaining those electrons (reduction), resulting in the change in the oxidation states of the elements involved in the reaction.
Ribonucleic Acid (RNA)	A molecule that plays a crucial role in protein synthesis and gene expression in living organisms. It is a single-stranded nucleic acid that carries genetic information from DNA to the cellular machinery responsible for protein synthesis, serving as a key intermediary in the process of converting genetic instructions into functional proteins.
Saccharide	A simple sugar or carbohydrate, which is a class of organic compounds composed of carbon, hydrogen, and oxygen atoms. Saccharides play essential roles in providing energy and structural support in living organisms and are the building blocks of more complex carbohydrates like starch, cellulose, and glycogen.
Schwann cell	A type of glial cell found in the peripheral nervous system (PNS). Their primary function is to produce myelin, a fatty substance that wraps around peripheral nerve axons, providing insulation and aiding in the rapid conduction of nerve impulses along these axons. Schwann cells also play a role in nerve regeneration and support the overall function and maintenance of peripheral nerves.
Self-assembled monolayers (SAM)	Single layers of molecules that spontaneously arrange themselves on a substrate in a well-ordered and uniform manner. SAMs are often used to modify surfaces, providing tailored properties like wettability, adhesion, and chemical reactivity, making them valuable in various fields, including nanotechnology, biosensors, and surface science.
Signal-to-noise ratio (SNR)	A measure of the strength of a desired signal compared to the level of background noise or interference in a system. A higher SNR indicates a more reliable and accurate signal, whereas a lower SNR suggests a higher likelihood of errors or inaccuracies in the signal detection or measurement.
Silicon nanowire (SiNW)	A nanoscale structure made of silicon, with a diameter in the nanometer range and a length much larger than its diameter. SiNWs have unique electrical and mechanical properties, making them attractive for various applications, including electronics, sensors, and biomedical devices.
Silicon photonics	An electro-optical silicon photonic integrated circuit platform built on silicon on insulator wafer technology.
Single-nucleotide polymorphism (SNP)	A variation in a DNA sequence where a single nucleotide (A, T, C, or G) differs among individuals in a population. SNPs are the most common type of genetic variation and can have various effects on traits, disease susceptibility, and drug responses, making them important markers in genetics and personalized medicine.

(Continues)

(*Continued*)

Context	Definition
Single-stranded DNA (ssDNA)	A type of nucleic acid consisting of a single, unpaired DNA strand. Unlike double-stranded DNA, which forms the iconic double helix structure, ssDNA exists as a single linear chain and is often involved in various biological processes, such as transcription, replication, and certain types of genetic research and diagnostics.
Solvent-polymer dynamics	the study of the interactions and movements of polymers in different solvent environments. Understanding these dynamics is crucial for various applications, such as in the development of new materials, drug delivery systems, and coatings, as the solvent's composition can significantly affect the behavior and properties of the polymer.
Soma	The cell body of a neuron, which contains the nucleus and other organelles that support the neuron's metabolic and structural functions. It plays a vital role in integrating incoming signals from dendrites and generating outgoing signals along the axon to communicate with other neurons or target cells.
Spatio-temporal resolution	The ability to precisely and accurately measure both the location (spatial) and the timing (temporal) of events or changes in a system. In various fields, such as imaging, neuroscience, and signal processing, high spatio-temporal resolution is crucial for capturing fine details and dynamics of phenomena with accuracy and precision.
Spin-coating	A thin-film deposition technique used to create uniform and smooth thin films on a substrate. It involves spinning the substrate at high speeds while applying a liquid solution or precursor, which spreads out and forms a thin layer due to centrifugal force, making it commonly used in microelectronics, optics, and other thin-film applications.
Sputtering	A physical vapor deposition (PVD) technique used to deposit thin films of materials onto a substrate. It involves bombarding a target material with energetic ions, causing atoms or molecules to be ejected from the target and then depositing on the substrate, enabling precise and controlled coating in various industries, including electronics, optics, and thin-film technology.
Stokes shift	Difference in the wavelength between the peak excitation and emission spectra.
Streptavidin	A protein derived from Streptomyces bacteria that has a strong affinity for biotin, a vitamin B7 derivative. Streptavidin is widely used in biotechnology and molecular biology as a valuable tool in various assays, purification, and detection methods due to its high binding specificity for biotinylated molecules.
Stromal molecule	Proteins or other biomolecules found in the stroma, the supportive connective tissue that surrounds and supports the functional cells of an organ or tissue. These molecules play essential roles in maintaining the structure, function, and

(*Continues*)

(*Continued*)

Context	Definition
	communication between different cell types within the tissue, contributing to tissue homeostasis and proper physiological function.
Sudan II	A synthetic dye with a reddish-orange color that is commonly used as a stain in histology and microscopy. It is often employed to colorize fat droplets in tissues for visualization and is widely utilized in various scientific and industrial applications.
Superconducting Quantum Interference Devices	Extremely sensitive magnetometers based on the quantum behavior of superconducting materials. They can measure extremely small magnetic fields and are used in various applications, including medical imaging (such as magnetoencephalography) and fundamental research in physics.
Surface grafting	The process of attaching or bonding functional molecules, such as biomolecules or recognition elements, onto the surface of a sensor substrate. This modification enhances the sensor's ability to interact with target molecules, improving sensitivity and selectivity. Surface grafting can involve various techniques like chemical reactions or physical adsorption, leading to a tailored sensor surface that efficiently captures and detects specific analytes, making it a crucial step in biosensor design.
Surface plasmon polariton (SPP)	Bound electromagnetic waves propagating at a dielectric-metal interface with exponential field decay normal to the interface.
Surface-enhanced Raman scattering	A vibrational spectroscopy technique with sensitivity down to the single molecule level that provides fine molecular fingerprints, allowing for direct identification of target analytes.
Synapse	A specialized junction between two neurons or between a neuron and a target cell, where electrical or chemical signals are transmitted. It allows for communication and information transfer between different cells in the nervous system, enabling the transmission of nerve impulses and the coordination of various physiological processes.
Synaptogenesis	The process of forming new synapses, or connections, between neurons in the developing nervous system. It is a fundamental aspect of brain development and plasticity, crucial for establishing neural circuits and networks that underlie various cognitive and sensory functions.
Syncytial	A type of cell or tissue structure formed by the fusion of multiple individual cells, resulting in a multinucleated cell or syncytium. Syncytial structures are commonly found in certain tissues, such as muscle and placenta, and are also observed in some viral infections, like respiratory syncytial virus (RSV), where infected cells fuse to form large multinucleated cells.

(*Continues*)

(*Continued*)

Context	Definition
Template immobilization	Attaching a specific biomolecule to a biosensor's surface. This biomolecule acts as a template, interacting selectively with target molecules. This interaction generates signals indicating target presence or concentration, enabling sensitive and specific detection in applications like medical diagnostics and environmental monitoring.
Tetramethyl ammonium hydroxide and water (TMAHW)	A mixture of the organic compound tetramethyl ammonium hydroxide (TMAH) and water. TMAH is a strong base commonly used in semiconductor manufacturing and microfabrication processes for developing photoresist patterns and etching silicon wafers. The addition of water to TMAH can help adjust its concentration and properties for specific applications.
Thalamocortical	The connection between the thalamus, a relay center in the brain, and the cerebral cortex, the outer layer of the brain responsible for higher cognitive functions. The thalamocortical pathway is essential for relaying sensory information from various parts of the body to the cerebral cortex, allowing for perception, awareness, and conscious processing of sensory stimuli.
Thermocouple	Thermocouple consisting of two dissimilar materials joined together at one end. When the junction of the two metals is heated, a voltage is generated that is proportional to the temperature.
Thermopile	Thermopile comprised of several thermocouples connected in series.
Thiol bond (also known as a sulfhydryl bond)	A type of covalent bond formed between the sulfur (S) atom of a thiol group (–SH) and another functional group or thiol. Thiol bonds are essential in various biochemical processes, including protein folding, enzyme catalysis, and the formation of disulfide bridges in proteins, contributing to the stability and structure of biomolecules.
Topographical mismatch	A situation where the surface features or patterns of two materials do not align or match properly when they are brought into contact. It often occurs in nanotechnology or microfabrication processes, where precise alignment is crucial for the successful integration of different components or layers.
Transducer	An electronic device that converts energy from one form to another.
Trypsin	A proteolytic enzyme produced in the pancreas and plays a key role in protein digestion. It cleaves peptide bonds specifically at the carboxyl side of the basic amino acids lysine and arginine, breaking down proteins into smaller peptides and amino acids for absorption in the small intestine. Additionally, trypsin is widely used in cell culture to detach adherent cells from culture surfaces for passaging and experimentation.

(*Continues*)

(*Continued*)

Context	Definition
Tumor necrosis factor-α (TNF-α)	A cytokine involved in the regulation of immune responses and inflammation. It plays a crucial role in various physiological processes, but excessive production of TNF-α can lead to harmful inflammation and is associated with certain autoimmune diseases and chronic inflammatory conditions.
Tunnel magnetoresistance	A quantum mechanical phenomenon observed in magnetic tunnel junctions, where the electrical resistance of a thin insulating barrier between two magnetic layers can vary significantly based on the relative orientation of their magnetization.
Ultraviolet (UV)/ ozone dry stripper	A surface treatment method that uses ultraviolet (UV) light in the presence of ozone to remove organic materials, such as photoresist, from substrates. It is commonly used in semiconductor manufacturing and microfabrication processes to clean and etch surfaces without the use of liquid chemicals or solvents.
Via holes	Small openings or holes in a semiconductor wafer or printed circuit board (PCB) that allow electrical connections between different layers of the device. They are filled with conductive material to establish vertical interconnections, enabling the flow of electrical signals and current between different parts of the circuit.
Waveguide	A structure that guides a light wave by constraining it to travel along a certain desired path. Typically, waveguides are hollow metal tubes (often rectangular or circular in cross section).
Well capacity	The maximum amount of charge or signal that a pixel or sensor element in a digital imaging device can hold before it becomes saturated. It's an important concept in photography and digital imaging, influencing the dynamic range and overall quality of the captured images.

List of frequently used acronyms and abbreviations

Explanation	Acronym/ Abbreviation
(3-glycidoxypropyl) trimethoxysilane	GOPS
11-mercaptoundecanonic acid	11-MUA
1-ethyl-3-(3-dimethylaminopropyl)-carbodiimide	EDC
2-(N-morpholino)ethanesulfonic acid	MES
3-Aminopropyltriethoxysilane	APTES
3-mercaptopropionic acid	MPA
4-(2-hydroxyethyl)-1-piperazineethanesulfonic acid	HEPES
4-carboxyphenylboronic acid	CPBA
Acetone	ACE
Action Potential	AP
Active Pixel Sensor	APS
Adenosine Triphosphate	ATP
Adsorptive Stripping Voltammetry	AdSV
Alternating Current	AC
Aluminum	Al
Aluminum and CMOS passivation layers	Al/pass
American Standard Code for Information Interchange	ASCII
Amplified Metal-Oxide-Semiconductor Imager	AMI
Analog Blocks	AB
Analog Front-End	AFE
Analog-to-Digital Converter	ADC
Anisotropic Magnetoresistance	AMR
Anodic Stripping Voltammetry	ASV
Antibody	Ab
Application-Specific Integrated Circuit	ASIC
Artificial Cerebrospinal Fluid	ACSF
Artificial Intelligence	AI
Auto-Balanced Bridge	ABB
Avian influenza virus	AIV
Back-End-Of-Line	BEOL
Base Stored Image Sensor	BASIS
Benzocyclobutene	BCB
Benzotriazole	BTA

(Continues)

(*Continued*)

Explanation	Acronym/ Abbreviation
Beta-Mercaptoethanol	BME
Biological CMOS	BioCMOS
Biological Field-Effect Transistor	BioFET
Biological/Chemical Field-Effect Transistor	Bio/ChemFET
Biorecognition Element	BRE
Bipolar CMOS	BCMOS
Bipolar Junction Transistor	BJT
Blood Oxygen Level Dependent	BOLD
Bovine Aortic Smooth Muscle Cell	BAOSMC
Bovine Capillary Endothelial	BCE
Bovine Serum Albumin	BSA
Bulk Charge Modulated Device	BCMD
Buried Oxide	BOX
Capacitance-to-Frequency Converter	CFC
Capacitance-to-Voltage Converter	CVC
Capacitive Coupled Device	CCD
Capacitive Trans-Impedance Amplifier	CTIA
Capacitively Coupled Instrumentation Amplifier	CCIA
Carbon Nanotube	CNT
Carbon Nanotube Field-Effect Transistor	CNT-FET
Cathodic Stripping Voltammetry	CSV
Central Nervous System	CNS
Cerium	Ce
Channel	Ch.
Channel Count	Ch. #
Charge Injection-induced Error-Free	CIEF
Charge Modulation Device	CMD
Charge Pump	CP
Charge Sharing	ChS
Charge-Based Capacitance Measurement	CBCM
Charge-Pump	CP
Charge-sensitive amplifier	CSA
Chemical Current Conveyor type 2	CCCII+
Chemical Field-Effect Transistor	ChemFET
Chemically Controlled Ring Oscillator	CCRO
Chromium	Cr
Chronoamperometry	CA
Colony-Forming Units per mL	cfu/mL
Common Mode Feedback	CMFB
Common Mode Rejection Ratio	CMRR
Complementary Metal-Oxide-Semiconductor	CMOS
Complementary to Absolute Temperature	CTAT
Computer-Aided Design	CAD
Constant-Voltage Constant-Current	CVCC
Continuous-Time Sigma-Delta Modulator	CTΣΔM
Correlated Double Sampling	CDS
Counter Electrode	CE

(*Continues*)

(*Continued*)

Explanation	**Acronym/ Abbreviation**
Current Conveyor	CC
Current Mirror	CMirr
Current Mode	CM
Current-Controlled Oscillator	CCO
Current-Controlled Stimulation	CCS
Current-to-Frequency	I-to-F
Current-to-Voltage	I-to-V
Cyanine-3	Cy3
Cyclic Voltammetry	CV
Dark Count Rate	DCR
DC Servo Loop	DSL
Deep Reactive Ion Etching	DRIE
Deionized water	DI water
Delay-Locked Loop	DLL
Sigma-Delta Modulator	ΣΔM
Deoxynucleotide Triphosphate	dNTP
Deoxyribonucleic Acid	DNA
Design Rule Check	DRC
Design-to-Manufacturing	DtM
Detection Dynamic Range	DDR
Diacrylate Bisphenol A	DABA
Dichloromethane	DCM
Dielectrophoresis	DEP
Differential	Diff.
Differential Pulse Voltammetry	DPV
Digital blocks	DB
Digital Frequency Locked Loop	DFLL
Digital Microfluidic	DMF
Digital Signal Processing	DSP
Digital-to-Analog Converter	DAC
Digoxigenin	Dig
Di-methyl-sulphoxide	DMSO
Direct Digital Synthesizer	DDS
Direct-Current	DC
Direct-Write Fabrication Process	DWFP
Discrete Fourier Transform	DFT
Dopamine	DA
Dry-Film Resist	DFR
Dual In-line Package	DIP
Duty-Cycle Resistor	DCR
Dynamic Element Matching	DEM
Dynamic Nuclear Polarization	DNP
Dynamic Range	DB
Dynamic Threshold MOSFET	DTMOST
Electro Thermal Filter	ETF
Electro Wetting-on-Dielectric	EWOD
Electrochemical Impedance Spectroscopy	EIS

(*Continues*)

(*Continued*)

Explanation	Acronym/ Abbreviation
Electroencephalogram	EEG
Electroless Nickel Immersion Gold	ENIG
Electrothermal Filter	ETF
Embryoid Body	EB
Embryonic Fibroblast Cells	mEF
Embryonic Stem Cells	mESC
Enhanced CBCM	ECBCM
Ethanol	EtOH
Ethylcellulose	EC
Ethylenediamine Pyrocatechol	EDP
Ethylenediaminetetraacetic Acid	EDTA
Exclusive OR	XOR
Extracellular Action Potential	EAP
Fast-Scan Cyclic Voltammetry	FSCV
Fast-Settling Miller Compensation	FSMC
Fetal Bovine Serum	FBS
Fiber-Optic Faceplate	FOF
Field-Effect Transistor	FET
Field-Programmable Gate Array	FPGA
Figure-of-Merit	FoM
Finite Difference Time-Domain	FDTD
Finite Impulse Response	FIR
Finite State Machine	FSM
Finite-Difference Time-Domain	FDTD
Fixed-Pattern Noise	FPN
Floating Diffusion	FD
Food and Drug Administration	FDA
Frames per second	fps
Frequency Control Register	FCR
Frequency Locked Loop	FLL
Frequency Response Analysis	FRA
Frequency-Division Multiplexing	FDM
Fully Differential	FDiff.
Gadolinium	Gd
Gated-Ring Oscillators	GRO
General Purpose Input/Output	GPIO
Genomic DNA	gDNA
Giant Magnetoimpedance	GMI
Giant Magnetoresistance	GMR
Glucokinase	GLK
Glutamate Oxidase	GOx
Glutaraldehyde	Glu
Gold	Au
G-Protein Coupled Receptor	GPCR
Graphene field-effect transistor	GFET
Graphical User Interface	GUI
Grounded Electrode	GndE

(Continues)

(*Continued*)

Explanation	Acronym/ Abbreviation
Hepatitis B Virus	HBV
High Density Microelectrode Array	HD-MEA
High-Frequency Interference Rejection	HFIR
High-Frequency Switch-Capacitor	HFSC
High-Pass Filter	HPF
Horizontal Cell Manipulation	HCM
Horseradish Peroxidase	HRP
Human Adenovirus C	AdvC2
Human Adenovirus E	AdvE
Human Embryonic Kidney	HEK
Human Immunodeficiency Virus	HIV
Human mammaglobin	hMAM
Human Parainfluenza Virus	HPIV
Human Serum Albumin	HSA
Immunoglobulin G	IgG
Inductor-Capacitor	LC
Influenza A	FluA
Influenza B	FluB
Injection Locked Frequency Divider	ILFD
Injection-Locked Oscillator	ILO
Inner Helmholtz Plane	IHP
In-phase	I
In-phase and Quadrature	IQ
Input Diode	ID
Input Dynamic Range	IDR
Input/Output	I/O
Instruction Multiple Data	SIMD
Instrumentation Amplifier	IA
Integrated Circuit	IC
Inter Layer Dielectric	ILD
Interdigitated Electrode	IDE
Interdigitated Microelectrode Array	IDMEA
Interferon-g	IFN-g
Intermediate Frequency	IF
Internal Register of PAC	PAR
Internet of Things	IoT
Inter-Pulse Charge Control	IPCC
Ion-Selective Field-Effect Transistor	ISFET
Isopropyl Alcohol	IPA
Isotropic Conductive Adhesive	ICA
Kelvin Connection DEM	KC-DEM
Lab-on-Chip	LoC
Layer-By-Layer	LBL
Lead	Pb
Light Emitting Diode	LED
Limit of Detection	LoD
Linear Feedback Shift Register	LFSR

(*Continues*)

(*Continued*)

Explanation	Acronym/ Abbreviation
Linear Scan Voltammetry	LSV
Local Field Potential	LFP
Local Oscillator	LO
Localized Backside Etching	LBE
Lookup Table	LUT
Loop Filter	LF
Loop-Mediated Isothermal Amplification	LAMP
Low Dropout	LDO
Lower Limit of Detection	LLoD
Low-Noise Amplifier	LNA
Low-pass Filter	LPF
Low-Temperature Co-fired Ceramic	LTCC
Luciferase	LUS
Lysogeny Broth	LB
Magnetic Correlated Double Sampling	MCDS
Magnetic Nanoparticle	MNP
Magnetic Resonance Imaging	MRI
Magnetorelaxometry	MRX
Magnetoresistive Correlated Double Sampling	MRCDS
Mesenchymal Stem Cell	MSC
Metal Oxide Semiconductor Field-Effect Transistors	MOSFET
Metal-Insulator-Metal	MIM
Metal-Organic Framework	MOF
Metal-Oxide-Semiconductor	MOS
Metal-Oxide-Semiconductor Field-Effect Transistor	MOSFET
Methanol	MeOH
Micro Lead Frame Package	MLP
Micro-controller Unit	MCU
Microelectrode Array	MEA
Micro-electromechanical System	MEMS
Microfluidics	μF
Micro-NMR	µNMR
Micro-Ring Resonators	μRR
Microwave	MW
Mixing Cell Manipulation	MCM
Molecularly Imprinted Polymer	MIP
Multi-Labs-on-a-Single Chip	MLoC
Multiple Input/Single Output	MISO
Multiply-accumulate (MAc)	MAc
Multi-walled carbon nanotube	MWCNT
Mycobacterium Avium Complex	MAC
Nano-Imprint Lithography	NIL
Nanoparticle	NP
Nanorod	NR
Nanowire	NW
Nanowire field-effect transistor	NW-FET
N-channel Metal-Oxide Semiconductor	NMOS

(*Continues*)

(*Continued*)

Explanation	**Acronym/ Abbreviation**
Near Field Communication	NFC
Near Infrared	Near-IR
Negative Dielectrophoresis	nDEP
Neonatal Rat Ventricular Myocyte	NRVM
New Delhi Metallo-β-lactamase	NDM
Nickel	Ni
Nicotinamide Adenine Dinucleotide + Hydrogen	NADH
Nitrilotriacetic Acid	NTA
Noise Efficiency Factor	NEF
Non-Selective	NS
Not Available	NA
N-terminal pro-Brain Natriuretic Peptide	NT-proBNP
Nuclear Magnetic Resonance	NMR
Numerically-Controlled Oscillator	NCO
Open-cavity Micro Lead Frame Package	MLP
Operational Amplifier	OpAmp
Operational Transconductance Amplifier	OTA
Optical Density	OD
Osmium-polyvinylpyridine gel polymer containing HRP	Os-HRP
Outer Helmholtz Plane	OHP
Oxide/Nitride/Oxide	ONO
Oxide-Semiconductor Field-Effect Transistor	OSFET
Parts per billion	ppb
parts per million	ppm
Passive Pixel Sensor	PPS
P-channel Metal-Oxide Semiconductor	PMOS
Peptide Nucleic Acid	PNA
Personal Computer	PC
Phase Domain Sigma-Delta Modulator	PΣΔM
Phase-Frequency Detector	PFD
Phase-Locked Loop	PLL
Phase-to-Amplitude Converter	PAC
Phosphate-Buffered Saline	PBS
Photomultiplier Tube	PMT
pH-to-Frequency Conversion	pH-FC
pH-to-Time Converter	pH-TC
pH-to-Time-to-Voltage converter	pH-TVC
Pixel array	PA
Plasma Enhanced Chemical Vapor Deposition	PECVD
Platinum Nanostructures	Pt-nanoS
Point-of-Care	PoC
Poly(3,4-ethylenedioxythiophene) doped with poly(styrenesulfonate)	PEDOT:PSS
Poly(allylamine) Hydrochloride	PAH
Poly(amidoamine)	PAMAM
Poly(cyanopropylmethylsiloxane)	PCM
Poly(dimethylsiloxane)	PDMS
Poly(epichlorohydrin)	PECH

(Continues)

(*Continued*)

Explanation	Acronym/ Abbreviation
Poly(ethylene glycol)	PEG
Poly(ethyleneimine)	PEI
Poly(methyl-methacrylate)	PMMA
Poly(N-isopropylacrylamide)	PNIPAM
Poly(sodium styrene sulfonate)	PSS
Polyelectrolyte Multilayer	PEM
Polyether urethane	PEUT
Polymerase Chain Reaction	PCR
Polyurethane	PUR
Polyvinylchloride	PVC
Positive Dielectrophoresis	pDEP
Power Amplifier	PA
Power Supply Rejection Ratio	PSRR
Printed Circuit Board	PCB
Process Voltage Temperature	PVT
Programmable Gain Amplifier	PGA
Programmable Gate	PG
Proportional Integral	PI
Proportional Integral Derivative	PID
Proportional To Absolute Temperature	PTAT
Propylene Glycol Monoether Acetate	PGMEA
Prostate-Specific Antigen	PSA
Pseudo Digital	pD
Pseudo-differential	PsDiff.
Pseudo-Reference Electrode	PRE
PTAT/Resistance	PTAT/R
Pulse Width Modulation	PWM
Quadrature	Q
Quadrature Voltage-controlled Oscillator	QVCO
Quantum dot	Qdot
Quantum Efficiency	QE
Quasi-Interdigitated Electrode	Q-IDE
Quiescent Current	IQ
Radio Frequency	RF
Radio Frequency Identification	RFID
Radio Frequency Silicon on Insulator	RFSOI
Reactive Ion Etching	RIE
Reactive Physical Vapor Deposition	RPVD
Readout Electrode	ReadE
Receiver	RX
Reference Electrode	RE
Reference Field-Effect Transistor	REFET
Refractive Index	RI
Refractive Index Change	RIU
Relaxation Oscillator	Rel.O
Resistive-TIA	RTIA
Respiratory Syncytial Virus	RSV

(*Continues*)

(*Continued*)

Explanation	**Acronym/ Abbreviation**
Return-to-Common-Mode	RCM
Revolutions Per Minute	RPM
Ribonucleic Acid	RNA
Ring Oscillator	RO
Ripple Rejection Loop	RRL
Saline Sodium Citrate	SSC
Scanning Electron Microscope	SEM
Secondary Alcohol Dehydrogenase	2° ADH
Serial Peripheral Interface	SPI
Sigma-Delta Analog-to-Digital Converter	ΣΔ ADC
Signal-to-Noise Ratio	SNR
Silicon	Si
Silicon Nanowire	SiNW
Single-Nucleotide Polymorphism	SNP
Single-Photon Avalanche Diode	SPAD
Single-stranded DNA	ssDNA
Single-Walled Carbon Nanotube	swCNT
Slave Select	ss
Small Outline Transistor	SOT
SNP-based LAMP	sbLAMP
Sodium Dodecyl Sulfate	SDS
Sodium Hydroxide	NaOH
Spermidine/Spermine N1 Acetyltransferase	SSAT
Spin-Valve	SV
Static Induction Transistor	SIT
Static Random-Access Memory	SRAM
Stereolithography	SLA
Subthreshold	Subth.
Successive Approximation Register	SAR
Superconducting Quantum Interference Device	SQUID
Surface Acoustic Wave	SAW
Surface Plasmon Polariton	SPP
Surface-Mounted Device	SMD
Switch Matrix	SM
Switch-Capacitor	SC
Switch-Capacitor Stimulation	SCS
Switched Capacitor Loop Filter	SCLF
System on Chip	SoC
Tape Automated Bonding	TAB
Temperature to Digital Converter	TDC
Temperature-to-Frequency Converter	TFC
Tetramethyl Ammonium Hydroxide and Water	TMAHW
The ground	GND
Three dimensional	3D
Time-Division Multiplexing	TDM
Time-to-Digital Converter	TDC
Titanium nitride	TiN

(*Continues*)

(*Continued*)

Explanation	Acronym/ Abbreviation
Toluene	TOL
Toluidine blue O	TBO
Topmost metal with one passivation layer	Al/1pass
Total Active Area	TAA
Total Sensing Area	TSA
Transimpedance Amplifier	TIA
Transmitter	TX
Transmitter	TX
Traveling wave Dielectrophoresis	twDEP
Triangular Voltage Analysis	TVA
Tryptic Soy Broth	TSB
Tumor Necrosis Factor	TNF
Tunnel Magnetoresistance	TMR
Two-dimensional	2D
Two-Polysilicon-Four-Metal	2P4M
Type II Positive Current Conveyor	CCII+
Ultraviolet	UV
Ultraviolet Light Emitting Diode	UV LED
Universal Serial Bus	USB
Utah Electrode Array	UEA
Variable Gain Amplifier	VGA
Velocity Saturation	VS
Vertical Cell Manipulation	VCM
Voltage Controlled Delay Line	VCDL
Voltage Gain Amplifier	VGA
Voltage Window Comparator	VWC
Voltage/Current-Controlled Stimulation	V/CCS
Voltage-Controlled Oscillator	VCO
Voltage-to-Current	V-to-I
Working Electrode	WE
Yeast Peptone Dextrose	YPD
Zero Crossing	ZC
Zinc	Zn
β-D-dodecyl-N-maltoside	DDM
β-lactoglobulin	β-LG
β-nicotinamide adenine dinucleotide phosphate	$NADP^+$

Parameters

Parameter	Definition
$(W/L)_{Mi}$	Aspect ratio of transistor M_i
x_H	The distance of the OHP from the metal electrode
Ψ_B	The potential of the bulk of the solution
Ψ_S	The potential of the oxide surface
Ψ_o	The insulator-electrolyte potential
χ^{sol}	The surface dipole potential of the solution
$\varnothing_{\text{PPF}}$	Phase shift of poly-phase filter
$\varnothing_{\text{WB}}$	Wien bridge phase shift
μ	Mobility
μ_{H}	Hall mobility
μ_{p}	The hole mobility coefficient
$1/f$	Flicker noise
A_{amp}	The gain of amplifier
A_{att}	Attenuation factor
A_{c}	Overlapping area of the electrodes of a capacitor
A_{CCO}	The gain of the CCO
A_{I}	The gain of current mirror
a_{IDE}	The width of the stripped area of interdigitated electrodes
A_{inv}	The gain of an inverter
A_{TIA}	The gain of a transimpedance amplifier
B	External magnetic field
B_0	Static magnetic field
B_1	Magnetic field orthogonal to B_0
B_{bead}	Induced magnetic field from the bead
B_{mag}	Generated magnetizing field
C_{bio}	The total capacitance of the surface functionalization layer, including the binding recognition elements, grafting molecules, and analyte binding, as well as any contribution from the Helmholtz layer
C_{cell}	Capacitance of cell layer
C_{Dif}	Differential double layer capacitance, or interfacial capacitance
C_{dl}	Electric double-layer capacitance
C_{E}	The spreading capacitance of the electrode to the electrolyte
C_{f}	Fringe capacitance
C_{fb}	Feedback capacitor
C_{gm}	The total capacitance of the conductor above the cell layer to all neighboring conductors (including voltage biases, power supply, and ground metal) except for the sensing electrodes

(Continues)

(*Continued*)

Parameter	Definition
C_{Gouy}	Diffuse layer capacitance
C_{gs}	Gate-source capacitance
C_{I}	The C_{dl} created at the electrode–electrolyte interface
C_{ins}	Insulating layer capacitance
C_{int}	Integrating capacitor
$C_{int\pm}$	The integrating capacitances used in a differential mode
C_{inter}	The capacitance of the dielectric interconnect
CLK	Clock
C_{os}	Offset capacitor
C_{ox}	Gate-oxide capacitance per unit area
C_{pi}	The parasitic capacitances of the node N_i
C_{PPF}	Capacitor of poly-phase filter
C_{R}	Reference capacitance
C_{S}	Sensing capacitance
C_{SC}	Capacitor of switched-capacitor circuit
C_{sensed}	Sensed capacitance
C_{sol}	The capacitance of the electrolyte solution
$C_{standing}$	Standing capacitance
C_{Stern}	Stern capacitance
C_{stray}	The stray capacitance of the working electrode terminal
C_{surf}	Surface capacitance
C_{tot}	Total capacitance
C_{WB}	Capacitor of Wien bridge sensor
D	Thermal diffusivity of the silicon
d_{c}	The distance between the plates of a capacitor
d_{IDE}	The distance between interdigitated electrodes' fingers
e_{n}	Current error
En_{i}	The *i*th command line
E_{ref}	The potential of the reference electrode
f	Frequency
f_{0}	Initial frequency
f_{CCO}	The output frequency of CCO
F_{CCO}	The average of the output pulse frequencies of CCO
f_{int}	Integrating frequency
f_{L}	Larmor frequency
f_{Out}	Frequency of the output pulse
f_{ref}	Reference frequency
f_{S}	Sampling frequency
G_{H}	Geometrical correcting factor for finite-size Hall devices
G_{m}	Transconductance
H_B^+	The proton in the bulk of the solution
I'_{R}	The average of the mirrored reference current
I'_{S}	The average of the mirrored sensing current
I_{B}	Bias current
i_{CCO}	The input current of CCO
I_{Charge}	Current of current generator
I_{inj}	Injection current

(Continues)

(*Continued*)

Parameter	Definition
I_{mag}	Current passing through a pair of metal wires on both sides of the N-well Hall-effect sensor
I_{osc}	The current of the oscillator
I_{PTAT}	PTAT current
i_R	Reference current
i_S	Sensing current
J_0	The zeroth Bessel function of the first kind
K_a	Acid dissociation constant
k_B	Boltzmann constant
K_b	Base dissociation constant
K_n	The dissociation constant of primary amine sites
K_{VT}	Ratio of the PTAT and V_{REF}
L_D	Debye length
L_H	Length of Hall plate
l_{IDE}	The length of interdigitated electrodes' fingers
m	Resonance order
n	The subthreshold slope parameter
n^0	The number concentration of each ion in the bulk
N_{Count}	Total number of pulses counted by the counter
N_{div}	The fixed frequency division ratio of a frequency divider
n_{eff}	Effective index of refraction
n_{IDE}	Number of interdigitated electrodes' fingers
n_{medium}	Refractive index dielectric medium
N_n	The surface density of primary amine sites
N_S	The surface density of silanol sites
$n_{waveguide}$	Refractive index of waveguide
q	Electronic charge
Q	Total charge
Q_B	The charges in the interface states
q_n	The output pulse of the comparator
Q_{ox}	The charges in the oxide
Q_{ss}	The charges in surface states
$R(T_0)$	Resistor at a reference temperature of T_0
R_d	The internal drain resistance
R_D	Drain current
$R_{discharge}$	Discharge resistor
R_E	The spreading resistance of the electrode to the electrolyte
R_{Isense}	Current-sense resistor
$R_N(T)$	Resistor with a negative temperature coefficient
R_P	Potentiometer
$R_P(T)$	Resistor with a positive temperature coefficient
$R_{PPF}(T)$	Resistor of poly-phase filter
R_s	The internal source resistance
R_{SC}	Average resistance of switched-capacitor circuit
R_{sol}	The resistance of the electrolyte solution
$R_{WB}(T)$	Resistor of Wien bridge sensor
S_V	Voltage-related sensitivity
T	Absolute temperature

(*Continues*)

(*Continued*)

Parameter	Definition
T_0	Nominal value of temperature
TC_1 and TC_2	First-order and second-order temperature coefficients
t_{IDE},	The thickness of interdigitated electrodes' fingers
T_{int}	Integration time
T_{osc}	Period of oscillator
T_S	Sampling period
V_{Am}	The voltage dropped over the ammeter
V_{BE}	Base-emitter voltage
$V_{bias,H}$	Bias voltage applied to the Hall plate
$V_{D,sat}$	The voltage point when the drain current saturates
V_{dd} and V_{ss}	Dual supply voltages
V_{DS}	Drain-source voltage
V_{GS}	Gate-source voltage
$V_{IN,PPF}$	Input voltage of poly-phase filter
$V_{IN,WB}$	Input voltage of Wien bridge sensor
V_{off}	Offset voltage
$V_{off\pm}$	The offset voltages in a differential mode
V_{osc}	Voltage of oscillator
V_{Out}	Output voltage
$V_{OUT,\ PPF}$	Output voltage of poly-phase filter
$V_{OUT,WB}$	Output voltage of Wien bridge sensor
$V_{Out\pm}$	Differential output voltages
$v_R(t)$	The instantaneous voltages across the reference capacitor
$v_S(t)$	The instantaneous voltages across the sensing capacitor
v_{sat}	The saturated carrier velocity
V_{st}	Stimulating voltage
V_T	Thermal voltage
V_{Th}	Threshold voltage
W_H	Width of Hall plate
w_{IDE}	The width of interdigitated electrodes' fingers
$w_{ss}(x,y)$	Displacement of the flexible electrode
Y_S	Sensing admittance
z	The valence of the ion
Z_{in}	Input impedance
Z_S	Sensing impedance
α_s	Seebeck coefficient
β_{int}	Intrinsic buffer capacity
β_p/β_n	The ratio of PMOS and NMOS aspect ratios
γ	Gyromagnetic ratio of the nuclei
δ	Conductivity
ΔC	Capacitance changes (ΔC=C_S-C_R)
ΔC_{Max}	The highest achievable capacitance change
ΔC_{Min}	The lowest achievable capacitance change
$\Delta C_{tot}(f_0)/C_{tot}$	The relative shift in the total capacitance with respect to the C_{tot}
Δf	Frequency shift
$\Delta f/f_0$	Relative frequency shift

(*Continues*)

(*Continued*)

Parameter	Definition
Δt_{int}	The charge integration time
ΔT_{SB}	Thermocouple voltage difference
ΔV_{BE}	The difference between base-emitter voltage
ΔV_{chem}	pH-related potential change
ΔV_{GS}	Gate-source voltage difference
ΔV_{SB}	Seebeck voltage
ε	Permittivity
ε_0	The vacuum permittivity
ε_r	The relative permittivity
θ	Angle of incidence of light entering the waveguide
θ_{B}	Tip angle
θ_C	Critical angle of incidence of light entering the waveguide
λ_{IDE}	IDE spatial wavelength or the periodicity of the fingers
λ_{res}	Resonant wavelength
$\sigma C(t)$	Time-varying sensing capacitance change
σ_{dl}	Double-layer charge
σ_{n}	Noise
σ_{s}	The charge in the semiconductor
τ	Time constant
τ_{d}	The delay generated by a comparator
$\mathbf{\Phi}_{\mathrm{f}}$	The potential difference between the Fermi levels of doped and intrinsic silicon
$\mathbf{\Phi}_i$	Clock pulse
$\mathbf{\Phi}_{\mathrm{Si}}$	Silicon electron work function
ω	Angular frequency
ω_{inj}	The angular frequency of the injection source
φ_{ETF}	Temperature-dependent phase shift

Chapter 1
Introduction

In recent years, complementary metal-oxide-semiconductor (CMOS) biosensors have emerged as a groundbreaking technology with immense potential in various fields, particularly in healthcare and diagnostics. The importance of CMOS biosensors lies in their ability to revolutionize the way we approach medical and diagnostic challenges. This chapter aims to highlight the significance of CMOS biosensors by exploring their unique attributes and the numerous advantages they bring to the table. CMOS biosensors offer several key benefits, including their compatibility with mass production, leading to cost-effective devices that can be used as single-use diagnostic tools. They rely on recognition elements to transduce biological activity into detectable physical changes, which can be accurately measured by CMOS technology. Throughout this book, we will delve into the diverse modalities employed by CMOS biosensors for detecting these physical changes. These modalities encompass electrical methods such as capacitive, impedance, or charge-based sensing, as well as optical, magnetic, and thermal techniques, providing versatile options for capturing and quantifying biological signals or analyte interactions. Additionally, this chapter will provide an overview of the organization of the book, outlining the different sections and their respective focus. The subsequent chapters will delve into specific applications, case studies, and cutting-edge research, providing a comprehensive understanding of the potential and impact of CMOS biosensors in the field of healthcare and diagnostics.

1.1 Why CMOS?

One of the key advantages of CMOS biosensors is their low cost. Traditional biosensing technologies often require expensive and complex fabrication processes, limiting their accessibility and widespread use. In contrast, CMOS biosensors leverage the mature CMOS platform, which has benefited from significant investments. This substantial investment has resulted in highly reliable and cost-effective manufacturing processes, enabling mass production of CMOS bio-sensors and actuators at a fraction of the cost of alternative technologies (Figure 1.1) [1].

Furthermore, CMOS biosensors exhibit exceptionally low power consumption. This characteristic is particularly advantageous for portable and point-of-care (PoC) applications. The ability to operate with minimal power requirements allows

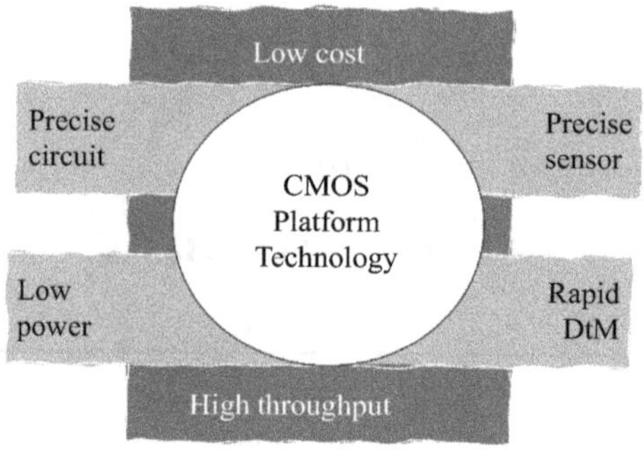

Figure 1.1 Advantages of CMOS DtM process

for extended battery life in handheld devices and enables the development of wearable biosensing systems that can continuously monitor vital signs and health parameters. Another notable advantage of CMOS biosensors is their rapid design-to-manufacturing (DtM) cycle. The CMOS platform's well-established infrastructure and standardized fabrication processes streamline the development process, significantly reducing the time required from initial design to final manufacturing. This accelerated timeline facilitates faster prototyping, optimization, and commercialization of CMOS biosensor devices, ultimately benefiting both researchers and end-users. Additionally, CMOS biosensors offer a novel feature that enables high-precision readout circuitry integration. The inherent compatibility of CMOS technology with sensor integration allows for the seamless integration of sensing elements with readout circuitry on the same chip. This integration enhances the overall performance and sensitivity of the biosensors while reducing noise and signal interference, ultimately leading to more accurate and reliable measurements.

It is noteworthy that CMOS circuits and sensors have been extensively documented, and innovative CMOS-based devices for life science applications have been developed, drawing upon these rich resources. A CMOS chip can be designed and implemented for biosensing or actuating purposes, requiring an interface board and graphical user interface (GUI), as depicted in Figure 1.2. The electronic engineer will take charge of integrated circuit (IC) design, hardware, and software requirements. On the other hand, the development of this device relies solely on classical knowledge of chemistry and biology for chemical and biological sensing/actuating purposes. The necessary chemical and biological protocols are typically designed and executed above the CMOS chip, in collaboration with life scientists. This promising and progressive field of biological CMOS (BioCMOS) chip design is rapidly advancing, aligned with the progress of the standard CMOS industry and the advancements in biology and chemical sciences. It benefits from talented ideas and interdisciplinary collaborations, driving innovation in the domain.

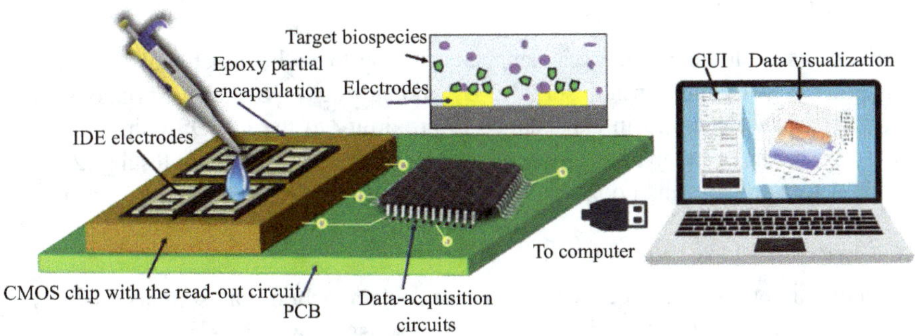

Figure 1.2 Schematic of a CMOS biochemical sensing system's architecture. The system comprises a CMOS chip with interdigitated electrodes (IDEs) placed on the topmost metal layer, partially encapsulated with epoxy to protect the pads from liquid samples. This CMOS chip is located on a Printed Circuit Board (PCB) alongside a microcontroller (MCU) responsible for control signals and data acquisition. The acquired data is transmitted to a custom-made GUI via a universal serial bus (USB), where it is visualized for analysis.

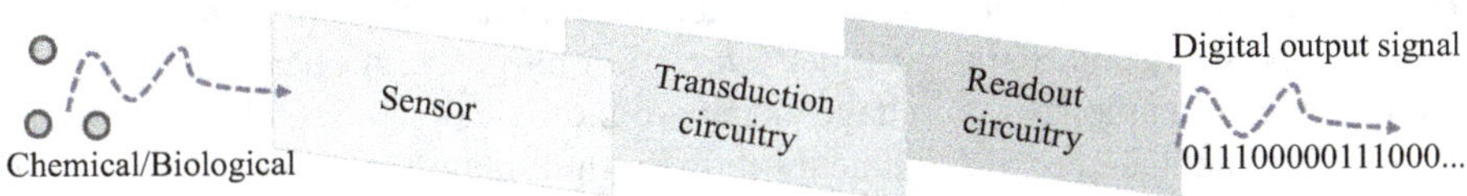

Figure 1.3 BioCMOS biosensor architecture with biological/chemical input and digital output

1.2 Sensors-on-CMOS

In the field of CMOS biosensors, various sensor types have been developed, each offering unique capabilities and applications. The following section will illustrate the CMOS biosensor architecture and describe different types of CMOS sensors, including capacitive, electrochemical sensors, optical, magnetic, nuclear magnetic resonance (NMR), biological/chemical field-effect transistor (Bio/ChemFET), and biopotential sensors. Additionally, advancements in the field have led to the development of mechanical sensors, such as cantilevers, and thermal sensors integrated with hotplates on CMOS chips. It is noteworthy that mechanical sensors have not been covered in this book.

The CMOS biosensor architecture (e.g., capacitive sensor), as depicted in Figure 1.3, typically consists of three main components: the sensing element, the transduction circuitry, and the readout circuitry. The sensing element is responsible for detecting the target analyte or biological interaction, while the transduction circuitry converts the measured physical or chemical signal into an electrical signal. The readout circuitry processes the electrical signal, amplifies it, and provides the output for analysis and interpretation.

1.2.1 Capacitive CMOS sensors

These sensors exploit changes in capacitance resulting from the target analyte's presence or interaction with the sensing surface. By monitoring capacitance variations, these sensors can detect changes in capacitance in proximity to the CMOS chip. Capacitive CMOS sensors have found applications in cell growth monitoring, drug testing [2,3], and bacteria detection [4]. This book introduces CMOS capacitive sensors in Chapter 4.

1.2.2 Electrochemical CMOS sensors

The main difference between capacitive and electrochemical sensors is the presence of an insulation layer above capacitive sensors, which is not found in electrochemical sensors (see Figure 1.4(a) and (b)). Electrochemical sensors [5,6] measure changes in impedance, current, or voltage resulting from the binding or interaction of the target analyte after applying sine wave voltage, direct-current (DC) voltage, or DC. These sensors provide label-free and real-time detection of biological and chemical entities. They hold significant value in monitoring biomolecular interactions, deoxyribonucleic acid (DNA) hybridization, and immunoassays, with diverse applications in medical diagnostics and drug discovery.

1.2.3 Biopotential sensors

The voltage generated by biological cells such as neuronal cells can be detected for neurobiology study and brain-computer interaction purposes. A charge amplifier is used to detect and amplify the voltage change of biological cells (see Figure 1.4(c)) [6]. CMOS enables the development of high throughput bipotential recording sensors and circuits for parallel neuronal signal recording and processing purposes.

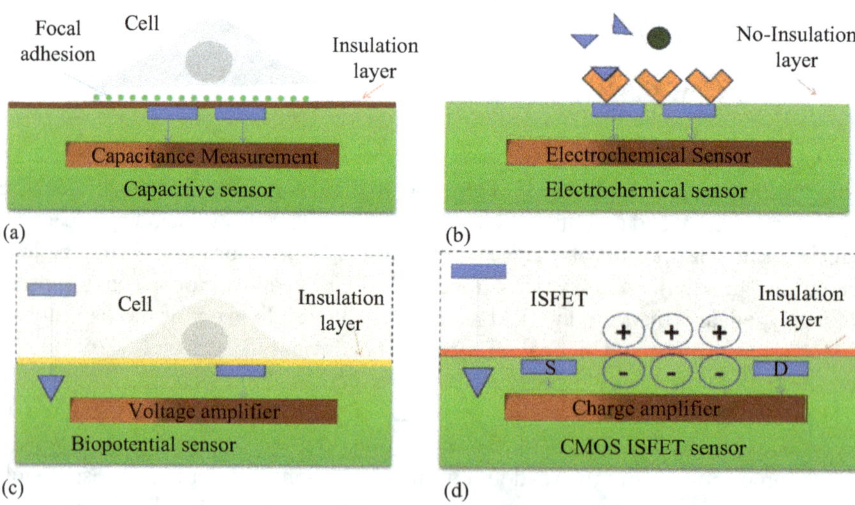

Figure 1.4 Illustration of CMOS: (a) Capacitive sensors, (b) electrochemical (voltammetric and impedimetric) methods, (c) biopotential recording methods, and (d) ISFET sensors

1.2.4 *Biological/chemical field-effect transistor (Bio/ChemFET) sensors*

Bio/ChemFETs encompass a range of transistor families designed for detecting field effects induced by charges. These transistors utilize diverse intermediate layers on their gates to serve as interfaces for biological or chemical interactions. Notable examples include ion-selective field-effect transistors (ISFETs), Enzyme-FETs, Gene-FETs, and Cell-FETs. These variants are utilized for detecting ions, enzyme reactions, DNA molecules, and cells, respectively. While the specific targets differ among these Bio/ChemFET-based sensors, they all operate based on the principles governing ISFETs' functionality.

The ISFETs are used to directly measure the charge without inserting current through the sample (see Figure 1.4(d)). This charge in the proximity of the CMOS chip increases the charge density and accordingly affects the transistor characteristics. The extra charge can increase the drain current as seen in Figure 1.4(d). These sensors have been widely used for pH change dedicated to life science applications such as DNA sequencing [7–9].

1.2.5 *Optical CMOS sensors*

Optical CMOS sensors utilize light-based detection methods to measure changes in absorbance, fluorescence, or refractive index (see Figure 1.5(a)). By integrating photodetectors and light sources on a CMOS chip, these sensors enable highly sensitive and miniaturized optical detection systems. They have found applications in

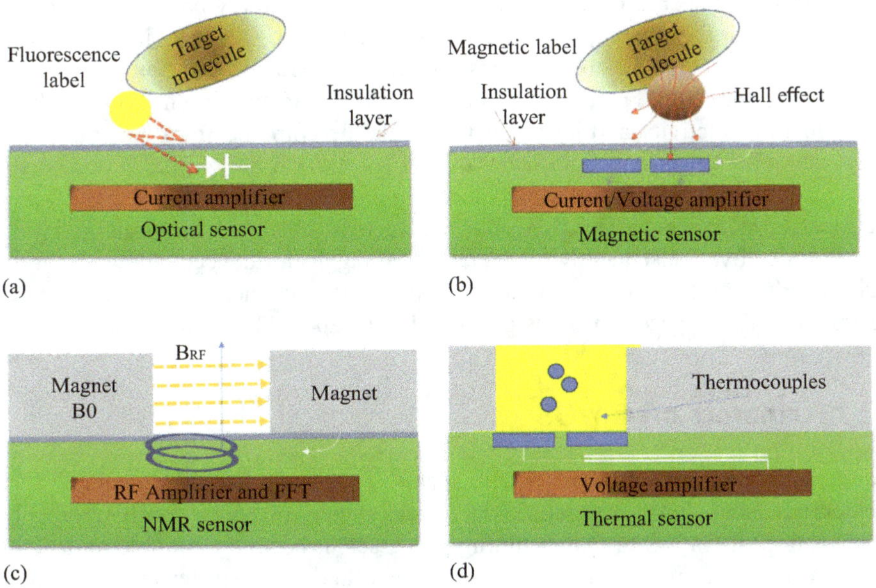

Figure 1.5 Illustration of CMOS: (a) optical, (b) magnetic, (c) NMR, and (d) thermal sensing methods

various fields, including biomedical imaging, DNA sequencing, and PoC diagnostics [10–13]. It is worth noting that the detection of target molecules using these sensors sometimes requires fluorescence labels for highly selective sensing purposes.

1.2.6 Magnetic CMOS sensors

Magnetic CMOS sensors utilize magnetic field variations to detect target analytes. By integrating magnetic sensing elements on the CMOS chip, these sensors can detect magnetic nanoparticles (MNPs) or changes in magnetic properties resulting from biological interactions (see Figure 1.5(b)). Magnetic micro- and nanoparticles act as magnetic labels, aiding in the detection process. Magnetic CMOS sensors find applications in bioassays, magnetic particle tracking, and molecular diagnostics [14,15].

1.2.7 Nuclear magnetic resonance (NMR) CMOS sensors

NMR is a highly precise sensing method used for the detection of molecules or atoms in a sample by observing resonance changes. The integration of NMR with a CMOS chip using a small magnet has opened avenues for detecting molecules at low resonance frequencies in miniaturized NMR systems (see Figure 1.5(c)) [16]. These integrated NMR CMOS sensors find extensive use in PoC disease diagnostics and environmental analysis due to their sensitivity and accuracy.

1.2.8 Thermal CMOS sensors

CMOS ICs have been effectively employed to develop highly precise temperature measurement sensors using parasitic metal-oxide-semiconductor (MOS) transistors. These sensors are capable of measuring the temperature of the environment in proximity to the CMOS chip (see Figure 1.5(d)) [17]. However, challenges related to thermal conductivity and heat transfer inside the chip and from the chip to the surrounding air exist. One significant area of research in the field of biosensors is focused on thermal thermocouples, which have garnered attention. These on-chip thermocouples can be affected by the presence of chemical materials and their reactions nearby. Consequently, the detected voltage is influenced, allowing for the quantification of chemical or biological materials and their reactions. Detailed information about these sensors is discussed in Chapter 11 of our study.

1.3 Actuators-on-CMOS

Standard CMOS technology offers the capability to manipulate biological cells and molecules using both magnetic and electrostatic methods. Furthermore, by employing post-CMOS micromachining processes, it becomes possible to develop other types of actuators utilizing mechanical and thermal techniques. These advancements open up exciting possibilities for various applications in the field of biotechnology and microfluidics.

1.3.1 Magnetic actuators

CMOS technology can be utilized to develop micro-coils that are connected to current sources. These micro-coils generate programmable magnetic fields, enabling the manipulation of magnetically labeled particles. By controlling the magnetic field, CMOS-based magnetic actuators can push, attract, or repel particles with high precision and efficiency. This capability has applications in magnetic particle separation, sorting, and manipulation (Figure 1.6 (a)) [18].

1.3.2 Electrostatic actuators

CMOS technology can generate electrical fields for realizing techniques like dielectrophoresis (DEP) or electrophoresis. Arrays of electrical electrodes can be implemented on the CMOS chip to precisely control the non-uniform electrical field [19]. This controlled field can manipulate the position and movement of cells or other materials, allowing for precise spatial manipulation and sorting (Figure 1.6(b)).

1.3.3 Other potential actuating techniques

1.3.3.1 Thermal actuators

CMOS technology can be leveraged to develop on-chip hotplates, which can significantly increase the temperature in localized areas. These hotplates enable the creation of thermal actuators and sensors in a compact and integrated form. By applying precise temperature changes, CMOS-based thermal actuators can induce controlled thermal effects for various applications such as microfluidics, chemical reactions, and micro-actuation [20].

1.3.3.2 Mechanical actuators

CMOS technology can be used to fabricate cantilevers that can be mechanically vibrated or actuated. These cantilevers, often integrated with sensing elements, can be utilized to manipulate or interact with biological materials or cells. By controlling the mechanical vibration or deflection of the cantilever, CMOS-based mechanical actuators can exert forces, perform sensing tasks, or induce specific responses in the target materials [20].

Mechanical and thermal actuators have not been covered in this book.

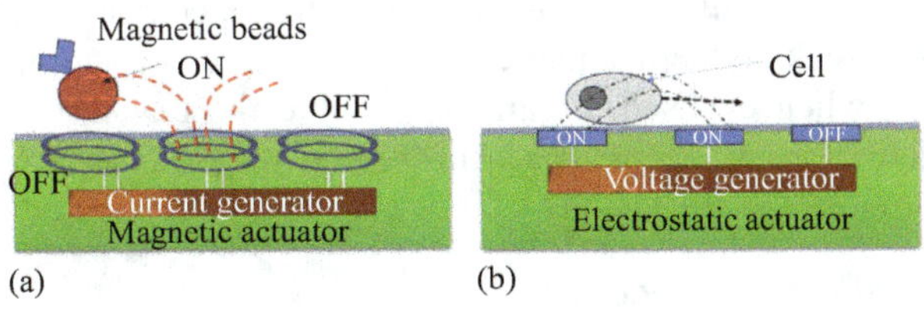

Figure 1.6 Illustration of CMOS: (a) magnetic, (b) electrostatic, actuating methods

1.4 CMOS interface circuitries

CMOS interface circuits play a critical role in CMOS biosensors as they facilitate the extraction and processing of signals from the sensors [21,22]. These interface circuits are designed using high-precision analog and digital circuitry, forming the core of the CMOS chip design process using tools like Cadence.

While numerous circuitries have been developed for various applications, the design of high-resolution, low-power consumption, and high-speed readouts or actuators remains a significant challenge. Each type of sensor and specific application may require a unique design approach to meet the desired performance requirements. In the upcoming sections of the book, we will delve into the intricacies of CMOS interface circuit designs for biosensors. We will explore different design strategies, methodologies, and techniques that have been proposed and developed to address the challenges associated with high resolution, low power consumption, and high-speed readout or actuation requirements. The book will encompass a comprehensive discussion of various CMOS interface circuit designs, considering different sensing modalities and applications. Examples of these designs may include high-resolution analog-to-digital converters (ADCs), low-noise amplifiers (LNAs), digital signal processing circuits, multiplexing techniques, and power management strategies.

By providing insights into these diverse circuit designs, readers will gain a deeper understanding of the intricacies and considerations involved in developing CMOS interface circuits for biosensors. This knowledge will enable researchers, engineers, and designers to explore innovative solutions tailored to specific sensor types and application requirements.

1.5 Microfluidic packaging and post-CMOS processes

The CMOS biosensor designers aim to design CMOS biosensors with minimal or no post-CMOS processing; however, in practice, some degree of post-processing is often necessary to optimize their performance. This section will discuss the need for post-CMOS processing and highlight specific processes commonly employed. Several references will be provided to support these discussions.

1.5.1 Specific metal deposition

In the development of CMOS biosensors, specific metal deposition, such as gold, is often required to create electrodes with desirable properties for sensor operation. Gold electrodes offer excellent biocompatibility, high electrical conductivity, and chemical stability, making them suitable for various biosensing applications. These metal electrodes, when exposed to samples, enable the development of electrochemical, capacitive, or impedimetric sensors that rely on the interaction of the analyte with the electrode surface [18].

1.5.2 Encapsulation process

To protect the bonding wires and prevent direct contact between the solution and wires, an encapsulation process is necessary. This process involves covering the bonding wires with a protective layer or encapsulant. By doing so, the sensors are shielded from external contaminants, ensuring long-term reliability and stability [19].

1.5.3 Insulation layer removal

Standard electrodes made of materials like aluminum may require direct contact with the solution for certain biosensing applications. In such cases, the insulation layer, typically composed of silicon oxide or silicon nitride, needs to be removed to expose the electrode surface. This can be achieved through an etching process, which selectively removes the insulation layer while preserving the integrity of the underlying sensor structure [20].

1.5.4 Other coatings

These post-CMOS processing steps are crucial for optimizing the functionality and performance of CMOS biosensors performance using, for instance, biorecognition elements (BREs). While the goal is to minimize post-processing requirements, these additional steps ensure compatibility with the specific sensing mechanisms and sample requirements. Additionally, soft material deposition plays a vital role in CMOS biosensors for various biosensing purposes as discussed in Chapter 14.

1.5.5 Microfluidic sample holder

The primary purpose of microfluidic packaging and hermetic bonding with the CMOS chip is twofold: to precisely direct the sample towards specific sensing sites and to prevent its contact with other areas. The sample holder can either be open-top or closed. However, the preference is for closed fluidic structures to avoid sample reduction caused by evaporation. This precaution ensures accurate and reliable results while maintaining the integrity of the sample throughout the analysis process [19].

1.6 Life science applications

As mentioned earlier, CMOS biosensors have emerged as highly promising candidates for the development of PoC diagnostics. A key advantage lies in the mass production capability of CMOS chips, enabling the manufacturing of cost-effective devices that can be utilized as single-use diagnostic tools for critical sensing applications. These biosensors rely on BREs to transduce biological activity into detectable physical changes, which can be accurately measured by CMOS technology. CMOS biosensors offer a wide range of modalities for detecting these physical changes, including electrical methods like capacitive, impedance, or charge-based sensing, as well as optical, magnetic, and thermal modalities, as mentioned in Section 1.2. This diversity of modalities provides various options for capturing and quantifying biological signals or analyte interactions effectively. It is worth noting that some applications, such as cell culture monitoring, focus not only on selectively detecting biomolecules or cells but also on monitoring the growth of cells, as described in Chapter 14 of our study. This monitoring is crucial for understanding cell behavior and optimizing cell culture processes in fields like biotechnology and medicine. Various BREs [23] are employed in CMOS biosensors to facilitate specific detection. One prominent group of BREs is the antibody-antigen mechanism, where antibodies are coated on the chip's surface using specific biochemical preparation methods. The main BREs are as follows.

1.6.1 Antibodies

Antibodies act as biological recognition elements (BREs) in CMOS biosensors, functioning through the antibody-antigen interaction. They are coated on the chip's surface using specific biochemical preparation methods. Antibodies offer highly selective and sensitive sensing capabilities, making them valuable for the detection of viruses and other target analytes.

1.6.2 DNA probes

DNA probes are immobilized on the CMOS chip's surface and designed to detect specific genes or genetic sequences. These probes enable precise genetic analysis and diagnostics by capturing DNA hybridization events.

1.6.3 Enzymes

Enzymes, such as glucose oxidase, can be immobilized on the chip's surface to detect target molecules like glucose. This enzymatic approach allows for the selective detection of specific analytes, making CMOS biosensors useful for applications in glucose monitoring, enzymatic assays, and more.

1.6.4 Aptamers

Aptamers are short, single-stranded DNA (ssDNA) or ribonucleic acid (RNA) sequences that can bind to specific target molecules with high affinity. They can be utilized as BREs in CMOS biosensors, offering selective detection capabilities for various analytes.

1.6.5 Molecularly Imprinted Polymers (MIPs)

Molecularly Imprinted Polymers (MIPs) are synthetic polymers designed to possess specific binding sites for target molecules. They can be integrated into CMOS biosensors as recognition elements, providing highly selective detection for a range of analytes.

1.6.6 Receptor proteins

Receptor proteins, such as G-protein coupled receptors (GPCRs) or ion channels, can be employed as recognition elements in CMOS biosensors. These proteins offer specific binding and transduction capabilities for target molecules, enabling their detection.

1.6.7 Aptamer-functionalized nanomaterials

Nanomaterials, such as nanoparticles (NPs) or nanowires (NWs), can be functionalized with aptamers to enhance the recognition and detection capabilities of CMOS biosensors. The combination of nanomaterials and aptamers provides amplified signal responses and improved sensitivity.

The utilization of these various recognition elements in CMOS biosensors allows for the specific detection of target analytes across a wide range of applications. By tailoring the BREs to the desired analyte or biomarker, CMOS biosensors offer customizable and adaptable solutions for PoC diagnostics and critical sensing applications.

1.7 Summary and organization of the book

In this introductory chapter, we discussed the importance of CMOS technology and provided an overview of the main contents of this book, which primarily focuses on sensors and actuators. Conventionally, CMOS chips have been utilized for electrical analog and digital signal processing. However, the emergence of BioCMOS chips has revolutionized the field by integrating microfluidic and biomaterial components. These advanced BioCMOS chips serve the purpose of biological and chemical sensing, offering both electrical and fluidic inlets/outlets for precise and efficient detection and analysis. The combination of CMOS technology with microfluidics and biomaterials opens new possibilities for a wide range of applications in the realm of biosensors and bioanalytical devices (see Figure 1.7). Moving forward, the next chapter will delve into design strategies relevant to this field. Chapter 3 introduces the sensing electrodes used in capacitive sensors, along with other electrochemical sensors and biopotential recording systems. Following that, Chapters 4–11 cover the design, implementation, and applications of various sensors, including capacitive, electrochemical, biopotential, ISFET, optical, magnetic, NMR, and thermal sensors, with a specific emphasis on life science applications. Chapter 12 concentrates on magnetic and electrostatic actuators integrated into the CMOS chip. These actuators play a crucial role in manipulating biological samples and enhancing sensor performance. Furthermore, Chapter 13 delves into post-CMOS and microfluidic packaging techniques, providing essential insights into the circuit design and packaging of CMOS-based biosensors. Lastly,

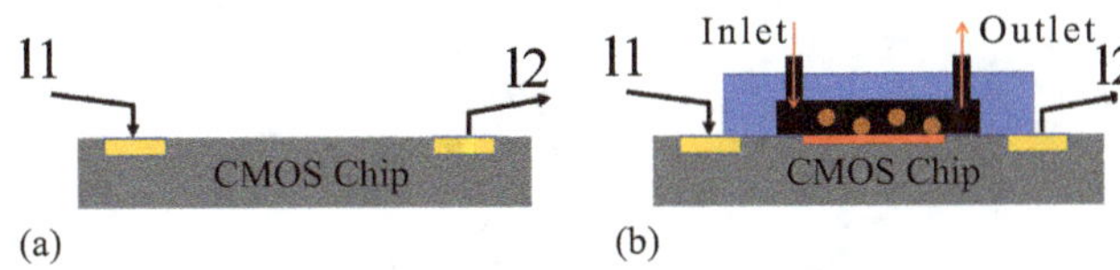

Figure 1.7 Illustration of (a) conventional CMOS with electrical input and outputs and (b) integrated CMOS microfluidic with fluidic inlet/outlet and electrical input/output

Chapter 14 focuses on life science applications, elucidating the recognition elements employed in biosensors. These elements play a pivotal role in the specificity and selectivity of biosensor detection. By organizing the book in this manner, we aim to provide a comprehensive understanding of CMOS biosensors and their applications across the life sciences. Each chapter will delve into specific aspects, fostering a deeper insight into this exciting and transformative technology.

Chapter 2
Design strategies

Complementary metal-oxide-semiconductor (CMOS) biosensor and actuator design is a truly multidisciplinary field that encompasses integrated circuit and sensor techniques, micromachining or microfluidic processes, and biochemical material coating protocols tailored for specific applications.

2.1 Design strategies

2.1.1 Design workflow

The design process for integrated circuits (ICs) used for sensing or actuating purposes can be effectively executed by following the steps depicted in the workflow diagrams of Figure 2.1(a) and (b). The starting point is the application requiring the sensing and/or actuating on the chip. For instance, let us consider a scenario where a sensor has been previously developed in the literature for the detection of biological materials, such as the N-gene associated with COVID-19, using a capacitive sensor built with off-the-shelf discrete electronic components or lab instruments like impedance spectroscopy [24]. Now, to capitalize on the advantages offered by CMOS technology, the aim is to justify the integration of this sensor into a CMOS chip. In this particular case, the application is COVID-19 detection, and the sensor needs to be developed on a CMOS platform to leverage the benefits offered by CMOS technology.

Similar to the off-chip capacitive sensor, the on-chip capacitive sensor is designed, characterized, and simulated using tools like Multiphysics COMSOL. These simulations help estimate the capacitance variation resulting from the presence and concentration range of the target molecule. Additionally, investigations are conducted into the sensor's dimensions and type, such as interdigitated or other shapes. The subsequent steps involve designing an interface electronic circuit capable of reading the variations in the input signal (e.g., capacitance changes) and converting the output into digital or analog signals. These signals can then be utilized for post-signal processing, storage, or display purposes.

According to the required design and schematic, post-layout simulations are conducted to estimate the layout shape and dimensions. Subsequently, the feasibility of testing the chip and the need for post-CMOS processes are investigated. For instance, if the chip requires the incorporation of gold electrodes, and these electrodes should be coated above the aluminum electrode and the topmost metal

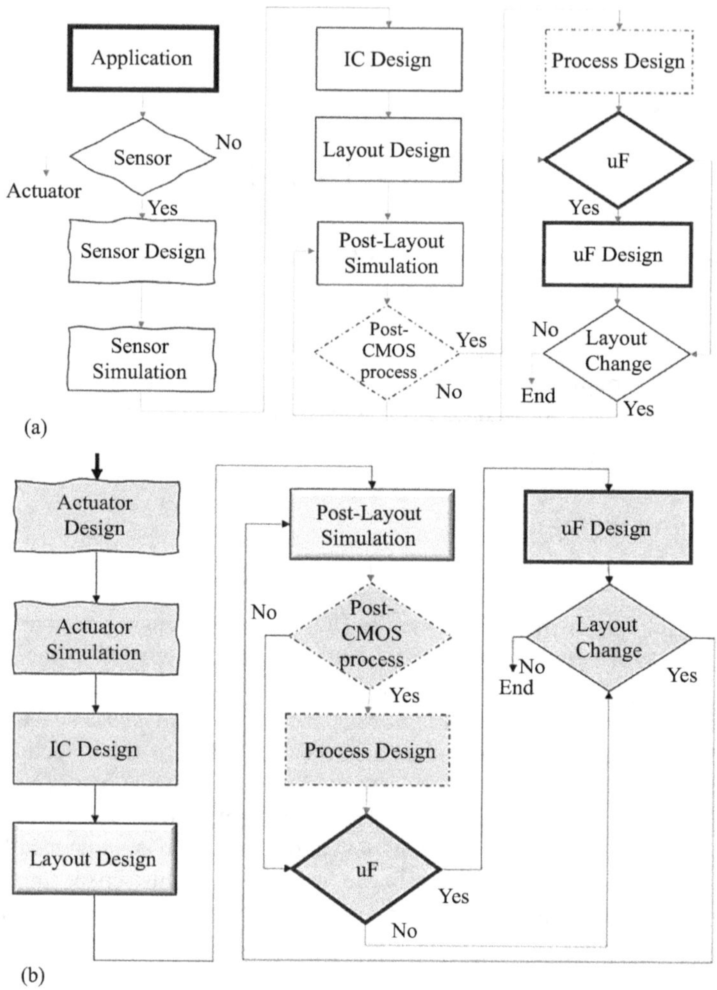

Figure 2.1 Design workflow for both integrated (a) sensor and (b) actuators for biological applications. (This workflow clearly outlines the sequential steps involved in designing a sensor or actuator on a CMOS chip. These steps encompass the design and simulation of the sensor or actuator, followed by the design and simulation of the chip. Subsequently, design layout and post-layout simulation are conducted, culminating in the integration of microfluidic and post-CMOS processes before coating the chip with biorecognition element (BRE) and conducting the biological experiments.)

layer, a micromachining process needs to be designed. This process may involve photolithography for achieving specific electrode sizes and implementing chemical procedures. To enable these post-CMOS processes, adjustments may be necessary in the layout. For instance, creating sufficient distance between the bonding pads and the sensor would be essential. This step ensures the facilitation of the subsequent post-CMOS processes to achieve the desired functionality and performance of the integrated sensor.

Regardless of any post-CMOS processes, the sensors and actuators on the chip need a protective design to prevent liquid from reaching the electronic components, rather than the sensing or actuating sites [25]. For this reason, we need to develop microfluidics (μF) packaging that applies some extra rules in the layout by increasing the dimensions to increase the surface of the chip that is bonded with microfluidics. These rules can result in some changes in the layout that require tuning the design and performing post-layout simulations.

2.1.2 Test strategies

In conventional IC design, the primary objective is to detect and analyze the electrical characteristics of the chip's output. This is typically achieved using either a probe station or a custom-made printed circuit board (PCB). Verifying the functionality of the chip through this process is the initial and crucial step. Following the electrical verification, the next essential step involves testing the functionality of the sensor in the presence of a small amount of liquid. To achieve this, the chip is exposed to a controlled nanoliter droplet, ensuring that the overall chip performance remains unaffected. This task is intricate and requires careful execution. To safeguard the chip from direct contact with the liquid droplet, it is advisable to encapsulate the chip bonding wires with epoxy. This protective measure helps prevent any adverse effects of the liquid on the chip's performance. Moving forward, the subsequent stages of the verification process involve describing the next steps based on the specific details outlined in the test plan.

The initial test serves to demonstrate the functionality of both the sensor and the circuit. Following this, the subsequent step involves completing the necessary post-CMOS chip and microfluidic packaging. Additionally, custom hardware and software must be developed to facilitate the connection to a computer for data analysis and control. Once all these stages are completed successfully, it is essential to repeat the dry and wet tests mentioned earlier to ensure the chip's robustness and reliability. These repeated tests are crucial in validating the chip's performance under different conditions. Upon satisfactory completion of the above steps and thorough verification, the chip will be deemed ready for dedicated biological testing. This critical phase involves evaluating the chip's performance and interaction with biological samples, where its effectiveness and accuracy in relevant applications can be assessed.

The development of biorecognition elements (BREs) on the chip and the handling of actual biological samples should be conducted within a specific laboratory environment with the appropriate biosafety permissions. The required biosafety permission class for the laboratory will depend on the level of risk associated with the target biological samples. For handling low-risk biological samples, a Biosafety Permission Class 1 may be sufficient. For samples presenting moderate risk, a Class 2 biosafety permission might be required. In cases involving higher-risk biological samples or potentially hazardous agents, a higher-level Biosafety Permission Class, such as Class 3 or Class 4, may be necessary.

Figure 2.2 illustrates the test workflow starting from the chip fabrication and electrical packaging stages until the biological test is conducted. This flowchart outlines the key steps involved in the process, leading to the evaluation of the chip's performance with biological samples.

Adhering to the appropriate biosafety regulations ensures the safety of laboratory personnel, the community, and the environment, while also safeguarding the integrity of the biological samples and the accuracy of the research results. It is imperative to follow these guidelines diligently to conduct responsible and secure experimentation with biological materials.

2.1.3 Life science applications

As aforementioned, the initial step in CMOS biosensor design is to investigate the application and determine why a CMOS IC should be used instead of discrete devices for the same purpose. As described in the last chapter, the primary motivation for utilizing CMOS technology is its low cost and the ability to integrate specific functions that are difficult to achieve with discrete devices. CMOS chips enable the development of point-of-care (PoC) disease diagnostic for example for infectious diseases [24] as demonstrated in Figure 2.3(a).

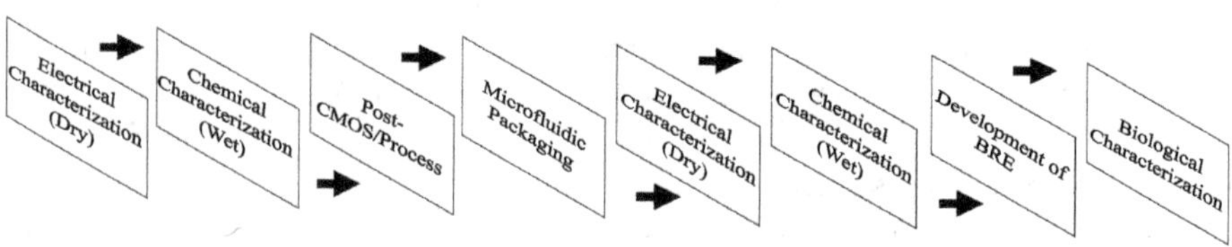

Figure 2.2 The test workflow

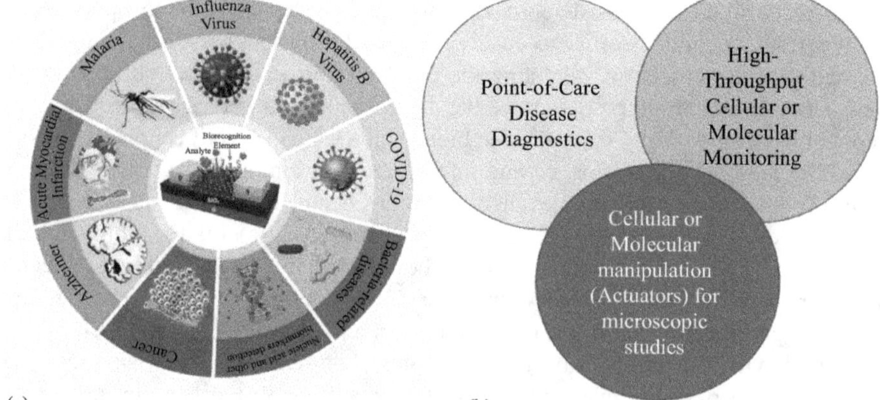

Figure 2.3 Life science applications: (a) infectious detection using FET on CMOS and (b) target technologies

Another notable advantage of CMOS for life science applications is the precise manipulation of cells or molecules in a non-invasive manner, enabling microscopic studies. This capability is described in the chapter on actuators and is well-illustrated in various experimental setups. CMOS also enables the development of high-throughput sensors with a large number of sensors on a single chip, which is not easily achievable with discrete components. This integration capability makes CMOS an attractive option for various applications. Therefore, the use of single-use CMOS chips must be well justified. For instance, if a CMOS chip for DNA sequencing costs hundreds of dollars, it must demonstrate clear advantages over competitive methods, such as whole genome sequencing using Ion Torrent or Oxford Nanopore, which are more cost-effective [25].

Even though the CMOS chip is disposable, it can still be suitable for the same application if it offers unique advantages. However, the reusability of CMOS chips remains a challenge in certain applications. One critical factor influencing this is the stringent regulatory requirements for clinical and biological uses. Securing approval from regulatory bodies such as the Food and Drug Administration (FDA) imposes strict limitations on reusability. Consequently, the removal of biochemical residues from the chip becomes imperative to enable multiple uses of a limited number of chips fabricated through the foundry process. Specific cleaning procedures are necessary to ensure that the cleaning process does not adversely affect the device's characteristics. While reusability poses difficulties, it is worth noting that in pathogen detection experiments like COVID-19 testing, there are often no viable alternatives but to use CMOS chips for single-use purposes. This is primarily due to the need for stringent biohazard control and the importance of preventing cross-contamination between samples. Therefore, striking a balance between reusability and adhering to biohazard control protocols is crucial, and the decision to use single-use or reusable CMOS chips should be made based on the specific requirements and safety considerations of each application.

Considering these factors, the design of CMOS biosensors should carefully evaluate the advantages and limitations of CMOS integration, cost-effectiveness, regulatory requirements, and the specific needs of the application. By addressing these considerations, designers can develop CMOS biosensors that meet the requirements of their intended applications while considering factors like cost, disposability, and regulatory compliance.

2.1.4 Recognition elements

Indeed, the selection of recognition elements or coating materials for CMOS biosensors is closely related to the specific application and the biomarkers or targets to be detected or actuated. The choice of materials on the surface of the CMOS chip plays a crucial role in creating a bio-friendly environment or facilitating selective detection.

For cell monitoring applications (see Figure 2.4(a)), where the focus is on cell growth and culture, it is important to design a surface that promotes cell adhesion and growth. The coating materials should create a favorable microenvironment for

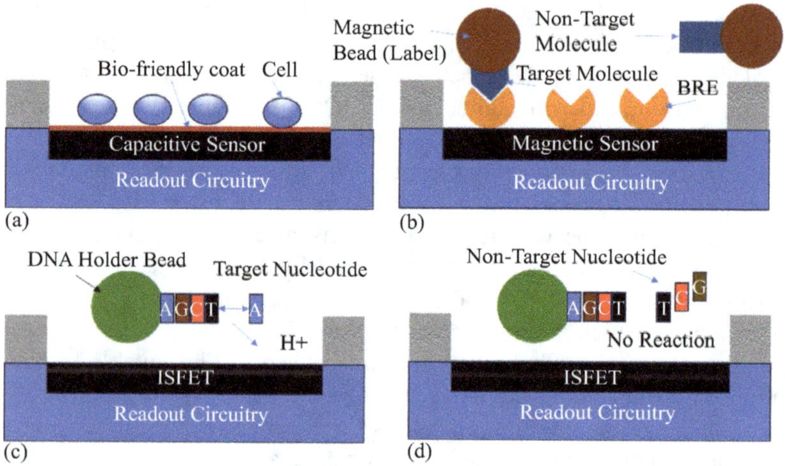

Figure 2.4 This figure illustrates the applications of bioelectrical impedance spectroscopy as follows: (a) non-specific cell monitoring/culture, (b) specific target molecule detection, and (c and d) specific nucleotide detection for sequencing purposes

cell attachment, proliferation, and maintenance of cell viability. Various biocompatible materials, such as polymers, hydrogels, or specific surface modifications, can be employed to achieve this.

On the other hand, if the objective is to selectively detect specific biomolecules like DNA or proteins such as antigens (see Figure 2.4(b)), the design and coating of BREs become essential. These BREs, such as DNA probes or antibodies, are immobilized on the surface of the CMOS chip. They allow for the specific binding and recognition of target biomolecules, enabling their detection or quantification. Numerous research papers have documented various biochemical surface modification methods for immobilizing recognition elements on CMOS chips. These methods facilitate the interaction between the surface material (such as SiO_2, aluminum, or other coatings) on one side of the chip and the recognition element on the other side. Collaboration with experts in biochemistry or biomaterials is highly valuable in optimizing the coating materials and ensuring the effective immobilization of recognition elements.

A groundbreaking non-specific sensing application utilizing CMOS was reported by Rothberg *et al.* [25]. This innovative application focuses on DNA sequencing, where the reaction of unknown DNA nucleotides (A, G, C, and T) is examined in the presence of a series of known nucleotides (A, G, C, and T). Detection of complementary matching between nucleotides (A-T, C-G) is achieved through an Ion-Sensitive Field-Effect Transistor (ISFET), which senses the pH change in the solution near the CMOS sensor (see Figure 2.4(c) and (d)). This pioneering approach shows great promise in advancing DNA sequencing technology.

The collaboration between engineers and biochemists is crucial to harness the synergistic expertise from both fields. Engineers provide knowledge of CMOS technology, chip design, and integration, while biochemists offer insights into the selection of appropriate recognition elements, surface modification techniques, and bio-functionalization methods.

By collaborating and leveraging the expertise of biochemists, designers of CMOS biosensors can develop effective surface coatings and immobilization strategies, enabling the selective detection of target biomarkers or the creation of a suitable microenvironment for cell-based applications.

2.2 Biological CMOS (BioCMOS) chip

In this section, we present the key considerations for designing a CMOS chip specifically tailored for life science applications.

2.2.1 CMOS sensor/actuators

2.2.1.1 Type of sensor/actuator

The type of sensor is chosen based on the dedicated recognition element. For example, if the recognition element is a specific antibody for COVID detection, various sensor modalities such as electrochemical, capacitive, or impedimetric can be used, without a specific label. Additionally, magnetic and optical sensors can be utilized if the sample involves magnetic beads or fluorescence labels, which can enhance the signal due to the presence of the label. The decision regarding which sensor to use depends on the complexity of the sensor and surface preparation factors. It is important to highlight that CMOS technology facilitates the design and implementation of diverse sensors (e.g., capacitive, optical, impedimetric, electrochemical) and actuators (e.g., magnetic, electric), which will be further elaborated in the subsequent sections.

2.2.1.2 Sensor/actuator array

Another important aspect in sensor/actuator design is the consideration of using an array of sensors or a single sensor, depending on the application requirements and the needs expressed by the end user. Furthermore, the size and shape of the sensor are additional factors that should be determined based on reports in the literature and other factors related to the application and biological considerations. The sensitivity and resolution required for the specific application should also be taken into account.

2.2.1.3 Multiphysics simulation

Simulating the sensor poses a significant challenge due to its dual nature. On one side, the sensor is connected to the readout system responsible for measuring the electrical changes caused by biological materials. On the other side, the sensor is exposed to the biomaterials being detected. This inherent complexity in biosensors makes simulation challenging, particularly when dealing with multiphysics simulation methods like

COMSOL, where the presence of the biomaterial or recognition element can introduce additional complexity. To overcome these challenges, it is often necessary to develop low-complexity models and then experimentally verify the functionality and applicability of the design. Alternatively, additional features can be incorporated into the circuitry design to address uncertainties associated with the effect of biological materials, such as the development of calibration methods. Combining simulation, experimental validation, and calibration techniques can help in designing robust CMOS biosensors and actuators, ensuring their functionality and reliability in detecting and measuring biological targets.

2.2.2 Interface circuit

Once the sensor/actuator type is selected and a low-complexity model is developed using multi-physics software like COMSOL, variations in the sensor's output can be obtained. These output values serve as input for the circuit design, with a specific focus on achieving high-performance reading capabilities. The primary emphasis of the book is on the design of IC interface circuits for CMOS biosensors and actuators. These interface circuits play a critical role in extracting, amplifying, and processing the sensor signals, enabling accurate and reliable measurement or actuation.

To effectively handle the variations in the sensor output (and actuators input) and ensure precise and reliable readout capabilities, the design of high-performance circuitry is paramount. Computer-aided design (CAD) tools such as Cadence can be employed to facilitate the circuit design process. Design considerations encompass noise analysis, gain stability, power management, high-speed design techniques, and optimization methods tailored to biosensor applications. By addressing these design considerations, engineers and researchers can develop IC interface circuits that efficiently interface with CMOS biosensors. The utilization of CAD tools streamlines the design process, ensuring that the circuitry can effectively handle the variations in the sensor output while delivering accurate and reliable measurement or actuation.

2.3 Microfluidic and postprocessing

Depending on the specific application and the required biochemical surface modifications, a design strategy should be developed for microfluidic structures and post-CMOS processing. For applications where a droplet of sample needs to be introduced to the chip, the design should incorporate a suitable sample holder. On the other hand, if an array of sensors needs to be coated with various sample recognition elements for the detection of multiple biomarkers, a microfluidic structure should be developed (Figure 2.5). Collaborating with microfluidic design groups can be beneficial in developing fabrication methods for such microfluidic structures.

Similarly, the post-processing methods should be designed and optimized. For instance, if DNA molecules are intended to be absorbed onto gold-coated surfaces

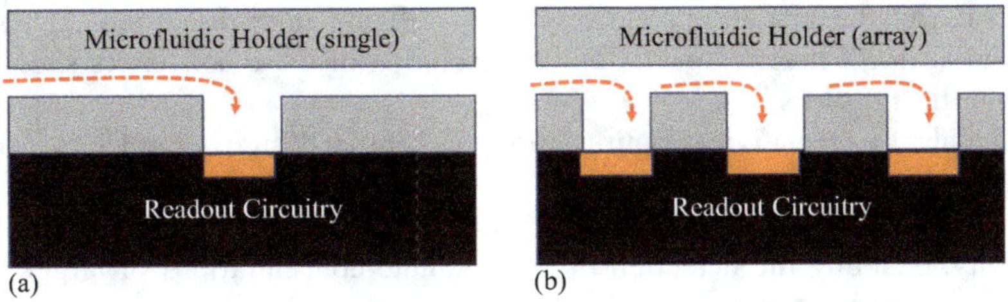

Figure 2.5 Microfluidic packaging: (a) sample holder for a single sensor and (b) sample holder for multiple sensors on the chip

functionalized with HS (thiol) groups, the chip's surface should be coated with gold. Photolithography methods, such as sputtering or other gold deposition techniques, can be applied in conjunction with CMOS processes. It is worth noting that depositing gold on a specific side of the die that has been separated from the wafer is a challenging task. However, if access to the whole wafer is possible, it becomes easier to utilize the wafer and adapt it to photolithography systems. Careful consideration and collaboration with experts in microfluidics, surface chemistry, and fabrication processes are crucial in developing the appropriate microfluidic structures and implementing the necessary post-processing methods. This ensures the successful integration of CMOS biosensors with microfluidic systems and facilitates the required biochemical surface modifications.

2.4 CMOS sensor/actuator performance factors

In this section, the main factors in the design of high-performance biological sensors and actuators are put forward.

2.4.1 Sensor

2.4.1.1 Sensitivity

Sensitivity is a crucial aspect in the design of CMOS biosensors and refers to their ability to detect very low concentrations of target molecules or cells in a sample. Achieving high sensitivity involves considering various factors, including the sensor geometry, materials used, and the design of the electronic interface readout circuitry.

The size, shape, and materials of the sensor play a significant role in determining the changes in electrical parameters and, subsequently, the measurable signal by the interface circuit. Smaller sensors tend to produce lower output signals. However, for applications requiring the detection of single cells or small particles, the sensor size should be comparable to the size of the target, such as a cell. The minimum variation in the sensor's output signal must be higher than the minimum detectable signal by the interface circuit. The resolution of the interface circuitry is determined by its ability to detect the smallest variation in the sensor's output signal. The presence of an analog-to-digital converter (ADC) can also impose limitations on the resolution. However, by utilizing high-resolution ADCs and digital-to-analog converters (DACs), the resolution of these converters should not pose significant issues.

To enhance sensitivity, optimization of sensor design parameters, careful selection of sensor materials, and the design of high-resolution interface circuitry are essential. By considering these factors, CMOS biosensors can achieve high sensitivity, enabling the detection of even minute concentrations of target molecules or cells in a sample.

2.4.1.2 In/out signal range

The input and output range of the interface circuit are indeed important considerations in sensor design. The input signal to the interface circuit is determined by the output signal of the sensor, while the output signal of the interface circuit can be digital and is constrained by the resolution of the ADC converter.

A wider variation in the input and output signals can be advantageous as it allows for the quantification of a larger range of the measured parameters. However, this wider range may also impose limitations on the required sensitivity to detect the smallest changes in the measurement. In other words, a broader input and output range may compromise the ability to detect subtle variations in the measured parameter. Therefore, finding the right balance between the input and output range of the interface circuit is crucial. It involves considering the dynamic range of the sensor's output, the desired resolution of the measurement, and the limitations imposed by the ADC converter. Optimizing the input and output range ensures that the interface circuit can effectively capture and process the sensor's output signal while maintaining the necessary sensitivity for accurate measurements.

By carefully designing and optimizing the input and output range of the interface circuit, CMOS biosensors can achieve a balance between quantifying a wide range of measurements and maintaining the required sensitivity for detecting small changes in the measurement.

2.4.1.3 Power consumption

Power consumption is a critical factor to consider in CMOS biosensor devices, particularly in applications such as PoC diagnostics where low power consumption is crucial. The goal is to design circuitry that minimizes power consumption, enabling the use of smaller batteries and longer device operation. CMOS technology inherently offers low power consumption due to its efficient transistor design and low leakage currents. To further reduce power consumption, additional design strategies can be implemented. These may include optimizing circuit architectures, utilizing low-power design techniques, and incorporating power management features. These measures help ensure that the biosensor device operates efficiently, consuming minimal power while maintaining accurate and reliable measurements.

Reducing power consumption is not only important for extending battery life but also for managing heat generation within the device. Excessive heat can affect the performance and accuracy of the biosensor by altering the characteristics of the sensor elements or interfering with biochemical reactions. Therefore, minimizing power consumption helps mitigate heat generation and maintains a stable and uniform operating temperature, enhancing the sensitivity and accuracy of the biosensor measurements.

In summary, achieving low power consumption in CMOS biosensors is critical for various reasons. It enables the use of smaller batteries, extends device operation time, and minimizes heat generation, thereby preserving the performance and accuracy of the biosensor. Through careful circuit design, optimization techniques, and power management strategies, CMOS biosensors can achieve efficient power utilization, making them suitable for PoC and other low-power applications.

2.4.2 Actuators

The primary circuit design features of on-chip actuators involve generating voltage and current to drive electrical or magnetic actuators. Moreover, key performance features include the response time for actuating the device and the ability to easily achieve multiplexing through CMOS IC techniques.

2.5 Graphical user interface (GUI) and data communication

The graphical user interface (GUI) plays a crucial role in collecting data from the sensor, displaying it, and performing further signal processing. Depending on the application and user requirements, the GUI can be designed as a mobile app, particularly when using a smartphone for data monitoring. In such cases, the communication protocol for data transfer, such as Bluetooth, Wi-Fi, or other methods, is selected accordingly.

Recently, near-field communication (NFC) has also gained popularity as a communication method, that can address power supply issues for the device. In laptop or desktop computers, wired communication protocols can be used to establish a connection with the sensor.

Moreover, the GUI or mobile app can facilitate data communication with a server or cloud infrastructure. This enables the development of artificial intelligence (AI) algorithms that can analyze sensed data and provide insights or make predictions. By leveraging cloud computing, the AI program can be trained using large datasets and deployed for real-time sensing purposes. The integration of AI with sensor data opens up possibilities for a variety of applications, enabling advanced data analysis, anomaly detection, or decision-making based on sensor measurements.

The GUI or mobile app serves as an interface for users to interact with the sensor system, visualize data, configure settings, and enable seamless communication between the sensor, local devices, and cloud-based services.

2.6 Summary

This chapter provides a comprehensive review of major design considerations, encompassing the design, implementation, and testing of sensor/actuators, circuits, microfluidic packaging, post-CMOS processes, BRE coating, and their applications in life science. An electrical engineer with a background in IC design can easily create and simulate the required circuitry for readout sensors or actuator drivers. Additionally, expertise in developing microfluidic packaging and post-CMOS processes is essential. Collaboration with researchers having micro-machining or microfluidic expertise can facilitate the necessary micro-machining processes on the CMOS chip. This multidisciplinary approach ensures a holistic design methodology. Moreover, the third group of expertise required for the successful implementation of biological sensors/actuators is chemical/biological procedures, necessary for coating the chip's surface or performing sample preparation before introduction onto the chip. By adhering to these generic considerations, the chip designer can develop the chip and proceed with other steps to verify the functionality and applicability of the proposed chip for life science applications.

Chapter 3
CMOS electrodes

In this chapter, we explore the various techniques employed for implementing sensing electrodes in diverse applications using standard complementary metal-oxide-semiconductor (CMOS) and post-CMOS processing structures. The topmost metal layer of the standard CMOS technology contains aluminum (Al) and silicon (Si) impurities. The 0.18 μm process has Al with 1% Si [26]. Passivation layers made of SiO_2, Si_3N_4, and polyimide are stacked on the topmost metal layer, as depicted in Figure 3.1(a).

3.1 Different on-chip electrodes

We categorize the different electrode structures for diverse applications into eight groups, namely passivated electrodes, high sensitivity passivated electrodes, quasi-interdigitated electrodes (Q-IDEs,) Al/Al_2O_3-electrodes, polymer-coated electrodes, gold (Au)-coated electrodes, platinum (Pt)-coated electrodes, and titanium nitride (TiN) electrodes, as compared in Table 3.1. It is worth noting that these electrodes can be utilized for various CMOS sensors, employing capacitive, impedimetric, voltammetry, amperometry, and biopotential recording techniques.

3.1.1 Passivated electrodes

Passivated electrode structure, as shown in Figure 3.1(a), is a common configuration where the passivation layers of the metal layers in CMOS technology are retained. This sensing structure has been utilized in various applications such as neurotransmitter dopamine (DA) detection [27] and monitoring cell growth [28–31]. Prakash *et al.* [29] used this sensing electrode type to track cell proliferation. The experiments conducted by this group, as presented in Table 3.1, employed sensing electrodes of different sizes, specifically $20 \times 20\ \mu m^2$, $30 \times 30\ \mu m^2$, and $40 \times 40\ \mu m^2$.

In their research, Lu *et al.* [27,32] utilized the Al layer (metal 3) present in 0.35 μm CMOS technology, specifically the two-polysilicon-four-metal (2P4M) variant, as the electrode material. To detect neurotransmitter DA, probe molecules were immobilized on the inter-metal SiO_2 located on metal 3 (see Table 3.1). The top passivation layers of this technology, consisting of Si_3N_4 and SiO_2, were removed through dry etching. Additionally, to enhance the sensitivity of the

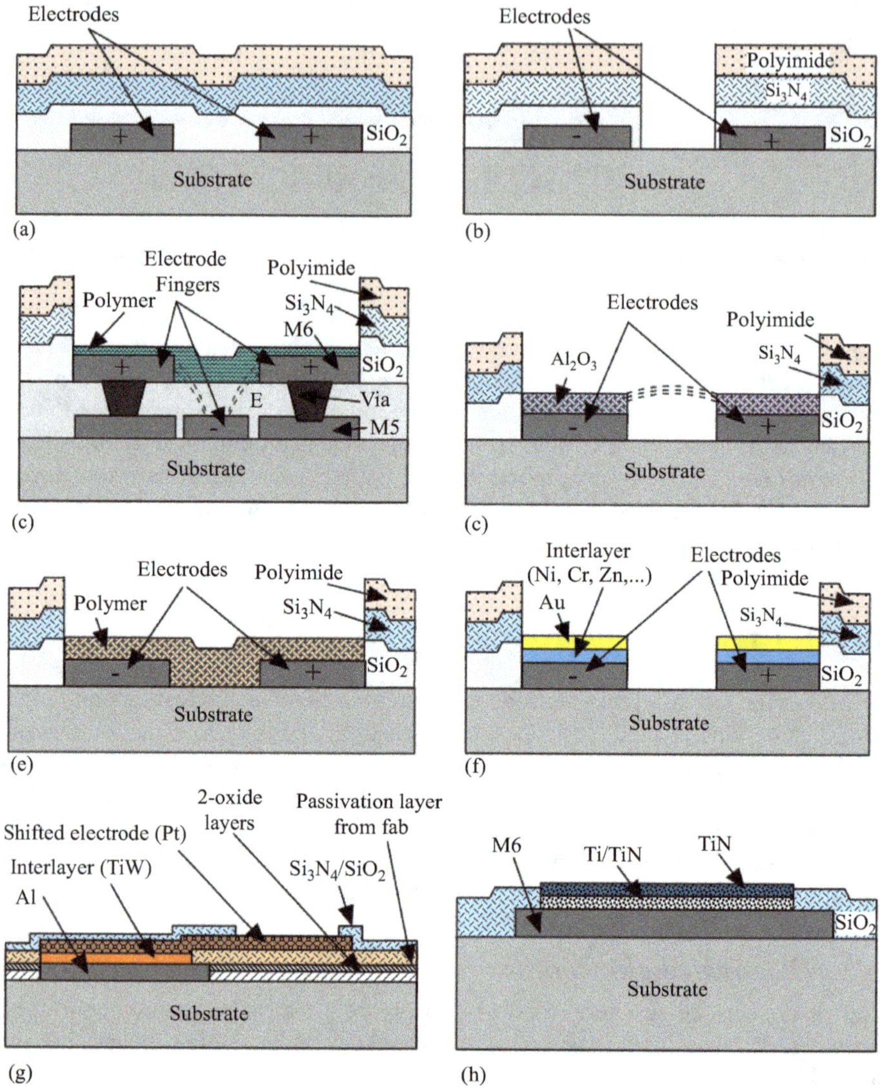

Figure 3.1 Cross-section of (a) a typical passivated electrode (IDE), (b) a passivated IDE with a window in between the fingers, (c) a Q-IDE, (d) a bare electrode, (e) a typical polymer-coated electrode, (f) a typical Au-coated electrode, (g) a shifted Pt electrode (in floating configuration), and (h) a TiN-electrode adapted

system, the remaining oxide thickness was thinned down by an optional dielectric reactive etching process.

The total capacitance measurement is affected by the equivalent parasitic capacitance of the coating passivation layer, which is positioned in series with the

Table 3.1 CMOS-based electrodes used for biosensing

Technique	Application	Target species	Sensor type	Array #	Sensing element material	Electrode size (μm^2)	Reference
0.5 μm 2P3M	Cell proliferation monitoring	Breast cancer cells	Capacitive	28	Al/1Pass[a]	20 × 20, 30 × 30, 40 × 40	[29]
0.35 μm 2P4M	Detection of the neurotransmitter DA	CPBA	Capacitive	5 × 5	Al/1Pass	20 × 27, 30 × 39, 50 × 67, 100 × 135, 150 × 203	[27]
0.25 μm	Bacteria detection	*S. epidermidis*	Capacitive	1	Al/Al_2O_3	220 × 230	[33]
0.25 μm	Detection of single bacterial cell	*S. epidermidis*	Capacitive	16 × 16	Al/Al_2O_3	14 × 16	[4]
0.18 μm 1P6M	Rapid identification of bacteria	*E. coli*	Capacitive	2	Al/Al_2O_3	~ 35 × 35	[34]
0.18 μm	Bacteria growth monitoring	*E. coli*	Capacitive	3	Op-fin	100 × 100	[35]
0.35 μm	Pathogenic bacteria detection	Pathogenic bacteria	Capacitive	30	Q-IDE	100.3 × 102.8	[36]
0.35 μm	Cell growth monitoring	Human lung carcinoma cell	Capacitive	8 × 8	Al/PDMS	50 × 50	[37]
0.35 μm	Cell culture monitoring and drug test	H1299 and Hek293 cells	Capacitive	8 × 8	Al/PEM	50 × 50	[2]
0.5 μm	Detection of DNA hybridization	DNA	Capacitive	8 × 16	Au	200 × 200	[38]
0.35 μm	Detection of DNA hybridization and protein	DNA, proteins	EIS	10 × 10	Au	40 × 40	[39]
90 nm	Cell monitoring	Insulating and conducting particles	High-frequency impedance spectroscopy	256 × 256	AuCu	0.025	[40]

(Continues)

Table 3.1 (Continued)

Technique	Application	Target species	Sensor type	Array #	Sensing element material	Electrode size (μm^2)	Reference
0.18 μm 2P6M	Neurotransmitter detection	Neurotransmitter	Stimulation, impedance measurement, and fast CV	332 × 180	Pt	3 × 7.5	[6]
0.35 μm	Intracellular electrophysiological imaging	Neonatal rat ventricular cardiomyocytes	Electrical recording/ stimulation	32 × 32	Pt	–	[41]
0.6 μm	Electrogenic cells	Chicken dorsal root ganglion neurons	Electrical recording/ stimulation	11,011	Pt	~ 24.5	[42]
0.35 μm	Electrogenic cells	Rat cortical neurons	Electrical recording/ stimulation	26,400	Pt	9.3 × 5.4	[43]
0.18 μm	Electrophysiological imaging of cells	Cardiac cells, brain slices	Electrical recording and impedance imaging	59,760	Pt	1 × 1,2 × 2,4 × 4,8 × 8	[44]
0.13 μm	Neuro-electrophysiological studies	Primary hippocampal neurons	Electrical recording and impedance imaging	16,384	TiN	11 × 11,7 × 7,4.5 × 4.5,2.5 × 3.5	[45]

[a]Al/1Pass: topmost metal with one passivation layer, Op-fin: passivated IDE with a window in between the fingers, EIS: electrochemical impedance spectroscopy, CPBA: 4-carboxyphenylboronic acid, CV: cyclic voltammetry.

sensing capacitance. If the passivation layer is made thinner, the parasitic capacitance will increase, thereby enhancing the sensitivity of the measurement. However, a specific process [30] is necessary to achieve a reduction in the thickness of the passivation layers. In certain situations where immobilization of specific bioparticles and biorecognition elements (BREs) is required above the electrodes, passivated electrodes may not suffice.

3.1.2 High sensitivity passivated electrodes

To increase sensitivity in between the conductors of electrodes, removing the passivation layer using a pad-etch process as depicted in Figure 3.1(b) can enhance the electric field. However, the electrodes have smaller dimensions than the pad, so the simulation of design rule check (DRC) violations is necessary for their analysis in Cadence software. Increasing the height of the sensing electrode by using a thicker topmost metal layer can improve sensor performance since the electric field generated in the space between parallel electrodes is greater than that formed above them. Ghafar-Zadeh *et al.* exploited this approach by removing the passivation layer in between the fingers of IDEs [46,47]. It is crucial to match the mask with the gap between the fingers. The lack of a stop layer above the substrate in these electrodes caused slight erosion in the substrate and deepened the gap. Using 0.18 μm CMOS technology, these electrodes were produced in a size of $100 \times 100\ \mu m^2$ (see Table 3.1). This structure was utilized by this group for detecting chemical solvents [48] and monitoring bacteria growth [35].

3.1.3 Quasi-interdigitated electrode (Q-IDE)

Figure 3.1(c) depicts the implementation of Q-IDEs in the two topmost metal layers without the passivation layers. The electrode beneath acts as the reference while the top electrodes serve as working electrodes (WEs), with the pad-etch process being used to remove the passivation layers on IDEs. This design approach [36,49–51] intensifies the electric field surrounding the electrode, improving its sensitivity. Several studies have used this structure as a CMOS capacitive gas sensor, as discussed in literature such as [50,52,53]. A polymeric thin film is commonly used to cover the chip's surface for gas sensing applications. For example, Kummer *et al.* [49] utilized a Q-IDE coated with a polymer to detect volatile organics in ambient air. Tashtoush *et al.* [36] employed this design for pathogenic bacteria detection by coating the sensing area with bacteriophage, designated as the BRE in Table 3.1, and using epoxy to cover unexposed regions.

3.1.4 Al/Al_2O_3 electrodes

Aluminum (Al) is highly susceptible to oxidation, which can be a problem in CMOS processes when attempting to expose the Al electrode directly to the sample. However, a native layer of aluminum oxide (Al_2O_3) can form above the bare Al electrode when it is exposed to air (as depicted in Figure 3.1(d)) [54]. This layer of Al_2O_3 acts as an insulating layer, shielding the electrodes from the biological and chemical environment, and also provides a suitable surface for functionalization purposes [55]. Bare electrodes have been used for a variety of applications, such as deoxyribonucleic acid (DNA) detection [55], bacteria growth [34], and chemical solvent monitoring [54]. One method of creating a sensing area between the electrodes is by using thick topmost metal layers and a large space between the electrodes, achieved by selecting a pad-etch mask [56]. For instance, Nikkhoo *et al.* [34] removed only the passivation layer above the topmost metal layer and implemented two electrodes measuring approximately $35 \times 35\ \mu m^2$, with a trench in between them for bacteria detection.

It is possible to intentionally add a layer of Al_2O_3 onto non-passivated Al electrodes using post-CMOS processes, as detailed in Reference [4]. In their work on bacteria detection, Couniot *et al.* [4,33] utilized such electrodes that lacked SiO_2 and Si_3N_4 passivation layers but instead had a thin layer of Al_2O3 deposited on the topmost metal layer using plasma-enhanced atomic layer deposition (PEALD) for corrosion protection (see Table 3.1).

3.1.5 Polymer-coated electrodes

Polymeric membranes are commonly used to coat sensing electrodes to enhance sensitivity for gaseous samples (as illustrated in Figure 3.1(e)), as reported in various studies [50,52,53]. Besides, polymer coatings have also been shown to improve biocompatibility, as evidenced by Nabovati *et al.* [37], who utilized a spin-coated poly(dimethylsiloxane) (PDMS) layer on bare Al electrodes. This technique was employed for cell growth screening. The same research group [2] also found polyelectrolyte multilayer (PEM) to be a useful intermediate layer for enhancing cell adhesion, particularly for applications like cell growth monitoring and drug tests. PEM was deposited on Al electrodes using the layer-by-layer (LBL) coating technique, and the functionality of these electrodes was demonstrated for life science applications. It is important to note that certain bioactive coating methods based on PEMs have also proven useful for glucose detection [57,58] and drug delivery [59], as highlighted in Table 3.1. Both [37] and [2] implemented an 8×8 array of these electrodes in their studies.

3.1.6 Gold-coated electrodes

The biocompatible and stable noble metal, gold (Au) [56,60], is widely used in various biosensing applications, including anti-Rubella [61] and DNA detection [38,62,63], owing to its good electron conductivity. Au can form self-assembled monolayers (SAMs) between electrodes [56], which facilitates the immobilization of recognition elements like glucose oxidase [64] and DNA oligonucleotides [65] on Au electrodes. However, post-CMOS processing steps are required for the fabrication of Au electrodes above the CMOS chip. Au can be deposited on different materials such as aluminum (Al) [66], platinum (Pt), copper (Cu), palladium (Pd), and nickel (Ni) [67]. In the case of Al, an interlayer of chromium (Cr), zinc (Zn), or Ni is often necessary to improve the adhesion of Au [60]. Figure 3.1(f) provides an illustration of the use of an interlayer, such as chromium, zinc, or nickel, to enhance the adhesion of Au on Al.

Several approaches have been reported for depositing Au on Al electrodes in CMOS technology. These include photolithography and electroless deposition methods, as described in References [61,66,68]. These techniques are explained in Appendix A.

Guiducci *et al.* [63] employed photolithography for capacitive DNA sensing and used tantalum (Ta) as the interlayer between Au and Al. They created an array of 128 Au electrodes for the same application, as listed in Table 3.1 [38]. However, aligning the photomask with the fabricated electrodes can be challenging with this technique.

Electroless nickel immersion gold (ENIG) plating procedure was used by Sun *et al.* [61] to create Au IDEs for a high-density 4,096-pixel electrochemical biosensor array. 3D trenches were formed between the two combs by removing the passivation layer which serves as nano-wells for cell culture. Similarly, Manickam *et al.* [39] employed ENIG to fabricate biocompatible Au microelectrodes for the detection of DNA and proteins.

Widdershoven *et al.* conducted research using nanoelectrodes for high-frequency impedance spectroscopy and imaging, as described in their publications [40,69,70]. These nanoelectrodes are linked to a vertical pillar formed by stacking four Cu vias of CMOS technology. The fourth via's Cu is alloyed with the Au deposited on the chip surface to create Au/Cu nanoelectrodes. This group reported the smallest electrode size in Table 3.1, which was utilized in a large 256×256 array compared to other works.

3.1.7 Pt-coated electrodes

Platinum (Pt) is an inert metal similar to gold (Au), but it is not as stable against oxidization. Moreover, the process for fabricating Pt electrodes requires post-CMOS processing. Typically, Pt is deposited and patterned using photolithography, and an adhesion layer of titanium-tungsten (TiW) is utilized between Al and Pt [60] (refer to Figure 3.1(g)). Pt electrodes are commonly used in applications such as stimulation, impedance measurement, and recording of electrogenic cells like cardiomyocytes or neurons [6,43,71,72]. In some studies [6,43,72], Pt is deposited onto Al electrodes by following a straightforward 2-mask post-processing procedure. First, the electrode is shifted, and the underlying Al is sealed with the electrode metal and a passivation stack. Next, TiW, an adhesion promoter, and Pt are sputter-deposited and structured using a lift-off process. Then, an alternating Si_3N_4/SiO_2 passivation stack is deposited using plasma-enhanced chemical vapor deposition (PECVD). The mixed-frequency PECVD process matches the stress with that of the underlying Si_3N_4 of the CMOS process. Finally, reactive ion etching (RIE) is used to etch openings in the nitride (the passivation layer) and define the shape and size of the electrodes. Dragas *et al.* [6] utilized 332×180 of these microelectrodes with a size of $3 \times 7.5\ \mu m^2$ in their system, occupying a large active area of $4.48 \times 2.43\ mm^2$. Similarly, Frey *et al.* [42] presented an array of 11,011 Pt electrodes, along with 126 channels for extracellular bidirectional communication with electrogenic cells. The same group [43] utilized a post-processed shifted-electrode plated with rat cortical neurons. As per Table 3.1, a 1,024-channel microelectrode array (MEA) with 26,400 of these Pt electrodes was created to record and stimulate electrogenic cells. Viswam *et al.* [44] adopted a similar approach to fabricate a densely packed 59,760-array of Pt-microelectrodes within a $4.5 \times 2.5\ mm^2$ sensing region.

3.1.8 TiN electrodes

Another type of electrode that has been used in biosensing applications is the TiN electrode (refer to Figure 3.1(h)) [45,73]. In one study by Lopez *et al.* [73], a chip containing 16,384 TiN-electrodes was developed on top of a Si substrate using a 0.13 μm Al CMOS technology with a six-metal-layer Al back-end-of-line (BEOL) stack. To fabricate the chip, the top layer of the last BEOL metal layer (M6) was passivated with SiO_2 using PECVD in the first step. Another layer of SiO_2 was added to achieve the appropriate passivation thickness. In the subsequent step, RIE was employed to open via holes in the passivation layer to reach the M6 contacts. Then, a Ti/TiN seed layer was added and covered by a TiN layer using reactive physical vapor deposition. The stack was patterned with RIE. The functionality of this chip for neuro-electrophysiological studies was demonstrated by the group [45]. Table 3.1 indicates four different sizes of electrodes presented on the chip, ranging from $2.5 \times 3.5\ \mu m^2$ to $11 \times 11\ \mu m^2$.

3.2 Summary

In this section, the focus is on various methodologies applied to integrate sensing electrodes above CMOS chips. The selection of electrode arrangement hinges on the biosensing technique employed. Crucially, these on-chip setups boast straightforward design and fabrication, underscoring a key advantage. However, constraints arise in opting for ultra-compact dimensions or materials directly exposed to the analyte. The advent of nano-scale CMOS processes has enabled the realization of expansive arrays of nano-scale electrodes. A balance must be struck between the electrodes' minimal features, sensor velocity, and the voltage or current applied, demanding thoughtful consideration.

Chapter 4

CMOS capacitive biosensors

This chapter is focused on reviewing and describing complementary metal-oxide-semiconductor (CMOS) capacitive sensors, their transducers, and interface circuits. As the electrodes used in biochemical sensors are in direct contact with the liquid analyte, we will first explore the related principles of such capacitive transducers in this chapter. Following that, we will review various techniques that have been reported for the readout of CMOS capacitive sensors.

4.1 Electrode–electrolyte interfaces

In the context of interfaces between electrodes and solutions, the capacitor consists of a working electrode (WE) as one of its plates, which is typically coated with a thin insulating layer. The WE is then placed in an electrolyte solution with an auxiliary electrode to create a two-electrode configuration [74]. By applying a polarization voltage, an electric double-layer (EDL) forms at the interface between the WE and the solution, which is characterized by an EDL capacitance (C_{dl}) (as illustrated in Figure 4.1). This type of capacitor is made up of two layers of ions with opposite polarities on the surface of the metal electrode and in the electrolyte solution, separated by a single layer of solvent molecules attached to the surface of the WE, which act as the dielectric of the capacitor.

The layer of Helmholtz or Stern is created by solvent molecules, and sometimes by specifically adsorbed ions or molecules in the solution. The outer Helmholtz plane (OHP) is formed by the closest solvated ions to the metal electrode, which is the collection of centers of the adsorbed ions on the insulator surface. These ions form pairs with the charged surface sites, creating the inner Helmholtz plane (IHP) [75]. The Stern capacitance, C_{Stern}, is described by [76]:

$$C_{\mathrm{Stern}} = \frac{\varepsilon_0 \varepsilon_{\mathrm{r}}}{x_{\mathrm{H}}} \tag{4.1}$$

where ε_0 and ε_{r} are the vacuum permittivity and the relative permittivity of the dielectric material, respectively. The distance of the OHP from the metal electrode is denoted by x_{H}. Additionally, a diffuse layer extending from the OHP into the solution bulk is created by the distribution of non-specifically adsorbed ions, which was first suggested by Gouy [77] and Chapman [78]. The thickness of the diffuse

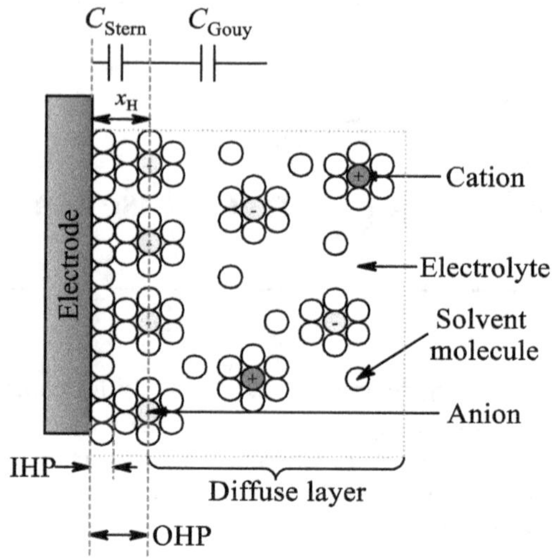

Figure 4.1 Electrode–electrolyte interface

layer, represented by Debye length (L_D), is influenced by the concentration of electrolyte in the solution, and therefore its conductivity. This diffuse layer can be modeled by a capacitor known as C_{Gouy} [76]:

$$C_{Gouy} = \frac{\varepsilon_0 \varepsilon_r}{L_D} \cosh\left(\frac{z\psi_0}{2V_T}\right) \tag{4.2}$$

The symbol V_T represents the thermal voltage that can be defined as $V_T = k_B T/q$, where k_B represents the Boltzmann constant, T is the temperature, and q is the electronic charge. The potential at the electrode is denoted as ψ_0. The ion's valence is denoted by the symbol z. C_{Dif}, the interfacial capacitance (or in other words differential double-layer was capacitance), is obtained by the combination of Helmholtz (Stern) layer and the diffuse layer, which are connected in series:

$$\frac{1}{C_{Dif}} = \frac{1}{C_{Stern}} + \frac{1}{C_{Gouy}} \tag{4.3}$$

Electrodes that have not undergone passivation exhibit lower impedance than their passivated counterparts. However, to minimize the faradaic current between the electrode and the electrolyte solution, non-passivated electrodes are often covered with a semi-passivated layer, such as a thin chemical or biological layer. This insulating layer has a capacitance known as C_{ins}. The thickness of the EDL depends on the thickness of the IHP, OHP, and diffuse layers. Upon modifying the electrode surface with an insulating layer and recognition layer, the distance between the plates increases. After specific probe-target binding, the distance between the plates further increases, making the electrode–electrolyte interface behave like a capacitor. C_{tot} is a combination of three capacitors in a series including C_{ins}, C_{bio}, and C_{Gouy}:

$$\frac{1}{C_{tot}} = \frac{1}{C_{ins}} + \frac{1}{C_{bio}} + \frac{1}{C_{Gouy}} \tag{4.4}$$

C_{bio} represents the total capacitance of the surface functionalization layer, including the binding recognition elements, grafting molecules, and analyte binding, as well as any contribution from the Helmholtz layer.

When the target biospecies binds with the recognition elements, it alters the value of C_{bio} and as a result, C_{tot} changes. The extent of the change in C_{tot} is determined by the capacitance with the lowest value. Therefore, to obtain larger values for C_{ins}, the insulating layer should be thin and have a high dielectric constant [79].

4.2 Capacitive transducers

The basic configuration of a capacitive sensor's capacitor involves two electrodes in parallel with a dielectric material sandwiched between them, and the capacitance value can be calculated by:

$$C = \varepsilon_0 \varepsilon_r \frac{A_c}{d_c} \tag{4.5}$$

The capacitance value of a capacitive sensor is calculated using the vacuum permittivity, ε_0, and the relative permittivity of the dielectric material, ε_r, along with the overlapping area of the electrodes, A_c, and the distance between the plates, d_c. Modifications to any of these parameters will affect the capacitance value. So far, no published research exists on utilizing changes in the overlapping area of electrodes for biochemical sensing purposes [79].

4.2.1 *Floating electrodes*

A floating electrode consists of a WE submerged in a liquid sample, which creates an electrode–electrolyte interface, and a reference electrode (RE) that is assumed to be grounded at an infinite distance. In CMOS technology, the WE is implemented using the top metal layer, and the RE is a grounded wire connected through the bonding pad. Living cell monitoring and cell proliferation tracking were performed by Prakash *et al.* [29] using this configuration, as depicted in Figure 4.2(a–d), which shows a floating electrode with a cell exposed to a low-frequency electric field. The cell-culture medium, acting as a conducting layer, covers the electrode, and the cells act as an insulation layer between the ionic medium and the metal electrode, leading to membrane polarization, as shown in Figure 4.2(a–c). When cells are in contact with the electrode, cell-substrate interaction occurs in three phases: sedimentation,

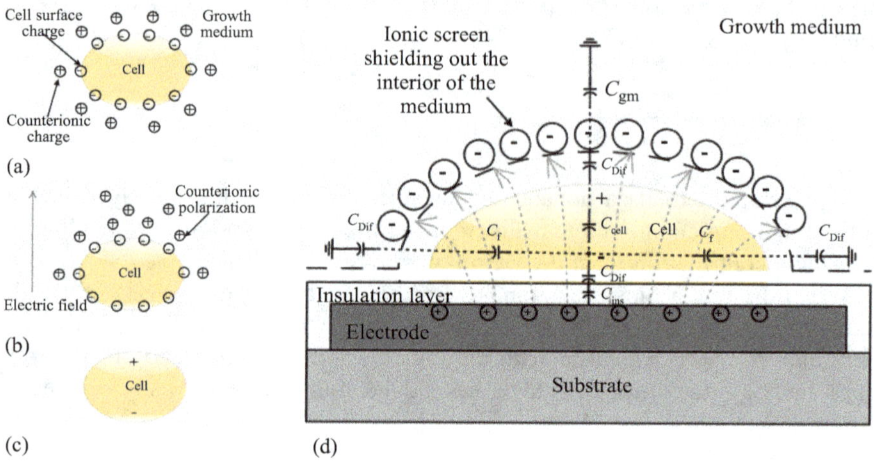

Figure 4.2 (a) Cell insulation suspended within the conductive medium, (b) electric field exposure triggers polarization, (c) creation of cell dipole moment, and (d) cell exposed to an electric field, connected to the floating capacitor (during adhesion and proliferation stages)

adhesion, and proliferation, as seen in Figure 4.2(d). The capacitance, C_{sensed}, can be calculated as follows:

$$\frac{1}{C_{sensed}} = \frac{1}{C_{Cell} + C_f} + \frac{1}{C_{ins}} + \frac{1}{C_{Dif}} + \frac{1}{C_{gm}} \tag{4.6}$$

The capacitance of the cell layer and insulation layer are represented by C_{cell} and C_{ins}, respectively. C_{Dif}, the interfacial capacitance, can be modeled using the Gouy-Chapman-Stern theory [80]. The fringe capacitance, C_f, arises from the lateral coupling of the electrode with neighboring metal lines, as shown in Figure 4.2(d). The growth medium creates an ionic screen that shields the interior from the sensing field. As cells adhere and proliferate, they become exposed to the sensing field within the ionic screen. The bulk of the growth medium can be considered an ideal ionic conductor that is electrically neutral under equilibrium conditions. C_{gm} is the total capacitance of this conductor above the cell layer to all neighboring conductors (including voltage biases, power supply, and ground metal) except for the sensing electrodes. Healthy cells adhere more tightly to the electrodes than dead or unhealthy cells, resulting in higher sensed capacitance values due to the stronger capacitive coupling between them and the underlying electrodes, as stated in Reference [29].

4.2.2 Interdigitated electrodes (IDEs)

The IDE operates by detecting changes in the dielectric constant, as depicted in Figure 4.3(a). By functionalizing the IDE with probe molecules, it is possible to

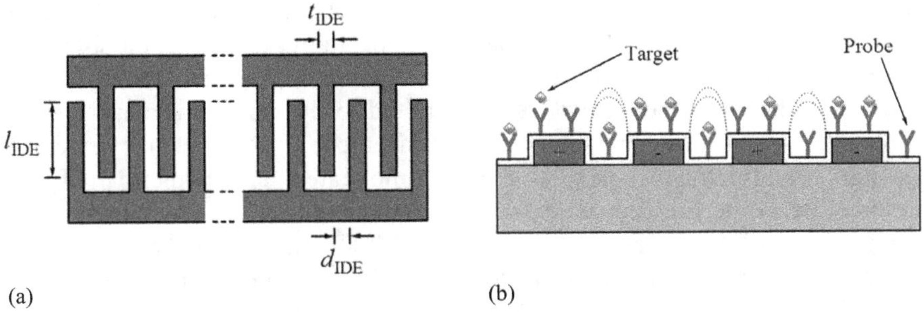

Figure 4.3 (a) Plan-view of an integrated sensing device and (b) Cutaway depiction of a modified integrated sensing device demonstrating probe-target binding

monitor reactions with target molecules and measure changes in capacitance or impedance, as shown in Figure 4.3(b). Equation (4.7) [79] can be used to simplify the capacitance of an IDE, assuming that the electrodes are thick and edge effects can be disregarded.

$$C = n_{IDE}\varepsilon \frac{l_{IDE}t_{IDE}}{d_{IDE}} \tag{4.7}$$

where n_{IDE}, t_{IDE}, and l_{IDE} represent the number, the thickness, and the length of IDE fingers, respectively. ε denotes the permittivity of the material that covers the electrodes, while d_{IDE} signifies the distance between them.

In their work, Den Otter [81] made an approximation of the capacitance of an IDE by assuming that the thickness of the fingers can be neglected compared to other dimensions. They considered a large number of parallel strips with uniform width (w_{IDE}) and space (d_{IDE}) between them. They defined λ_{IDE} as the IDE spatial wavelength or the periodicity of the fingers, given by $\lambda_{IDE} = 2(w_{IDE} + d_{IDE})$. Based on this, they demonstrated that the following equation holds true:

$$C = l_{IDE}a_{IDE}\frac{8}{\pi\lambda_{IDE}}\varepsilon\sum_{n=1}^{\infty}\left[\frac{1}{2n_{IDE}-1}J_0^2\left(\frac{(2n_{IDE}-1)\pi d_{IDE}}{\lambda_{IDE}}\right)\right] \tag{4.8}$$

In this equation, J_0 refers to the zeroth Bessel function of the first kind, while a_{IDE} denotes the width of the stripped area. It should be noted that this approximation does not discriminate between the locations of charges, whether between or on the strips. Interestingly, (4.7) demonstrates that although the capacitance of IDE does not change linearly with variations in its area, there exists a linear correlation between capacitance and the permittivity of the dielectric.

IDEs, which are smaller than traditional electrodes, offer advantages such as higher sensitivity and consuming smaller samples, making them more compact, simple, and portable. However, the limitation in producing short and reproducible distances between fingers hinders their performance [82]. For an IDE-based sensor to perform optimally, its features must be comparable in size to the target biospecies [83]. When IDEs are nano-scaled, more complex modules and expensive production facilities are required. Gerwen *et al.* [84] found that the electric field between the electrodes is restricted to a layer thinner than the IDE wavelength, λ_{IDE}. Analytical models, such as those used to characterize IDE capacitance and frequency response, are complicated, and numerical approximations like Finite Element models [85] have been developed to simulate IDE behavior [86–90]. Software programs like COMSOL and Sonnet can be used for simulation and analysis. Igreja *et al.* [87,88,91] demonstrated that the capacitance of the IDE saturates at a layer thickness of half the IDE wavelength above the surface, $\lambda_{IDE}/2$.

IDE structures for capacitive and impedimetric biosensing come in both planar [92] and three-dimensional (3D) versions [93,94], utilizing insulating barriers between adjacent electrode digits. An analysis of impedance in [95] showed that planar IDEs have lower impedance but higher sensitivity to impedance changes, whereas 3D structures offer a higher interdigit surface area and more space for molecular immobilization. However, the model used in this study did not account for surface-charge-governed properties [96–98]. To address this, Bäcker *et al.* [99] compared planar and 3D structures using a model that considers these properties, revealing that both IDE structures exhibit non-linear behavior in low-ionic strength solutions due to surface resistance caused by surface charge. Bäcker *et al.* [99] assembled a charged polyelectrolyte layer on the barrier surface to affect surface resistance, showing that 3D IDEs exhibit higher surface-charge sensitivity than planar IDEs with the same fingerprint.

IDEs, which refer to coplanar electrodes made up of two planar electrodes, are commonly produced on the highest metal layer in CMOS technology. These electrodes have a broad range of applications, including but not limited to, the detection of the neurotransmitter dopamine (DA) [27], drug testing [2], and monitoring bacterial growth [35].

In Figure 4.4, a basic electrical model of a coplanar IDE is illustrated. This model comprises several components including C_{stray}, C_{inter}, C_I, R_{sol}, and C_{sol}. C_{stray} denotes the stray capacitance of the WE terminal, while C_{inter} refers to the capacitance of the dielectric interconnect. C_I represents the C_{dl} that is created at the electrode–electrolyte interface, while R_{sol} refers to the resistance of the electrolyte solution. Lastly, C_{sol} denotes the capacitance of the electrolyte solution.

4.2.3 Surface stress-based devices

In recent years, surface stress-based devices like capacitive membranes and cantilevers have gained significant attention. These devices consist of a flexible and a rigid electrode, with the former being functionalized with probe molecules. The operation of these devices is based on alterations in the distance between the plates. Upon interaction of the probe molecules with the target biospecies, surface stress is induced, leading to a change in the deflection of the flexible structure and the distance between the two capacitor plates (as illustrated in Figure 4.5(a) and (b)).

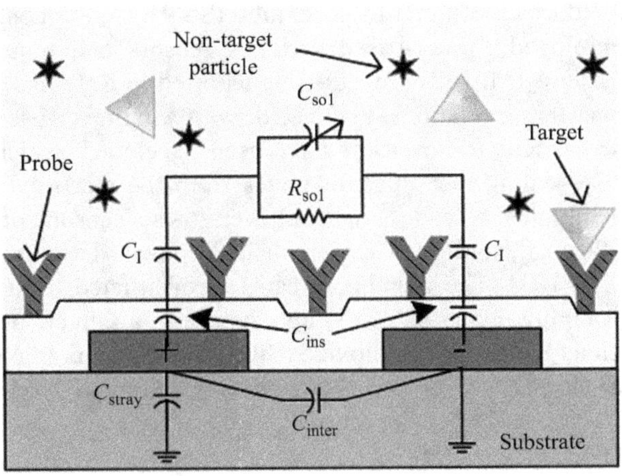

Figure 4.4 An IDE arranged on the same plane, fixed sensing elements, and target biological species

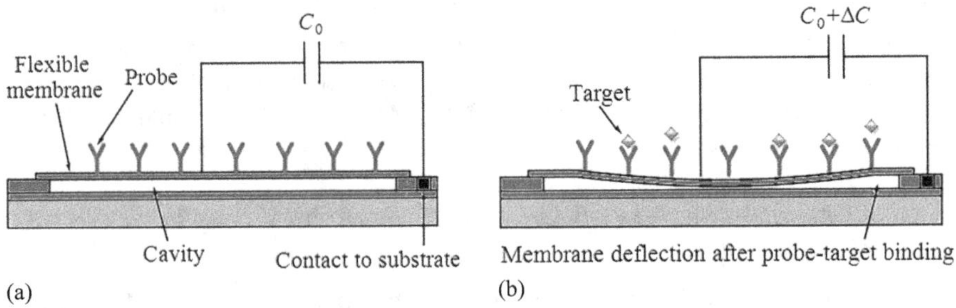

Figure 4.5 Functioning of capacitive membrane devices based on surface stress changes: (a) Pre-binding state and (b) Post probe-target binding state

The displacement of the flexible electrode, $w_{ss}(x,y)$, can be represented by (4.9), as modified in [79]:

$$C = \varepsilon_0 \varepsilon_r \iint_{A_c} \frac{1}{(d_c - w_{ss}(x,y))}. \tag{4.9}$$

Micro-cantilevers have demonstrated exceptional sensitivity, and they can be arranged in an array structure with different functionalizations. Optical and piezoresistive techniques are commonly used to detect cantilever bending, but optical methods are expensive and cumbersome, and they cannot be utilized in opaque liquids like blood. Piezoresistive techniques, on the other hand, are less sensitive and susceptible to temperature variations. Capacitive sensing offers highly sensitive detection, but the presence of electrolyte solutions leads to faradaic currents between the plates of cantilever biosensors. This makes capacitive detection challenging and reduces the signal-to-noise ratio (SNR). Consequently, cantilevers are primarily employed for air-based and gas sensors and require microelectromechanical systems (MEMS) processes, as reported in References [100,101].

To overcome the limitations associated with cantilever-based biosensors, micro-membrane capacitive biosensors have been developed, which provide reliable detection. By sealing the capacitor plates from the electrolyte solution and employing a membrane instead of a cantilever, these sensors offer sensitivity comparable to that of cantilever sensors using optical detection, as noted in References [74,102,103]. The membrane can be constructed using a conductive material, such as highly doped Si, or through the use of a thin conductive layer on an insulator, such as Si_3N_4, silicon dioxide (SiO_2), or polymers like parylene, poly (methyl-methacrylate) (PMMA), and poly(dimethylsiloxane) (PDMS), as stated in Reference [74].

4.3 Capacitive interface circuits

Here are six main categories of CMOS capacitive interface circuits, which include: (1) capacitance-to-frequency converters (CFCs) based on oscillators; (2) voltage-controlled oscillator (VCO)-based capacitive sensors; (3) lock-in detection; (4) triangular voltage analysis (TVA); (5) charge redistribution techniques; and (6) charge-based capacitance measurement (CBCM). Table 4.1 compares different CMOS capacitive biosensors.

4.3.1 Oscillator-based capacitance-to-frequency converters (CFCs)

One common approach for operating this type of capacitive sensor is through the use of relaxation oscillators or ring oscillators (ROs).

4.3.1.1 Capacitance-to-frequency converters (CFCs) using a comparator-based relaxation oscillator

Relaxation oscillators are used in certain types of sensors to modulate the output frequency in response to changes in the C_S oscillator [27,38,104]. A CFC is presented in Figure 4.6 as a straightforward example of this process. The sensing capacitor, C_S, is charged and discharged by a reference current (I_B). Next, a comparator compares the voltage across the capacitor with a threshold voltage (V_{th}), and when it reaches V_{th}, the comparator's output signal controls a switch that resets the voltage to zero. The frequency of the output pulse (f_{Out}) is inversely proportional to C_S [27], as shown in the following equation:

$$f_{Out} = \left(\frac{C_S.V_{th}}{I_B} + \tau_d\right)^{-1} \tag{4.10}$$

where the term τ_d represents the delay generated by the comparator, which is assumed to be negligible.

Table 4.1 CMOS interface circuits of capacitive biosensors

Application	Method	D/ND	Digital/ Analog	Technique	Array #	Resolution	Sensitivity	IDR	Area (mm^2)	Power (V_{dd})	Reference
Detection of single bacterial cell	CVC by ChS	ND	Analog	0.25 μm	16 × 16	450 aF	55 mV/fF	0.45 fF-57 fF	0.1(TA)	29 μW (2.5 V)	[4]
Cell proliferation monitoring	CVC by ChS	ND	Analog	0.5 μm	28	5 fF	NA	NA	2.25 (TA), 0.16 (EA)	NA (3 V)	[29]
Localization of bioparticles	CVC by CSA	D	Analog	0.35 μm	320 ×320	21 aF	345 mV/fF	0.42 fF	NA	NA (3.3 V)	[105]
Cancer biomarker detection	SC+ SAR	ND	Digital	0.18 μm	9	4.5 fF	NA	16.137 pF	0.06 (AA)	2.1 μW (0.9 V/1 V)	[106]
Detection of neurotransmitter DA	CFC by Rel.O	ND	Digital	0.35 μm	5 × 5	NA	21 kHz/fF @ 4.9 MHz @ 123 fF	12 fF-700 fF	NA	NA	[27]
Detection of DNA	CFC by Rel.O	D	Digital	0.5 μm	8 × 16	NA	23 Hz/pF @ 7.5 kHz @ 330 pF	330 pF-10 nF	28.8 (TA)	NA (5 V)	[38]
Cell volume growth monitoring	CFC by RO	D	Digital	0.13 μm	3 × 4	10 aF	235 mV/fF	+/−100 fF	2.25 (TA),1 (AA)	NA	[107]
Real-time measurement of cell proliferation	CFC by RO	ND	Digital	0.35 μm	4 × 4	14.4 aF	590 kHz/fF	12 fF	9 (TA), 0.0144 (EA)	8 mW (3.3 V)	[3]
Single cell analysis	CFC by RO	D	Digital	0.35 μm	1	14 aF	570 kHz/fF@ 1.37 GHz @ 500 fF	NA	4 (TA)	NA	[108]
Detection of cells and biomolecules (using magnetic bead detection)	CFC by RO	ND	Digital	0.35 μm	8 × 8	2.5 fF	223 kHz/fF @ 5.2 MHz @ 23 fF	NA	NA	NA	[104]

(Continues)

Table 4.1 (Continued)

Application	Method	D/ND	Digital/ Analog	Technique	Array #	Resolution	Sensitivity	IDR	Area (mm^2)	Power (V_{dd})	Reference
Bacteria detection	CFC by RO	ND	Digital	0.25 μm	1	10 fF	11 kHz/fF @ 254 MHz @ 17.5 fF	NA	0.05 (AA), 0.0506 (EA)	29 mW (2.5 V)	[33]
Detection of nano- and micro-particles	CVC by CBCM	ND	Analog	0.18 μm	3 × 3	1 fF	10 fF/mV	NA	1.8 (TA)	−(−1.2 to 1.3 V)	[109]
Cell monitoring	CVC by CBCM	D	Analog	0.5 μm	6 × 6	15 aF	200 mV/fF	25 fF	3	165 μW (3 V)	[110]
Bacteria growth monitoring	CBCM+ΣΔ ADC	ND	Digital	0.18 μm	3	10 aF	250 mV/fF	~ 2.7 fF	2 (TA)	NA (1.8 V)	[35]
Cell growth monitoring	CBCM+ΣΔ ADC	D	Digital	0.35 μm	8 × 8	10 aF	350 mV/fF	10 fF	9 (DA)	NA (3.3 V)	[37]
High-frequency impedance spectroscopy and imaging	CBCM+ADC	ND	Digital	90 nm	256 × 256	1 aF	NA	4 fF	2,525,253 (TA), (Radius of NE = 90 nm)	15 mW (1.2 V)	[40]
NA	CBCM+CCO	ND	Digital	0.18 μm	1	873 aF ^{NP}10 aF ^{P}	138 pulses/fF	70 fF	0.0645 (TA), 0.007224 (AA), 0.0236 (EA)	103 μW (1.8 V)	[111]
Detection of DNA hybridization	TVA+ SC+ ADC	ND	Digital	0.35 μm	16 × 8	1 pF	NA	10 pF-10nF	10 (TA)	NA	[112]
Airborne particle detector	Lock-in + ADC	D	Digital	0.35 μm	32	0.065 aF	NA	Less than 1 fF	6 (TA), 1.15 (EA)	84 mW (3.3 V)	[113]
Droplet analyses	CBCM+CCO	ND	Digital	0.35 μm	16 × 16	150 aF	NA	100	2.8 × 2.5 (TA)	84 μW/Ch. (3.3 V)	[114]
Chemical solvent sensing	CBCM+CCO	ND	Digital	0.35 μm	2	416 aF	2.46 pulses/fF	>400 fF	1 × 2.5 (TA)	29.91 mW (3.3 V)	[115]

ChS: Charge sharing, ND: Non-differential, Ch.: Channel, TA: Total area, EA: Electrode(s) area, CSA: Charge-sensitive amplifier, D: Differential, AA: Active area, Rel.O: Relaxation oscillator, RO: Ring oscillator, NE: Nanoelectrodes, NP: Without pre-distortion, P: With pre-distortion, TVA: Triangular voltage analysis.

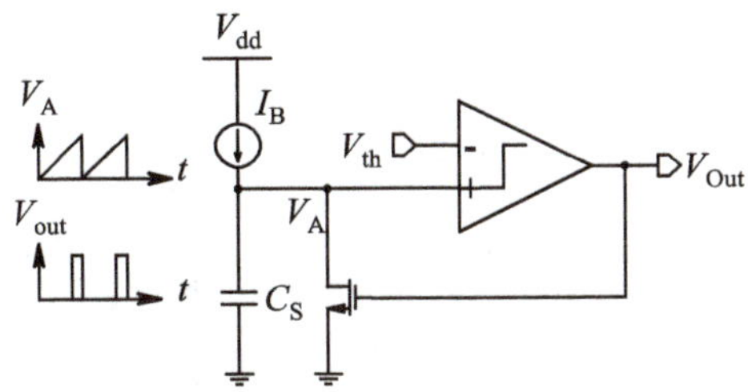

Figure 4.6 Illustration of a basic configuration for comparator-driven relaxation in CFC systems

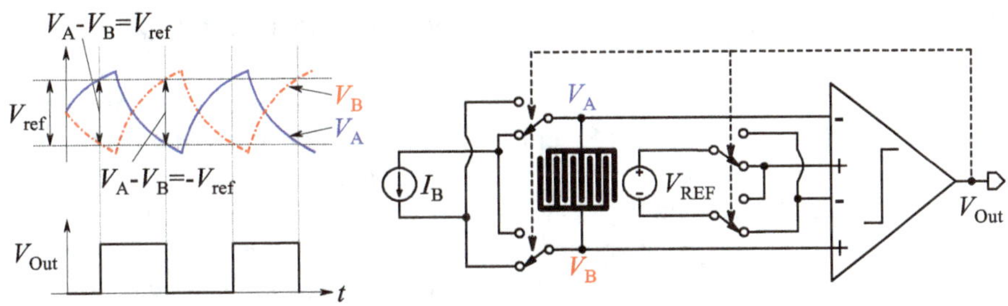

Figure 4.7 A CFC system utilizing a relaxation oscillator, featuring a pair of symmetrical threshold voltages

An alternative configuration for charging and discharging a capacitor, C_S, involves the use of two identical reference currents. One of the reference currents acts as a source while the other acts as a sink. To switch between the two currents, a comparator with hysteresis, such as a Schmitt trigger, can be used. Alternatively, as depicted in Figure 4.7 [38], a single reference current can be used along with reversible current direction switches. This setup was utilized in an 8×16-array for deoxyribonucleic acid (DNA) hybridization detection. The circuit uses a comparator with a single threshold voltage, but the switching strategy produces transient voltage waveforms, V_A and V_B, on the sensing electrode's two plates. Additionally, it generates two symmetrical threshold voltages during two phases, as shown in Figure 4.7. The time constant of these waveforms is primarily determined by the electrode–electrolyte impedance. If V_A-V_B exceeds V_{ref} or falls below -V_{ref}, the switches will reverse the direction of the reference voltage and current simultaneously. If the solution resistance is negligible, the squared waveform's frequency output can be determined using the following equation:

$$f_{Out} = \left(2RC\ln\left(1/\left(1-\frac{V_{ref}}{I_B R}\right)\right)\right)^{-1} \tag{4.11}$$

where R and C are the contributions of the resistance and capacitance of both electrode and solution. If the frequency is not too low, (4.11) can be simplified to

(4.12) by the first-order Taylor approximation:

$$f_{\mathrm{Out}} = \frac{I_{\mathrm{B}}}{2V_{\mathrm{ref}}C} \tag{4.12}$$

The counter can be utilized to gauge the input capacitance, which has an inverse correlation with the output frequency as indicated by (4.12).

The accuracy of current references and comparators is a crucial consideration in the design of comparator-based relaxation oscillators. In addition, the output frequency of such oscillators can be influenced by various factors, including parasitic capacitances, process-variable parameters (e.g., resistors), and temperature-variable parameters (e.g., transistor transconductance).

4.3.1.2 Ring oscillator (RO)-based capacitance-to-frequency converters (CFCs)

ROs can also be utilized to transform sensing capacitance into frequency [3,33,107,108,116–119]. Figure 4.8 depicts a CFC that contains three delay cells, whose delays depend on the charge of their load capacitors, facilitated through the current I_{B} to a specific voltage V_{osc}. One of the delay cells is considered the sensing capacitor, C_{S}, which can be expressed as $C_{\mathrm{S}} = C_{\mathrm{R}} + \Delta C$, where C_{R} denotes the reference capacitance and ΔC represents capacitance changes. It can be proven that the period of an M-stage RO can be determined by the following equation [107]:

$$T_{\mathrm{osc}} = 2\left[\frac{C_{\mathrm{S}}.V_{\mathrm{osc}}}{I_{\mathrm{B}}} + (M-1)\frac{C_{\mathrm{R}}.V_{\mathrm{osc}}}{I_{\mathrm{B}}}\right] \tag{4.13}$$

A differential structure for RO-based sensors is illustrated in Figure 4.9, as reported by Mohammad *et al.* [116]. This design incorporates two separate ROs, namely RO1 and RO2, with the former linked to the sensing electrode in a microfluidic channel. The XOR gate is used to detect the frequency difference between these two ROs, producing a pulse-width modulated at the difference frequency. A low-pass filter (LPF) is then utilized to remove high-frequency components, while an amplifier buffers the intermediate frequency (IF), and a frequency counter is employed to determine the difference frequency. The output IF frequency of the sensor can be adjusted using two variable capacitors.

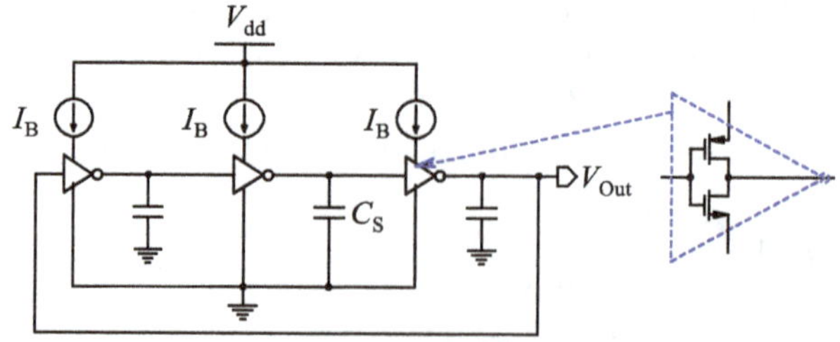

Figure 4.8 A three-stage RO-based CFC

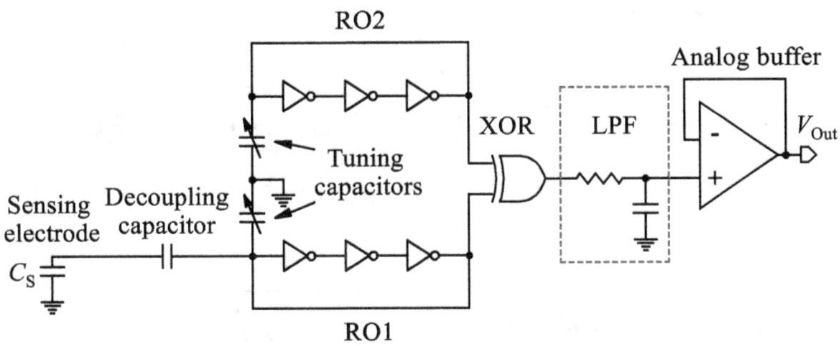

Figure 4.9 A differential RO-based CFC

The output frequencies of ROs can be susceptible to various factors such as process variations, temperature fluctuations, supply voltage fluctuations, and device sizes. Despite this, the key advantage of all CFCs is that their output is semi-digital, which can be easily converted to digital form using a counter without the need for additional analog-to-digital converters (ADCs). This approach reduces the complexity and power consumption of the system.

4.3.2 Voltage-controlled oscillator (VCO)-based capacitive sensors

VCO-based capacitive sensors are commonly employed for measuring complex permittivity and conducting dielectric spectroscopy in frequency ranges spanning from mega- to giga-Hertz. This has been highlighted in previous studies [120–128].

In capacitive sensing, VCO-based sensors are often employed for dielectric spectroscopy and complex permittivity sensing in the frequency range of mega- to giga-Hertz [120,123]. A VCO-based sensor consists of a sensing VCO and a frequency detector, as shown in Figure 4.10(a). An example of an inductor-capacitor (LC) VCO-based sensor is depicted in Figure 4.11, where C_S is a component of the LC tank [120,123]. The presence of an analyte causes electric field perturbations, leading to a frequency shift in the VCO. The relative frequency shift with respect to f_0, the LC oscillator's frequency before exposure, can be obtained by [128]:

$$\frac{\Delta f}{f_0} = \frac{1}{\sqrt{1 + \Delta C_{tot}(f_0)/C_{tot}}} - 1 \tag{4.14}$$

where C_{tot} represents the total capacitance, which includes C_S, parasitic capacitors, and varactors, and $\Delta C_{tot}(f_0)/C_{tot}$ indicates the relative shift in the total capacitance with respect to C_{tot}.

Single VCO-based sensors have a major drawback in that they are highly sensitive to low-frequency temperature and environmental effects. To overcome this issue, a differential technique such as the one demonstrated in Figure 4.10(b) can be employed to enhance the performance of the sensors. This technique involves utilizing two VCOs—a sensing VCO and a reference VCO—and

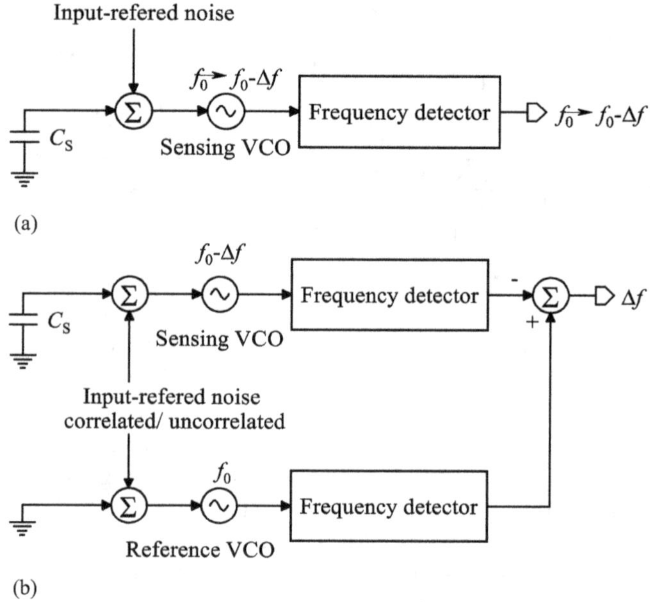

Figure 4.10 (a) Configuration of a solitary sensor based on a VCO and (b) Setup involving a VCO-centered sensor integrating both a sensing and a referencing VCO

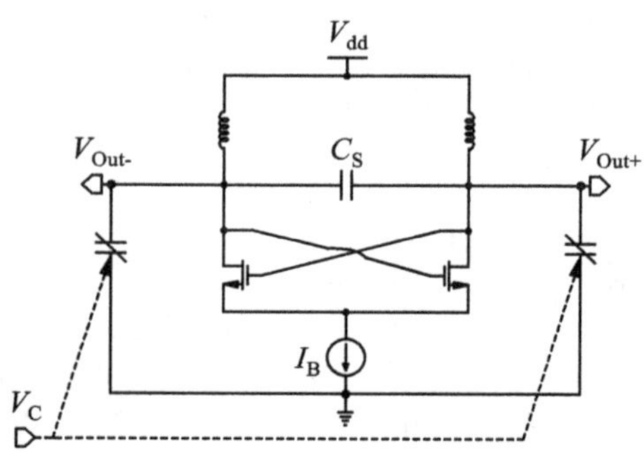

Figure 4.11 A capacitive sensor utilizing an LC VCO

differentiating their output frequencies. However, it is important to note that the two VCOs need to be as similar as possible for optimal performance. To prevent VCO frequency pulling and filter out correlated low-frequency noise, only one VCO should operate at a time, with periodic activation of the two VCOs. While a frequency counter can be used to detect frequency, it may require a long measurement time. Furthermore, the presence of an open-loop system containing the VCO can result in absolute oscillator frequency drifts, thus leading to imprecise measurements [123].

To measure the capacitance at a specific frequency with improved speed, a closed-loop frequency detector such as a phase-locked loop (PLL) architecture can be employed. This method allows for the tuning and stabilization of the frequency shift, resulting in a faster measurement time. The control voltage (V_C) can be used to represent variations in the free-running frequency of the VCO and capacitive sensor (C_S). This V_C can be digitized by an ADC [120,121]. A readout circuit for a dielectric sensor is shown in Figure 4.12, which includes an ADC and a frequency synthesizer loop [120]. The frequency synthesizer loop consists of a VCO, a frequency divider with a fixed frequency division ratio (N_{div}), a phase and frequency detector, a loop filter, and a charge pump. This loop tunes the VCO control voltage (V_C) such that the output frequency of the VCO (f_{Out}) is equal to N_{div} multiplied by the reference frequency (f_{ref}). If N_{div} and f_{ref} remain constant, then f_{Out} remains constant. Therefore, any change in the sensing capacitor is compensated for by a corresponding change in V_C. The voltage across the capacitor C_1 in the loop filter changes according to V_C and is then fed to the ADC for further processing and compensation. V_C is digitized by the ADC and used to control either a variable f_{ref} or a variable N_{div}. It is important to note that the resolution of the ADC can limit the accuracy of the frequency shift detection. Therefore, care must be taken to ensure that the ADC is of sufficient resolution to accurately detect any changes in V_C.

Elhadidy *et al.* [128] have introduced an RO-based PLL with a semi-digital output, which incorporates a sensing VCO and a reference VCO. The sensing VCO is designed to detect changes in capacitance and conductance of the sensing capacitor loads of a delay cell. Meanwhile, the reference VCO is immune to the analyte and is controlled by an amplitude-locked loop (ALL).

According to some studies [125–127], the usage of quadrature voltage-controlled oscillators (QVCO) can enhance precision and SNR in various applications. These circuits are made up of LC tanks that are part of injection-locked oscillators (ILOs). For instance, in their work, Chien *et al.* [125] employed an ILO-based dual-channel interferometer capacitive sensor for high-speed flow cytometry. Figure 4.13 shows a single sensing channel of this sensor, which comprises a

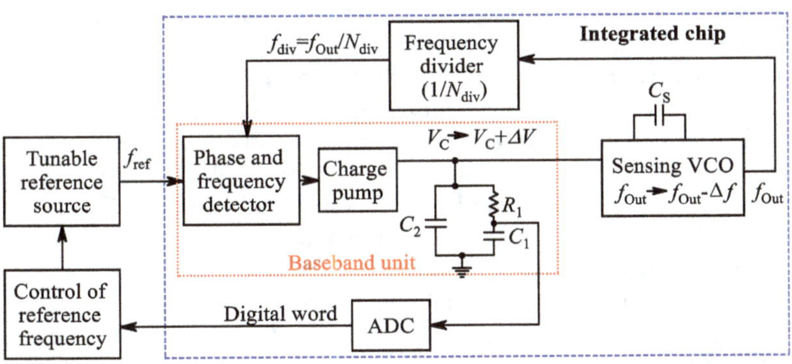

Figure 4.12 A sensor utilizing a VCO with a frequency detection system comprising a PLL and an ADC

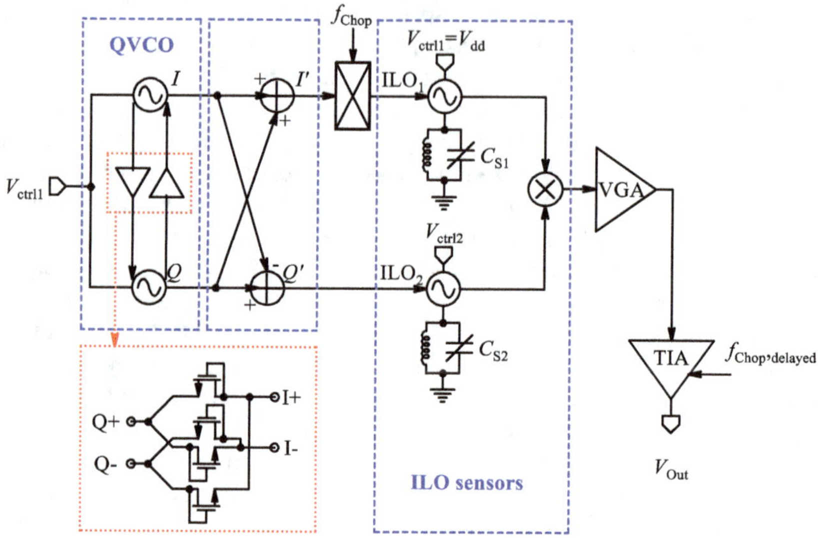

Figure 4.13 A sensing channel of a dual-channel CMOS capacitive sensor utilizing an ILO

QVCO, a mixer used as a phase detector, two ILOs, and an analog front-end for signal conditioning. In this circuit, the sensing oscillators (ILO) are locked by an injection source at a similar frequency. The amount of phase shift between the input and output is proportional to the frequency changes and can be calculated using the Barkhausen criteria [125]:

$$\sin\theta \approx \frac{2Q_{LC}}{\omega_0}\frac{I_{osc}}{I_{inj}}\left(\omega_0 - \omega_{inj}\right) \tag{4.15}$$

where ω_{inj} denotes the injection source frequency, which deviates from the free-running frequency of the frequency-locked oscillator, ω_0. Q_{LC} refers to the quality factor of the LC tank, and I_{osc} and I_{inj} represent the currents of the oscillator and injection source, respectively. The ILO acts as a phase amplifier that precedes the phase detector and has a gain that can be regulated via I_{osc}/I_{inj}. A diode-coupled QVCO is used for injection locking of the sensing oscillators. To reduce the effects of flicker noise, chopper stabilization is utilized in one of the signal injection paths. The mixer gain can be maximized by adding and subtracting the I and Q signals with/from each other before injection and removing the quadrature phase error to the first order. After the mixer, a variable gain amplifier (VGA), a baseband amplifier, and a transimpedance amplifier (TIA) are used for demodulation. In another study, the same group [129] designed a CMOS/microfluidic platform that utilized a similar circuit in an arrayed format and distributed four sensing channels at 6.5, 11, 17.5, and 30 GHz with the frequencies centered at 18 GHz (around the water relaxation frequency).

According to a study [120], VCO-based sensors require a costly tunable frequency source and the LC tanks' inductors are typically bulky, requiring significant separation to prevent substrate or electromagnetic coupling, resulting in a larger occupied area. To address these issues, Nagulapalli *et al.* [130] proposed the use of relaxation oscillators in place of VCOs to enhance the sensitivity of frequency deviation-based sensors.

4.3.3 Lock-in detection

In the field of impedance spectroscopy, a technique widely used for this purpose is discussed in Section 5.3.2.1 in Chapter 5 [6,39,131,132]. However, this technique can also be applied to capacitive sensors. For instance, Ciccarella and colleagues [113,133] utilized this technique to achieve a resolution better than 100 zepto-Farad (zF) for counting single micrometer-sized airborne particles in a 32-channel sensor [105,106]. Nonetheless, this approach presents some limitations, such as increased complexity and issues with mismatch and synchronization in the sensor. The schematic of each channel of this sensor, as shown in Figure 4.14, includes a charge preamplifier with adjustable high-pass filtering, a first-order g_m-C tunable LPF, a square-wave mixer, and a digital network. The C_{up} and C_{dw} electrodes share a common electrode connected to the charge preamplifier. The other two sides of the electrodes are driven by counter-phased square wave voltages produced by two on-chip clocks, Φ_1 and Φ_2. The amplitude of these voltages is V_{Ref}, equivalent to the

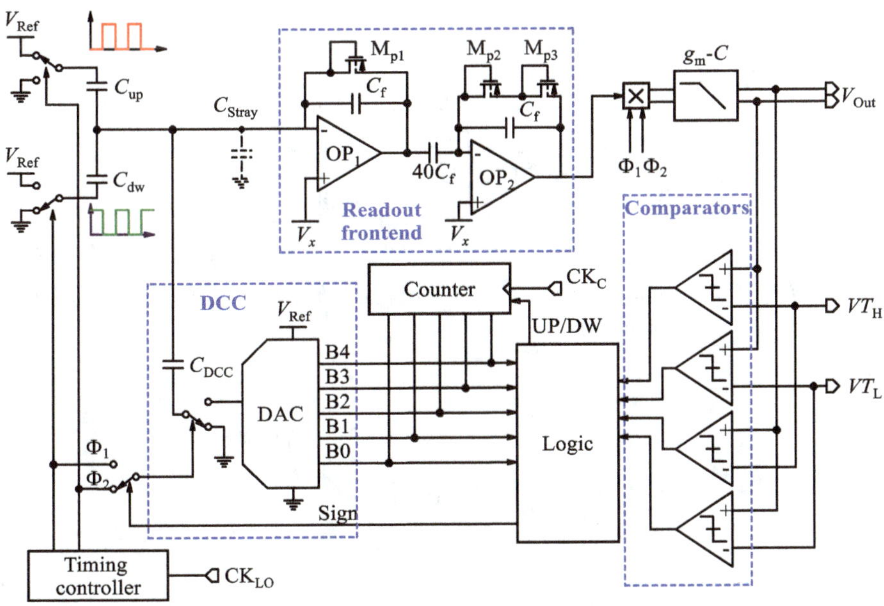

Figure 4.14 Each channel of a 32-channel CMOS capacitive sensor including a charge preamplifier with adjustable high-pass filtering, a first-order gm-C tunable LPF, a square-wave mixer, and a digital network

supply voltage. The use of a rail-to-rail square waveform as the excitation voltage, instead of a sinusoidal signal, maximizes the charge injected in the preamplifier. The charge preamplifier reads the current of the sensing capacitor with the help of its capacitive feedback. Ideally, the difference in charge on one of the two electrodes is measured. The differential structure of the sensor allows for the elimination of the large capacitance of the sets of electrodes and the noise of V_{Ref} and capacitance fluctuations, resulting in a high gain for the input amplifier. The lock-in demodulation is performed by a non-return-to-zero (NRZ) passive mixer and the low-pass g_m-C filter. The voltage amplifier placed between the lock-in demodulator and the charge preamplifier limits the noise of the g_m-C filter. Additionally, to maintain a balanced condition of the differential input capacitance and handle the saturation of the front-end amplifier, a digital-to-capacitance converter (DCC) was employed.

4.3.4 Triangular voltage analysis (TVA)

It is feasible to measure the capacitance and resistance of an analyte individually through a specific technique. This involves using a triangular voltage signal generated by a pulse current stimulator, as demonstrated in the first scenario presented in Figure 4.15. Alternatively, the same technique can be used to stimulate the analyte, as shown in the second scenario in Figure 4.16.

The calculation for obtaining CS in the first scenario, as demonstrated in Figure 4.15, involves the use of a current-controlled oscillator (CCO) for the

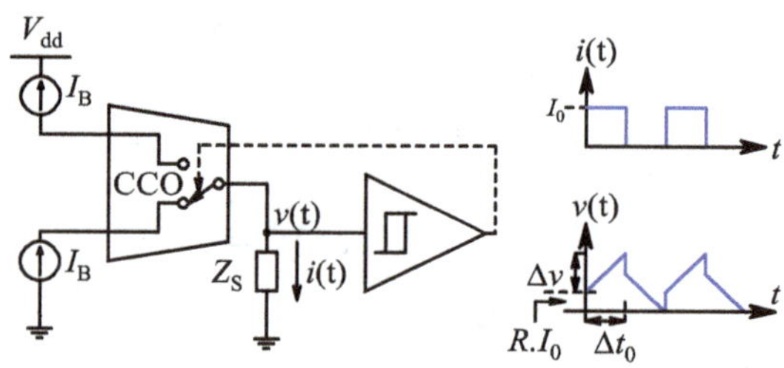

Figure 4.15 The first scenario of the TVA approach with an output triangular voltage waveform

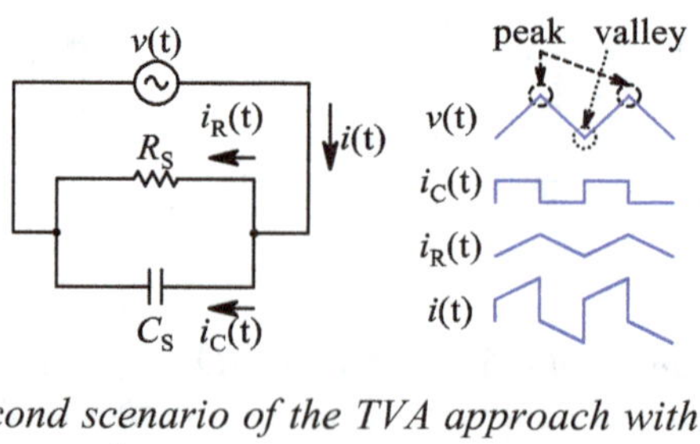

Figure 4.16 The second scenario of the TVA approach with an input triangular voltage waveform

charging and discharging of C_S, thereby eliminating the need for a variable signal source to stimulate the circuit. This method utilizes the slope of each rising or falling segment of the output voltage, which is represented by the ratio of $\Delta v/\Delta t_0$. According to the research conducted by Druart *et al.* [134], the formula for C_S can be expressed as $C_S = (I_0.\Delta t_0)/\Delta v$.

In the second scenario depicted in Figure 4.16, a triangular voltage waveform is utilized to stimulate the analyte, leading to the generation of a square current that flows through the sensing capacitor (C_S). The slope of the stimulus signal determines the amplitude of this current. The "valley" and "peak" regions of the input signal, which have non-zero second derivatives, are valuable for sensing capacitance measurement. After the output current reaches a steady-state value, the output can be evaluated. Lee *et al.* [112] utilized a switch-capacitor (SC) circuit in a 16 × 8 array to process the output current, for example. Direct measurement of impedance from the output current or voltage is feasible with TVA, without the need to evaluate the amplitude and phase shift of the input impedance [112].

4.3.5 Charge redistribution techniques

There are three subcategories into which these techniques can be classified: (1) those based on the charge-sharing principle, (2) those based on charge-sensitive amplifiers (CSAs), and (3) SC circuits.

4.3.5.1 Charge sharing principle

The operational principle of this method is founded on the manipulation of the charge and discharge cycles of multiple capacitors. A basic arrangement of this approach is depicted in Figure 4.17 [29,30]. In this particular sensing system, the charging and discharging processes of the capacitors located at N_1 and N_2 are governed by the switches $M_{1\text{-}3}$.

The circuit is subjected to two distinct phases, namely the reset phase and the evaluation phase, by controlling clock pulses. In the reset phase, the nodes N_1 and N_2's capacitances are charged with V_{dd} and V_{ss}, respectively. In contrast, during the evaluation phase, the charge distribution between the capacitances of the nodes N_1

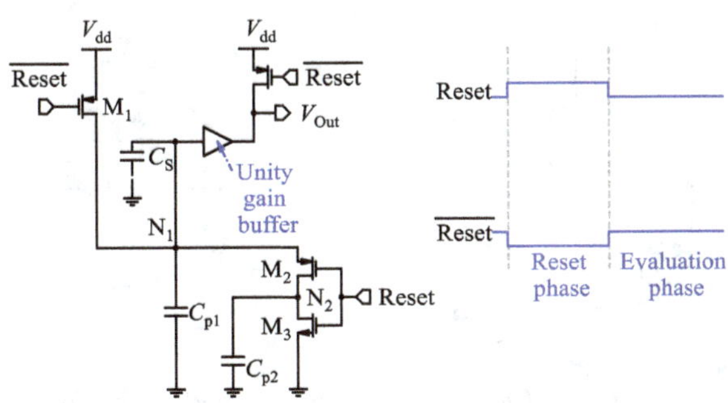

Figure 4.17 A basic charge-sharing-based sensor

and N_2 occurs. The output voltage's value is determined by C_S and its mathematical expression can be given by:

$$V_{Out} = \frac{(C_{p1} + C_S).V_{dd} + C_{p2}.V_{ss}}{C_{p1} + C_{p2} + C_S} \tag{4.16}$$

The notation C_{p1} and C_{p2} are utilized to represent the parasitic capacitances of the nodes N_1 and N_2, respectively.

Couniot *et al.* [4] conducted a study on a capacitive sensor array using the charge-sharing principle. The array consists of 16 × 16 pixels, and Figure 4.18 displays one pixel of the array. To perform the sensing operation, four distinct phases are employed: initialization, reset, integration, and readout. During the initialization phase, both V_g and V_S voltages are grounded. In the reset phase, the switch M_3 sets V_{pix} to its reset value. Subsequently, M_1 is activated to load C_S to V_{ref}, and M_2 is enabled, which causes the charge distribution between C_S and the fixed capacitance $C_D + C_g$ (where C_g denotes the capacitance of the node V_g). As a result, V_g is a function of C_S and is calculated using the equation mentioned below:

$$V_g = \frac{C_S.V_{ss}}{C_D + C_g + C_S} \tag{4.17}$$

In the third phase, the value of V_g remains constant. When operating in the subthreshold region, M_4 exponentially amplifies V_g. This amplification leads to the progressive discharge of V_{pix} by M_4. Figure 4.18 does not depict the column amplifiers, which store the initial and final values of V_{pix} during the integration period. In the fourth phase, these values are transmitted by the sensor to the output stage. Subsequently, the output analog voltage can be converted to digital using an off-chip ADC. This sensor offers advantages over References [29,30], including an additional subthreshold gain stage, a more linear response, and the ability to calibrate with V_{ref}.

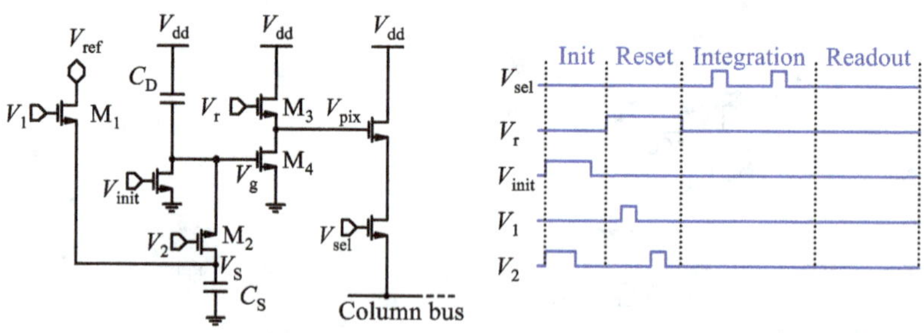

Figure 4.18 A single pixel of a charge-sharing-based sensor array with four distinct phases

4.3.5.2 Charge-sensitive amplifier (CSA)-based techniques

An alternative method known as the CSA technique is utilized for charge redistribution purposes. The configuration depicted in Figure 4.19 demonstrates a straightforward implementation of this technique [105,135,136]. In this particular circuit, a voltage pulse denoted as **Φ** is applied to C_S. Once this pulse concludes, the charge stored in C_S is redistributed into a reference capacitor, C_R. Subsequently, a capacitive feedback amplifier integrates the voltage across C_R. To prevent amplifier saturation, a reset switch is incorporated into the feedback path. By comparing the charge on C_S and C_R and completing the integration process, it can be easily established that the output voltage of the integrator is directly proportional to the disparity between C_S and C_R [136]:

$$V_{Out} = V_A \frac{C_S - C_R}{C_{fb}} \tag{4.18}$$

V_A represents the amplitude of the **Φ** signal in this context. Although the periods of the reset signal and **Φ** are equal, the duty cycle of the reset signal is lower than that of **Φ**. Whenever the reset signal becomes high, it causes the feedback capacitor, C_{fb}, to discharge. If the non-inverting input of the amplifier is linked to V_{ref} (rather than ground), then the output voltage is increased by V_{ref}.

De Venuto *et al.* [135] employed a technique that utilized a CSA for measuring the output voltage without the use of a reset switch for discharging the feedback capacitor (C_{fb}). Instead, a resistor ($R_{discharge}$) was employed for the same purpose. The schematic diagram depicting this method is shown in Figure 4.19 [136]. The mathematical expression for the output voltage can be derived through:

$$V_{Out} = -\frac{Q}{C_{fb}} \exp\left(-\frac{t}{\tau}\right) \tag{4.19}$$

The feedback capacitor charge integration during each period of **Φ** is denoted by Q, where $\tau = C_{fb}R_{discharge}$. However, the exponential discharging time poses a challenge in achieving linear discrimination of the pulse width containing the necessary information. Therefore, it is preferable to incorporate a feedback loop with a constant discharging current. When a voltage step is applied at the input, a positive and negative pulse is produced at the output. To exclusively detect capacitive value from positive pulses, De Venuto *et al.* suggested the utilization of Schottky diodes in the feedback loop [135].

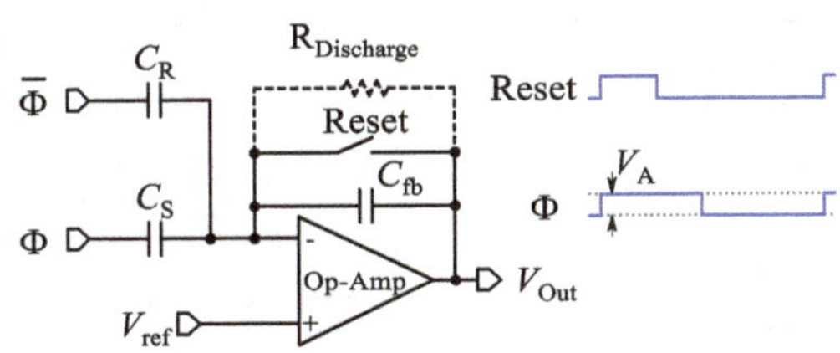

Figure 4.19 Schematic of a CSA

4.3.5.3 Switch-Capacitor (SC) circuits

SC circuits bear a resemblance to CSAs in their operation. However, they differ in that their input accepts a direct-current (DC) voltage rather than a voltage pulse. As demonstrated in Figure 4.20, multiple switches facilitate the charging and discharging phases of the sensing and reference capacitors. This design feature helps to counteract parallel parasitic capacitances by keeping the input impedance low [136].

Frequency modulation of a sawtooth waveform output can be achieved by employing an amplifier and a counter. However, if the output voltage is in the form of a level, an ADC is required to convert it into a digital signal [137,138]. Various types of ADCs can be utilized, such as the successive-approximation register (SAR) ADC, the successive sigma-delta analog-to-digital converter (ΣΔ ADC), and the pipeline ADC, etc. [136]. These ADCs can be implemented with SC circuits, which can convert the input signal into a digital format either directly or after undergoing voltage conversion [106,139,140].

For instance, Alhoshany *et al.* [106] used an SC circuit in conjunction with an SAR ADC. Figure 4.21 shows the circuit's components, which include a binary-weighted

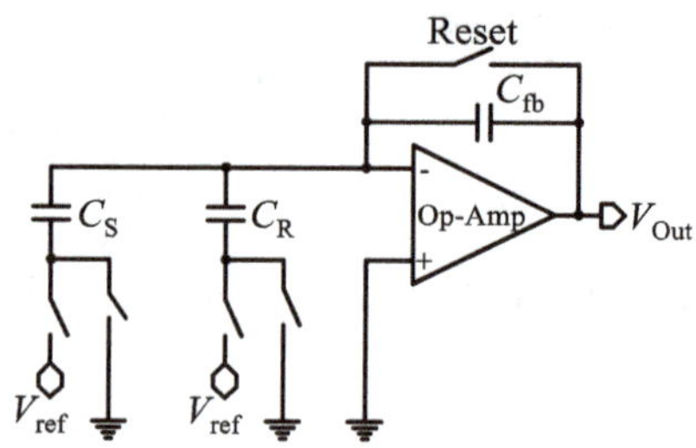

Figure 4.20 A SC circuit

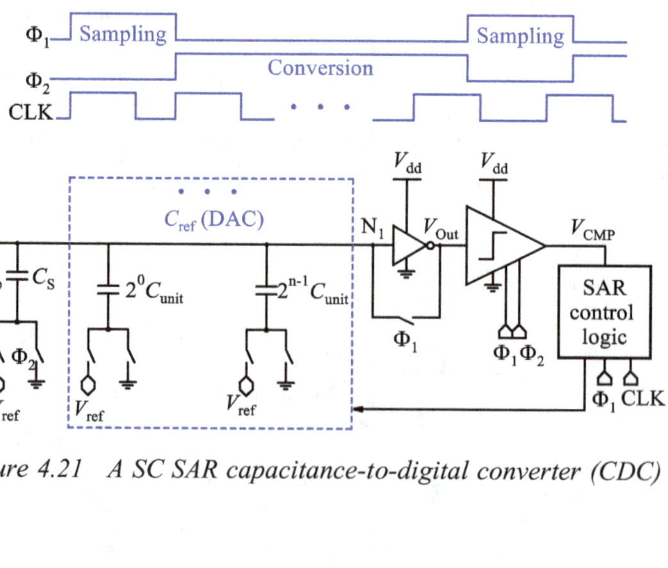

Figure 4.21 A SC SAR capacitance-to-digital converter (CDC)

digital-to-analog converter (DAC), a single-ended comparator, control SAR logic, and a feedback amplifier. This sensor operates in two phases: (1) sampling, and (2) conversion, based on clock pulses Φ_1, Φ_2, and a SAR clock (*CLK*).

During the sampling phase, Φ_1 becomes high, and the feedback switch is activated, forming a unity-gain amplifier. The binary-weighted DAC, C_S, and parasitic capacitance, C_p, are then charged using the switching threshold voltage (V_{N1}). The bottom plate of C_S is connected to V_{ref}, while the binary-weighted DAC is connected to the ground. Turning off the comparator reduces the average power consumption.

In the second phase of the process, the signal Φ_1 experiences a low state which causes the feedback switch to turn off, creating an open-loop amplifier. At this point, the *n*-bit SAR logic output determines the connection of the bottom plates of the binary-weighted DAC to either ground or V_{ref}. The comparator is then activated, resulting in cascaded amplification. By comparing the input capacitance with the digital output of the DAC, the amplifier produces an output equal to:

$$V_{Out} = V_{N1} + \Delta V_{Out} = V_{N1} + \frac{\left(C_S - C_{ref,ON}\right) V_{ref}}{C_S + C_{ref} + C_p} A_{inv} \tag{4.20}$$

The second expression represents the fluctuations in output voltage, with A_{inv} representing the limited gain of the inverter. The comparator's output voltage can be mathematically represented as:

$$V_{CMP} = \begin{cases} 0, \Delta V_{Out} < 0 \\ 1, \Delta V_{Out} > 0 \end{cases} \tag{4.21}$$

The algorithm known as successive approximation aims to modify the reference voltage $C_{ref,ON}$ until it corresponds with the sensed voltage C_S, with an error margin of one least significant bit. The determination of V_{CMP} is based on the difference between C_S and $C_{ref,ON}$, without regard for ΔV_{Out}'s absolute value. As a result, changes in V_{ref} and C_p do not affect the sensor's sensitivity.

In general, there are certain issues that arise with SC circuits, including the addition of offsets to the circuit due to switches, as well as charge injection. Techniques such as auto-zeroing and double sampling are commonly employed to address these issues. However, these techniques can lead to higher power consumption and increased complexity of the circuit as a whole. It is worth noting that these problems are commonly encountered in various fields where SC circuits are used, including but not limited to electronics and telecommunications. To alleviate these problems, it is necessary to develop more advanced techniques that can reduce power consumption while maintaining the desired performance of the circuit.

4.3.6 Charge-based capacitance measurement (CBCM)

The CBCM method utilizes a reference capacitor (C_R) and a sensing capacitor (C_S) along with a bio-interface layer to detect variations in the analyte. Pseudo-inverters consisting of pull-up and pull-down switches are used to charge and discharge the

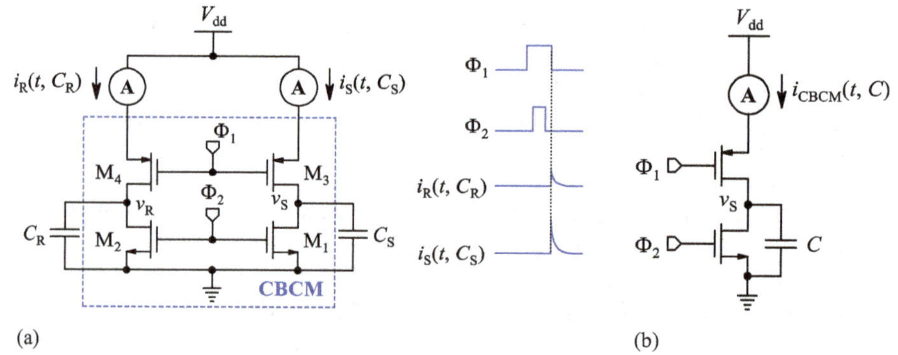

Figure 4.22 (a) The core of CBCM with two electrodes and (b) The core of CBCM with one electrode

capacitors in a controlled manner (see Figure 4.22(a)). Clock pulses Φ_1 and Φ_2 are used to control these switches and prevent short-circuit current [141]. The capacitors are charged up to a known voltage level via pull-up switches (M_3 and M_4), and then discharged via pull-down switches (M_1 and M_2). The resultant instantaneous currents $i_S(C_S,t)$ and $i_R(C_R,t)$ flowing through the CBCM core branches are time-dependent sharp exponential currents, as shown in Figure 4.22(a). The difference between the capacitances of the sensing and reference capacitors, $\Delta C = C_S - C_R$, results in a variation in the analyte. The reference node can be left unconnected or connected to insensitive electrodes. The voltage dropped over the ammeters is denoted by V_{Am}, and the supply voltage is represented by V_{dd}:

$$i_S(t) = C_S \frac{dv_S(t)}{dt} \tag{4.22}$$

$$i_R(t) = C_R \frac{dv_R(t)}{dt} \tag{4.23}$$

The instantaneous voltages across the sensing capacitor (C_S) and reference capacitor (C_R) are represented by $v_S(t)$ and $v_R(t)$, respectively. The averages of sharp exponential currents flowing through the CBCM core branches, $i_S(C_S,t)$ and $i_R(C_R,t)$, throughout one period of clock pulses Φ_1 and Φ_2 (denoted by T_S) are directly proportional to the capacitance of each branch, as indicated by (4.24) and (4.25):

$$\begin{aligned} I_S &= \frac{1}{T_s}\int_0^{T_s} C_S \frac{\mathrm{d}v_s(t)}{\mathrm{d}t}\mathrm{d}t \\ &= \frac{C_S}{T_s}\int_0^{V_{dd}-V_{Am}} \mathrm{d}v_s \\ &= \frac{C_S(V_{dd}-V_{Am})}{T_s} \end{aligned} \tag{4.24}$$

$$I_R = \frac{1}{T_s}\int_0^{T_s} C_R \frac{dv_R(t)}{dt} dt = \frac{C_R}{T_s}\int_0^{V_{dd}-V_{Am}} dv_s = \frac{C_R(V_{dd} - V_{Am})}{T_s} \tag{4.25}$$

The average currents $i_S(C_S,t)$ and $i_R(C_R,t)$ are denoted by I_S and I_R, respectively. The difference between the capacitances of the sensing and reference capacitors, $\Delta C = C_S - C_R$, can be determined by subtracting the average current of the reference branch from that of the sensing branch, i.e., $I_S - I_R$:

$$I_S - I_R = f_S(V_{dd} - V_{Am}).\Delta C \tag{4.26}$$

The proposed method enables the extraction of ΔC from C_S with high precision by utilizing the frequency of Φ_1 and Φ_2 ($f_S = 1/T_S$), as stated in previous sections. It is worth noting that the method can also be implemented by using half of the circuit depicted in Figure 4.22(b), which consists of only one electrode and one pseudo-inverter. However, measurements must be carried out in two separate steps. In the first step, the electrode is shielded from the biochemical sample, and the output current is solely influenced by the parasitic capacitances of the electrode and switches. In the second step, the biochemical sample alters the capacitance under measurement from its initial value [56,142]. Recent advancements in core-CBCM capacitive sensors have incorporated integrated readout circuits to process CBCM currents instead of relying on ammeters and off-chip elements, which will be elaborated on in subsequent sections.

4.3.6.1 Some non-idealities of core-CBCM circuits

Errors in the results of CBCM circuits can be caused by various factors such as leakage current [143], charge injection [143–146], noise [147] mismatch [22,110,147], and parasitic capacitances [148].

Leakage current

Leakage current is primarily caused by subthreshold and junctions' leakage. Notably, leakage current can also occur in the standby mode, even when there is no switching, due to the presence of unselected devices [149].

Charge injection

In addition to leakage current, charge injection is a non-ideal factor that can lead to errors, particularly in sub-micron CMOS technologies lower than 90 nm. During clock Φ_1's high state (as shown in Figure 4.22(a)), a current flows through the gate-source capacitance (C_{gs}) of the P-channel metal-oxide semiconductor (PMOS) switches onto the measurement bus [143]. This injection is related to the edge rate of Φ_1 and can be described by $I = C.dv/dt$, where dv/dt represents the edge rate. Figure 4.23 depicts the origin of this error, which is caused by the current injection. It is important to note that charge injection can significantly impact circuit

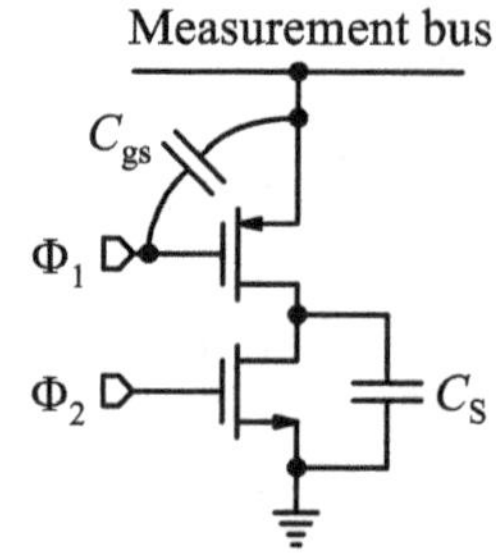

Figure 4.23 Current injection through C_{gs}

performance and must be considered when designing CBCM circuits, particularly in advanced CMOS technologies with smaller dimensions (lower than 90 nm).

Noise

One of the factors that can have an impact on the performance of CBCMs is noise coupling from other devices, which can affect the supply voltage and substrate. To address this, conventional approaches include using guard rings and distinct supply voltages for mixed-signal circuits, as referenced in [147].

Mismatch

Process and device mismatches can also contribute to the imbalance of the two branches of the circuit. This can be caused by mismatches of switches and transistors in readout circuits, as well as parasitic capacitances. Additionally, the presence of bio-particles on the electrodes, temperature fluctuations, and other unpredictable factors can result in an offset current that prevents i_S and i_R from being equal when $\Delta C = 0$. It is important to calibrate this offset based on the readout circuit's structure and mitigate sensor variability. In voltage-mode circuits, especially those using low-supply voltage technologies, the offset current is converted to an offset voltage, which can limit the dynamic range and resolution of the circuit. One approach to reducing the impact of mismatches is to use larger devices, but this increases the chip footprint in larger sensor arrays. These issues are discussed in more detail in [22,110].

Parasitic capacitances

Considering a cross-section view, Figure 4.24 showcases a simplified model representing the sensing and reference nodes (C_{IDE} and C_R, respectively). These nodes are implemented on the highest metal layer of CMOS technology. In Figure 4.24, the reference node is unconnected. In some configurations, it might be connected to an electrode identical to the sensing electrode. The sensing electrode can be represented by an equivalent circuit denoted as C_{Sensed}, which can be described using the following equation:

$$C_{Sensed} = C_{Stray} + C_{Standing} + \Delta C(t) + \sigma C(t) \tag{4.27}$$

where C_{Stray} accounts for the parasitic capacitance arising between the electrode and the underlying interconnect layers or substrate. $C_{Standing}$ represents the parasitic

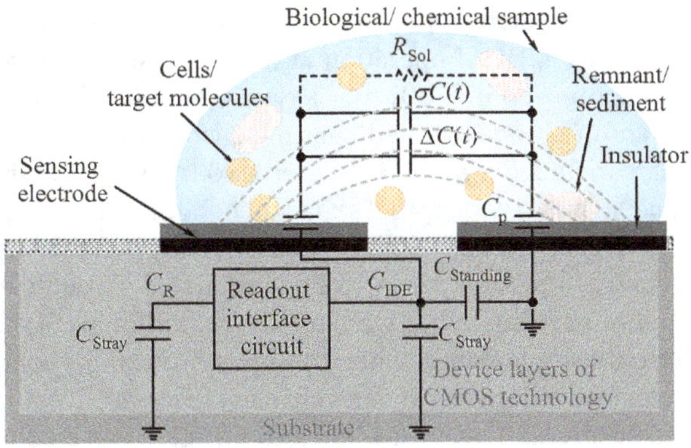

Figure 4.24 Depiction of a capacitive model for cell–electrode interaction alongside parasitic capacitances linked to the IDE placed atop the CMOS chip. Alterations in dielectric properties and resultant capacitance shift in the IDE arise from the binding of cells or target molecules onto the electrodes. This modification can be attributed to the attachment of entities like cells (σC(t)) or sedimentary residues (ΔC(t)).

capacitance between the fingers of the IDEs. $\Delta C(t)$ corresponds to the time-varying offset capacitance, which can originate from various factors, such as residual deposits from the fluidic sample on the sensing electrodes or chemical reactions leading to the formation of solid particles (e.g., salt crystals) or other by-products in proximity to or on top of the electrodes. Lastly, $\sigma C(t)$ denotes the time-varying sensing capacitance change, which emerges due to the presence and interaction of target molecules (e.g., cancer biomarkers, bacteria, or cells) with the electrode surface.

4.3.6.2 Some modifications to reduce the non-idealities

In the literature, various modifications to the basic CBCM block have been proposed for industrial applications. These approaches may also prove useful for life science applications. One such proposal by Bach *et al.* [143] involves the use of thick-oxide transistors and high sampling frequency to increase the measured current and reduce the impact of gate leakage current. This method ensures that the leakage current between the reference and sensing branch remains constant, allowing for accurate measurement of the difference between I_S and I_R and the corresponding capacitance value. Bach *et al.* [143] also suggest minimizing the gate-source capacitance by reducing transistor width and increasing the transistor length to lower charge injection and decrease the edge rate of the clock Φ_1, respectively.

To mitigate charge injection effects, Vendrame *et al.* [146] utilized transmission gates, while Ning *et al.* [149] proposed a leakage suppressor and a transmission gate scheme to improve CBCM. Additionally, charge injection-induced error-free (CIEF)

CBCM circuits were presented in [144,145] to suppress charge injection-induced errors. To improve CIEF CBCM operation at high frequencies, a self-differential CBCM method was proposed in [150]. For industrial applications operating in the giga-Hertz range, an on-chip self-timed control scheme circuitry was proposed in [147]. This circuitry included a pseudo inverter, two transmission gates connected to C_S and C_R, and a non-overlapping signal generator. Thick-oxide transistors were used to reduce gate leakage current. By utilizing a single pseudo-inverter instead of two, charge-injection errors were eliminated through two-pass measurements, and transistor mismatches were minimized. Silicon (Si) calibration was also performed to reduce mismatch, whereby C_S with an empty load and C_R were used to measure and calculate the mismatch parameter, $\Delta C_0 = C_S - C_R$. Guard rings were used to surround the CBCM block to mitigate noise coupling from the substrate. Yu *et al.* [151] proposed an on-chip signal generator for CBCM measurement circuits to reduce the mismatch between the two clock signals $\mathbf{\Phi}_1$ and $\mathbf{\Phi}_2$. Additionally, various calibration techniques were reviewed in the following subsections.

Until recently, there have been four compensation schemes utilized to counteract the impact of parasitic and offset capacitances in core-CBCM circuits. The initial scheme (CSc1), as described in [110], involved leaving the reference capacitor node unconnected to the electrodes, resulting in a reference node capacitance equal to the stray capacitance ($C_R = C_{Stray}$). However, this technique failed to compensate for the standing capacitance ($C_{Standing}$), thereby restricting the circuit's input dynamic range (IDR) due to the offset introduced by $C_{Standing}$. Despite this limitation, CSc1 was preferred for creating smaller pixels in arrays.

The second scheme (CSc2), introduced in [152] aimed to compensate for the standing capacitance. It involved replicating the sensing IDE on the reference side, where the RE remained either insensitive to or not in contact with the analyte. In an ideal scenario with complete circuit symmetry, the capacitance at the reference node would be $C_R = C_{Stray} + C_{Standing}$. This technique offered more precise compensation for the time-invariant offset capacitance. However, practical device mismatch resulted in compensation errors. Compared to CSc1, CSc2 provided higher IDR but required larger pixel sizes.

The third scheme (CSc3), similar to [22,35,37], incorporated tunable current mirrors and potentiometers on the reference side. This scheme enabled compensation of offset capacitances ($C_{Stray} + C_{Standing}$) through the reference capacitor and the ability to adjust the gain of the current mirror to compensate for device mismatch. The adjustment replicated the effects of electrode mismatch, resulting in $C_R = C_{Stray} + C_{Standing} + C_{os}$, where C_{os} represented the equivalent value of the reference capacitance after proportional adjustment. The advantage of this technique was more precise compensation for the time-invariant offset capacitance. However, it required configuring the optimum gain of the digitally programmable current mirror before each experiment and for each pixel, making it cumbersome for chips with large arrays of electrodes.

The fourth scheme (CSc4), employed in [110], compensated for offset capacitances without replicating the sensing structure. It involved implementing an offsetting capacitor (C_{os}) through interlayer dielectric capacitances and adding it to the reference node. Consequently, C_R could be expressed as $C_R = C_{Stray} + C_{os}$. While this technique closely resembled our work, the fixed value of the reference capacitor limited its ability to compensate for large offset capacitance caused by sedimentation and remnant effects, thereby constraining the IDR of the sensor.

Most previously reported CBCM circuits in the literature, such as in [22,35,37,110], were voltage-mode circuits that utilized an integrating capacitor in the analog domain to average the differential current. Despite efforts to develop fully differential topologies like those presented in [110] and [37] to increase the IDR, these voltage-mode CBCM circuits still faced limitations due to restricted voltage headroom, especially in low-voltage technologies. To enhance the IDR of CBCM circuits, a current-mode topology was proposed in [111,153]. This approach involved integrating the differential needle-shaped current (i_{out} in Figure 4.22(a)) in the digital domain using a fast CCO and an up/down counter based on the structure of Fibonacci linear feedback shift registers (LFSRs). Adjustable current mirrors and the up/down counter provided finer calibration, and a pre-distortion technique employing a look-up table and a modeled error was employed to mitigate non-linearity in the output response.

Additionally, a calibration-free method is introduced in [148] that involves sweeping the reference capacitance. This technique effectively generates a distinct and informative pattern of the specimen, eliminating the necessity for compensation or calibration.

Further elaboration on a subset of these methods will be provided in the subsequent sections.

4.3.6.3 Primary core-CBCM capacitive sensors

In 1996, the CBCM method was initially introduced for studying the crosstalk of integrated circuits (ICs), using DC ammeters to measure the charging and discharging current of the CBCM circuit [141]. Since then, numerous attempts have been made to develop CBCM for various applications, particularly in the field of life sciences, by integrating the core-CBCM capacitive sensors with other necessary components like electrodes and readout interface circuitry onto a single chip. A review of several important core-CBCM sensors is presented in this section. The CBCM method has demonstrated the capability of achieving a resolution of tens of atto-Farad while being characterized by its low complexity, low power consumption, and small area requirements. These features have facilitated the design of large arrays of core-CBCM sensors, such as the 246 × 256-array proposed in [40], which was specifically developed for high-frequency impedance spectroscopy.

Core-CBCM technique using discrete components

The CBCM technique was employed by Stagni and Guiducci and their colleagues for detecting DNA hybridization in several papers, including References [63,142, 154–156]. In their original research, this group demonstrated the effectiveness of their method for DNA detection using a non-integrated prototype setup with discrete components [142,154]. Instead of the traditional three-electrode setup, which necessitates a RE and a potentiostat, they utilized a simple two-electrode electrochemical system without a RE to form an electrode–electrolyte interface, as shown in Figure 4.25. The utilized gold electrodes were 12 mm in diameter. This system is more suitable for integration on a single chip. In Figure 4.25, the WE is functionalized for capturing single-stranded DNA (ssDNA). The capacitance of the electrode–electrolyte interface is affected by the physicochemical characteristics of the interface, which is measured in two steps, before and after exposure to ssDNA. A capacitance-to-current converter based on the CBCM technique is used, and the average current flowing through the functionalized electrode is measured using a multimeter.

The group behind a previous study proposed an alternative to using laboratory instruments by introducing a microcontroller-based system [155]. In this setup, a microcontroller determines the maximum frequency that can be utilized and then utilizes a DAC to generate the necessary input pulse waveforms. The electrode's output current is transformed into voltage with the aid of a resistor, and an operational amplifier amplifies the signal. The microcontroller then receives the signal from an ADC and processes the data by averaging them at various frequencies to calculate the equivalent capacitance using the curve of average voltage versus frequency. The same group took the first step towards developing a fully integrated capacitive biosensor for DNA detection by combining Au microelectrodes with a silicon CMOS standard process [63]. The ionic aqueous condition can add a resistive path to the sensing capacitor and cause imperfect behavior, which the group addressed by showing the potential of using the CBCM method in ionic conditions in a qualitative study [142]. The group proposed a completely electronic DNA sensor chip [156], which integrated the CBCM structure and pairs of Au electrodes with CMOS technology, but the readout circuit was implemented off-chip. Figure 4.26 depicts a single cell from the capacitive sensor array used in [156]. An external I/V converter circuit transforms the output transient current of the chip into a voltage signal, which is then sampled and processed by a personal computer (PC) using the trapezoidal integration method to provide the capacitance value. Carrara and Stagni *et al.* utilized the same measurement system as the

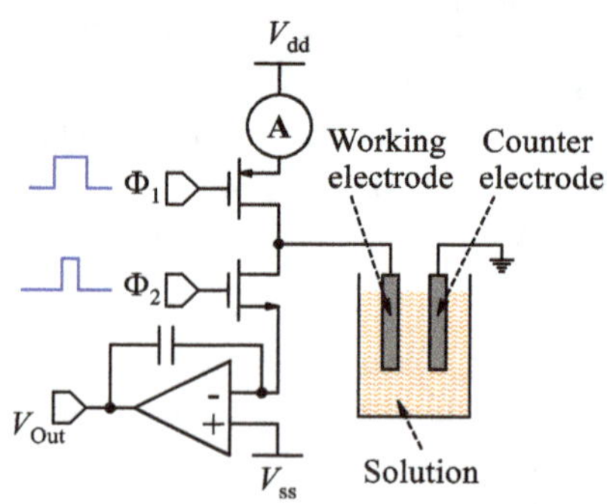

Figure 4.25 Configuration of a measurement arrangement with two electrodes

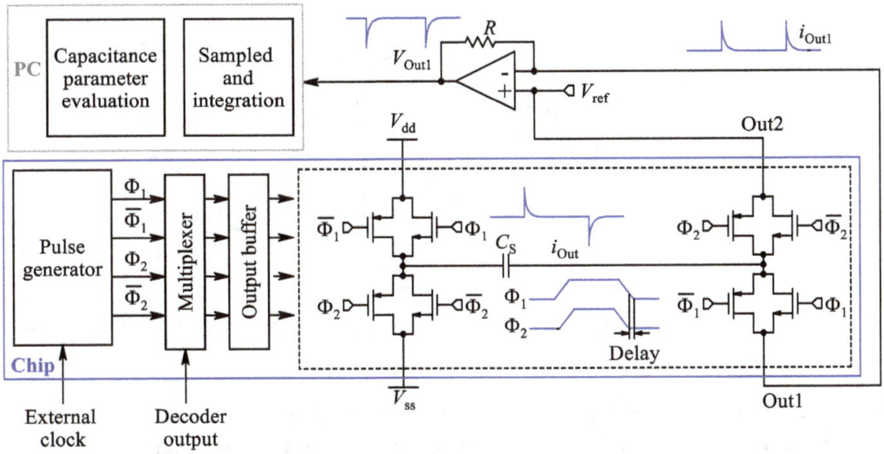

Figure 4.26 The electrical circuit linked with each individual cell within the capacitive biosensor array

previous group but improved it by developing new immobilization strategies for DNA detection, immunosensing, and cancer biomarkers detection [157–159].

Integrated sensors can utilize IDEs as electrodes, replacing the conventional three-electrode configuration with a simpler two-electrode setup. An IDE is comprised of two sections, one of which is grounded while the other acts as a WE. Sampson *et al.* [160] demonstrated the application of an IDE and the fundamental principle of capacitively coupled contactless conductivity detection with a single pseudo-inverter in detecting paraffinophilic Mycobacterium Avium Complex (MAC), a hazardous pathogen. They deposited paraffin wax on Cr/Au interdigitated array microelectrodes designed in both conventional and square spiral configurations. Consumption of the dielectric layer by the bacteria altered the sensing capacitance. To implement the measurement setup, dual N- and P-channel pairs of metal-oxide-semiconductor field-effect transistors (MOSFETs) were integrated onto a chip. Commercial capacitors were used to calibrate the circuit's operation by comparing the measured values.

Core-CBCM technique using integrated current mirrors

To enable on-chip measurement of the CBCM current, a straightforward alternative to an off-chip ammeter involves using a current mirror-like (M_3, M_4) shown in Figure 4.27. This current mirror senses and amplifies the CBCM current, providing a gain equal to A_I. During rapid charging of C_S when Φ_1 is low, the drain current of M_4 (I'_S) is A_I times greater than the drain current of M_3 (I_S). Hence, $I'_S = A_I.I_S$. Neglecting the on/off resistances of the switches M_1 and M_2, their relationship can be expressed as:

$$I'_S = A_I C_S \frac{dV_S}{dt} = A_I \mu_p C_{ox} \left(V_{sg3,4} - V_{thp}\right)^2 \tag{4.28}$$

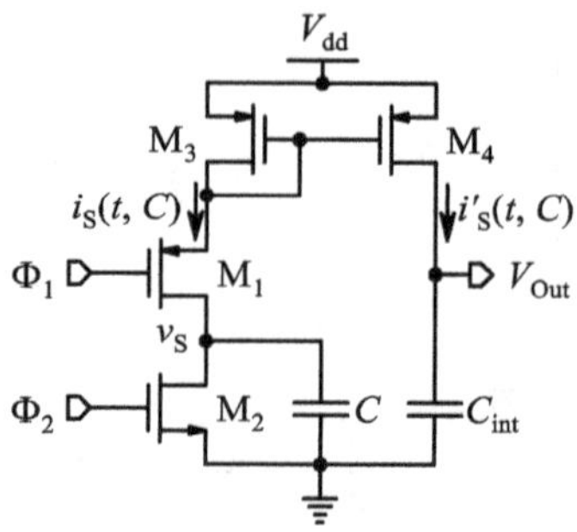

Figure 4.27 Utilization of an integrated current mirror for on-chip core-CBCM CVC

The expression for the relationship between switches M_1 and M_2 can be represented in terms of μ_p and C_{ox}, where μ_p denotes the entire mobility coefficient and C_{ox} stands for the gate capacitance per unit area. Furthermore, the source-gate voltage and the threshold voltage of M_3 and M_4, $V_{sg3,4}$ and V_{thp} respectively, are also incorporated in the equation.

To determine the value of V_S, considering $V_{sg3,4} = V_{dd} - V_S$, where V_S denotes the source voltage, becomes necessary. Assuming that $V_S = 0$ at $t = 0$ and that Φ_1 and Φ_2 are high while C_S is discharging, the value of V_S can be calculated using the following expression:

$$V_S = (V_{dd} - V_{thp}) - \frac{(V_{dd} - V_{thp})C_S}{\mu_p C_{ox}(V_{dd} - V_{thp})t + C_S} \tag{4.29}$$

By consolidating the aforementioned equations, it can be inferred that I'_S, a function of both C_S and time t, can be expressed as:

$$I'_S(C_S, t) \approx \frac{\left[(V_{dd} - V_{thp})C_S\right]^2 \mu_p C_{ox} A_I}{\left[\mu_p C_{ox}(V_{dd} - V_{thp})t + C_S\right]^2} \tag{4.30}$$

The output voltage can be obtained by integrating the current, $I'_S(C_S, t)$, with the aid of an integrating capacitor, C_{int}. The formula representing the resultant voltage output is provided as follows:

$$I'_S = C_{int}\frac{dV_{Out}}{dt} \tag{4.31}$$

In the limit of t being much greater than zero, the expression for V_{Out} can be ideally represented as:

$$V_S = \frac{C_S}{C_{int}}\frac{W_4}{W_3}\frac{L_3}{L_4}(V_{dd} - V_{thp}) \tag{4.32}$$

The width and length of M_3 and M_4 transistors are denoted by W_3 and W_4 and L_3 and L_4, respectively. The capacitance of C_S is the sum of the parasitic capacitance

C_R and the variable capacitance ΔC. However, since the value of ΔC is much greater than that of C_R, the dynamic range is limited [56]. To measure the capacitance changes ΔC, the signals related to C_S and C_R such as current, voltage, frequency, etc. can be subtracted either before or after averaging. This subtraction can be achieved through differential voltage- and differential current-based techniques, which will be discussed in the following sections.

4.3.6.4 Voltage-mode core-CBCM capacitive sensors

Non-differential core-CBCM capacitance-to-voltage converters (CVCs)

To fully integrate core-CBCM capacitive sensors, an integrator circuit or capacitor is often used to average their sharp exponential current signal (refer to Figure 4.22 (a)). This process converts the current signal to a voltage signal proportional to the measured capacitance. A capacitance tomography system based on an adapted CBCM, also known as charge transfer, was proposed by York *et al.* [161] (refer to Figure 4.28). The system mirrors the charging current of an unknown capacitor to a known integrating capacitor, using a gain MOSFET that operates in an active region. To increase the measurement range without reducing sensitivity, two integrating capacitors are utilized, along with six programmable offset capacitors (C_{os}) that are used in parallel for three purposes. These include reducing the measurement range, providing the biasing current needed to turn on the transistors for the first stage current mirror, and providing discrete test capacitors in the femto-Farad range for initial testing. By using a single transistor in a negative feedback loop, a virtual ground node can be obtained, which enhances the circuit stray immunity and reduces charge injection effects from the switching transistors. To achieve high sensitivity (80 mV/fF in this case), the transistors of the first stage operate in weak inversion, resulting in a high transconductance.

Non-differential core-CBCM capacitance-to-digital converters (CDCs)

The use of single arrays of nanoelectrodes in voltammetric or amperometric electrochemical assays has been extensively studied in the literature. However, when it comes

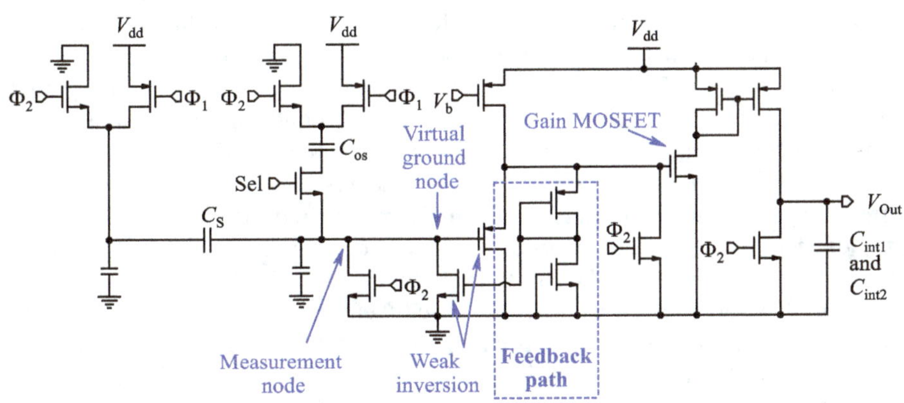

Figure 4.28 Tomography system employing the core-CBCM capacitive method

to electrochemical impedance spectroscopy (EIS), the nano-scale has not received much attention. Impedance spectroscopy is typically performed at low frequencies, around 10 kHz, and in the absence of faradaic contribution, the EDL has enough time to be rearranged. This results in an exponential decay of the instantaneous electric field from the electrode surface, which extends to the equilibrium Debye screening length L_D. As a result, target analytes can only be detected when they are within a few nanometers or less from the electrode, which is in the same order as L_D. Therefore, if a target analyte is located beyond the EDL formed by screening salt ions, even at a short distance from the sensor, it cannot be detected at low frequencies. However, certain applications require the ability to probe beyond the EDL under physiological salt conditions, as documented in References [40,70,162].

An effective approach to address the issue of perturbations caused by the EDL when detecting small adsorbates using nanoelectrodes is to increase the frequency of the alternating current (AC) excitation. By doing so, the ions in the solution will not be able to rearrange quickly enough to disturb the electric field. This results in the electric field being transmitted radially from the electrode to the solution at a distance that is proportional to the smallest electrode dimension. This method offers improved selectivity and sensitivity for supra- or macro-molecular entities. However, the sensitivity to small adsorbates within the EDL decreases due to the reduced perturbations caused by the electric field. The detection beyond the EDL is achieved by modulating the frequency above the salt-concentration-dependent cut-off frequency, which is equivalent to:

$$f_{\text{cut-off}} = \frac{1}{2\pi R_{\text{sol}}(C_{\text{dl}} + C_{\text{sol}})} \tag{4.33}$$

where R_{sol} and C_{sol} represent the solution resistance and capacitance, respectively. To effectively measure the properties of biomolecules using nanoelectrodes, a high frequency is required, which can only be achieved by lowering the electrolyte ionic strength. However, this may lead to changes or denaturation of the biomolecule properties. Another option is to use a large number of nanoelectrodes in parallel to increase the total admittance, which enables probing a small volume beyond the EDL. Nevertheless, this comes at the expense of missing individual detection at each electrode. To address this issue, an array of nanoelectrodes can be integrated with the required readout circuits to minimize background capacitance and enable high-frequency operation at physiological salt concentrations [70,162].

The CBCM technique was employed by Widdershoven and colleagues [69] to perform capacitive measurements on a densely packed and individually addressable array of 256 $\times$ 256 nanoelectrodes at high frequencies of up to 200 MHz. A temperature sensor integrated on the chip allowed for the analysis of temperature-induced variations. The same research group [162] used their massively parallel biosensor to conduct high-frequency impedance spectroscopy and real-time imaging of the dynamic variations of BEAS, THP1, and MCF7 cancer cells in a growth medium. With an atto-Farad resolution on the sub-micrometer scale, their nano-capacitor biosensor was capable of imaging micro-particles and living cells under physiological salt conditions.

The group expanded its biosensing platform with microfluidics and calibration systems [40]. Multi-frequency measurements, which can be calibrated, were used to record capacitance spectra, revealing novel information about the analyte. Figure 4.29(a) depicts the core-CBCM readout circuit, while Figure 4.29(c) illustrates the array architecture of the biosensor proposed by Widdershoven *et al.* [40,70].

The N-channel metal-oxide semiconductor (NMOS) transistors are employed as switches in the CBCM core of this particular biosensor, which facilitate the charging and discharging of the sensing capacitor to V_1 and V_2, correspondingly. The summation of charges transferred by one sensor cell is determined after N cycles of Φ_1 and Φ_2, which can be expressed mathematically as:

$$Q = N(V_1 - V_2)(C_S + C_p) \tag{4.34}$$

where C_p is the parasitic capacitance of the nanoelectrode. The detection of the analyte leads to a modification in capacitance, which subsequently causes a corresponding alteration in Q value:

$$\Delta Q = N(V_1 - V_2)\Delta C \tag{4.35}$$

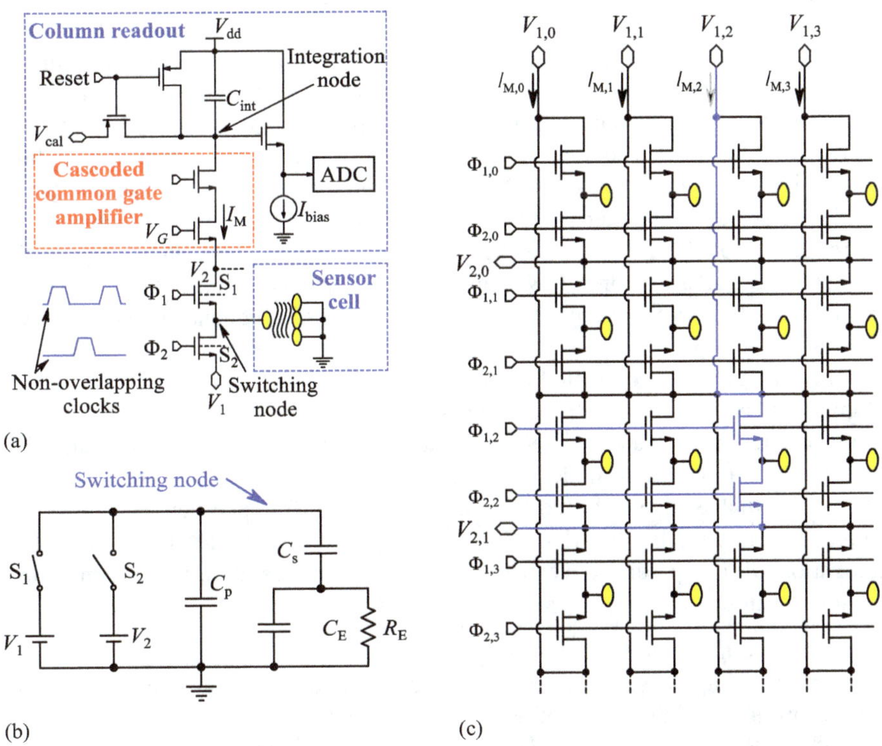

Figure 4.29 (a) Structure of the array, (b) Switching model's equivalent circuit and (c) Readout circuit for the core-CBCM biosensor, utilized for both real-time imaging and impedance spectroscopy

Flicker noise is not generated by the circuit. However, the signal is subject to transistor reset noise during the $2N$ switching steps, with an equivalent value provided in:

$$\sigma_n^2 = 2Nk_BT(C_S + C_p) \tag{4.36}$$

where k_B is Boltzmann's constant and T stands for the absolute temperature. Hence, the intrinsic SNR can be expressed as:

$$\left(\frac{\Delta Q}{\sigma_n}\right)^2 = \frac{N(V_1 - V_2)^2 \Delta C^2}{2k_BT(C_S + C_p)} \tag{4.37}$$

The use of small nanoelectrodes can enhance the SNR by increasing the differential capacitance (ΔC) and reducing the sum of the parasitic capacitance (C_S) and the double-layer capacitance (C_p). However, achieving small C_S requires non-CMOS technologies such as electron-beam lithography, while sub-micrometer dense integration in modern CMOS technology is necessary for small C_p [70].

To measure one row of the array architecture simultaneously, Φ_1 and Φ_2 of Figure 4.29(a) can be applied. The non-selected nanoelectrodes can be disabled by setting a low potential on their respective Φ_1 and then connecting them to AC grounded discharge lines using high voltages of Φ_2. This creates a reconfigurable on-chip counter electrode, which prevents unwanted electrochemical modifications of the electrode surfaces and provides a low-impedance AC return path from the electrolyte to the chip. The simplified equivalent circuit of the proposed switching model is illustrated in Figure 4.29(b), where C_p represents the parasitic capacitance, C_{surf} is the surface capacitance, and C_E and R_E denote the spreading capacitance and resistance of the electrode to the electrolyte. The C_{dl} dominates in ionically conductive solutions and/or at low frequencies and depends on the effective area of the electrode. In low-conductive solutions and/or at high frequencies, the C_{dl} is less dominant.

The voltage V_1 is controlled by the cascoded common-gate amplifier, and the column current is averaged by the parasitic capacitance and passed to the integrating capacitor C_{int}, as depicted in Figure 4.29(a). After integration, an ADC samples the voltage at the node V_{int}, which is then compared with the calibration voltage V_{cal} that is also sampled by the ADC. The difference between these two samples is converted by the ADC. Eight on-chip 13-bit ADCs are used for parallel digitization of V_{int}. Each ADC processes 32 columns consecutively using time-division multiplexing.

Differential capacitance-to-voltage conversion

Differential voltage-based methods involve averaging the currents of the core CBCM through an integrator immediately after extraction. Another approach, as reported in References [152,163], uses two separate integrating capacitors to average the currents i_S and i_R. The voltages of these integrating capacitors are then amplified and subtracted by a differential amplifier, sacrificing resolution and dynamic range for increased sensitivity (see Figure 4.30). This approach has a

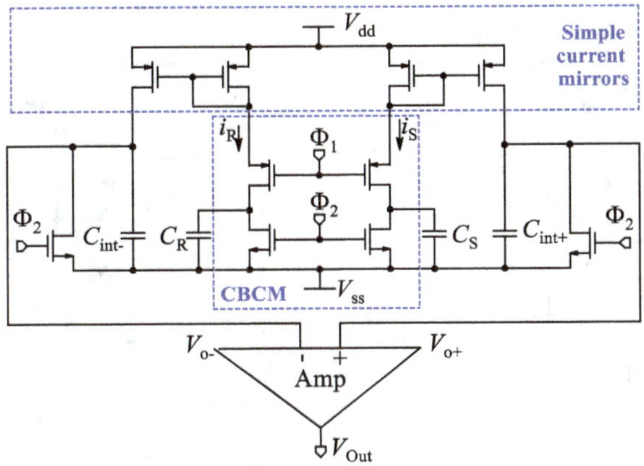

Figure 4.30 A core-CBCM CVC approach, where integration of CBCM currents occurs before the subtraction process

resolution of around 10 aF and was used for particle detection by York and Evans *et al.* [152,164], who utilized an IDE consisting of Al fingers on a Si chip surface. The standing capacitance and capacitance changes in the presence of small particles were determined through 3D finite element modeling [164].

Tanskanen *et al.* [163] used the same idea with novel current–voltage mirrors for MEMS applications, as shown in Figure 4.31. These mirrors consist of a two-transistor configuration of a simple current mirror (M_6, M_7) (or (M_{12}, M_{13})) to mirror their input and output currents and a four-transistor configuration (M_5, M_8, M_9, and M_{10}) (or M_{11}, M_{14}, M_{15}, and M_{16}) to mirror their input and output voltages. A virtual short between their input and output caused by the transistors (M_9, M_{10}) (or (M_{15}, M_{16})) helps to equalize them. Utilizing these new current–voltage mirrors, CBCM currents are mirrored more accurately, although the output voltages are limited due to the reduced output voltage swing. The sensitivity of this circuit in 0.35 μm CMOS technology is reduced from 47.73 mV/fF to 28.7 mV/fF after replacing the simple current mirrors with the new ones.

In their study, Tanskanen *et al.* [163] introduced a new approach using current–voltage mirrors with the aim of enhancing MEMS applications (refer to Figure 4.31). The current–voltage mirrors consist of a two-transistor current mirror (M_6, M_7) (or (M_{12}, M_{13})) for mirroring input and output currents, and a four-transistor configuration (M_5, M_8, M_9, and M_{10}) (or M_{11}, M_{14}, M_{15}, and M_{16}) for mirroring input and output voltages. The input and output transistors (M_9, M_{10}) (or (M_{15}, M_{16})) create a virtual short, which ensures equalization between the input and output of the circuit. By setting $V_{sg10} = 0$ (or $V_{sg16} = 0$), subthreshold current flows through (M_9, M_{10}) (or (M_{15}, M_{16})), resulting in a virtually shorted source-gate voltage for M_9 (or M_{15}), $V_{sg9} = 0$ (or $V_{sg15} = 0$). This feature allows the circuit to mirror both input and output voltages and currents more accurately. The novel current–voltage mirrors improve the accuracy of CBCM currents at the cost of reducing the output voltages due to limited output voltage swings. In 0.35 μm

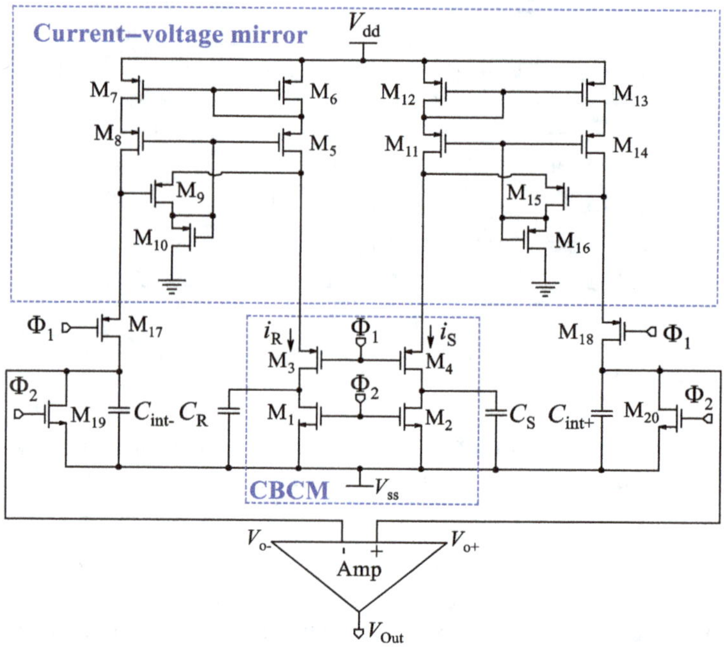

Figure 4.31 A current–voltage mirror integrated into a capacitive sensor designed for MEMS applications

CMOS technology, the sensitivity of the circuit decreased from 47.73 mV/fF to 28.7 mV/fF after the simple current mirrors were replaced with the new ones.

In their study, Yusof and colleagues [165] have introduced a fully differential capacitance-to-voltage converters (CVCs) circuit based on CBCM that is capable of measuring the EDL capacitances at the interface of the electrode–electrolyte. The circuit diagram of the proposed system is shown in Figure 4.32. To minimize the negative impacts of charge injection and clock feedthrough effects, the researchers have employed transmission gates as switches. Assuming the ideal performance of switches S_1 and S_2 and considering only the non-faradaic impedance, the mean value of the transient current across sensing capacitor C_S and reference capacitor C_R can be estimated by averaging the current over the voltage steps applied to the switches.

$$I_{S,R} \approx f_S C_{S,R}(V_{ref1} - V_{cm}) \tag{4.38}$$

In the circuit shown in Figure 4.32, two distinct reference voltages, V_{ref1} and V_{ref2}, are used alongside a common-mode voltage (V_{cm}) which is equal to the average of the supply voltages ((V_{dd} + V_{ss})/2). The transient currents across sensing capacitor C_S and reference capacitor C_R are integrated by capacitors C_{int} and then averaged. The resulting voltages are amplified by a differential amplifier with a gain of A_{amp}. At a stable state, the differential output voltage is

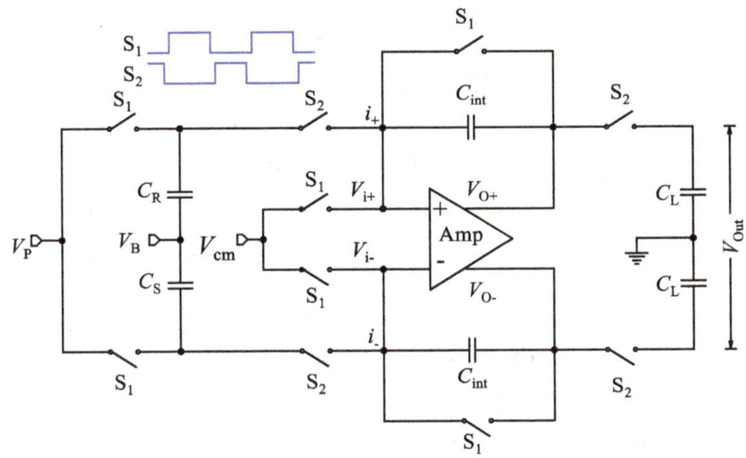

Figure 4.32 A fully differential CVC based on CBCM method

obtained by a calculation involving various circuit parameters, as given in the following formula:

$$V_{\mathrm{Out}} = \frac{\Delta C(V_{\mathrm{ref1}} - V_{\mathrm{cm}})A_{\mathrm{amp}}}{C_{\mathrm{int}}(1 - A_{\mathrm{amp}})} \tag{4.39}$$

The operational capabilities of the sensor are determined by C_{int}, the integrating capacitor value. This sensor's performance is assessed in solutions containing sodium chloride (NaCl) with an ionic concentration spanning from 0.1 mM to 1 M, encompassing a wide frequency range. Furthermore, when exposed to DNA hybridization in a phosphate-buffered solution, a 20% capacitance shift (10 fF) was observed.

The wireless biosensor developed by Kim *et al.* [166] utilized a core-CBCM sensor, a radio frequency identification (RFID) circuit, a spiral inductor tag antenna, and a ΣΔ ADC. Nakazato [167,168], on the other hand, employed the same core-CBCM sensor in a multi-modal sensor array that includes amperometric and potentiometric sensors for DNA detection. In another study [169], Nakazato's team presented a 4 × 4 array of this sensor in Table 4.1, where the differential CBCM cell was arranged in a matrix structure, and the operational amplifier (OpAmp) and SC integrator were shared among cells. Moreover, they designed N-type and P-type array-clock controllers to facilitate the selection of array cells.

Differential capacitance-to-current and current-to-voltage conversion

Ghafar-Zadeh and colleagues [170] have proposed a novel solution to tackle the issue of limited dynamic range in differential voltage mode. Figure 4.33 depicts their approach, which involves the subtraction of CBCM (common-base common-mode) currents before averaging. In this method, M_1 and M_4 are turned on while M_2 and M_3 are turned off, resulting in the discharging of currents (Δi) from the output node (V_{Out}).

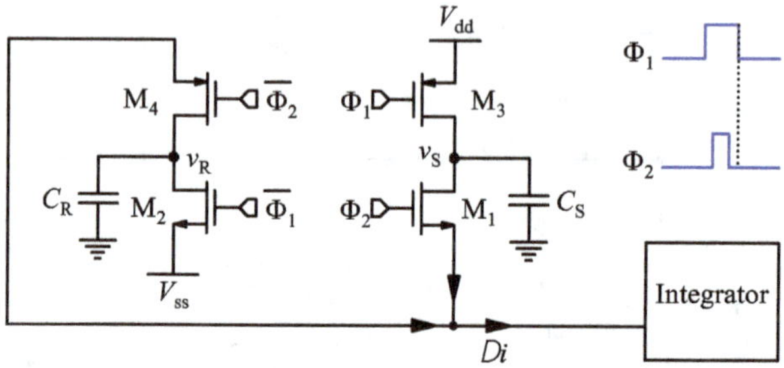

Figure 4.33 A fundamental core-CBCM capacitive sensor in which subtraction of CBCM currents precedes the integration process

The next step involves turning on M_1 and M_4 while turning off M_2 and M_3, which prevents current flow through the output node. The resulting differential current (Δi) can then be integrated by an integrating capacitor or an integrator circuit. The primary cause of error in this circuit is the mismatch between PMOS and NMOS transistors.

The circuit designed by Ghafar-Zadeh and colleagues is depicted in Figure 4.34 [22]. Their method involves subtracting and amplifying the currents i_S and i_R using current mirrors, followed by injection of the differential current ($\Delta i = i_S - i_R$) into an integrating capacitor.

The circuit response of a biosensor is analyzed in [22] for solutions with low and high ion conductivity placed on the electrodes. The parasitics generated on top of the electrodes are represented by an equivalent circuit, as shown in Figure 4.35, where R_{sol} is the resistivity of the analyte and C_{ins} and C_1 denote the capacitance of the passivation layer and the analyte capacitance. C_R remains the same for solutions with high or low conductivity. In the case of a low-conductive analyte, R_{sol} is very large, resulting in $C_{eq} = C_1||C_2||C_3$, where $C_{ins}>>C_1$. For an ionically conductive solution, R_{sol} is negligible, and $C_{eq} = C_2||C_3||C_{ins}$ [22]. When the solution is non-conductive, the output voltage is given by:

$$V_{Out} \approx \frac{C_{T1}}{C_{int}} A_{I1}\left(V_{dd} - V_{thp}\right) - \frac{C_{T2}}{C_{int}} A_{I2}\left(V_{dd} - V_{thp}\right) + V_{off} \tag{4.40}$$

where V_{off} is the output offset voltage. The capacitance of the solution is equal to the sum of C_1, C_2, and C_3 ($C_S = C_1+C_2+C_3$) [22].

The switches $M_{1\text{-}4}$ are assumed to have negligible on/off resistances. Parasitic capacitances C_A and C_B created by $M_{1\text{-}4}$ are incorporated into the equivalent capacitances C_{T1} and C_{T2}, respectively. C_{T1} is the sum of C_A and C_S, while C_{T2} is the sum of C_B and C_R. The gain of current mirrors A_{I1} and A_{I2}, as indicated in Figure 4.34, is equal to [22]:

$$A_{I1} = \frac{(W/L)_7}{(W/L)_5} \tag{4.41}$$

$$A_{I2} = \frac{(W/L)_8}{(W/L)_6} \times \frac{(W/L)_9}{(W/L)_{10}} \tag{4.42}$$

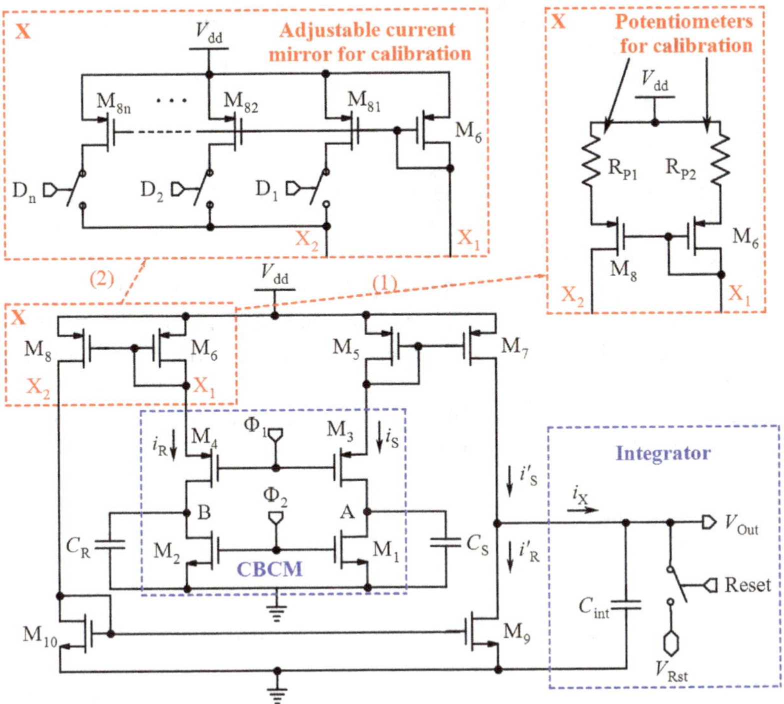

Figure 4.34 In the core-CBCM CVC configuration, the components within rectangle X can be substituted with either (1) potentiometers, or (2) an adjustable current mirror, offering calibration capabilities

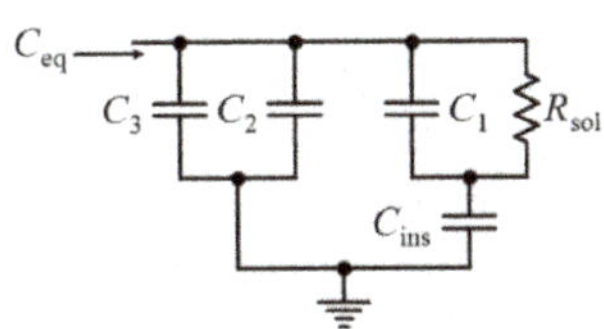

Figure 4.35 The circuitry representing the parasitic elements generated at the electrodes

Equation (4.41) can be expressed in an alternate form, as shown in [22]:

$$V_{\text{Out}} \approx \frac{C_{\text{T1}} - C_{\text{T2}}}{C_{\text{int}}} A_{\text{I1}} \left(V_{\text{dd}} - V_{\text{thp}}\right) + V'_{\text{off}} \tag{4.43}$$

The term V'_{off} denotes the output offset voltage equivalent, and its value is stated as [22]:

$$V'_{\text{off}} = \frac{C_{\text{T2}}}{C_{\text{int}}} (A_{\text{I1}} - A_{\text{I2}}) \left(V_{\text{dd}} - V_{\text{thp}}\right) + V_{\text{off}} \tag{4.44}$$

If we assume that C_A is equal to C_B, the sensor's sensitivity concerning the change in capacitance, ΔC, which is defined as C_{T1} minus C_{T2}, can be expressed as [22]:

$$S_{\Delta V/\Delta C} = \frac{\Delta V_{Out}}{\Delta C} = \frac{A_{I1}\left(V_{dd} - V_{thp}\right)}{C_{int}} \tag{4.45}$$

The sensitivity of a system can be increased by increasing the value of A_{I1} and decreasing the value of C_{int}, where ΔV_{Out} represents the change in output voltage. However, the minimum value of C_{int} cannot be less than the parasitic capacitance at node A. Multiple-stage amplifiers can offer higher sensitivity and A_{I1}. Although larger electrodes can produce larger changes in capacitance (ΔC), using an array of small sensing electrodes is preferred because it reduces the impact of parasitic capacitances.

According to Reference [22], it is possible to acquire the V_{Out} for $t > 0$ in solutions that have high conductivity:

$$V_{Out} = \frac{C_{ins}}{C_{int}} A_{I1}\left(V_{dd} - V_{thp}\right)\left[1 - \frac{C_1}{C_1 + C_{ins}} \exp\left(\frac{-t}{R_{sol}(C_1 + C_{ins})}\right)\right] + V''_{off} \tag{4.46}$$

Reference [22] provides the equation for V''_{off}, which can be expressed as follows:

$$V''_{off} = \frac{\left(V_{dd} - V_{thp}\right)}{C_{int}}\left[(C_2 + C_3 + C_A)A_{I1} - (C_R + C_B)A_{I2}\right] + V_{off} \tag{4.47}$$

The capacitance C_{ins} is much larger than the capacitance C_S, and the time constant $\tau = R_{sol}C_{ins}$ is smaller than the sampling time. Fluids demonstrate a particular behavior during low-frequency clock pulses [22]:

$$V_{Out} \approx \frac{C_{ins}}{C_{int}} A_{I1}\left(V_{dd} - V_{thp}\right) + V''_{off} \tag{4.48}$$

Applying clock pulses with low frequency to R_{sol} enables the output voltage of V_{Out} to remain constant, rendering it insensitive to changes in C_S. As a result, the sensor can detect conductive fluids that have similar characteristics. To measure C_S in series with C_{ins}, the appropriate circuitry and differential measurement techniques must be employed, as described in References [35,171,172].

A differential measurement technique can be utilized to compare two distinct analytes by directing them toward the sensing and reference capacitors.

To calibrate the offset voltage, V_{off}, a solution proposed by the authors involved using two potentiometers, R_{P1} and R_{P2} (as shown in Figure 4.34, state (1)) along with an adjustable current gain (as shown in Figure 4.34, state (2)), in conjunction with current mirrors. However, the relationship between changes in the values of the potentiometers and variations in V_{Out} is non-linear, making it only possible to perform a coarse adjustment using R_{P1} or R_{P2}. To achieve a high resolution, digital calibration circuitry is employed to adjust the current gain and reduce the offset voltage.

$M_{81\text{-}8n}$ generates I_R based on the aspect ratios corresponding to each of the digital input signals depicted in Figure 4.34, state (2), namely $D_1, \ldots, D_n$:

$$\left(\frac{W}{L}\right)_6 = 2^1\left(\frac{W}{L}\right)_{81} = 2^2\left(\frac{W}{L}\right)_{82} = \ldots = 2^n\left(\frac{W}{L}\right)_{8n} \tag{4.49}$$

The aspect ratio of transistor M_i (where $i = 6$ and 81, 82, ..., $8n$) is denoted by $(W/L)_{Mi}$. The digital inputs, $D_1, \ldots, D_n$, can be produced either by an on-chip or an off-chip programmable unit. The residual offset, which remains after offset cancellation, is removed automatically through the subtraction of two subsequent measurements taken before and after analyte injections.

In accordance with the information provided in Figure 4.34, it can be noted that the sensitivity of the sensor can be adjusted to achieve a value of approximately 250 mV/fF. This adjustment can be made using a 1.8 V power supply voltage and 0.18 μm technology. By doing so, it is possible to obtain a resolution of about 10 aF, along with an IDR of approximately 2.7 fF. These figures are presented in Table 4.1 for reference.

Ghafar-Zadeh and colleagues utilized the direct-write fabrication process (DWFP) method [21] to create microfluidics made of polymers on top of the CMOS chip and fluidic packaging. They integrated microfluidics with a basic circuit proposed in [22] to demonstrate the idea. In [46], the feasibility and applicability of this novel technique for CMOS-based applications were demonstrated using a polyelectrolyte hydrogel and dichloromethane (DCM) as ionically conductive and non-conductive analytes, respectively, for on-chip measurement of parasitic capacitance.

Yao and colleagues [173] employed a circuit that bears resemblance to the one illustrated in Figure 4.36 to identify lethal bacteria such as *Salmonella* and *E. coli*

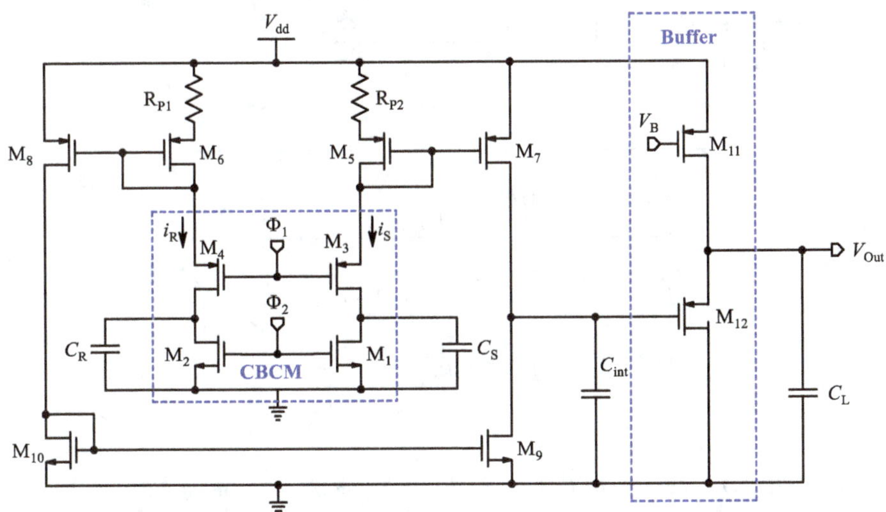

Figure 4.36 The core-CBCM capacitive sensor using resistors for balancing the offset

(refer to Table 4.1). In Figure 4.36, the resistors R_{P1} and R_{P2} were utilized to balance the offset, while the load capacitor was driven as an LPF by the source follower that is composed of M_{11} and M_{12}.

Differential capacitance-to-current, current-to-voltage, and voltage-to-digital conversion

Ghafar-Zadeh *et al.* built on their previous research in [22] and introduced a novel low-complexity ΣΔ modulator designed specifically for the differential circuit depicted in Figure 4.34 of [171]. This modulator converts the voltage across the integrating capacitor into digital data using a voltage comparator, as illustrated in Figure 4.37(a). The design employs adjustable current mirrors for calibration purposes. While conventional ΣΔ modulators have the benefits of high resolution,

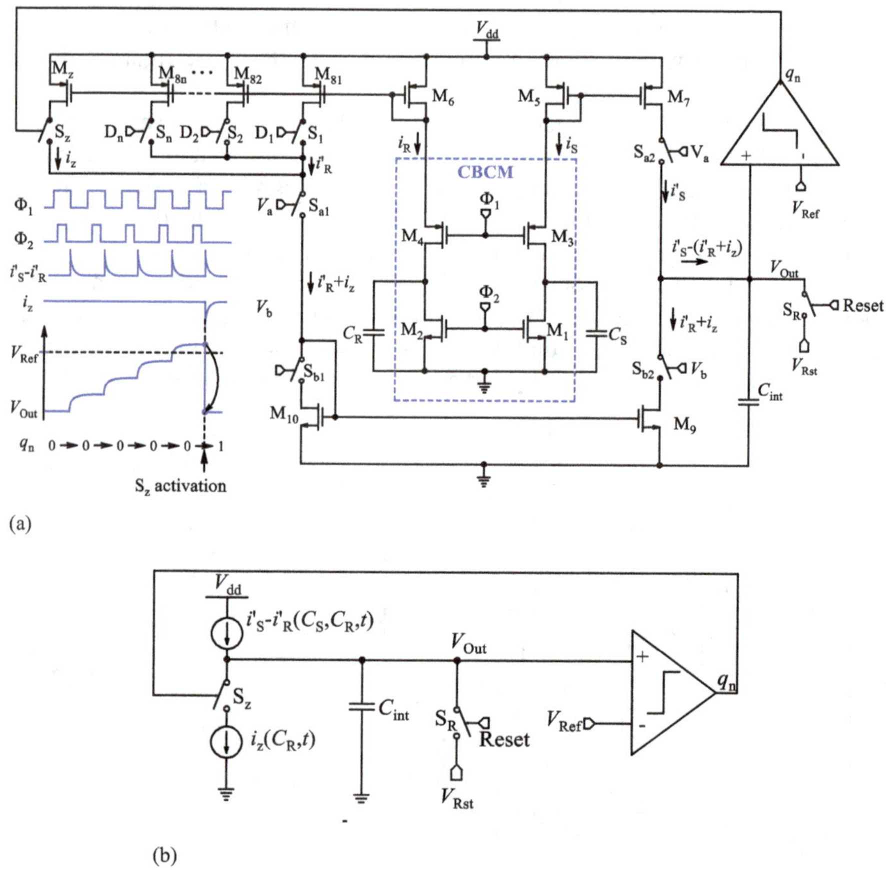

Figure 4.37 (a) Circuitry of the core-CBCM ΣΔ capacitive sensor with corresponding clock signals, differential current input, ΣΔ current, sensor output voltage, and the resultant bitstream and (b) Simplified representation of the ΣΔ modulator model

simple architecture, low power consumption, small size, and easy implementation, the modulator proposed by Ghafar-Zadeh *et al.* differs in that it uses a step waveform as input voltage, rather than a ramp.

A simple model for the ΣΔ modulator proposed by Ghafar-Zadeh *et al.* is depicted in Figure 4.37(b). In this model, the voltage on the integrating capacitor, C_{int}, is compared to the reference voltage, V_{ref}, using a voltage comparator. The output pulse of the comparator, q_{n}, is then applied to the switch $\mathrm{S_z}$, which generates a current, $i_{\mathrm{z}}(C_{\mathrm{R}}, t)$, via the transistor $\mathrm{M_z}$. This current causes a change in the DC voltage of the output, V_{Out}, by charging C_{int}. The interactions associated with C_{S} cause V_{Out} to increase above V_{ref}, which in turn closes the switch $\mathrm{S_z}$ and causes the value of V_{Out} to decrease to a level below V_{ref}, as shown in Figure 4.37(a). The generated current, i_{S}, in each period of $\mathbf{\Phi}_1$ and $\mathbf{\Phi}_2$ incrementally increases the voltage on C_{int}, which serves as a delay integrator [56].

The first order ΣΔ ADC in Figure 4.37(a) utilizes a 1-bit DAC composed of switch $\mathrm{S_z}$ and the current mirror ($\mathrm{M_6}$, $\mathrm{M_z}$) to convert the digital input to i_{z}. The amplified currents of i_{S} and i_{R} are i'_{S} and i'_{R}, respectively, and their difference generates a current error, e_n, given by $e_n = i'_{\mathrm{S}} - (i'_{\mathrm{R}} + i_{\mathrm{z}})$. This current error is transformed into a voltage error, which is then added to the capacitor output voltage. A track and latch voltage comparator, followed by an RS flip-flop, is responsible for the quantization process. Since i'_{S}-i'_{R} is a linear function of C_{S}, the digital output of the comparator, q_{n}, is proportional to C_{S}.

In their work, the research group has evaluated the practicality of a hybrid CMOS-microfluidics sensor that utilizes the ΣΔ modulator and DWFP-based microfluidics. To test the sensor, they utilized polyelectrolyte solutions and chemical solvents with recognized dielectric constants, as outlined in References [47,171,174]. Moreover, they have explored its utility for monitoring bacterial growth, as mentioned in References [35,172]. The team has also proposed a new algorithm for decoding first-order ΣΔ modulator data, which is relevant for lab-on-chip (LoC) applications, as presented in Reference [175].

In their study, Miled and colleagues [176] proposed a novel approach to enhance the sensitivity and decrease the power consumption of the CBCM technique. The authors suggested utilizing the subthreshold conduction of current mirror transistors instead of relying on the saturation region to amplify the CBCM current. This method aims to achieve improved performance by exploiting the unique characteristics of the subthreshold region.

The sensitivity of the drain current in a current mirror transistor, such as $\mathrm{M_5}$ and $\mathrm{M_7}$, to its gate-source voltage, $V_{\mathrm{sg5,7}}$, is linear when it operates in the saturation region. This is expressed by:

$$\frac{di_5}{dv_{sg5}} = \mu_{\mathrm{p}} C_{ox} \frac{W_5}{L_5} \left(V_{sg5} - V_{thp} \right) \tag{4.50}$$

In Figure 4.34, transistors $\mathrm{M_5}$ and $\mathrm{M_7}$ are intentionally operated at the subthreshold region, where their gate-source voltage is lower than their threshold voltage ($V_{\mathrm{sg5,7}} < V_{\mathrm{thp}}$). This makes their drain current more sensitive to variations in V_{sg5}. The sensitivity of i_{D5} to V_{sg5} in the subthreshold region is described by:

$$\frac{di_5}{dv_{sg5}} = \frac{W_5}{L_5} I_{D0} \frac{V_{sg5}}{nV_T} \exp\left(\frac{V_{sg5}}{nV_T}\right) \tag{4.51}$$

which includes an exponential term due to the nature of the subthreshold operation. Comparing (4.51) with (4.50) shows that the drain current of the transistor in the subthreshold region is more sensitive to its source-gate voltage. This means that small changes in C_S can cause significant variations in current with low power consumption. However, the behavior of transistors in the subthreshold region is not well-known and it can result in a limited dynamic range.

The aforementioned group proposed a micro- and nano-particle manipulation dielectrophoresis (DEP) device that utilizes an electric field. The device was integrated with an array of electrodes and utilized a modified version of the circuit displayed in Figure 4.34. The current mirror transistors in the circuit operate in the subthreshold region [109,177,178]. The circuit was evaluated using a 3 × 3 array of capacitors, which can be activated line by line, and measure three distinct capacitances simultaneously (refer to Figure 4.38 and Table 4.1 for details).

The three command lines, namely En_0, En_1, and En_2, depicted in Figure 4.38, activate a single line that comprises three unique capacitive sensor cells. These cells facilitate the simultaneous measurement of capacitances at different locations within the microchannels. A sensing current circuit is linked to each column of the

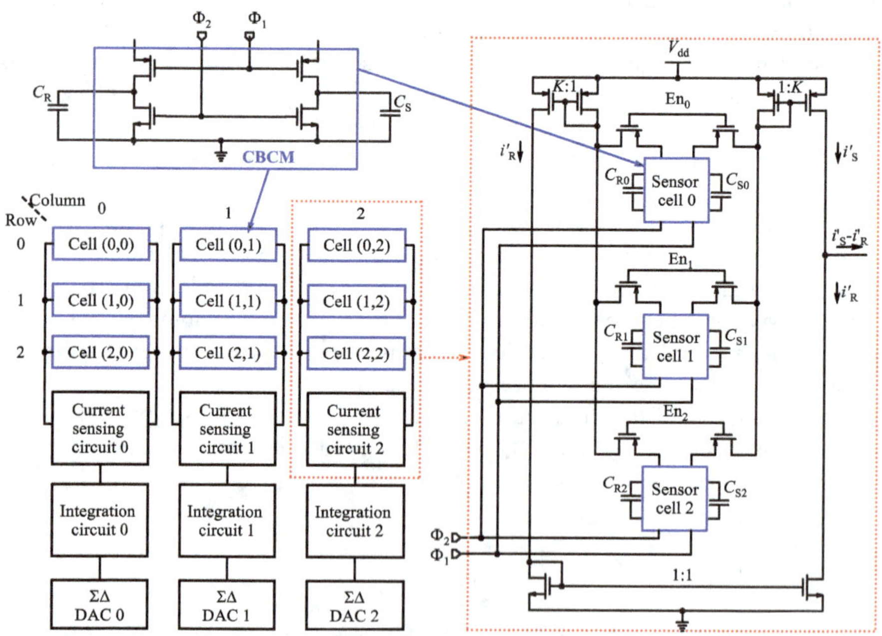

Figure 4.38 Configuration of a 3 × 3 array of sensors designed for concurrent assessment of three distinct capacitance values

array, enabling it to detect any current alterations in the two branches of the sensor cells in that specific column. This process results in an output differential current, which is then transformed into voltage by an integration circuit. Finally, the voltage is digitized by a ΣΔ ADC, completing the measurement process.

Fully differential capacitance-to-current and current-to-voltage conversion

The articles referenced in [110,179] describe a fully differential rail-to-rail CVC designed for on-chip cell monitoring, which is illustrated in Figure 4.39. This converter addresses some of the issues present in single-ended structures caused by parasitic capacitances and common-mode noise. By utilizing a fully differential circuit, the capacitance measurement's dynamic range is doubled.

In this circuit, the integrating capacitors are all equal to each other ($C_{int+} = C_{int-} = C_{int}$). The currents Δi_+ and Δi_- are produced by subtracting the currents $\Delta i_+ = i_S - i_R$, and $\Delta i_- = i_R - i_S$, respectively. These two currents are then amplified by a gain of $A_I = A_{I1} = A_{I2}$ using current mirrors. The output voltages of the circuit are denoted by $V_{Out\pm}$ and can be calculated using the following equation:

$$V_{Out\pm} = \pm A_I.\Delta C.\frac{(V_{dd} - V_{ss} - |V_{thp}|).f_S}{C_{int}.f_{int}} + V_{off\pm} + V_{cm} \quad (4.52)$$

Here, V_{dd} and V_{ss} represent the dual supply voltages, V_{cm} is the common mode voltage at the output nodes, and $V_{off\pm}$ denotes the offset voltage after integrating Δi_+ and Δi_-. The integrating frequency (f_{int}) is equal to the sampling frequency (f_S). The difference between the two output voltages, ΔV_{Out}, is given by [179]:

$$\begin{aligned}\Delta V_{Out} &= V_{Out+} - V_{Out-}\\ &= 2.A_I.\Delta C.\frac{(V_{dd} - V_{ss} - |V_{thp}|).f_S}{C_{int}.f_{int}} + \Delta V_{off}\end{aligned} \quad (4.53)$$

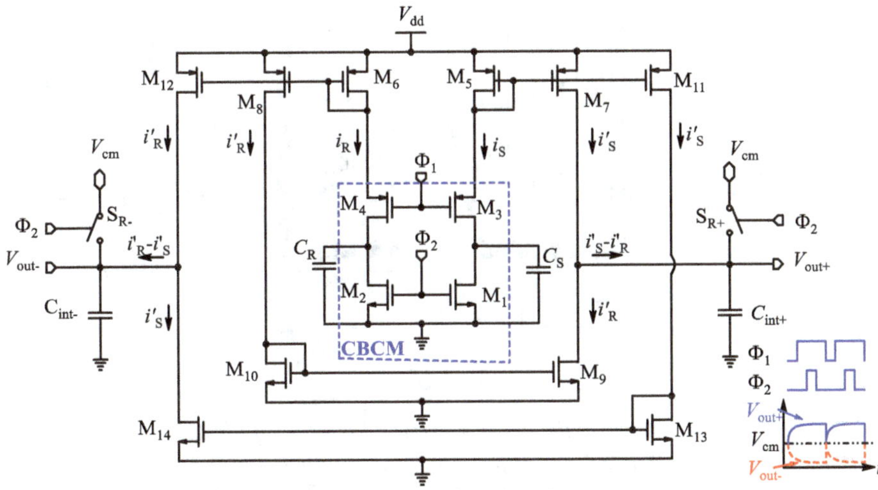

Figure 4.39 A core-CBCM capacitive sensor with a fully differential configuration

In an ideal scenario, ΔV_{off} (the difference between V_{off+} and V_{off-}) is zero. However, practical mismatches in the transistors of the circuit can lead to $\Delta V_{off} \neq 0$.

The fully differential capacitance sensor array utilizes current bus shielding in [110,179] to safeguard the array from noise and junction leakage while also simplifying calibration by avoiding individual pixel calibration. As illustrated in Figure 4.40(a) and (b), the current bus lines are an extension of the nodes N_+ and N_-, which enables a common single-sensor evaluation module (SEM) block including current mirrors, subtractors, integration capacitors, and output buffers to be shared with a column of CBCM pixel units based on a column-parallel architecture. A larger metal shield in a lower metal layer can effectively isolate the bus line from the substrate, improving the immunity of the current bus to substrate noise. Furthermore, the buffer amplifiers, shield line, and individual n-wells of the PMOS switches in each CBCM pixel are driven with a potential that tracks the bus line potential, eliminating the effects of source-to-bulk junction leakage and bus-to-shield capacitance.

The use of floating gate transistors in current mirrors has been explored in [110,179] to address voltage offset issues in fully differential architectures. This calibration technique involves a combination of electron injection and tunneling and offers several benefits such as long-term retention, programming capability, and compatibility with standard CMOS fabrication processes. In the circuit depicted in Figure 4.39, the current subtractors have been modified by replacing transistors M_7 and M_{12} with floating-gate transistors. This allows for modification of the gate offset voltage and cancelation of the corresponding offset current. Figure 4.40(c) illustrates the floating-gate transistor, which features a high-quality insulator (SiO_2) that isolates the gate and provides a nonvolatile charge storage node. The poly 1 and 2 of the CMOS process are used to create the floating gate and the control gate, respectively, forming capacitor C_{cont}. Additionally, a PMOS capacitor called C_{tun} is connected to the floating gate, with its gate serving as the

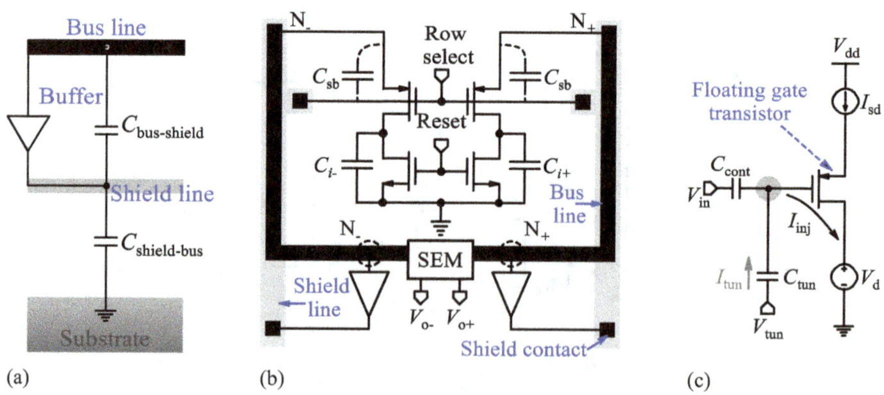

Figure 4.40 (a) Side perspective of the shielded current pathway, (b) Individual units within a fully differential capacitance sensor array exhibiting column-parallel architecture utilizing shielded current routing and (c) Conceptual depiction of a floating-gate transistor

tunneling terminal. Two different programming mechanisms are employed for the floating-gate transistor: impact-ionized hot electron injection and Fowler–Nordheim tunneling. In the former approach, a high source-to-drain voltage generates a negative shift in the gate offset voltage, causing electrons to be generated at the drain edge of the drain-to-channel depletion region via hot-hole impact ionization. The rate of injection current, I_{inj}, is controlled by the source-to-drain voltage and the source-to-drain current. In the latter mechanism, a positive shift in the gate offset voltage is generated using Fowler–Nordheim tunneling, which involves electric field-assisted tunneling by applying a very high electric field on C_{tun}.

Prakash and Abshire [110] have demonstrated a dynamic range of several tens of femto-Farad with atto-Farad resolution, using a supply voltage of ±3V as illustrated in Table 4.1. This achievement makes their sensor technology suitable for on-chip cell monitoring applications. The authors have also explored different array architectures, electrode configurations, and mismatch compensation techniques in their study [180].

In their research study, Jun-Rui and colleagues [181] utilized a fully differential core-CBCM sensor, similar to the one depicted in Figure 4.39, for detecting low-energy biomarkers. They replaced the simple current mirrors with cascade current mirrors, which resulted in achieving a higher output impedance that did not affect the output current. This modification allowed for an accurate study of both the steady-state and the transient current. They quantitatively explored the potential of applying the CBCM method in ionic conditions. The functionalized Au IDEs' binding with biomarkers caused a change in both real and imaginary components of the sensing impedance. They employed a parallel resistance-capacitance model for the sensing and REs ($R_S||C_S$ and $R_R||C_R$, respectively) and defined two variables as $\Delta R = R_S - R_R$ and $\Delta C = C_S - C_R$. They studied the behavior of the circuit in two different states: (1) with $\Delta R = 0$ and sweeping ΔC, (2) with $\Delta C = 0$ and sweeping ΔR. In the first case, ΔC affected the transient response of the circuit and the peak value of the output voltage, V_{Out}. In the second case, ΔR changed the final value of V_{Out}. By analyzing the peak and final values of the transient response, they could extract the real and imaginary parts of the sensing impedance, respectively.

The fully differential CVC described in Reference [110] uses output voltages that are amplified by a differential voltage amplifier in Figure 4.41 in [182] to achieve high sensitivities. The amplifier gain, A_{amp}, increases the sensor's sensitivity to 1.42 V/fF. The formula for the sensitivity is

$$S_{\frac{dV_{Out}}{dC_S}} = 2.A_{amp}.A_I.\Delta C.\frac{\left(V_{dd} - V_{ss} - \left|V_{thp}\right|\right).N}{C_{int}} \tag{4.54}$$

Despite its high resolution, the limited output voltage swing restricts the output dynamic range to about 300 aF. To address this issue, a common mode feedback (CMFB) circuit is proposed to regulate the output common-mode voltage of the first fully differential stage to half of the supply voltage. This results in more symmetrical responses at the two differential output nodes of the first stage (V_{Out+} and V_{Out-}) and prevents overshooting of these voltages.

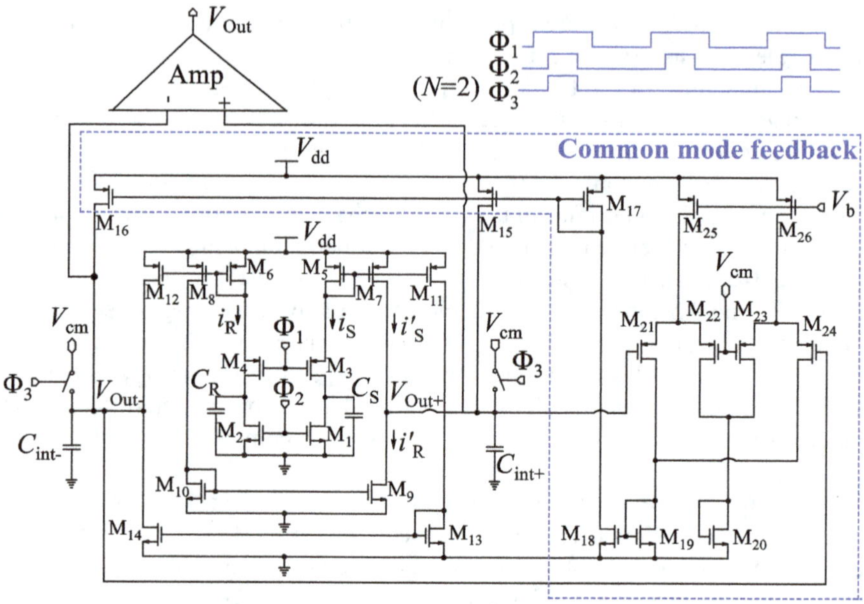

Figure 4.41 Fully differential capacitive sensor utilizing core-CBCM technology, where V_{cm} represents the necessary common-mode voltage, equal to half of V_{dd}

Fully differential capacitance-to-current, current-to-voltage, and voltage-to-digital conversion

In their research paper [54], Nabovati *et al.* propose a fully differential CDC architecture that utilizes a fully differential CVC and a ΣΔ modulator specifically designed for this structure. The ΣΔ modulator used in this CDC is an extended version of the single-ended modulator proposed by Ghafar-Zadeh *et al.* [171].

The proposed CDC, as depicted in Figure 4.42(a)–(c), comprises a comparator, a quantizer, and a DAC. The first-order ΣΔ modulator in the CDC consists of integrating capacitors C_{int+} and C_{int-}, which create a summing node and an LPF. In each quantizer's sampling period, the voltages of these capacitors are compared with reference voltages. If $V_{Out+} - V_{Out-} > V_{Ref+} - V_{Ref-}$, the output signals produced by the comparator activate switches S_{z+} and S_{z-}, injecting currents i_{z+} and i_{z-} (dependent on C_R) into the nodes V_{Out-} and V_{Out+}, respectively. This causes the positive current $i'_S - i'_R$ to have i_{z-} subtracted from it, while the negative current $i'_R - i'_S$ has i_{z+} added to it. The switches S_{z+} and S_{z-} and the current sources of i_{z+} and i_{z-} (M_{z+} and M_{z-}) act as a DAC in the first-order ΣΔ modulator.

The modulator's simple model is illustrated in Figure 4.42(b), and the corresponding signal diagrams are shown in Figure 4.42(c). The fully differential structure's sensitivity is improved by integrating the CBCM currents over N cycles

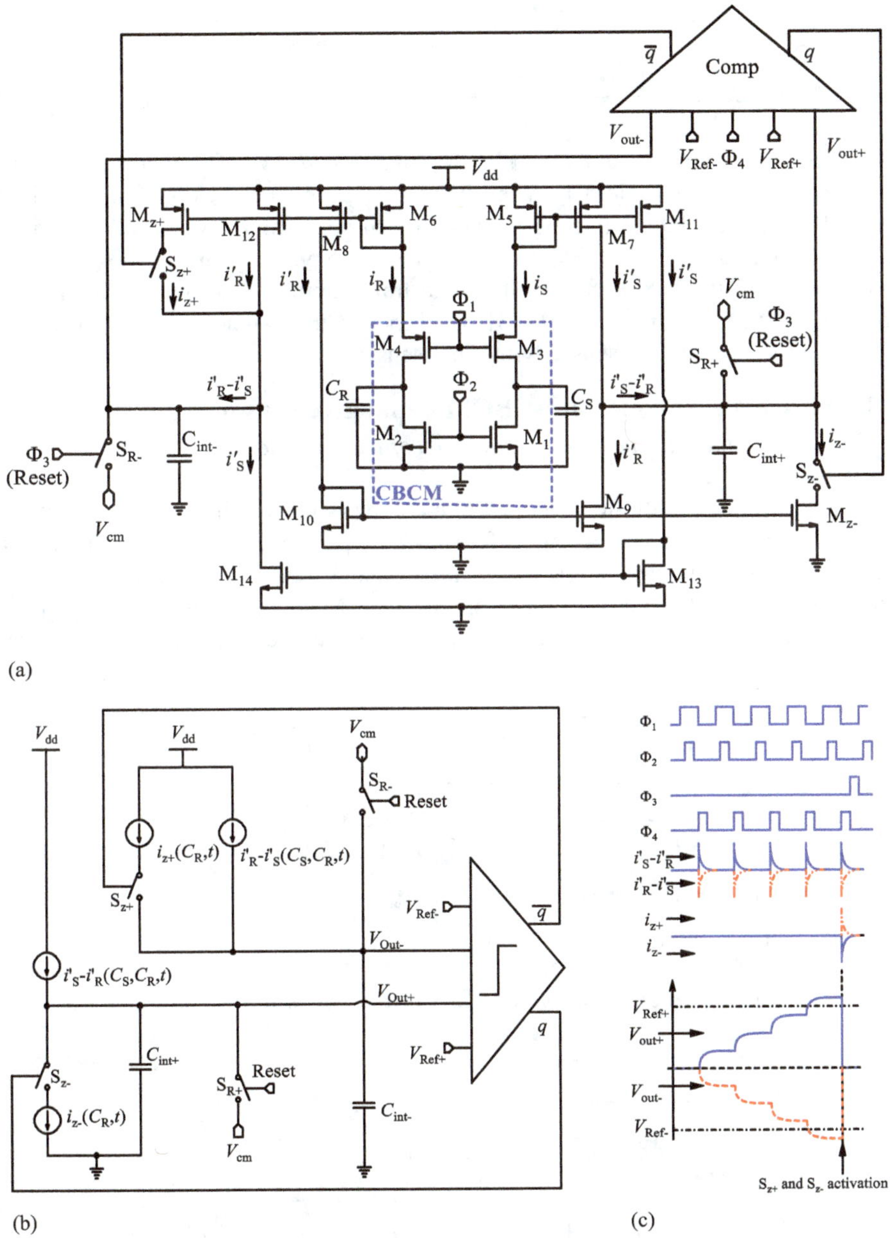

Figure 4.42 (a) Illustration of a fully differential core-CBCM ΣΔ CDC, (b) Adaptation of a ΣΔ modulator for a fully differential core-CBCM capacitive sensor and (c) Corresponding signal diagrams

of the periods of Φ_1 and Φ_2 ($f_S = Nf_{int}$). The sensitivity is then given by:

$$S_{\frac{dV_{Out}}{dC_S}} = 2.A_I.\Delta C.\frac{(V_{dd} - V_{ss} - |V_{thp}|).N}{C_{int}} \tag{4.55}$$

This modulator was employed by Nabovati *et al.* [54] for chemical solvent detection, as seen in Table 4.1. When integrated over 10 cycles of sampling time ($N = 10$), the modulator achieves a sensitivity of approximately 350 mV/fF and a dynamic range of 10 fF in 0.35 μm CMOS technology. The same group [37] used the same CDC in an 8 × 8-array for high-throughput cell growth screening. In another study, the same group [183] developed a multi-functional biosensor by combining the capacitive sensor with a pH sensor.

4.3.6.5 Current-mode core-CBCM capacitive sensors

Differential capacitance-to-current, current-to-frequency conversion

The utilization of both CBCM and CFC methods is present in certain sensors, as demonstrated in Figure 4.43 of Reference [184]. The circuit proposed in this reference uses an integrating capacitor in the analog domain to average the sharp exponential currents of each branch in the CBCM block. This voltage is then converted into a frequency that corresponds to the average current. Two structures,

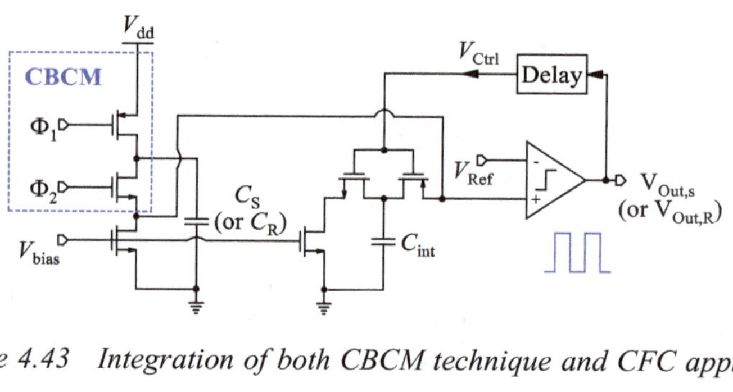

Figure 4.43 Integration of both CBCM technique and CFC approach

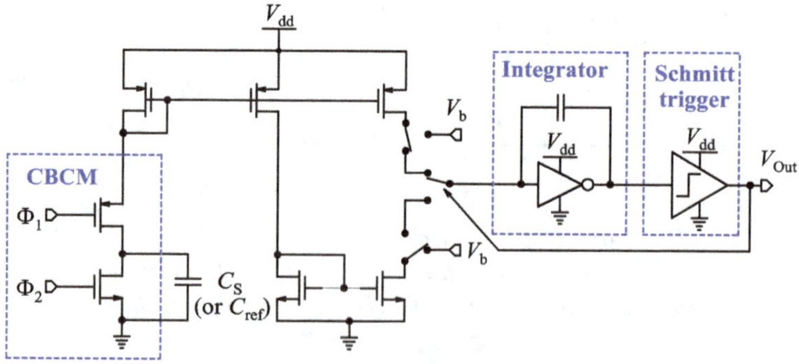

Figure 4.44 A core-CBCM CFC using current mirrors, an integrator and a Schmitt trigger

similar to the one used for the sensing capacitor, are used for the reference capacitor to convert their capacitance into frequency. By subtracting the outputs of these two structures, the change in capacitance (ΔC) is proportional to the difference in frequency. However, this sensor requires a large integrating capacitor to achieve high accuracy and its IDR is in the femto- to pico-Farad range.

Another approach that combines the CBCM and CFC methods is proposed in [185]. As seen in Figure 4.44, this sensor uses an integrator and a Schmitt trigger to accumulate and convert the CBCM current into a square signal. The sensing signal is then fed to a digital block containing a counter that compares its period with a reference signal obtained from a symmetrical circuit. The counter stops counting the sensing signal when the reference signal reaches a known number (2^8). The counter value at this point provides information about the ratio of the sensing and reference capacitances (C_S/C_R). However, this sensor suffers from low resolution.

Tashtoush [36,186] utilized a CBCM sensor that was previously suggested in [173], alongside a comparator-based signal processing circuit, in a multi-labs-on-a-single chip (MLoC) system as indicated in Table 4.1. The output of the CBCM-based capacitive sensor can be read as a voltage signal, but the signal processing circuit converts it to a frequency. Apart from the CBCM-based capacitive sensor that can detect bacterial pathogens at femtomolar levels, the MLoC also incorporates an optical CMOS biosensor, an electrochemical spectroscopy biosensor, and a CMOS micro-coil that is integrated with an interdigitated microelectrode array (IDMEA) and microfluidic system.

The previous core-CBCM sensors faced limitations in dynamic range due to their use of voltage mode. These circuits converted the output current of the CBCM block to the voltage before digitization, resulting in integration in the analog domain, which restricted the swing of the integrating capacitor voltage due to supply voltage limitations. To overcome this, a current-mode circuit was introduced in [187] for low-supply voltage CMOS technologies, which featured a core-CBCM CFC. Figure 4.45(a) illustrates the proposed concept through a simple circuit. In this approach, the amplified currents of the CBCM core are directed to a CCO after subtraction. The CCO can modulate the sharp exponential currents of the CBCM block to pulse frequency and follow their variations effectively. The output frequency is averaged using a counter, as the average of the differential current is proportional to ΔC. Unlike previous works, this sensor performs the required averaging operation part by part using the CCO in the analog domain, with the counter accumulating the total integrated parts, each having a unit value equal to A_u, in the digital domain.

In Figure 4.45(b), a basic design of a CCO is presented to explain the performance of the circuit shown in Figure 4.45(a). The CCO takes the input current i_{CCO} and alternately charges nodes V_1 and V_2, resulting in variations of the output voltages V_3 and V_4, which are inversions of each other due to the latch. The output frequency is determined by the intervals (T_m) during which V_1 and V_2 reach a toggling voltage (V_{tog}) and change the state of the latch. The integration of i_{CCO} during each interval is calculated by:

$$\int_{\left(\sum_{z=1}^{m} T_z\right)-T_m}^{\sum_{z=1}^{m} T_z} i_{CCO}(t, C_S, C_R)dt = C_{t1} V_{tog}, \; m = 1, 2, \ldots, J \tag{4.56}$$

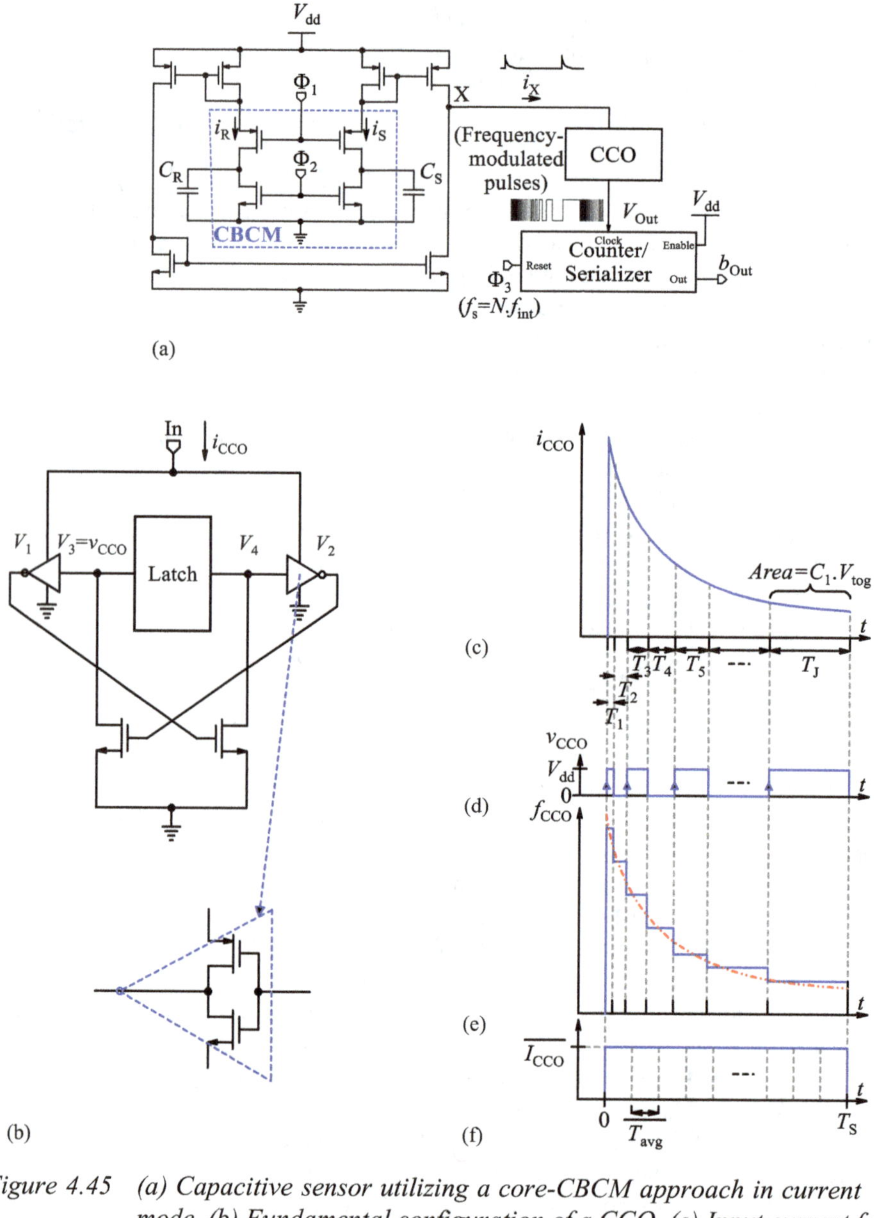

Figure 4.45 *(a) Capacitive sensor utilizing a core-CBCM approach in current mode, (b) Fundamental configuration of a CCO, (c) Input current for the CCO, (d) Output voltage of the CCO, (e) Frequency of CCO's output pulses along with their enveloping and (f) Average input current of CCO determined via the mean value theorem (represented with exaggeration)*

Assuming that $C_{t1}V_{tog}$ is constant, the smaller amplitude of i_{CCO} leads to longer intervals (T_m), while the larger amplitude leads to shorter intervals. Therefore, the total area under $i_{CCO}(t)$ during a certain time is divided into very small blocks, each with an area of $C_{t1}V_{tog}$ g. These blocks are called unit blocks, and J unit blocks are obtained during the sampling time ($T_S = 1/f_S$), as shown in Figure 4.45(c)–(f). After each T_m, the state of the output voltage (v_{CCO}) changes from high to low or low to high, as depicted in Figure 4.45(d). The integration of i_{CCO} during the total integration time $T_{int} = 1/f_{int}$ is obtained by counting these unit blocks by a counter:

$$\int_0^{T_{int}} i_{CCO}(t, C_S, C_R)dt = N\sum_{m=1}^{J}\left(\int_{\left(\sum_{z=1}^{m} T_z\right)-T_m}^{\sum_{z=1}^{m} T_z} i_{CCO}(t, C_S, C_R)dt\right) = NJC_{t1}V_{tog} \tag{4.57}$$

The integration process is continued and completed in the digital domain. If the rising edges of the output pulses are considered, the total number of pulses counted by the counter (N_{Count}) during T_{int} is equal to $NJ/2$. Therefore, it can be proved that N_{Count} is linearly proportional to ΔC. The equation for N_{Count} is given by:

$$N_{Count} = \frac{NA_I\left(V_{dd} - V_{thp}\right)\Delta C}{2C_{t1}V_{tog}} \tag{4.58}$$

where N is an even number.

The CCO circuit, illustrated in Figure 4.45(e), is capable of converting the input capacitance (i_{CCO}) to frequency, following the envelope of the discrete variable frequency of the output pulses. The output frequency (f_{CCO}) is time-variant and can be expressed as:

$$f_{CCO}(t, C_S, C_R) = A_{CCO}.i_{CCO}(t, C_S, C_R) \tag{4.59}$$

where A_{CCO} represents the gain of the CCO. The input current i_X can be expressed as:

$$i_X(t, C_S, C_R) = A_I\Delta C\frac{dv}{dt} \tag{4.60}$$

By combining (4.59) and (4.60) and averaging the results over the sampling time (T_S), we can obtain the average of the variable output pulse frequencies, F_{CCO}, over T_S, as given by:

$$F_{CCO}(\Delta C) = \frac{A_{CCO}.A_I.\Delta C.\left(V_{dd} - V_{thp}\right)}{T_S} \tag{4.61}$$

Furthermore, averaging over the total integration time (T_{int}) results in:

$$F_{CCO}(\Delta C).T_{int} = A_{CCO}.N.A_I.\Delta C.\left(V_{dd} - V_{thp}\right) \tag{4.62}$$

where N_{count} represents the counter number, as shown in Figure 4.45(d):

$$N_{Count} = \int_0^{T_{int}} f_{CCO}(t, C_S, C_R).dt = F_{CCO}(\Delta C).T_{int} = \frac{NJ}{2} \tag{4.63}$$

The sensitivity of the sensor is determined by three parameters: A_{CCO}, A_I, and N, where N is the number of cycles of Φ_1 and Φ_2, which can be controlled off-chip:

$$S_{\frac{d(Numb)}{d(\Delta C)}} = A_{CCO}.N.A_I.\left(V_{dd} - V_{thp}\right) \tag{4.64}$$

Increasing the gain of the current mirror (A_I) can result in greater sensitivity, but it may be limited by the IDR of the CCO and the size of the counter. Increasing the number of cycles (N) leads to a longer measurement time.

To summarize, the CCO circuit integrates the CBCM current signals in two steps: analog integration by the CCO over the time of each unit block, followed by digital counting of the number of unit blocks. Each unit block generates an individual output frequency, which together produces a string of output pulses with discrete frequencies that follow the envelope of the instantaneous frequency-modulated current response. The original circuit presented in [111] is calibrated using a fast up-down counter and is designed to operate in five phases: initialization, down counting, saving, up counting, and cycling. The LFSR is adapted to count bi-directionally and works as a serial-input serial-output shift register. A digital buffer is also required before the counter to mitigate the loading effect. Integration over N cycles of the sampling period ($f_s = N.f_{int}$) increases the sensitivity of the sensor. The non-linearity of the CCO is the primary cause of the limited resolution of this sensor, which can be improved using the pre-distortion technique. The IDR of the CCO and the size of the counter are the limiting factors for the sensor's IDR. Table 4.1 shows that this method results in a dynamic range of about 70 fF in 0.18 μm CMOS technology, which is high compared to previous core-CBCM sensors.

Capacitance-to-frequency conversion based on enhanced CBCM (ECBCM) circuit

In their research on core-CBCM capacitance-to-current converters, Osouli *et al.* [115] introduced an innovative core-CBCM current-mode CFC with a wide input capacitance range. This converter includes a calibration circuitry, which incorporates an array of switched capacitors and IDEs implemented on the topmost metal layer. A current-controlled 300 MHz oscillator and a counter-serializer are also integrated to generate digital output. The chip's output is linked to a microcontroller, facilitating the transmission of data to a tailored graphical user interface (GUI).

The provided circuit diagram in Figure 4.34(a) illustrates a typical configuration of a core-CBCM capacitance-to-current converter which converts the differential input capacitance ($dC = C_S - C_R$) into the differential current. Figure 4.46(a) shows a sensor using this circuit. If an integrating capacitor serves as the load, the drain-source voltage of N_3 (V_D) fluctuates during the charge and discharge cycles of the integrating capacitor, while the drain-source voltage of N_4 (V_C) remains constant for a given C_R. Hence, the currents I_1 and i_R are also affected by the drain-source voltage due to the channel length modulation effect. Consequently, I_1, which ideally should equal i_R, may deviate significantly (where λ represents the

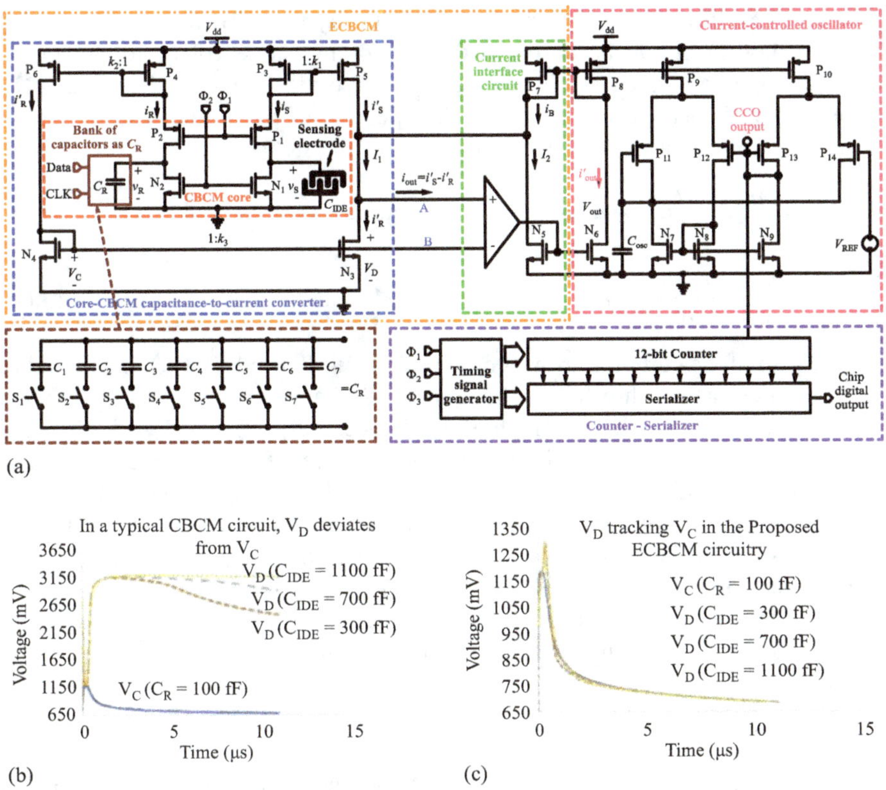

Figure 4.46 *(a) ECBCM-driven CFC configuration, (b) Graphs depicting V_C (solid line) and V_D (dashed line) variations across different C_{IDE} values, with an integrating capacitor in the output path [115] and (c) V_C (solid line) and V_D (dashed line) fluctuations across different C_{IDE} values in ECBCM with a feedback loop engaged [115]*

channel-length modulation parameter, which is non-zero for MOSFETs). Although using long-channel devices for N_3 and N_4 can partially mitigate the adverse effects of λ, it does not completely eliminate this systematic deviation. Cascode current mirrors have been proposed as a means to alleviate the impact of channel length modulation by employing an additional stage to shield the current mirrors from voltage variations of the load [188]. Despite improvements, the cascode topology still faces limitations due to restricted voltage headroom. The research of Osouli and colleagues [115] focuses on demonstrating the efficacy of a feedback loop for CBCM circuits, aimed at reducing the aforementioned tracking error across a wide range of input differential capacitance values. They proposed enhanced CBCM (ECBCM) topology, depicted in black in Figure 4.46(a), which incorporates a feedback path comprising N_5, N_3, and an OpAmp to enforce V_D to track V_C, regardless of the value of C_S. Simulation results in Figure 4.46(b) illustrate the deviation of V_D from V_C for different C_S values in a typical CBCM circuit with C_{int}. In Figure 4.46(c), it can be observed that the drain-source voltage of N_3 in the ECBCM topology successfully follows the drain-source voltage of N_4. Consequently, I_1 becomes equal to i'_R, irrespective of the value of C_S. The following equations express the transfer function of the ECBCM circuit:

$$i_{out} = \frac{W_{N6}}{W_{N5}} \times \frac{L_{N5}}{L_{N6}} \times \frac{1 + \lambda V_{Out}}{1 + \lambda V_D} \times (A_I \times (i_S - i_R) - i_B) \tag{4.65}$$

$$i_B = \frac{A'_I}{2} \times \frac{W_{P7}}{L_{P7}} \times (V_{dd} - V_D - V_{ThP})^2 \times (1 + \lambda(V_{dd} - V_D)) \tag{4.66}$$

The gain of the current mirrors, composed of (P_3, P_5) and (P_4–P_6), is denoted as A_I. V_{ThP} represents the threshold voltage of the P-channel MOSFET device, while W and L refer to the width and length of the devices, respectively. V_{dd} signifies the supply voltage, and i_B, as given in (4.66), represents the offset current from P_7. The equation is inherently nonlinear due to the terms $1 + \lambda V_{out}/1 + \lambda V_D$ and i_B. To mitigate this nonlinearity, long-channel transistors are employed for N_5 and N_6 in the second stage. This choice helps satisfy the conditions $\lambda V_{out} \ll 1$ and $\lambda V_D \ll 1$, allowing the nonlinear term $1 + \lambda V_{out}/1 + \lambda V_D$ to be approximated as unity. To minimize the impact of i_B, the aspect ratio W/L of P_7 is minimized. As a result, $A_I \times (i_S - i_R) \approx i_B$, reducing (4.66) to a linear equation given by:

$$i_{out} = \frac{W_{N4}}{W_{N3}} \times \frac{L_{N3}}{L_{N4}} \times A_I \times (i_S - i_R) \tag{4.67}$$

It should be noted that the channel-length modulation effect of P_5 and P_6 is not considered in (4.65) for simplicity purposes. Additionally, P_7 is introduced to the topology to provide a bias current proportional to the reference capacitance value, serving as a baseline. This proves beneficial when the CCO exhibits nonlinearity for small currents (less than 1 μA in [115]). In the evaluation of this circuit's performance using ECBCM compared to CBCM, Figure 4.46(b) illustrates the variations of V_C (solid line) and V_D (dashed line) for different C_{IDE} values when the output is connected to an integrating capacitor. On the other hand, Figure 4.46(c) depicts the same simulation for the ECBCM with the feedback loop. An external system consisting of microcontrollers (MCUs) is used to read the digital output and generate the clock pulses as described in Appendix B.

Based on their simulation results, the input capacitance range is approximately 200 fF (ranging from 100 fF to about 300 fF) for the CBCM circuitry, around 600 fF (ranging from 100 fF to 700 fF) for the CBCM with cascode current mirrors, and roughly 1,800 fF (ranging from 100 fF to 1,900 fF) for the ECBCM. These numbers may vary slightly based on different sizing configurations, but they generally reflect the effectiveness of the proposed feedback loop.

The presence of mismatch-induced offset in capacitance measurements is a time-invariant phenomenon that can be mitigated through calibration. However, when working with biological and chemical samples, the accumulation of sediments can introduce significant offset changes during the experiment [4,56].

Consequently, a simple pre-exposure calibration of the offset capacitance based on pre-analyte conditions may not effectively eliminate such instantaneous offset variations occurring during the experiment. To address this challenge, this group [148] proposed a calibration-free approach in their ECBCM-based CFC by using a bank of variable reference capacitors. By leveraging the wide input capacitance range and differential measurement capabilities of this sensor, the proposed method enables rapid sweeping of the variable reference capacitor value at periodic intervals. This enables the real-time detection of instantaneous and gradual changes in capacitance across a broad range during the experiment, without the need for sensor calibration. The resulting two-dimensional (2D) data, when plotted over time, forms an informative 3D graph. This graph reveals the presence of non-target molecules, including sedimentation such as dead cells.

Assuming the constant nature of $\Delta C(t)$, which is the time-variant offset capacitance due to the remnants and/or other by-products, etc. (as explained in Section 4.3.6.1), over a time interval between t_1 and t_2, where $t_2 - t_1 > T_S$, it is possible to utilize a reference capacitance C_R in the CBCM interface circuit to counteract the impact of ΔC. In the initial time period T_{init}, the offset capacitance of an IDE, C_S, can be nullified by ΔC_{Min} (part (1) of Figure 4.47(a)), ensuring that the average of the differential output current, i_{out}, in the CBCM circuit is solely proportional to the change in $\sigma C(t)$. However, during the experiment, remnants present on the IDE can introduce additional offset capacitance, denoted as ΔC_1 and ΔC_2 (see parts (2) and (3) of Figure 4.47(a)). Consequently, the same C_R cannot guarantee the cancellation of $\Delta C(t)$ throughout the entire duration of the experiment. If any of the calibration schemes described in Section 4.3.6.2 were employed, C_R would be maintained at ΔC_{Min}, potentially causing i_{out} to quickly reach the saturation region or leaving limited room for the measurement of the target sensing capacitances, $\sigma C(t)$. To address this issue, Osouli *et al.* [148] proposed the use of a variable C_R that is swept across its complete dynamic range, encompassing values between ΔC_{Min} and ΔC_{Max}. Here, ΔC_{Min} and ΔC_{Max} represent the lowest and highest achievable capacitances, respectively, with the implemented bank of capacitors. The sweeping of C_R enables the detection of sudden and gradual changes in capacitance caused by remnants above the electrodes, $\Delta C(t)$, as well as the binding of target molecules, $\sigma C(t)$. Figure 4.47(b) illustrates i_{out} as a function of ($C_S - C_R$) and time, presenting a 3D representation. This depiction builds upon the foundation of Figure 4.47(a), with the addition of the C_R axis. Each plate in the figure corresponds to a baseline concerning the value of $\Delta C(t)$. Consequently, Figure 4.47(b) extends Figure 4.47(a) by incorporating the C_R axis. By acquiring consecutive curves, a 3D representation encompassing the entire duration of the experiment can be obtained, revealing both $\Delta C(t)$ and $\sigma C(t)$. This approach establishes a calibration-free technique for monitoring a wide range of capacitance changes in a solution, encompassing abrupt changes resulting from remnants, $\Delta C(t)$, as well as gradual changes due to the interaction of target molecules with the electrode, $\sigma C(t)$.

In Figure 4.47(c), a 3D graph showcases the experimental results obtained using Methanol droplets, known for their higher dielectric property ($\varepsilon_r = 30$) and slightly elevated conductivity ($\delta = 0.3$ mS/cm). The graph depicts the behavior of

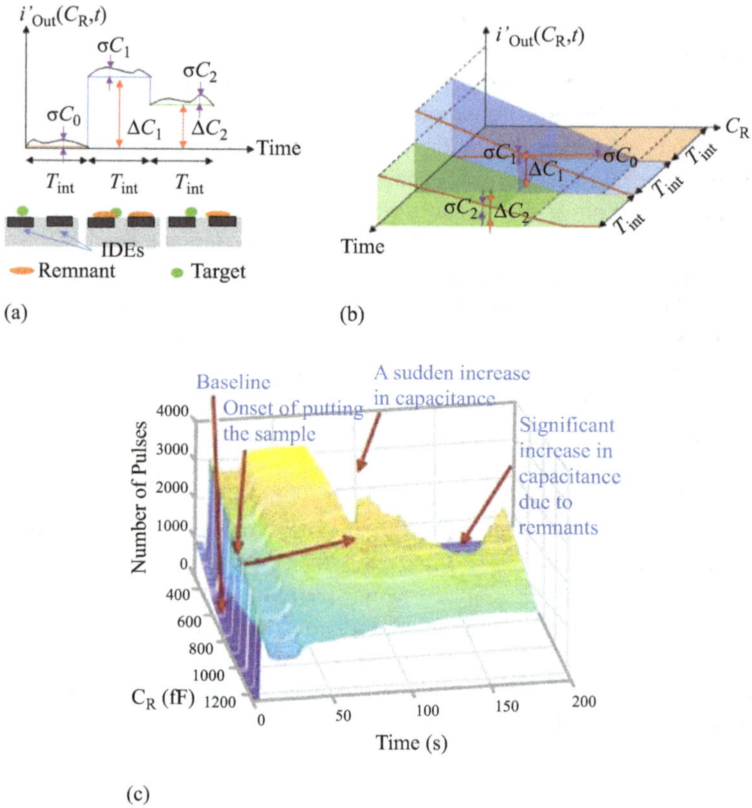

Figure 4.47 (a) Conceptual view of a 2D graph of CBCM's output signal, (b) Conceptual view of a 3D graph of CBCM's output signal and (c) A measurement result showing the 3D signature for evaporating a 2μL-droplet of MeOH [148]

capacitance over time, starting with a baseline and exhibiting a sudden increase. Around the 100-second mark, the curves undergo a noticeable change, showing consistently higher capacitance until the conclusion of the experiment, when the droplet fully evaporates. Notably, once the sample dries, the curves do not revert to the initial baseline. This particular observation can be attributed to the significant sedimentation of salty crystals present in the original sample.

In another effort, this group [114] extended their core-CBCM sensor to a dual 8×16 capacitive array using a novel multiplexing scheme. The proposed circuit configuration, as depicted in Figure 4.48, showcases the arrangement of an electrode array. In this particular setup, a set of 16×8 IDEs are multiplexed and linked to a single ECBCM circuit. Within this multiplexing arrangement, the in-pixel devices consist solely of P(i,j) and N(i,j), where i and j denote the row and column, respectively. The gates of each P(i,j)-N(i,j) transistor pair are connected to a network of multiplexed clock

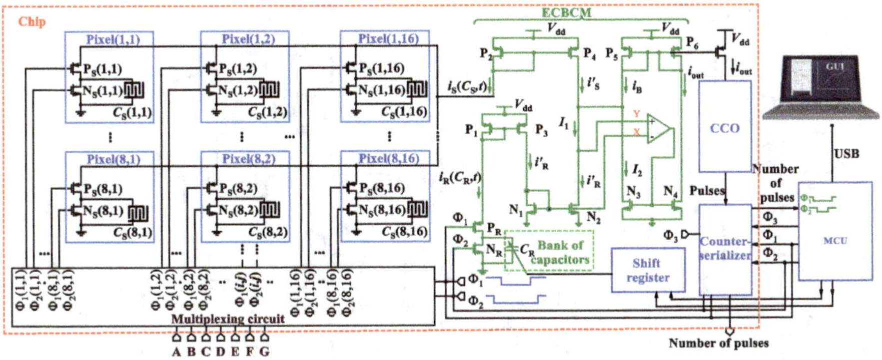

Figure 4.48 Capacitive sensor array reported in [114]

signals, namely $\Phi_1(i,j)$ and $\Phi_2(i,j)$ illustrated in Figure 4.49(a). This digital circuitry incorporates 3-bit and 2-bit multiplexers with control signals A, B, C, D, E, F, and G, where G represents the most significant bit of the 7-bit number. Figure 4.49(b) and (c) provides timing signals exemplifying the generation of $\Phi_1(i,j)$ and $\Phi_2(i,j)$ in the multiplexing circuit for specific rows and columns, such as $i = j = 1$, $i = 2, j = 1$, and $i = 8$, $j = 16$. These examples demonstrate how the multiplexing circuit operates. In this configuration, all PMOS switches connected to the IDEs, responsible for charging, remain off except for the Active IDE under evaluation. For instance, as shown in Figure 4.49(b), when the input of the multiplexers (GFEDCBA) is 0000000, P(1,1) and N(1,1) receive the Φ_1 and Φ_2 signals, while the PMOS devices of the remaining IDEs ($P(i,j)$ for $i = 2,\ldots 8, j = 1,2,\ldots 16$) remain off. Notably, the IDEs charge during the falling edge of $\Phi_1(i,j)$ and discharge during the rising edge of $\Phi_2(i,j)$. The clock pulse signals for the reference side, supplied to P_7 and N_5 (refer to Figure 4.48), are not multiplexed as the reference is shared among all the IDEs. Figure 4.48 also illustrates that, in this multiplexing configuration, all non-active IDEs are connected to the ground through NMOS devices. This offers a significant advantage in electrochemical tests by creating a large counter electrode and ensuring that the solution's voltage remains uniformly close to zero. The IDEs in this study are implemented as 2-finger IDEs on the topmost metal layer, with dimensions of 35 μm by 30 μm. The array employed both passivated and non-passivated electrodes. Through simulation and experimentation, it was demonstrated that the sensor possesses the capability to differentiate between the dielectric constants of various liquids, determine the precise location of a droplet in the sensing area, and measure the time it takes for the droplet to evaporate from the chip surface. An external system consisting of an microcontroller (MCU) is used to read digital output and generate the clock pulses as described in Appendix C.

4.3.7 Comparison of different capacitive sensors

Table 4.1 presents a comparison of various techniques for capacitive biosensors in terms of their sensitivity, power consumption, area, number of arrays, and interdigitated capacitance resolution. The lock-in technique in Reference [113] achieves

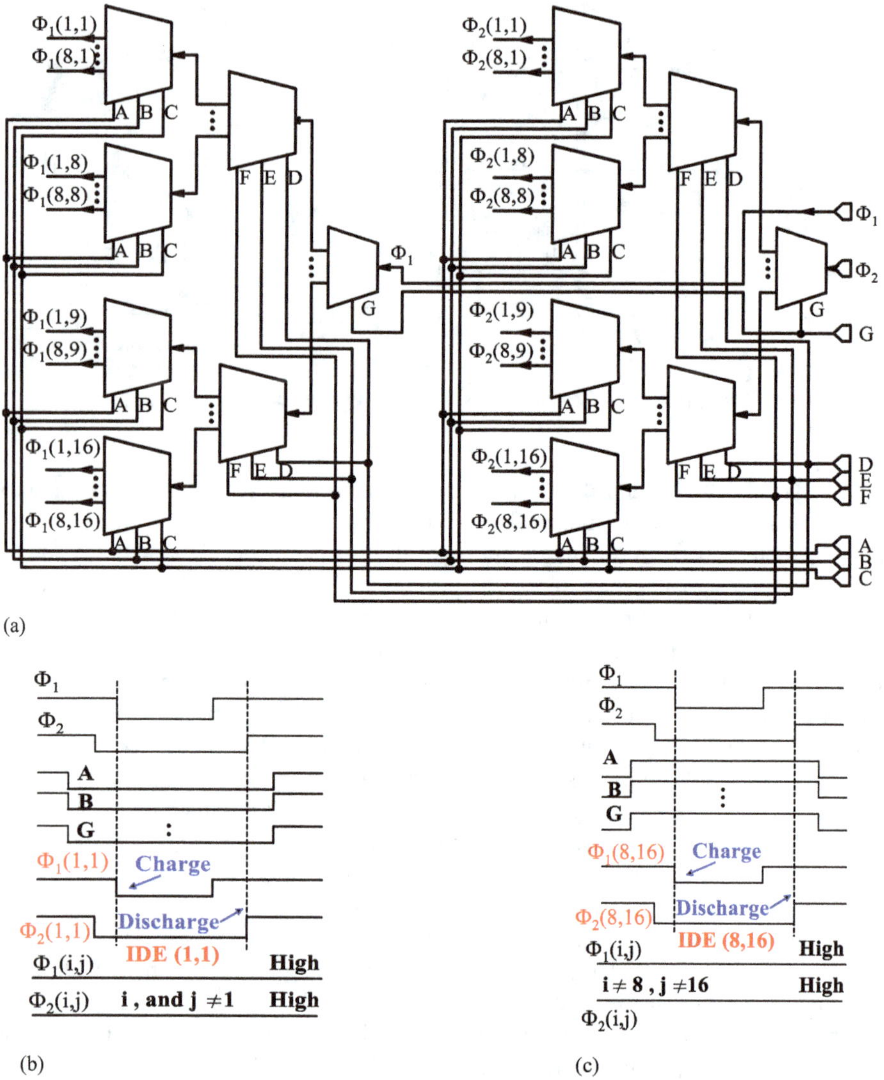

Figure 4.49 (a) The multiplexing circuitry, (b and c) Example timing diagrams of the multiplexed clock pulses: (a) when all multiplexing signals are zero, IDE (1,1) becomes active and receives the $\mathbf{\Phi}_1$ *and* $\mathbf{\Phi}_2$ *clock pulses. PMOS and NMOS devices of all other IDEs are high, making the IDE connected to the ground through the NMOS device, (b) IDE (8, 16) is active when all multiplexing signals are high*

a high resolution down to 0.065 aF, but its power consumption is high and its IDR is limited to 1 fF. In contrast, the combination of TVA and SC techniques proposed in [112] achieves a resolution of about 1 pF, although its accuracy depends on the accuracy of the triangular voltage and output signal processing.

CMOS-based techniques such as CFCs, CVCs, and charge redistribution-based techniques, using ROs or charge-sharing methods, have also been reported. RO-based sensors in [104] and [33] have resolutions of 2.5 fF and 10 fF, respectively, but using ROs in the CFCs proposed in [3,107,108] provides higher sensitivities and resolutions of tens of atto-Farad with an IDR in the tens of femto-Farad range. The comparator-based relaxation oscillator in [27] demonstrates an IDR between 12 fF to 700 fF in a 5 × 5 array, while the largest array of CFCs in Table 4.1 is the 8 × 16-array presented in [38], which provides an IDR between 330 pF to 10 nF.

The techniques of charge sharing, CSA, and SC-based methods rely on charge redistribution and have demonstrated high sensitivities in certain cases. The 28-array sensors utilizing the charge-sharing principle, as reported in Reference [29], have a resolution of only 5 fF. However, the charge-sharing-based sensor proposed by Couniot *et al.* [4] has a resolution of approximately 450 aF, with low complexity, resulting in reasonable power consumption of about 29 μW and a total area of around 0.1 mm^2. The CSAs described in [105] offer a higher resolution of approximately 21 aF and an IDR of 0.42 fF. Although this circuit was used in a large 320 × 320 sensor array, its outputs are not digital and require external ADC conversion. The SC technique presented in [106], adapted to SAR ADC, provides a resolution of about 4.5 fF, which is not as fine as other charge redistribution-based methods but offers the lowest power consumption among the sensors reported in Table 4.1, at a total of approximately 2.1 μW for a 9-array.

In the design of capacitive biosensors for life science applications, accuracy is a crucial factor because biological signals are typically small. Although circuit simplicity allows for high-throughput measurements, the accuracy of resistive-oxide-based sensors is often not as high as the CBCM method. Furthermore, RO-based sensors are susceptible to variations in supply voltage, process, and temperature. In contrast, CBCM offers both high accuracy and low power consumption, making it an ideal candidate for such applications. Table 4.1 indicates that CBCM can achieve high accuracy in the range of atto-Farad while consuming only micro-watts of power. Therefore, the development procedure of core-CBCM circuits is the focus of the rest of this chapter. Array structures based on CBCM and CFC techniques are popular due to their low complexity, but CBCM stands out in terms of accuracy and power consumption.

4.4 Summary

In this segment, an outline is presented encompassing diverse CMOS capacitive interface circuits. These circuits encompass charge redistribution methodologies such as the charge-sharing technique, CSA-driven configurations, and SC circuits. Additionally, CFCs utilizing comparator-based relaxation oscillators, ROs, VCO-based sensors, lock-in detection, triangular voltage analysis, and CBCM method, specifically tailored for biosensing applications, are addressed. Among these methodologies, lock-in detection, RO-centered CFCs, and CBCM demonstrate elevated precision. Nevertheless, the intricacy of lock-in detection often leads to heightened power consumption and a larger circuit footprint. Conversely, RO-based CFCs are frequently susceptible to variations in temperature, device dimensions, fabrication process, and supply voltage. Given the accuracy and streamlined nature of core-CBCM biosensors, these sensors are singled out for comprehensive elucidation in this chapter. Subsequently, the developmental trajectory of core-CBCM circuits is delved into. Despite the considerable interest garnered by CMOS capacitive biosensors, sustained endeavors remain imperative to conceive versatile capacitive biosensors housing an extensive array of sensing electrodes tailored suitably for diverse biological applications.

Chapter 5
CMOS voltammetric and impedimetric biosensors

Voltammetric and impedimetric sensors work based on electron transfer. So, in these sensors, a current or voltage is created in the output due to the reactions in the electrochemical cell.

The sensing element of these sensors are electrodes that can be fabricated by complementary metal-oxide-semiconductor (CMOS) technology using the same techniques discussed in Chapter 3. However, it is noteworthy that the required electrodes are not coated with insulating layers because the insulation layer will prevent electron transfer which is the basis of the operation of these types of sensors.

Voltammetric sensors involve the application of a potential that changes over time to the electrochemical cell, followed by the measurement of the resulting current in relation to that applied potential [189]. Based on how the applied potential changes, when the current is measured, and whether the solution is stirred, voltammetric techniques can be categorized into different types such as polarography, hydrodynamic voltammetry, stripping voltammetry (including anodic stripping voltammetry (ASV), cathodic stripping voltammetry (CSV), and adsorptive stripping voltammetry (AdSV), differential pulse voltammetry (DPV), cyclic voltammetry (CV), amperometry, and chronoamperometry (CA). Among these techniques, the most reported ones for integrating with the CMOS process are: (1) Amperometry, which is a type of voltammetry, in which a constant potential (see Figure 5.1(a)) is applied to the working electrode (WE) and the current is measured as a function of time, (2) CA which is based on applying a step potential (see Figure 5.1(b)) to the electrochemical cell, and (3) CV in which the applied potential is ramped linearly versus time and after reaching a set potential, it is ramped in the opposite direction to return to the initial potential (see Figure 5.1(c)).

Impedimetric sensors measure the impedance of the sensing electrode as a function of frequency and work based on the application of Ohm's law by applying a small sinusoidal stimulus voltage (or current) of variable frequency (such as Figure 5.2) and evaluating the output current (or voltage) [190]. Like voltammetric sensors, there are electron transfers in impedimetric measurements. One of the merits offered by the impedimetric approach is that the stimulus signal is small and does not damage the bio-recognition layer [191].

Electrochemical cells might be configured by two electrodes (Figure 5.3(a)) or three electrodes (Figure 5.3(b)). Two-electrode configuration [39,44,132,192–195]

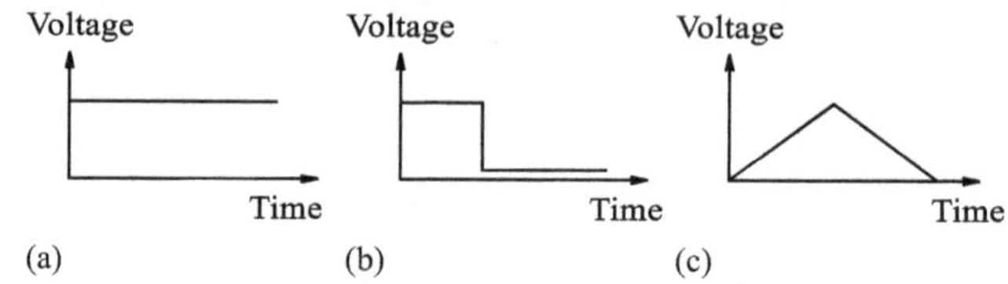

Figure 5.1 *Electrode voltage waveform for (a) amperometry, (b) CA, (c) CV*

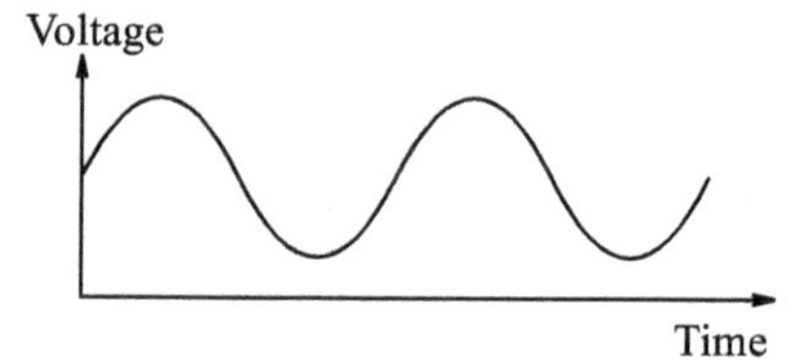

Figure 5.2 *Electrode voltage waveform for impedance spectroscopy*

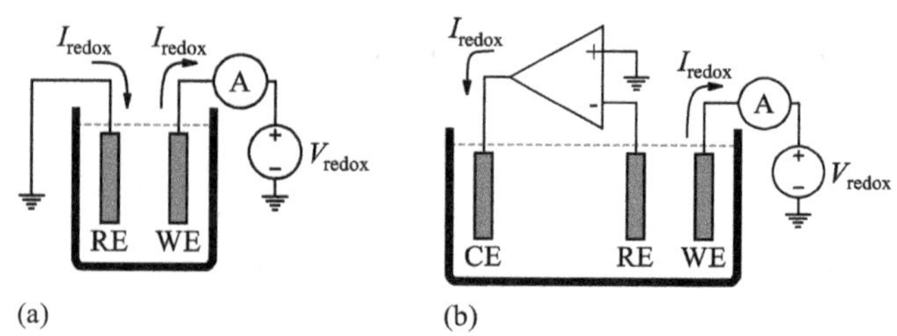

Figure 5.3 *(a) A two-electrode electrochemical system. (b) A three-electrode electrochemical system.*

includes a WE and a reference electrode (RE). The chemistry of interest occurs at the WE while the RE completes the circuit and maintains a constant interfacial potential. After applying the potential between the two electrodes, the current flowing between the two electrodes can be measured from each of them. However, since none of the electrode potentials is fixed (as a reference potential), it is very difficult to maintain a constant reference voltage while the current is flowing, and, under most conditions, it is impossible to say at which voltage an electrochemical reaction occurs. That is the reason why a three-electrode configuration using an auxiliary or counter electrode (CE) is more popular [196,197].

Figure 5.4 illustrates the structure of the electrochemical instrumentation employed for voltammetric and impedimetric sensors including signal generator, potentiostat, readout circuit, and signal processing. The signal generator generates the required stimulus signal shapes which can be a constant voltage, pulse, triangle, or saw waveform based on the desired sensing approach. The potentiostat is used for controlling the electrochemical cell and current acquisition. The readout circuit

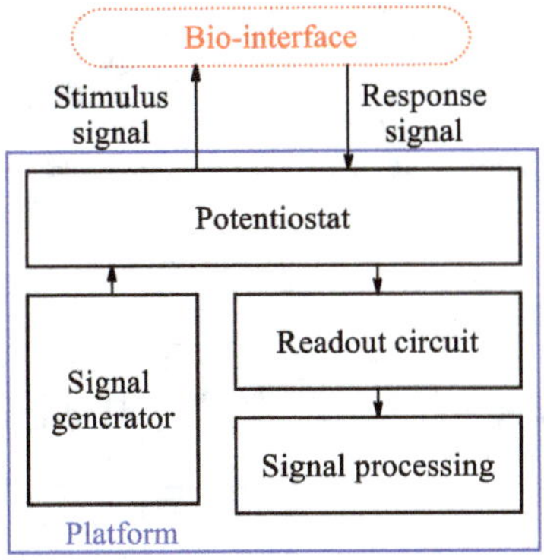

Figure 5.4 The structure of a potentiostat-based electrochemical sensor

is required for the amplification and conditioning of the sensor's response signal. The signal processing block improves data received from the readout circuit by filtering and alike. Since the signal generator and signal processing blocks do not directly connect to the electrode-electrolyte interface and can be implemented by general circuits, the main focus of this section will be on the other two blocks. So, in this section, after a brief review of the integrated signal generators reported for these biosensors, different types of CMOS potentiostats and the readout circuits for reading direct currents (DC currents) and alternating currents (AC currents), respectively obtained from voltammetric and impedimetric sensors, are reviewed.

5.1 Signal generator

As mentioned earlier, voltammetric and impedimetric sensors use different types of excitation signals. In some cases, these signals are generated off-chip. But, to fully integrate these sensors, a signal generator can be integrated with the other parts of the sensor. A low-noise signal generator is essential for high-resolution applications. In this section, a few examples of these signal generators will be demonstrated.

5.1.1 Signal generators for voltammetric sensors

Synthesizers are among those circuits that can be used to generate the required signals for voltammetric and impedimetric sensors. For example, Ghoreishizadeh *et al.* [198] designed a direct digital synthesizer (DDS) by mixed mode method to produce sub-Hertz triangular waveforms with different slopes for cyclic voltammetry in a multi-target sensor designed for CV and CA (Figure 5.5). This circuit includes a digital up/down counter and a digital-to-analog converter (DAC) implemented in a resistor ladder scheme. The slope of the waveform is set by the up/down counter step size controlled by a signal called "slope control."

In multi-mode signal generators, programmable analog circuits can be used to generate continuous signals of various shapes, frequencies, and amplitudes. However,

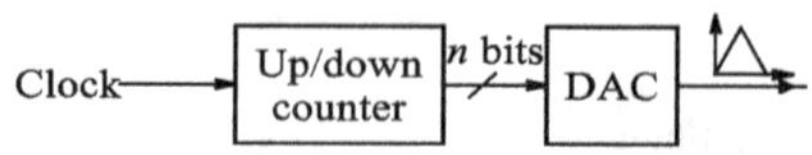

Figure 5.5 Schematic of a triangular waveform generator

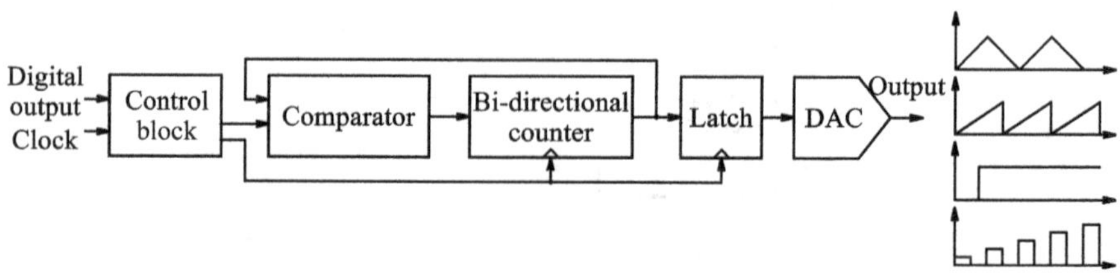

Figure 5.6 Schematic of a multi-mode voltage waveform generator

each desired signal requires a specific analog circuit resulting in increased power and chip area. So, digital circuits are more popular because they provide a lower complexity using a DAC and digital controller. However, the complexity of the controller could be significant [199]. In an effort, a programmable multi-mode voltage waveform generator was designed by Li *et al.* [199] to support a variety of voltammetric techniques including constant potential, linear sweep, cyclic, and pulse voltammetry. This circuit which is shown in Figure 5.6(a) consists of a digital control block, a comparator, a bidirectional counter, and a DAC. The control block produces the required signals to reset the generator, choose up/down count, and select voltammetry mode. After selecting the desired frequency as well as the maximum and minimum values of the waveform amplitude by the control block, the counter counts based on the comparator's output in either direction as required to generate a given waveform. The final analog voltage is generated by the DAC receiving the counter-binary bits.

5.1.2 Variable frequency signal generators for impedimetric sensors

Electrochemical impedance spectroscopy (EIS) needs sinusoidal signals with low noise, precise frequency (usually between 0.1 Hz and 10 MHz), high harmonic quality, and high stability against environmental factors such as temperature [200]. A frequency synthesizer can be used to generate a variety of output frequencies in the required range from a single reference frequency. In this section, after a brief review of reference frequency generators, a few frequency synthesizers reported for impedimetric sensors will be discussed.

5.1.2.1 Reference frequency generator

The reference frequency can be generated by: (1) non-linear (also known as resistive or relaxation) oscillators such as ring oscillators which usually output a square signal; or (2) linear (or so-called resonant, harmonic, tuned, or tank) oscillators such as inductor-capacitor (LC) and crystal oscillators which usually output a sinusoidal signal [200].

Although ring oscillators occupy less silicon area in comparison to resonant oscillators like LC and crystal oscillators, they have considerably higher phase noise, and, as a result, they are not very popular in impedance spectroscopy [200].

LC oscillators can generate highly pure sinusoidal waves, achieving easily a quality factor of 60. But the tolerances of the inductor and capacitors used in these oscillators are in the best case 0.1%, which makes it difficult to match a targeted frequency [200].

Crystal oscillators are more precise resonant oscillators that can provide matched frequencies with tolerances measured in parts per million (ppm) and with the addition of a minimum quality factor of 10,000 [200].

5.1.2.2 Frequency synthesizer

Frequency synthesizers may employ the techniques of frequency division, frequency multiplication, frequency mixing, DDS, and phase-locked loop (PLL) to generate variable frequency signals from their reference input frequency.

Figure 5.7 illustrates a block diagram of a PLL whose output frequency, f_{out}, is equal to Nf_{ref} and comprises a phase-frequency detector (PFD), a charge-pump (CP), a loop filter (LF), a voltage-controlled oscillator (VCO) and a frequency multiplier, N. PLL is prone to closed-loop instability. This can be avoided by the appropriate designing of an LF which can be freely changed by the designer based on the PFD and CP gain, VCO gain, and N. This flexibility of resonant VCO-based PLLs makes them suitable for EIS systems [200].

The configuration known as DDS offers an alternative means of generating a signal with a variable frequency, derived from a reference frequency. Figure 5.8 presents the schematic representation of a direct digital synthesis setup, comprising

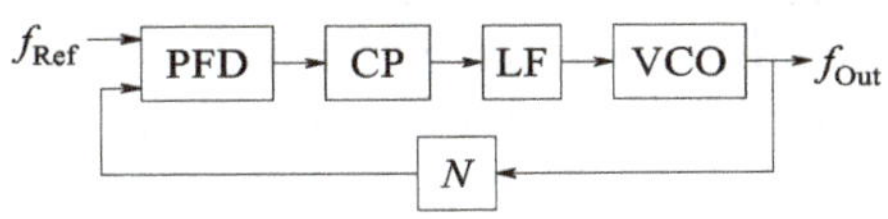

Figure 5.7 Block diagram of a phase-locked loop

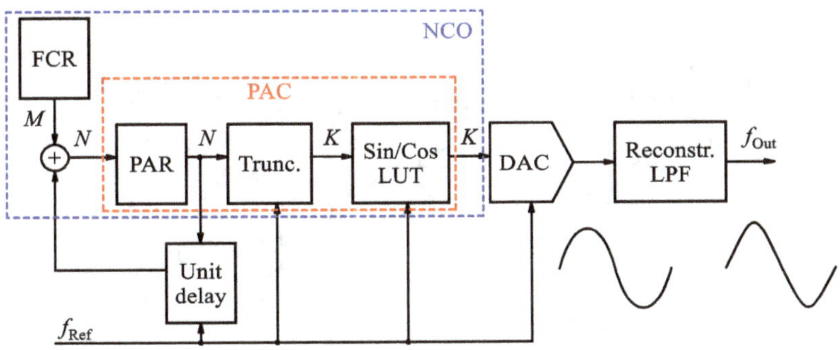

Figure 5.8 Block diagram of a direct digital synthesis

a numerically controlled oscillator (NCO), a DAC, and a low-pass filter (LPF). Within the NCO, there exists a frequency control register (FCR) alongside a phase-to-amplitude converter (PAC). This PAC involves an internal N-bit register (PAR) and also incorporates lookup tables (LUTs) housing a specific number of evenly distributed values for a single oscillation cycle. The system's resulting output frequency is determined by $f_{out} = f_{ref}\,(FCW/2^N)$, where FCW corresponds to the value stored within the FCR [200].

DDS requires simpler circuits in comparison to PLLs, consumes low power, occupies a small area, and can provide a flexible frequency sweep with a reference frequency. These merits make it more popular for use in EIS [200].

5.2 Potentiostats

An electronic device, called potentiostat, is usually required to control the potential of the electrochemical cell and for current acquisition, especially for constant voltage amperometry, CV, and impedance spectroscopy in a three-electrode configuration.

Three-electrode systems can be implemented in three configurations: (1) grounded WE (Figure 5.9(a)) which has been the most popular one [201]; (2) grounded RE (Figure 5.9(b)) which has been used mainly in multi-electrode systems [198]; (3) grounded CE (Figure 5.9(c)) which has received little attention. Since the first two configurations are electrically the same, there are actually only two different configurations [202].

Figure 5.9(a) illustrates a basic grounded-WE configuration of a potentiostat. When the WE is connected to the true ground, the system is not sensitive to interference and noise. However, in this configuration, the CE is placed in the control feedback loop and the connection of the output of the control amplifier and the CE might create a large capacitor resulting in the instability of the control amplifier.

The grounded-CE configuration is usually more complex than the two other configurations and needs more components and consequently, it is more vulnerable

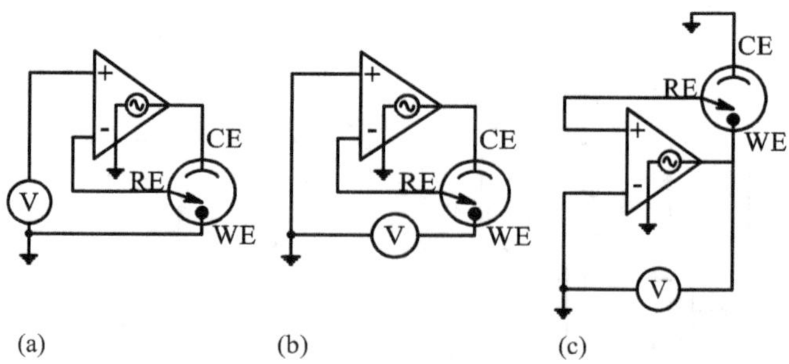

Figure 5.9 Potential control using (a) the grounded-WE configuration, (b) the grounded-RE, (c) the grounded-CE configuration

to component mismatches. However, when it is difficult to implement the shielding and screening of the WE connection from external electromagnetic interference, this configuration may help to improve the current measurement [202].

The output current is usually read through WE or CE. When there is no direct connection between WE and true ground, environmental noise and some form of interference might be created. So, it is better to connect the WE directly to the ground and the current is read through CE. However, in some works, the current is read through WE while it is virtually grounded.

There are two common types of current acquisition circuits for potentiostats including: (1) a transimpedance amplifier (TIA) (Figure 5.10), and (2) a current conveyor (CC) (Figure 5.17). After a brief review of different potentiostats in this section, different applications of this element will be discussed along with the readout circuits in Section 5.3.

5.2.1 Transimpedance amplifier (TIA)-based current acquisition

In most applications, it is required to source and sink the reaction (like redox) current. A simple approach is the direct conversion of the bidirectional current to a voltage which can be done by capacitive or resistive feedback across an operational transconductance amplifier (OTA), known as a TIA [192].

Since the voltages of WE and RE (versus the ground) change together, it is essential to make sure that the voltage difference between WE and RE is equal to the stimulating voltage (V_{st}). In many TIA-based potentiostats, a buffer is used to make sure that no current flows through the RE and apply a certain voltage difference between RE and WE. In an effort, Li *et al.* [199] used a voltage level shifter in the feedback loop. In this potentiostat which is depicted in Figure 5.11, OP_1 level shifts the output of the waveform generator and matches it with the input range of OP_2. For microscale sensors, the potential of CE and RE can be assumed the same due to the negligible solution resistance between RE and CE. RE is connected to the input of OP_3 through which no current flows. CE is connected to the output of the OP_2 which is buffered through negative feedback utilizing OP_3. The input stimulus voltage at CE results in a current that can flow only between CE and WE.

The low power supply trend of the CMOS process limits the functionality of single-ended potentiostats because biosensors' redox potentials can be higher than the power supply. In addition, the dynamic range is degraded and these circuits are not able to remove common mode noise and interference. To address these issues,

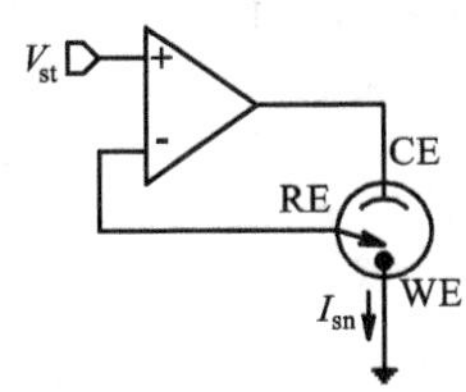

Figure 5.10 A three-electrode TIA-based system

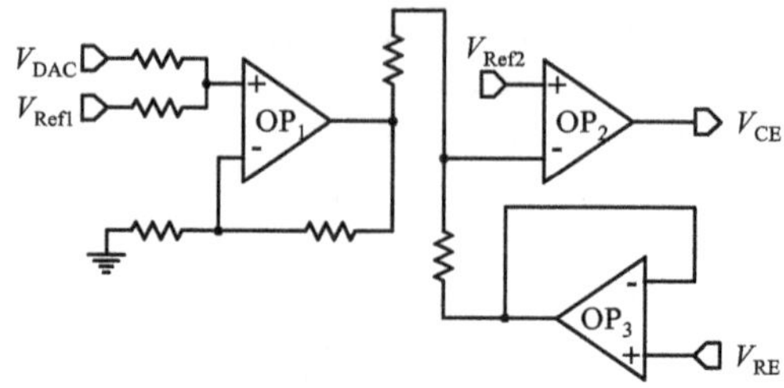

Figure 5.11 Schematic of a single-ended potentiostat using a level shifter in a three-electrode configuration

fully differential potentiostats were introduced [203–209]. Figure 5.12(a) and (b) show, respectively, the block diagram of a standard single-ended potentiostat and the fully differential potentiostat proposed by Martin *et al.* [210]. This developed fully differential potentiostat doubles the V_{cell} output range in single-ended potentiostat and enables chemical micro-transducers to detect a large suite of analytes even when implemented in high-performance, low-voltage processes. However, it still falls short in terms of sensitivity and accuracy. Ghodsevali *et al.* [196] used an error cancelation loop composed of high gain and rejection ratio amplifiers to subtract the voltage drop created by the sense resistor, thus enabling the use of high resistance values for low current sensing with no adverse effect on the biosensor's dynamic range; as a result, the sensitivity of the circuit is considerably improved (Figure 5.12(c)).

To reduce the effects of charging currents and the currents due to undesired analyte, pseudo-differential amperometric readout circuits were also developed [211]. The schematic of a pseudo-differential readout is presented in Figure 5.13, where the reading time is controlled by switches S_n and S_p. By subtraction of the currents from WE_n and WE_p, nearly all parasitic currents can be canceled while preserving the analytical signal of interest. But, two WEs are used instead of one [212].

Figure 5.14 shows another pseudo-differential potentiostat reported by Martin *et al.* [211]. In this circuit, when the switches S_p, S_n, and S_1 are closed, the feedback around OP_3 and OP_4 forces WE_p and WE_n to a virtual ground, and the feedback around OP_1 and OP_2 ensures V_{cell} tracks the desired potential V_{src}. The WE currents flow through R_f and create a pseudo-differential output voltage. Based on the timing diagram shown in Figure 5.14, first, S_p is closed allowing current to flow through W_p, and then, S_1 is closed enabling the potentiostat to set the potential of the electrochemical cell to the desired voltage, V_{src}. Then, this potential changes to E_d, and the analyte is preconcentrated on WE_p by electrochemical deposition. After S_n is closed, preconcentration begins on WE_n. t_d and t_{ds} show the deposition time for WE_p and WE_{ds}, respectively. The subtraction of the currents from WE_p and WE_n during the voltammetric sweep can almost cancel the parasitic current and preserve the signal of interest.

Differential and pseudo-differential topologies are also reported for array sensors. Figure 5.15 illustrates a pixel of a 5 × 10-array with a pseudo-differential

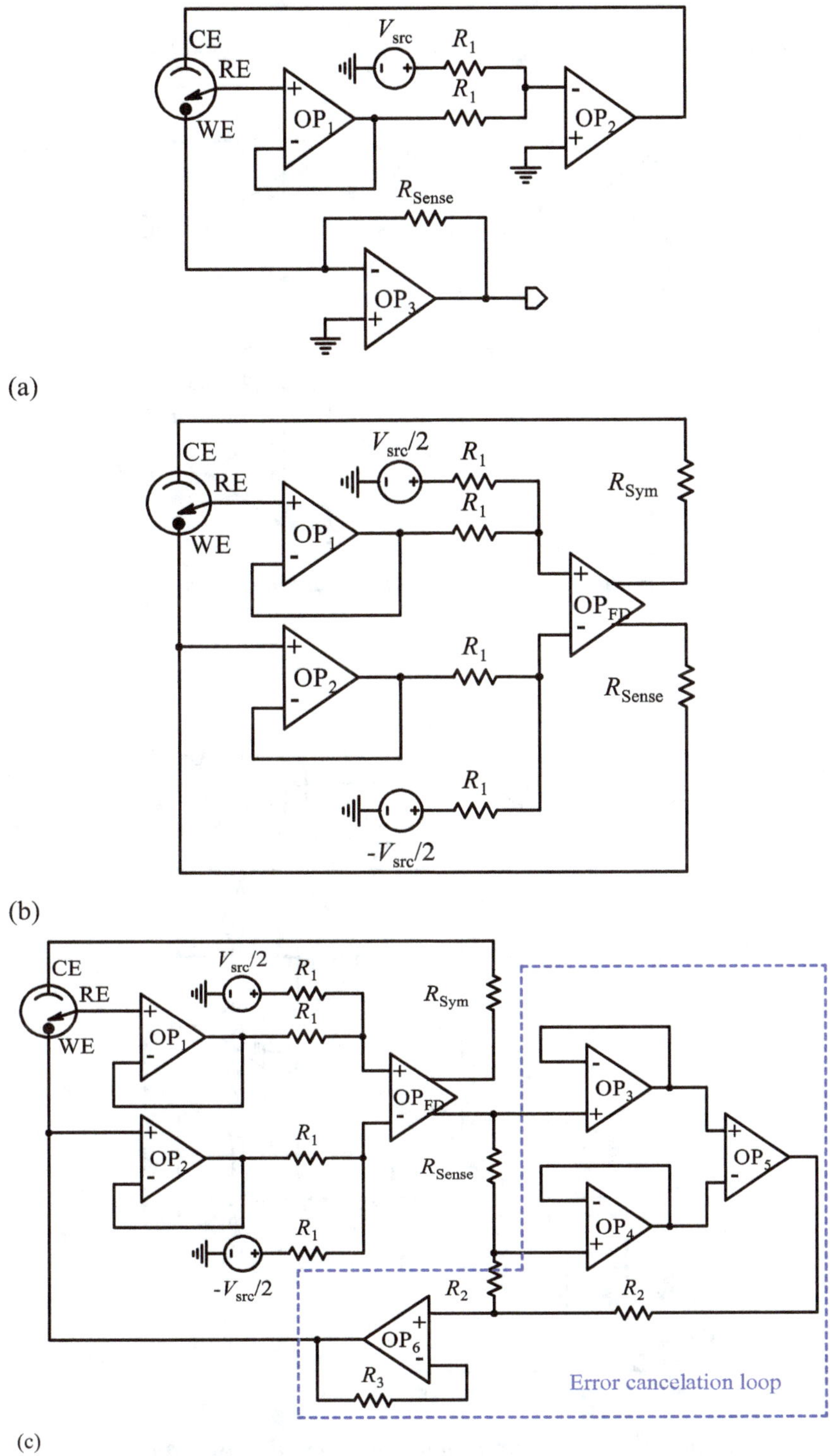

Figure 5.12 (a) Single-ended potentiostat diagram. (b) A fully differential potentiostat. (c) The architecture of fully-differential potentiostat where OP_3, OP_4, OP_5, and OP_6 form an error cancelation loop.

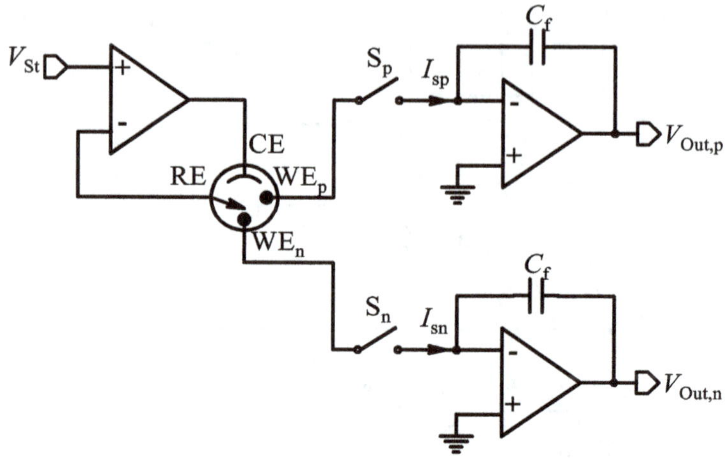

Figure 5.13 The structure of the pseudo-differential amperometry readout circuit

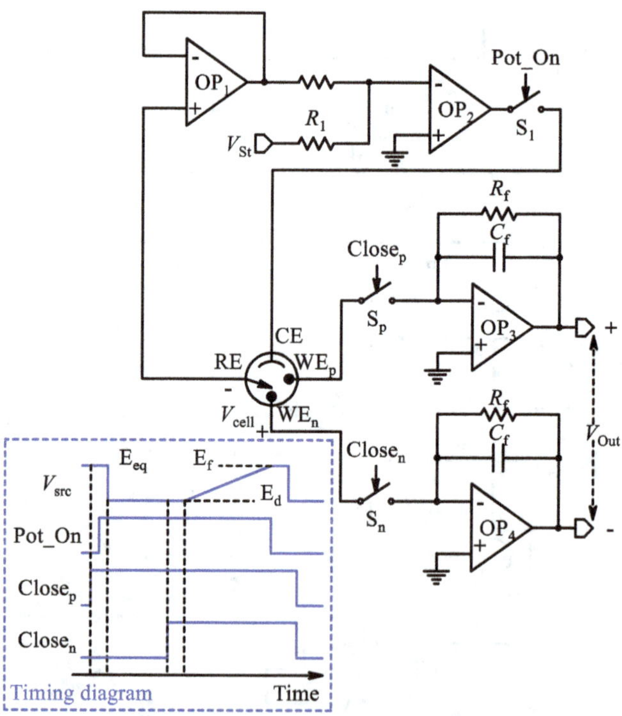

Figure 5.14 Schematic and control timing diagram of a pseudo-differential potentiostat along with a capacitive-resistive readout circuit in a three-electrode configuration, and two WEs

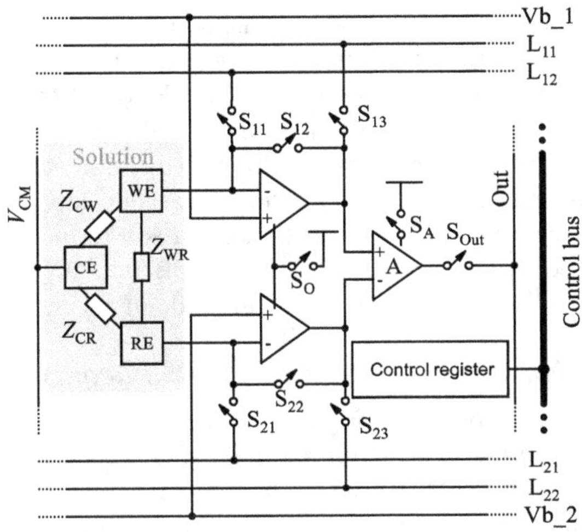

Figure 5.15 *A pseudo-differential potentiostat including two OpAmps, a differential amplifier, and nine transmission gate switches for different operation modes*

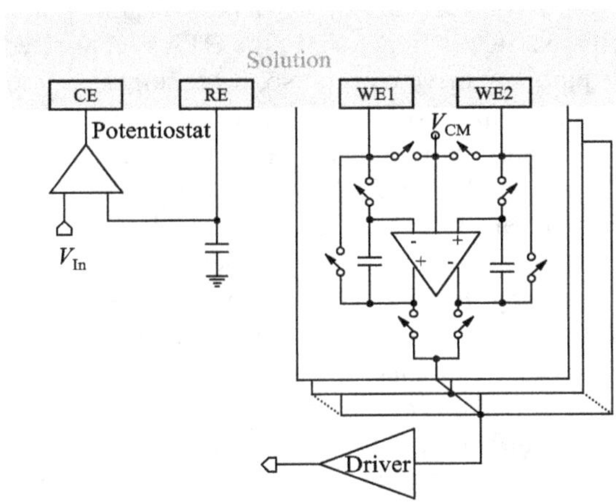

Figure 5.16 *A differential potentiostat using a fully differential integrator within the pixel*

architecture proposed by Hassibi *et al.* [26]. The topology of this sensor is configurable and provides different operation modes including impedance spectroscopy, voltammetry, potentiometry, and field-effect sensing. Each pixel of this sensor has two operational amplifiers (OpAmps), a differential amplifier, and nine transmission gate switches, as depicted in Figure 5.15. This circuit, as a pseudo-differential amplifier, has two identical TIAs with off-chip components in their feedback loop. In another work, Augustyniak *et al.* [213] proposed the circuit shown in Figure 5.16 for an array sensor including an array of 24 × 16 WEs grouped in 12 × 16 differential pixels and a potentiostat controlling and adjusting the electrochemical cell potential through RE and CE. While the WEs are held at a constant potential, the electrolyte voltage changes step-like. A fully differential integrator within the pixel collects the resulting charge and accommodates differential and common mode loops using a folded-cascode topology with approximately rail-to-rail output voltage swing. In comparison to the pseudo-differential architecture reported by Hassibi *et al.* [26], this differential potentiostat required TIAs with smaller active areas for an identical grade of accuracy.

5.2.2 Current conveyor (CC)-based current acquisition

Another common circuit for measuring the reaction current is a CC which is like a current buffer isolating the sensor from the following stages. In this approach, the reaction current is conveyed from WE, which is held at a virtual potential, to a high-impedance output node, and then it is converted to a voltage, time, or frequency [192]. This technique is mostly reported in two-electrode configurations and for measuring one directional current measurement. This technique is more attractive for high-density on-chip arrays due to their low power and area. CCs can be as simple as an OTA and one additional transistor, as shown in Figure 5.17 without the need for large resistors or capacitors resulting in low power consumption and small area. In this figure, since the OTA is connected to the transistor gate, it does not require any drive current; so, it consumes very low power [214].

Bidirectional measurement of the currents is also reported in the literature. This can be done by using two symmetrical circuits [215,216] or adding offset current to the input of the CC [192]. Adding offset current raises two problems. First, since the bias current noise can be directly added to the input current, it increases the possibility of error occurrence. Second, the analog-to-digital converter (ADC) that is fed by the resulting signal must have a high resolution. In an effort, Jafari *et al.* [192] reported a bidirectional CC-based circuit comprising an N-channel metal-oxide semiconductor (NMOS) and a P-channel metal-oxide semiconductor (PMOS) transistor, M_n and M_P, connected to the feedback of an OTA without the need for adding a DC offset current to source and sink the input current. As shown in Figure 5.18, the currents of the transistors, M_n and M_p, are mirrored to the output of the CC and integrated on the load capacitor and sampled. OTA chopping helps to decrease the effects of flicker noise.

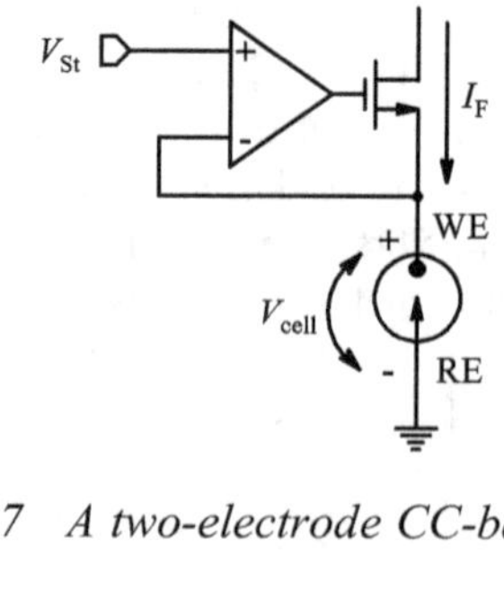

Figure 5.17 A two-electrode CC-based system

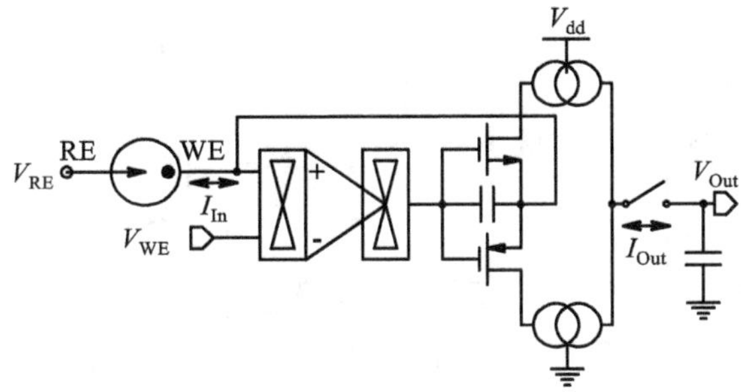

Figure 5.18 Chopper-stabilized bidirectional current conveyer architecture in a two-electrode configuration

In [217], a Wilson current mirror is used along with a cascode amplifier to decrease the input resistance in addition to mitigating the error due to current mirror mismatches and also bidirectional current measurement. Decreasing the input resistance makes it possible to achieve the desired bandwidth. In [218–221], three-electrode CCs were used along with current mirrors. One of the advantages of three-electrode CCs is that, in these systems, WE can be grounded and the current can be read from the CE. However, current mirror circuits suffer from limited saturation voltage and nonlinearity of the drain-source resistance of transistors resulting in decreased linearity of the potentiostat.

CCs can also be used in array structures. For instance, Kim *et al.* [193] used a CC-based potentiostat in a two-electrode configuration along with an integrating capacitor and multiplexer in a 10 × 10 array. This group used a time-multiplexing technique in such a way that while one amplifier in a column is being read out, all the other potentiostats in that column continue to integrate their input currents, and all outputs in a column are serially presented at a single output.

Using amplifiers has demerits for high-density array recording due to their large area and high power consumption. As a result, in some array structures using CCs [214,222,223], half of the amplifier circuit is used as a common part for all the WEs of a row of the array resulting in a smaller area and lower power consumption. For example, White *et al.* [223,224] adapted a half-shared operational amplifier design to enable the integration of an array of 32 × 32 amplifiers and electrodes on a single chip with lower power consumption, and cell area. As seen in Figure 5.19(a), the non-inverting input transistor, M_1, is in the half OpAmp and the inverting transistor, M_{3N}, is in the individual TIA. The voltage at electrode N is regulated by the negative feedback through M_{6N} to be the same as V_{ref} at the non-inverting input. Since a two-stage amplifier is employed in the TIA structure, the phase margin must be considered. So, a compensation capacitor, C_{compN}, is utilized to raise the phase margin of the TIA. In the array architecture, one non-inverting half OpAmp is shared among four TIAs and each TIA includes a 1-bit static random-access memory (SRAM), as depicted in Figure 5.19(b), which is fully addressable by column and row decoders.

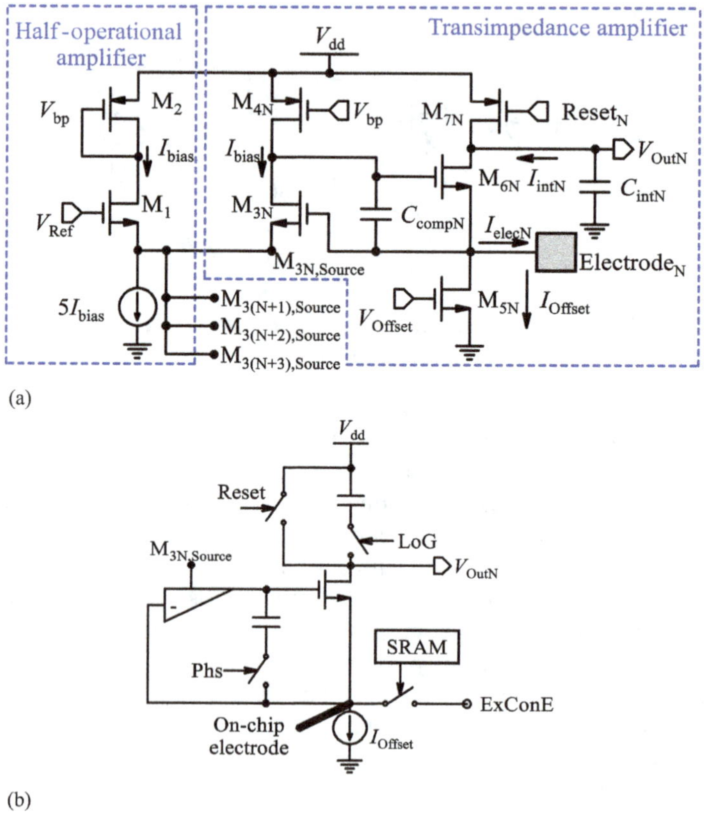

Figure 5.19 (a) A half-shared OpAmp and TIA designs for a two-electrode configuration and (b) each TIA including 1-bit SRAM

Ayers *et al.* [222] also took advantage of the shared amplifier scheme shown in Figure 5.20 to minimize the transistor count per detector in a 5 × 5 array. In this circuit, a common half circuit is shared among all regulation amplifiers in a row. The potentiostat employs a regulated cascode stage featuring a high-gain amplifier that maintains the electrode voltage via negative feedback. In another effort, Li *et al.* [214] used shared circuit resources in the multi-channel system to reduce the total circuit area and power. To provide an accurate gain, good noise performance, and sufficient bandwidth with a small transistor count, this group designed a two-stage CC (Figure 5.21(a)) for bidirectional current sensing in a group-cluster system architecture (Figure 5.21(b)). The input bias voltage V_{WE} is set externally. The feedback of OTA_1 held the WE voltage at V_{WE}. The output current of the first stage is mirrored and amplified by a gain equal to *AB* using high-precision current mirrors. This circuit, as a continuous-time technique, does not inject charge current into the electrode-solution interface. The voltage of V_{o1} and V_{o2} must be held at a

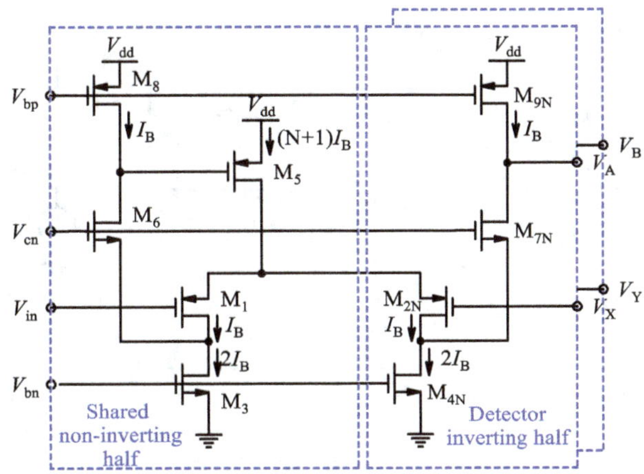

Figure 5.20 A shared half amplifier scheme in a two-electrode configuration based on a folded-cascode OpAmp topology

DC voltage by their respective following stages. Compared to single-stage CCs, a two-stage structure can boost the current gain. Since the positive terminals of the OTAs are connected to the same voltage, V_{WE}, it is possible to share a segment of the OTAs by all channels and connect it to a common bias voltage, as shown in Figure 5.21(b). This way, only a pixel segment should be repeated for each channel. The shared segment concept is used in both stages. However, a single second stage, or so-called gain stage, is shared by multiple pixel stage channel amplifiers by employing a multiplexer in an interleaved structure. Each channel includes M pixels, as a cluster. Each cluster is connected to N-shared gain amplifiers. N gain amplifiers with their associated pixel clusters from a group including $M \times N$ shared-segment pixel amplifiers and N shared-segment gain amplifiers. The whole system is composed of Q groups connected to a Q-channel signal processing stage. In total, there are $M \times N \times Q$ recording channels in this group-cluster architecture.

In another work, Huang *et al.* [225] reported a CC-based potentiostat using a sharing technique in a 16-channel, 64-electrode-per-channel chip (arranged in a 32 × 32-format) for bidirectional current measurement required for CV (see Figure 5.22). The second half-shared amplifier helps to ensure the current mirror transistors, M_2 and M_3, have the same drain voltage. To reduce the number of transistors used in each detector pixel, the second half-shared amplifier was replaced by a peripheral cascode bias circuit shared by the entire row of electrodes which is composed of a half-shared amplifier, a current source, and an NMOS transistor, M_0. Both positive and negative current is converted to a positive voltage by the integrating capacitor, C_{int}, showing the absolute value of the measured current. To indicate the polarity of the current, a sign binary output circuit, including M_6 and M_7, is also added to each detector pixel which compares two bias points in the circuit. When the current is positive, this circuit generates a high voltage and when the current is negative, this output will be low. The current

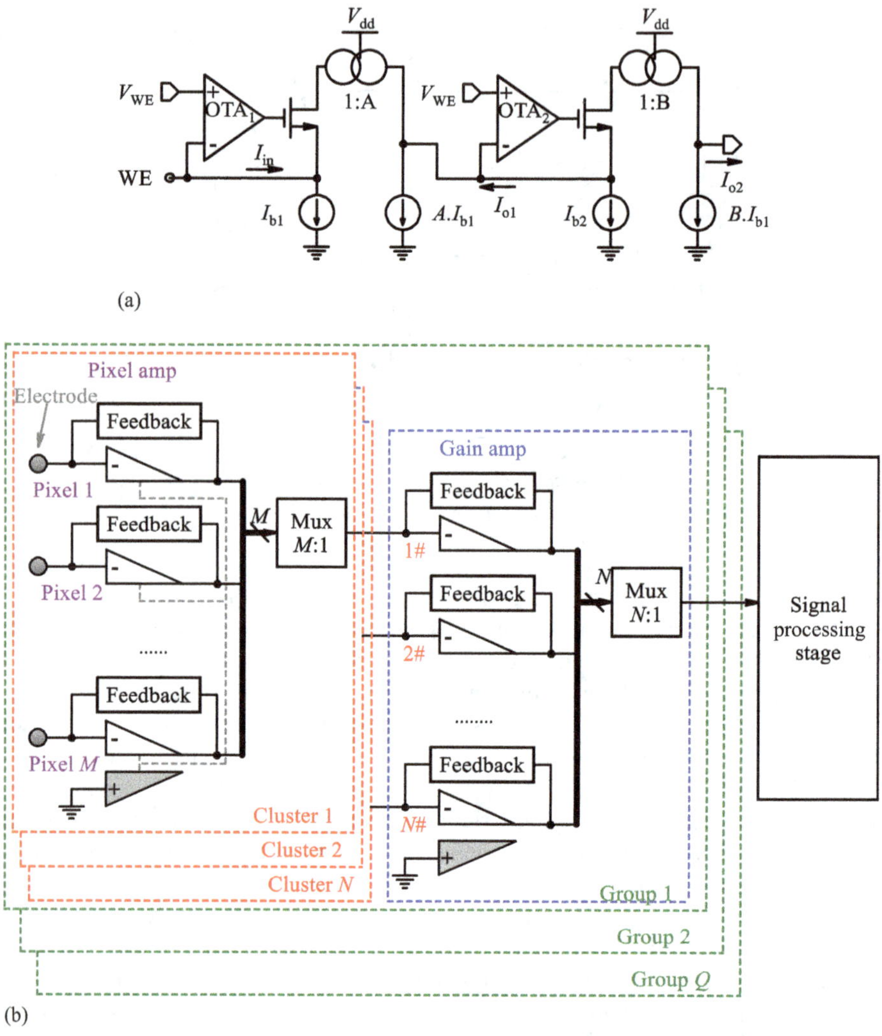

Figure 5.21 (a) Block of a two-stage CC for bidirectional current sensing in a two-electrode configuration. (b) Block diagram of a group-cluster architecture.

mirrors B_P and B_N are, in turn, for the calibration of positive, I_P, and negative current, I_N. Sixty-four electrodes of each channel, which are arranged in two 32-electrode columns, share one output buffer and the 32 electrodes in the same row. Each detection pixel of this sensor includes only 11 transistors for CV measurements and two transistors for calibration.

Huang *et al.* [226] proposed the circuit shown in Figure 5.23 to convert oxidation current to a voltage. In this circuit, the oxidation current is first amplified by a wide-swing cascode current mirror and then converted to a voltage employing an

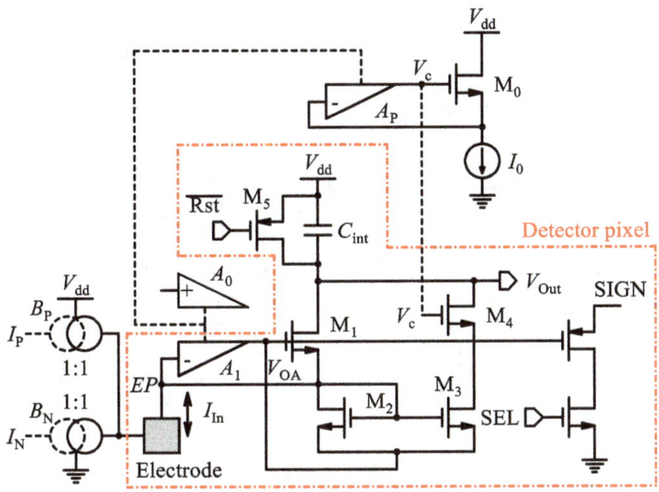

Figure 5.22 A current-conveyor-based potentiostat for bidirectional current measurement (the gray box shows the single detector pixel) in a two-electrode configuration

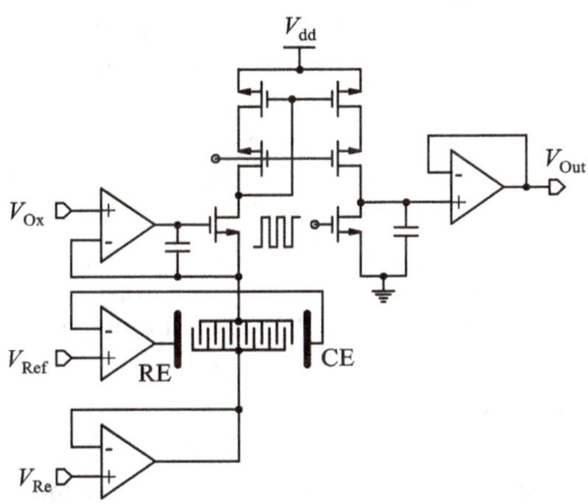

Figure 5.23 A CC-based potentiostat along with a wide-swing cascode current mirror and a capacitor-based current-to-voltage converter

integrating capacitor. The integration time is controlled periodically by an external clock that discharges the capacitors through a switch. Then, the output voltage is buffered. As seen in Figure 5.23, the RE and CE are biased through an operational amplifier.

CC-based potentiostats can also be designed by taking advantage of differential techniques. Figure 5.24(a) illustrates a differential form of a potentiostat circuit using CCs [227–229], while Figure 5.24(b) shows a single-ended version of the same circuit with the lumped electrode model and the shorting unity feedback

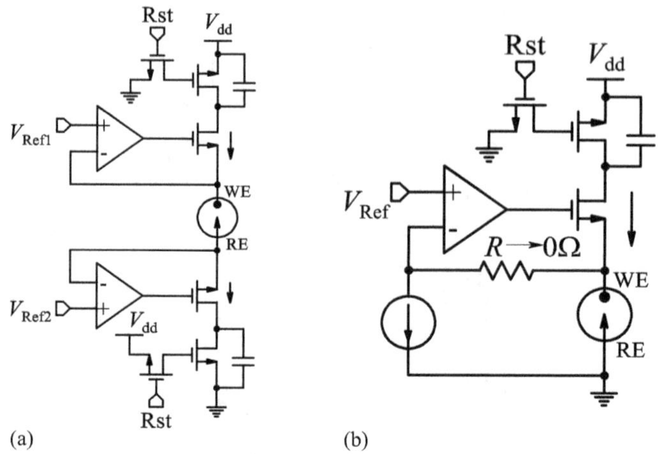

Figure 5.24 (a) Differential current sensing potentiostat. (b) Single-ended current sensing potentiostat using CC

path expressed with a resistor (for facilitating feedback circuit analysis). In an effort, Wang *et al.* [203] presented two potentiostats with similar structures, one in a single-ended configuration (Figure 5.25(a)) and the other one in a fully differential topology (Figure 5.25(b)) to avoid the deviation of electrochemical cell voltage and converting the cell current to voltage using resistive feedback (see Section 5.3.1.1). When the sensor current level is low (for example in the range of pA and μA), the harmonic distortion quantity increases which results in a decrease in the linearity of the single-ended sensor, and consequently limits the detectable current and dynamic range. In comparison to the single-ended potentiostat, the fully differential potentiostat can provide a wider output voltage swing and dynamic range and performs relatively better in terms of linearity.

5.3 Readout circuits

This section is dedicated to discussing different types of readout circuits along with the potentiostats reported for reading DC or AC output currents in voltammetric and impedimetric sensors (Table 5.1).

5.3.1 DC current readout

In voltammetric sensors, the output signal is usually a DC current. Different techniques have been reported to convert the reaction current directly to an output voltage using a TIA. To do so, the TIA can be configured with a feedback resistor or/and capacitor or implemented as a switch-capacitor (SC) circuit. The output current of CC-based potentiostats can also be integrated using capacitors. There are also some current-to-frequency (I-to-F) conversion or digitization techniques to convert the reaction current to a digital signal.

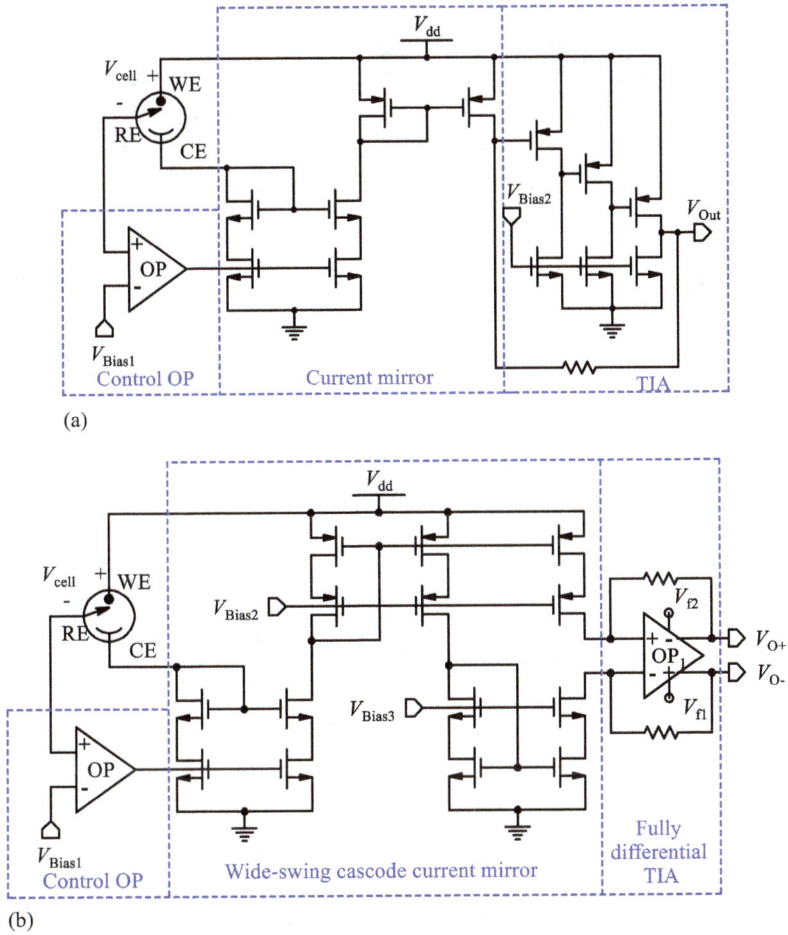

Figure 5.25 The building blocks of (a) single-ended potentiostat, (b) fully-differential potentiostat proposed in

5.3.1.1 Resistive readout circuits

In an effort, Ghoreishizadeh *et al.* [198] proposed a CV and CA sensor on a single chip along with the required signal generator circuit. In CV mode, a triangular signal is applied to the WE of interest through a TIA and a fixed voltage is applied to the RE through a potentiostat allowing the sweeping of both positive and negative voltages of V_{WE}-V_{RE}. Moreover, the TIA used in the readout circuit of the CV mode reads the bidirectional current, I_{WE}. The resulting current flows through a resistor, R_M, and is converted to a voltage which is buffered through two operational amplifiers, OP_2, and OP_3. Then, the buffered voltage across R_m is amplified by a common difference amplifier. Furthermore, a DC voltage is added by the last amplifier to the signal to keep it within the supply voltage range.

Table 5.1 CMOS amperometric and voltammetric biosensors

Application	Technique	Array #	Power (Supply voltage)	Area (mm^2)	DR	Sensitivity	Resolution	Output (D/A/ pD?)	Technique	Reference
Amperometry, electrochemical imaging	0.6 μm	16064 WEs 4016 WEs	59.6 μW 39.6 μW	19 × 19	4 μM to 512 μM	4.7 pA/μM	NA	A	TIA, R/C	[230]
CV CA	0.18 μm	5 WEs,1 RE,1 CE	220 uW (1.8 v), CV:117 uW	2.2×15	500 to 3,000 μM for glucose and 50 to 200 μM for lactate	Glucose: 27.2±5.2 Lactate: 286±42.9 $\left(\frac{mA}{mM \times cm^2}\right)$	NA	CV: A, CA: pD	TIA, R CC, I-to-F	[198]
ASV	0.5 μm	68 channel (2 WEs, 1 RE, 1 CE For each channel)	16 mW/Ch. (5V)	36 mm^2/Ch.	97.7 dB	NA	36.6 pA	A	PsDiff., TIA, R/C	[211]
Voltammetry	0.35 μm	NA	CMirr: 36 nW, OpAmp: 0.9 mW	370 × 280	0.2 nA to 0.4 μA, 66 dB	2.89 nA/μM 1.73 nA/μM	0.1 μM	A	CC, C	[226]
CV	0.35 μm 2P4M	1024	NA	32 × 32	−700 pA to1968 pA	NA	NA	A	Half-shared OPA and TIA, C	[223]
CV	0.25 μm	48 (Ch. #: 4×4) (1 RE, 4 CEs, 16 WEs)	160 mW, 10 mW/Ch. (2.5 V)	5×3	60 dB100pA-250nA	NA	240 pA	D	TIA, SC, dual-slope ADC	[231]
CA	0.5 μm	16 × 8 (1RE, 1CE)	NA	6.4 × 4.5 mm^2	1pA-100nA,	NA	1 pA	D	CC, SC, ADC	[232]
CV, DNA detection CA	0.25 μm 1 M	4 × 4 1WE,1RE, 1CE	(2.5 V) 25 μW/Ch.	5 × 3 6.44 mm^2/Ch.	NA 2-32 μA	NA NA	NA 2 μA	D D	TIA, SC, ADC ΣΔ ADC	[233] [201]
CV	0.5 μm	8 × 8 32 × 32	73.5 μW @ 0.1 μA, 132 μW @ 10 μA	NA	−0.4 to 0.4 μA	NA	NA	A	TIA	[234]
Linear sweep, CA, CV, PV	0.5 μm	2 × 2 (WE, WE, CE)	22.5 mW (5 V)	400 μm^2/Ch.	120 dB	NA	±500 fA	A	TIA + SC	[199]
Amperometry	NA	10 × 10	NA	3 × 3	3.5 to 105 μM	1.2 pA/μM	~pA	A	CC, C	[193]

CV	0.13 μm	1 RE, 1WE	3 μW/Ch.) (1.2 V)	4,800 μm^2/Ch.	10 pA-350 nA	NA	135 fA	D	TIA, SC, ADC	[192]
			4 μW/Ch. (1.2 V)	10,000 μm^2/ Ch.			8.6 pA		CC, SC, ADC	
CV	0.6 μm	2 × 2 (4WE, 1CE,2 quasi-RE)	3 mW (5 V)	5.06	6 pA-10 μA	0.62 $\mu A.mM^{-1}.cm^{-2}$	6 pA	A	TIA, SC	[235]
Simultaneous detection of O_2 and H_2O_2	0.18 μm	20 × 20	NA	10.4 × 10.4	NA	NA	NA	A	TIA, SC	[236]
Single-cell recording of vesicle release from human neuroblastoma cells	0.35 2P4M	32 ×32	12.5 mW	5 × 5	NA	NA	NA	A	CC, C	[224]
Chronoamperometry	0.35 μm	NA	9.3 mW (3.3 V)	0.6	156 dB	NA	0.47 pA	D	Diff., TIA, SC	[237]

D: Digital, A: Analog, pD: Pseudo Digital, NA: Not available, PsDiff.: Pseudo-differential, Diff.: Differential, SC: Switch Capacitor, CC: Current Conveyor, Ch.: Channel, R: Resistive readout, C: Capacitive readout.

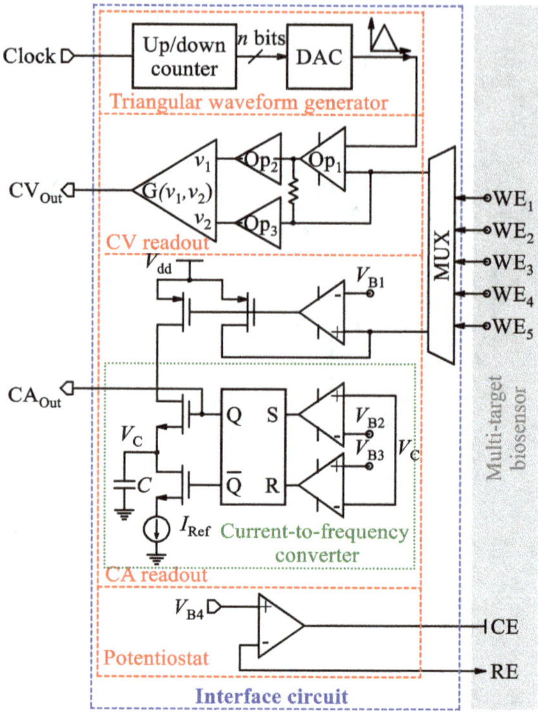

Figure 5.26 Schematic of an electrochemical sensor driver/readout interface including a waveform generator, CA readout, CV readout, and potentiostat

Resistors can inject thermal noise into the biosensor [192]. Moreover, it may employ a large resistor to achieve high gain and signal-to-noise ratio (SNR) which occupies a large area. Although an active feedback structure has also been used instead of resistive feedback to decrease the area, the total area is still so large that only eight channels can occupy an area of about 9 mm^2 [238,239]. Moreover, resistive feedback TIAs in multi-channel readout circuits require gain calibration since the gain variations are large which is because of the resistance mismatch due to process variations [214].

5.3.1.2 Capacitive readout circuits

As aforementioned, one approach to convert the reaction current to voltage is using TIA with capacitive feedback and SC techniques. A SC feedback TIA plays the role of an integrator capable of measuring small currents. Capacitive feedback occupies a small area and, due to its averaging behavior, acts as an LPF which helps to remove high-frequency noise [192]. As an example, Jafari *et al.* [192] used a TIA-based readout circuit in a two-electrode configuration which is composed of a chopper-stabilized OTA and capacitive feedback, C_F. Figure 5.27 illustrates this circuit which works in two phases based on the switch, S, being open or closed. When it is open, the input current is integrated over C_F. The negative terminal of OTA is at WE

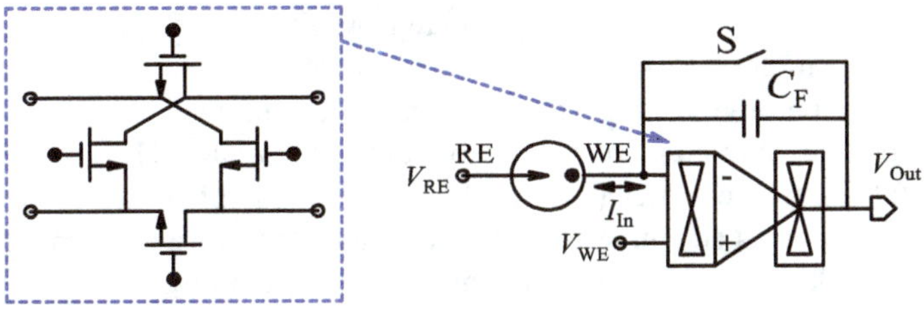

Figure 5.27 Chopper-stabilized bidirectional-input TIA in a two-electrode configuration

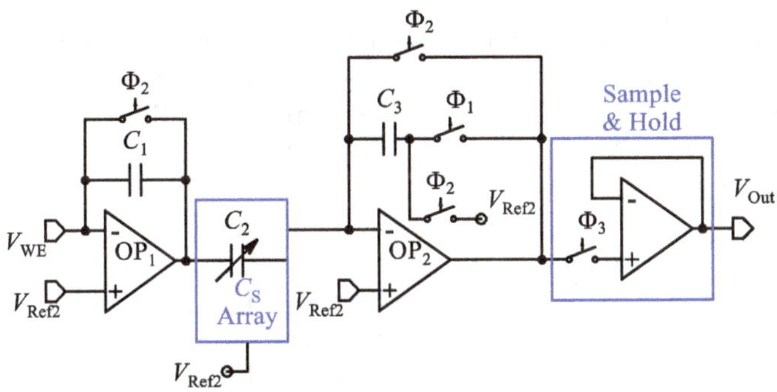

Figure 5.28 Schematic of an SC readout circuit

potential due to the OTA feedback. When the switch is closed, the feedback capacitor is discharged. As a result, any residual on C_F is removed at the end of one cycle. A fully differential integrator was also introduced in Figure 5.16.

In another effort, Li *et al.* [199] designed a 2 × 2 array of electrochemical sensors using a multi-mode signal generator (Figure 5.6, see Section 5.1.1), a three-electrode single-ended potentiostat (Figure 5.11) and a SC readout circuit (Figure 5.28(b)). The scan range is controlled by a DC offset control input, V_{ref2}. An SC-based readout circuit converts the output current of the potentiostat to voltage using an integrating capacitor, C_1. Correlated double sampling (CDS) was used to decrease the $1/f$ noise. The output voltage of the SC integrator is fed to a PGA with the gain of C_2/C_3 (C_2 is a digitally programmable on-chip capacitor array) and then sampled and held. The generated voltage of this current readout circuit is equivalent to:

$$V_{out} = \frac{I_{WE}}{f_s.C_1} \cdot \frac{C_2}{C_3} \tag{5.1}$$

where I_{WE} denotes the current at WE, and f_s stands for the frequency of switch phi_2 (which is generated by an on-chip clock generator). After passing through an LPF, the output is fed to an ADC.

Another SC readout circuit was presented by Yang *et al.* [235] for reading a 4 × 4-array of on-chip WEs. Figure 5.29 illustrates an electrode potential drive block, the SC amperometric readout block, and a current switching matrix (MUX). The SC charge integrator converts the sensor current to voltage. Next, it goes through a programmable gain amplifier (PGA) which is then sampled, held, and fed to an ADC. The circuit takes advantage of CDS to reduce the noise and amplifier offset. Although SC feedback TIA occupies less area compared to resistive feedback TIA, a long integration time compels low signal bandwidth. Moreover, this technique adds switching noise issues and noise aliases to the system. In addition, injecting charge into the sensing electrodes due to the switching noise can disturb the charge balance and the electrode-solution interface which will be more critical when the concentration of the reaction is low. So, this approach needs calibration too [214].

As aforementioned in Section 5.2.2, one of the types of potentiostats is based on the structures of CCs which can be simply employed along with an integrator to convert the reaction current to voltage. For example, in the 32 × 32-array proposed by White *et al.* [223,224] shown in Figure 5.19(a), the output current, I_{elecN}, is integrated by an integrating capacitor, C_{intN}.

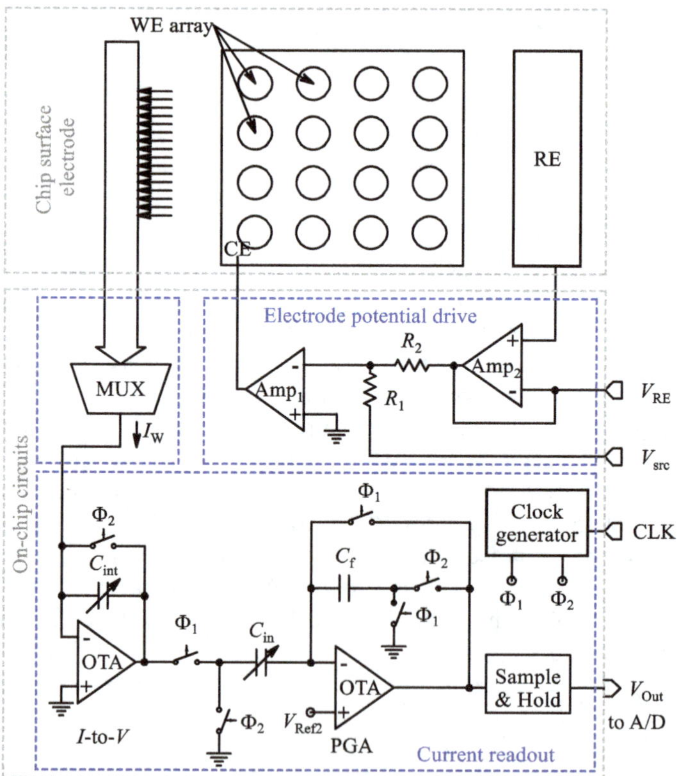

Figure 5.29 Schematic of an amperometric sensor in a three-electrode configuration including a SC readout block

Dorta-Quiñones *et al.* [240] also took advantage of the shared amplifier scheme and integrating capacitor, as shown in Figure 5.30(a). This circuit is composed of inverting half amplifiers, A_1 and A_2, cascade transistors M_1 and M_4,

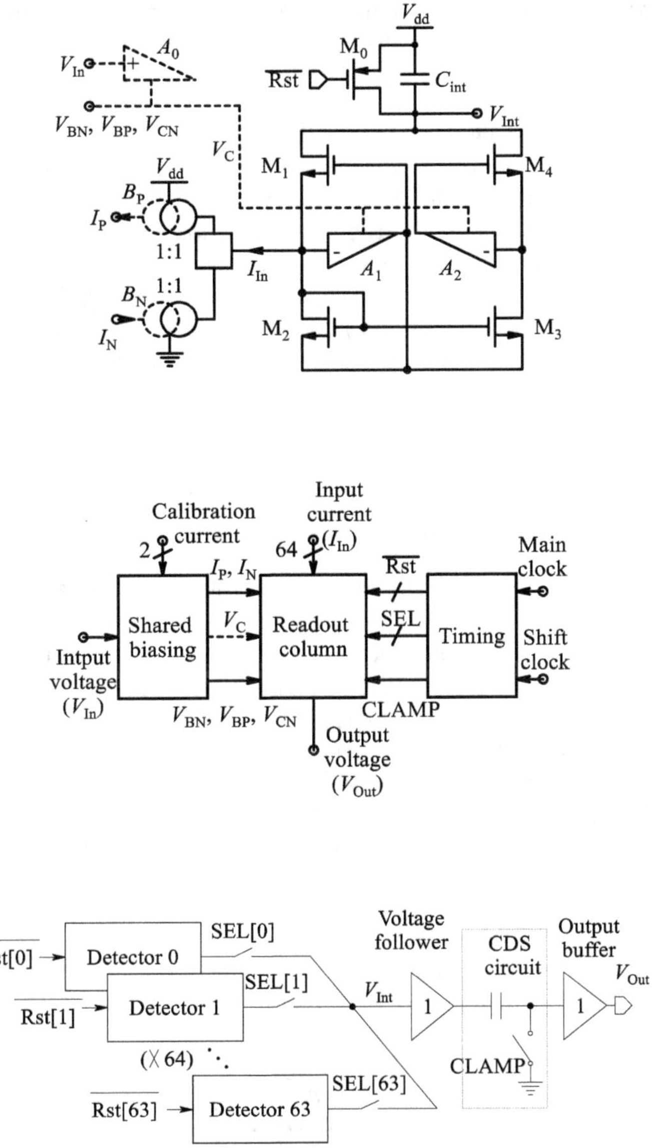

(a) (b) (c)

Figure 5.30 (a) Bidirectional-current potentiostat unit in a two-electrode configuration using a shared amplifier scheme and integrating capacitor, (b) the system of an FSCV detector array, and (c) the readout column of (b) including 64 detectors sharing a common output stage

mirror transistors M_2 and M_3, an integrating capacitor C_{int} and switch M_0. This topology is used in a 64-channel fast-scan cyclic voltammetry (FSCV) detector array where each pixel includes 14 transistors and an integrating capacitor. Figure 5.30(b) shows the system of the FSCV detector array and its readout column is shown in Figure 5.30(c) which shares a common output stage including a voltage follower, a CDS circuit, and an output buffer.

5.3.1.3 Resistive/capacitive readout circuits

Another approach for current-to-voltage (I-to-V) conversion is using both capacitors and resistors. For example, in the circuit shown in Figure 5.14, Martin *et al.* [211] used a capacitive-resistive readout circuit along with a pseudo-differential potentiostat. In another work, Tedjo *et al.* [230] employed a TIA with a T-network resistive feedback along with a capacitor for the I-to-V conversion of 16,064 channels. As shown in Figure 5.31, the OP_{TIA} with $R_{1\text{-}3}$ constitutes each of the TIA read channels. The resulting voltage from this circuit is given by:

$$V_{out} = -I_{in}R_1\left(1 + \frac{R_3}{R_2}\right) \tag{5.2}$$

It is noteworthy that the T-network configuration amplifies both gain and noise in comparison to the single resistor TIA.

5.3.1.4 Readout circuits with digital and pseudo-digital outputs

Different circuits can convert DC output currents to digital, or frequency-modulated signals. Since the output signals of electrochemical sensors are usually current-domain signals, and low-voltage compatibility, high noise immunity, and area efficiency are among the important factors for designing electrochemical sensors, current-mode circuits such as current-controlled oscillators (CCOs) and CCO-based quantizers can be suitable to be integrated into sensor arrays. For example, Laifi *et al.* [241] reported a low-voltage fully differential current-mode

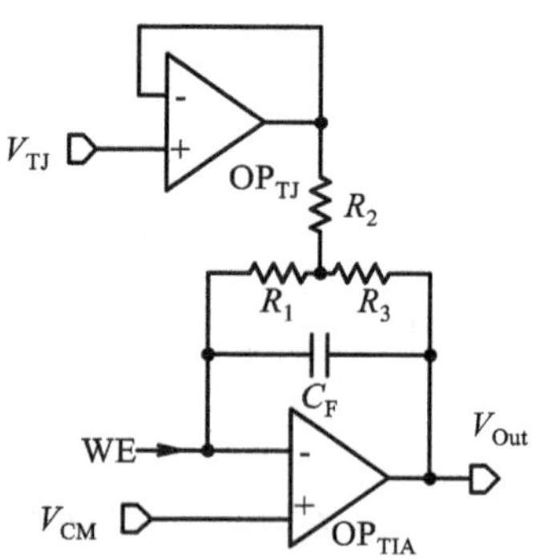

Figure 5.31 A TIA read channel with high gain T-network resistive feedback in a three-electrode configuration

continuous-time ΣΔ modulator based on CCOs. Moreover, a switched-shaped current feedback DAC in this circuit addresses the clock-jitter sensitivity.

In another effort, Nazari *et al.* [195] used a CC-based potentiostat along with an ADC which is a combination of an I-to-F and the single-slope ADC. Figure 5.32 illustrates one channel of the 8 × 12-array reported by this group which is capable of measuring bidirectional currents. In another effort, Schienle *et al.* [232] also used in-pixel ADCs composed of an I-to-F converter and a counter/shift register in a 16 × 8 sensor array chip. In the multi-target sensor reported by Ghoreishizadeh *et al.* [198], which was designed for both CV and CA measurement using five WEs (Figure 5.26), a three-electrode CC was used along with an I-to-F for the CA mode. In this mode, a fixed voltage is applied to the WE of interest, and a unidirectional current is copied by a current mirror. Then, a current-to-pulse width converter generates a pulse width inversely proportional to this resulting current. The output of these sensors can be easily transmitted outside the chip without the need for an ADC.

Levine *et al.* [231,233] reported a 4 × 4 array of sensors each pixel of which includes a square WE (with different sizes in each row), and dual-slope ADC with local digital control circuitry (see Figure 5.33). A control amplifier drives each row of four WEs. An array using the SC technique and ADC was also designed by Kruppa *et al.* [234] which includes 24 × 16 sensor sites. Figure 5.34 shows the sensor architecture including a potentiostat, an array of analog pixel circuits, ADC, and a finite state machine (FSM). Realizing a pixel integrator for each WE paves the way for multi-analysis in parallel. To enable the integration of a high input dynamic range (IDR) sensor into a small pixel area, in comparison to dual-slope integration, successive-approximation register (SAR), or direct integration, Manickam *et al.* [39] reported a 32 × 32-array of 100 μm × 100 μm pixels each of which consists of WE, RE, and CE connected to a voltammetry circuit underneath them. As shown in Figure 5.35, a first-order current sensing ΣΔ modulator is connected to the WE whose current is first fed into an integrator and then quantized by an ADC to generate a digital output. The differentiation operation is controlled by this output driving a DAC.

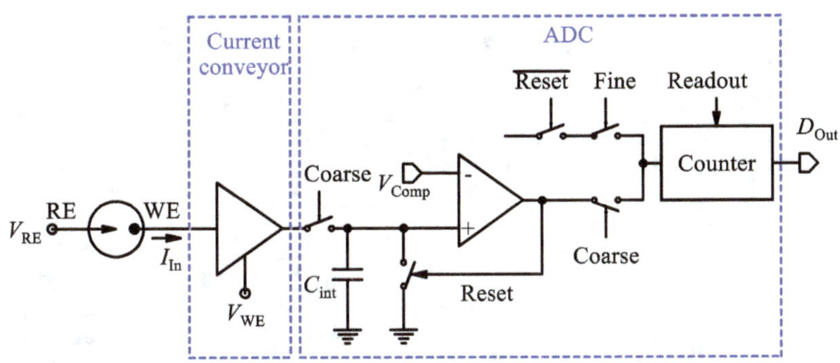

Figure 5.32 One channel of an array sensor capable of measuring bidirectional currents in a two-electrode configuration

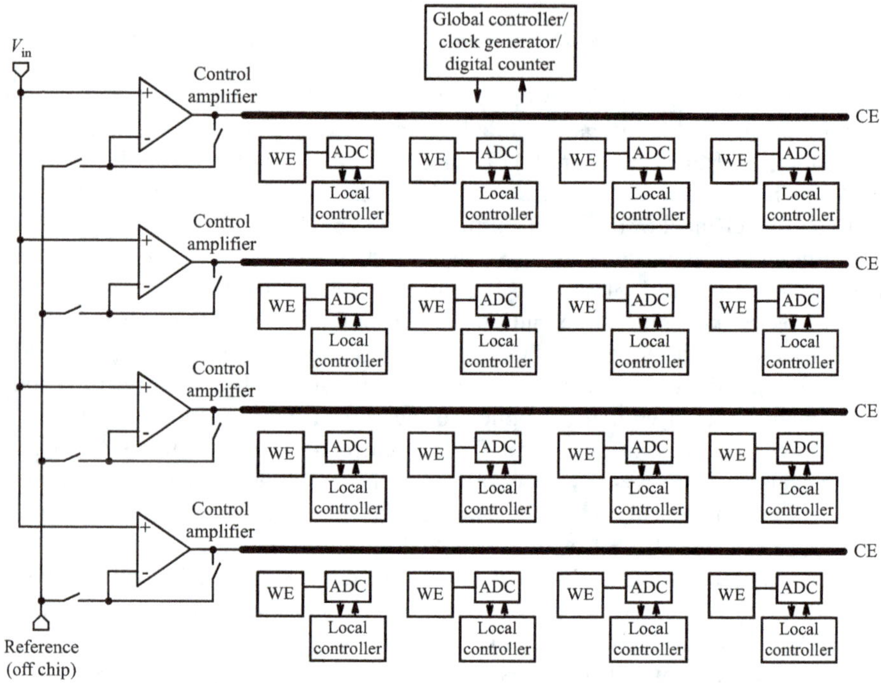

Figure 5.33 A sensor array including a square WE, and dual-slope ADC with local digital control circuitry

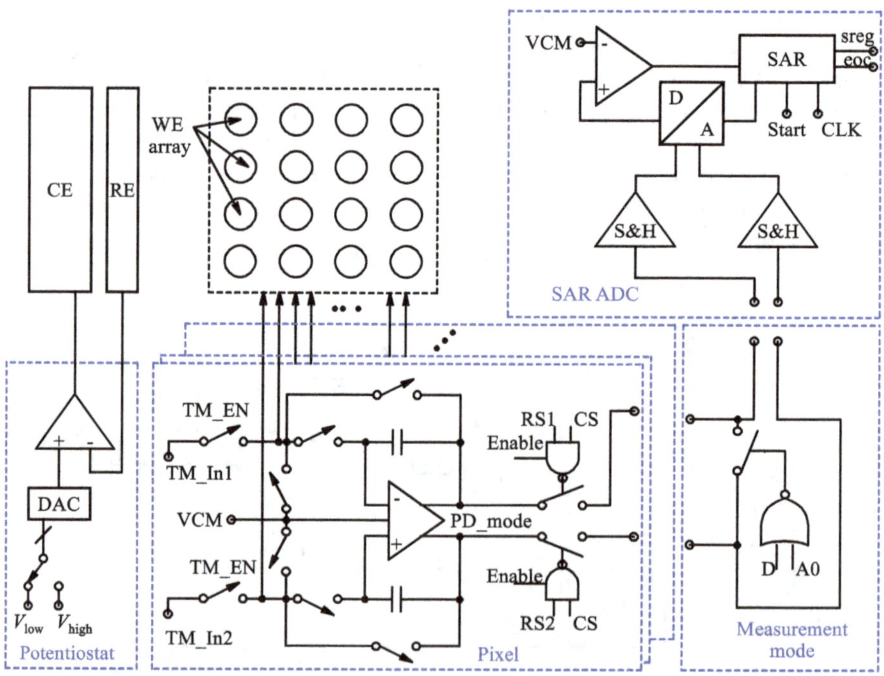

Figure 5.34 A sensor platform in a three-electrode configuration using a potentiostat, an array of analog pixel circuits, SAR ADC, and an FSM

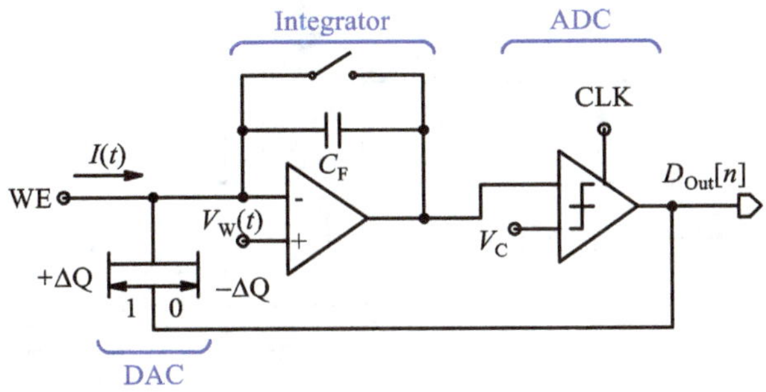

Figure 5.35 Each pixel of the circuit for a three-electrode configuration using a first-order current sensing ΣΔ modulator

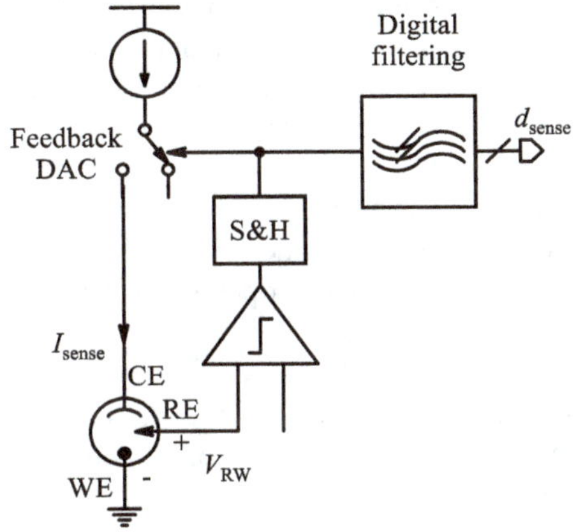

Figure 5.36 A potentiostat including a ΣΔ modulation loop

As an innovative approach, Sutula *et al.* [201] proposed to use the typical long-time constant of an electrochemical cell to implement a small-size and low-power electrochemical ΣΔ modulator interface circuit by eliminating the need for large on-chip capacitors. As shown in Figure 5.36, the time constant of the electrochemical cell is reused in a loop to implement a continuous-time first-order noise shaper. The other parts of the system including a single-bit quantizer, a sample-and-hold, and a feedback DAC can be completely implemented in the electronic domain. Due to the closed loop of the modulator, V_{rsw} is kept at the desired potential V_{ref}. The output current I_{sens} is digitally modulated to q_{mod} which is available at the input of DAC. The analog-to-digital conversion is completed by an LPF which scales the sampling frequency down to the Nyquist rate.

5.3.2 AC current readout

Impedance readout circuits calculate the sensing impedance by comparing the excitation and response signal. Conventional non-CMOS impedance spectroscopy measurement

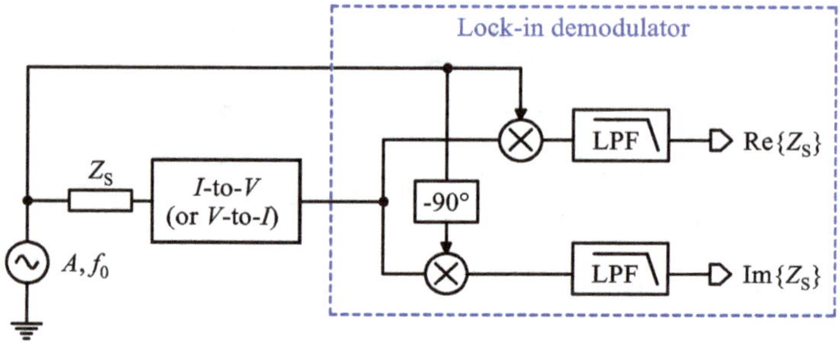

Figure 5.37 The block diagram of a simple lock-in architecture

techniques are usually based on an auto-balanced bridge (ABB), or a DDS along with an ADC, and a discrete Fourier transform (DFT) multiply-accumulate (MAc) core (like impedance analyzers AD5933), as well as lock-in approach (Figure 5.37). ABB bridge-based systems are usually used for frequencies up to 100 kHz. But, in most EIS applications, frequencies below 100 kHz are used which is compatible with usual OpAmps allowing for low-cost impedance analyzers such as AD5933. Lock-in techniques are used for measuring impedance within the same range without the need for fast sampling. Among these techniques, the latter is the most popular technique for implementing on a single chip using CMOS technology [200].

Herein, we categorized the CMOS impedimetric biosensors into three groups: (1) lock-in detection, (2) phase/amplitude technique, and (3) silicon cochlea-based sensor. However, to the best of our knowledge, as shown in Table 5.2, the first approach is mostly reported for the CMOS impedimetric biosensors which are tested experimentally and using on-chip electrodes.

5.3.2.1 Lock-in approach

Figure 5.37 illustrates the principle of the lock-in approach. In this technique, the sensing impedance along with low-frequency noise and offset are modulated to a high frequency by a specific signal. Based on the type of the approach which can be voltage mode or current mode, an I-to-V or voltage-to-current (V-to-I) conversion and amplification are required before down conversion. The original signal can be retrieved by synchronous demodulation and using LPFs. A quadrature demodulator can help extract the real and imaginary part of the sensing impedance (Z_S) by multiplying the signal with the in-phase (I) and quadrature (Q) signal. Although this technique is widely reported for impedance spectroscopy [6,39,131,132], some studies are using this technique for ultra-wideband dielectric spectroscopy and complex permittivity sensing [243–245].

Manickam *et al.* [39,132] reported the first fully integrated CMOS impedimetric biosensor composed of an array of 10 × 10 on-chip electrodes (Figure 5.38(a)). Figure 5.38(b) shows each pixel of this array comprising a coherent demodulator consisting of two mixers and a low-noise TIA. The TIA amplifies the current of the sensing impedance, $Z_{S,}$ and then the output of the TIA is multiplied by orthogonal sinusoidal signals (I and Q) at the angular frequency ω. The DC values of the output

Table 5.2 CMOS impedimetric biosensors

Application	WE, RE, CE	Technique	Sensing Element size (μm^2)	Array #	Power (Supply voltage)	Area (mm^2)	DR	Sensitivity	Resolution	Output(D/A/pD?)	Readout circuit technique	Reference
DNA analyzing	WE, RE, CE	0.13 μm	55 × 55, 5 × 5, two 2 × 2	64 (Ch. #: 4 × 4)	0.35 mW/Ch., 42 μW (1.2 V)	1.05 × 1.6 (300 × 200 μm^2/Ch.)	140 dB (3-mode)	NA	320 fA	D	lock-in approach, dual-slope ADC	[242]
Impedance spectroscopy and electrophysiological imaging of cells	WE, RE	0.18 μm 6M1P	1 × 1,2 × 2, 4 × 4,8 × 8	59,760	412 μW/Ch. (3.3 V)	12 × 8.9 (TSA:4.5 × 2.4 mm^2) (0.1 mm^2/Ch.)	~102 dB (10 kΩ -10 GΩ)	NA	(spatial resolution: 13.5 μm)	D	lock-in approach	[44]
Detection of DNA hybridization Detection of protein attachment	WE, RE	0.35 μm	40 × 40	10 × 10	84.8 mW @ 100 kHz	2 × 2 (Pixel size: 100 × 100 μm^2)	97 dB (BW: 10 Hz)	(Current sensitivity: 330 pA)	10^{-8} Ω^{-1}	A	Lock-in approach (10 Hz-50 MHz, Scan rate: 0.55^{-1} @ BW=10 Hz)	[39, 132]

D: Digital, A: Analog, NA: Not available, Ch.: Channel, Ch. #: Channel count

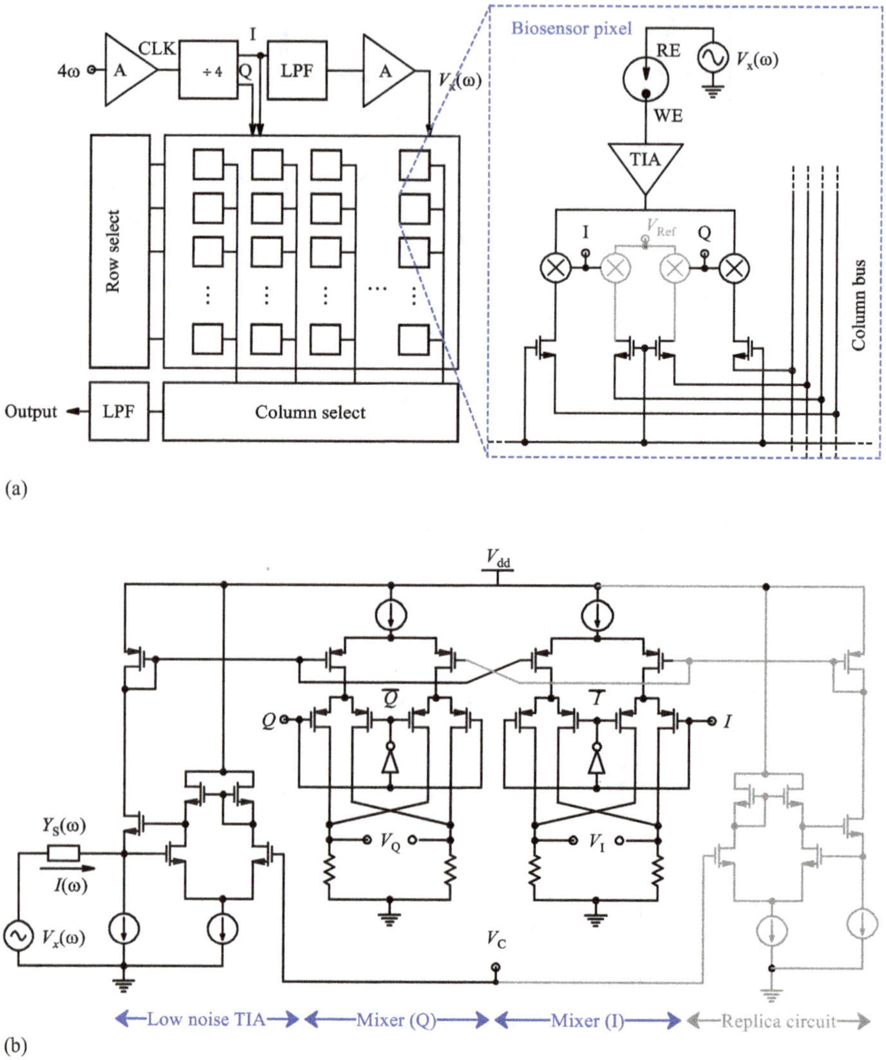

Figure 5.38 (a) An array of EIS biosensors based on a lock-in approach. (b) One pixel of the circuit shown in (a) (without bias circuits).

voltages of the two mixers, V_I and V_Q, give us the amplitude and phase of the sensing admittance, $Y_S(\omega) = Z_S(\omega)^{-1}$, as expressed in (5.3) and (5.4).

$$|Y_S(\omega)| = \frac{\sqrt{V_I^2 + V_Q^2}}{A_{TIA} \cdot |V_x(\omega)|} \tag{5.3}$$

$$\angle Y_S(\omega) = \tan^{-1}\left(\frac{V_Q}{V_I}\right) \tag{5.4}$$

where $V_x(\omega)$ is the sinusoidal excitation voltage signal and ATIA denotes the gain of the TIA. In this work, the quadrature signals, I and Q, and $V_x(\omega)$ are generated by an external reference clock operating at 4ω. The quadrature signals are generated after a divide-by-4 digital counter fed by this clock and $V_x(\omega)$ is derived from the I signal after going through an LPF and an amplifier, as shown in Figure 5.38(a).

Mazhab-Jafari *et al.* [242] reported a 16-channel DNA analyzer using frequency response analysis (FRA) for impedance spectroscopy. To minimize the area and power consumption due to the multiplication and integration operations required for the FRA algorithm, they used an in-channel dual slope multiplying ADC (see Figure 5.39(a)). To avoid the complexity of the implementation of the

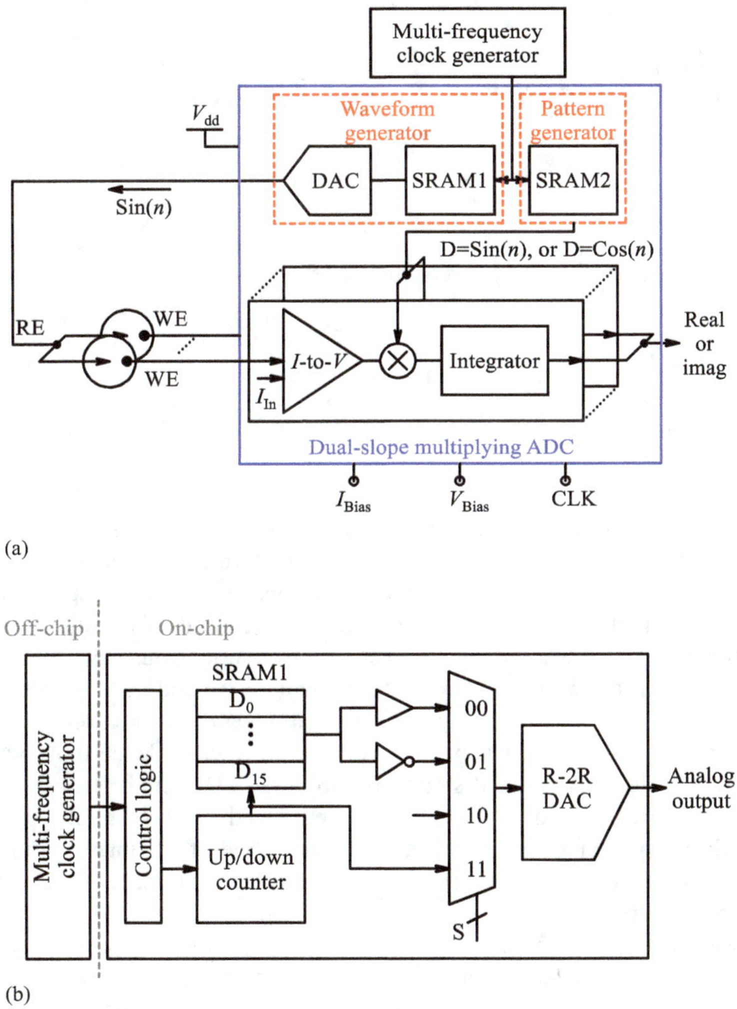

Figure 5.39 Block diagram of (a) an impedance spectroscopy microsystem, (b) the waveform generator used in (a)

signal generators producing sine waveforms and also reduce the systematic errors due to higher order harmonics existing in rectangular waveforms, this group employed a programmable analog waveform generator to produce the stepwise approximation of the interrogation signal (Figure 5.39(b)).

As seen in (5.5) and (5.6), in a frequency response analyzer, if the sensor response, $A\sin(\omega t+\theta)$, is multiplied by $\sin(\omega t)$ and $\cos(\omega t)$, and the result is integrated over one cycle of the interrogation signal, the real and the imaginary components of the sensor impedance will be separated.

$$\begin{aligned} Real &= \int_0^T A\sin(\omega t+\theta)\times\sin(\omega t)\mathrm{d}t \\ &= \int_0^T 0.5\times(A\cos(\theta)+\cos(2\omega t+\theta))\mathrm{d}t \\ &= \frac{0.5T}{\pi}A\cos(\theta) \end{aligned} \tag{5.5}$$

$$\begin{aligned} Imag &= \int_0^T A\sin(\omega t+\theta)\times\cos(\omega t)\mathrm{d}t \\ &= \int_0^T 0.5\times(\sin(\theta)+\sin(2\omega t+\theta))\mathrm{d}t \\ &= \frac{0.5T}{\pi}A\sin(\theta) \end{aligned} \tag{5.6}$$

The sensor depicted in Figure 5.39(a) consists of: (1) a programmable analog waveform generator, (2) a programmable digital pattern generator, and (3) an array of impedance extraction units. The interrogation waveform driving the RE is produced by the waveform generator. Digital multiplication coefficients representing either sine or cosine waves synchronized with analog sinusoidal interrogation voltage on the RE are produced by the digital pattern generator. Moreover, a dual-slope multiplying ADC is employed in each impedance extraction unit and an analog integrator converts the input current at the low-impedance input node to a controlled potential. The dual-slope multiplying ADC multiplies the sensor response with the set of digital coefficients produced by the digital pattern generator and then accumulates the results over one cycle of the interrogation signal using a counter as a digital integrator to extract the real and imaginary components of the sensing impedance.

Viswam *et al.* [44] presented an impedance spectrometry sensor including 59,760 platinum microelectrodes and 32 on-chip lock-in amplifiers to measure a selected subset of the electrodes in the array. This system includes 2,048 integrated extracellular-action-potential (EAP) readout channels (Figure 5.40(a) and (b)). A stepwise approximation of a sinusoidal voltage with a frequency ranging from 1 Hz

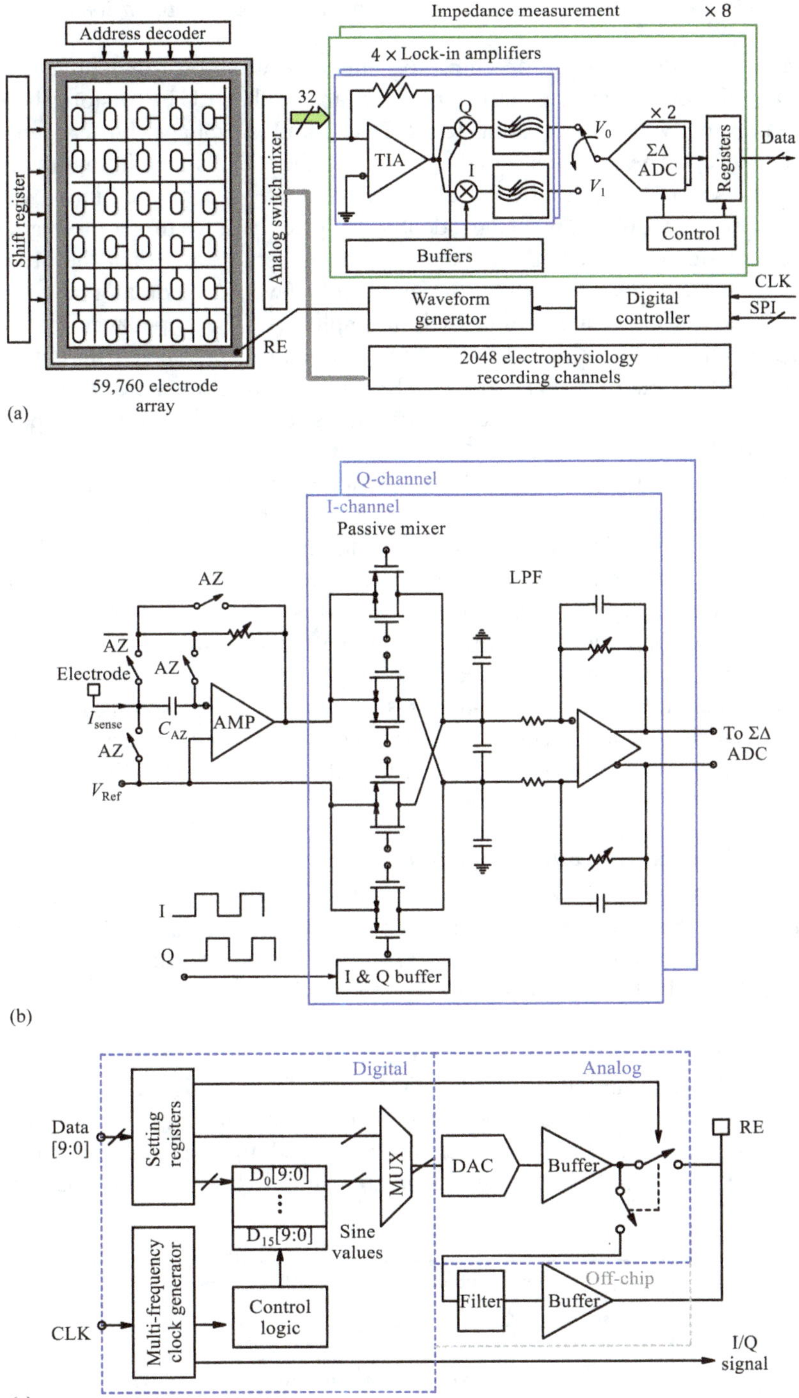

Figure 5.40 Block diagram of (a) a high-density microelectrode array impedance spectroscopy microsystem, (b) the lock-in amplifier used in (a), (c) the waveform generator used in (a)

to 1 MHz is produced by an on-chip programmable waveform including a multi-frequency clock generator, a setting register, an on-chip LUT (to store the sine-wave sampling values), a 10-bit R-2R DAC, and a voltage buffer generator (Figure 5.40(c)). The stimulus signal is applied to the RE. As seen in Figure 5.40(a) and (b), the sensing current arriving at the selected WE is converted into a voltage by a TIA and demodulated by using a double-balanced passive mixer mixing it with the I and Q signals generated by the waveform generator and then the produced signals go through an active fully differential LPF. Two ΣΔ modulators digitize the output signals of every set of four lock-in amplifiers. The impedance magnitude and phase can be then extracted from the in-phase and quadrature data (similar to what is explained in (5.3) and (5.4). The TIA with the programmable capacitive and resistive feedback shown in Figure 5.40(b) allows for providing a wide impedance range from 10 kΩ to 10 GΩ.

The lock-in detection approach is a synchronous method that suffers from matching, inherent DC offsets at the output, and synchronization errors. Matching issues are challenging for simultaneous real and imaginary measurements in multi-channel and/or multi-frequency systems. These circuits require precise synchronization between the injecting and measuring stages. Moreover, offset cancelation techniques like chopping are needed to remove the DC offsets. However, these solutions can increase circuit complexity [246].

In this technique, the signal must pass through an LPF resulting in increased measurement time at low frequencies. Moreover, since each frequency is applied one at a time, the measurement speed will be limited for high-frequency resolution [190].

5.3.2.2 Phase/amplitude approach

As an alternative to the lock-in approach, the magnitude and phase of the sensing impedance can be measured by replacing the multipliers of the lock-in approach with modulator, comparator, XOR, and integrator circuitries, as seen in Figure 5.41. To the best of our knowledge, this technique is barely reported along with on-chip electrodes.

The integrated phase/amplitude circuit proposed by Kassanos *et al.* [246] can be mentioned as an example of this technique, in which a single frequency at a time is applied. However, the electrodes of this sensor are off-chip. In the proposed technique by this group, the magnitude is obtained through multiplication. As shown in Figure 5.42, two instrumentation amplifiers measure the voltage

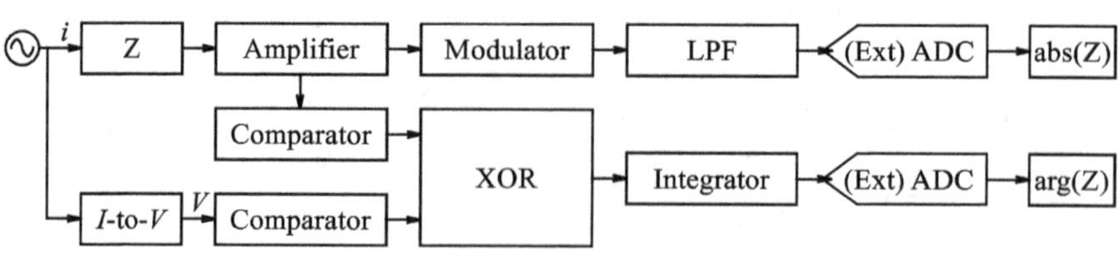

Figure 5.41 Block diagram of the phase/magnitude approach

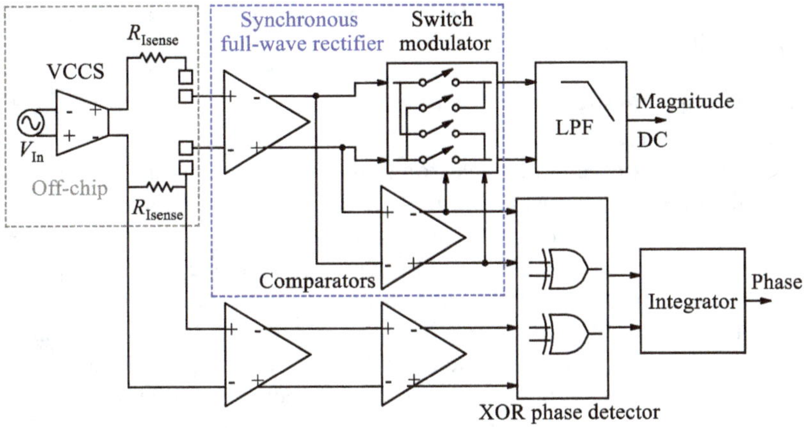

Figure 5.42 Block diagram of a phase/magnitude approach

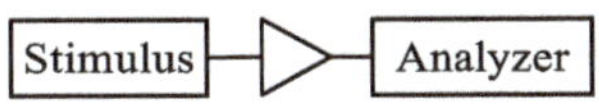

Figure 5.43 Simultaneous frequencies

differences across the known current-sense resistor (R_{Isense}) and the sensing impedance. Then, the two clockless comparators convert the outputs of the amplifiers into square waves. Next, an XOR gate generates a pulse signal whose width represents the phase delay introduced by the sensing impedance at that specific frequency. Two XOR gates produce a differential signal which is applied to a G_m-C integrator. A switch across the capacitor driven by one of the XOR outputs helps to generate a symmetrical signal by synchronously discharging the capacitor at the end of every pulse. As a result, a sawtooth wave is produced whose peak is proportional to the pulse duration and subsequently the phase. The computation of the phase detector's output is determined using the following equation:

$$\theta = 360.\Delta t.f \tag{5.7}$$

where θ is the phase of the sensing impedance. Δt stands for the pulse width and f denotes the frequency of the applied signal. The square wave produced to calculate the phase of the sensing impedance can serve as the clock signal for a switching modulator, facilitating synchronous rectification of the measured signal. Next, the signal is low-pass filtered to achieve the full-wave rectified signal DC average voltage. It can be proved that the magnitude (M) of the sensing impedance can be expressed as:

$$M = \frac{\pi V_{DC}}{2.A_v.I} \tag{5.8}$$

where Av is the amplifier gain, I stands for the amplitude of the injected current, and VDC denotes the filter output.

5.3.2.3 Silicon cochlea-based sensor

By stimulating the electrodes with a multi-frequency signal, the system measurement time can be improved in comparison to the lock-in technique. Silicon cochlea can be used to generate sinusoidal stimuli at the frequencies spaced periodically. After these stimuli are summed, the analyzer divides the output into the related frequency to find a relationship between frequency and impedance [190].

Hamilton *et al.* [247] presented such an impedance sensor which is depicted in Figure 5.44(a). They used silicon cochleae instead of a filter bank of band-pass filters due to their steeper high-frequency cut-offs. This makes it possible to reject out-of-band frequencies without the need for a higher-order filter with high complexity, area, and power consumption. In the circuit illustrated in Figure 5.44(a), the stimulus cochlea, whose structure is shown in Figure 5.44(b), generates sinusoidal stimuli at logarithmically spaced frequencies. The stimuli are summed and injected into the test structure. The analyzer cochlea divides the output of the test structure into the contributing frequencies to establish the impedance-frequency relationship which is (shown in Figure 5.44(a)):

$$Z(i) = \frac{I_{\text{out}}(i)}{I_{\text{ref}}(i)} \times scale \tag{5.9}$$

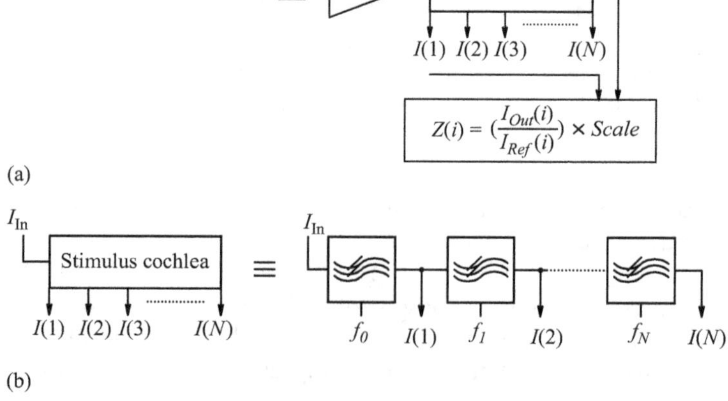

Figure 5.44 (a) An impedance measurement system based on silicon cochlea. (b) The silicon cochlea cascade structure.

where $I_{out}(i)$ denotes the *i*th current from the analyzer cochlea, $I_{ref}(i)$ stands for the *i*th current from the stimulea cochlea, and *Scale* identifies a constant determined by the V-to-I converter.

5.4 Summary

In this chapter, electron transfer-based biosensors were discussed and categorized into two general groups including impedimetric and voltammetric. While the former uses an AC signal for stimulation, the latter requires DC stimuli. Electrochemical cells are usually configured by using two or three electrodes. These sensors need a signal generator to produce the required stimulus signal with the desired shape, a potentiostat to control the electrochemical cell and current acquisition, and a readout circuit to amplify and read the output signal. Here, after a brief review of the structures of integrated signal generators, different potentiostats were reviewed in two categories of TIA-based and CC-based potentiostats. Then, various types of CMOS-based readout circuits for DC and AC currents obtained, respectively, by voltammetric and impedimetric biosensors were discussed.

Chapter 6

Bio-voltage recording

Functional analysis of brain activity necessitates the utilization of high-throughput and high-resolution tools to observe and manipulate neural processes. The bio-electric activity propagated along a neuron's axon can be measured either intra-cellularly or extracellularly using a microelectrode. External action potentials (APs) are generated by depolarizations of the neuron's membrane and manifest as waveforms between 100 Hz and 10 kHz, lasting a few milliseconds. The occurrence rate of APs ranges from 10 to 120 per second. Intracellular APs exhibit amplitudes around 70 mVpp [248], while extracellular APs have weaker amplitudes of 50 μVpp to 500 μVpp, depending on the distance between the active neuron and the recording electrode [249]. Intracellular APs are measured by penetrating the neuron's membrane using a sharp glass micropipette filled with saline solution. A metal electrode connected to the pipette's electrolyte links it to an amplification circuit and computer interface, facilitating the measurement of strong voltage signals. However, this method leads to cell death within a few hours after recordings. Another type of bioelectrical signal, called local field potentials (LFPs), consists of lower-frequency neural waveforms ranging from mHz to 200 Hz, with amplitudes spanning 500 μVpp to 5 mVpp. LFPs are also detected using extracellular microelectrodes. They likely contain valuable complementary information on neural network dynamics and find utility in brain interfacing applications [250,251].

Table 6.1 presents an overview of the neural recording parameters across various signal modalities. The results indicate that intracellular AP measurements exhibit robust signals but are limited in their long-term applicability. Conversely, extracellular AP and LFP measurements capture weaker signals yet hold significant potential for chronic utilization. Electroencephalogram (EEG) measurements share similar parameters with LFPs; however, the latter is considerably more invasive compared to the former. Carbon fiber [252,253] or iridium oxide microelectrodes [254] can be employed for neurotransmitter measurement using techniques like amperometry or voltammetry. Fast-scan cyclic voltammetry (FSCV) is often preferred due to its ability to monitor electrochemical species in real time on a shorter time scale. By understanding the preferred redox potential and associated redox current of a specific species, it becomes possible to identify the neurochemical molecule and determine its concentration. Chemical measurements involve minute redox currents but offer substantial prospects for long-term utilization. In contrast,

Table 6.1 Neural recording parameters across various signal modalities [255]

Neural signal modality	Measurement technique	Amplitude	Bandwidth	Transducers	Potential for chronic utilization
Extracellular APs	Voltage amplification	50 to 500 μVpp	100 Hz to 10 kHz	Metal/Silicon microelectrode	High
Intracellular APs	Voltage amplification	10 to 70 mVpp	100 Hz to 10 kHz	Glass micropipette	Very low
LFP	Voltage amplification	0.5 to 5 mVpp	1 mHz to 200 Hz	Metal/Silicon microelectrode	High
EEG	Voltage amplification	1 to 10 mVpp	1 mHz to 200 Hz	Surface electrode	Very high
Ionic current	Patch clamping	1 to 10 nA	1 mHz to 10 kHz	Glass micropipette	Low
Redox current	Amperometry	100 fA to 10 μA	1 mHz to 100 Hz	Iridium oxide/ carbon fiber microelectrode	High
	FSCV		1 mHz to 10 kHz		

patch-clamping (see Section 6.1) measurements handle larger currents but possess a relatively lower capacity for chronic utilization [255].

6.1 Measurement methods of electrogenic cells' electrical activity

The measurement of electrical activity in cultured cells can be conducted through direct methods like patch-clamp [256] or microelectrode array (MEA) [257–259], or indirect methods such as optical techniques [260]. Direct measurement of electrogenic cells' electrical activity involves two main approaches: (1) transmembrane measurements, and (2) extracellular recordings. Achieving cellular resolution in extracellular recordings is achieved through three distinct approaches [42]: (1) constraining cells in specific electrode positions [261,262], (2) adapting electrode layout and biological structure [263], and (3) employing high-density arrays capable of simultaneous recordings from all electrodes [257,258].

Patch-clamp is a precise technique for investigating cell electrophysiology, enabling accurate intracellular access. This method captures minute changes in membrane potential by inserting an electrode into the cell or measures the currents flowing through individual ion channels. However, patch-clamp is time-consuming, has low throughput, and is invasive [264].

In extracellular recordings, cells are directly cultured on micro-transducers. When an AP occurs in a cell near the micro-transducers, the ions crossing the cell membranes generate an electric field or voltage that influences the open-gate region of field-effect transistors (FETs) [265], or metallic microelectrodes [257–259,266]. These methods are non-invasive, suitable for long-term and multi-site measurements, including recording and stimulation [266]. Complementary metal–oxide–semiconductor (CMOS) technology facilitates the utilization of open-gate FETs and metallic electrodes for *in vitro* measurements. For instance, Eversmann *et al.* [258] developed a FET-based, 128 × 128-MEA that adopts a matrix readout scheme for the extracellular recording of neural activities. The array integrates on-chip amplifiers, multiplexers, and buffers. This array operates based on the oxide-semiconductor field-effect transistor (OSFET) principle, where the potential of the cleft formed between the cell and sensor surface modulates a transistor's drain current.

The MEA platform, when combined with cell culture or brain slice, serves as an effective means to study neural information processing *in vitro*. To enable simultaneous long-term recording of neural signals from multiple cells, the preferred technique is extracellular neural recording, which relies on subcellular-scale microelectrodes, dense arrays, and flexible substrates. Conventional MEAs, although they measure population activity averages, fail to capture signals from the majority of cells in the sample. In contrast, high-density microelectrode arrays (HD-MEAs) are increasingly employed for the observation and manipulation of neurons and networks *in vitro*. The use of HD-MEAs allows for the assessment of individual cell activity with sub-cellular precision, providing a significantly greater level of detail for each cell. The ability to study neurons and their network functionality, as well as observe activities in large neuronal populations, relies on the inclusion of a large-scale component. Additionally, long-term recording capabilities are essential for investigating plasticity in networks within the context of cellular embodiment or statistical coherence. In this chapter, we present the advancements in neural recording and stimulation, especially CMOS HD-MEA technology.

6.2 High-density microelectrode array (HD-MEA) for neuroscience applications

Contemporary neuroscience emphasizes the collaborative efforts of neurons in executing major brain functions, leading to inquiries regarding the intricate nature of their activities. This necessitates the development of advanced experimental methods and analytical approaches [267]. As mentioned earlier, MEAs provide a solution by facilitating label-free, non-invasive, and long-term measurements of extracellular action potentials (EAPs). MEAs find applications in diverse areas such as neuropharmacology [268], neurotoxicity assessments [269,270], evaluation of functional compatibility for new materials [271,272], and the investigation of neural network plasticity [273,274].

Current clinical practices *in vivo* primarily rely on the use of millimeter-scale electrodes to record and stimulate deep brain regions [275]. However, the implantation process itself poses significant challenges, including electrode damage, tissue inflammation, and corrosion, which affect the long-term stability of these penetrations [276]. The precise monitoring of rapid intracellular dynamics *in vivo* remains technically difficult due to the lack of high-resolution tools [277,278], especially in the presence of complex intracellular events such as calcium signaling, energy homeostasis, and protein redistribution [279,280]. Although recent advancements in two-photon imaging have enabled the recording of activity-dependent calcium dynamics *in vivo*, their utility is limited by their configuration [281,282]. For instance, imaging calcium dynamics in axons or pre-synaptic boutons within the living brain is challenging and experimentally demanding due to the difficulty in identifying isolated axons and synapses within intricate neural networks [278,283,284]. Additionally, correlating calcium dynamics with specific electrical patterns in identified neuronal compartments like axons or dendrites is a complex task to systematize [285]. Such experimental setups typically involve patching neurons *in vivo* or in acute slices while simultaneously imaging projecting axons or target neurons that are located several hundred micrometers away from the cell body [286,287]. Furthermore, pharmacological interventions targeting specific neuronal compartments, such as soma, dendrites, axons, or synapses, are particularly difficult due to the uncontrolled diffusion of compounds. Consequently, researchers are constrained to use time-consuming alternatives such as viral injections or local infusions using nearby micropipettes [277,288]. As a result, the precise molecular mechanisms governing activity-dependent subcellular calcium dynamics within axons and synapses in neuronal networks are not yet fully understood [273].

To overcome the limitations mentioned, many cell biology studies employ *in vitro* primary cultures of neurons, which offer greater accessibility and compatibility with high-resolution video microscopy.

The development of *in vitro* MEA technology has enabled researchers to delve into the intricate details of cultured neural networks. Neurons are cultivated on a planar MEA capable of detecting electrical signals from individual neurons or neural circuits [267].

The integration of cultured neuronal networks with electronic systems has found utility as *in vitro* model systems across various neuroscience disciplines. *In vitro*, MEAs are employed with dissociated neuronal preparations or brain slices, and numerous studies have explored their potential applications in drug screening, biosensing [289,290], rapid medical diagnosis [291], pharmaceutical testing [292], as well as military and environmental fields, such as the detection of biological or chemical agents [293]. They have also been utilized to investigate synaptogenesis [294], axon guidance [295], nerve regeneration [296], and neural plasticity [297,298]. Moreover, specific methodologies like compartmental chambers [299], micro-island cultures [300], cell patterning [301], and organotypic slice culture [302] have been developed to study the mechanisms of neural information processing in neural networks [303].

The success of each experiment and the overall effectiveness of MEA technology greatly depends on the quality of neural recordings. Unlike traditional MEAs with 64 electrodes, CMOS-based MEAs have up to 16,384 electrodes integrated onto the same chip along with the necessary addressing circuitry [258,304,305]. One crucial design consideration for large arrays is multiplexing, which allows for a higher number of electrodes without the limitation of bond pads. Recent advancements in CMOS technology have made it possible to incorporate recording electrodes with integrated multiplexing circuits [306]. This integration enables access to a greater number of closely spaced electrodes, resulting in larger sensing areas with high spatial resolution [307]. On-chip multiplexing is a significant advantage of using CMOS, as it improves connectivity and signal quality, and enables the conditioning of small signals through dedicated circuitry units like filters and amplifiers. CMOS technology has demonstrated its capability to achieve high spatiotemporal resolution, connectivity, signal quality, and low-noise recordings from various biological preparations, such as brain slices and cultured cells. The integration of on-chip microelectronics provided by CMOS technology enhances system capabilities. For instance, on-chip signal conditioning ensures faithful recording of weak neural signals [308]. Furthermore, integrating recording amplifiers and analog-to-digital converters (ADCs) on the same substrate as the electrodes eliminates off-chip parasitics and interference while allowing for a large number of recording channels with minimal connections [43]. Achieving high signal-to-noise ratios (SNRs) is possible by placing on-chip amplifiers directly beneath the sensing electrodes, reducing the signal path length [72,259]. CMOS technology also enables direct interfacing with electrogenic cells, such as brain cells *in vitro*. In this approach, cells are cultured directly on top of CMOS chips that undergo specific post-processing to establish a reliable bidirectional interface using noble metal microelectrodes or high-k dielectrics.

However, a drawback of CMOS chips is their lack of transparency to visible light, unlike standard cell culture substrates used in biology. Additionally, the operation and prolonged exposure of the chip or its components to salt water can cause corrosion. This issue becomes more prominent as CMOS chips are typically not designed for long-term immersion in saline solutions spanning weeks or months. Furthermore, when stimulation features are implemented on-chip, significant problems can arise from electrochemistry during the delivery of current or voltage pulses and due to the varying redox potentials of exposed materials. These factors can generate local electrochemical elements and lead to extensive corrosion, including the complete dissolution of CMOS metal aluminum. Hence, an effective packaging solution is necessary to protect the chip from metabolic products and chemicals in the cell culture, while also preventing the cells from being affected by toxic materials released by the chip, such as dissolved CMOS metal aluminum (or copper in modern processes) in saline solution.

HD-MEAs implemented in standard CMOS technology offer the potential for recording at single-cell or subcellular resolution. However, these devices must satisfy two conflicting requirements: achieving a high SNR and simultaneously maintaining high spatial and temporal resolution measurements. The challenge lies in the fact that these two requirements contradict each other in terms of technical implementation [264].

To achieve high spatial resolution, smaller electrodes are employed, which unfortunately exhibit higher thermal noise. Furthermore, due to limited space, only a few and small circuitry elements can be integrated for addressing and signal amplification per electrode. The reduction in electrode and circuitry size amplifies the noise, as the noise of a transistor scales inversely with its size. Additionally, the realization of filters or amplifiers requiring large capacitors becomes unfeasible. Consequently, the circuitry necessary for reading out such high-density arrays compromises the overall signal quality [264].

On the other hand, using larger electrodes offers advantages such as increased signal strength and more space for *in situ* circuits, resulting in a better SNR. However, addressing signals at the single-cell level becomes challenging when larger electrodes are employed [264].

In recent studies, MEAs with different patterns of cell adhesion have been utilized to examine the behaviors of neural systems with diverse morphological structures *in vitro*. The key to success lies in achieving both biological activity in the neuronal networks and effective measurement of bioelectric signals by the sensors.

6.3 Readout schemes for HD-MEAs

In recent electrophysiological studies focused on neuronal signals and network behavior, CMOS-based HD-MEAs have gained significant popularity [43]. These devices enable the measurement of extracellular electrical activity *in vitro*, capturing the local changes in charge distribution resulting from neuronal activity. This activity generates a potential difference between the microelectrode and the reference bath potential, provided by a reference electrodc [309].

In the literature, three main types of readout schemes have been utilized: passive pixel sensor (PPS), active pixel sensor (APS), and switch matrix (SM) sensors.

Passive pixel arrays consist of an array of electrodes without any active electronic signal conditioning components. Hence, extensive signal processing is necessary to enable in-depth analysis [264] (Figure 6.1(a)).

On the other hand, active pixel arrays incorporate an amplifier within each pixel as well as on-chip multiplexers (Figure 6.1(b)).

To attain high spatial resolution and accommodate complex morphologies, certain sensors employ an analog SM approach. In this configuration, each pixel of the array is comprised solely of static random-access memory (SRAM) cells and switches [43]. The SM approach incorporates transistors within the array to serve as switches, facilitating the routing of signals from electrodes to the readout circuitry positioned outside the electrode array. The routing mechanism in the SM operates in a static mode, meaning that electrode selection is set by opening or closing the switches, after which recordings are initiated without altering the electrode selection (see Figure 6.1(c)).

Table 6.2 provides a summary of various *in vitro* MEAs, highlighting recent developments and facilitating comparisons. The table reveals the adoption of different electrode materials, shapes, and sizes to improve recording capabilities. Electrodes are being scaled down to subcellular resolution, allowing for higher packing densities. Nanoscale fabrication techniques are being employed to enhance sensitivity and spatial selectivity. Moreover, there is growing interest in using flexible substrates, in addition to rigid ones, for MEAs [310].

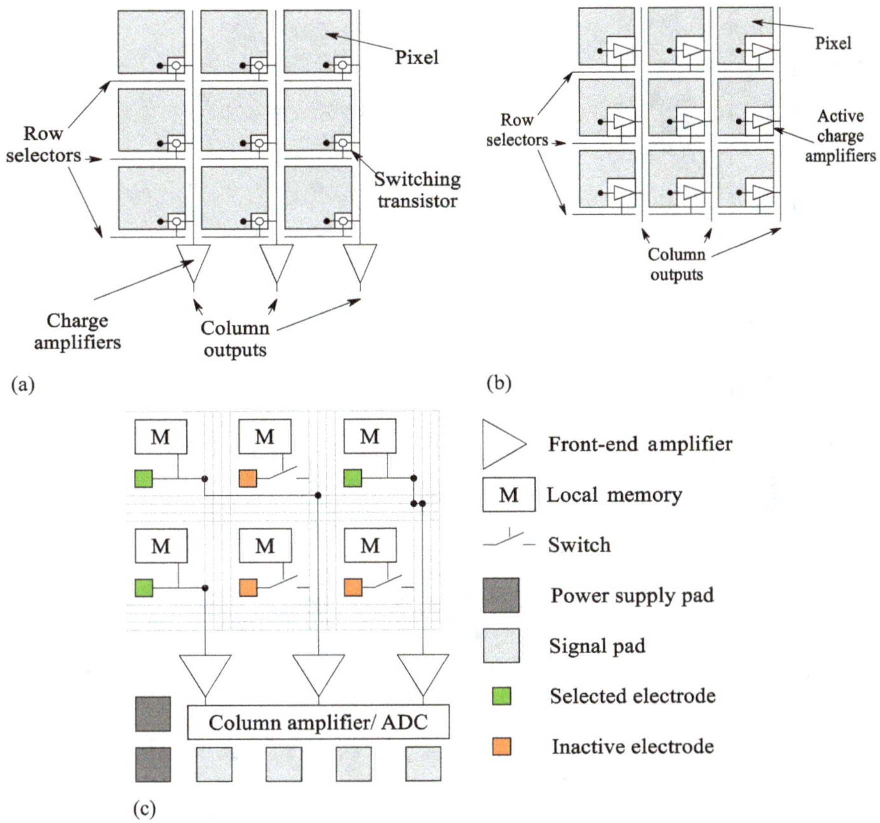

Figure 6.1 *(a) Passive pixel CMOS arrays, (b) active pixel CMOS arrays, and (c) SM*

6.3.1 Passive pixel CMOS arrays

The initial CMOS arrays, known as passive pixel CMOS arrays, were characterized by fixed wiring and a passive design, lacking active circuit elements such as amplifiers. Each electrode in a conventional passive CMOS array (refer to Figure 6.2) directly connects to a signal pad located outside the array through a wire. These pads are then linked to external equipment for signal conditioning [311]. The fabrication of passive MEAs is relatively straightforward, allowing for the utilization of various substrates and electrode materials. Additionally, users have direct access to all electrodes simultaneously.

Nevertheless, passive MEAs, which commonly include metal electrodes on a glass substrate, come with restrictions concerning the number of electrodes (generally fewer than 300) and spatial precision (typically around 30 μm). These limitations arise due to wiring constraints and electrode geometry [312].

The presence of a glass ring around the passive electrodes serves to contain the cell culture and the medium. The medium contains essential nutrients to sustain the

Table 6.2 Recent advances of CMOS-based in vitro *MEAs*

PPS/ Aps/ SM	Technology (μm)	Die Size	Number of channels	Number of transducers	Area/Ch.	Sensing area (mm^2)	Electrode size	Electrode material	Electrode shape	Target application	Reference
APS	0.35	3.5×4.5 mm^2	16×16	16×16	200×200 μm^2	NA	$L = 1{,}500$ μm	Pt	3D UEA	Intact mouse hippocampus	[319]
APS	0.5 (2p, 3M)	5.4×6.5 mm^2	128×128	128×128	21×21 μm^2	1×1	$D = 4.5$ μm*	Au	Circular	Snail neuron	[258]
APS	0.6 (2P, 3M, HR)	6.5×6.5 mm^2	8×16	128	NA	2×4	$D = 30$, and 40 μm	Pt	Circular	Dissociated cortical tissue of fertilized chicken eggs	[72]
APS	0.18	2.2×1.7 mm^2	1120	1120	NA	2.2×1.7	50×50 μm^2	Pt	Rectangular	Mouse olfactory bulb slice	[320]
APS	0.13	1.6×1.6 mm^2	768	768 electrodes, 2,048 photopixels	NA	1.6×1.6	50×50 μm^2	Pt	Rectangular	Mouse olfactory bulb slice	[321]
APS	0.35	21×21 μm^2	640 (culture), 1046 (Slice), 1558 (Retina)	64×64	NA	2.67×2.67, 5.12×5.12	21×21	NA	Rectangular	Neuronal cell cultures, brain slices, and mouse retina preparation	[322]
APS	0.18	NA	4,225 recording, 1,024 stimulation	4,225 recording, 1,024 stimulation	NA	1×1, 2×2	$D = 8$ μm	TiN/ Ti-Zr oxide	Octagon	Neural tissue	[323]
APS	0.5	9.5×9.5	256×128	256×128	NA	2.28×1.14	NA	Pt/TiO_2	Rectangular	Cultured cells from the pond snail	[324]
APS	0.35	NA	64×64	64×64	NA	2.67×2.67	21×21 μm^2	NA	Rectangular	Hippocampal cells	[305]
SM	0.6 (3M 2P)	7.5×6.1 mm^2	126	11,011	0.07 mm^2	2×1.75	NA	Pt-Black	NA	Neonatal rat cardiomyocytes	[42]
SM	0.18	12×8.9 mm^2	2,048	59,760	0.1 mm^2	4.48×2.43	3×7.5	Pt	Rectangular	Mouse cerebellar slice	[6]
SM	0.35	7.6×10.1 mm^2	1,024	26,400	NA	3.85×2.10	9.3×5.4	Pt	Rectangular	Rat cortical neurons	[43]

NA: Not available, Active area, D: Diameter, L: Length, UEA: Utah electrode array, *: Typical cell diameters are between 10 and 50 μm

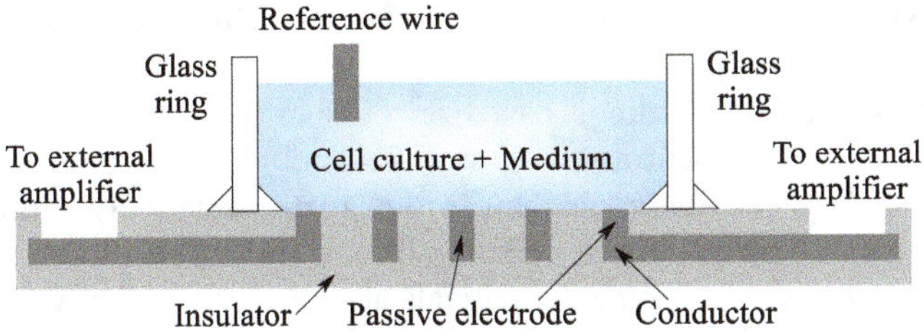

Figure 6.2 Schematic representation of the passive MEAs main elements

viability of the culture. The external amplifier measures an extracellular potential signal, which exhibits an extremely low amplitude. A reference wire is utilized to provide a reference potential for external purposes. These passive MEAs typically comprise 30–160 electrodes with an inter-electrode spacing ranging from 100 to 500 μm, as offered by Multichannel Systems, Panasonic, and Ayanda. However, the signal amplitude is insufficient for direct multiplexing, thereby highlighting the significance of achieving connectivity between the electrode and the amplifier [312].

Various examples of passive MEAs have been developed and utilized for both *in vivo* and *in vitro* neural recordings. Robinson introduced the first metal microelectrode in 1968 [313], and in 1972, Thomas *et al.* presented the initial MEA [314]. This MEA enabled the recording of LFP signals from dissociated chick myocytes using a 2 × 15 array of planar gold electrodes (100 μm pitch), coated with platinum black to reduce impedance. However, the device was unable to capture activity signals from individual cells. In 1977, G. Gross proposed the use of multielectrode arrays comprising 36 gold electrodes (10 μm in diameter, 100 or 200 μm pitch), fabricated through laser etching. Gross *et al.* employed these arrays to record single-APs (up to 3 mV p–p) from snail ganglion cells [315]. Three years later, J. Pine achieved the first recording from rat-dissociated neurons cultured on-chip [316]. The multielectrode array designed by Pine and colleagues featured 16 platinized microelectrodes (250 μm pitch), insulated with silicon dioxide and featuring a square electrode opening of 10 μm [317]. This array, known as the "Pine array," was employed in various experiments conducted by the Pine lab and other research groups.

In 1989, Regehr *et al.* [318] utilized a 61-electrode array initially fabricated by Gilbert and Pine to study invertebrate neurons. The arrangement was crafted to facilitate the recording of neurons from a sparsely populated culture positioned on its surface, building upon the insights derived from Pine's experiments in 1980. To facilitate the observation of cells at high magnification with a short working distance lens and an inverted microscope, the array was fabricated on a thin glass cover slip. The experimenters aimed to explore these possibilities by recording from a range of invertebrate neurons, including snails, aplysia, and leeches.

In the late 1990s, research groups in Japan, specifically at the Matsushita and NTT laboratories under the leadership of Taketani and Kawana, developed

64-electrode MEAs for use in slice experiments and with cultures of dissociated cortical neurons [325,326].

While extracellular recordings based on passive MEAs enable the simultaneous access of multiple neurons with sufficient temporal resolution to discern individual APs, they encounter challenges due to the relatively large electrode spacing (>30 μm). Consequently, it is often difficult to capture all neurons within a population, resolve overlapping neuron signals accurately, or reliably detect axonal signals [327]. Despite the potential for small pixel size and a large fill factor, passive MEAs suffer from low sensitivity and high noise attributed to the considerable capacitance of the columns in relation to the pixels [328].

6.3.2 *Active pixel sensor arrays (APS-MEA) technology*

APS technology, characterized by its low cost, lower power consumption, and high dynamic range pixels, holds great promise for researchers and clinicians. This technology, built upon CMOS compatibility, has emerged as an electronic imaging device, surpassing the capabilities of capacitive coupled devices (CCDs). APS offers advantages such as low-noise readout, improved scalability for large array formats, and faster readout speeds compared to PPSs. The distinguishing feature of an active pixel image sensor is the inclusion of one or more active transistors within each pixel unit cell [329].

The APS integrated circuit (IC) is incorporated or etched onto the substrate of the MEA, with the need for protective measures to shield it from the surrounding medium, as depicted in Figure 6.3. The primary objective is to minimize the distance between the measurement electrode (pixel) and the preamplifier [330]. By adopting the APS approach, the bottleneck for electrode density shifts from connectivity to the size of the amplifier. In other words, reducing the amplifier's area per channel allows for increased electrode density.

The origin of the APS technique can be traced back to image-sensing devices [330]. Various APS technologies have been developed, including the amplified metal-oxide-semiconductor imager (AMI) [308], charge modulation device (CMD) [308], bulk charge modulated device (BCMD) [331], base stored image sensor (BASIS) [42], and static induction transistor (SIT) [332]. When these techniques are applied to MEAs, they result in the formation of active electrodes or pixels, as depicted in Figure 6.3.

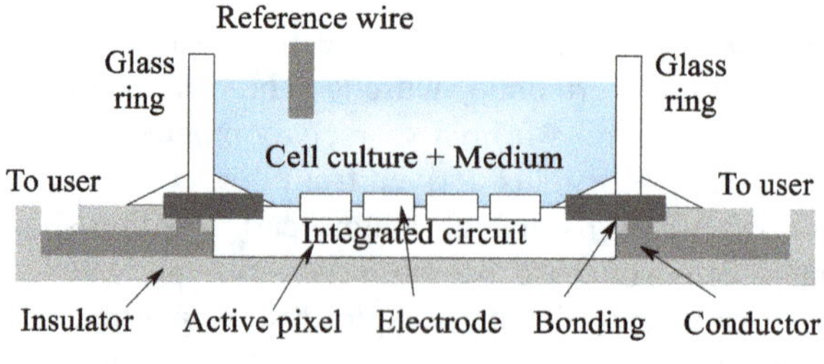

Figure 6.3 MEA schematic: the signal is amplified as close as possible to the culture, and is delivered to the user with an optimal SNR

In APS MEAs, all electrodes or pixels can be sampled at high speeds using full-frame readout. Alternatively, rectangular sub-arrays can be selected as regions of interest and sampled at even faster rates than full-frame readouts. For full-frame readout, the front-end amplification and filtering must occur before the multiplexing stage. This necessitates the placement of the front-end amplifier within each pixel itself. The reason behind this arrangement is that the electrodes exhibit high impedance, and without amplification, they cannot drive multiplexed readout lines at the required speed.

However, the small area available near each electrode poses a challenge when designing low-noise circuitry for APS MEAs. This is because small-sized amplifiers inherently generate higher levels of noise. Consequently, although all electrodes can be simultaneously recorded, only relatively large signals can be reliably detected above the noise level.

In 2001, the APS-MEA platform was introduced for *in vitro* neural recording [333]. This groundbreaking platform facilitated simultaneous recordings from a large-area array, spanning several square millimeters, containing 4,096 microelectrodes with small separations on the order of tens of microns. The APS-MEA utilizes a highly scalable circuit architecture to implement an array of active electrodes. The platform consists of disposable chips, a real-time hardware unit, and a software application for the recording, visualization, and analysis of acquired signals.

APS MEAs have been extensively developed and utilized for both *in vivo* and *in vitro* neural recording (Aziz *et al.* [319], Berdondini [317], Heer [72], Bertotti [323], Berdondini [333], Eversmann [258,324], Johnson [320,321], Maccione [322]). The latest generation of high-density APS MEAs, known as HD-MEAs, aims to enhance the total electrode count and spatial resolution to meet the demands of increasingly challenging applications. Notably, all electrodes within the array can be read out during each frame.

Nevertheless, the design of on-chip signal conditioning plays a vital role in attaining high-quality signals. The power consumption limitations and restricted pixel area inherent in APS-based HD-MEAs often result in higher levels of noise [334,335]. Furthermore, APS pixel size tends to be larger to accommodate the necessary circuitry, leading to lower spatial resolution [309].

6.3.3 Multiplexed array with switch matrix (SM) concept

The concept of the SM involves the use of transistors to create switches within the array, enabling the routing of signals from electrodes to readout circuitry positioned outside the electrode array itself. In the SM configuration, as depicted in Figure 6.4, these routing mechanisms operate in a static mode, which means that specific electrodes are selected by opening or closing the switches, and subsequent recordings are conducted without altering the electrode selection. Typically, not all electrodes exhibit activity during an MEA experiment, allowing for the selection of a subset of "interesting" electrodes. A common procedure involves initially scanning all the electrodes in consecutive recordings to identify which electrodes should be continuously monitored throughout the experiment. The benefit of this concept is that it

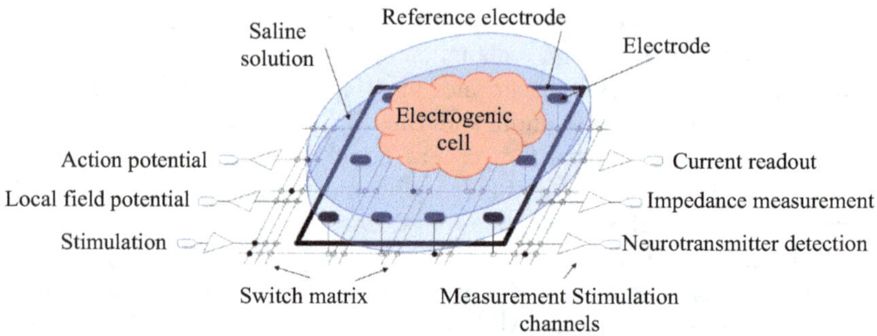

Figure 6.4 Illustration of the multi-functional, SM-based HD-MEA system featuring AP and LFP readout, current readout, impedance measurement, neurotransmitter detection, and stimulation channels

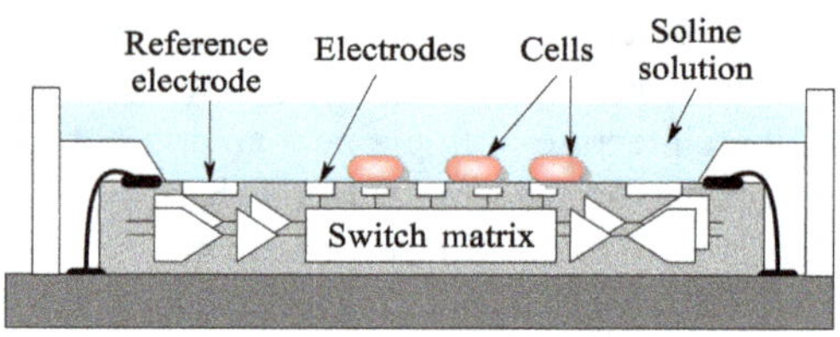

Figure 6.5 Diagram of the packaged MEA chip (device concept)

allows for the implementation of large, low-noise amplifiers outside the electrode array, thereby optimizing the amplifiers to achieve the best possible SNR [336].

An innovative approach for achieving a multiplexed array with flexible addressing involved incorporating additional routing resources within the array [337]. In a study by Ballini *et al.* [43], a CMOS-based MEA SM system was developed. This system effectively maintained sub-cellular spatial resolution across a large sensing area of 8.09 mm and featured 1,024 channels for high-temporal-resolution recording at a rate of 20 kS/s. Despite an eight-fold increase in channel count compared to a previous work in [42], the state-of-the-art noise performance was successfully achieved, thanks to an efficient design of the circuitry for amplification and analog-to-digital conversion, which effectively utilized area and power resources. Furthermore, the design addressed issues such as cross-over distortions and channel-length modulation effects that were observed in the stimulation unit of a different study [338], leading to their significant reduction. To create a device that could be easily handled like a standard MEA dish, post-CMOS fabrication techniques were employed for electrode fabrication, as well as biocompatible die-bonding and encapsulation processes, as depicted in Figure 6.5.

Various implementations of SM MEAs have demonstrated different levels of flexibility in their routing capabilities. Simple row and column-based selectability have been utilized [264]. For *in vivo* probes, increased degrees of freedom in electrode selection have been achieved [339,340]. By employing a large number of wires, switches, and local memory, more complex routing paths have been established to flexibly connect specific subsets of electrodes to readout and stimulation channels. Frey *et al.* [42] achieved fairly complex routing with an average of 1.2 memory cells per electrode. Ballini *et al.* [43] significantly expanded the possibilities by using 2.2 memory cells per electrode, enabling more extensive subset selections. Viswam *et al.* [341] further enhanced the flexibility by increasing the number of bits per electrode to over three, thereby virtually allowing arbitrary subset selections. Notable SM MEAs have been developed in the literature by Dragas [6], Seidl *et al.* [342], Lopez *et al.* [339], Frey *et al.* [42], and Ballini [43], with their key specifications summarized in Table 6.2.

6.4 Neural stimulation and recording circuitries

This section gives an overview of the CMOS circuits reported for both *in vitro* and *in vivo* neural stimulation and recording.

6.4.1 Neural stimulation

Charge delivery is the fundamental aspect of electrical stimulation, whereby the accumulation of charge in tissues reaches a certain threshold, triggering an AP and subsequent muscle contraction.

6.4.1.1 Voltage/current-controlled stimulation (V/CCS)

The prevailing method of electrical stimulation is current-controlled stimulation (CCS) [343–345], known for its controllable charge and high integration advantages. Figure 6.6(a) depicts the conventional bidirectional current stimulation scheme, consisting of well-matched current sources (I_p and I_n), an electrode for stimulating charge transfer (V_E represents the electrode voltage), and a reference voltage electrode (V_{Ref}) [346]. The electrode-tissue interface can be modeled as a combination of capacitance and resistance. Cathodic stimulation triggers the AP, while anodic stimulation compensates for the accumulated charge. In Figure 6.6(b), an intermediate delay ensures AP transfer and high-matching bidirectional current is crucial to prevent charge accumulation and potential nerve tissue damage [347].

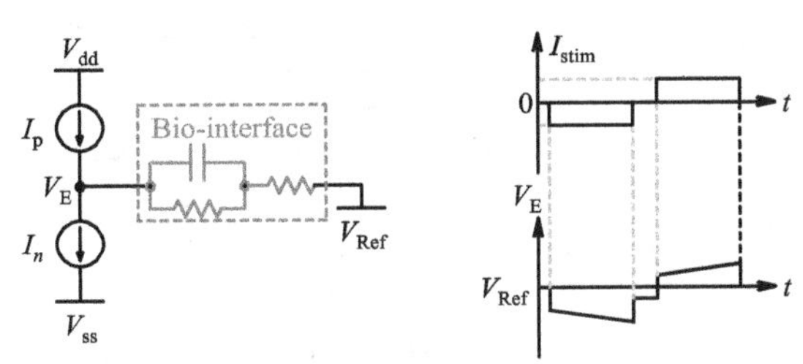

Figure 6.6 (a) Conventional configuration of a bidirectional current stimulator and (b) Waveforms depicting the current and voltage behavior of the bidirectional current stimulator.

In certain applications, voltage/current-controlled stimulation (V/CCS) is employed to regulate both the output current and voltage simultaneously, as highlighted in studies [6,43]. The utilization of CCS allows for precise charge delivery, independent of the electrode impedance, ensuring accurate stimulation. On the other hand, controlling the stimulation voltage becomes essential to prevent tissue damage when the stimulation range exceeds safety limits. Figure 6.7 [6] depicts a buffer circuit that demonstrates these functionalities. The core component of this circuit is a type II positive current conveyor (CCII+), responsible for converting digital-to-analog converter (DAC) voltages into electrode currents. By closing S_1 and S_2, the default current range can be expanded, amplifying the gain of the current mirror in the output stage. To enable voltage control, two operational transconductance amplifiers, OTA_{UP} and OTA_{DN}, are incorporated, which deactivate either the positive or negative output current respectively, once the output voltage reaches predetermined values of V_{UP} or V_{DN}. For instances where voltage-only controlled stimulation is desired, OTA_{VolBuf} operates in a unity-gain feedback configuration and is directly connected to the output. OTA_{VolBuf} possesses a dominant pole positioned at the output node, making it suitable for driving predominantly capacitive electrodes.

6.4.1.2 Switch-capacitor stimulation (SCS)

While CCS has become the most prevalent method, electrode impedance significantly affects the stimulation voltage, especially in multi-channel stimulation scenarios. To achieve the required power efficiency, the supply voltage must ensure a minimum applicable voltage level for various loads. Alternative stimulation techniques, such as SCS [348–350], provide better control over stimulus charges and higher power efficiency. However, integrating large capacitors into the chip proves challenging. Recent developments have demonstrated the feasibility of high-frequency stimulation [349], leading to the proposal of a high-frequency switch-capacitor stimulation (HFSC) method in [348]. Figure 6.8(a) presents a

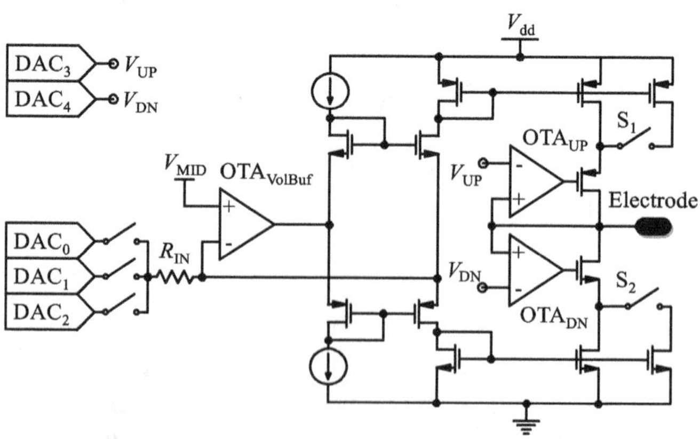

Figure 6.7 Schematic of a V/CCS buffer

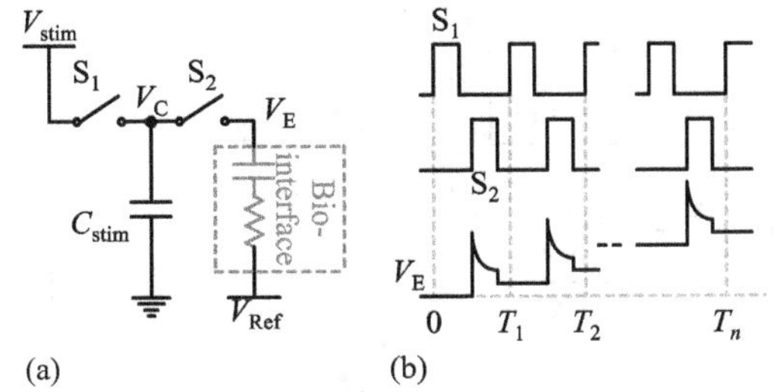

Figure 6.8 (a) Structure of an HFSC and (b) Voltage waveform on the electrode with the stimulator's switch control signals.

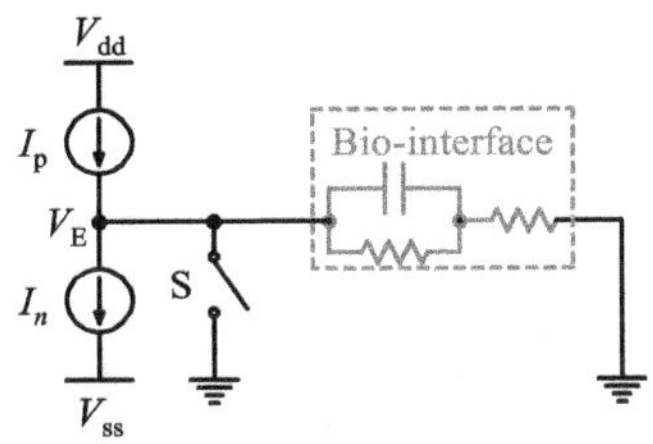

Figure 6.9 Stimulator structure employing an electrode short-circuit technique

high-frequency switching approach for stimulation. Due to the small charge transferred in each switching process, the required capacitance (C_{stim}) can be exponentially reduced, facilitating on-chip integration, and minimizing chip area and cost. Figure 6.8(b) illustrates the switching timing and voltage waveform of the electrode (V_E). Introducing a phase difference between S_1 and S_2 eliminates the dead zone. V_E increases with the number of switches until the charge accumulation reaches the threshold, triggering an AP [347].

6.4.1.3 Addressing non-idealities in neural stimulation circuits

Balancing accumulated charge

While undergoing stimulation, the two-way current fails to achieve full alignment, resulting in remaining charge within the nerve tissue. The accumulation of residual charge can cause irreversible damage to the nerves. To ensure neural stimulation safety, the designed stimulator aims to minimize residual charge within a single cycle and promptly remove accumulated charge after multiple cycles. Real-time monitoring of V_E is crucial to eliminate residual charge when the voltage fails to return to the reference level at the end of the stimulation cycle [351]. Various methods for balancing accumulated charge are introduced, including electrode short-circuit technology [352] and short-time pulse insertion technology [346,353,354]. In Figure 6.9, the electrode short-circuit technology utilizes switch S to connect the electrode to the ground. This method introduces unpredictable discharge times, contingent upon the electrode impedance. The short-time pulse insertion technique enables controllable compensation. As shown in Figure 6.10, when V_E deviates from the reference voltage

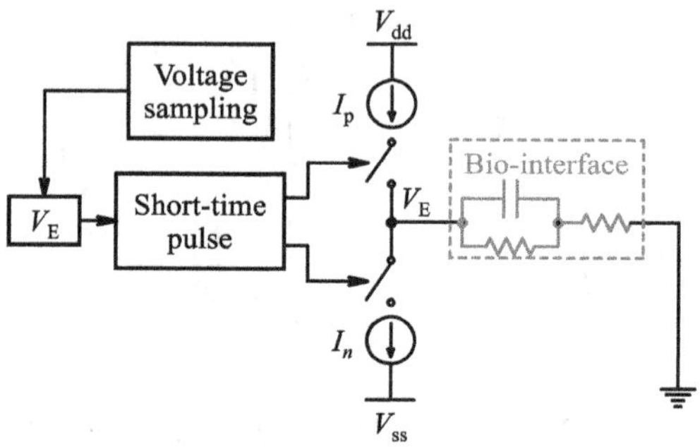

Figure 6.10 Stimulator structure utilizing the short-time pulse insertion technique

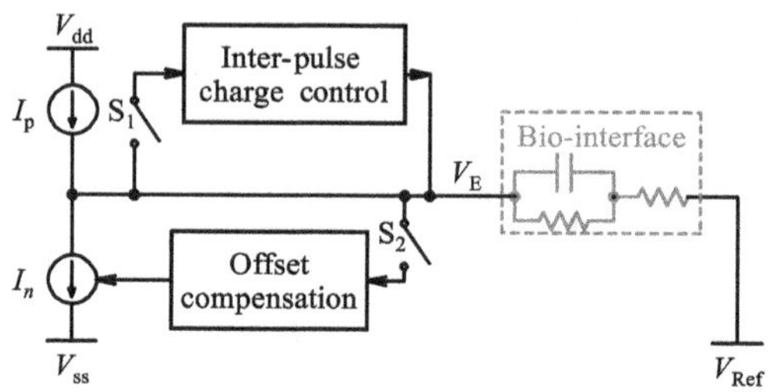

Figure 6.11 Structure of the stimulator incorporating IPCC and offset compensation

range at the end of each stimulation cycle, a series of short-time pulses regulates the switch to achieve charge compensation and restore the V_E voltage level [355]. However, frequent short-time pulse stimulation introduces additional switching noise and reduces the SNR of the recorded signal [347].

A cooperative approach to compensation is proposed by Butz *et al.* [344] to protect nerve tissue from damage during long-term stimulation. This method utilizes cause-based and consequence-based systems, with the stimulation mode being CCS, as shown in Figure 6.11. The outcome-oriented mechanism, referred to as inter-pulse charge control (IPCC), offers immediate adjustment between successive stimulations. When the voltage V_E experiences significant changes at the end of the stimulation cycle, the high-voltage output stage of IPCC generates a constant compensation current to offset the residual charge until V_E returns to the reference voltage. On the other hand, the cause-based compensation method, called offset compensation, incorporates a stable feedback system through proportional integral (PI) control. The compensation is performed in the subsequent stimulation cycle, where the PI control adds additional bias current to the cathode current. This approach ensures that the accumulated charge is counterbalanced through an enhanced bidirectional current following each bidirectional stimulation. The two compensation methods offset compensation and IPCC, work independently using S_1 and S_2 control, respectively, to avoid disturbance caused by simultaneous sampling. It is important to note that voltage sampling should be completed before IPCC starts working since offset compensation has a shorter duration [347].

Power consumption

The design of neural stimulation circuits also considers power efficiency to minimize thermal power consumption. For example, Ha *et al.* [356] adopt a new type of adiabatic current-controlled stimulator architecture that ensures better power efficiency and enables a complete wireless power supply. The adiabatic stimulator effectively tracks the change of electrode voltage and minimizes voltage drop across the current source. On-chip resonant coils provide adiabatic waveforms, which are synthesized through cascading and folding auxiliary rectification stages based on the stimulating voltage requirements. Additionally, the circuit implements better energy-saving performance by recovering electric charge from the nerve tissue. Figure 6.12 illustrates this process, where the stimulation is initially supplied by $V_{DD\text{-stm}}$ and $V_{DD\text{-stm}}$. After the AP is triggered, reverse current compensation takes place. In the second stage, the charge flows into V_{DD} and V_{SS}, thereby returning the energy back to the supply rails. This approach prevents energy loss during the stimulation process and improves power efficiency [347].

Table 6.3 presents a comparison of different neural stimulation circuit designs. Notably, CCS remains the predominant stimulation type for neural stimulators due to its ability to control the injected charge into the tissue. Implantable stimulators require

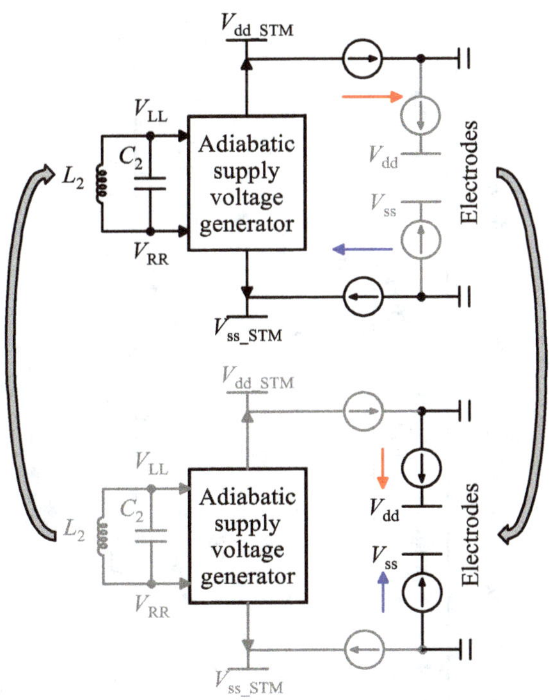

Figure 6.12 Configuration of an adiabatic stimulator

Table 6.3 Comparison of neural stimulation circuits

Technology (μm)	Stimulation type	Supply voltage	Peak efficiency (%)	Maximum stimulation signal	Power/Ch. (mW)	Area/Ch. (mm^2)	Application	Reference
0.18	HFSC	5	49	NA	0.063	0.035	*In vivo*	[348]
0.35	CCS	22	NA	5.12 mA	11	1.5	*In vivo*	[344]
0.35	CCS	20	62	1 mA	1.16	0.2	*In vivo*	[343]
0.35	SCS	4	80.4	4 mA	NA	3	*In vivo*	[350]
0.5	CCS	5	68	2.48 mA	3.75	0.3	*In vivo*	[345]
0.35	V/CCS	3.3	NA	NA	0.073	NA	*In vitro*	[43]
0.18	V/CCS	3.3	NA	±100 μA,1.65 V	0.251 (CCS), 0.306 (VCS)	0.046	*In vitro* (Multi-functional)	[6]

NA: Not available, V/CCS: Voltage/Current-Controlled Stimulation

a wireless power supply with an inductive link, while external wearable stimulators utilize batteries. Ensuring stimulation safety and energy efficiency is crucial in neural stimulator design. Voltage detection circuits and pulse injection circuits, or alternative circuits with better current matching, are necessary for timely monitoring and removal of residual charge from the tissue. Peak efficiency performance is measured by the ratio of maximum output power to the power supply, with the highest peak efficiency observed as 80% among the referenced works. The maximum stimulation current represents the strength of neural or muscular stimulation, with most stimulators providing a maximum output current of at least 1 mA [347].

6.4.2 Neural recording

In a closed-loop neural system, neural recording is an essential component alongside the stimulator responsible for triggering APs. Together, these components form an integrated system that enables feedback and control based on the recorded neural signals. The neural recording aspect serves as the digital backend of the entire closed-loop system. Unlike the stimulation signals generated by the stimulator, neural recordings involve the sampling of LFPs. LFPs are characterized by their low amplitude (measured in microvolts, μV) and low frequency (ranging from 1 to 200 Hz) electric potentials in the extracellular space surrounding neurons [347]. To capture these weak analog signals produced by extracellular electrodes, specialized circuits called analog front ends come into play. The extracellular electrodes pick up the neural activity, but the signals they generate require amplification, filtering, and digitization to be compatible with computing systems for further processing and storage [255]. The analog front ends are designed to perform these functions efficiently, enabling the neural recordings to be processed and analyzed effectively within the closed-loop system.

6.4.2.1 Amplification

The primary component of the front ends used in neural signal recording is the low-noise amplifier (LNA) [43,357], also called a neural amplifier. Its role is to amplify and filter the neural waveforms, eliminating any direct-current (DC) offset observed across a pair of differential electrodes. The LNA must deliver adequate gain, appropriate bandwidth, high SNR, excellent linearity, and high common mode and power supply rejection ratios (CMRR and PSRR) to ensure the desired signal quality. In a multi-channel setup, each electrode requires its own sensory circuit. Therefore, a neural amplifier must possess low power consumption, compact size, and scalability for multiple parallel channels. Additionally, it is crucial to employ a dedicated circuit design methodology, such as the one described in [358,359], to optimize the amplifier for extremely low-power operation. The noise efficiency factor (NEF) is widely used as the primary performance metric for evaluating neural amplifiers and comparing different existing topologies [255].

6.4.2.2 DC offset suppressions

Achieving both low-noise and low-power characteristics while suppressing the DC input voltage, and preserving the broadband neural signal within the frequency range of 1 mHz to 10 kHz, poses a significant challenge in the design of integrated devices. This

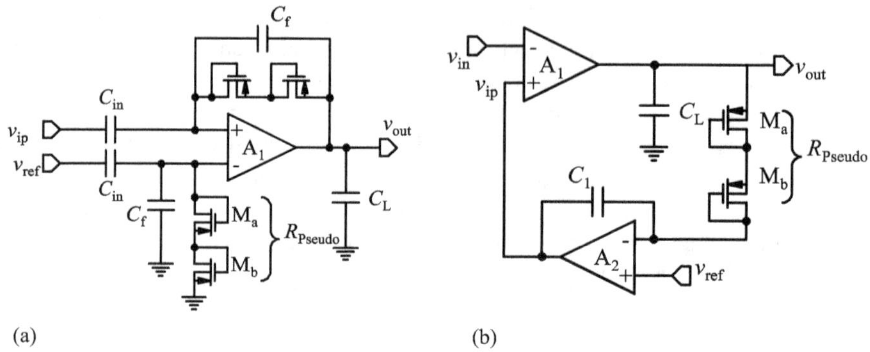

Figure 6.13 The schematic of two neural amplifiers featuring (a) capacitive feedback along with an ADC, and (b) active feedback

challenge arises due to the need for implementing an extremely low-cutoff frequency. While certain designs focus on addressing extremely low-power [360] or very low-noise requirements [361], the development of reliable integrated neural amplifiers has transitioned from open-loop to closed-loop topologies. Initially, techniques involved loading the recording site with a high-value resistor, which could be implemented using a reverse-biased diode [362] or by utilizing subthreshold transistors to create a high-pass filter (HPF) in conjunction with the electrode capacitance [363].

In recent devices, closed-loop DC suppression schemes have replaced open-loop approaches. One commonly employed approach, illustrated in Figure 6.13(a), utilizes a capacitive feedback network and a highly resistive metal-oxide-semiconductor (MOS)-bipolar element to establish a closed-loop gain with a low-frequency pole [358]. MOS-bipolar elements can possess equivalent resistors in the range of several thousands of giga-Ohms for small voltages, enabling the use of small integrated capacitors (typically in the tens of pico-Farads range) in the feedback network to implement a low-frequency cutoff in the mHz range. The feedback loop typically incorporates a low-noise OTA whose implementation may vary [255] (see Section 6.4.2.3). The common approach for reliable quantization of microvolt-level nerve signals involves integrating an LNA at the front end of the recording process, followed by digital quantization circuits [43,364–366]. This combined setup ensures accurate signal capture and conversion. To eliminate any DC offset in neural signals and prevent recording saturation, an additional HPF with a low cutoff frequency, formed by input capacitors, is employed. By incorporating these components in the front end, the system can effectively capture and process microvolt-level nerve signals with minimized noise and distortion, allowing for reliable data acquisition and analysis [347].

6.4.2.3 Low-noise operational transconductance amplifiers (OTAs)

The prevailing configurations for low-noise OTAs include the current mirror OTA [367–370] (depicted in Figure 6.14(a)) and the two-stage operational amplifier

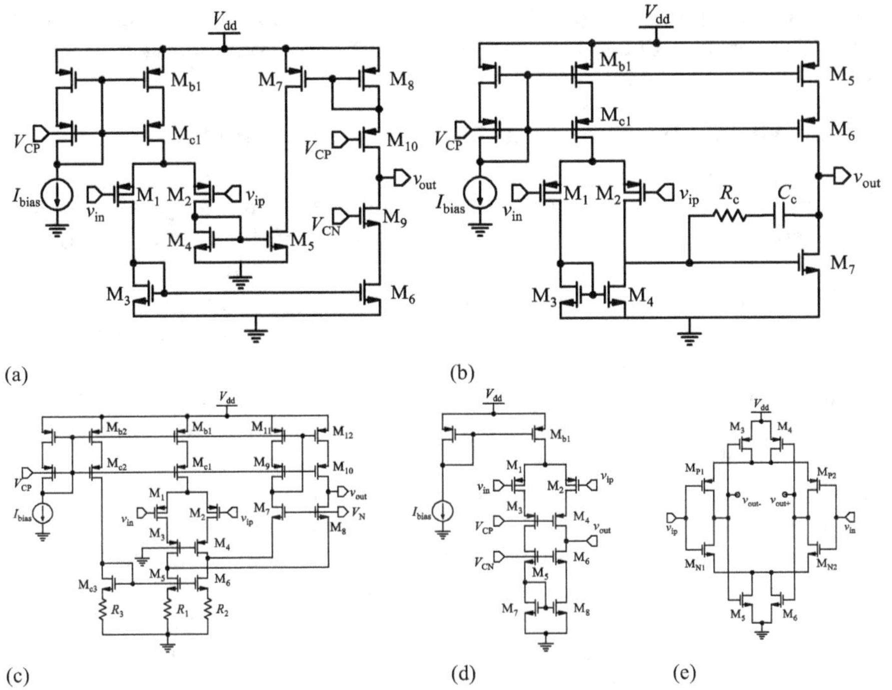

Figure 6.14 (a) The current-mirror OTA, (b) The Miller OTA, (c) The folded OTA with source degeneration, (d) The telescopic OTA and (e) The self-biased OTA

(OpAmp) [364,371,372] (illustrated in Figure 6.14(b)). To minimize power overhead and reduce noise generated by the second stage, alternative approaches have been employed. These include folded cascode OpAmps [46] (shown in Figure 6.14 (c)), telescopic cascode amplifiers [29,47] (represented in Figure 6.14(d)), and fully-differential self-biased OTAs [48] (depicted in Figure 6.14(e)). An interesting capacitive network-based topology proposed in [373] (schematic in Figure 6.14(c)) enhances the NEF by utilizing source degeneration resistors within the active load of the differential pair M_1–M_2. Through the utilization of a source-degenerated load (M_5–M_6 and R_1–R_2) in place of a standard active load, the bias current within the folded branch becomes a minor portion of the current in the differential pair. As a result, the noise contribution from this branch becomes negligible compared to that of M_1–M_2.

Furthermore, the dominant source of noise in the source-degenerated load arises primarily from the degeneration resistors. This noise is primarily of thermal nature. In contrast, active loads employing MOSFETs contribute a significant amount of $1/f$ noise. The expressions provided include the open-loop gain and input referred noise density. The open-loop gain of an OTA is determined by the product of the overall transconductance and the output resistance of the OTA. Comparing different OTA topologies, the Miller OTA and telescopic cascode OTA exhibit lower noise densities due to a reduced number of noise-contributing transistors in their implementation [255].

6.4.2.4 Alternative neural amplifier and signal quantization designs

To enhance the performance of closed-loop amplifiers, Shen *et al.* [374] proposed a novel approach. They introduced a configuration that involves stacking inverters and dividing the capacitor feedback network. This innovative design resulted in a six-fold increase in current reuse, effectively boosting the transconductance and reducing noise levels. Remarkably, these improvements were achieved without any additional current consumption.

Moreover, the demand for achieving a compact silicon area in neural amplifiers has become increasingly crucial due to the continuous growth in the number of recording channels. Nonetheless, capacitive feedback networks often occupy a significant amount of space. An alternative solution presented in [375] suggests a different design (Figure 6.13(b)). This particular design incorporates an active feedback loop utilizing a Miller integrator that employs a MOS-bipolar component and a small-integrated capacitor. The purpose of this configuration is to eliminate any systematic offset and drift at the electrode-tissue interface. Not only does this approach offer a competitive NEF and consume low power, but it also enables a remarkably small chip area. This advantage is due to the usage of just one small integrated capacitor, as opposed to the capacitive-feedback network-based configuration which necessitates two larger capacitors and two smaller capacitors.

Modified versions of the neural amplifier designs mentioned earlier have gained widespread usage for extracting lower-frequency signal modalities like EEG and LFPs [358]. Some designs employ active techniques, such as chopper stabilization, to reduce $1/f$ noise [376,377], which dominates the low-frequency range (1 mHz to a few hundred Hz) in CMOS-integrated devices. These circuits have also demonstrated the ability to capture multimodal information by accommodating different frequency ranges [249,367,372,373,378]. Various methods are employed to achieve different bandwidth settings: (1) tuning resistive values in a filter, (2) selecting different capacitor values from an array, or (3) changing the operating points of a circuit. In certain designs, the high cutoff is adjusted by varying the gate voltage of a pseudo-resistor [373,379] or a weakly inverted MOSFET [363,380], while the low cutoff is modified by directly adjusting the operating point of the low-noise OTA [358] or by changing its capacitive load through the selection of different output capacitors [381].

However, many of these designs suffer from significant distortion and variations in the high cutoff due to the voltage-dependent resistance of MOS devices. MOS resistors exhibit asymmetric and nonlinear resistance when the voltage across them fluctuates, and they are also highly sensitive to process variations. To address this, approaches have been proposed to linearize MOS resistors by implementing dedicated linearization circuits that appropriately bias the MOS device gates [367,382]. These approaches have been further enhanced in [383] by replacing current mirrors and source followers with closed-loop OpAmps, enabling high linearity above 74 dB within ±200 mV and process-independent frequency cutoff values. In this work, the pseudo-resistor of a low-pass filter (LPF) is linearized by feeding back any voltage variations in the input voltage directly to its gate with unity gain. This maintains a constant gate voltage, canceling out non-linearity while providing the necessary DC biasing to set the desired cutoff frequency within a preferred range. In an effort, a switch-capacitor (SC) neural amplifier with tunable characteristics was demonstrated in [384] as an alternative method for multimodal measurements. This amplifier, apart from offering low noise and satisfactory gain, can accommodate LFPs and APs by tuning its clocking frequency, allowing for straightforward implementation of different low- and high-cutoff frequencies [255].

In conventional approaches, the DC offset is eliminated using an alternating current (AC)-coupled configuration with an HPF. Nevertheless, the presence of low-frequency information in nerve signals requires the HPF to have a large input capacitance, consequently lowering the input impedance of the neural recording circuit. Jeon *et al.* [385] proposed a neural-recording IC utilizing a voltage-controlled oscillator (VCO) for direct signal quantization and achieving a high input dynamic range (IDR). The circuit structure of this recording method is shown in Figure 6.15. It converts the voltage into differential currents using an AC-coupled input transconductance circuit. These currents are then fed into current-controlled oscillators (CCOs) CCO_P and CCO_N. The quantizer generates a digital output, D_{out}, based on the phase difference between CCO_P and CCO_N. Through negative feedback controlled by a DAC, the input resistance is adjusted to reduce the difference between I_{in+} and I_{in-}. This negative feedback ensures that even with large input voltages, I_{in+} and I_{in-} assume nearly identical values, resulting in a wide input linear range. Moreover, by directly connecting the circuit input to the metal-oxide-semiconductor field-effect transistor (MOSFET) gate, which exhibits a large input impedance, the recording stability is improved [347].

The choice of an ADC in neural-recording front ends often favors the power-efficient successive approximation register (SAR-ADC) [6,42,72,320,339]. However, when the input swing is large, a high gain in the instrumentation

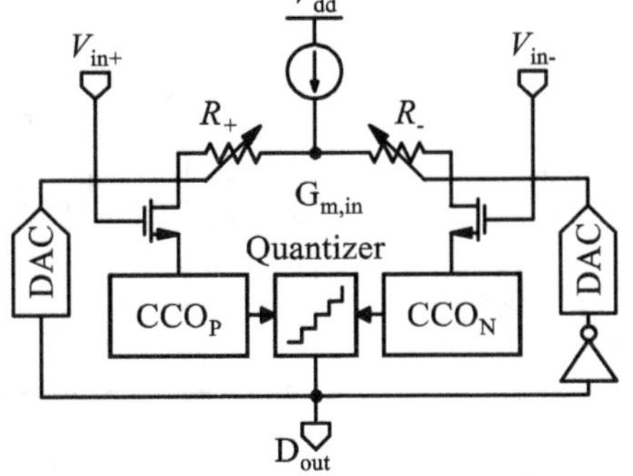

Figure 6.15 Configuration of the neural recording device utilizing a VCO

amplifier (IA) can lead to saturation at the front end. This creates a trade-off between the required input range and the voltage gain in voltage/current domain systems. While circuits designed for low input referred noise have a limited input range, large stimulation artifacts can saturate the IA, making neural signal recovery impossible. The commonly used SAR-ADC is not ideal for high resolution due to capacitor matching limitations in modern CMOS processes.

In an attempt to address the input range/gain trade-off in the voltage domain, some approaches have been explored. In [386], a frequency-shaping amplifier architecture attenuates the gain at low frequencies, enabling a larger dynamic range without increasing ADC complexity. However, this results in a sharp increase in input-referred noise power spectral density at low frequencies, and high-frequency components of artifacts may still saturate the front end. Although the circuit reported in [386] consumes high power, its noise and linearity performance is insufficient.

Another possibility involves reducing the front-end gain while designing an oversampled ADC with higher effective resolution. However, this places the linearity/noise requirements on the feedback DAC and typically necessitates a precise OpAmp, resulting in significantly higher power consumption compared to SAR-ADC.

In [357], an equalization scheme is combined with an oversampled Δ-modulated ΣΔ ADC, which can be implemented in a more compact area compared to a SAR-ADC. Alternatively, Jiang *et al.* [387] propose a different approach where the input signal is directly converted from the voltage domain to the frequency domain using a VCO. The forward path consists of a broadband IA and a VCO-based ADC [388].

Another novel architecture employs a continuous-time sigma-delta modulator (CTΣΔM) as the recording front end. Researchers have developed a structure based on a second-order CTΣΔM [389], as illustrated in Figure 6.16. This strategy involves a sequence of an integrator and a feedforward compensation design, featuring distributed input paths and a one-bit quantizer. The first integration stage

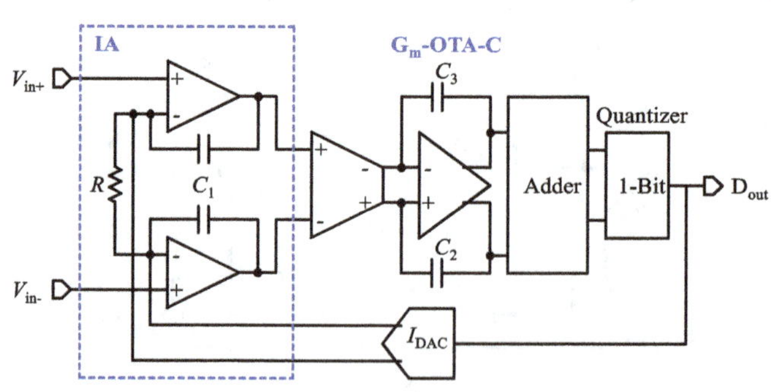

Figure 6.16 Configuration of the neural recording circuit based on CTΣΔM

Table 6.4 Recent advances in neural amplifiers and recording circuits

Technology (nm)	Supply voltage (V)	Target application	Number of channels	Number of transducers	Input referred noise (μV_{rms})	Total Gain (dB)	Bandwidth (Hz)	Power/ Ch. (μW)	Total power	Area/Ch.	ADC	Application	Reference
180	1.8	LFP	NA	NA	2.3	NA	250	23	NA	0.694 mm^2	IA-CTΣΔM	*In vivo* (implantable)	[389]
180	1.0	AP, and LFP	NA	NA	5.5	25.6	4–10 k	0.025	0.25 μW	0.29 mm^2	NA	NA	[374]
180	0.5/1.0	AP, and LFP	128	1,024	3.32	37.5–52.9	0.4–10.9 k	1.22	NA	0.05 mm^2	Δ-ΔΣ	NA	[357]
65	0.5	LFP	64	64	1.3	NA	1–500	2.3	225 μW	0.025 mm^2	VCO-based	*In vivo* (implantable)	[388]
40	0.45/1.2	LFP	NA	NA	5.2	NA	1–200	7	NA	0.135 mm^2	VCO-based	*In vitro*	[387]
350	3.3	AP, and LFP	1,024	26,400	5.9	70	1–10 k	73	75	NA	Single-slope	*In vitro*	[43]
0.35	3	AP, and LFP	16 × 16	NA	7	48–68	1–5 k	15	5.04	200 × 200 μm^2	Off-chip	*In vitro*	[319]
0.5 (2p, 3M)	5	AP, and LFP	128 × 128	128 × 128	NA	NA	10–1 k	NA	656	21 × 21 μm^2	Off-chip	*In vitro*	[258]
0.6 (2P, 3M, HR)	5	AP, and LFP	8 ×16	128	11.7	10–20	0.1–100 k	156.25	120	NA	SAR	*In vitro*	[72]
0.18	1.8	AP, and LFP	1,120	1,120	4.3	18–26	20–9 k	12.6	14.1 mW	NA	SAR	*In vitro*	[320]
0.13	1.5	AP, and LFP	768	768 electrodes, 2,048 photopixels	4.1	43.5–43.9	3.6 k	NA	13.3 mW	NA	Ramp	*In vitro*	[321]
0.35	NA	AP, and LFP	640 (culture), 1,046 (Slice), 1,558 (Retina)	64 × 64	26	60	0.1–5 k	NA	NA	NA	NA	*In vitro*	[322]
0.18	1.8	AP, and LFP	4,225 recording sites, 1,024 stimulation site	4,225 recording, 1,024 stimulation	<300	NA	1–10 k	NA	NA	NA	Off-chip	*In vitro*	[323]

(Continues)

Table 6.4 (Continued)

Technology (nm)	Supply voltage (V)	Target application	Number of channels	Number of transducers	Input referred noise (μV_{rms})	Total Gain (dB)	Bandwidth (Hz)	Power/Ch. (μW)	Total power	Area/Ch.	ADC	Application	Reference
0.5	5 V	AP	256 × 128	256 × 128	NA	NA	2.4 k	NA	4 W (max.)	NA	Pipeline	*In vitro*	[324]
0.35	3.3	NA	64 × 64	64 × 64	11	NA	7.8 k–125 k	NA	132 mW	NA	Off-chip	*In vitro*	[305]
0.6 (3M 2P)	5 analog, 3.3 digital	AP, and LFP	126	11,011	2.4	0–80	1–100 k	160	135 mW	0.07 mm^2	SAR	*In vitro*	[42]
0.18	1.8	AP, and LFP	52	455	3.2	30–72	0.5–200, and300–6 k	27.84	949.8 μW	0.19 mm^2	SAR	*In vivo* (implantable)	[339]
0.18	3.3	AP, and LFP	2,048	59,760	5.4 (AP channel), 4.3 (LFP channel)	29–77	1–10 k	412	86 mW	NA	SAR	*In vitro* (Multi-functional)	[6]

NA: Not available

utilizes an improved IA, while the second integrator is implemented using a G_m-C-OTA circuit. The feedforward branches are summed using an SC adder, and digital signals are generated by the quantizer. The feedback current source (I_{DAC}) adjusts the bias current of the IA, enhancing the stability of the first stage. Similar to the VCO method, the input of the recording circuit is directly connected to the MOSFET gate, exhibiting a large input impedance. Both the VCO and CTΣΔM employ quantizer output and feedback DAC modulation, thereby improving the circuit's linear input range. The differential recording method offers significant common-mode rejection for large-signal artifacts. With the increasing demand for neural recordings, such as monitoring signals from numerous electrodes simultaneously, it becomes essential to develop intelligent data acquisition systems with low power consumption and a small footprint. Hence, a high input impedance front end for neural recording is necessary due to varying interface impedances resulting from electrode size and materials. The conventional recording technique involving the gain stage and ADC is progressively being substituted with direct conversion to the analog front end. This approach offers advantages such as high input impedance, low power consumption, and a compact footprint, making it a promising direction for future development [347].

Table 6.4 presents a comparison of neural recording circuits, considering the different bandwidth requirements based on the recorded signals. LFPs occupy a low-frequency band of 1–200 Hz, while APs occupy a higher frequency range of 200 Hz to 10 kHz. The input-referred noise directly affects the quality of the neural recording, and reducing input-referred noise improves the SNR. To preserve neural signal integrity, the input impedance (Z_{in}) of the neural recording circuit must be significantly higher than the electrode impedance ($Z_{in} > 1$ GΩ), with the electrode's DC current limited to 100 pA. However, a high-gain LNA (>40 dB) may lead to poor artifact tolerance, as large-scale artifacts can saturate the amplifier [347].

6.5 Summary

This chapter discusses the need for advanced tools to analyze brain activity and explores methods for measuring bioelectric activity along neurons' axons. Intracellular and extracellular approaches using microelectrodes capture external APs and LFPs. It highlights trade-offs between robust intracellular signals and long-term applicability, favoring extracellular measurements. The chapter covers different techniques, especially MEAs for electrical activity measurement in cultured cells. It emphasizes the utility of CMOS technology, particularly in HD-MEAs for detailed neural network studies. The importance of achieving effective cell adhesion patterns is underscored, and CMOS-based HD-MEAs' popularity in electrophysiological studies is highlighted. It mentions various readout schemes, including PPS, APS, and SM sensors, for capturing extracellular electrical activity *in vitro*.

Advancements in fabrication and semiconductor technologies have enabled the miniaturization of modern biosensor devices. This progress has involved integrating active electronic components on the same substrate as the electrodes in MEAs.

These recent MEAs represent the outcomes of research primarily focused on optimizing electrodes and recording performance [262,390]. This has involved exploring different electrode materials such as Au, Pt, black Pt, IrO_x, and TiN, as well as investigating planar or three-dimensional (3D) electrode morphologies and their interface properties [391,392]. More recent approaches have involved the use of nanomaterials like carbon nanotubes (CNTs) to enhance the recording and stimulation capabilities of microelectrodes [336,393–395]. CNTs exhibit biocompatibility and offer a suitable surface for neuronal cell adhesion and growth, regardless of surface uniformity [396,397]. Their minute nature is similar to that of bio-membranes, making them suitable for electrode fabrication [398].

Additionally, new materials such as nanoparticles (NPs), nanowires (NWs), organic semiconductors, conducting polymers, and graphene have been developed [399]. These materials offer unique functionalities, such as chemical sensitivity, which are valuable for various sensor applications. Novel fabrication techniques like inkjet printing have also enabled the patterning of biologically active materials for tissue engineering purposes [400,401]. With the increasing interest in bioelectronics, wearable electronics, and flexible electronics for sensor applications, significant progress has been made in applying flexible electronics technology to neural interfaces. Examples include silk-based conformal MEA, optical stimulators for optogenetics, flexible MEA for bladder control, and more [402,403]. Flexible electronics technology holds promise as a key technology for future ultra-small, ultra-dense, and ultra-soft *in vitro* neural interfaces.

The design of the electrode arrays has been adapted to suit various experimental models, ranging from cultured networks to brain tissues. Leveraging developments in microfluidics and biosensors, additional on-chip functionality has been incorporated to enhance the capabilities of *in vitro* recordings from cultured neurons [404] and brain slices [405]. These advancements have also contributed to organizing the topology of cultured neuronal networks [406–408].

Chapter 7

CMOS Bio/Chem field-effect transistor (Bio/ChemFET)-based biosensors

Field-effect transistors (FETs) have been receiving much attention due to their superior sensing characteristics. Bio/Chem field effect transistors (Bio/ChemFETs) refer to all families of FETs dedicated to sensing charge-induced field effects using various intermediate layers as biological/chemical interface on their gates such as ion-sensitive FETs (ISFETs), Enzyme-FETs, Gene-FETs, and Cell-FETs which are, respectively, employed for sensing ions, enzyme reactions, deoxyribonucleic acids (DNAs), and cells. Although the biological or chemical targets in these Bio/ChemFET-based sensors are different, their sensing mechanism is based on ISFETs' operation.

For the first time, ISFETs were introduced in the early 1970s by removing the metal layer on top of the oxide layer of metal-oxide-semiconductor FETs (MOSFETs) [409]. Primarily, it was believed that by this subtle variation, an electrolyte solution can be considered as the fluidic gate while the surface potential of the oxide layer can be manipulated by changing the voltage using a reference electrode. Since the invention of ISFETs in 1976, various structures of ISFETs have been introduced for different biological and chemical samples, leading to the development of various Bio/ChemFET structures [410].

In general, all Bio/ChemFETs include an electrolyte containing the target species in an ionic solution (KCl, PBS, NaCl, etc.) buffer. The presence of the target species and their concentration/quantity induces a charge on the sensor which results in a change in the surface potential leading to a detectable electron current in the conductive channel of the Bio/ChemFET-based sensor. The output signal can be processed after physical sensing of the activities on the surface [411].

7.1 Structures and fabrication of Bio/ChemFET-based sensors

In 1972, Piet Bergveld introduced the ISFET technology and tested it for pH measurements of NaCl solution. Within the next 50 years, various geometries have been developed that can be categorized into six groups: (1) conventional oxide-electrolyte structures, (2) floating gate structures, (3) extended gate structures, (4) double gate structures, (5) top gated structures, and (6) standard complementary metal-oxide semiconductor (CMOS) BioChemFETs. These structures have been the basis of other BioChemFETs such as carbon nanotube-FETs (CNT-FETs), graphene FETs (GFETs), nanowire-FETs (NW-FETs), or other new sensing materials like metal-organic frameworks (MOFs) and MoS_2 [412–414]. Here, the focus is on CMOS Bio/ChemFETs. More information about the other structures of Bio/ChemFETs can be found in [415].

The development of ISFETs using the well-matured CMOS technology was a smart step toward the development of standard ISFETs [416,417]. Employing unmodified CMOS structures for the fabrication of ISFETs makes this structure comparative compared to other Bio/ChemFET structures because CMOS adaption makes this type of sensor very low-powered and scalable as two main features which are important for implementing lab-on-chip (LoC) platforms.

In the CMOS ISFET structure shown in Figure 7.1, the elongated gate of a transistor is connected to the solution which is perfectly separated from the sensor area. Using the metal connectors buried in the oxide layer of CMOS technology, the gates can be connected to ion-sensitive layers such as SiO_2, Si_3N_4, and alike [416] (Figure 7.2). For example, Mulgrew *et al.* [417], proposed a 16 × 16-array of these unchanged CMOS sensors for direct extracellular imaging (see Figure 7.3).

Some CMOS chemical FETs (ChemFETs) have also been reported in the literature used for chemical sensing. An instance of this is demonstrated in the work of Errachid *et al.* [418], where they introduced a needle-shaped ISFET sensor designed to explore minute regions within biological and chemical samples. The sensitivity of this sensor was measured to be 45 mV/pH. Georgiou and Toumazou [419] also modified a

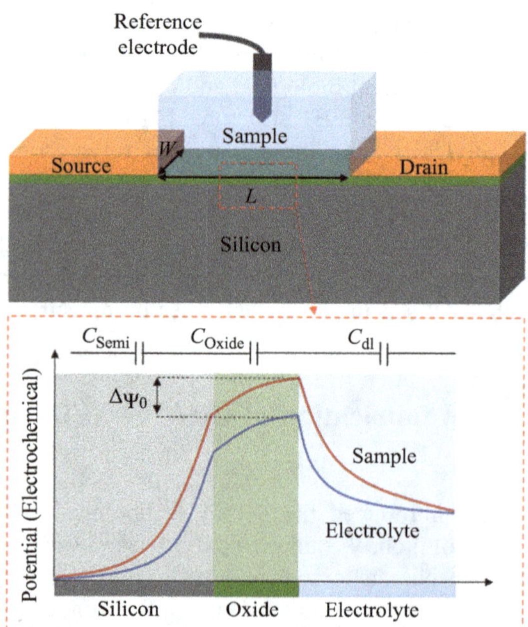

Figure 7.1 Standard structure of a Bio/Chem-FET

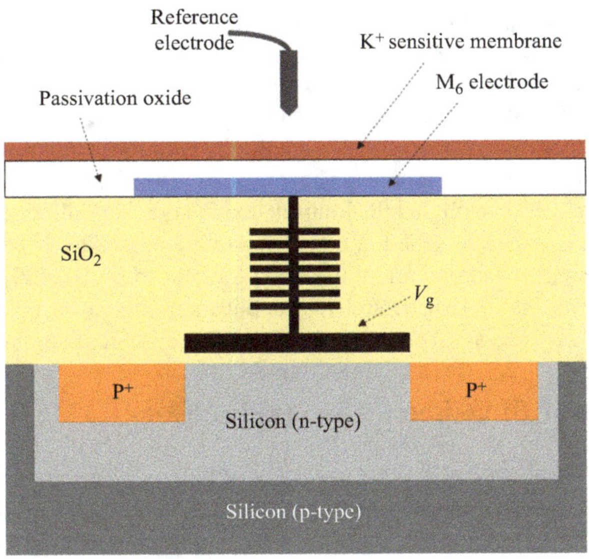

Figure 7.2 The standard CMOS ISFET structure coated with a potassium-sensitive membrane

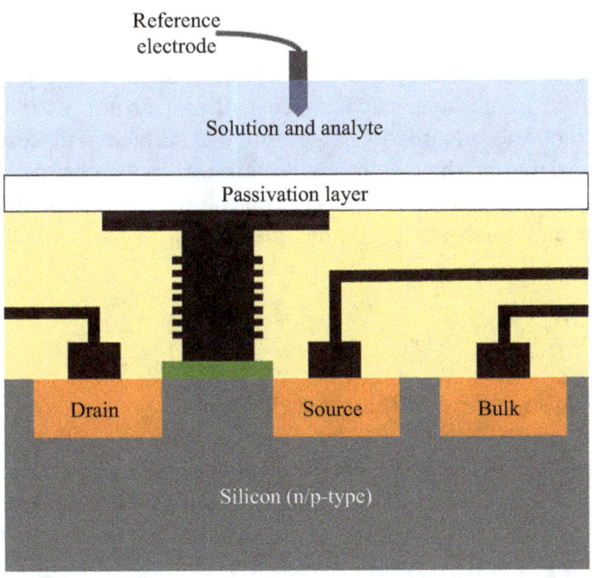

Figure 7.3 A standard CMOS ISFET for direct extracellular imaging

commercial 0.25 μm CMOS technology as a standard ISFET for pH sensing. To find important geometrical parameters affecting the operation of CMOS-based ISFETs, Sohbati *et al.* [420] presented the dimensional and shape analysis in another study. A common feature in CMOS-based ISFETs is using Si_3N_4 or SiO_2 or both which affects the ISFET sensitivity and parasitic errors due to the additional capacitance formed by these layers above the silicon channel. The unwanted capacitance due to the coverage of all parts of the sensor with this thick nitride/oxide layer can influence the threshold voltage of the sensor because the ion complexation/trapping sites on the nitrite/oxide surface will affect the electron current in silicon for next tries [419,420]. CMOS-based ISFETs have demonstrated successful real applications based on pH measurements such as genome sequencing [421], and detection of nucleic acid amplification [422].

7.2 Physics of Bio/ChemFET-based sensors

Adapting Bio/ChemFETs to CMOS technology helps to miniature them together with the required readout circuits including multiplexers, signal amplifiers, filters, and analog-to-digital converters (ADCs) on a single chip which offers the great advantages of CMOS-based biosensors. In 1999, Bausells *et al.* [423] showed the possibility of the fabrication of ISFETs in an unmodified CMOS process (standard CMOS process) which follows the same steps required for MOSFET fabrication. Physico-chemical models were also developed as SPICE built-in models to characterize ISFETs on which a layer of the ion-sensitive membrane was in interaction with the analyte [424–429].

As illustrated in Figure 7.1, the interaction of the oxide layer and the ions causes the double-layer capacitance (C_{dl}) to be created in the proximity of the oxide. The built-in oxide layer capacitance will exit on top of the channel. The depletion effect of the adsorbed charges on the surface will result in another capacitance modulated by the electron/hole transport in the channels as well as the surface charge density. The charge effects on the oxide layer surface lead to the drain-source current which can be expressed as (7.1) for the non-saturated (linear) region ($V_{DS} < V_{GS} - V_{Th(ISFET)}$):

$$I_D = \mu C_{ox} \frac{W}{L} \left\{ \left[V_{GS} - V_{th(ISFET)} \right] V_{DS} - \frac{1}{2} {V_{DS}}^2 \right\} \quad (7.1)$$

where μ is average mobility in the channel; W and L, respectively, stand for the gate's width and length; V_{DS} and V_{GS}, respectively, denote the drain-source voltage and the gate-source voltage; C_{ox} is the capacitance of the gate oxide. $V_{Th(ISFET)}$ identifies the threshold voltage of the ISFET, which can be expressed as:

$$V_{th(ISFET)} = E_{ref} - \Psi_o + \chi^{sol} - \frac{\Phi_{Si}}{q} - \frac{Q_{ox} + Q_{ss}}{C_{ox}} - \frac{Q_B}{C_{ox}} + 2\Phi_f \quad (7.2)$$

where E_{ref} denotes the potential of the reference electrode; χ^{sol} is the surface dipole potential of the solution; Φ_{Si} stands for silicon electron work function; q is the elementary charge; Q_{ox}, Q_{ss}, and Q_B denote the charges in the oxide, charges in surface states and interface states, and the depletion charge, respectively; and Φ_f identifies the potential difference between the Fermi levels of doped and intrinsic silicon. Ψ_o identifies the insulator-electrolyte potential which is the difference between the potential of the oxide surface Ψ_S and the bulk of the solution Ψ_B:

$$\Psi_o = \Psi_S - \Psi_B \tag{7.3}$$

All of the parameters mentioned in (7.1) and (7.2) are constant values except Ψ_o which demonstrates the effect of surface potential on the oxide layer and affects the ISFET's threshold voltage ($V_{Th(ISFET)}$). The ionic solution and biomolecular content of the solution change the threshold voltage and consequently the current in the channel. According to the side-binding model [430], which describes the mechanism of oxide surface charge, the equilibrium between the amphoteric MOH surface sites and the H^+ ions in the solution can be described as follows:

$$MOH \leftrightarrow MO^- + H_B^+ \tag{7.4}$$

$$MOH_2^+ \leftrightarrow MOH + H_B^+ \tag{7.5}$$

where H_B^+ is the proton in the bulk of the solution and each reaction has an associated dissociation constant identified, respectively, by K_a and K_b. The relationship between the activity of protons in the vicinity of the insulator surface and bulk protons can be derived employing the Poisson–Boltzmann equation [431]:

$$a_{H_S^+} = a_{H_B^+} \exp(-q\Psi_o/KT) \tag{7.6}$$

The surface charge density can be obtained by [431]:

$$\begin{aligned} \sigma_o &= qN_S\left(\frac{a_{H_S^+}{}^2 - K_aK_b}{K_aK_b + K_b a_{H_S^+} + a_{H_S^+}{}^2}\right) + qN_S\left(\frac{a_{H_S^+}}{a_{H_S^+} + K_n}\right) \\ &= qN_S f_a(\psi_o, pH) + qN_S f_b(\psi_o, pH) \end{aligned} \tag{7.7}$$

where N_S and N_n show the surface densities of silanol sites and primary amine sites and K_n expresses the dissociation constant of primary amine sites. The charges and capacitances in the electrolyte can be described using the Gouy–Chapman-Stern model. An equal and opposite charge in the electrolyte balances the surface charge resulting in a double-layer charge, σ_{dl}, and differential double-layer capacitance, C_{Dif}, which is the resultant of the diffuse charge layer (Gouy–Chapman layer) and the Stern capacitance (Helmholtz layer). The Stern layer can be divided into two parts for SiO2 as the sensing layer including the outer Helmholtz plane (OHP) and inner Helmholtz plane (IHP) [431]. It can be written that:

$$\frac{\partial \sigma_0}{\partial pH_s} = -q\beta_{\text{int}} \tag{7.8}$$

where β_{int} represents intrinsic buffer capacity expressing the effect of a small change in proton activity on the surface charge density because of pH variation.

To develop a pH-sensitive FET model using the built-in MOSFET model, the electronic stage, like MOSFET, and the electrochemical stage, including the electrolyte-insulator interface, should be decoupled from each other. So, the charge neutrality of the structure is considered as follows:

$$\sigma_o + \sigma_{dl} + \sigma_S = 0 \tag{7.9}$$

where σ_s denotes charge in the semiconductor. Considering to be constant with pH and $\sigma_s << \sigma_o$ and $\sigma_s << \sigma_{dl}$ and σ_s, the equation can be reduced to (7.6) which enables us to decouple the electronic and electrochemical stages:

$$\sigma_o + \sigma_{dl} = 0 \tag{7.10}$$

According to the Gouy–Chapman-Stern model, the charge in diffuse can be expressed as:

$$\sigma_{dl} = -\sigma_o = -C_{Dif}\psi_o \tag{7.11}$$

The potential difference across Stern capacitance is $\Psi_2 - \Psi_0$. The potential Ψ_2 can be found as:

$$\psi_2 = \psi_0 - \frac{\sigma_0}{C_{\text{Stern}}} = \psi_0 - \frac{\sigma_0 x_H}{\varepsilon_r \varepsilon_0} \tag{7.12}$$

where x_H is the distance of the stern layer from the surface. Differential double-layer capacitance shows the ability of the double layer to store charge in response to a small variation in potential and can be expressed as:

$$C_{\text{Dif}} = \frac{\partial \sigma_o}{\partial \psi_o} = -\frac{\partial \sigma_{dl}}{\partial \psi_o} \tag{7.13}$$

It can be shown that C_{Dif} is equal to [431,432]:

$$C_{\text{Dif}} = \frac{\left(2\varepsilon_r\varepsilon_0 z^2 q^2 n^0 / k_B T\right)^{1/2} \cosh\left(zq\psi_2/2k_B T\right)}{1 + (x_2/\varepsilon_r\varepsilon_0)\left(2\varepsilon_r\varepsilon_0 z^2 q^2 n^0 / k_B T\right)^{1/2} \cosh\left(zq\psi_2/2k_B T\right)} \tag{7.14}$$

where ε_0 and ε_r are the vacuum and relative permittivity, respectively. n^0 identifies the number concentration of each ion in the bulk and z stands for the magnitude of the charge on the ions.

Combination of both sides of the double layer, (7.8) and (7.14), yields:

$$\frac{\partial \Psi_o}{\partial pH_s} = \frac{\partial \Psi_o}{\partial \sigma_0} \frac{\partial \sigma_0}{\partial pH_s} = \frac{-q\beta_{\text{int}}}{C_{\text{Dif}}} \tag{7.15}$$

$$\frac{\partial \Psi_o}{\partial pH_s} = -2.3 \frac{K_B T}{q} \alpha \tag{7.16}$$

where α denotes a dimensionless sensitivity parameter and is equal to $\alpha = 1/\left(\left(\frac{2.3K_B T C_{\text{Dif}}}{q^2 \beta_{\text{int}}}\right) + 1\right)$. α identifies a constant parameter between 0 and 1 and highly depends on the differential capacitance and the intrinsic buffer capacity. K_B stands for Boltzmann constant, and T is the absolute temperature [432].

A small variation in the electrolyte ion concentration, due to either a chemical reaction in a biological sample or the existence of charged molecules in the solution, can change the surface potential. Generally speaking, the addition of a specific concentration of an analyte changes the pH of the solution, which can be correlated to the surface potential using (7.15). As a result, the drain-source current in the conduction channel is affected. Thus, the threshold voltage of an ISFET can be expressed as:

$$V_{\text{th(ISFET)}} = V_{\text{th(MOS)}} + (\gamma + 2.3\alpha V_T pH) \tag{7.17}$$

where γ denotes a pH-independent grouping of chemical potentials; $V_T = K_B T/q$ is thermal voltage; and $V_{\text{Th(MOS)}}$ stands for the threshold voltage of a conventional MOSFET.

Regardless of the channel-gates arrangement and geometry, these principles can be applied to semiconductor-oxide-electrolyte Bio/ChemFETs. Nonetheless, the principle might not work for some specific nanomaterial-based Bio/ChemFETs which use the sensing properties of nanomaterials such as carbon nanotubes (CNTs) and graphene because site-bonding theory might not be able to describe the behavior of the ion-complexation on these surfaces. Due to difficulties of mathematical modeling of the molecular interaction at the oxide or nitride surface, it is impossible to predict surface potential like what is provided by site-bonding theory. Recently, in terms of modeling the interaction in Bio/ChemFET surface complex systems, atomistic molecular dynamics simulation has been employed to analyze the ion complexation on the surface and evaluate the surface potential on the channel [433–435]. Molecular analysis has opened new horizons to get insight into the molecular interactions at the channel surface and the double-layer capacitance considering the interaction of oxide, nanomaterials, ionic solution, and proteins on the surface, which is not practically possible with current models.

7.3 Operational regions of ion-sensitive field-effect transistors (ISFETs)

Different operational regions of the ISFETs can be categorized into four groups: (1) linear or triode region (Equation (7.1)), (2) saturation or strong inversion, (3) or subthreshold or weak inversion; and (4) velocity saturation. As above mentioned, (7.1) expresses the drain current in the linear or triode region. When an ISFET is in the saturation region ($V_{DS} > V_{GS} - V_{\text{Th(ISFET)}}$), its drain current can be given by:

$$I_D = \mu C_{ox} \frac{W}{L} \left(V_{GS} - V_{\text{Th(ISFET)}}\right)^2 (1 + \lambda V_{DS}) \tag{7.18}$$

where λ denotes the channel length modulation factor. If I_D and V_{DS} both are held constant, V_{GS} will change to compensate for any variations in $V_{\text{Th(ISFET)}}$.

By considering a zero bulk-source voltage and $V_{DS} > 4V_T$, the operation of an ISFET in the weak inversion region can be expressed by [436]:

$$I_D = I_0 \frac{W}{L} \exp\left(\frac{V_{GS} - V_{Th(ISFET)}}{nV_T}\right) \tag{7.19}$$

where I_0 stands for a positive constant current and n denotes the subthreshold slope parameter.

In the region of velocity saturation, the behavior of the device can be characterized as stated in [437]:

$$I_D = v_{sat} C_{ox} W \left(V_{GS} - V_{Th(ISFET)} - V_{D,sat}\right) \tag{7.20}$$

where v_{sat} and $V_{D,sat}$ are, respectively, the saturated carrier velocity and the point when the drain current saturates.

7.4 Readout circuits of Bio/ChemFET-based sensors

The parameters of Bio/ChemFETs can be controlled and measured using electrical devices which might be off-chip instruments, such as a potentiostat and a semiconductor parameter analyzer [438–441], or on-chip integrated circuits [7,25,416,442,443]. In 1999, Bausells *et al.* [423] showed that it is possible to integrate ISFET sensors including both sensing and electronic functions in a standard CMOS process and proposed an integrated ISFET-amplifier by using standard cells of this process. Table 7.1 compares different readout circuits for this type of electrochemical sensor. Most of the circuits reviewed in this table are designed for ISFETs because many efforts have been made to fabricate this type of Bio/ChemFETs in CMOS technology.

The expression for the output current of the ISFET in the linear region was given by (7.1), where the threshold voltage can be determined using (7.17). However, the actual drain-source voltage and the gate-source voltage of the ISFET are, respectively, equal to $V_{DS,ISFET} = V_{DS}\text{-}(R_s + R_d)I_D$ and $V_{GS,ISFET} = V_{GS}\text{-}R_sI_D$ where R_s and R_d are the internal source and drain resistances formed based on the actual geometry of the device which is in contact with electrolyte with no metal in its vicinity. So, the sensitivity is practically declined [444]. Shortening the length of the source and drain diffusions is one of the technological approaches to mitigate this issue. But it implies new technologies while it is easier to design readout circuits insensitive to series resistances. If an ISFET is biased at a constant I_D with a constant drain-source voltage, V_{DS}, any change in threshold voltage because of pH variation will be compensated by adjusting the gate-source voltage, V_{GS} [444].

As seen in Table 7.1, depending on the operation region of Bio/ChemFETs, various readout circuits can be designed. Linear operation mode is usually reported for current-mode readout circuits. In triode ISFETs with a low V_{DS}, the output current, shown in (7.1), can be approximated by [437]:

$$I_D = \mu C_{ox} \frac{W}{L} \left[\left(V_{GS} - V_{th(ISFET)}\right) V_{DS}\right] \tag{7.21}$$

Table 7.1 Different readout circuits for Bio/ChemFET sensors

CMOS Tech.	Array #	Diff.	Operational region of FET	Configuration	Output signal	Resolution	|Sensitivity|	IDR	Sensing area (μm²)	Pixel Size (μm²)	Total area (mm²)	Power (mW) (Supply voltage)	Ref.
1.6 μm	1	No	Sat.	Wheatstone-bridge, variable V_{Ref}	V	NA	58 mV/pH	NA	NA	NA	4 × 4	NA	[458]
NA	1	No	Tr.	CVCC	V	NA	NA	NA	NA	NA	NA	NA	[444]
0.35 μm	1.5 M 7.2 M 13 M	NA	NA	CVCC	NA	NA	NA	NA	NA	NA	10.6 × 10.9 17.5 × 17.5 17.5 × 17.5	NA	[25]
1.0 μm	1	No	NA	CVCC, Feedback	V	NA	47 mV/pH	pH: 2.5 to 9.2	NA	NA	NA	NA	[423]
2.5 μm	NA	No	Tr.	CVCC	V	NA	58 mV/pH	pH: 3 to 11	NA	NA	< 0.25	10 (9 V)	[459]
0.18 μm	6	No	Tr.	CVCC	V	NA	NA	NA	NA	NA	1.5 × 0.6	NA	[416]
0.35 μm	64 × 64	No	NA	CVCC, SPT	V	NA	20 mV/pH	pH: 4 to 10	NA	10.2 × 10.2	0.7158 × 0.7158	NA	[460]
0.35 μm	64 × 64	No	Tr.	CVCC, APS	D	NA	9.23 mV/pH	NA	NA	96	0.56	NA	[442]
5 μm	10 × 10	No	NA	CVCC, APS	V	NA	229 mV/pH	pH: 4 to 9.1	2,000 × 2,000 (total)	200 × 200	5.1 × 5.1	NA	[461]
0.15 μm	256 × 256	No	Sat.	APS	D	0.06 pH	55.1 mV/pH	pH: 3 to 10	1.42 × 1.42	NA	5.2 × 5.2	271 mW (3.3 V)	[462]
1P1M	32 × 32	No	NA	APS	V	NA	450 mV/pH	NA	42.5 ×37.5	NA	7.4 × 8.0	NA	[463, 464]
2 μm	1024 × 1024	No	NA	CVCC, Charge transfer APS	V	NA	29.8 mV/pH	pH: 2 to 10	NA	23.55 × 23.55	14.8 × 14.8	NA	[465]
0.35 μm	3 × 11	No	Sub.	CVCC, integrate-and-fire topology, AER	D	NA	7.73 dB/pH	pH: 1 to 14	57.5 × 57.5	80 × 100	NA	(EP: 157 nW) (3.3 V)	[466]
0.35 μm	8 × 4	No	NA	CVCC, Feedback to the gate	V	60.3 mpH	42.1 mV/pH	pH: 1 to 14	NA	60 × 70	2 × 2.5	SFE:4.841 × 10^{-4} (3.3 V)	[450]
0.35 μm	1	No	Sat.	CVCC, Feedback, PG	V	NA	200 mV/pH	NA	30 × 100	NA	0.6 × 0.5	NA	[449]
1 μm	4 × 16	Yes	NA	Bio-amplifier	V	NA	50 mV/pH (2.8 to 54 mV)	pH: 5 to 11 (40 dB from 1.1 to 110 ng/μL)	90 × 90	NA	16.5 × 16.5	40 (5 V)	[467]
0.35 μm	1	No	NA	CVCC, VCO	F	NA	78 kHz/pH	pH: 0 to 7	NA	NA	0.045	0.12	[468]
0.35 μm	64 × 128	No	Tr.	CVCC, APS	D	NA	9.23 mV/pH	NA	9.3 × 10.3	NA	2×4	NA	[469]
	64 × 200		VS	I-M			1.033 μA/pH		6.5 × 7.775				
	64 × 200		VS	I-M, PG			0.717 μA/pH		6.5 × 7.775				

(Continues)

Table 7.1 (Continued)

CMOS Tech.	Array #	Diff.	Operational region of FET	Configuration	Output signal	Resolution	|Sensitivity|	IDR	Sensing area (μm^2)	Pixel Size (μm^2)	Total area (mm^2)	Power (mW) (Supply voltage)	Ref.
0.18 μm	16 × 8	No	Sat.	I-M, P-N complementary source follower, SC	D	0.014 and 0.017 pH	30 mV/pH	NA	NA	50 × 50	1.7 × 2	0.98 (AB: 3.3 V, DB: 3.3, 1.8 V)	[470]
NA	1	No	Sub.	I-M, Current feedback	V	NA	49.4 mV/pH	pH: 4 to 9	NA	NA	NA	NA	[471]
0.35 μm	3 × 3	No	Sub	I-M, PG, RO	F	0.008 pH	6 to 8 kHz/pH	pH: 5 to 7	55 × 65	64 × 54	0.1089	6 × 10^{-3}	[472]
0.35 μm	128 × 128	No	Tr.	I-M, CC, Auto-zeroing, S/H	D	0.24 pH (@1000 fps) 0.45 pH (@3000 fps)	50 LSBs/pH	pH: 4 to 10	NA	18 × 12.5	2.6 × 2.2	376 (3.3 V)	[473]
0.35 μm	8 × 8	No	Sub.	V-M, PG, Optic., MM	D	NA	57 mV/pH	pH: 4 to 10	NA	NA	NA	NA	[474]
0.18 μm	64 × 64	No	NA	pH-TC, Optic.	D	NA	26.2 mV/pH (G = 1) −103.8 mV/pH(G = 4)	pH: 1 to 14	Chem.:22.3Opt.:20.1	10 × 10	2.5 × 5	105.6	[475]
65 nm	512 × 128	No	Sub.	pH-TVC	D	0.01 pH	123.8 mV/pH	pH: 2.5 to 11.5	15	4.4. × 4.4	512 × 128	PA: 80.6 (2.8 V), AB: 108.4 (2.8 V), DB: 6.5 (1.2 V)	[7]
0.35 μm	8 × 8	No	Sub.	pH-TC, APS, PWM	t	33 mpH	6.1 μs/pH	NA	NA	16.5 × 16.25	6.7	8.3 (3.3 V)	[476]
0.35 μm	1	No	Tr.	pH-TC (Sawtooth oscillator)	t	0.1604	2.74 %pH	(0.4860 V)	NA	110 × 130	NA	6.899 (I_{bias}=1 μA)	[477]
				pH-FC (CCRO)	F	0.036	2.76 %pH	(0.5038 V)		75 × 75		655 × 10^{-6} (0.7 V)	
0.18 μm	3 × 3	No	NA	pH-TC	D	0.028 pH	27ns/pH	NA	10 × 10	NA	0.036	0.23 (1.8 V)	[478]
0.18 μm	32 × 32	No	Sub.	pH-TC	D	0.013 pH	30 mV/pH	NA	22 × 22	26 × 26	0.9	11.286 @1 kHz (1.8 V)	[479]
0.35 μm	64	No	NA	pH-FC (Oscillator)	F	(Near pM level)	53mV/pH	1 pM to 10 nM	30 × 30	NA	NA	NA	[480]
0.18 μm	8 × 8	No	Sub.	pH-FC	F	NA	37 mV/pH	pH: 4 to 10	NA	NA	2.6	0.076 (1.8 V/ 3.3 V)	[481]
0.35 μm	NA	No	Near-threshold	pH-FC (Comparator)	F	NA	32 mV/pH	NA	W/L=25 μm/0.5 μm, 50 μm/0.5 μm	NA	NA	NA	[482]
0.35 μm	8	No	Tr.	pH-FC (I-to-F converter+ ΣΔ ADC)	D	NA	NA	pH: 3 to 10	NA	NA	3.4 × 4.4	I2F: 0.228, ΣΔ ADC: 1.2	[483]

0.35 μm	1	No	Sub.	ISFET logic	V	0.5 pH	3,700 mV/pH	pH: 3.7 to 10.95	95 × 200	NA	NA	NA	[484]
0.35 μm	8 × 8	No	Sat.	ISFET logic	V	0.5 pH	50 mV/pH	pH: 1 to 14	NA	NA	NA	NA	[485]
0.35 μm	4 pairs	Yes	Sat.	CVCC, PG, Feedback to the gate	D	NA	100 mV/pH	pH: 5 to 9	30 × 100	120 × 120	0.65 × 0.5	NA	[486]
0.35 μm	2 × 2	Yes	Tr.	ISFET/REFET diff., CVCC	D	NA	40 mV/pH	NA	11.6 × 11.6	NA	1.4 × 2.6	15 (3.3 V)	[487]
0.35 μm	16 × 16	Yes	Tr.	ISFET/REFET diff., CVCC	D	NA	46 mV/pH	pH: 3.28 to 7.22	11.6 × 11.6	12.8 × 12.8	NA	60 (3.3 V)	[417]
0.6 μm	1	Yes	Sat.	ISFET/REFET diff., CVCC	V	NA	400 mV/pH	6 pH	NA	NA	18.225	2.1 (3 V)	[488]
NA	1	Yes	NA	ISFET/REFET diff.	V	0.01 pH	40 to 43 mV/pH	pH: 4 to 9	NA	1650 × 2600	4.9 × 3.9	NA	[451]
2.5 μm	1	Yes.	Tr.	ISFET/REFET diff., CVCC	C	0.15 pH	0.3875 μA/pH	pH: 3 to 11	NA	NA	NA	NA	[489]
0.35 μm	1	Yes	Tr.	ISFET/ISFET diff., I-M	D	0.1 pH	800 mV/pH	pH: 5 to 9	NA	NA	NA	NA	[490]
0.35 μm	1	Yes	Sub.	Diff, Gilbert gain cell, I-M, Translinear	C	NA-	45 mV/pH	pH: 5 to 9	34 × 100	NA	2.5 × 2.81	1.65×10^{-4} (3.3 V)	[491]
0.35 μm	1	Yes	NA	Diff, direct charge accumulation	D	36 μV	NA	88.3 dB	71 × 71	1000 × 1640	4 × 5	NA	[443]

NA: Not available, V-M: Voltage-mode, I-M: Current-mode, VS: velocity saturation, PG: Programmable gate, Tr.: Triode, Sat.: Saturation, PA: Pixel array, AB: Analog blocks, DB: Digital blocks, pH-TVC: pH-to-time-to-voltage converter, pH-TC: pH-to-time converter, pH-FC: pH-to-frequency converter, Sub.: Subthreshold, IDR: Input dynamic range, F: Frequency, SFE: Single front end., CVCC: Constant-voltage constant-current, CC: Current conveyor, SPT: Standard pixel topology, APS: Active pixel sensor, AER: address event representation, V: Voltage, D: Digital, C: Current, T: Time, EP: Each pixel, Opt.: Optical sensor, Chem.: Chemical, RO: Ring oscillator, S/H: Sample and hold, MM: Magnetic manipulation, G: Gain, W: Width of Bio/ChemFET, L: Length of Bio/ChemFET, SC: Switched capacitor.

where $V_{Th(ISFET)}$ is dependent on pH based on (7.17). Thus, I_D shows a linear sensitivity to pH [437]:

$$\frac{dI_D}{dpH} = g_m(pH) = -2.3\alpha\mu C_{ox}\frac{W}{L}V_T V_{DS} \tag{7.22}$$

The gain requirements determine the width of the channel (W) biased in this region but there is a compromise between the gain and the occupied area, especially in array structures where smaller footprints are more desirable to achieve a suitable array density. On the other hand, the length of the transistor (L) should be long enough to decrease the short-channel effects and the second-order sources of non-linearity (like mobility degradation).

The technique for the assessment of the threshold voltage of saturated ISFETs is similar to triode ones. Moreover, the ISFETs operating in weak inversion are suitable for designing low-voltage low-power circuits where ISFETs are usually used as a translinear element.

Velocity saturation, often regarded as a short-channel effect, can also be viewed as a linear mode of operation, making it suitable for the design of current-mode readout circuits. If $V_{Th(ISFET)}$ follows (7.17), the ISFET shows a linear sensitivity to pH [437]:

$$\frac{dI_D}{dpH} = g_m(pH) = -2.3\alpha V_T v_{sat} C_{ox} W \tag{7.23}$$

In comparison to (7.22), it can be seen that in contrast to the triode region, the operation of the transistor in velocity saturation is insensitive to mobility (μ), length (L) mismatches, and V_{DS}, especially when the minimum length is employed. In velocity saturation, the transistor is typically designed as a short-channel device, and it is advisable to employ a narrow width to restrict the biasing current, as recommended in Reference [437].

Some non-idealities in Bio/ChemFETs should be considered during a design, including:

1) Trapped charges
   One of the non-idealities that can cause a large unpredictable variation in the threshold voltage of ISFETs is the accumulation of trapped charge on the gate and passivation layer that happens during the fabrication of floating gate ISFETs in the standard CMOS process. The most convenient approaches to remove the excess charge are ultraviolet (UV) radiation and bulk substrate biasing. UV radiation can penetrate the inner polysilicon by utilizing a small aperture positioned on top of the ISFET. This UV exposure stimulates the trapped charge, causing it to migrate into the bulk region. In this method, it is not required to use any additional circuitry. This technique can be utilized for matching the electrical characteristics of ISFET pixels in an array architecture. Elimination of the mismatch between pixels' threshold voltage makes it possible to use a single reference electrode for biasing the entire array [445].

   Hot-electron injection is another method that can be adapted for threshold voltage programming. By adding the electron to the floating gate and increasing the threshold voltage, the operating point of ISFET can be calibrated [446]. However, this method cannot decrease the threshold voltage. To eliminate electrons from the floating gate of the ISFET, Al-Ahdal *et al.* [447] suggested using electron tunneling.

Another approach was proposed by Georgiou *et al.* [448] to program the floating gate of the ISFET by employing a capacitively coupled floating gate. This technique allows the tenability of the operation of the device and the operating point within a tolerable gate voltage range by adjusting the operating point. In a study conducted by Yan *et al.* [449], a feedback configuration was employed, utilizing two additional programmable nodes. Utilizing a switch for extra connections to the floating gate of the ISFET in a feedback loop is another technique that is reported by Hu *et al.* [450].

2) Temperature sensitivity

Temperature sensitivity is another factor affecting BioChemFET's performance. One solution involves using circuit compensation techniques to mitigate the effects of temperature changes.

A differential configuration of an operational amplifier (OpAmp) can be used to match the ISFET sensor with a MOSFET and remove the temperature sensitivity [451]. Equalizing the bias of two devices to ensure they exhibit the same temperature characteristics can be challenging, especially when dealing with a differential pair comprising a MOSFET and an ISFET on the same chip. This difficulty arises from variations in their threshold voltages, leading to potential differences in bias conditions [444].

Selecting a proper bias condition and empirical determination of a thermal operating point is another approach that can be used to cancel the ISFET temperature coefficient which is dependent on the operating point [452]. Simultaneous detection and employing the parameters of the measuring device causing the temperature drift is an unusual approach in conventional electronics. In an effort, Bergveld *et al.* [444] proposed to utilize a feedback system and inject a sinusoidal signal by using a transformer connected between the source and the bulk of the ISFET to tune a certain bias condition and generate a sensitivity independent of $\beta = \mu C_{ox}(W/L)$ and internal drain and source resistances. In another effort, a periodic gate reset was used by Sohbati *et al.* [453] to accurately tune the direct-current (DC) operating point of the ISFET.

To cancel out the positive temperature coefficient of the ISFET, Chin *et al.* [454] suggested employing a summing circuit and a P-N diode as a temperature sensor that has a negative temperature coefficient. Another technique was also reported by Chung *et al.* [455] to compensate for the ISFET thermal characteristic which is based on using a threshold voltage extractor circuit.

3) Noise

Another non-ideality affecting the sensor's resolution and accuracy is the noise performance of the ISFETs. Flicker noise is the dominant low-frequency noise. The measured noise characteristics in [456] demonstrated that the Si/SiO_2 interface dominates the noise behavior and the interface between the solution and the gate insulator does not contribute to measurable noise. To decrease the noise, the ISFETs based on p-MOSFET can be employed rather than n-MOSFET, because it is well established that the noise of p-MOSFET is lower by nearly two orders of magnitude [457]. However, n-type ISFETs show better drift performance and higher mobility.

The rest of this section is dedicated to different types of readout circuits and the techniques employed in each circuit for the amelioration of the above-mentioned non-idealities.

7.4.1 Non-differential measurements

Herein, the non-differential circuits are categorized into five groups including: (1) variable voltage reference in feedback mode, (2) constant-voltage constant-current (CVCC), (3) current-mode readout circuits, (4) ISFET logic, (5) pH-to-time conversion (pH-TC), and (6) pH-to-frequency conversion (pH-FC).

7.4.1.1 Variable voltage reference in feedback mode

In this technique, the variation in I_D is controlled through a feedback loop, and the reference electrode's potential is adjusted automatically with respect to the source and drain potentials. Thus, V_{Th} variations will be compensated by changing V_{gs} via the reference voltage of the reference electrode, V_{Ref}.

Figure 7.4 illustrates an automatically balanced bridge circuit [458] as an example of such a circuit that compensates the variations of V_{Th} by changing V_{gs} via the reference electrode. However, this approach forces the analyte to a specific voltage while it may not be grounded. The asymmetrical impedance of the drain and source of the device causes the system to be sensitive to the electric field and static electricity [444]. Moreover, this technique is not appropriate for measuring several ISFETs with one RE [492].

Figure 7.5 demonstrates a readout circuit with a programmable gate (PG)-ISFET [493]. The feedback loop of this circuit helps to compare and set the ISFET's current with a reference current, I_{Ref}, and the effective gate-source voltage of the ISFET adjusts V_{Out}. Another technique is to set the voltage reference in a feedback loop, as reported by Jamasb *et al.* [494].

This configuration is not suitable for array structures, because a distinct reference electrode is required for each pixel due to the inconstant biasing which increases the complexity of the sensor [495].

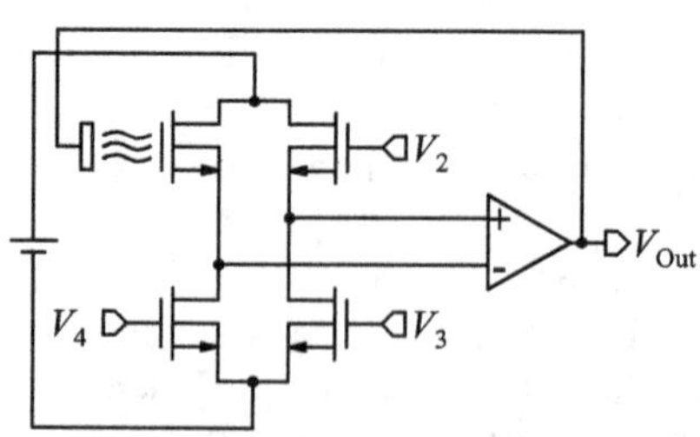

Figure 7.4 Direct Wheatstone-bridge

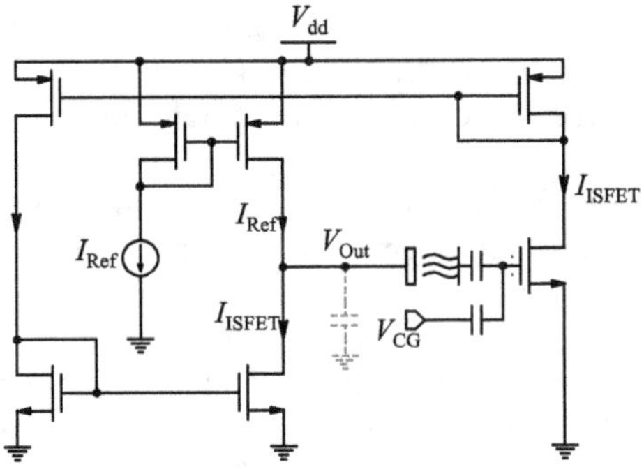

Figure 7.5 The circuit employed to assess the ISFET drift in the feedback mode

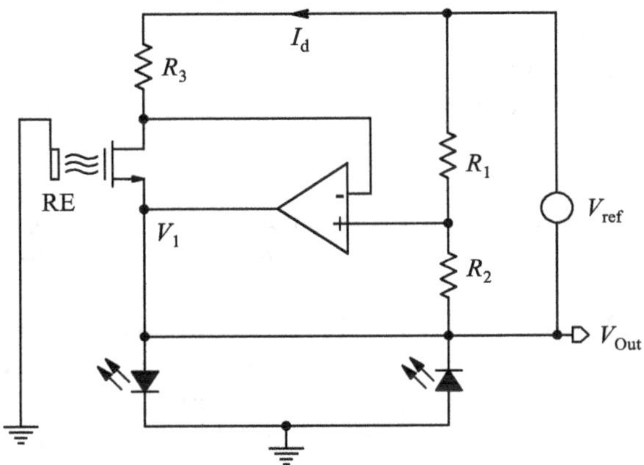

Figure 7.6 A source/drain follower readout circuit

7.4.1.2 Constant-voltage constant-current (CVCC)

This is the most popular configuration for the readout circuits of Bio/ChemFET sensors. In this technique which was proposed by Bergveld [444], drain voltage and current are constant and the source voltage changes with pH. In this approach, the source/drain potentials are automatically adjusted with respect to a constant liquid protentional. The best option is to ground the liquid using a grounded reference. One of the most popular circuits to provide these conditions and detect the changes in ISFET threshold voltage is a source/drain follower circuit [444]. The floating gate of the ISFET in this configuration is immune to the capacitive scaling of the pH. Furthermore, the array sensors based on the CVCC technique are compact and only require an ISFET and a switch for each pixel [495].

Figure 7.6 illustrates a simple source/drain follower which employs unsaturated ISFET [492]. Since the input voltages of the OpAmp are equal to each other and, V_{R1} and V_{R2} are parts of constant V_{ref}, $V_{R1} = V_{R3}$, and $V_{R2} = V_{ds}$ which are constant. Thus, $I_d = V_{R3}/R_3$ is also constant. So, V_{gs} automatically adjusts the source voltage to compensate for the variation in ISFET threshold voltage, V_{Th}, so that the output voltage accurately reflects the change in oxide surface potential. When the system is out of range, the diodes at the output light up, providing a limitation on V_{Out} and thus the source voltage.

Another source/drain follower is depicted in Figure 7.7 which uses the unsaturated ISFET and instrumentation amplifier (IA) concept [444]. The inputs of the IA are connected to a fixed potential equal to IR_1 and there is an inverse relationship between its output and the channel resistance of the ISFET. The difference between V_{ref}, which is an adjustable reference voltage, and the output voltage of the IA is amplified by an OpAmp. This amplifier injects a feedback current, I_f, into the resistor R_2 and provides a constant current and drain/source voltage $V_{ds} = IR_1$ by controlling the source voltage (V_{R2}) and drain voltage (V_{R1+R2}). An adjustable resistor R_9 measures I_f. The drain and source voltage (with respect to ground) follow a change in the effective input voltage V_1 which is amplified across R_9 and $V_{Out} = (R_9/R_2)V_1$. The reference voltage should be independent of temperature and adjusted to provide the conditions that $I_f = 0$ when the ISFET is in a neutral buffer (pH = 7). To measure the temperature and compensate for the temperature drift in $V_{gs}(T)$ for constant I_d, a circuit was also added to this readout circuit.

Nikkhoo *et al.* [416] integrated a source/drain follower along with potassium-sensitive FET employing CMOS technology for bacteria detection. They used the popular source/drain follower which includes a feedback loop composed of a fixed resistor and two OpAmps to provide the bias condition of CVCC (like what is shown in Figure 7.8). Considering both drain current and drain-source voltage are fixed and a fixed DC voltage is applied to the solution, the source voltage, as the output, follows the variations of the threshold voltage of ISFET. This output voltage can be measured using a buffered readout circuit and converted to digital. To

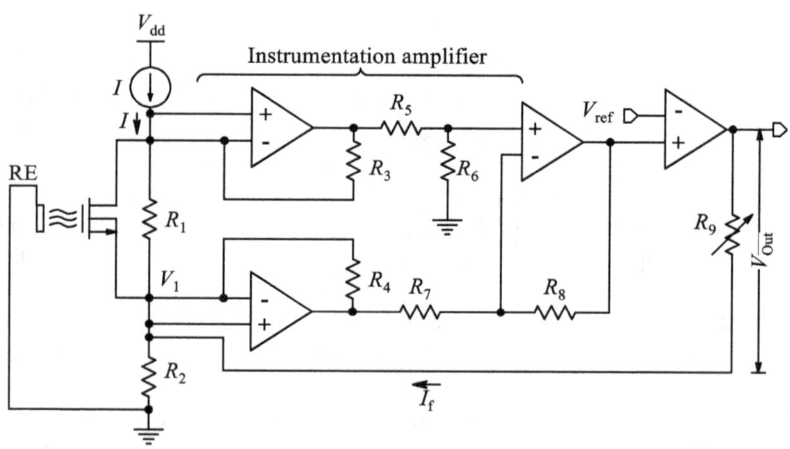

Figure 7.7 An ISFET amplifier with an additional circuit for temperature compensation

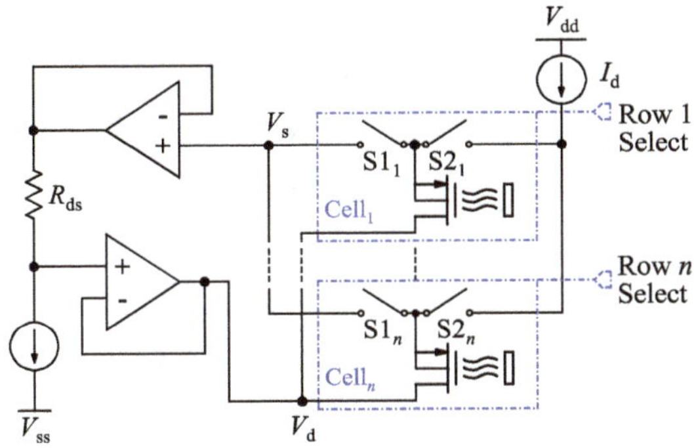

Figure 7.8 A dual-pixel schematic of an array based on a source/drain follower including a feedback loop composed of a fixed resistor and two OpAmps

provide the ISFET drain current with low dependence on its V_{ds}, Ravezzi *et al.* [459] proposed the use of cascode current mirrors. In their design, two OpAmps were used to bias the ISFET at the triode region, and a third OpAmp was employed as a follower to de-couple the output from the biasing circuit.

The input impedance of a source/drain follower prevents the circuit from loading the system during measurement. These circuits achieve excellent linearity by maintaining constant biasing and effectively suppressing parasitic capacitance from the drain, source, and bulk nodes to the floating gate. As a result, the output voltage remains closely aligned with the input measurement voltage [479]. However, using OpAmps results in a larger occupied area and high power consumption. This will be more critical when the readout circuit is going to be used for an array of sensors with a large number of pixels. An alternative in array structures is to share the readout circuit between pixels in the same row or column [417,487,488,496,497]. For example, a single column of n half-cells in an $n \times n$ array is illustrated in Figure 7.8. A 16×16-array of ISFET sensors with a similar readout architecture is reported by Milgrew *et al.* [417] for direct extracellular imaging. To achieve dense arrays with small pixels and high accuracy and sensitivity, Rothberg *et al.* [497] improved FET pixels in array structures and proposed different pixel designs based on both n-channel ISFETs and p-channel ISFETs. This group [25] reported a chip including 1.5M, 7.2M, and 13M ISFETs for simultaneous detection of independent DNA sequencing of different bacterial genomes (see Table 7.1).

Duan *et al.* [470] proposed a floating CMOS ISFET array with high offset distribution tolerance, cost-effective auto-compensation, and a high resolution of 0.017 pH. The readout circuit is based on a P-N complementary source follower which helps to extend the input range and obtain high offset distribution tolerance (Figure 7.9). Moreover, digitally common-mode voltage adjustment, which is performed before amplification and quantization, helps to compensate for the signal offsets from a wide range. Combining an amplifier and an ADC paved the way to achieve a high resolution.

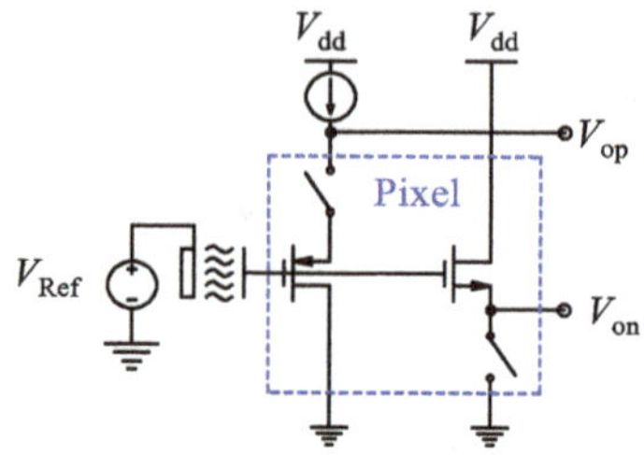

Figure 7.9 An ISFET front end based on complementary source follower

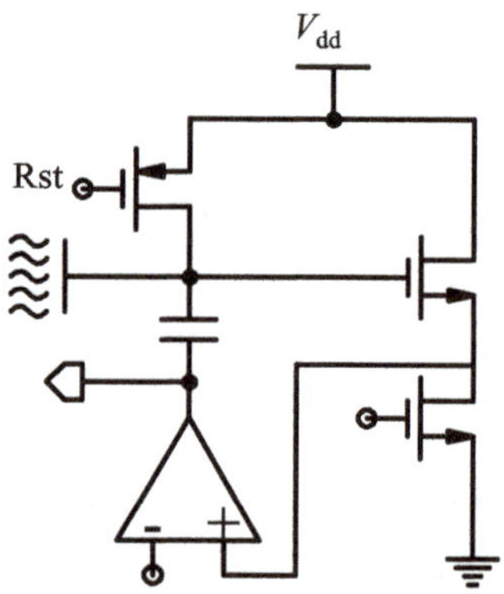

Figure 7.10 Closed-loop readout with ISFETs within the loop

In some other designs based on the CVCC method, the output node of the readout circuit is connected to the floating gate of an ISFET through a capacitor or switch. Since capacitive division and DC offset can result in trapped charge and transconductance reduction, Hu *et al.* [450] proposed a circuit using capacitive feedback with large open-loop gain and low-leakage switching scheme to suppress the variation on the floating gate and provide gain to the signal with high linearity (see Figure 7.10). In addition to the power and speed limitations imposed by the internal amplifiers used in this circuit, offset compensation is required before establishing the feedback loop.

7.4.1.3 Current-mode readout circuits

Several readout circuits [436,437,471,498,499] have been reported based on current-mode readout techniques to measure the output current of ISFET which is compatible with velocity saturation, as well as linear and weak inversion regions [495]. Moreover, many of these current-mode circuits take advantage of current conveyors (CCs) which play the role of buffer in the voltage-mode circuits.

The range of the drain current provided by the FETs operating in the weak inversion region is in the order of nanoamperes and below which makes them a suitable option for designing low-power sensors. Furthermore, this region can give us the maximum intrinsic voltage gain because it has the highest transconductance-to-current ratio [436]. The other beneficial features of this region include low required V_{ds}, low gate voltage, and minimum gate capacitance [419]. In addition, novel designs can be developed by using translinear circuits and bipolar circuit techniques such as log-domain filters due to the exponential relationship of V_{gs}-I_d.

For instance, Shepherd *et al.* [436,500] reported a hydrogen cell (H_{cell}) based on a translinear operation with an output current that is directly proportional to pH (see Figure 7.11). By considering $K_{chem} = \exp(-\gamma/nV_T)$ and (7.19) and setting $I_{b1} = I_{b2}$, the output current can be expressed as [436]:

$$I_{Out} = I_{b1}\exp\left(\frac{-2V_{ref}}{nV_T}\right)K_{chem}^2[H^+] \tag{7.24}$$

In another effort, a log domain filter with an ISFET used in its input, which was biased in weak inversion, was proposed by Georgiou *et al.* [499,501]. Figure 7.12(a) illustrates a first-order log-domain filter with this feature. Another configuration is depicted in Figure 7.12(b) where the drain current of the MOSFET operating in weak inversion is replaced by a chemically modulated signal. The filter parameters such as cut-off frequency and gain can be dependent on the ion concentration of the analyte.

As another example, Miscourides *et al.* [437] reported an array configuration whose pixels can be accessible by directly controlling the V_{ds} of each device. A generic CC is used in each drain row resulting in a constant voltage for that row (V_{DR}) guaranteeing that the generated output current depends only on V_{gs} during the readout of each pixel. They used the ISFETs biased in the velocity saturation region in an array where only one pixel is turned on at any time using a pixel switch. According to (7.22) and (7.23), the drain currents of the transistors operating in the velocity saturation region are higher compared to the currents in the triode region. Thus, it is unsustainable to output the currents of all column pixels at the same time. The same group [502] extended this current-mode circuit to dedicate a PG inside each pixel to compensate for the offset. As a continuation of these works, this group [469] designed the circuit shown in Figure 7.13 including a

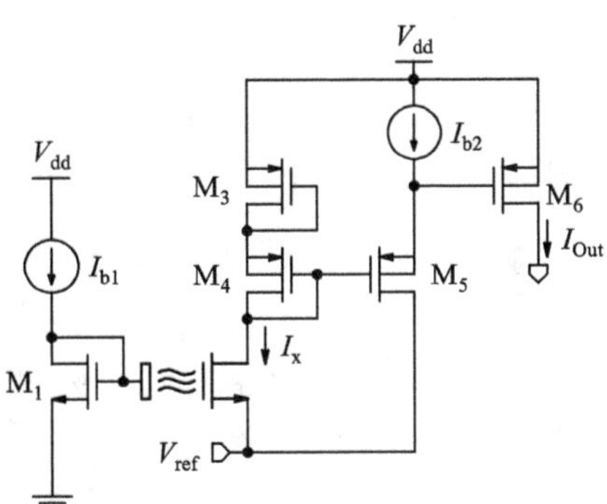

Figure 7.11 HCell with the translinear operation

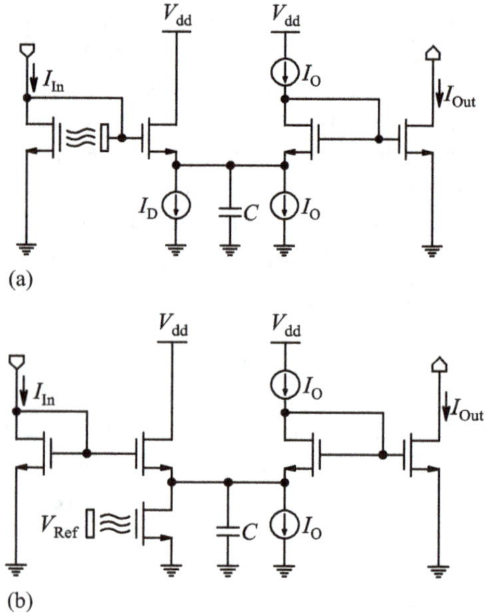

Figure 7.12 (a) A first-order log-domain filter with an ISFET employed at its input. (b) Chemically tunable log-domain filter

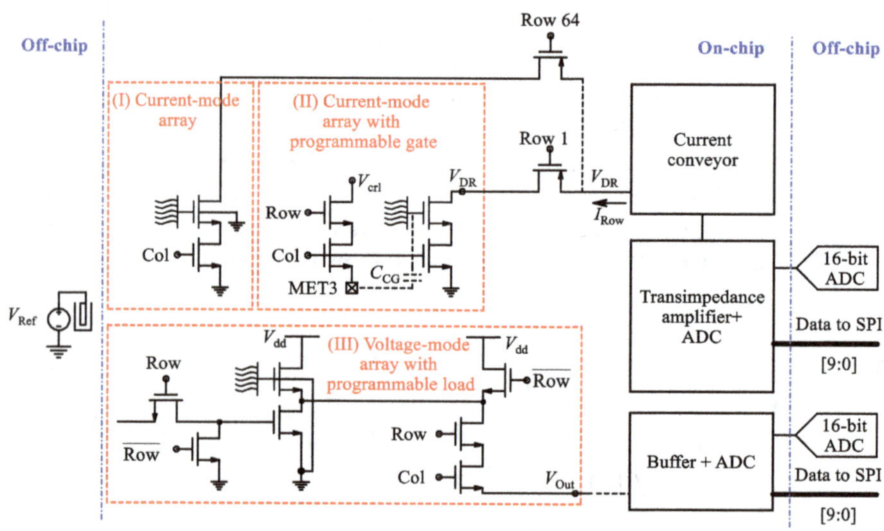

Figure 7.13 Schematic of a system comprising three configurations: (I) Current-mode configuration in a 64 × 200 array. (II) Current-mode PG pixel architecture in a 64 × 200-array. (III) Voltage-mode pixel structure in a 64 × 128-array

current-mode structure in a 64 × 200 array, current-mode PG pixel configuration in a 64 × 200 array, and voltage-mode pixel architecture in a 64 × 128 array to compare them. The two current-mode circuits have a smaller footprint, better linearity in comparison to the voltage-mode one, and provide the best overall performance. In another effort [503], they used the 64 × 200-array 2-transistor current-mode structure with a global reference electrode real-time ion imaging and detection of DNA amplification.

In another design, Toumazou *et al.* [504,505] described a chemical current conveyor type 2 (CCCII+) incorporating an input ISFET pH sensor transistor in the Y-node circuitry of the current conveyor type 2 (CCII+) (see Figure 7.14(a) and (b)). Figure 7.14(a) shows the operation of a CCII+ by its matrix. The matrix shows that there is a unity voltage gain between the input voltage at the Y-node and the output voltage at the X-node. Moreover, there is a unity inverting current gain between the input current into the X-node and the output current from the Z-node. This circuit exploits the g_m enhancement technique proposed in [506]. In another effort, Zeng *et al.* [473] proposed a 128 × 128 current-mode ISFET array with in-pixel calibration using a CC, auto-zeroing circuit, and sample and hold, as shown in Figure 7.15. The CC sets a fixed drain-source voltage across each ISFET pixel and mirrors the pixel current to subsequent stages.

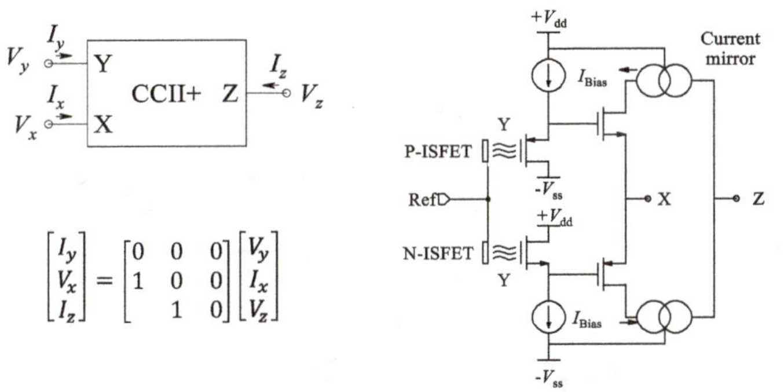

Figure 7.14 (a) A CCII+ and the hybrid matrix. (b) A CCCII+ using a P-ISFET and an N-ISFET

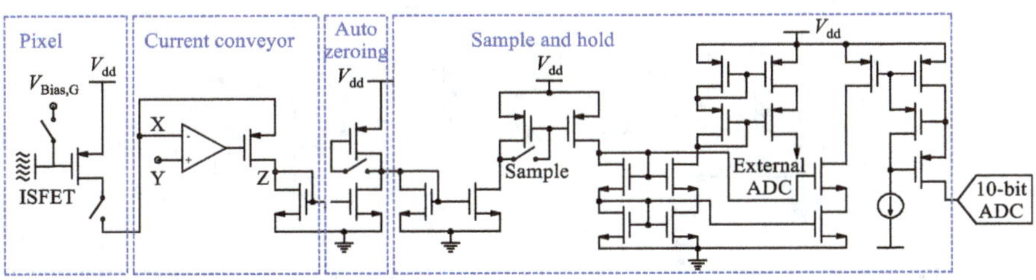

Figure 7.15 The circuit of the complete signal pipeline

To stabilize DC operation, current feedback is employed in some current-mode circuits [471,507,508]. For example, Figure 7.16 illustrates a current-mode readout with global feedback reported by Premanode *et al.* [471].

Another circuit is proposed by Juffali *et al.* [509] which works based on translinear operation using a division circuit and two ISFETs. Figure 7.17(a) shows the ISFET-based translinear divider. In another work, Kalofonou *et al.* [510] reported a ratiometric technique of DNA methylation by employing ISFET in a translinear circuit, shown in Figure 7.17(b).

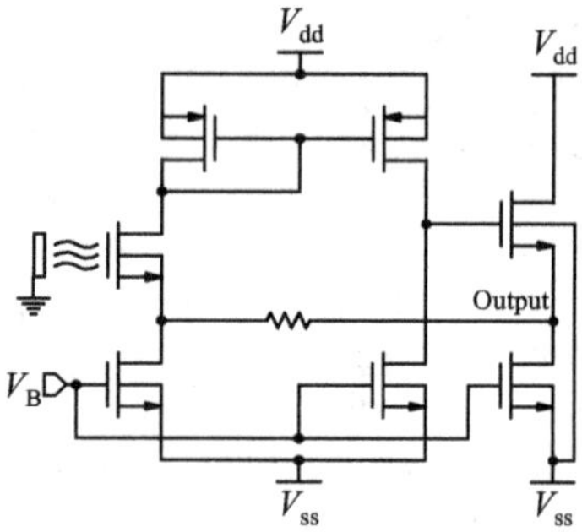

Figure 7.16 A current-mode readout with current feedback

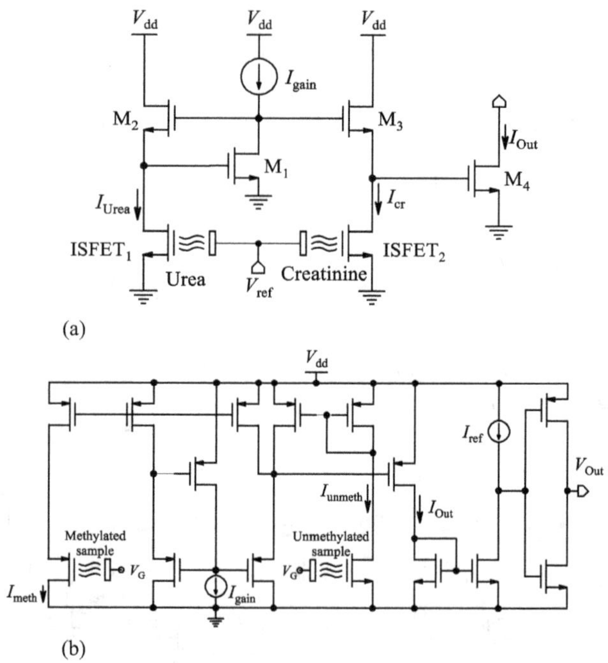

Figure 7.17 Ratiometric measurement: (a) an ISFET-based translinear divider and (b) a translinear circuit

7.4.1.4 ISFET logic

Discrete ISFET logic was proposed by Wong *et al.* [511] for the detection of chemical reaction thresholds without the need to measure the exact value of pH. For instance, the switching threshold of an inverter can provide a yes/no answer based on the pH changes. Other usable components as comparators include a Schmitt trigger [512], a NAND gate [40], or a switch [484,513].

Three kinds of configurations of an ISFET inverter are shown in Figure 7.18 (a–c) using only one p-type or n-type ISFET or a pair of both. The sensitivity of the threshold voltage of the three inverters shown in Figure 7.18(a), (b), and (c) are, respectively equal to [485]:

$$\frac{\partial V_{th,a}}{\partial pH} = \frac{\left(1+\sqrt{\frac{\beta_p}{\beta_n}}\right)2.303\alpha V_T}{\left(1+\sqrt{\frac{\beta_p}{\beta_n}}\right)} = 2.303\alpha V_T \tag{7.25}$$

$$\frac{\partial V_{th,b}}{\partial pH} = \frac{\sqrt{\frac{\beta_p}{\beta_n}}2.303\alpha V_T}{\left(1+\sqrt{\frac{\beta_p}{\beta_n}}\right)} \tag{7.26}$$

$$\frac{\partial V_{th,c}}{\partial pH} = \frac{2.303\alpha V_T}{\left(1+\sqrt{\frac{\beta_p}{\beta_n}}\right)} \tag{7.27}$$

where β_p/β_n stands for the ratio of P-channel metal-oxide semiconductor (PMOS) and N-channel metal-oxide semiconductor (NMOS) aspect ratios.

For instance, a complimentary pair of n-type and p-type ISFETs sharing the same ion-sensitive membrane is proposed by Al-Ahdal *et al.* [484,513], as an ISFET-based chemical switch which is depicted in Figure 7.18(a). In another work, these switches were used by Nabovati *et al.* [485] in each pixel of an 8 × 8 array, and a sensitivity of 50 mV/pH and a wide pH range from 1 to 14 was achieved (see Table 7.1).

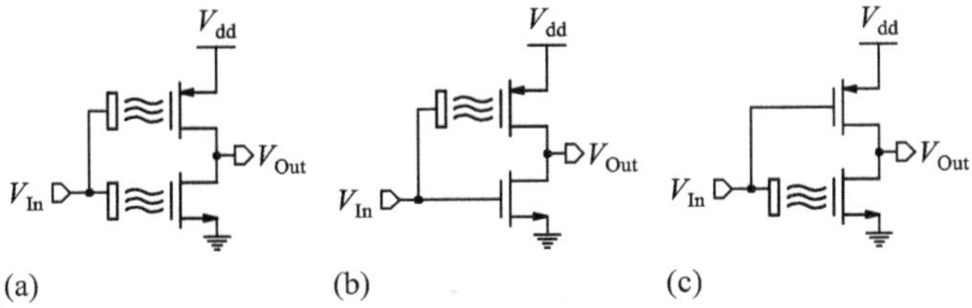

Figure 7.18 Three configurations of ISFET inverters: (a) n-type and p-type ISFET sharing the same membrane, (b) n-type MOSFET and p-type ISFET with membrane, (c) p-type MOSFET and n-type ISFET with membrane

7.4.1.5 pH-to-time conversion (pH-TC) and pH-to-frequency conversion (pH-FC)

One of the techniques reported for reading the output of ISFETs is converting pH to time and pulse-width modulation (PWM) using the aforementioned inverters based on complementary ISFETS as the core of a pH-to-time converter capacitively coupled to a triangular wave whose switching point and, consequently, its output pulse width is modulated with pH [514,515]. A completely time-based readout makes it possible to achieve a maximally-compact readout and overcome speed limitations.

PWM-based techniques are appropriate candidates for in-pixel quantization to improve the SNR [495]. However, special consideration is required for readout synchronization, and parasitics in a large array [479]. Figure 7.19 shows a PWM-based interface reported by Liu *et al.* [514]. This circuit needs extra amplification for data conversion, such that the noise in the following stages is directly referred to the chemical signal. This group [479] used in-pixel inverter-based sensors controlled by a triangular waveform in a 32 × 32-array to convert pH to time. An asynchronous column-wise time-to-digital converter (TDC) enables fast asynchronous conversion.

In an effort, an ISFET pH-TC is reported by Wang *et al.* [478] which is integrated with a TDC as the ADC with a thermometer-to-binary coder and a delay line. As another example, an active pixel sensor (APS)-based pH-TC was proposed by Moser *et al.* [476,516,517] which includes current discharging output nodes, as illustrated in Figure 7.20. A periodic reset scheme along with discharging a capacitor through the ISFET is used to modulate pH to pulse width.

Jiang *et al.* [7,518] used the subthreshold region of ISFETs for pH-to-time-to-voltage conversion as depicted in Figure 7.21. Digital double sampling and pixel-level calibration were used to decrease the non-ideal effects, such as trapped charge, parasitic capacitance, passivation attenuation, etc. In Figure 7.21, M_{n0} and M_{n1} are, respectively, an ISFET and a row-selected transistor. The other parts show the shared circuit by the pixels of each column. M_{p0} and S_0 are, in turn, a pre-charge switch and a transfer switch. An integrator, a buffer, and a biasing circuit are

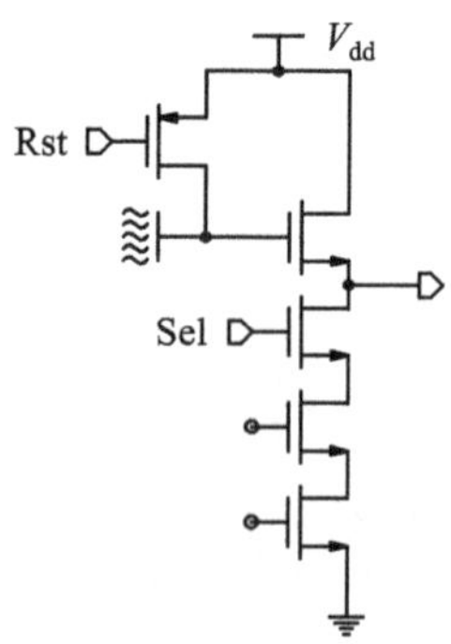

Figure 7.19 PWM-based interface with time output

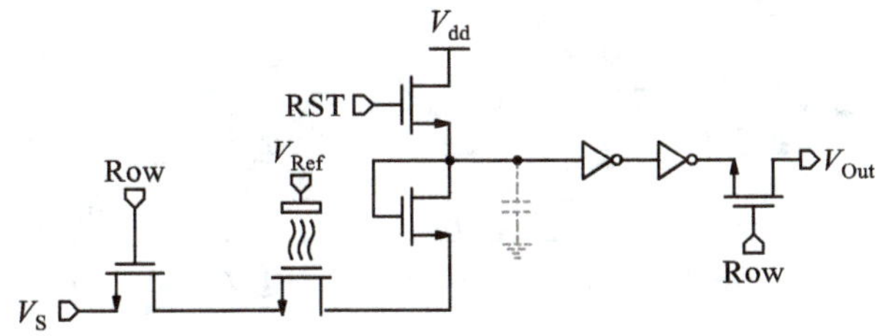

Figure 7.20 An APS-based pH-to-time converter including current discharging output nodes

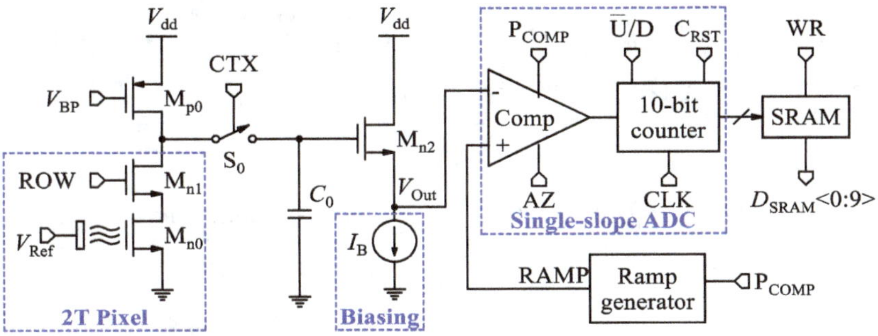

Figure 7.21 A readout circuit including a 10-bit voltage digital converter and a pH-TVC amplifier

also employed which are implemented by the integrating capacitor C_0, the source follower M_{n2} and the current source I_B shown in Figure 7.21. The pH-TVC readout circuit also includes a single-slope ADC, a ramp generator, and a static random-access memory (SRAM) group. pH variations cause a shift in the V_{Out} which is equal to:

$$\Delta V_{Out} = \left(\frac{\Delta t_{int}}{C_0}\right).G_m.A_{att}.\Delta V_{chem}, \tag{7.28}$$

where G_m, ΔV_{chem}, A_{att}, and Δt_{int} denote the ISFET transistor transconductance, pH-related potential change, attenuation factor, and the charge integration time, respectively. $(\Delta t_{int}/C_0)G_m$ is defined as the amplification factor which compensates for the passivation sensitivity attenuation effect. This group [7] utilized this sensor for *E. coli* detection, and as seen in Table 7.1, achieved the resolution and sensitivity of 0.01 pH and 118.7 mV/pH, respectively.

Cacho-Soblechero *et al.* [477] reported two oscillator-based sensors that convert the measured ion concentration into a chemically controlled duty cycle using PWM. One of these sensors, shown in Figure 7.22(a), includes a sawtooth oscillator front end which is suitable for compensation and robustness on safety-critical diagnostic platforms. The other one, depicted in Figure 7.22(b), takes advantage of

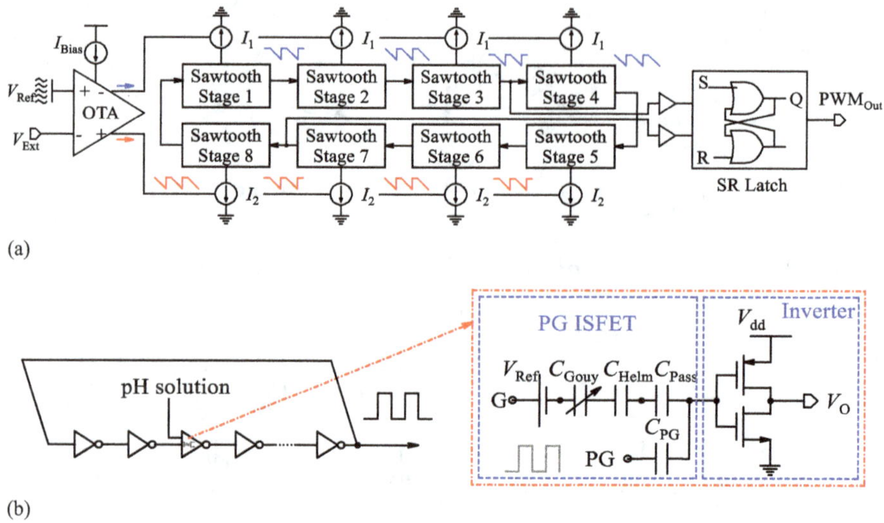

Figure 7.22 A pH-TCs based on: (a) sawtooth oscillator and (b) CCRO

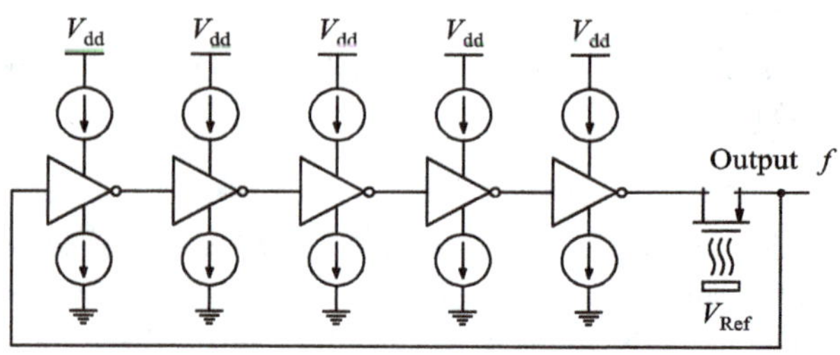

Figure 7.23 One pixel of a pH-FC with 64 sensors of four designs

a chemically controlled ring oscillator (CCRO) which offers the benefits of low power consumption and node scalability appropriate for point-of-care (PoC) devices. The difference in the propagation delay generated by the ISFET delay cell modulates both the oscillation frequency and duty cycle of the output signal. Duty cycle modulation offers the advantages of temperature insensitivity and linearity whereas frequency modulation suffers from correlation with temperature and non-linearity. Moreover, duty cycle modulation makes it possible to control the sensor's sensitivity by adjusting the voltage supply.

As above mentioned, some Bio/ChemFET readout circuits are designed based on the frequency modulation of their inputs. As another example, Lee *et al.* [480] reported a pH-FC using CMOS ISFETs with a capacitor structure in the oscillator readout circuit. As seen in Figure 7.23, this chip includes four designs, each of which has 16 ISFET sensors and operators from 1 MHz to nearly 100 MHz, respectively. During the measurement, only one sensor is activated by controlling the row and column decoders to avoid inaccuracy and signal drift due to cross-coupling and on-chip heating. The readout circuit consists of a five-stage ring oscillator with a p-type metal-oxide-semiconductor (MOS) capacitor as the load which controls the output frequency. Choosing various bias currents and inverter sizes leads to achieving a range of operating frequencies from the nominal low to high. The output of this circuit is connected to the drain and source terminals of MOS capacitor instead of the external reference electrode to prevent excessive loading from the sample impedance. All outputs of the sensors are processed through a shared frequency divider.

As part of their research, Georgiou *et al.* [466] introduced an integrated-and-fire configuration combined with address event representation (AER) to transform the pH-sensitive current into a frequency output. Although this circuit consumes low power (157 nW at pH = 7), the signal processing utilizing ADCs in the spike domain results in higher complexity. For robust measurement, the same group [472] employed a similar current-mode pixel architecture with an averaging mirror. Current-to-frequency conversion of this sensor was also performed by using a ring oscillator (see Figure 7.24).

Chang *et al.* [482] proposed another pH-FC including 16 ISFET sensors whose outputs are measured sequentially through a buffer with the timing controlled by row and column decoders. Figure 7.25 shows the readout circuit of each sensor where the drain-source current, which is varied by the interfacial potential change, is first integrated by C_{int}. After the resulting ramp signal is compared with the switching level of the comparator (V_{comp}), a reset signal is generated to discharge the capacitor through an n-type transistor. A delay stage including two inverters helps to provide sufficient pulse width for reset. Therefore, the frequency of reset pulses indicates the interfacial potential change.

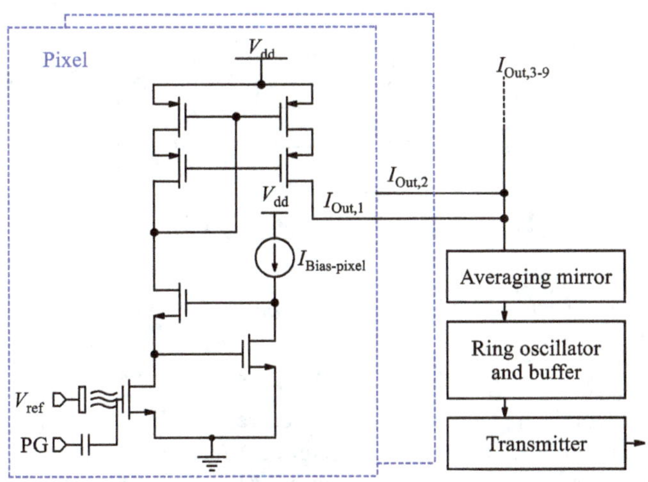

Figure 7.24 A 3 × 3 array using a current-mode pixel architecture with an averaging mirror and a ring oscillator-based current-to-frequency converter

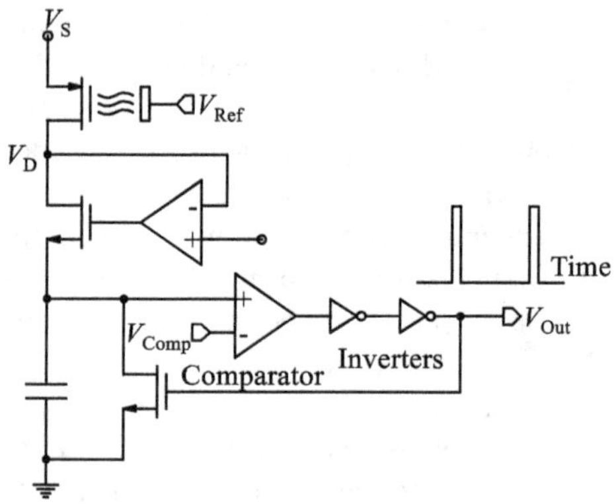

Figure 7.25 A pH-FC array sensor with sequential measurement using a comparator

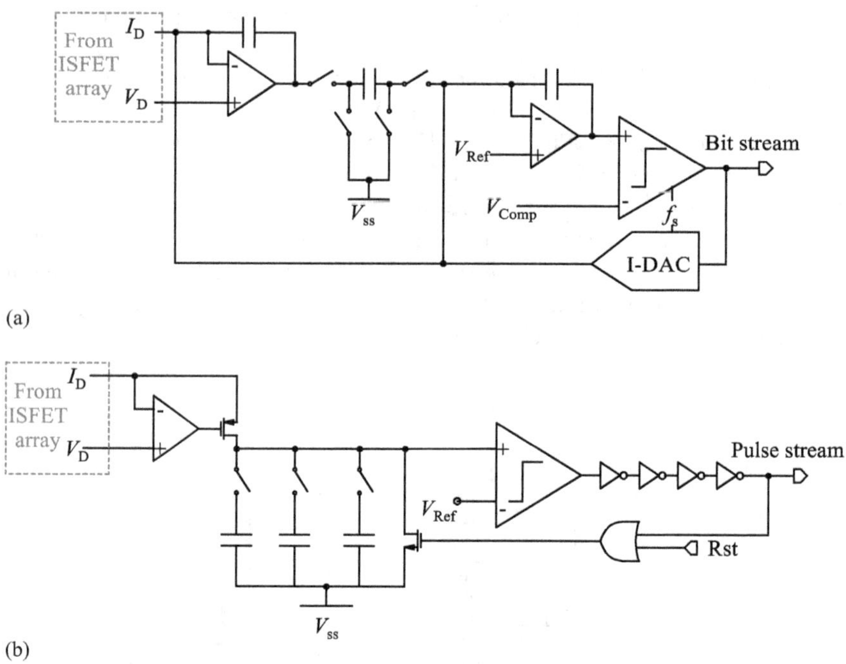

Figure 7.26 pH-FCs based on: (a) a second-order ΣΔ converter and (b) an I-to-F converter

In another work, a pH-FC including 16 current readout units was reported by Livi *et al.* [483] which was composed of eight ΣΔ modulators (Figure 7.26(a)) and eight current-to-frequency (I-to-F) converters (Figure 7.26(b)) and digitize the analog signal on the chip. This sensor contains four sub-arrays with 12 silicon nanowire (NW)-based ISFET with a shared source contact. Sixteen voltage buffers were used to determine the voltage at the source terminal of each NW. Each ΣΔ modulator, as seen in Figure 7.26(a), is composed of a second-order architecture with two switched-capacitor integrators. The current-mode digital-to-analog converter (DAC) is based on the switched-capacitor technique. This topology helps to reduce dead zones and increase the linearity and immunity to parasitic capacitances. Each I-to-F converter, shown in Figure 7.26(b), includes an input CC, an integrating capacitor, and a continuous time comparator. The I-to-F converter together with an external counter forms a complete ADC. The current flowing from the drain of the ISFET is buffered by the CC and then integrated by a capacitor and converted to a voltage which is compared to a reference voltage by the comparator. Whenever this voltage crosses the reference voltage of the comparator, the integrating capacitor is discharged and after resetting, the current integration starts again. This process creates a sequence of output pulses whose frequency encodes the input-measured signal.

Chan *et al.* [481] proposed using an averaging array working based on a global negative current feedback method to regulate the undefined threshold voltages of ISFETs (see Figure 7.27(a)). In this sensor, a log domain voltage-controlled oscillator (VCO) using the weak inversion region is utilized whose Q value and oscillation frequency are adjusted independently by two bias currents. The oscillator can be linearly tuned by the pH. Figure 7.27(b) illustrates this VCO in which the ISFETs averaging array and the external resistor constitute a voltage-to-current (V-to-I) converter and determine the input signal of the connected log domain CCO (I_0).

7.4.2 Active pixel sensors (APSs)

Some sensors include integrated ISFET-based APS by using ISFETs as standard buffers [474,475] or replacing photodiodes with the ISFETs [461,465] (by employing a charge transfer technique). Common source-based circuits convert the chemical voltage to current and thereafter output a voltage by discharging the load capacitors [517] or output a current via CCs [460]. These circuits need constant biasing and the capacitor ratio in the floating gate causes non-linearity and signal reduction. The biasing current and power consumption directly affect their noise performance. Furthermore, since such open-loop techniques do not lead to sufficient linearity, calibration and pH-to-voltage mapping are required. Some structures of array sensors and their pixels have been discussed in previous sections. Here, we discuss a few more examples of APSs in array sensors.

In some sensor arrays [460,519], a standard pixel readout like what is depicted in Figure 7.28 including an ISFET and two switches is used which is similar to CVCC technique. In another effort, in the 64 $\times$ 64-array of pH-to-voltage converter designed by Malpartida-Cardenas *et al.* [442], a source follower configuration was employed in the pixels.

A dual-mode sensor has been proposed by Huang *et al.* [475] which integrates a CMOS image sensor with the ISFET as standard buffers. One of the pixels of this sensor is shown in Figure 7.29 where the gray part is the added circuit to provide optical sensing in addition to pH measurement. The poly-gate of ISFET (the device

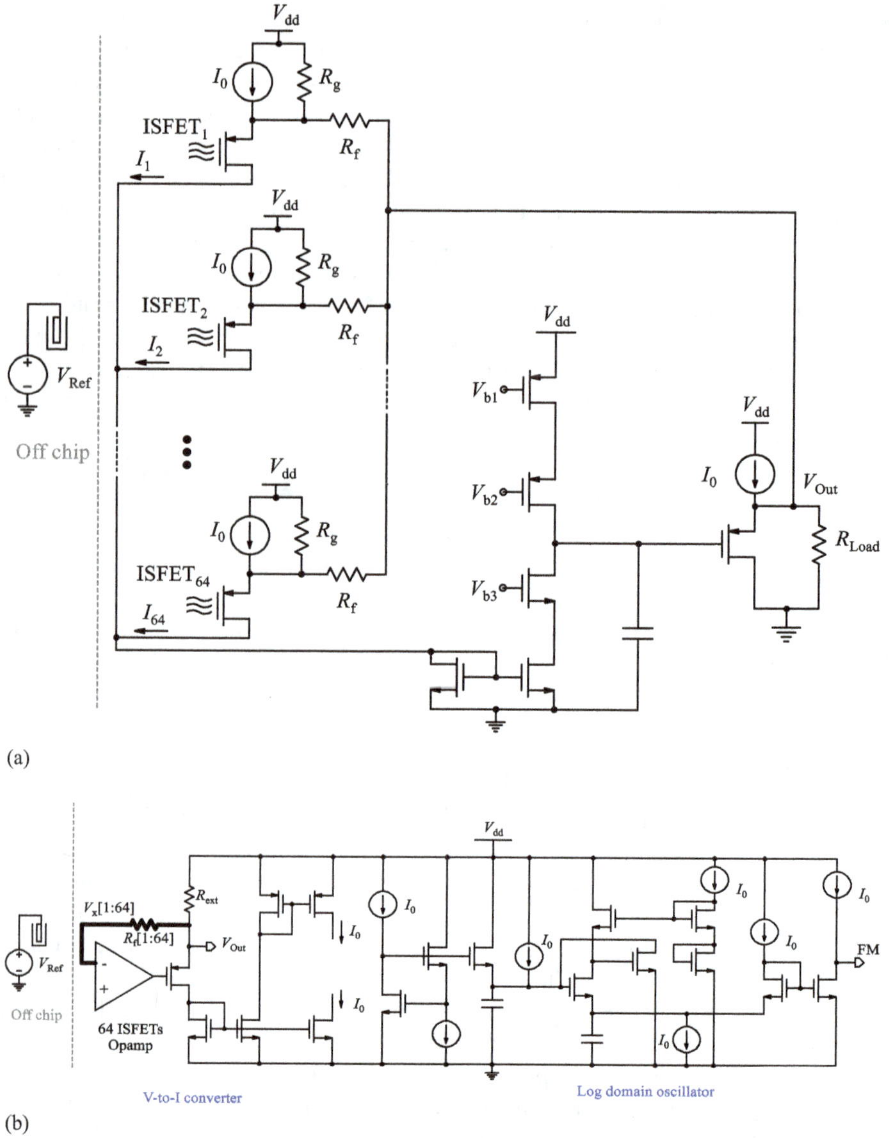

Figure 7.27 (a) ISFET averaging array using global current feedback. (b) A log domain VCO including a V-to-I converter and a CCO

named SF) is connected to the top metal and Si_3N_4 passivation layer which plays the role of the ion-sensitive membrane of ISFET. Correlated double sampling was also utilized to reduce pixel-to-pixel non-uniformity such as threshold voltage mismatch.

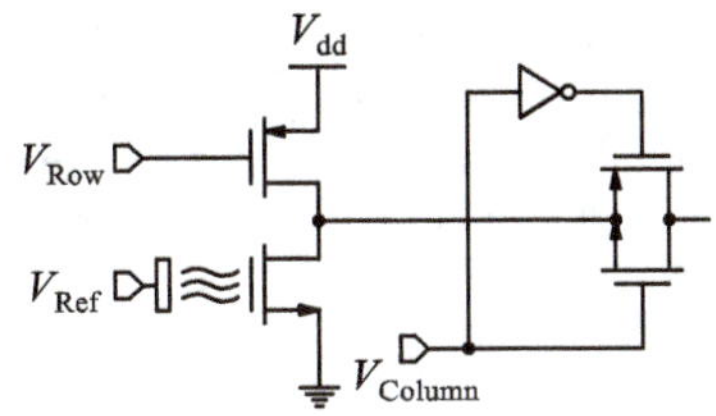

Figure 7.28 A standard ISFET pixel readout

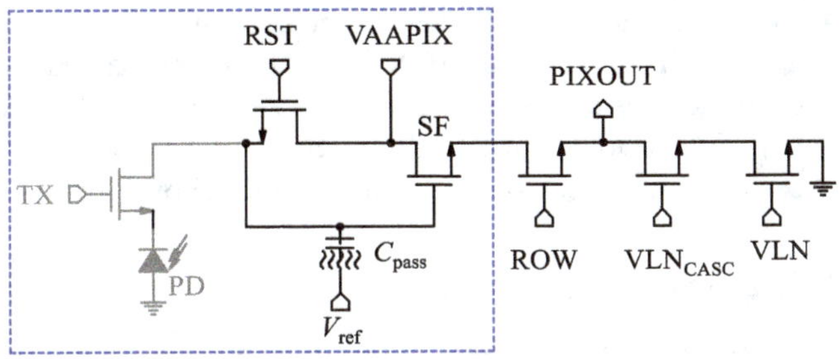

Figure 7.29 A dual-mode circuit comprising a source follower and a CMOS image sensor

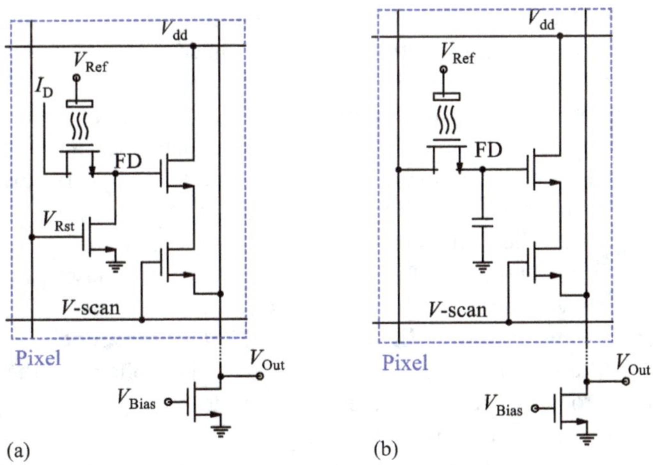

Figure 7.30 The APS topologies with (a) four transistors and (b) three transistors

In another work, Sawada *et al.* [461,463–465] proposed an accumulation method for ion sensing which is based on charge transfer techniques. They [461,465] designed the four-transistor APS, illustrated in Figure 7.30(a). The image sensor includes an ion sensing region, a floating diffusion (FD) region, a reset switch, and a source follower as well as an input diode (ID), an input control gate, and a transfer gate. In this sensor, the surface potential variation is transformed into an electrical charge packet by charge transfer and then it is converted into a voltage through the FD amplifier and source follower. The pH-dependent electrical signal is amplified by the following factor:

$$A = \frac{C_i}{C_{FD}} A_{FS} \tag{7.29}$$

where A_{FS}, C_i, and C_{FD} stand for the gain of the source follower circuit, the capacitance of the insulator (including the ion-sensitive membrane), and the capacitance of the FD region, respectively. To diminish the pixel size and increase the spatial resolution, the same group [462] proposed a simpler APS with three transistors (see Figure 7.30(b)). In this circuit, by setting ID to the ground (GND), electrons are charged into FD. Then, the electrons, except the ones corresponding to the potential level of the sensing area, are discharged to the ID due to the raised voltage level of the supply voltage. Next, the potential at the sensing area is read out through a source follower circuit.

7.4.3 Differential measurements

These techniques are based on measuring the difference in the output signals of two ISFETs with the same reference electrodes. Differential measurement helps to eliminate common-mode noise, compensate for temperature and long-term drift, and tackle the problem of the unstable potential of the electrolyte-reference electrode interface and the instability imposed by the membrane layers of the ISFETs.

One of the issues of Bio/ChemFET-based sensors is the limited stable lifetime of the liquid-filled reference electrode that they need. Therefore, a rather large reference electrode is used together with a small sensor probe which limits the possibility of measuring the small analyte volumes. As a result, differential measurement techniques between an ISFET and an identical reference FET (REFET) have attracted the attention of many researchers due to their very low sensitivity (ideally zero) to the sample ion concentration. In this case, a metal wire is employed to ground the sample solution [492]. Thus, a differential structure including an ISFET and a REFET can be biased by a common counter-electrode, the so-called pseudo-reference electrode (PRE), which is used instead of the reference electrode, as shown in Figure 7.31 [492]. This electrode is usually made of noble metal like platinum or gold, to determine the voltage of the electrolyte without the need for any external reference electrode.

To create low-sensitive REFET, the gate oxide of the ISFET is usually covered with an additional ion-insensitive membrane like chemical modification of the original gate oxide surface sites [520], or a polymer [521,522]. Source and bulk connections are usually short-circuited because of practical reasons [522]. In the pioneering studies conducted by Bergveld *et al.* [523], bulk feedback was employed to regulate the depletion charge in the bulk region. Some of the issues of this technique are drift and different electrical properties of REFET and ISFET [496,524,525]. To achieve identical electrical behavior of REFET and ISFET, their DC bias should be the same.

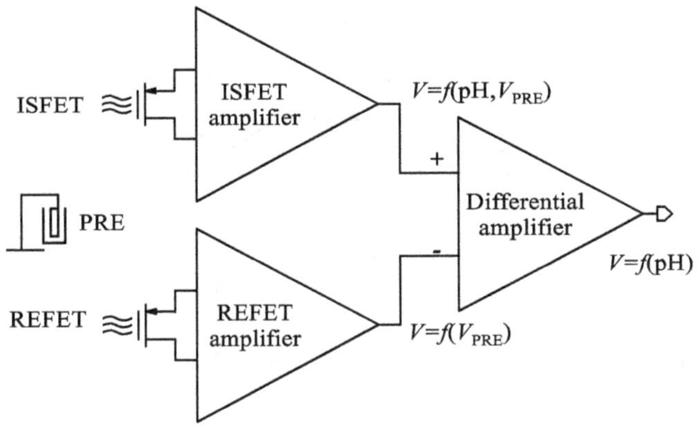

Figure 7.31 A differential ISFET/REFET measuring system

A modified ISFET was reported by Chudy *et al.* [526] including additional membranes where the outer membrane contains a buffer component for the ion measurement. In a differential system including the modified ChemFET and REFET, the electrically identical REFET with the same membrane does not include such a buffer.

Wong and White [451] presented a CMOS-integrated operational amplifier that integrated an ISFET as an input and incorporated a feedback loop to a MOSFET for pH measurement. The outputs of two ISFET OpAmps with a PRE were amplified differentially. To obtain different sensitivities, the gates of ISFETs were coated with special layers, including Ta_2O_5/SiO_2 and $SiO_xN_y/Si_3N_4/SiO_2$ whose sensitivities are respectively 58–59 mV/pH and 18–20 mV/pH. Moreover, an ISFET, playing the role of a reference device, was submerged in a buffer solution of known pH. The common temperature effects and drift in both ISFETs were canceled in this device and a sensitivity of 40–43 mV/pH was reported for the whole device.

In another effort, a differential architecture was proposed by Chodavarapu *et al.* [527] comprising a gold PRE and a pair of ISFET operational transconductance amplifiers with identical ISFETs as the inputs of each amplifier. However, the different sizes of p-MOSFET loads employed in the amplifiers make variations in the amplifiers' sensitivities (50 mV/pH and 30 mV/pH). They reported a sensitivity of 40–45 mV/pH for the whole device.

Hammond *et al.* [496] proposed a digital pH meter for use in a wireless diagnostic capsule which consists of on-chip programmable voltage references and multiplexers controlled by a microcontroller. This device has shown a sensitivity of 37 bits/pH within an operating range of 7 pH units. The analog part of this differential pH measurement device is illustrated in Figure 7.32. As seen in this figure, the circuit includes a common PRE and two source/drain followers for the REFET and ISFET whose outputs are fed to an IA. A similar differential structure was also reported by Palan *et al.* [489] which includes two sources/drain followers and a V-to-I converter. In this sensor, $ISFET_1$ and $ISFET_2$ included Si_3N_4 and SiO_2 layers

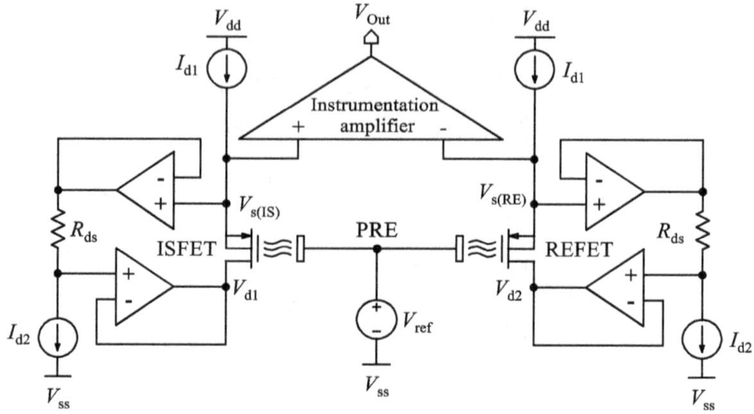

Figure 7.32 A differential pH-measurement pixel with ISFET and REFET for array structure

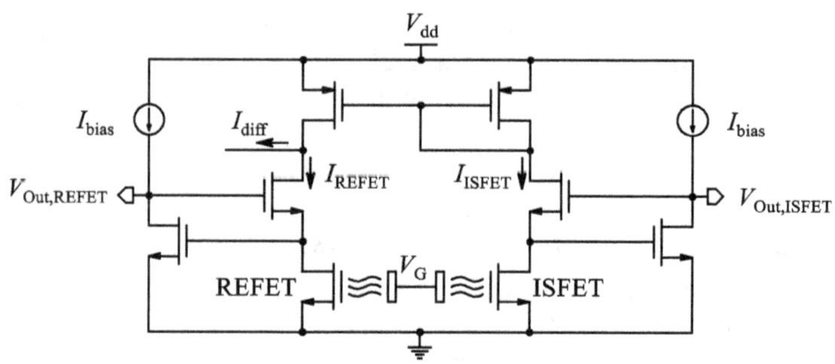

Figure 7.33 ISFET/REFET voltage-clamped configuration

with a sensitivity of 52–58 mV/pH and 24–36 mV/pH, respectively. According to their simulation results, this interface circuit has a precision of ± 0.15 pH.

Figure 7.33 shows a voltage-clamped circuit as an example of a pseudo-differential ISFET/REFET topology [528] which can operate in either strong or weak inversion and provide a fixed electrical bias. In another effort, Liu *et al.* [486] reported a differential CVCC-based topology controlled with feedback to the floating gate of ISFET.

Nabovati *et al.* [490] proposed a differential current-mode ΣΔ ISFET readout circuit by employing the ISFETs operating in the linear region (see Figure 7.34). IS_2 is a REFET whose pH value is 7. This sensor works in three phases: (1) reset phase activated by RST and $\overline{\text{RST}}$ during which the value of C_{int} is set to the common-mode voltage, V_{cm}; (2) evaluation phase during which the clock Φ_1 is activated and C_{int} is charged by the ISFET current; (3) sampling phase in which the circuit is connected to the output for sampling. The output voltage is converted to digital by using a charge-to-digital converter. The results verified that the current differential technique increased sensitivity.

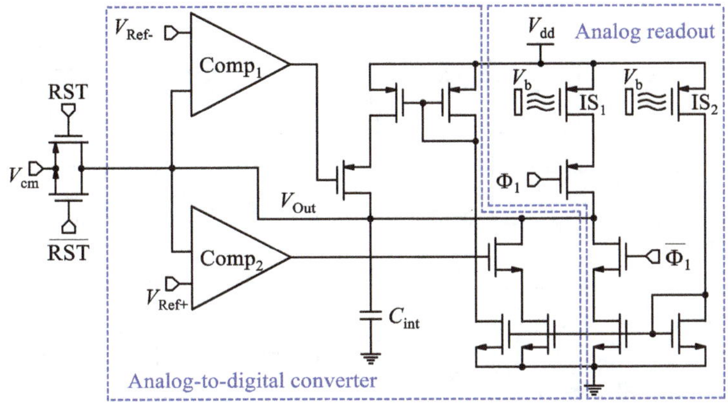

Figure 7.34 A ΣΔ ISFET readout circuit based on differential measurement

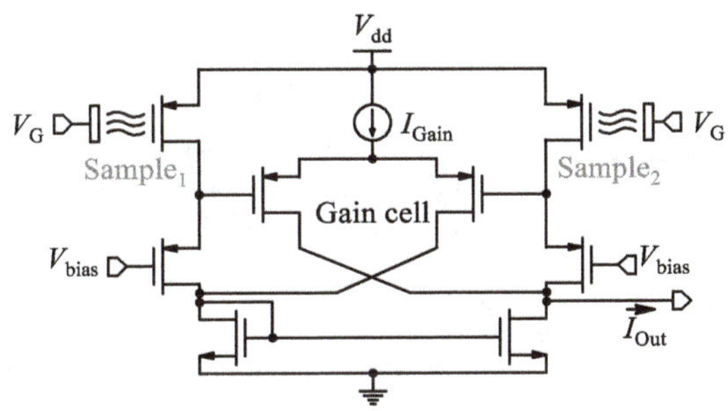

Figure 7.35 A chemical Gilbert cell

A chemical Gilbert cell can also be used for the readout of ISFETs. Figure 7.35 illustrates such a circuit where the two ISFETs used in the differential pair can be considered a chemical translinear circuit [491]. This circuit is suitable for differential monitoring in several chambers and with small volumes, such as analyzing more than one gene in DNA detection.

In another work, a CMOS device was proposed by Lee *et al.* [443] which directly measures the induced charges without any conversion loss and with improved signal-to-noise ratio (SNR). In this circuit, two working electrodes (Wes) are connected to a voltage buffer circuit with high input impedance which transfers the induced charges in these electrodes to a fully differential switched-capacitor integrator and then to an ADC for signal accumulation and digitization.

In the 4 × 16 array designed by Seong-Jin *et al.* [467], shown in Figure 7.36, four lines of an array including 1 × 16 pixel sensors are integrated inside a corresponding chamber. So, there are four chambers, one as the reference chamber to suppress any possible signal drift and background noise in the electrode interface and electrolyte, and three chambers for biosamples which can be of different types. In this circuit, a FET, in each pixel, was connected to a sensing electrode as the

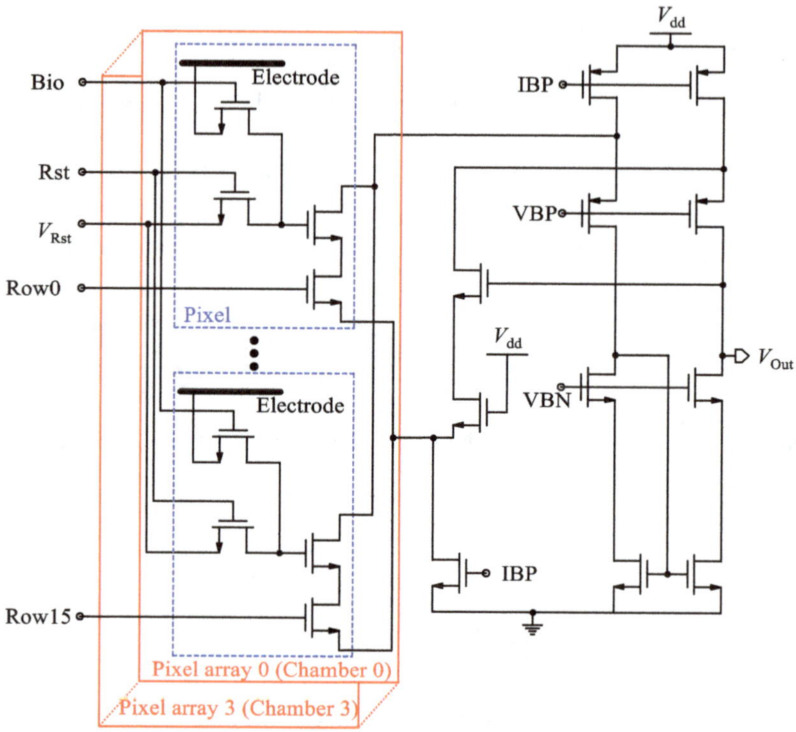

Figure 7.36 A Bio-amplifier with reduced gain FPN in a small pixel area

positive input transistor of the differential input stage of a folded cascode amplifier with a shared output stage and negative input transistor among all pixels to configure a voltage-follower, or so-called Bio-amplifier, with reduced gain fixed-pattern noise (FPN) in a small pixel area. In this work, which was reported for real-time monitoring of DNA concentrations without immobilization on the sensing surface, the charges of DNA molecules in the sample directly affect the potential of the sensing electrode by capacitive-coupling through the floating gate of FETs of each pixel and the surface potential of the sensing electrode is represented by the output voltage of the bio-amplifier.

7.5 Summary

In this chapter, Bio/ChemFET was introduced as a sensing element whose structure is based on MOSFET structure in CMOS technology. Since the focus of this book is on CMOS-based sensors, the structure of ISFETs as the first Bio/ChemFET, which has been developed by this technology, was discussed and the theory of its operation was outlined. After introducing different non-idealities of these sensors, different readout circuits reported for them were reviewed.

Chapter 8
CMOS-based optical biosensors

Optical biosensors present good performance in the real-time monitoring of biomolecular events and promote significant advances for many applications such as health care [529–532], food diagnostics [533–535], and environmental monitoring [536–539]. Although promising reports of optical biosensors are being presented, there are not many reports of optical biosensors for point-of-care (PoC) platforms with the potential of being integrated into a chip using complementary metal-oxide-semiconductor (CMOS) technology. The objective of this chapter is to provide a comprehensive review of CMOS-based optical biosensors by investigating different structures of optical biosensors, and various methods used to enhance their sensing properties and specifically looking into the deployment of a complete PoC biosensing system using CMOS technology.

Optical biosensors present great advantages over conventional analytical techniques, for their good performance in detecting biological species, including high sensitivity, robustness, reliability, real-time analysis, cost-effectiveness, and potential to be integrated into one chip [540–544]. These biosensors detect biological interactions by evaluating the variations induced in the light properties, such as intensity [545,546], wavelength [13,547,548], or refractive index (RI) [549–551], which will be proportional to the concentration of a specific analyte. Different structures of optical biosensors with potential applicability in clinical diagnosis are reported in the literature. These biosensors are fluorescence-based [551–558], bioluminescence-based [559–561], absorption photometry-based [562–564], silicon photonics-based [565–571], surface-enhanced Raman scattering-based [572,573], and plasmon-based biosensors [574–578].

While remarkable advancements in optical biosensors have been documented, there is a scarcity of reports detailing the successful integration of CMOS optical biosensors into practical lab-on-chip (LoC) systems. This chapter provides an overview of recent advances in CMOS-based optical biosensors and focuses on monolithic different parts of optical biosensors such as photodiodes, filters, and readout circuits. We will describe their important features and discuss challenges for their further development for PoC applications.

8.1 Principles operation of optical sensing methods

The most widely used optical reorganization methods to detect the binding between probes and targets, which are more capable of being implemented in CMOS

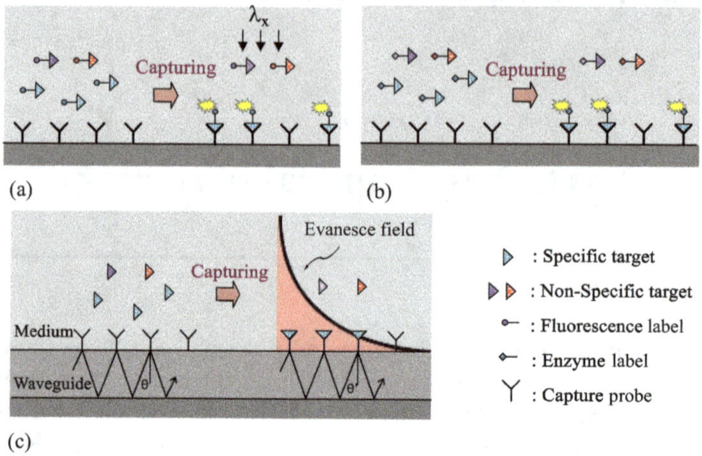

Figure 8.1 Detection principle, (a) fluorescence, (b) bioluminescence, and (c) evanescence field

technology, include fluorescence-based, bioluminescence-based, and evanescent wave-based. Each of these biosensors uses a different way to generate an optical signal based on analytes as shown in Figure 8.1 and will be briefly discussed in the following sections.

8.1.1 Recognition principles of fluorescence-based biosensors

In fluorescence-based biosensors, as shown in Figure 8.1(a), fluorescent molecules, also known as fluorophores, are used as molecular labels for the targets, while capturing probes are immobilized on a surface. The difference in the wavelength between the peak excitation and emission spectra is known as the Stokes shift. The most known fluorophores are quantum dots (Qdots) [579–581] and cyanine-3 (Cy3) fluorophores [582,583]. The principle for the operation of a fluorescence-based biosensor is that the fluorophores absorb excitation light photons at a particular wavelength (generated by an excitation source) and later emit photons in a different and longer wavelength [584]. Fluorescence labels provide the information created by the reactions between the probe and target and transduction into optical signals with high sensitivity. This optical signal, in most applications of fluorescing, requires a precise filter to reject the background excitation light and pass a desirable signal, which finally will be detected by photosensitive elements like photodiodes.

Fluorescence sensing has been one of the key enabling technologies in biomolecular sensing, which offers sensitive, specific, and robust sensing modalities in diagnostic and medical research [584]. However, such a fluorescence sensing approach often requires external elements such as an excitation source to stimulate photoemission of the fluorophores and a filter to reliably discern the generated photoemission amidst the high background. Furthermore, integrating a filter with a good rejection ratio is one of the most important challenges of this type of biosensor. On the other hand, labeling can be an expensive and time-consuming process sometimes interfering with the molecular event examined [585].

8.1.2 Recognition principles of bioluminescence-based biosensors

Chemiluminescence is a phenomenon where light is emitted by a chemical reaction. Bioluminescence, as shown in Figure 8.1(b), is a form of chemiluminescence where the light-yielding reaction is derived from a naturally catalyzed process, such as the enzymatic reactions found in luminous fireflies, jellyfish, bacteria, and others. A well-known bioluminescence example is a reaction of luciferin with adenosine triphosphate (ATP) in the presence of the luciferase label-based enzyme [586], which occurs within fireflies to produce their light. In bioluminescence-based biosensors, signal photoemission is triggered simply by a chemical reaction, without much-associated background emission. Hence, in contrast to fluorescence methods, an excitation source or filter is not required. So, bioluminescence can be cost-effective and more amenable to LoC integration than fluorescence methods.

8.1.3 Recognition principles of evanescent field-based biosensors

One of the promising alternatives for affinity-based biosensors is label-free sensing [587]. In this method, instead of labels, the generated light signal is the result of a change in the intrinsic physical properties of the target analyte when it binds to the receptor molecules [588]. Most of the integrated optical biosensors make use of the evanescent field detection principle for sensing in label-free schemes [589–595]. In the evanescent field sensing detection, a receptor layer is immobilized onto the surface of the waveguide as shown in Figure 8.1(c). When light is launched into a waveguide placed into a dielectric medium of lower RI ($n_{waveguide} > n_{medium}$), all the light will reflect within the waveguide if the angle of incidence of light entering the waveguide, θ, is greater than the critical angle, θ_c, as defined by $\theta_c = \sin^{-1}(n_{medium}/n_{waveguide})$ [592,594]. Although the incident light beam totally reflects, the electromagnetic field called the "evanescent wave" still penetrates at a small distance from the interface into the lower index medium.

The binding between the receptor and analyte induces a local change in the optical properties of the biological layer. This change is detected via the evanescent field of the guided light. Because the evanescent field distance is so short, the biosensor only interacts with the material in direct proximity to its surface, such as proteins that are chemically attached to the surface, while being unaffected by unbound materials suspended in solution. Since there is no need for any complex labeling process, this technique can be cost-effective and time-efficient. At the same time, label-free sensing allows monitoring the molecular interactions in real time providing useful information about the affinity rate of the reaction under test and the dynamic interplay of molecules [588].

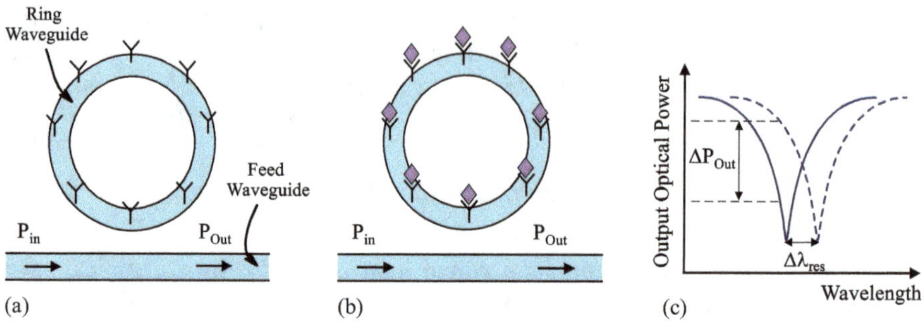

Figure 8.2 Basic principle of a ring resonator biosensor, (a) before sensing, (b) after sensing, and (c) shift in resonance induces a fluctuation in the transmitted optical power

Most employed evanescent-field biosensors for analyses with potential applicability in clinical diagnosis and versatile applications and scalable production prospects by CMOS fabrication methods are silicon photonic biosensors which makes them promising candidates for LoC solutions [588,591]. Silicon photonic biosensors can be distinguished in two areas, including resonant architecture and interferometric architecture. Each of these device architectures has advantages and disadvantages in terms of sensitivity, limit of detection (LoD), compactness, and system complexity. A widely used photonic structure is the ring resonator [588,596–599]. As illustrated in Figure 8.2, a ring resonator consists of a feed waveguide coupled to a ring waveguide. For resonant sensors, the circulating optical mode is sensitive to refractive changes within the confines of the evanescent field. When the RI near the ring resonator surface is modified due to the capture of target molecules on the surface, n_{eff} (effective index of refraction) will change. The change in n_{eff} is detected by monitoring the resonances, which occur when L (ring circumference) of the ring waveguide is equal to an integer multiple (m, resonance order) of the feed waveguide's wavelength (see (8.1)); that makes a negative peak in the spectrum of light leaving the ring.

$$m = n_{\text{eff}}\, L/\lambda_{\text{res}} \tag{8.1}$$

where λ_{res} is the resonant wavelength. The n_{eff} variations lead to a shift in the λ_{res} [568]. This wavelength shift is a quantitative measure of the number of binding events near the surface [600]. Ring resonator biosensors require a tunable laser and external readout equipment for resonance tuning and readout processing. The most challenging issue of this method is the co-integration of a planar ring resonator with on-chip electronics to introduce a fully integrated electronic-photonic biosensing platform in a CMOS process.

8.2 Photodetectors

In optical biosensors, the analyte-bioreceptor interaction varies the light properties, which can be detected by optical measurements [601]. The core of the sensing element of optical biosensors is the photosensitive element. In the past, a photomultiplier tube (PMT) has been widely used as a sensor element for light detection, which makes the system complex, bulky, and expensive, and limits its application. Replacing the PMT with optical transducers in CMOS is essential for realizing handheld and portable devices for optical sensor applications.

Optical transducers, in other words, photodetectors in CMOS, are limited by the silicon substrate in which devices are fabricated. Silicon is suitable for high-performance visible range (like 400–800 nm wavelength range) detectors [1]. Optical signals (in the visible range) can reach the CMOS transducers from the top and bottom of the chip. Notably, silicon oxide and the passivation layers at the top of the chip are transparent in the visible range, but silicon is not. Therefore, when light is coupled from the bottom, it becomes necessary to decrease the thickness of the silicon substrate through post-fabrication methods. In applications that use the evanescent field for detection, the silicon substrate should be removed by the process detailed in [602] to confine light in the waveguide (which is constructed in buried oxide (BOX) layers of SiO_2) and reduce evanescent optical leakage.

Suitable alternative CMOS photodetectors are phototransistors and photodiode which pave the way for the development of miniaturized sensing systems through integrated optical components. In the following section, photodiodes as the most employed photosensitive element will be discussed.

8.2.1 CMOS-based photodiodes

The basic structure of photodiodes consists of a p-type and an n-type semiconductor that converts photons (or light) into electrical current. These diodes are specifically engineered to operate under reverse bias conditions, creating a depletion region between the N and P layers where no free carriers are present. When light is made to illuminate the diode, a photon with enough energy can enter the depletion region and hit the atom. This results in a pair of mobile electron-hole pairs (charge carriers) that induces the photocurrent in the photodiode.

Various types of photodiodes have been reported [603–608]. Although the basic operation of these diodes is the same, some have better performance compared to others. The types of photodiodes can be classified as PN photodiode, PIN photodiode, and avalanche photodiode which are shown in Figure 8.3. Each type of this photodiode offers some benefits and drawbacks. The key performance specifications of photodiodes that characterize the effectiveness of detection for a specific application are response time, quantum efficiency (QE), responsivity, breakdown voltage, dark current, noise, wavelength, reverse bias constraints, and gain [606]. Among these parameters, dark current and QE are the most investigated parameters in CMOS optical biosensors. So, brief definitions of these parameters are expressed here.

Dark current is the minimum amount of reverse current presented in the absence of light illumination. It is an unwanted current, resulting from thermally excited electrons or leakage of current along the current path. Dark current is one of the main sources of noise in the photodiode. QE is defined as the probability of an

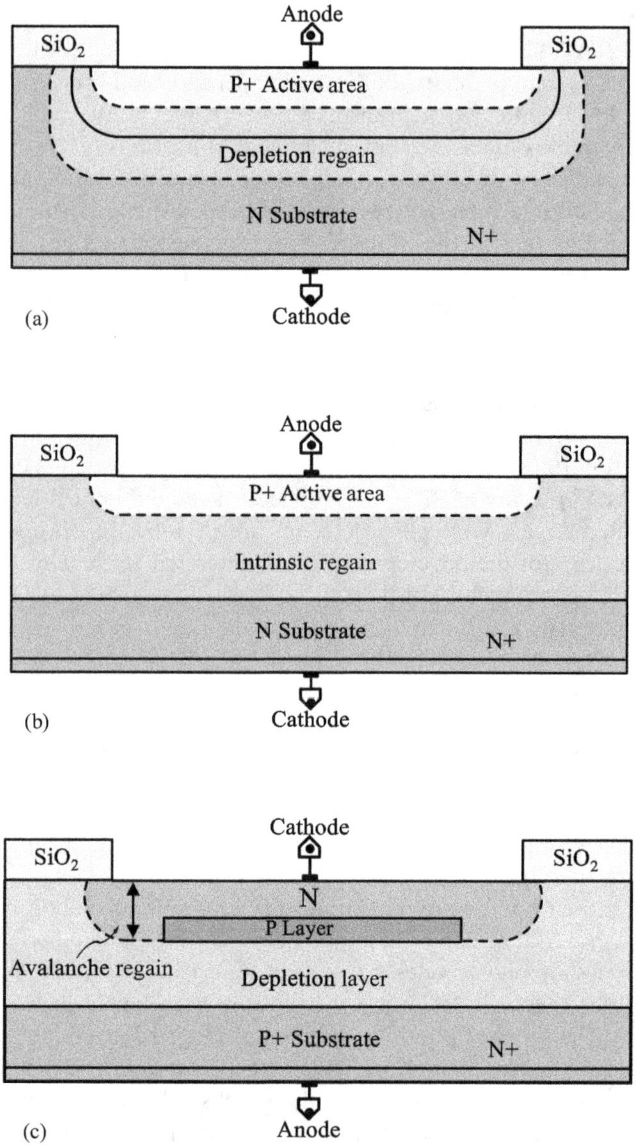

Figure 8.3 Cross-section of (a) PN Photodiode, (b) PIN photodiode, (c) avalanche photodiode

incident photon generating an electron-hole pair contributing to the detection signal, which is affected by material properties and device geometry. Table 8.1 summarizes the various structures of photodiodes, which compare area, dark signal, and QE.

Table 8.1 Comparison of photodiodes for optical biosensors

Photodiode	**Area (μm^2)**	**Dark signal**	**QE**	**Process**	**Ref.**
$P^+/N/P_{sub}$	150 × 150	2.6 mV/s	0.4 @560 nm	0.18 μm	[586]
N-well/P_{sub}	100× 100	2.31 pA	0.45 @635 nm	0.25 μm	[581]
N-well/P_{sub}	50 × 50	20 fA-5.5 pA@30-90°C	> 0.36@480 nm-800 nm	0.25 μm	[555]
P-well/N-well/P_{sub}	9.14 × 123	15.5 fA	7	65 nm	[580]
N-well/P_{sub}	100 × 70	450 fA	0.28 @562 nm	0.18 μm	[611]
P^+/deep-N-well (Avalanche)	NA	1,059 Hz @1 V	NA	0.35 μm	[618]
P^+/deep-N-well (Avalanche)	140 × 70	3,000 count/s @4 V	NA	0.35 μm	[617]
P^+/deep-N-well (Avalanche)	15 × 15	100 Hz @1 V	NA	0.13 μm	[615]
P-well/deep N-well (Avalanche)	6 × 6	500 count/s @ 1 V	NA	28 nm	[603]

NA: Not available

8.2.1.1 PN photodiode

To the best of our knowledge, as overviewed in Table 8.1, the most employed type of photodiode for CMOS optical biosensors is the PN type [560,579–583,604,609–612]. For instance, Eltoukhy *et al.* [586], have used a $P^+/N/P_{sub}$ photodiode as a photon-to-electron transducer in an integrated bioluminescence detection LoC-based element. Due to the fully depleted nature of the intermediate N-region, the $P^+/N/P_{sub}$ doping profile makes a very low photodiode junction capacitance, which helps to achieve a high conversion gain. The P^+ layer helps passivate surface-state traps to prevent the formation of interface states on the photodiode surface, which diminishes the dark current. Also, some techniques with lower photodiode reset voltage have been utilized to reduce the dark current. In these techniques, the dark current increases as the reset voltage increases. In this design [586], the measured dark signal at a reset voltage of 0.1 V is reported around 2.6 mV/s. The use of a very low-dark current photodiode reduces background noise and thus increases signal detectability. As a result, emission rates below 4×10^5 photons/cm^2 over 30 s of integration time at room temperature are achieved. The reported QE at λ ~560 nm is about 0.4.

In another design, Patounakis *et al.* [581] employed N-well/P_{sub} photodiodes in a CMOS image sensor array, which contains an N-well guard ring to collect carriers generated by neighboring pixels. The reported measured dark current in the photodiode is approximately 2.31 pA. The external QE versus wavelength (400–800 nm) is measured. The peak of QE in the range of 0.45–0.5 occurs at wavelengths between 600–700 nm. In another effort, Singh *et al.* [611], reported that the N-well/P_{sub} photodiode offers the highest overall internal QE at $\lambda \sim 562$ nm and the least dark current compared to the other possible photodiodes available in 0.18 μm CMOS technology. The reported QE and dark current are 0.28 and 450 fA, respectively. Also, in [609,610], the photosensor element is an N-well/P_{sub} photodiode for the design of a fully integrated CMOS fluorescence biochip for deoxyribonucleic acid (DNA)/ribonucleic acid (RNA) testing in molecular diagnostics. External QE of greater than 0.36 in the spectrum of 480–800 nm, and dark currents of 20 fA and 5.5 pA at 30 °C and 90 °C are, respectively, reported in this work.

In another work, four different P/N photodiode structures with different doping are designed, fabricated, and measured by Hong *et al.* [580] to investigate photodiode performance. These photodiodes are N-well/P_{sub} (D_1), N-well/ P_{sub} with a larger active region (D_2), P-well/N-well (D_3), and N^+/P-well (D_4) and are shown in Figure 8.4(a–d), respectively. Responsivity, which is defined as the ratio of photocurrent generated from incident light to that incident light power for all photodiodes at 780 nm, was measured. The reported responsivity for D_1 and D_3 was almost 2–3 times higher than that for D_2 and D_4. This is attributed to the higher optical absorption in the highly doped active regions and shallow depletion regions of the photodiodes. In this work [580], the utilized diode in measurement is D_3 as each sensing site comprises a differential photodiode laid out alternatively covered by an integrated filter and metal shield to minimize the differential dark current as 15.5 fA and the reported QE is 7.

8.2.1.2 PIN photodiode

The PIN photodiode is composed of an intrinsic layer placed between the P and N layers. This intrinsic layer increases the depletion region leading to a decrease in the capacitance of the junction. The intrinsic layer also allows for a larger volume of photon-to-electron-hole conversion and higher quantum efficiency [607,613]. PIN photodiode in some structures of biosensors is used as an a-Si:H photosensor, which is an n-type/intrinsic/p-type structure deposited by plasma-enhanced chemical vapor deposition (PECVD) process [605,614].

8.2.1.3 Avalanche photodiode

Avalanche photodiodes are p–n junction diodes reverse-biased above their breakdown voltage, which use impact ionization (avalanche effect) to amplify the photogenerated current [606,607,613]. The main application of this diode is for the detection of weak optical signals such as single photon events. So, it is also known as single-photon avalanche diodes (SPADs). SPADs can resolve photon arrival times with sub-nanosecond precision. Therefore, it is more employed to measure bioluminescence intensity and fluorescence lifetime in time-gated approaches [603,615–618], which eliminates optical filters. As aforementioned, this kind of diode works with a reverse bias voltage over its breakdown voltage. Applying an excess bias voltage produces an avalanche more readily, leading to more sensitivity of the detector. However, the maximum practical excess voltage is limited by the dark count rate (DCR) [619]. In an effort, Schwartz *et al.* [618] proposed a 64 × 64-SPAD array for fluorescence lifetime imaging microscopy, where the average DCR

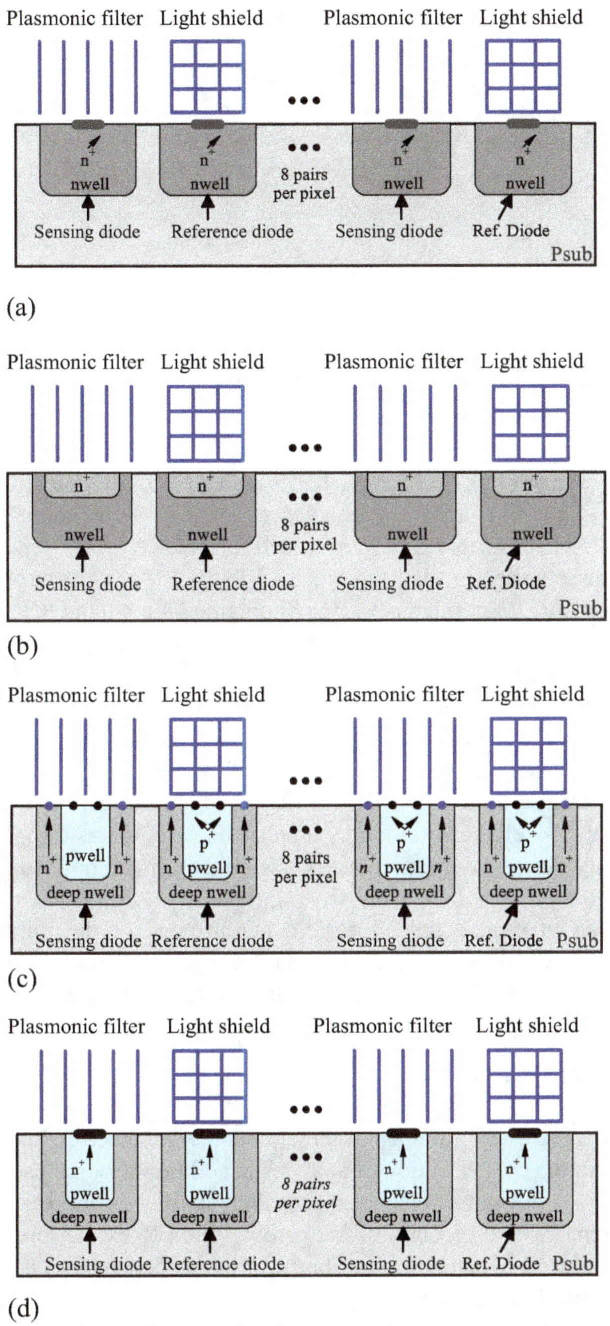

Figure 8.4 *(a) D_1: nwell-psub (small n+), (b) D_2: nwell-psub (large n+), (c) D_3: nwell-pwell (isolated), (d) D_4: pwell-n+ (isolated)*

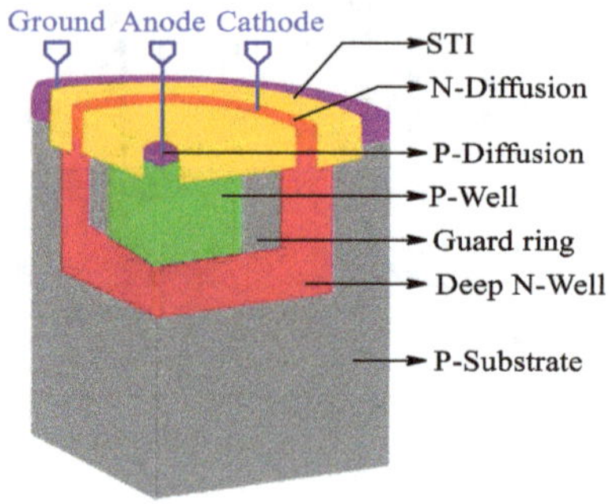

Figure 8.5 Cross-sectional view of the fabricated SPAD

at excess bias voltage (V_E) of 2.7 V is 1,059 Hz or 1,502 photons/μm^2. In another design, Stoppa *et al.* [617] achieved an average DCR of 3,000 count/s at V_E of 4 V for a CMOS SPAD sensor for time-resolved fluorescence measurements. Also, the breakdown voltage varies in the range of 27.5 to 28.4 V. For an array of a single-photon image sensor, Veerappan *et al.* [616] obtained a median DCR, as a function of excess bias voltage and temperature, which rises with an increase in both bias voltage and temperature. For instance, at the temperature of 20 °C, by varying excess bias voltage in the range of 0.2 V to 1.2 V, DCR is achieved between around 20 to 85 Hz. Also, the median DCR at 25 °C and V_E of 0.73 V is 50 Hz.

Figure 8.5 shows the cross-sectional view of the SPAD that is employed in the design of a microwave-optical biosensor [603]. The multiplication region is formed with P-well/deep N-well, surrounded by a native P_{sub} region as the guard ring. It is noted that highly doped layers increase DCR and even promote tunneling breakdown due to the increased defect density; so, it is not used. A lower doped guard ring is employed to reduce the edge field at the edge of the junction to avoid premature breakdown [620]. The SPAD has a DCR of 500 count/s at V_E of 1 V and a breakdown voltage of 15.3 V is reported at room temperature.

8.3 Optical filtering

In optical biosensors, after biological interactions, the produced light signals contain information about the concentration of analytes in addition to some background light. In many different optical biosensing approaches, a filter is used as a part of the biosensing system for the rejection of background light to extract precise biological information. The need for such a contributing filter is dependent on the method of signal transduction. For instance, in chemiluminescence and bioluminescence sensing, light is emitted as a result of a chemical reaction, and negligible background light is produced. Hence, there is no need for an optical filter [560,621–623].

As aforementioned, in fluorescence-based sensing, when fluorophores are excited by an external light source at a particular wavelength, secondary light is emitted at a higher wavelength. Alongside the emitted light, the background light or the reflected excitation light will be produced [624–626]. So, in the fluorescence method, the excitation sources are required to stimulate photoemission. Furthermore, precise optical filters must be employed to block the shorter wavelength excitation light of the external source and allow the longer wavelength emission light to pass through and be sensed by the photodetector. Also, it is needed to attenuate the background light to consequently achieve a high dynamic range.

There have been reported efforts to integrate a monolithic filter as an on-chip element for fluorescence biosensors. But, it requires complicated post-processing steps to fabricate the optical filter structures, which increases the system's cost and complexity [582,583,609,610]. For instance, Jang *et al.* [553] used a multilayer thin-film dielectric long-pass (edge) optical filter for an integrated fluorescent-based biosensor microarray system. This filter is externally grown and rejects incident photon flux by 98 dB at 532 nm, with approximately 1 dB loss in the optical passband. It is noted that not only should the filter provide the functionality of filtering but also it serves as a planar surface platform for immobilized capturing probes [1]. In this design, both the emission filter and the fiber optical faceplate (FOF) are used to present the aforementioned functionalities. In another design reported by Singh *et al.* [582], for a fluorescence contact imaging microsystem, an optical filter was externally added as a diced optical filter attached to the CMOS die, which again requires post-CMOS processing. In this design, a Cy3 was used as a fluorophore. It is noted that Cy3 is a widely adopted fluorophore for DNA analysis and is conventionally used as a fluorescent label. Cy3 fluorophore has a small difference in the wavelength between the peak excitation and emission spectra; so, the emission filter design is challenging so that it sufficiently blocks the excitation light and transmits the emission light. The attenuation property of an optical filter is characterized by its optical density (OD), which is defined as the ratio of the intensity of light incident on the filter to the intensity of the light passing through it. In this design, to provide a high OD, multilayer dielectric interference filters with ODs greater than 6 at the wavelength of 575 nm are used, which provide 60 dB excitation rejections to the 532-nm laser.

To the best of our knowledge, an integrated nanoplasmonic filter was presented, for the first time, by Hong *et al.* [627] and realized in 65 nm CMOS process to enable angle and scattering insensitive filtering and open the door toward implementing fully integrated fluorescence-based biosensors with no post-processing and external optical elements (except for an excitation source). The filter are realized with copper-based bottom five metal layers and four via layers. It can reject short wavelengths and pass through long wavelengths as a result of the surface plasmon polaritons (SPPs) at the SiO_2-Copper interface [628]. The plasmonic filters guide the fluorescence signals from the Qdot fluorophores (Qdot 800) on the chip surface to the on-chip detectors, while an extinction ratio of greater than 51 dB (63 dB by finite-difference time-domain (FDTD) simulation) is achieved [580]. Qdot fluorophores have a larger Stokes shift, which relaxes the requirements on the steepness of the filter cutoff [581] but are not typically utilized in applications such as DNA detection [582].

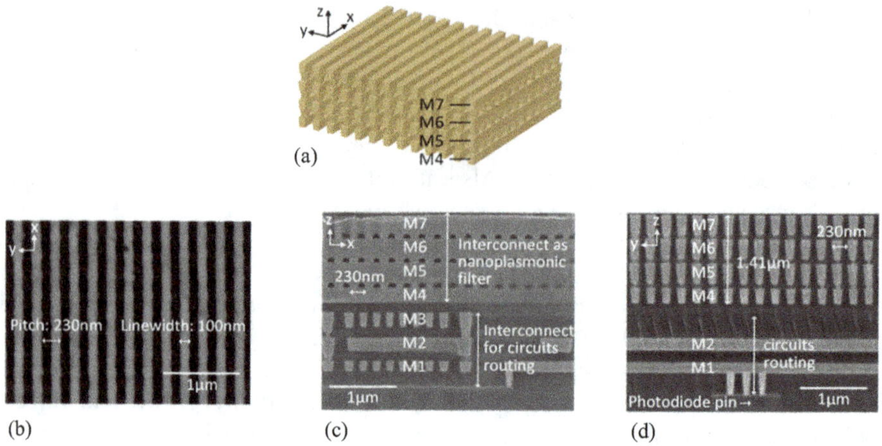

Figure 8.6 *(a) Integrated nanoplasmonic filter, (b–d) SEM images of the interconnect layers [579], https://pubs.acs.org/doi/full/10.1021/acsphotonics.8b00440. (Readers are advised to direct any inquiries regarding additional permissions related to the excerpted material to the ACS.)*

The system presented by Hong *et al.* [632], in addition to enabling angle and scattering insensitive filtering, exploits the metal and via layers between M_4 and M_7 in modern CMOS processes to guide and diffract light in a controlled fashion to be detected and processed for spectrum information between 500 nm and 830 nm. In another effort, the same group [579,629,630] implemented a 96-microarray fluorescence detection system with integrated nanoplasmonic filters co-packaged with an ultraviolet light emitting diode (UV LED) source. The structure of the integrated nanoplasmonic filter is shown in Figure 8.6. The interconnect layers serve the dual functionalities of integrated optical filtering and electronic routing for the photon-detection circuitry underneath the filter. The routing of the circuitry is achieved with the lowest three metal layers, while the nanoplasmonic filter is realized with the 4th to 7th metal layers. As a result, the combination of optical filtering of around 47 dB with additional electronic filtering of 30 dB, leads to the detection capability of fluorescence signals which are 77 dB below excitation. Similarly, Jiang *et al.* [631] presented a CMOS image sensor with nano-filter integration by employing the same technique reported by Hong *et al.* [627,632] for implementing the filter. The filter is realized by stacked and periodically patterned nano-scale copper lines such as M_5 and M_2. The nano-filter achieved a simulated rejection ratio of 51 dB (a measured value of 24.2dB) for 467 nm excitation light and 820 nm emission light. The comparison between different structures of optical filters is summarized in Table 8.2.

Traditional fluorescence sensors are mainly based on measuring the intensity of luminescence at their characteristic wavelength. Beyond intensity, other

Table 8.2 Comparison of different optical filters

Filter type	Rejection ratio (dB)	Excitation light (nm)	Emission light	Fluorophore	OD	Implement type	CMOS process	Ref.
Multilayer thin-film dielectric long-pass filter ($ZnS+Na_3AlF_6$)	98	532	NA	Cy3	NA	Require Post-processing	NA	[553]
Long-pass multi-dielectric emission filter	70	490	NA	Quencher molecules (BHQ2)	3.6	Require Post-processing	0.25 μm	[555]
Multilayer thin-film interference filter	60	532	575 nm	Cy3	6 >	Require Post-processing	0.35 μm	[582]
Nano-plasmonic filter	51	405	780 nm	Qdot 800	NA	On-Chip	65 nm	[627]
Nano-plasmonic filter + electronic filter	77	405	800 nm	Qdot 800	NA	On-Chip	65 nm	[632]
Nano-plasmonic filter	24.2	467	820 nm	PbS QDs	NA	On-Chip	65 nm	[631]

NA: Not available

characteristics of the emission light, such as the lifetime of emission, are also measured. Lifetime is the average time a fluorophore spends in the excited state before returning to the ground state. Time-resolved fluorescence is a technique that is used to measure the lifetime of emission, which contains the information of the reacting analyte [633]. Such a time-resolved fluorescence detection technique provides distinguished advantages in improved background rejection even in the absence of optical filtering [581,617,634–638]. Hence, these methods provide filterless fluorescence biosensing.

8.4 CMOS-based readout circuits of optical biosensors

Current sense circuits that have been reported for most optical biosensors can be classified into two groups including: (1) direct current integration, and (2) TIA. Each of these front ends has a constellation of performance tradeoffs, including bandwidth, noise, power consumption, input impedance, dynamic range, and sensitivity among others. Choosing and designing the appropriate front ends for readout circuits are critical for achieving high performance from any sensor [639,640]. In array structures of optical sensors, most pixels are composed of photodiodes, current readout circuits, and data converters. Since photodiodes do not have gain, an accurate current measurement, in the front end of the readout circuit, is crucial to detect the low-current signal generated by the photodiode.

8.4.1 Direct current integration

In imaging applications, an integrated signal from the photodiode is often preferred [606]. Most CMOS image sensors use direct integration [560,581,582,641–647]. In this structure, as shown in Figure 8.7, the photocurrent generated by photons incident is directly integrated across the photodiode capacitance C_D within a set period of time and converted to charge Q. The amount of charge is limited by the

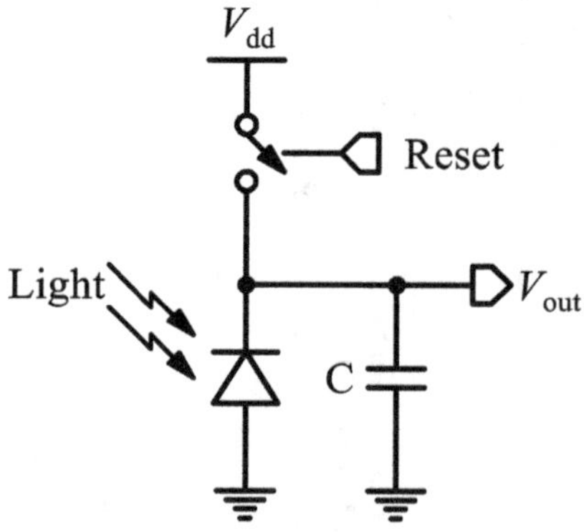

Figure 8.7 Direct current integration principle

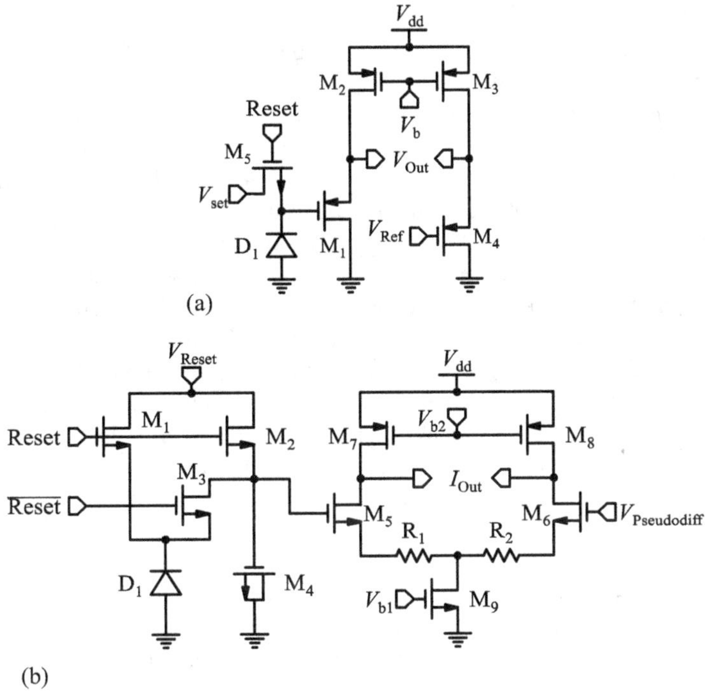

Figure 8.8 Direct current structure, (a) pseudo-differential, (b) current differential

well-capacity Q_{max}. At the end of the integration time T_{int}, the charge is read out as a voltage signal V_{Out}, which is related to Q by the conversion gain [622].

Figure 8.8(a) shows a pseudo-differential circuit by employing the direct current integration concept for implementing a readout circuit [586]. The pseudo-differential circuit consists of an NMOS reset gate and two PMOS source follower amplifiers. To further reduce dark current and to achieve high linearity a PMOS follower instead of the NMOS follower is employed. The readout circuit of this biosensor comprises an 8 × 16 pseudo-differential circuit array, a 128-channel 13-bit 2-step analog-to-digital converter (ADC), and a column-level digital signal processing (DSP) single instruction multiple data (SIMD) array, which all integrates with 0.18 μm CMOS image sensor process. Each pseudo-differential circuit is connected to its own ADC and to achieve low noise and high immunity to common-mode disturbances, a differential architecture is used for both pseudo amplifier and ADC. This structure is customized for applications such as determining gene expression using reporter gene assays, pathogen detection, determining intracellular ATP, and sequencing DNA. It involves some specifications such as the capability of detecting emission rates below 10^{-6} lux (4×10^5 photons/cm^2) over 30 s of integration time at room temperature, the intrinsic dynamic range of 61 dB, and static power consuming of 26 mW.

Patounakis *et al.* [581] used the direct current integration principle for fluorescence-based assays that enable time-gated, time-resolved fluorescence spectroscopy. The readout circuit contains an 8×4-array that is time-multiplexed into four current-mode ADCs and the digital results are stored in an on-chip static random access memory (SRAM). Each readout front end, as illustrated in Figure 8.8(b), contains two reset transistors (M_1 and M_2), an isolation device (M_3), a storage capacitor (M_4), and a transconductor ($M_{5\text{-}9}$). The transconductor converts the voltage across the storage capacitor, which results from the integrated photocurrent, into a differential current for subsequent current-mode data conversion. Also, R_1 and R_2 are utilized to linearize the transconductor through source degeneration. The sensitivity of the array to photon densities is as low as 1.15×10^8 photons/cm^2, and it has a dynamic range of over 74 dB.

8.4.2 Transimpedance amplifier (TIA)

Transimpedance amplifier (TIA) with negative feedback has low input impedance, which makes it a suitable structure for employment in optical sensing to convert a weak current signal to a measurable voltage signal [568,580,583,604,632, 639,640,648–650]. TIA employs resistors or capacitors as the elements of negative feedback, known as resistive-TIA (RTIA) and capacitive-TIA (CTIA), respectively.

8.4.2.1 Resistive transimpedance amplifier (RTIA)

In a conventional RTIA, a resistor (R_F), is connected in feedback between the inverting and output nodes of an amplifier, as shown in Figure 8.9(a). Also, a

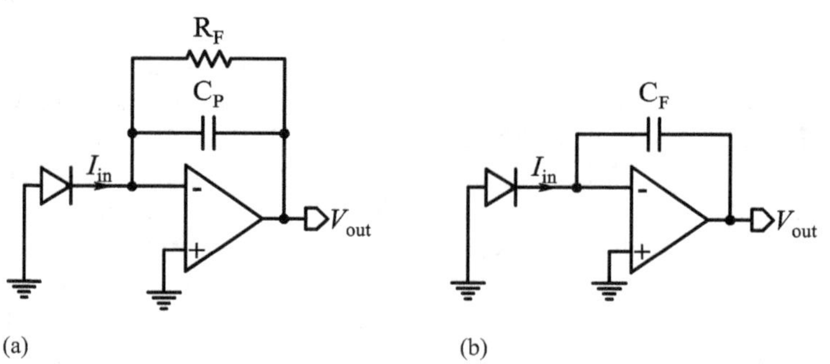

Figure 8.9 Schematic of a TIA with (a) resistive feedback considering the C_p, (b) capacitance feedback

feedback capacitor C_P (parasitic capacitor) is connected in parallel with R_F. The output voltage can be expressed as (8.2), where I_{in} is the output current of the biosensor (input current of RTIA):

$$V_{out} = I_{in} R_F \frac{1}{1 + S R_F C_P} \tag{8.2}$$

It can be deducted from (8.2) that to gain reasonable output voltage amplitude from a low input-current signal, a large R_F is needed, while a large resistor restricts the bandwidth, and in addition, requires a large area in CMOS implementations [640]. This structure to achieve high gain and low noise performance will suffer from poor bandwidth [640]. So, RTIA is not so desirable in biosensor applications. In the structure of the electronic-photonic biochip proposed by Adamopoulos *et al.* [549,568], the photocurrent is amplified by a programmable-gain RTIA. This gain programmability, accomplished by using a digitally controlled large feedback resistance (50–800 kΩ), offers a very wide sensor dynamic range and can be used to adjust between different sensing modes. However, increasing the feedback resistance decreases the noise term and improves LoD, as long as the bandwidth of the TIA remains sufficient for the application in question. To the best of our knowledge, this work presented the world's first fully integrated electronic-photonic, real-time, and label-free RI PoC sensor, utilizing on-chip electronics and 10 nm micro-ring resonators (μRRs) in a commercial CMOS 45 nm radio frequency silicon on insulator (RFSOI) platform. The minimum reported bulk RI as LoD is 3.5×10^{-5}.

8.4.2.2 Capacitive transimpedance amplifier (CTIA)

As mentioned earlier, a large R_F of RTIA affects the bandwidth and consumes a large area. CTIA is an effective solution to eliminating such limitations by replacing R_F with a capacitor, which makes it a popular topology for current sensing in optical biosensors [579,580,583,604,610,651] (see Figure 8.9(b)). For instance, in the sensor reported by Jang *et al.* [553], a CTIA is used to create the photocurrent integrator for a fully integrated fluorescent-based biosensor for DNA microarrays. In this design, a CTIA structure is preferred to a direct integrator because the CTIA linearity is not limited by the photodiode junction capacitance-voltage dependency and also it has a diode-independent well capacity set by the feedback capacitor. The current readout circuit comprises CTIA, chopper-stabilized preamp, and 14-bit ADC that is shown in Figure 8.10. ADC compares the CTIA's output with the voltage of V_{R2}, to measure the time when the ramp reaches the voltage of V_{R2}. The role of the chopper-stabilized preamplifier is suppressing the offset of the ADC's comparator.

In other sensors reported by Hong *et al.* [580], the sensor architecture includes the sensing modality with a sensing site and a control site. This allows the laser light leakage into the sensors to appear as common-mode perturbation and to be suppressed electronically down the chain which results in obtaining a

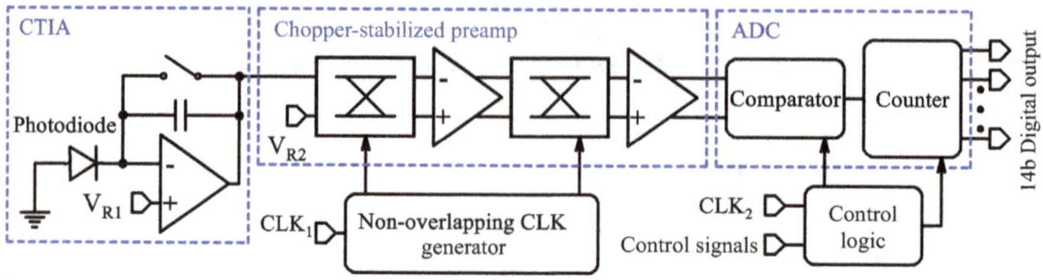

Figure 8.10 Readout circuit with CTIA, chopper, and ADC

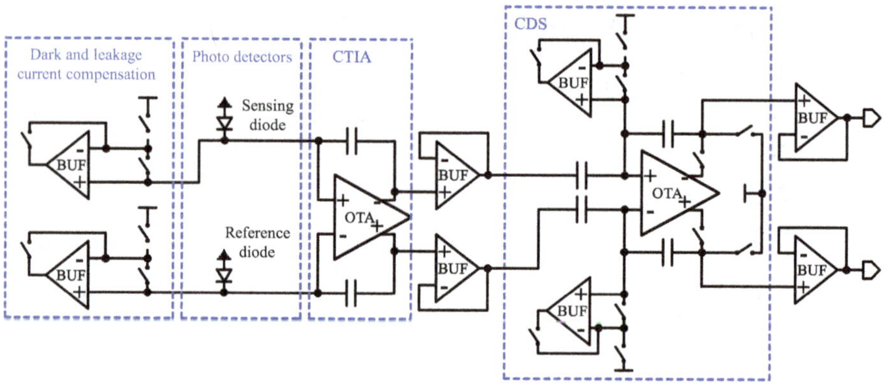

Figure 8.11 Readout circuit with CTIA and CDS [580]

net background suppression of orders of magnitude higher than what can be achieved purely optically. As shown in Figure 8.11, each sensing site comprises a differential photodiode, which is amplified by a CTIA, and is processed through a differential correlated double sampling (CDS) circuit to suppress offsets and low-frequency correlated noise. The signals are finally acquired by a 12-bit data acquisition and digitizer board. This sensing site is employed to design a fully integrated fluorescence-based bio-molecular sensor in 65 nm CMOS with integrated nanoplasmonic waveguide-based filters capable of more than 50 dB of rejection ratio. The sensor is capable of detecting 70 molecule/μm^2 (48 zeptomoles) on the surface without any post-fabrication, external optical filters, lenses, or collimators, which creates the opportunity to enable large multiplexed assays on a single chip and a wide variety of applications, from *in vitro* to *in vivo*.

In some designs [610,611,652], CTIA is used as an integrator in the loop of sigma-delta ($\Sigma\Delta$) current sensing structure. As shown in Figure 8.12(a), the main concept of $\Sigma\Delta$ current sensing is that the reverse-biased current of the photodiode

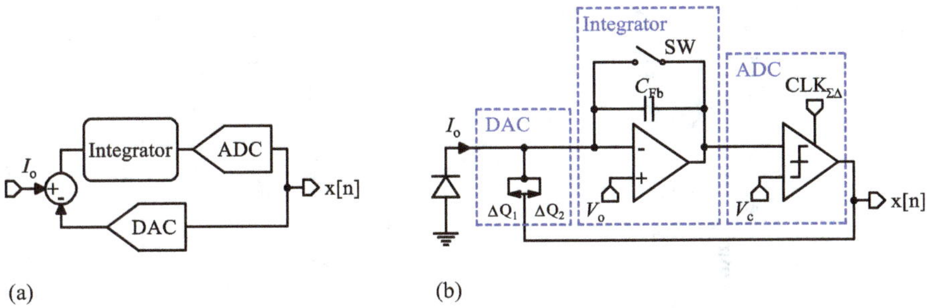

Figure 8.12 ΣΔ current sensing, (a) block diagram, (b) circuit diagram [555]

(I_D) as input signal feeds into a high DC gain integrator (Σ operator) followed by an ADC (quantizer). The digital output of the ADC also controls a digital-to-analog converter (DAC) to subtract an output-dependent current from the input (Δ operator). In the sensor reported by Manickam *et al.* [555], which is illustrated in Figure 8.12(b), the integrator is a CTIA that integrates I_o on its feedback capacitor, C_{Fb}, while pinning the reverse-bias voltage of the photodiode to V_o. The ADC is a clocked comparator, comparing the output of the CTIA with V_C, at every oversampling clock ($CLK_{\Sigma\Delta}$). The current DAC is implemented as a charge packet subtractor which subtracts ΔQ_1 or ΔQ_2 from the CTIA input, depending on the comparator output being 1 or 0, respectively. Some advantages of adopting this architecture are reported in [555]. First, it can achieve a very high detection dynamic range (DDR) (>100 dB) compared to direct current integration architectures used in CMOS image sensors [560,582]. Second, it provides superior noise performance due to ΣΔ operation, and third, the architecture has a digital output that can be easily read sequentially (scanned) in an array format. The important part of the reader electronics is the Spartan 6 XC6SLX16 field-programmable gate array (FPGA).

This ΣΔ current structure is the sensing part of a fully integrated CMOS fluorescence biochip for DNA and RNA testing in molecular diagnostics. The biochip includes a 32 × 32-array of continuous wave fluorescence detection biosensing elements, which is capable of DDR of about 116 dB (10 fA–10 nA) over the 25 °C–100 °C temperature range. The module of this chip contains a CMOS IC, integrated emission filter, DNA probes, and an optically transparent fluidic cap with an inlet and outlet. This module was designed at InSilixa, beside a reader which is used as a handheld biochip system for PoC DNA analysis [609,653]. Table 8.3 summarizes the various readout techniques discussed in this section. One of the most critical figure-of-merits (FoMs) for a biosensor is the LoD. For the different optical detection methods, LoD can be expressed with various quantities such as minimum photoemission rates (photon/cm^2), minimum bulk refractive index unit (RIU), minimum analyte concentration (ng/ml), minimum detectable molecular mass (g), etc.

Table 8.3 Comparison of CMOS optical biosensors

Array #	Current sensing method	LoD	Dynamic Range (dB)	Power Con. (mW)	Application	ADC	Die size (mm^2)	Tech.	Ref.
8 × 16	Direct current Integration	4×10^5 photons/cm^2	61	26	Pathogen detection, determining gene expression intracellular ATP, DNA sequencing.	128 channel 2-step ADC 13-bit	5 × 5	0.18 μm	[586]
8 × 4	Direct current Integration	1.15×10^8 photon/cm^2	74	NA	DNA Analysis	4th current-mode ADC	5 × 5	0.25 μm	[581]
128 × 128	Direct current Integration	5,000 fluorophore / μm^2	NA	26.2	DNA Analysis	128 channel ADC	4.4 × 2.9	0.35 μm	[582]
1	CTIA	70 molecule/ μm^2(48 zeptomoles)	NA	66	*In vitro* and *in vivo* applications	NA	0.4 × 0.8	65 nm	[580]
12 × 8	CTIA	1 molecule/ μm^2	NA	NA	*In vitro* and *in vivo* applications	NA	NA	65 nm	[579]
2 × 2	RTIA	3.5×10^{-5} (RIU)	NA	NA	Streptavidin target	SAR ADC 9-bit	5.53 × 3.03	45 nm	[549, 568]
32 × 32	ΔΣ current sensing	NA	116	118	DNA & RNA in Molecular Diagnostics	NA	7 × 9	0.25 μm	[555]

NA: Not available, RIU: Refractive index change

8.5 Summary

This chapter provides a review of different types of CMOS optical biosensors along with working principles and constructions. Miniaturization of an entire optical sensing system into a single chip can enable new modalities for sensing in PoC diagnosis. Although many studies have been published on the development of optical biosensors for integration with CMOS technology, most of them are CMOS-compatible or need some post-processing. Only a few counts of the developed optical biosensors are fully integrated, which are capable of opening the pathway to the LoC sensing system on a single die.

There are various technical difficulties in implementing optical biosensors but still, more research efforts are required for offering better alternatives. Some of them which should be considered to achieve better performance in feature design are as follows. Low photocurrent measurement, especially in the presence of background, is one of the important detection challenges. One of the issues with implementing fully integrated optical biosensors is integrating optical filters in CMOS technology. Some challenges of this issue, without requiring additional post-fabrication processes, have been addressed in recent years, particularly in the form of nanoplasmonic filters. However, the excitation source has not been integrated yet, which prevents the fully integrated biosensor for LoC platforms. Therefore, more attempts are required.

Chapter 9
CMOS-based magnetic biosensors

Magnetic biosensors have attracted significant interest in biological and medical applications, which benefit from the inherently negligible magnetic background in biological samples that leads to high-sensitivity measurements. In parallel, CMOS compatibility makes magnetic sensors attractive for Point-of-Care (PoC) applications [654]. Since biological entities have a non-magnetic nature, and most of the bio-molecules have zero magnetic susceptibility, the magnetic particles (usually magnetic nanoparticles (MNPs)) are used as labels in affinity-based assays for biological detection, in which the magnetic particles are functionalized with specific probes that can selectively bind to target analytes, including nucleic acids, proteins, viruses, cells, and bacteria [655–657]. An underlying magnetically responsive sensor detects the presence of the MNPs to measure the number of surface-bound biomarkers quantitatively. Moreover, magnetic signals are stable in biological solutions, which provide robust detection and long shelf life. In this chapter, we will focus on recent advances in various CMOS magnetic biosensors with a perspective on PoC settings for biological diagnostics.

Many techniques are available for detecting magnetic labels such as Hall-effect [658–665], inductive [666–676], giant magnetoresistance (GMR) [677–690], superconducting quantum interference devices (SQUID) [691,692], tunnel magnetoresistance (TMR) [693–696], anisotropic magnetoresistance (AMR) [697–699], and giant magnetoimpedance (GMI) [700–704]. These schemes present different trade-offs among various sensor performance metrics [705]. Among different kinds of magnetic field sensors, this chapter will focus on the advances of sensors that are more compatible with CMOS technology including Hall-effect, inductive, and GMR magnetic biosensors for biological applications with an emphasis on implementing them by CMOS technology for PoC platforms.

9.1 Principles of operation of magnetic sensing

A crucial component of magnetic biosensors is the magnetic particle that is used as a label or magnetic tag in the sensing mechanism. In the presence of an external magnetic field, which can be produced by the influence of electrical current and magnetic material, magnetic particles induce a net magnetic moment. The magnetic particles can be studied from different aspects such as magnetism, size, and assay configuration which are briefly discussed here.

9.1.1 Classes of magnetic particles

Magnetic particles according to their properties and their response to the externally applied magnetic field can be classified into six basic types of magnetism including diamagnetism, paramagnetism, ferromagnetism, ferrimagnetism (ferrite), anti-ferromagnetism, and superparamagnetism. Diamagnetism is a very weak form of magnetism that is induced by a change in the orbital motion of electrons due to an applied magnetic field. The direction of the formed magnetism is opposite to that of the applied field [706]. This magnetism is nonpermanent and persists with very small magnetic only in the presence of an external field [654]. Common examples of diamagnetic materials are gold, water, mercury, bismuth, copper, etc.

Paramagnetic materials have a permanent magnetic moment and can cause a net magnetic field to be produced in the direction of the applied field. However, by removing the applied field, the materials tend to lose their magnetism [707]. Examples of paramagnetic materials are iron oxide, oxygen, magnesium, molybdenum, titanium, lithium, aluminum, transition metal complexes, etc. Ferromagnetic materials have a permanent magnetic moment. In the presence of an external magnetic field, the domains in the direction of the external field tend to expand by utilizing other domains. So, a strong internal moment will be generated and remain in the material, which cannot be canceled even if the external field is removed. Ferromagnetic materials are usually metals such as iron, cobalt, nickel, or their alloys [708].

Ferrimagnetic materials are metal oxides and the most important examples of these materials are magnetite (Fe_3O_4) and maghemite (Fe_2O_3, γ-Fe_2O_3). The key difference between ferromagnetism and ferrimagnetism is that the curie temperature of ferromagnetic materials is higher than that of ferrimagnetic materials [709]. Besides the wide range of ferromagnetic materials, iron oxide is mostly used for bioanalytical applications [710]. Antiferromagnetism is a type of magnetic order that occurs when the magnetic moments of adjacent atoms or ions align in opposite directions, resulting in a net magnetic moment of zero. Antiferromagnetic materials typically exhibit magnetic ordering below a specific temperature called the Néel temperature. Above this temperature, the material becomes paramagnetic, losing its antiferromagnetic properties. Some examples of antiferromagnetic materials include hematite, metals such as chromium, alloys such as iron manganese (FeMn), and oxides such as nickel oxide (NiO) [711]. Superparamagnetism is a form of magnetism that appears in small ferromagnetic or ferrimagnetic nanoparticles, where in the absence of a magnetic field their total average magnetism in any direction appears to be zero. Once in the presence of an external magnetic field, they become magnetized, with their magnetic moments aligning with the field [712].

9.1.2 Size of magnetic particles

The sizes of the magnetic particles used as labels for various biomedical applications vary from a few micrometers to tens of nanometers for biosensing. Typically, micron-sized magnetic particles comprise a large number of MNPs in a silica or polymer matrix, as shown in Figure 9.1. They exhibit a large magnetic moment and provide sensitive analyte detection, even down to a single particle, which makes them an

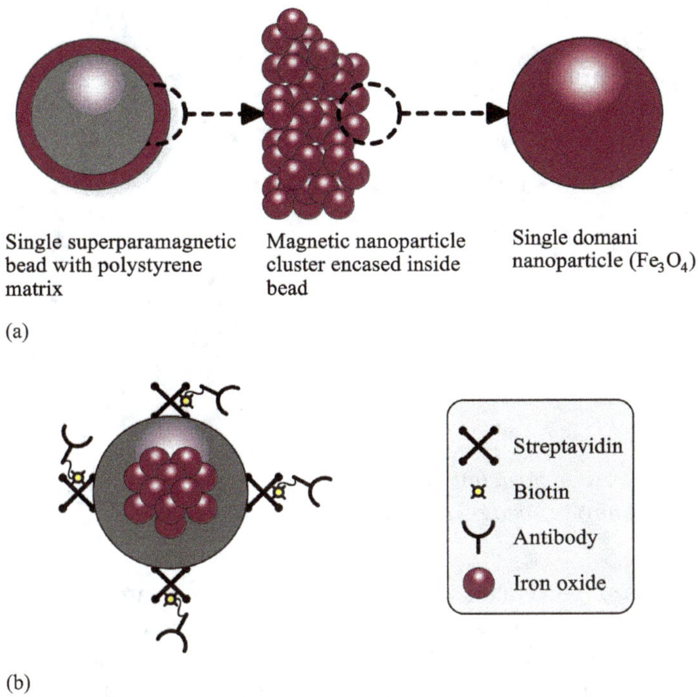

Figure 9.1 *(a) Construction of magnetic bead from iron oxide nanoparticles encased in a non-magnetic polystyrene matrix. (b) The magnetic tag comprises a dozen iron-oxide cores embedded in a dextran polymer*

attractive choice as a label [713,714]. On the other hand, large-sized labels cannot be used in experiments in which the kinetics of the binding reactions is studied since the labels significantly degrade the mobility of the antibodies. Smaller magnetic particles offer more reproducible detection, with minimal nonspecific binding and higher colloidal stability. Another benefit of using small tags is that the unbound tags do not need to be removed from the assay (by washing) because they remain in solution and are not in enough proximity to the sensor to be detected [687]. So, a proper label size will be selected with respect to the different magnetic biosensor types and design specifications.

9.1.3 Assay configurations of magnetic particles

Utilizing magnetic particles as the labels in magnetic sensing techniques can be performed in different assay configurations, including clustering assay, direct labeling assay, magnetic amplification assay, and sandwich assay [715], which are shortly described here.

In clustering assay type, as shown in Figure 9.2(a), magnetic particles with multivalent affinity ligands bind simultaneously to a single target molecule, which results in particle aggregation. Clustering assay is a wash-free assay procedure capable of detecting small molecules, nucleic acids, and proteins in suspension

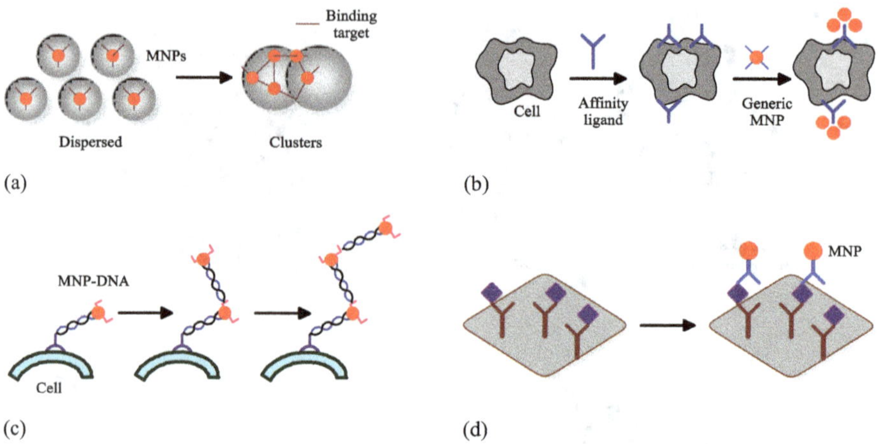

Figure 9.2 Molecular targeting and labeling strategies: (a) clustering, (b) direct labeling (c,) magnetic amplification, (d) sandwich

[64]. Also, it is a good choice for rapid analyses compared to surface-based detection as the reaction takes place in the entire sample volume.

For large biological entities that are not amenable to surface immobilization, the whole target can be labeled with magnetic particles. In this method, which is known as a direct labeling assay (Figure 9.2(b)), the unbound particles can be removed due to the size difference between target cells and magnetic particles leading to increases in the overall detection sensitivity [716,717].

In magnetic amplification assay, as shown in Figure 9.2(c), multiple magnetic layers can be formed by sequentially applying magnetic particles conjugated with orthogonal binding partners [718]. Magnetic signals can be amplified to detect rare molecular targets, so the overall detection sensitivity will be improved [719].

The standard magnetic assay, which is used for the majority of magnetic sensors, is an affinity-based sandwich assay whose working principle is shown in Figure 9.2(d). In this approach, the surface of the sensor is functionalized by capturing probes of desired targets. Then, the sample is introduced, and the target analytes bind to the capturing probes, while other analytes do not. Next, the surface must be washed to eliminate the nonspecific targets from the substrate. Then, the magnetic particles, as labels, are fed into the system and bind the target analytes via conjugated capturing probes specific to another section of the target analyte. After washing the surface, the unbound targets are removed. Finally, generated magnetic signals from magnetic particles, which are correlated with the presence of the target analytes in the samples, are detected by the underlying magnetic sensors.

These washing steps can have a profound impact on the quality and reproducibility of the result. So, the standard sandwich assays can cause limitations for PoC applications. The washing steps can be eliminated by reorganizing the sequence of events in the standard magnetic assay, as illustrated in Figure 9.3 [720]. Before explaining washing-free assay, it is worth mentioning that typically in magnetic assays,

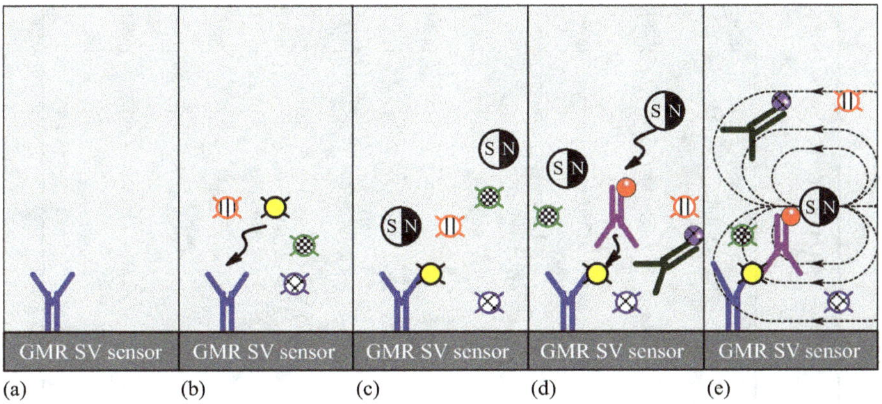

Figure 9.3 Wash-free sandwich assay: (a) capturing antibody is immobilized on the sensor surface, (b) the sample is added to the assay and incubated, (c) the magnetic nanotags are added, (d) the addition of detection antibodies results in the binding of magnetic nanotags to the analyte, (e) completed version of the wash-free magnetic assay

the magnetic particles are decorated with streptavidin. Streptavidin has an extremely high-affinity bond with biotin, which is one of the strongest noncovalent bonds known in nature [721]. The detection antibodies are biotinylated meaning that they have attached biotin which allows for a facile connection of magnetic particles. So, it is preferred to continue the assay description by a streptavidin/biotin attachment.

Free washing is similar to the standard sandwich assay with the difference of adding the magnetic particles after introducing samples (Figure 9.3(c)). Next, the biotinylated detection antibodies are added to the assay (Figure 9.3(d)). The biotin on the detection antibodies binds to the streptavidin on the magnetic particles and tethers them to the surface of the sensors (Figure 9.3(e)). It can be deducted from Figure 9.3(e) that all of the unreacted analyte, detection antibodies, and magnetic particles remain in the assay, as there are no washing steps to remove them. This often results in a slightly higher value of the lower limit of detection due to additional nonspecific binding [722]. In addition to the reviewed assays, other schemas have also been reported such as antibody assay, competitive antibody assay, reverse phase assay, and biotin BSA assay whose more detailed information regarding these assays can be found in [722]. Furthermore, different applications of magnetic sensors will be discussed in Chapter 14.

9.2 CMOS sensor chips of magnetic biosensors

As mentioned earlier, various types of magnetic sensors have been developed to measure the magnetic signal from the labeled biological objects with magnetic particles. In this section, we will review the theory for the operation and CMOS implementation of magnetic sensors including Hall-effect, inductive, and GMR magnetic sensors. Comparisons of reviewed CMOS magnetic sensors are summarized in Table 9.1.

Table 9.1 Comparison of different CMOS magnetic sensor

Type	Sensitivity	PN@1MHz /Frequency operation	Min. DMP*/ Readout time	Array	Power (mW)	Application	Post Proc.	External Magnetic	Fluidic	Area of chip/Each sensor	Process	Ref.
Hall-effect (Relaxation)	S_V: 50 mV/T	NA	2.8 μm, 64 s	256	138	Immunoassay	Yes	No	No	2.5 × 2.5 mm^2, 8.5 × 6 μm^2	0.18 μm	[659]
Hall-effect (Relaxation)	S_V: 50 mV/T	NA	<1 μm, 8.2 s	10240	300	Immunoassay	Yes	Yes	No	5.1 × 3.5 mm^2, 8 × 6 μm^2	0.18 μm	[728]
Inductive (LC)	$\Delta f/f$: 0.23 ppm 1 nM (DNA)	−135 dBc/Hz 1 GHz	1 μm, <90 s	8	165	DNA Analyses	No	No	PDMS	2.95 × 2.56 mm^2, NA	0.13 μm	[730]
Inductive (LC)	$\Delta f/f$: 63 ppm	−135.3 dBc/Hz 1.01 GHz	1 μm	64	80	DNA/RNA/Bacteria Detection	No	No	No	1.44 × 2.07 mm^2, NA	65 nm	[673, 675]
Inductive (LC)	$\Delta f/f \sim$ 1 ppm @ 1.1 GHz	−133.7 dBc/Hz for 1.4 GHz −124.8 dBc/Hz for 3.3 GHz 1.1-3.3 GHz	1 μm	4	1.8-6	Magnetic Spectrometer Protein /DNA Detection	No	No	No	1.2 × 1.2 mm^2, 260 × 260 μm^2	65 nm	[671]
Inductive (LC)	2.17 pM	1.4 GHz	200 nm	14	42.1	Immunoassay	No	No	PDMS	2.3 × 1.9 mm^2, NA	0.18 μm	[667]
Inductive (Transformer)	NA	1.2 GHz	4.5 μm, 20 ms	1	61.2	Cell (mEF) Flow Cytometer	No	No	PDMS	1.1 × 1.3 mm^2, NA	0.18 μm	[670]
Inductive (Transformer)	0.14 V/mT/ GHz	0.9 GHz 2.6 GHz	4.5 μm, 10 ms	2	-	Flow Cytometer breast cancer	No	No	PDMS	1.5 × 0.45 mm^2, NA	65 nm	[15]
Inductive (Transformer)	0.3 ppm	−131.8 dBc/Hz for 1.4 GHz −122.8 dBc/Hz for 3.6 GHz	4.5 μm	4	5	DNA Analyses	No	No	No	NA, 250 × 250 μm^2	65 nm	[738]
Inductive (Bridge)	NA	770 – 1450 MHz	4.5 μm, 30 min	1	19.7	Immunoassay Cytometer	No	No	No	1 × 1.5 mm^2, 20 × 200 μm^2	65 nm	[740]
GMR (SV)	NA	NA	50 nm, 250 ms	1,008	NA	DNA Analyses	Yes	Yes	No	2 × 2 mm^2, NA	0.25 μm	[688]
GMR (SV)	0.01 pM (protein)	NA	50 nm, 4 s	256	50.4	Protein detection	Yes	Yes	No	2.7 × 2.7 mm^2, NA	0.18 μm	[687]
GMR (SV-MRX)	4.4 ppm	NA	50 nm, 11 ms	1	1.39	Immunoassays	Yes	Yes	No	2.3 × 0.9 mm^2, NA	0.18 μm	[679]
GMR (MRX)	NA	NA	40 nm, 704 ms	1	4.32	Immunoassays	Yes	Yes	No	2.31 × 1.65 mm^2, NA	0.18 μm	[677]
GMR (SV)	NA	NA	250 nm, 192 s	192	6.5	DNA analyses	Yes	Yes	No	7.59 mm^2, NA	0.35 μm	[682]

*Minimum Detectable Magnetic Particles Size, NA: Not available

9.2.1 Hall-effect magnetic biosensors

9.2.1.1 Principles of operation of Hall-effect magnetic biosensors

Hall-effect was discovered by Edwin Hall in [723]. When a current-carrying conductor (Hall plate) is placed in an external magnetic field, B_{EXT}, the charge carriers move away from the nominal current direction. As a result, a transverse electrical field is created which eventually stops the carrier drift and can be measured as a transverse electrical potential, called Hall voltage, V_H. This phenomenon is known as the Hall effect and is a fundamental principle in designing Hall-effect sensors to detect magnetic particles. A generated Hall voltage is perpendicular to both the current and the magnetic field as described by (9.1):

$$V_H = G_H\,\mu_H B_{EXT}\,V_{bias.H}\frac{W_H}{L_H} \tag{9.1}$$

where W_H and L_H are the width and length of the Hall plate, μ_H is the Hall mobility of the majority carriers, $V_{bias,H}$ is the bias voltage applied to the Hall plate, and G_H is the geometrical correcting factor for finite-size Hall devices. As seen in (9.1), V_H is linear with the applied magnetic field, B, and bias voltage, $V_{bias,H}$. Therefore, the sensitivity parameter for a Hall-effect sensor, which is known as voltage-related sensitivity can be defined as:

$$S_V = \left|\frac{V_H}{V_{bias.H}\,B_{EXT}}\right| = G_H\,\mu_H\frac{W_H}{L_H} \tag{9.2}$$

According to (9.2), S_V is proportional to carrier mobility. Typical values of S_V for semiconductor Hall devices are less than 0.1 T^{-1} in silicon [724]. Magnetic Hall sensors are suitable for integration and scaling in CMOS technology as Hall plates can be realized either with a diffusion layer (e.g., N-well) or the channel region of a transistor. Generally, for Hall-effect sensors, N-well prototypes are chosen due to their higher carrier mobility.

A cross-section of the entire detection platform of the Hall-effect magnetic bead detector is shown in Figure 9.4 [713]. A current I_{mag} passing through a pair of metal wires on both sides of the N-well Hall-effect sensor generates the magnetizing field B_{mag}. The induced magnetic field from the bead, B_{bead}, is detected by

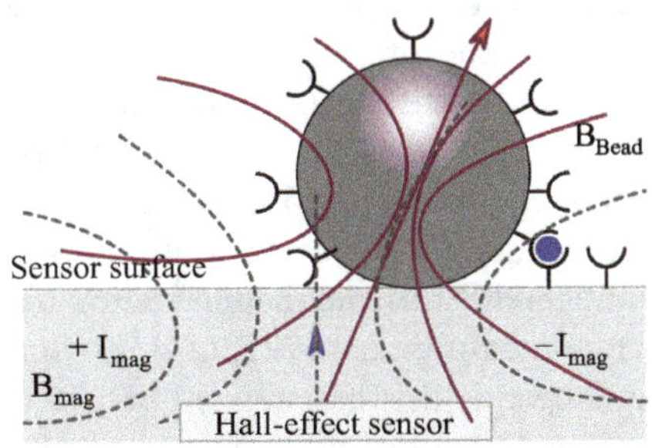

Figure 9.4 Cross section of a CMOS Hall-effect magnetic label detector

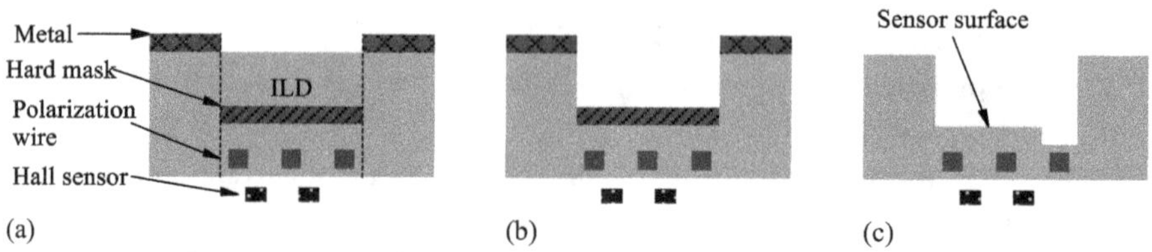

Figure 9.5 Post-process etching steps performed after CMOS chip fabrication to reduce the distance between beads and Hall sensors. (a) Chip is fabricated with a metal hard mask defining the sensor region. (b) Reactive ion etching is used to remove interlayer dielectric over the sensor region. (c) Exposed metal layer over the sensor surface is removed with a wet etch

the embedded Hall-effect sensor and converted to an electrical signal for processing [725].

Hall-effect sensors also detect the stray field from the polarized magnetic particles. So, the distance between the magnetic particles and the sensor surface must be minimized to achieve measurable signals. In typical multilayer metal processes, the chip surface is above the active devices and the magnetic signal from the bead decreases with the third power of distance. Therefore, a simple post-processing step is required to remove the passivation layer and most of the top interlayer dielectric, and consequently, the magnetic particles will be closer to the sensor surface and the sensitivity will increase. For instance, in the work reported by Skucha *et al.* [658], the distance between the surface and sensors was reduced from 9 μm to 3 μm by etching the interlayer dielectric above the sensor area, which resulted in a 10x–20x signal strength increase for the bead sizes on the order of 1 μm. Figure 9.5 shows the post-processing flow [713]. Notably, post-processing steps are cost-effective and could also compromise the yield [726].

9.2.1.2 Implementation of Hall-effect magnetic biosensors

During the past decades, Hall-effect magnetic sensors with various bead detection methods have been reported, which were integrated with standard CMOS technology [658–665]. In some bead detection methods such as the ones reported by Florescu *et al.* [663] and Aytur *et al.* [664], the authors attempted to resolve the magnetic field from a bead in the presence of a much larger magnetizing field baseline. Since the baseline is sensitive to environmental variations, these solutions, generally, require reference sensors, baseline calibration, and temperature stabilization, which can add significant penalties on-chip area, power consumption, and detection time. One approach to overcome these problems is based on relaxation measurement that extracts the intrinsic magnetization dynamics of the labels to take an ideally baseline-free measurement [725,727]. In this measurement method, as shown in Figure 9.6, the bead is first magnetized by a large field. Afterward, the magnetizing field is removed rapidly, and measurement starts; thus, the large baseline is eliminated, and the decaying magnetic field is detected from the bead.

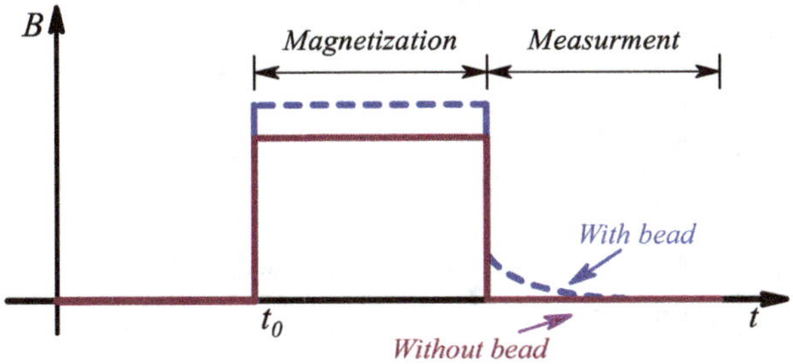

Figure 9.6 Detection method based on bead relaxation

Liu *et al.* [662] reported a fully integrated Hall-effect magnetic sensor based on magnetorelaxometry (MRX) which includes four 8 × 8-arrays implemented in a 0.18 μm CMOS process without requiring any baseline calibration or reference sensor. The sensor relies on the wires embedded in the chip to generate a local magnetizing field. Thus, not only does it eliminate the need for any external magnet but also it provides rapid switching off to observe the relaxation field of the magnetic particles. Due to the small magnitude of the bead signal, non-idealities such as stray magnetic field, DC offset, and thermal effects of the Hall-effect sensors and the CMOS readout circuit must be suppressed. In this work, modulation of the magnetizing field and correlated double sampling (CDS) were used for rejecting low-frequency errors and thermal effects.

As an extension of this work [662], this group [659,725] also used CDS and magnetic field modulation, combined with offset servo loops technique to suppress the detector's offset to sub-μT. The offset cancelation in the servo loop is realized by 4-bit calibration digital to analog (DAC) in each of the four programmable gain amplifiers (PGAs) that process the sensor's output. The block diagram of the magnetic bead detector chip is shown in Figure 9.7(a). Since the micron-sized beads generate only a localized field and therefore cannot be detected by a single sensor with a large area, the magnetic sensor is organized into four banks of 64 individual Hall-effect devices (256-element Hall-sensor array). In this scheme, the sensor's outputs are processed serially through a multiplexed readout channel that performs offset rejection and signal amplification. Also, each sensor element consists of two access transistors and a Hall plate modeled by a four-resistor bridge as shown in Figure 9.7(b). In the relaxation method, the delay between switching off the magnetizing field and detecting the magnetic particle's field is known as the dead zone, which usually is several hundred microseconds. They achieved the readout time of 64 ns by the integration of an electromagnet and readout channel. The minimum detectable bead is 2.5 μm with S_V of 0.03 T^{-1} (50 mV/T) [725].

Although Liu *et al.* [659] indicated the outstanding features of relaxation measurement to reduce baseline infection, its 256-element array was too small to provide clinically relevant results [728]. Also, as seen in Figure 9.7(a), this sensor uses a single-channel readout chain that limits the speed at which the array can be measured. To address these challenges, this group [728] proposed another

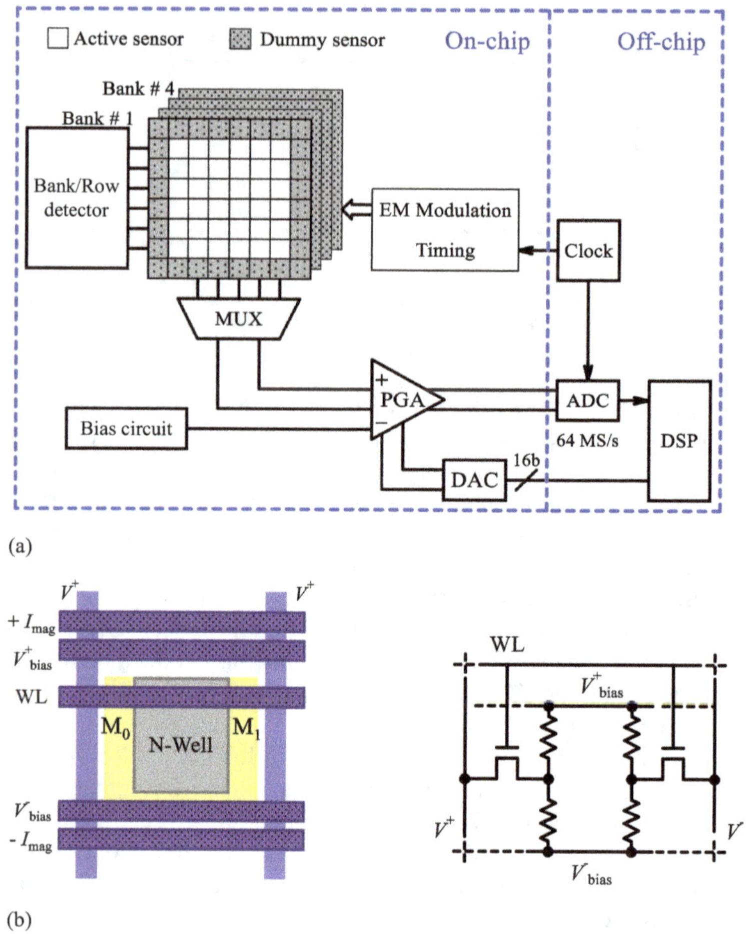

Figure 9.7 (a) Chip block diagram of a magnetic bead detector and (b) Hall-effect sensor element (equivalent circuit and layout)

Hall-sensor design based on relaxation measurement fabricated by 0.18 μm CMOS process. This sensor is composed of 10,240 sensing sites with column parallel readout electronics that combine auto-zeroing and nested chopping to enable low-offset and power-efficient signal acquisition. A block diagram of the entire system is shown in Figure 9.8(a). The chip is subdivided into two sub-arrays, each comprising a magnetization current generator, a timing generation block, and two blocks of Hall sensors with associated readout electronics. Each readout chain serves 64 sensor cells and contains low-noise amplification, averaging, and digitization. The sensor structure is also a modification of the one discussed in [659] and is shown in Figure 9.8(b). Each Hall-effect pixel contains a 3 μm × 4 μm Hall-effect device implemented with an N-well resistor to maximize mobility and hence

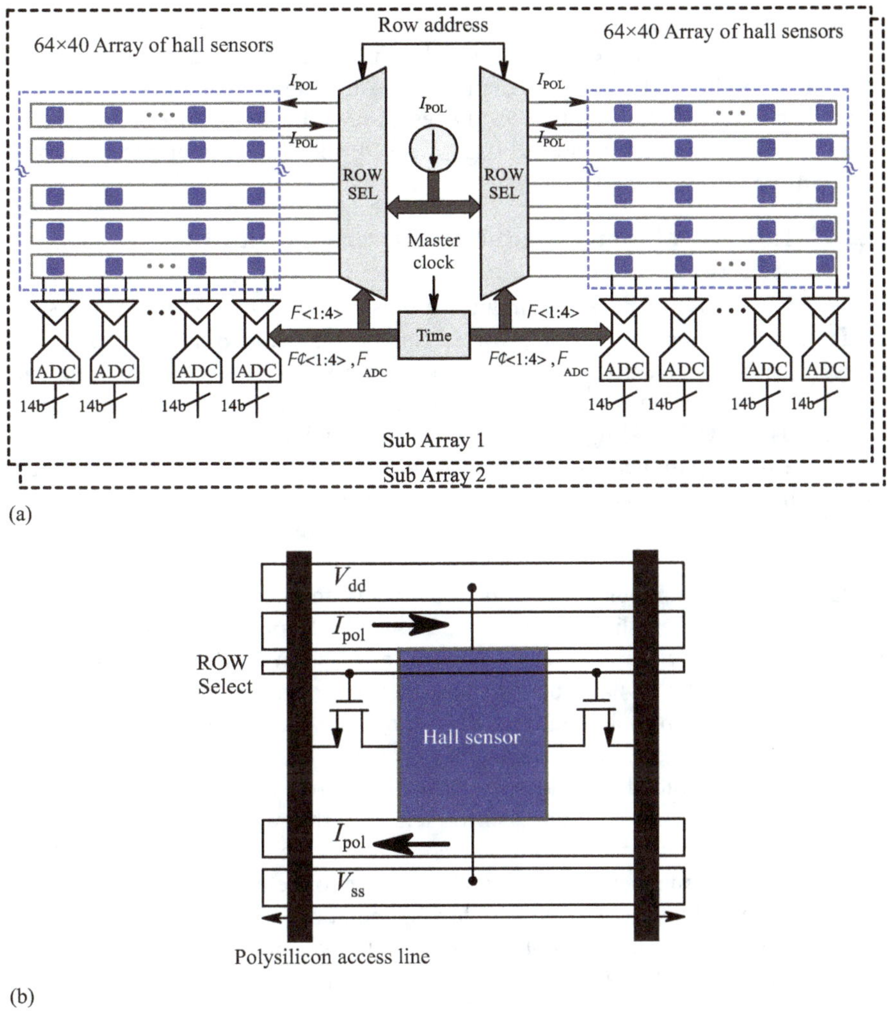

Figure 9.8 (a) Array and electronics block diagram of the entire system and (b) annotated Hall-effect sensor element

transduction gain. The sensor successfully detects magnetic particles whose sizes are less than 1 μm within a readout time of 8 s. As reported in Table 9.1, S_v of both structures is 50 mV/T [659,660].

9.2.2 *Inductive magnetic biosensors*

Magnetic detection schemes such as the Hall effect generally need external biasing magnetic fields to bias the immobilized magnetic labels during their sensing processes. In addition, fabrications of magnetic Hall sensors require expensive post-processing steps to improve sensitivity and biocompatibility. Inductive magnetic biosensors are completely compatible with CMOS processes without post-process modifications and do not need any external biasing magnetic field, which simplifies and diminishes the cost of system implementation. Numerous inductive magnetic sensors implemented through CMOS processes have been documented in the literature, including inductor-capacitor (LC)-based and transformer-based sensors, which will be briefly reviewed next.

9.2.2.1 Implementation of inductive biosensors with LC structure

One promising class of magnetic inductive biosensing scheme is frequency-shift-based magnetic resonant sensors, which can be realized with standard CMOS processes with no need for costly post-processing steps or any electrical or permanent external biasing magnets [672–676,729]. The core of the sensor of these structures as shown in Figure 9.9 is an on-chip low-noise LC resonator, the current through the on-chip inductor generates a magnetic field, and polarizes the magnetic labels on the sensor surface during the detection process. This increases the total magnetic energy and thus the effective inductance of the resonator that subsequently leads to a resonant frequency down-shift of the LC tank, which can be correlated with the number of beads near the resonator.

The first frequency shift sensor was designed by Wang *et al.* [730] as a single chip containing eight parallel sensor cells in a 0.13 μm CMOS process. A poly dimethylsiloxane (PDMS) microfluidic structure was also fabricated and bonded to the CMOS sensor chip to facilitate the sample delivery. It is noteworthy to mention that the oscillation frequency of an on-chip oscillator is sensitive to environmental perturbations, which leads to undesired frequency drifts and impacts the detection sensitivity on oscillation frequency shift. Hence, in this work [730], a differential frequency sensing scheme was introduced that was composed of two matched oscillators, one for magnetic sensing and the other for reference track, that shares the same supply, bias, and local on-chip temperature controller. In addition, a simple frequency counter was implemented next to the sensor oscillators to achieve digital outputs of the oscillation frequencies for further processing. Sensor

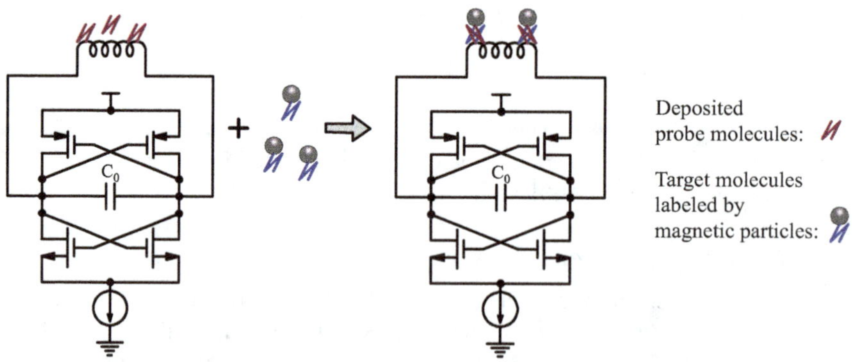

Figure 9.9 Concept of the frequency-shift magnetic sensor

frequency-shift response ($\Delta f/f_0$) to 1 nano-molar deoxyribonucleic acid (DNA) sample labeled by MNPs (diameter = 50 nm) is 2.8 parts per million (ppm). More details about this structure can be found in [731].

In another design reported by Wang *et al.* [675], a novel noise suppression technique, correlated double counting (CDC), was proposed to further enhance the sensitivity of the frequency-shift magnetic sensor, which was implemented with 64 arrays in 65 nm CMOS process. A simplified CDC configuration for the frequency-shift magnetic sensor is shown in Figure 9.10. Two LC resonators are implemented as differential sensing sites that share the same oscillator active core and the switches S_1–S_4 control which tank is connected to the oscillator active core at any given time for proper operation. Since the oscillator's $1/f^3$ phase noise is mainly due to the up-conversion of the active device flicker noise, the active core sharing directly establishes the correlation between the $1/f^3$ phase noise of the differential sensing and the reference counting results. To verify the magnetic sensing capability of the sensor, DynaBeads®MyOne™ Carboxylic Acid (diameter = 1 μm), is used as a test sample [675]. The achieved sensor frequency-shift responses ($\Delta f/f_0$) at 1.01 GHz with and without CDS are 63 ppm and 125 ppm, respectively. Thereby, the CDC scheme achieves an additional 6-dB noise suppression. Since the $\Delta f/f_0$ for a given magnetic particle is proportional to the particle's local polarization magnetic field strength normalized by the sensing inductor current, for conventional on-chip inductor geometries, this normalized magnetic field strength varies spatially and results in a spatially non-uniform transducer gain on the sensor area. To address this issue, this group [674] optimized the sensor inductor geometry, so that the polarization

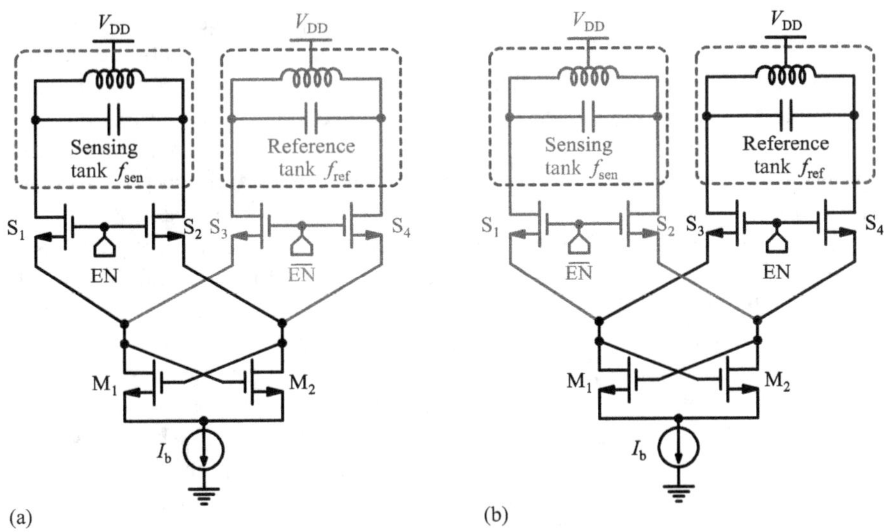

Figure 9.10 The inductive-based magnetic biosensor with a CDC configuration for a frequency-shift magnetic sensor: (a) sensing tank is enabled and (b) references tank is enabled

magnetic field changes and the spatial uniformity of the transducer gain across the sensing area is improved, and directly the system dynamic range increases. Frequency-shift magnetic sensors are capable of configuring as cell-based sensors, which utilize living cells or tissues as an integral part of the sensor front end [732,733]. They [673] used the frequency-shift magnetic sensor of [675] alongside cardiac progenitor cells that are differentiated directly on-chip to develop a magnetic cell-based sensing platform. In this design, also, a PDMS reservoir on top of the CMOS sensor chip was constructed to hold the sensing cells and the test chemicals.

The frequency-shift magnetic sensors, such as [673–676], despite their cost and sensitivity advantages, cannot diagnose different particle types. To solve this issue, Sideris and Hajimiri [672] proposed a magnetic spectroscopy concept, which is based on considering frequency-dependent response in the magnetic susceptibility [734], as different types of magnetic particles exhibit different resonance frequencies in terms of their induced magnetizations when being polarized. If the polarization magnetic field is at or beyond the resonance frequency, the paramagnetic particles become nonmagnetic or even diamagnetic, which results in a zero or negative magnetic susceptibility and, in turn, leads to a zero or positive frequency shift. Therefore, if the magnetic sensor can perform sensing across a wide frequency range, and effectively measure the magnetic spectrum, it can infer both the types and the quantities of the different magnetic particles present. The magnetic spectroscopy of this work [672] allows for the characterization of the magnetic susceptibility of magnetic beads across a broad frequency range of 1.1 GHz to 3.3 GHz that is achieved with a thermometer-coded switched capacitor (SC) bank, allowing for 16 unique frequencies of operation. This enables the differentiation and quantification of multiple beads of varying types concurrently present in the active volume of a sensor's surface [671]. The frequency response of three different kinds of beads, 1-μm and 4.5-μm Dynabeads magnetic beads from Invitrogen and 1-μm beads from Bangs Laboratories were measured. The results demonstrate successful direct magnetic particle detections and classifications.

The outstanding features of frequency-shift magnetic sensors such as extremely low-cost manufacturing, robust, easily integrated, and magnetic particle multiplexing capabilities make them ideal for handheld PoC applications. Pai *et al.* [735] presented another design that provided modifications to the frequency-shift magnetic sensor scheme reported by Wang *et al.* [675] for PoC viability. In this work, a handheld reusable reader and a disposable open-well cartridge for the detection of nucleic acids and antigens were introduced. The diagnostic system utilized a novel "magnetic freezing" technique to reduce measurement time, obviate baseline measurement before or during the biological assay, and reduce sensor noise [735]. In another work, Geng *et al.* [667] reported a high-sensitivity oscillator-based CMOS magnetic biosensor array to detect clinical biomarkers in human samples and provide accurate information to diagnose human diseases in a 0.18 μm radio frequency (RF) CMOS process. In this design, to increase the detection sensitivity, dynamic range, and target number, a novel inductor structure and a general methodology for designing the on-chip inductor are proposed, which is fully compatible with the standard CMOS processes.

9.2.2.2 Implementation of inductive biosensors with transformer structure

Flow cytometers are instruments that can detect and classify specific cells in a biological sample, which are widely used in medical diagnostics for hematology, immunology, chemotherapy, human immunodeficiency virus (HIV) detection, etc. [736]. Numerous research studies have explored label-free chip-scale flow cytometers, yet their practicality is constrained by a lack of specificity, limiting their potential applications. Substituting magnetic labels avoids this limitation. It is important to note here that magnetic labels with superparamagnetic nanoparticles embedded in a polymer matrix exhibit frequency-dependent complex susceptibility [670]. At low frequencies, the susceptibility depends both on size and material composition, but at frequencies beyond 1 GHz, the material properties dominate [737]. For example, at 2 GHz, the phase of the susceptibility of iron oxide and cobalt differs by about 60 degrees. Therefore, variations of the complex susceptibility of superparamagnetic nanoparticles with different material compositions can be used to distinguish between label classes independent of label count or signal strength [670]. The phase difference can be used for classifying magnetic labels. This concept is used to design magnetic flow cytometry. Despite all the advantages, the previously reviewed structures, such as [671–673,675], do not satisfy the bandwidth and sensitivity required for flow cytometry.

In one effort, Murali *et al.* [670] presented a magnetic flow cytometer integrated using a 0.18 μm standard CMOS technology assembled in a single-use microfluidic cartridge fabricated with PDMS. As shown in Figure 9.11, the magnetic sensor consists of a primary excitation coil to generate magnetic flux with embedded secondary pick-up coils. Magnetic labels passing over either of them modulate the flux coupling between the primary and secondary coils, generating a pulse at the output. The archived results indicate that the phase permits unambiguous separation of labels including FeO_x and Co. The nominal excitation frequency is 1.2 GHz. The active die area is 1.1 × 1.3 mm^2, sufficiently small for use in a disposable cartridge. It dissipates 61.2 mW from a 1.8 V supply.

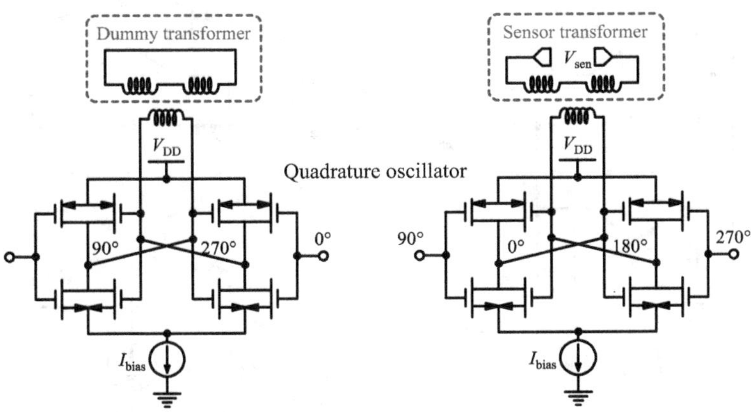

Figure 9.11 The inductive-based magnetic biosensors with a quadrature oscillator and the sensor transformer

Another magnetic flow cytometer was also presented by Murali *et al.* [15] that benefits from multi-frequency interrogation which improves the classification accuracy of different magnetic label classes and is fabricated in 65 nm CMOS technology. The CMOS chip is integrated within the microfluidic channel using the hot-embossing technique. Encapsulation of the multi-frequency CMOS magnetic sensor chip into a microfluidic cartridge enables the detection of multiple label classes. In this design, a weak magnet (10 mT) beneath the CMOS chip is used to pull the beads closer to the sensor surface. The chip can detect SKBR-3 breast cancer cells specifically tagged with magnetic labels ($MnFe_2O_4$, Fe_2O_3, and $CoFe_2O_4$) by demodulating the signal from a pick-up coil enclosing a differential excitation coil operating at 0.9 GHz and 2.6 GHz of a differential spiral transformer.

As mentioned earlier, one way to achieve magnetic multiplexing is to design a multi-frequency structure [672]. In such topologies, like the one reported by Sideris and Hajimiri in [672], an SC structure provides two oscillation frequencies, but it is a poor approach for accurately tracking the drift of the lower frequency. Also, switching structures can add a significant amount of extra parasitic capacitance and contribute more noise to the system. To solve these problems, a transformer-based approach can be considered as proposed by Sideris *et al.* in [666]. In this design, as shown in Figure 9.12, the fourth-order transformer-based system is designed with two resonance frequencies of 1.44 GHz and 3.65 GHz in a standard CMOS 65 nm process. A 2×2-array of the transformer-based drift-canceling magnetic sensor enables sub-ppm signal detection without the need for any thermal compensation. The approach is self-referencing, leading to significant savings in the chip area by removing the need for replica reference cells. The measured phase noise was −130 dBc/Hz at a 1-MHz offset for the 1.44-GHz frequency and −122.8 dBc/Hz at 1 MHz offset at 3.65 GHz. The

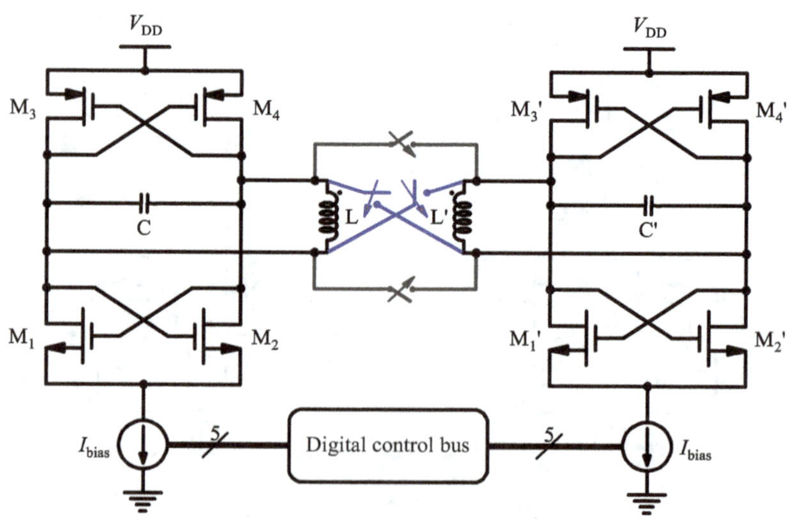

Figure 9.12 Dual-resonance transformer-based magnetic sensor

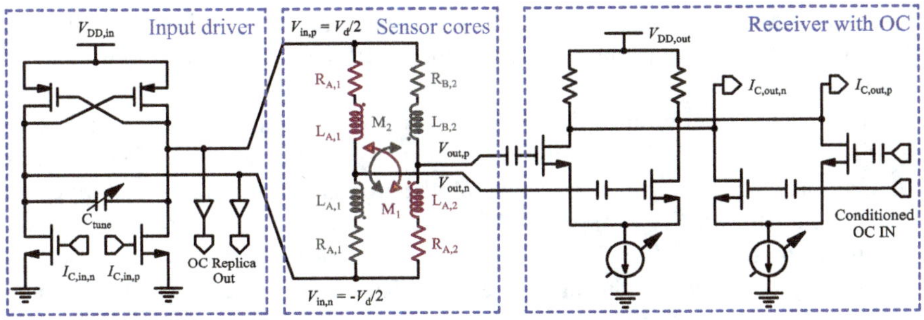

Figure 9.13 The inductive-based magnetic biosensors with the sensor circuit diagram, simplified driver, and receiver reported by Gal-Katziri and Hajimiri

results of an initial DNA bio-detection experiment are also achieved which confirms the capability of the sensor to be used for realistic bioassay [738].

Most reviewed CMOS-integrated inductor-based magnetic sensors provide a solution for point-of-use diagnostics with different trade-offs, as indicated in Table 9.1. For instance, frequency shift oscillators [730] are compact and have the highest sensitivity, but suffer from oscillator long-term drift [738]. On the other hand, integrated transformers [668,670] are more suitable for cell sorting applications but require additional tuning inductors to achieve their required single-bead sensitivity. To address the mentioned issues, Gal-Katziri and Hajimiri [739] proposed a novel sensor topology based on a compact differential coupled inductive bridge in a 65 nm bulk CMOS process. The sensor is shown in Figure 9.13, which is based on a differentially driven, differential AC (alternating current) bridge, and unlike standard bridge structures, it is composed of two pairs of coupled inductors [740]. It offers a response time similar to transformer-based sensors and sensitivity similar to frequency shift oscillators that may enable it to serve as a single solution for a wide range of applications. It can operate between 770 MHz and 1,450 MHz. The new bridge sensor demonstrates a reliable and continuous detection of 4.5 μm iron-oxide magnetic beads over time periods longer than 30 min.

9.2.3 GMR magnetic biosensors

9.2.3.1 Principles of operation of GMR magnetic biosensors

Magnetoresistance is the property of a material to change its electrical resistance in response to a magnetic field. GMR, AMR, and TMR are different types of magnetoresistance. The widespread applications of the seminal discovery of the GMR effect by Fert *et al.* [741] and Grunberg *et al.* [742] won them a well-deserved Nobel Prize in Physics in 2007 [743]. GMR is a quantum mechanical magnetoresistance effect observed in multilayer film stacks composed of two or more ferromagnetic layers separated by a thin non-ferromagnetic spacer. The resistance depends on whether the magnetization of adjacent ferromagnetic layers is in parallel or an antiparallel alignment resulting in low or high resistance, respectively

[743]. The GMR sensors utilize the giant magnetoresistive effect for magnetic detections. In this section, the focus is on the GMR sensor integrated with CMOS technology.

Although the operation of GMR sensors is based on multilayer film stacks, its operation can be simply described by three key layers, including the free, the pinned, and non-magnetic spacer layer, which the free layer and the pinned layer are ferromagnetic (Figure 9.14(a)) [678]. After fabrication, the pinned layer has a fixed magnetization, but the free layer magnetization can rotate freely owing to the external magnetic field H_A during the sensing. The angle of the magnetic moments between the free layer and the pinned determines the sensor resistance as shown in Figure 9.14(b). With no applied field, the sensor has a nominal resistance, R_0. When H_A biases the free layer to be in parallel with the pinned layer, carriers' orientation both of the free layer and the pinned layer will be the same, which results in resistance decrement. Contrariwise, the anti-paralleling of the free layer with the pinned layer leads to opposite carriers' orientation, consequently, resistance will increase. Hence, it can be deduced that the external field strength is approximately linear with the effective resistance of the sensor.

One drawback of the basic multilayer GMR film stack is the high magnetic field required to fully switch the magnetization from the antiparallel to the parallel state which tends to cause the sensors to have poor low-field sensitivity [683] and limits their utility as biosensors. One widely used GMR sensor configuration implemented in CMOS processes that overcome this drawback is GMR spin-valve (SV) sensors by introducing a synthetic antiferromagnet to the film stack [677,678,680,684,685,687,689,743–746]. This antiferromagnet formed by the CoFe/Ru/CoFe tri-layer structure pins the magnetization with exchange coupling. The free layer rotates freely with the applied magnetic field and a pinning layer, typically PtMn or IrMn, defines the magnetization of the synthetic antiferromagnetic and is also used to direct the microstructural texture of the subsequent thin films. A thin biocompatible passivation layer, such as a gold or silicon dioxide layer, can be used to create a noncorrosive sensor surface. The GMR SV structure of the two reported works is shown in Figure 9.15 [684,688].

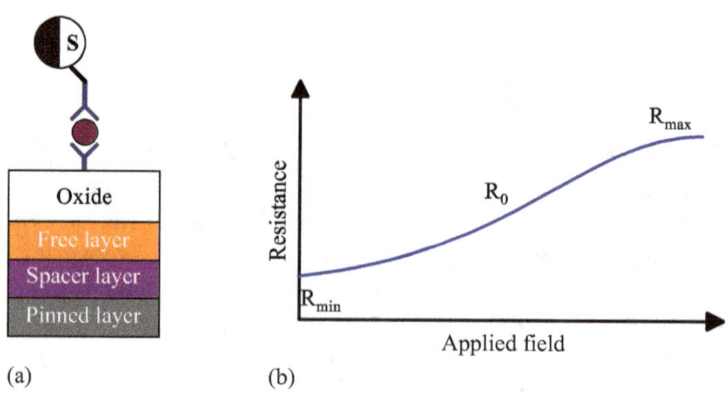

Figure 9.14 Operation principle of basic GMR sensor. (a) Simplified layer stack and (b) sensor resistance vs. applied field

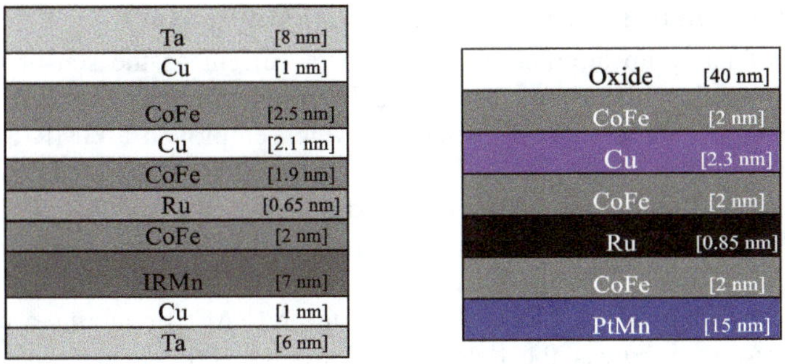

Figure 9.15 GMR SV sensor film stack annotated with thicknesses

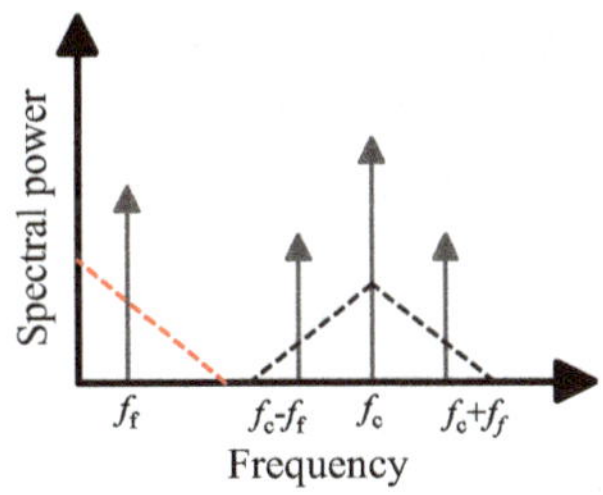

Figure 9.16 Exemplary spectrum using the double modulation scheme; the black dashed line shows the flicker noise of the GMR SV sensor, and the red dashed line depicts the flicker noise of the interface electronics

Notably, the operation of the SV sensor requires two external magnetic fields including the excitation field and the biasing field. Although the biasing field may be optional in certain sensing modes, the external magnetic field is generally needed and requires extra off-chip hardware and dedicated alignments, directly increasing the system's complexity, cost, and power consumption [726]. The GMR SV sensor will be integrated with CMOS integrated circuits (ICs) through post-process steps. The details about the fabrication process are reported in [682,747].

There are several challenges in designing a sensor interface for GMR SV biosensors. One of the most important challenges is that these sensors exhibit high flicker noise at low frequencies. Each magnetic particle can only induce a small resistance change, due to its small size. So, in the presence of the sensor's flicker noise ($1/f$) detection, such minute resistance changes will degrade the sensitivity of the entire system. To reduce flicker noise contribution, a double modulation technique is utilized to modulate the signal from the magnetic particles to a higher frequency where the sensor has lower flicker noise [677–680,684,685,687–690]. This is accomplished by applying a time-varying magnetic field (at a frequency f_f) and a time-varying excitation voltage (at a frequency f_c) to the sensor [655]. As shown in Figure 9.16, modulating the sensor excitation moves the signal band to a higher frequency, away from the flicker noise of the interface electronics, while modulating the magnetic field reduces the flicker noise of the sensor.

9.2.3.2 Implementation of GMR magnetic biosensors

The choice of using an alternating voltage or a current as the sensor excitation signal has spectral implications. The original implementations reported in the literature [688,690] used a current excitation, which employed a GMR SV CMOS sensor in a 0.25 μm bipolar CMOS (BICMOS) process that provided a very large-scaled array with 1,008 sensing sites (16 subarrays per chip and 7 × 9 sensors per subarray). To reduce the read-out time of this high-density array, both frequency-division multiplexing (FDM) and time-division multiplexing (TDM) schemes are applied. The building block of the analog front end (AFE) contains a mixer and PGA. The system achieves the detection of around 100 magnetic particles per sensor in DNA hybridization assays at the DNA concentration of 10 nM.

As another current excitation implementation, Costa *et al.* [682,748] GMR SV sensors in AMS 0.35 μm technology as a CMOS AFE design which is optimized to have lower noise and power consumption, which are 16 $nV/\sqrt{Hz}$ and 6.5 mW, respectively and appropriate its usage at the PoC platform. The full system can detect 250 nm MNPs with a readout time of 192 s. Since this design only uses one amplifier and does not employ FDM, the readout time of the proposed work is not so appropriate; however, readout time optimization was outside the scope of this work in terms of input-referred circuit noise.

Hall *et al.* [684,685,722] presented a voltage excitation implementation of the GMR SV sensor in a 0.18 μm CMOS process, which facilitates multiplexing the sensors by summing their currents without the need for a dedicated amplifier per sensor and shows protein detection of 0.01 pM. The sensor count increased from 16 to 256 compared to the current excitation implication [688]. Moreover, this work has over an order of magnitude lower readout time per sensor (250 ms) due to the combination of FDM and TDM with parallel readout channels. Architecturally, this system is arranged like an image, as shown in Figure 9.17. The AFE consists of two pseudo-differential transimpedance amplifiers (TIAs) and a fully differential ADC driver followed by a sigma-delta modulator [684].

Magnetic particles could be distinguished by their magnetic relaxation characteristics or MRX, which was employed for implementing Hall-effect sensors as described in Section 9.2.1. In MRX, the magnetic particles are first magnetized. Then, the temporal response is monitored after removing the magnetic field. Hence, this sensing modality is insensitive to the magnetic field homogeneity and requires comparably simpler readout electronics and signal processing making it more amenable to low-power portable applications [681]. In an effort, Hall *et al.* [681,746] reported, for the first time, a GMR-based time-domain MRX bioassay. The system consists of an 8 × 10-GMR sensor array, a Helmholtz coil, an electromagnet driver, and an integrator-based AFE. The block diagram of the MRX detector is shown in Figure 9.18. A personal computer (PC) provides a user-defined magnetization time and magnetic field amplitude to the PC-FPGA (PC-Field-Programmable Gate Array) interface. The FPGA controls the timing of the electromagnet and readout circuitry. The power amplifier (PA) provides a DC current, I_0, to the electromagnet driver that delivers a pulsed current, I_{Coil}, to the Helmholtz coil, generating a pulsed magnetic

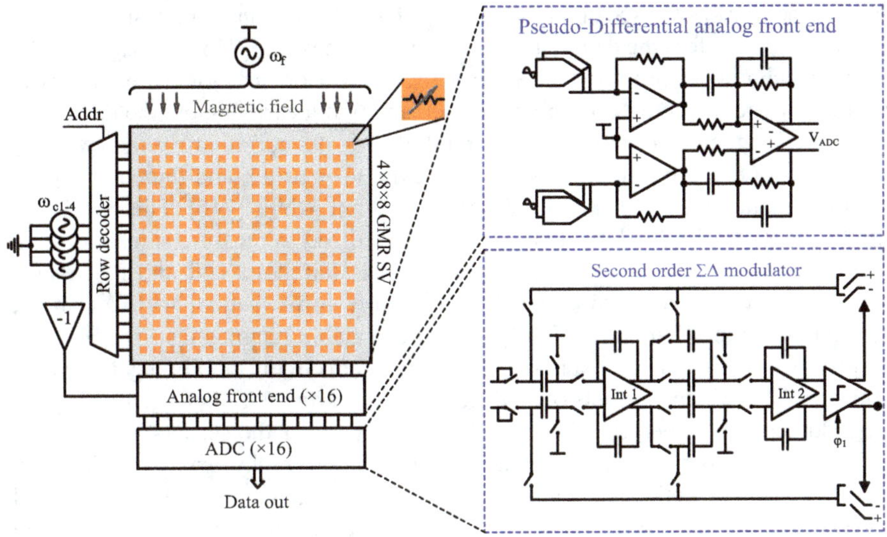

Figure 9.17 A 256 Pixel GMR sensor

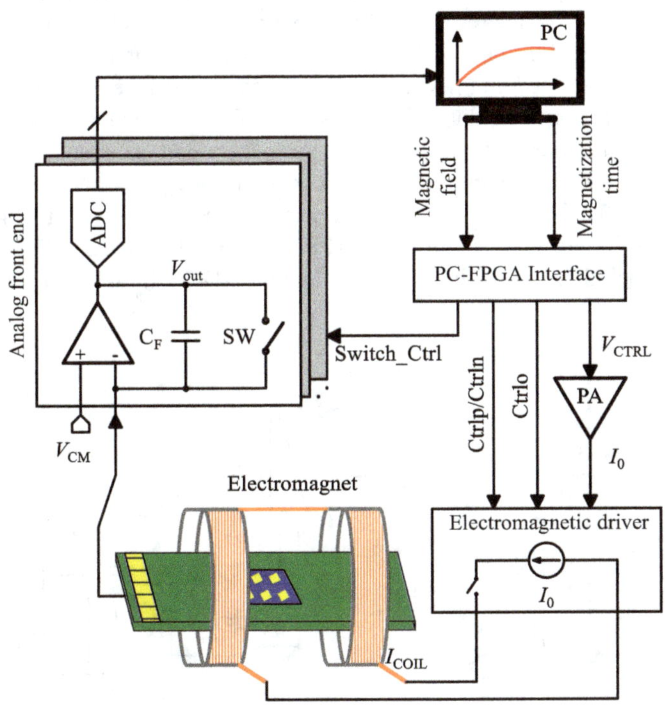

Figure 9.18 The MRX detector

field for the sensor chip located in the center. The AFE captures the resulting sensor response and transfers the data to the PC for post-processing [746]. A magnetic CDS (MCDS) technique is proposed to reduce sensor-to-sensor variation by 99.98% while also reducing temperature drift, circuit offset, and nonlinearity below the noise level. A custom-designed electromagnet driver and Helmholtz coil improve the switch-off speed to >5 Oe/μs, limiting the dead zone time to <10 μs, and thus enabling the system to monitor fast relaxation processes of 30 nm MNPs [746]. Therefore, time-domain analysis with MCDS can be effectively utilized to measure signals from the sensor arrays. However, the usage of an external Helmholtz coil and PA requires significant power consumption as well as a large form factor [749]. As another design architecture, Kim *et al.* [749] present a scalable magnetoresistive biosensor array with on-chip pulsed excitation and a readout circuit employing the MCDS technique. The system consists of a GMR SV sensor array with field-generating strip line inductors embedded on the same chip directly under the sensors and a custom AFE. The on-chip strip lines eliminate the need for a bulky external coil and its corresponding PA while also making this design much more scalable. Also, a fluidic reservoir was created on top of the GMR sensor array using a ¼ piece of Tygon tubing (57547, Tygon) adhered with two-part epoxy.

The large baseline-to-signal ratio and sensor-to-sensor mismatch in GMR biosensors severely complicate the design of the AFE due to the high dynamic range and lower noise required. To address these issues several techniques have been proposed by Zhou *et al.* [679,680]. In their design, a GMR sensor fabricated in a 0.18 μm CMOS process has been reported and new architectural and circuit-level techniques were proposed including fast settling duty-cycle resistors (DCRs) to reduce readout time and a high-frequency interference rejection (HFIR) sampling technique embedded in the ADC to relax the dynamic range requirement. Moreover, this design uses a reference sensor to cancel most of the baseline signals and reject common-mode interference, such as temperature drift. The architecture of the AFE, as shown in Figure 9.19, contains three main blocks including sensor bias, a PGA, and an ADC with a decimation filter and

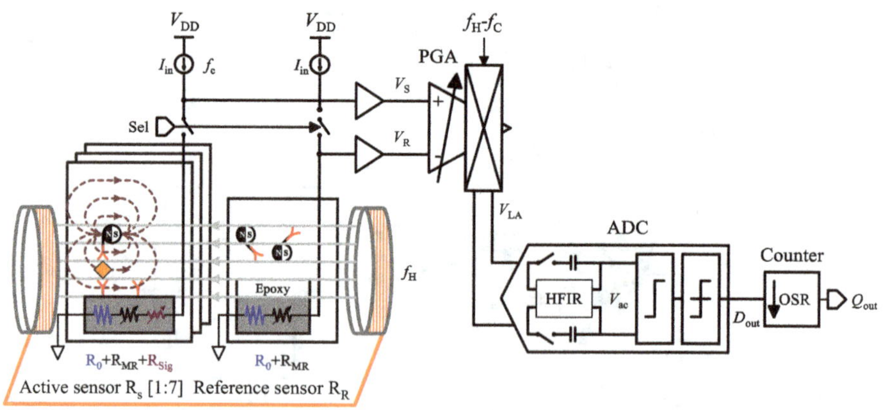

Figure 9.19 AFE of a GMR sensor with mismatch-tolerance and sub-ppm sensitivity

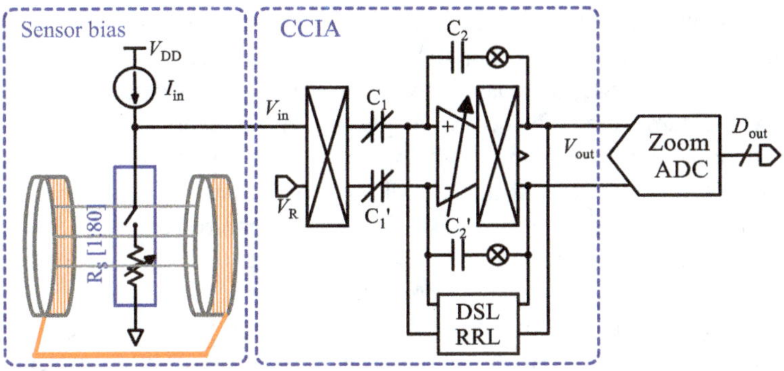

Figure 9.20 AFE of a GMR sensor to detect MNP relaxation

HFIR sampling. It should be noted that mismatches in the PGA and ADC are eliminated by chopping and CDS, respectively. The presented AFE achieves an input-referred noise of 46.4 nT/$\sqrt{}$Hz, an input-referred baseline of less than 0.235 mT, a sensitivity of 0.98 ppm, and a readout time of 11 ms while consuming just 1.39 mW (per channel).

In another effort, Zhou *et al.* [677] presented another MRX implementation which was a low-noise magnetic sensor front end in a 0.18 μm CMOS process to detect MNP relaxation. As shown in Figure 9.20, the proposed sensor AFE consists of a sensor bias block, a capacitively coupled instrumentation amplifier (CCIA), and an 18-bit Zoom ADC. A low-noise DC bias current, I_{in}, flows into the selected sensor, R_S, which is magnetically biased by an external Helmholtz coil. The resulting voltage, V_{in}, is amplified by the CCIA, which contains a DC reference input, V_R, and a DC servo loop to continuously cancel the R_0 baseline. A ripple rejection loop rejects the chopping ripple on V_{Out}, which is quantized by the ADC. The Zoom ADC consists of a 6-bit SAR and a 13-bit Σ modulator for coarse and fine quantization, respectively. In this design, several techniques, such as dynamic element matching (DEM) and magnetoresistive correlated double sampling (MRCDS) are presented to remove the sensor and AFE 1/*f* noise while a fast-settling miller compensation (FSMC) technique is proposed to reduce the amplifier power. The proposed MRCDS techniques remove one step from the previous MCDS technique. The system uses either MRCDS or MCDS. While MCDS is better for baseline rejection and temperature drift cancelation, MRCDS provides better 1/*f* noise rejection and takes less readout time. Collectively, these results in state-of-the-art input-referred noise performance (9.7 nT_{rms}) and a figure-of-merit (FoM) that is 6.6× and 210× better than previously reported magnetic sensor and relaxation-based AFEs, respectively. This design consumes 4.32 mW from a 1.8 V supply.

9.3 Summary

Integrated magnetic biosensor presents several outstanding advantages, especially in terms of the sensitivity and implementation form factor. Sine samples naturally lack any detectable magnetic content, providing a sensing platform with a very low background signal, which leads magnetic detection to be highly selective even in complex biological media. A variety of magnetic sensors have been employed as detectors for magnetic labels. Here, we summarize advances in magnetic sensing technologies including Hall-effect-based, inductive-based, and GMR-based, with an emphasis on implementing CMOS fully integrated system spatially dedicated to PoC settings.

Hall-effect sensors due to the proximity sensing nature require post-processing to minimize the distances between the magnetic particles and the Hall sensors, which is cost-effective and could also compromise the yield. Some detection methods resolve the magnetic field from magnetic particles in the presence of a much larger magnetizing field baseline. Since the baseline is sensitive to environmental variations, these solutions generally require an extra controlling circuit, which consumes more area, and power consumption that can affect detection time. One promising approach to overcome these problems is based on relaxation measurement without baseline requirement. Moreover, wires can be embedded in the chip to generate a local magnetizing field which eliminates the need for any external magnet. Also, several techniques, such as parallel readout electronics, CDS, and magnetic field modulation have been presented. Combining these techniques will provide a magnetic sensor with numerous sensing sites and improved performance.

The inductive magnetic biosensor is another scheme investigated in this chapter, which has two major advantages compared to Hall-effect biosensors. Implementation of inductive-based biosensors does not need any post-process modifications and external biasing magnetic field, which makes it a good choice for low-cost, compact, and robust design for PoC applications. One widespread type of magnetic inductive biosensing scheme is frequency-shift-based magnetic resonant sensors that were reported with different configurations dedicated to various applications. Inductive sensors offer good sensitivity by utilizing quality-factor amplification but require extensive calibration due to high baseline signals and potential interference from non-magnetic conductors. In other words, it requires circuit design optimization techniques to provide high performance of the frequency-shift magnetic sensor. As a suggestion to improve design specifications, current-reuse LC architecture [750] that consumes power as half that of conventional cross-coupled structures combined with some adaptive controlling techniques [751] can be employed as a resonator architecture for inductive-based sensors, which will provide low-power, low-voltage, and good-phase noise performance. Also, to achieve robust performance and reliable design against process voltage temperature, inductive biosensors can be accomplished in feedback structures such as phase-locked loops or be designed with a compensation circuit [752].

GMR sensors provide excellent sensitivity and support a large-scale sensor array but require external biasing magnetic fields and post-processing steps, as indicated in Table 9.1. There have been a lot of reports to improve AFE performance through different approaches. Several promising techniques have been employed for GMR sensors such as FDM, TDM, double modulation, time-domain MRX, paralleling readout channels, DEM, MCDS, MRCDS, and FSMC. In conclusion, integrated magnetic sensors present excellent potential for LoC biological diagnosis applications as they provide low form factor, system cost, high sensitivity, and present highly stable signals.

Chapter 10
CMOS-based NMR sensors

Nuclear magnetic resonance (NMR) is a phenomenon in which nuclei absorb and re-emit radio frequency (RF) electromagnetic signal in the presence of a static magnetic field (B_0) that allows explorations of the molecular information of specific nuclei [753]. Among various bio-sensing methods, NMR sensing stands out due to its exceptional characteristics, including non-destructive operation, the absence of direct contact with the sample, and the ability to employ probe-decorated magnetic nanoparticles (MNPs) for selective target quantification. These attributes make it highly suitable for bio-molecular sensing applications. The MNP provides inherent signal amplification to NMR since a single MNP cluster can affect billions of adjacent water molecules [754]. Furthermore, NMR sensing is free from probe immobilization on the surface, allowing various assays on different targets with the same readout circuit for higher throughput [755]. Hence, NMR-based detection has become a powerful analytical tool for a wide range of fields such as chemistry [756–758], biosensing [759–763], medical imaging [764,765], agriculture [766], etc.

The fundamental components of an NMR system are a magnet to produce a static magnetic field for polarization nuclear spins, a RF coil surrounding a sample, and an RF transceiver to generate excitation RF magnetic fields and monitor the resonance. The sensitivity and resolution of NMR analysis heavily rely on a powerful magnetic field, necessitating the use of a large and unwieldy system for conducting NMR experiments. Recently, pioneering NMR systems aiming for weight and size miniaturization have been demonstrated in CMOS which attempt to design NMR-on-a-chip devices [767–771]. In this chapter, the discussion about the miniaturization of NMR systems will be around devices with target applications in the biomedical with a focus on the progress in electronic aspects of the systems.

10.1 Principles of operation of NMR sensing methods

NMR experiments can be done with various atomic nuclei. Hydrogen contains a positively charged proton that has spin and electric charge, which act like tiny bar magnets due to their spins [759]. So, protons, which are the nuclei of hydrogen atoms (^{1}H), are used in the NMR experiment with water. Due to spin, the proton has an intrinsic magnetic moment. The principle of NMR is shown in Figure 10.1

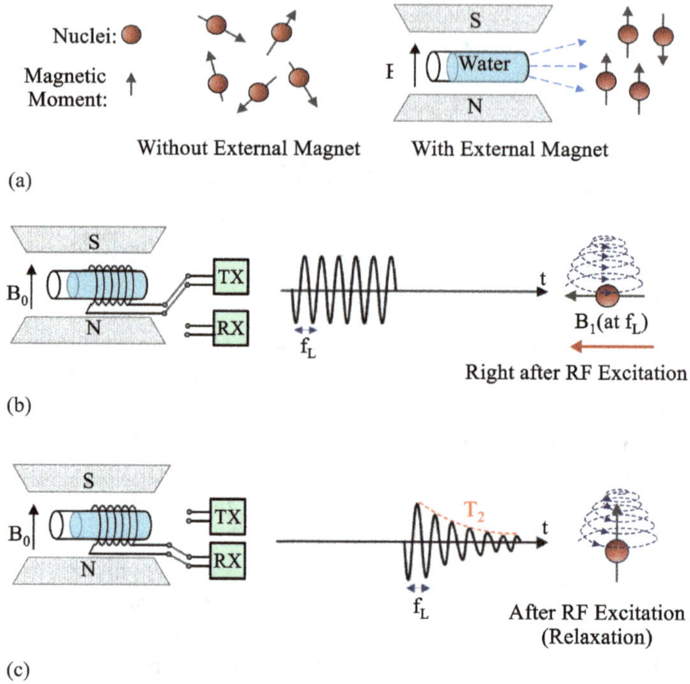

Figure 10.1 The principle of NMR operation with a macroscopic view of the spin nuclei. (a) B0 applied to the nuclei. (b) RF magnetic field excitation the nucleus at f_L*. (c) RF magnetic field is removed and the nuclei still resonate and return to the equilibrium*

[755,772]. When B_0 is applied to the nuclei, some of them will align with B_0 and have a spin-up state while others do not align (Figure 10.1(a)). Then, the transmitter (TX) transmits an excitation RF signal with a specific frequency known as a Larmor frequency (f_L) to the coil to generate a magnetic field B_1 orthogonal to B_0 (Figure 10.1(b)). The nuclei will start to process around B_0 to exchange resonant energy between RF magnetic fields and nuclear spins, and as a result, the bulk magnetization will deviate from B_0 with a tip angle of θ_B. This resonant energy exchange between the RF field and nuclear spins is NMR. Notably, the nuclei only accept RF excitation at Larmor frequency, defined by (7.1).

$$f_L = \gamma B_0 \tag{10.1}$$

where γ is the gyromagnetic ratio of the nuclei. After a certain period, the nuclei are tipped away from B_0 with θ of 90°, and the excitation is turned off by switching the coil to the receiver (RX). As shown in Figure 10.1(c), the nuclei will still resonate at f_L and decay exponentially with spin–spin relaxation time, called T_2. This T_2 parameter and the resonance frequency depend on the material specifications, so NMR can be used to sense biomolecules.

NMR experiments are categorized into relaxometry, spectroscopy, and magnetic resonance imaging (MRI) [771]. In the case of relaxometry, the primary focus is on retrieving the T_2, as the rate of phase decoherence varies with material species containing ^1H protons. In a spectrometer, the primary interest is in measuring and identifying the spectrum and its components. MRI is a variant of relaxometry in which T_2 measurements at different positions across a body part produce a spatial heat map of T_2 [771].

Consider, for example for relaxometry (T_2) measurements, a biological sample containing a large number of water molecules (thus a large number of ^1H protons) and MNPs with surfaces coated with antibodies (biotin) that will specifically bind to target proteins (avidin) are introduced into the sample (Figure 10.2) [773]. In Figure 10.2(a), water with no magnetic particles yields due to the basic proton–proton interactions. Figure 10.2(b) shows in the absence of the target proteins, the magnetic particles stay monodispersed. These magnetic particles due to Brownian motion produce fluctuating local magnetic fields, which disturb the precessions of the proton spins and due to the basic spin–spin interactions cause more phase noise beyond. Thus, the phase coherence is lost more rapidly and T_2 is reduced. As shown in Figure 10.2(c), in the presence of target molecules, the coated magnetic particles self-assemble into local clusters, which are even more efficient in accelerating the dephasing of ^1H proton precessions, yielding the smallest T_2. By monitoring T_2 of ^1H NMR signal, target molecules can be detected.

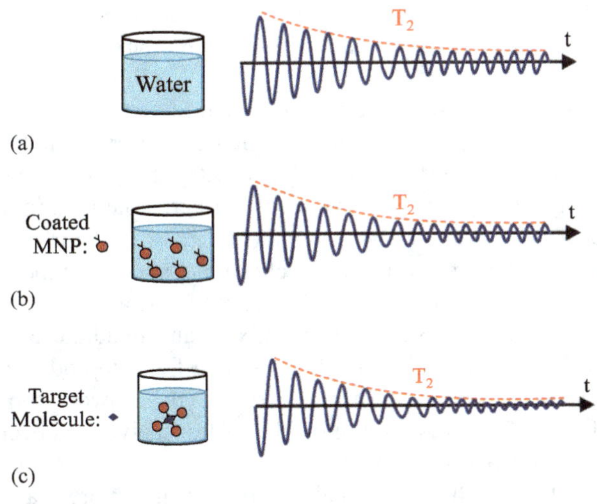

Figure 10.2 Conceptual of NMR biosensor (relaxometry). (a) The spin–spin relaxation time (T_2) for water molecules without test samples. (b) T_2 is reduced with monodispersed magnetic particles. (c) The target molecules, lead to the aggregation of magnetic particles to form clusters, which further decrease the T_2 of the water molecules

10.2 Implementation of NMR systems

Over the decades, a lot of researchers have attempted the miniaturization of NMR magnets, coils, and electronics to prepare these devices for point-of-care (PoC) scenarios [759,760,762,763,767,769–771,774–782]. A large-sized magnet produces a stronger NMR signal and relaxes the sensitivity requirement of the transceiver, however, it results in a bulky and heavy design, which limits the applications of NMR sensors for the PoC platforms. For miniaturization NMR-based sensors, a small magnet is necessary, but with decreasing magnet size, a weak NMR signal will be produced, so the NMR electronics play an increasingly important role in determining the overall system performance. To address this challenge, a highly sensitive RX integrated with CMOS technology is required. In this section, the recent miniaturized CMOS-based NMR systems will be reviewed, highlighting their application in relaxometry, spectroscopy, and MRI and the comparisons of them will be summarized in Table 10.1.

10.2.1 Relaxometry functionality

In one effort, an NMR system was designed by Lee *et al.* [754] that consisted of in-house fabricated 500 nH planar copper microcoils on a glass substrate, microfluidic channels, and a portable magnet. The microcoils were arranged in an array format to provide parallel and multiplexed detection. The microfluidic channels generate maximal NMR signal with small sample volumes (5–10 μl), and due to the small size of sensing components, portable permanent magnets are used to generate polarizing magnetic fields (B_0 = 0.1–0.5 T). But, besides the mentioned specifications, the RF transceiver which is built with discrete components, left the system bulky [754].

In another implementation, Liu *et al.* [772] and Sun *et al.* [759] adopt the same microcoil and magnet of [754] but integrated the RF transceiver in 0.18 μm CMOS process, substantially a portable NMR relaxometry system is achieved for the detection of biological objects such as cancer marker proteins. Figure 10.3 shows the architecture of the RF transceiver integrated circuit (IC) for the CMOS mini-NMR system. In the RX part, a weak spin-echo signal appears at the front-end node 2 and is amplified by a low-noise amplifier (LNA) and a variable gain amplifier (VGA), and then is down-converted using mixers and quadrature local oscillator signals. In the TX part, the excitation RF magnetic field is produced by the same quadrature local oscillators used in the RX and the power amplifier (PA) is implemented off chip. The noise figure of the RF transceiver has been measured at 0.7 dB that this high sensitivity enabled the construction of a 2 kg NMR system, which is 60 times lighter, 40 times smaller, yet 60 times more mass sensitive (20 fmol) than 120 kg state-of-the-art commercial benchtop system [783]. This experiment uses water proton spins and B_0 = 0.5T, so the NMR frequency is around 21.3 MHz. Notably, the pronounced field inhomogeneity causes the NMR signal to be damped far faster than T_2, which makes it hard to know the true T_2, even with high RX sensitivity. To suppress this event, the TX employs a digital pulse

Table 10.1 Comparison of different NMR CMOS sensor

Functionality	Sensitivity/Spin sensitivity (Spin/$\sqrt{HZ}$)	Larmor frequency/ Operation frequency (MHz)	Input referred noise (nV)/ $\sqrt{HZ}$)/NF (dB)	Power (mW)	Application	Magnetic (T)/ System weight (kg)	Chip area (mm^2)	Process (μm)	Ref.
Relaxometry	20 fmol (4nM)/NA	21.3/NA	2.5/0.7	NA	Cancer marker, Proteins	0.5/2	3.8	0.18	[759, 772]
Relaxometry	1 Avidin in 600 million water molecules/NA	23.9/NA	1.26/0.9	NA	Avidin detection	0.56/0.1	1.96	0.18	[784, 785]
Relaxometry	0.05 mM/NA	20.9/NA	1.26/2.2	NA	Proton, hCG cancer, Bladder cancer detection	0.49/2	11.25	0.18	[784, 785]
Relaxometry	2 nM ($CuSO_4$) 0.2 pM (Avidin)/NA	20/2-33	0.92/NA	30.3	$CuSO_4$/Avidin	0.46/1.25	2.08	0.18	[755, 762, 791]
Relaxometry	<100 pM/NA	19.6/NA	1/NA	81.5	DNA/Human IgG	0.46/1.4	7.6	0.18	[794, 795]
Relaxometry	35 μl/1.1×10^{14}	15.3/NA	0.61/NA	262	Body liquids	0.36/5.5	0.99	0.13	[770]
Relaxometry Spectroscopy	NA/NA	21.8/NA	0.82/0.23	NA	Aspirin, L-alanine, Glucose, etc.	0.51/7.3	4	0.18	[804]
MRI	NA/NA	300/64-450	NA/0.9	< 240	*In vivo* imaging (human wrist)	1.5-10.5/NA	22	0.13	[775]
Relaxometry Spectroscopy MRI	9.8 nL	600/30-700	1.26/0.7	20	Probing brain physiology	14.1/NA	NA	0.13	[769]
Relaxometry Spectroscopy MRI	NA	21.8/10-40	0.82/NA	NA	Biotin, Biomolecular	0.51/7.3	6	0.18	[776]
Relaxometry Spectroscopy MRI	2.1×10^{15}/NA	42/10-60	0.78/NA	603	Ethyl acetate	0.51/NA	5	0.18	[771]

* NA: Not available

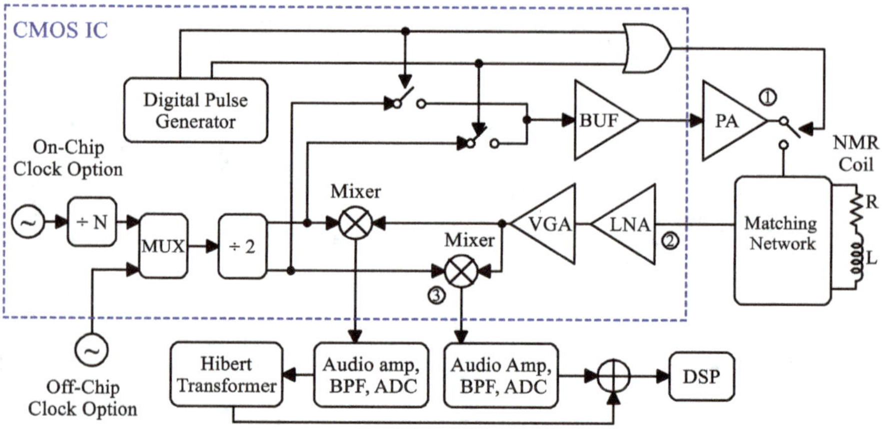

Figure 10.3 CMOS RF transceiver architecture of the mini-NMR system

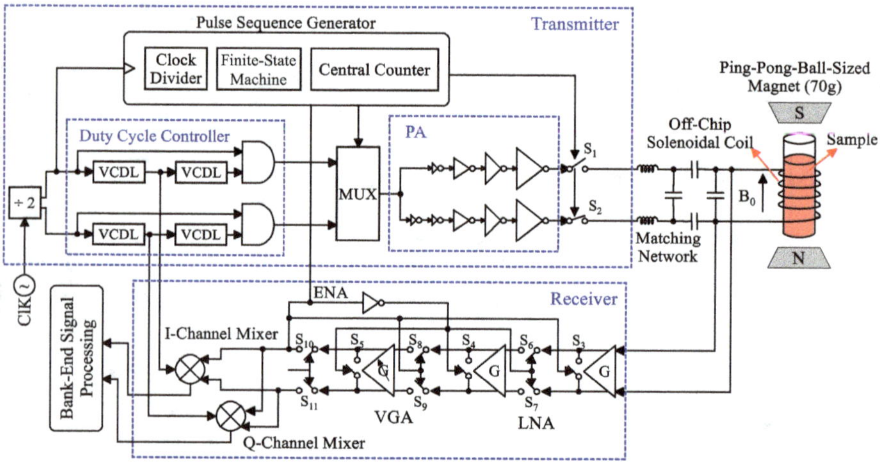

Figure 10.4 CMOS RF transceiver architecture of the palm system

generator, which transmits the RF signal in a sequence devised. This sequence effectively offsets field inhomogeneity, and the resultant provides true T_2 measurement. The absence of avidin results in a measured T2 value of 73 ms, while its presence causes T2 to decrease to 31 ms, signifying the detection of avidin.

In another work, Sun *et al.* [784,785] reported the refined versions of the NMR system [759,772], that successfully integrate the coil on-chip, offering a compact NMR sensor solution with lab-on-chip (LoC) capabilities, and also representing orders-of-magnitude size reduction. The structures of these NMR systems known as a palm NMR system and a 1-chip NMR system are shown in Figures 10.4 and 10.5,

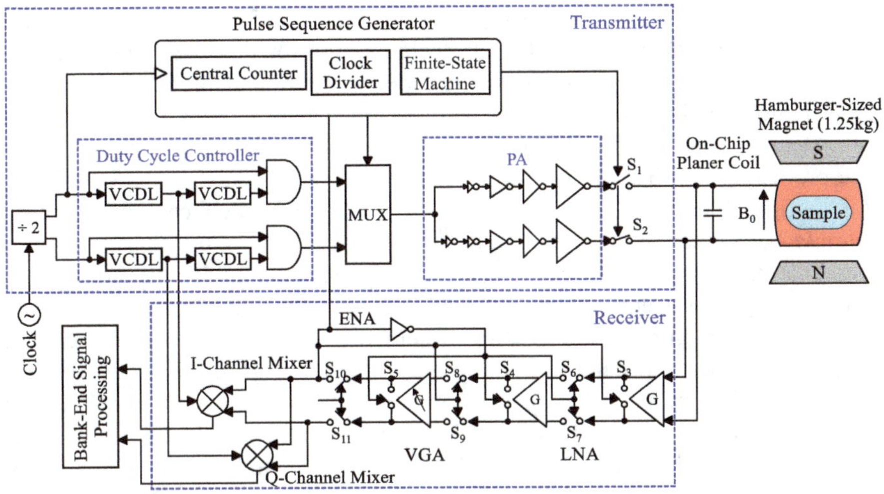

Figure 10.5 CMOS RF transceiver architecture of the 1-chip system with lab-on-a-chip capability

respectively. The transceiver architecture is essentially the same between the two systems, but the transceiver-coil matching networks are different. In these designs compared to previously reviewed NMR systems, the integration level is increased, which incorporates a PA on the chip. For power tuning of PA, the duty cycle is tuned by using voltage-controlled delay lines (VCDLs) and logic gates. The 0.1 kg palm NMR system uses a tiny magnet only the size of a ping-pong ball which weighs only 0.07 kg. Due to the use of the small magnet, the NMR signal is weakened, so this "palm" system uses a high-quality (Q = 28) off-chip solenoidal coil, in which a sample is placed inside the solenoidal coil. Moreover, the high-performance RF transceiver integrated circuit (IC) with noise figure (NF) of 0.9 dB is employed to further improve the sensitivity. The palm system is 1,200 times lighter and 1,200 times smaller than the aforementioned commercial system [783].

The 1-chip system integrates the planar coil and RF transceiver in the same CMOS 0.18 μm. The 1-chip system has the same spin-mass sensitivity as the work of [759,772], while 60 times more spin-mass sensitive than the commercial system [783]. The transceiver's performance with NF of 2.2 dB allows the use of the lossy on-chip coil. On the other hand, in order not to further weaken the signal, this system operates with a hamburger-sized magnet (1.25 kg). The main point of the 1-chip system is LoC operation, which is provided by introducing biological samples directly on the coil of the chip. The palm and 1-chip NMR systems detected a range of biological molecules and compounds. For instance, the palm system detects 1 avidin molecule in 600 million water molecules, as in the absence of avidin T_2 is 48 ms, and in the presence of avidin T_2 is reduced to 40 ms. For another example, a 1-chip system capable of detection of human bladder cancer cells down to around 18 cells per μL as in the absence of cancer cells, T_2 is 506 ms, but in the presence of the cancer cells, T_2 decreases to 440 ms.

The previous reviewed works [754,759,772,784,785], attempt to develop micro-NMR (μNMR) systems in CMOS technology through weight and size miniaturization. Regrettably, it is inflexible to pipeline multiple samples to the μNMR sensing region for higher throughput and real-time result comparison due to the small inner space and limited NMR sensing region of a portable magnet [786]. One of the promising approaches to address this issue is employing the biocompatible and reconfigurable digital microfluidic (DMF) [786–790]. DMF can cope with the 2D droplet-sample operation (e.g., transportation, mixing, and splitting) accurately by modulating the surface tension of the DMF electrodes under the principle of electro-wetting-on-dielectric (EWOD) [755].

The implementation of the μNMR system with a DMF device for chemical/biological assay was presented by Lei *et al.* [755,762,791]. In this study, an integrated transceiver system that incorporates a 2D multi-electrode DMF device within a compact magnet (0.46 T, 1.25 kg) is developed. This setup enables efficient multi-sample handling for μNMR measurements and facilitates electronic automation. When compared with the conventional channel microfluidic devices, DMF can avoid separate and complex networks of connections, and laboratory gadgets such as pumps and valves and the device is reconfigurable in such a way that the droplet can move freely over a surface of electrode matrix [755]. The electrical schematics of the system fabricated in 0.18 μm CMOS are illustrated in Figure 10.6, which includes a butterfly coil, a CMOS μNMR transceiver, a DMF device with its DMF electronic, and an FPGA with analog-to-digital converters (ADCs) that link up the system with a computer for easy execution and upgrade of protocols written in software. The measurement results demonstrate the feasibility

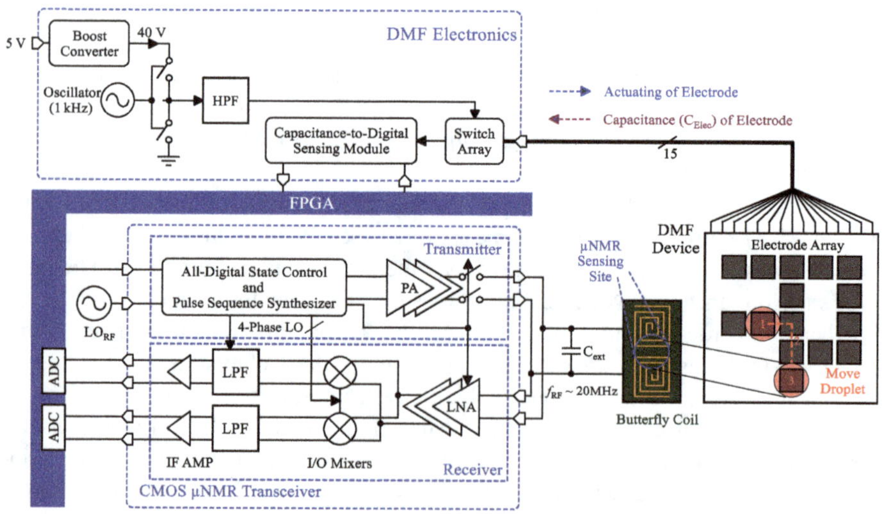

Figure 10.6 Architecture of the NMR system with a DMF device, and a DMF electronic

of electronic-automated biological (avidin) and chemical ($CuSO_4$) assays achieving a detection limit on avidin of 0.2 μmol.

Another important challenge to implementing a CMOS NMR system with a portable magnet is the variation of the B_0 field. Since f_L shifts with the temperature-dependent B_0, a calibration scheme to enhance the robustness of the system for PoC application is required [792,793]. In the presence of large fluctuates of B_0-field, conventional frequency stabilization techniques may not work properly. In other reported work by Lei *et al.* [794,795], to overcome this challenge, a B_0-field stabilization scheme for a handheld μNMR CMOS platform utilizing a portable 0.46 T magnet is proposed. As illustrated in Figure 10.7, the main component of the system is an NMR-on-a-chip transceiver application-specific integrated circuit (ASIC), which integrates a quadrature RX with baseband filters and amplifiers as well as a PA with a 4-phase pulse sequence synthesizer. A Hall sensor with a low-noise readout circuit fabricated in 0.18 μm CMOS is embedded with a μNMR transceiver to achieve better robustness. The current driver stabilizes the B_0 field of the magnet against ambient variation. The stabilized B_0 avoids the need for a frequency synthesizer to tune the local oscillator (LO), and an un-tuned LO can be generated directly by the crystal oscillator. To minimize the B_0-field offset error between the sample and the Hall sensor, the samples under assay are loaded on the on-chip planar coil by a handheld pipette. The sensing

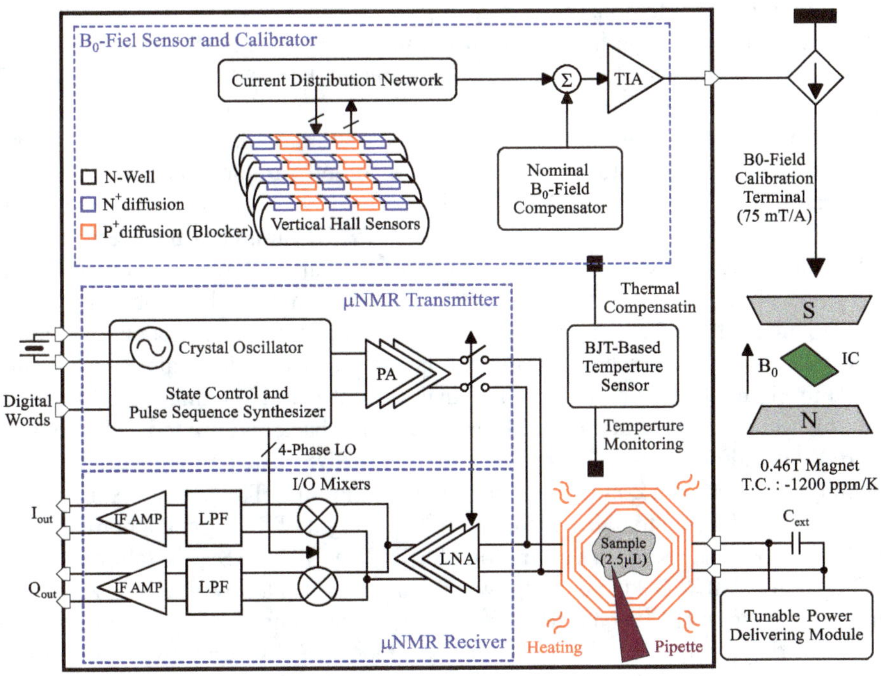

Figure 10.7 The bock diagram of the micro-NMR platform with B0-field stabilization and LoC feasibility for multi-type biological/chemical assays

coil also serves as a sample heater for thermal profiling. Operating from a battery supply, the presented system does not require any external instrumentation apart from a computer for data analysis and display. Further, benefiting from the versatility of the NMR assays, this handheld tool unifies multi-type assays. The system is capable of detecting <100 pM of *Enterococcus faecalis*-derived deoxyribonucleic acid (DNA) from a 2.5 μL sample. Compared to a commercial NMR-assay product [783], this platform greatly reduces the sample consumption (120×), hardware volume (175×), and weight (96×). The overall system size of 14 × 6 × 11 cm^3 and a weight of only 1.4 kg. Other specifications of this system are summarized in Table 10.1

As we mentioned earlier, the benefits of portable μNMR systems come at the price of low NMR signal sensitivity. One of the promising techniques to address the issues is the dynamic nuclear polarization (DNP) strategy [777,782,796,797]. DNP is a strongly hyperpolarizing effect that was first proposed by Overhauser [798] and soon afterward confirmed by Carver *et al.* [799]. DNP enables the transferring of the large magnetic moment of electron spins under microwave (MW) irradiation to nuclear spins. Therefore, enhancement of the NMR signal intensity by up to two orders of magnitude will be achieved [800]. Kiss *et al.* [782] presented a miniaturized probe head based on Overhauser DNP. It contains an MW resonator, a microfluidic chip, electrical shims, and an RF transceiver coil, which is mounted inside a palm-sized 0.5 T permanent magnet. The experiment results demonstrated that a signal enhancement of ~40-fold was achieved in the detection of 130 nL of deionized (DI) water. So, this work indicates a promising way to overcome instrumental barriers to the use of NMR detection for PoC applications.

Notably, NMR systems achieved increasing attention as a tool for the quantitative analysis and pathological assessment of blood samples [770,801–803]. As one example, in Dreyer *et al.* [770], a miniaturized NMR relaxometry system has been presented which centered on a BiCMOS NMR-on-a-chip transceiver ASIC and a direct-digital synthesizer (DDS)-based arbitrary reference frequency generator and phase control. This design focuses on an optimized performance in the determination of the different relaxation time components in large sample volumes corresponding to human blood samples that its electrical block diagram are presented in Figure 10.8. It consists of the analog front end, including the NMR on-a-chip transceiver ASIC and the tuned sensor coil, the back end printed circuit boards (PCBs), including anti-aliasing filters and the DDS-based reference frequency generator, as well as a commercial data acquisition unit. The magnet is a custom-designed 0.36 T benchtop NMR magnet, with a weight of 5.5 kg and a homogeneity of 30 ppm over the sample volume of 35 μL. The NMR ASIC is fabricated in a 0.13 μm SiGe BiCMOS technology, which co-integrates a low-IF RX, a PA, and a phase-locked loop (PLL) on a footprint of 1,100 × 900 μm^2. The TX path consists of an H-bridge output stage and its drivers, fed by a divide-by-two-based quadrature generation, which is in turn fed by the on-chip integer-N-based frequency synthesizer. The RX path consists of an LNA, a protection switch, which disconnects the LNA from the NMR coil during TX, a quadrature down-conversion mixer fed by the same in-phase and quadrature (IQ) generation module as the TX

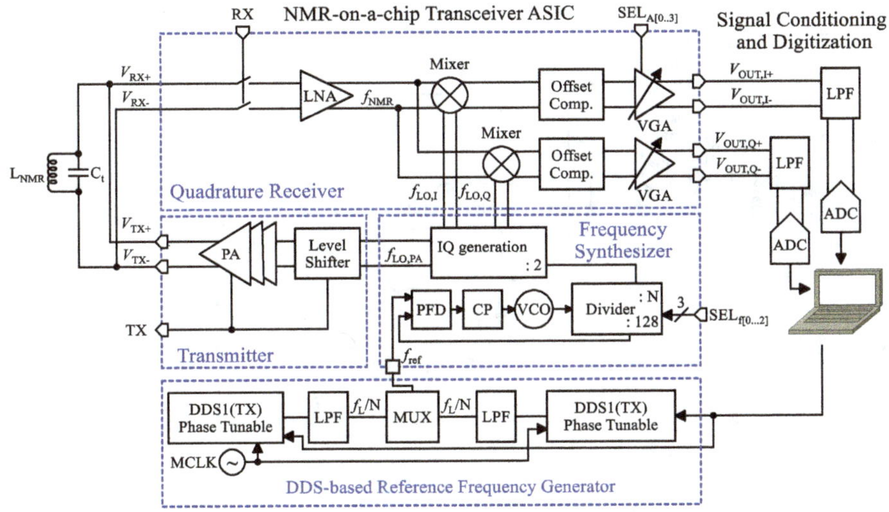

Figure 10.8 The bock diagram of the NMR relaxometry system

path, and a variable gain amplifier, including an on-chip offset compensation block. The arbitrary TX and RX phase control of this system enables the use of a modified water suppression CPMG sequence which enables the reliable detection of lipid and protein components of the T_2 distributions in samples with a strong water peak.

10.2.2 Spectroscopy functionality

The first μNMR system capable of performing 1D and 2D spectroscopy with customized silicon chips was reported by Ha *et al.* [804]. The structure of this NMR spectrometer is shown in Figure 10.9. The utilized magnet is a custom-designed, 0.51 T magnet with a six-direction electrical shim system and a homogeneity of 0.13 ppm. An external NMR coil with a diameter of 1 mm to improve concentration sensitivity is used. The integrated spectrometer electronics chip consists of three main parts, an RF RX, TX, and an arbitrary pulse sequencer. The RX amplifies the voltage signal across the coil induced by nuclear spin precessions using an LNA and down-converts the amplified signal into two phase-sensitive audio-frequency signals through mixers driven by quadrature local oscillators. The TX consists of a multiphase generator and a PA to excite nuclear spins and manipulate their motions. The delay-locked loop-based on-chip multi-phase generator provides choosing 1 out of 32 equidistantly phase-shifted excitation signals. The amplitude of the PA can be switched with a 5-bit resolution. So, by controlling the PA and multiphase generator, this structure allows for versatile use of the NMR system, covering a broad range of applications, including 1D and 2D NMR spectroscopy as well as NMR relaxometry. The NMR spectra demonstrate this homogeneity is worked for the analysis of several molecules such as ethylbenzene, aspirin, L-alanine, and glucose.

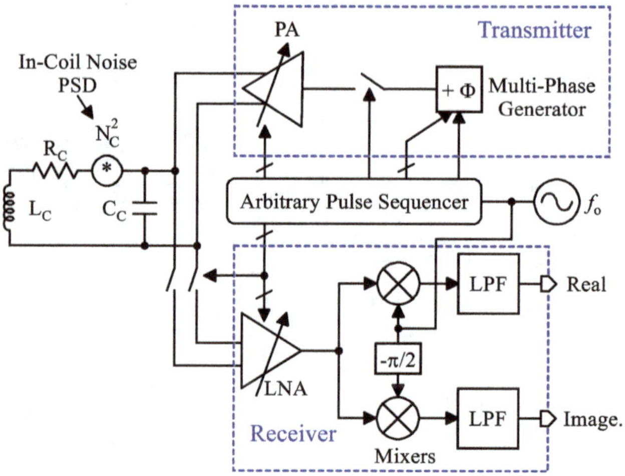

Figure 10.9 CMOS RF transceiver architecture for NMR spectrometer

In another effort, the first fully integrated multi-channel system with multiple detection coils was introduced by Anders *et al.* [805] for parallel spectroscopy at 300 MHz, with on-chip matching circuitry and detection coils occupying an area of 520 μm × 520 μm. In another work, Grisi *et al.* [806] presented a broadband single-chip CMOS transceiver that can be interfaced with on-chip integrated microcoils or external LC resonators and its use for multi-nuclear spectroscopy has been demonstrated. To improve sensitivity for NMR spectroscopy applications, Cherifi *et al.* [807] proposed an in-field preamplifier associated with receiver microcoil.

10.2.3 MRI, relaxometry, and spectroscopy functionality

In addition to NMR spectroscopy and NMR relaxometry, recently, MRI has started to gain the attention of researchers [769,771,775,776,780,808]. For instance, Lei *et al.* [776] proposed the first portable high-resolution multidimensional NMR with throughput-accelerating parallelism. This configuration occupies one shared NMR coil and is connected to a single silicon chip, simultaneously measures the relaxation time of 18 samples, and is imaged all at once multiple times.

To the best of our knowledge, a fully integrated MRI RX designed for on-coil operation in scanners for the first time was presented by Sporrer *et al.* [775]. The system features a complete quadrature down-conversion RX which fetches and digitizes the signal of two coils simultaneously and is suitable for MRI scanners operating up to 10.5 T. The IC was verified through *in vivo* images of a human wrist acquired in a 3 T scanner from the two receive channels, which indicate a comparable quality against commercially available systems while consuming 4× lower power.

MRI and spectroscopy are versatile methods for probing brain physiology, but their intrinsically low sensitivity limits the achievable spatial and temporal resolution [769]. Handwerker *et al.* [769] presented an implantable, needle-shaped NMR sensor in a 0.13 μm CMOS technology for *in vivo* studies of brain physiology inside a 14 T small animal magnet. This structure combines an ultrasensitive 300 μm NMR coil with a complete NMR transceiver that provides a 100-fold improvement in temporal resolution in detecting the BOLD effect in response to paw stimuli. It provides *in vivo* measurements of blood oxygenation and flows in nanoliter volumes at a sampling rate of 200 Hz. As shown in Figure 10.10, the RX path contains a complete quadrature RX, containing a PLL as a robust frequency synthesizer and protection switches for the LNA, and the overall reported noise figure is about 0.7 dB. The TX path features an H-bridge PA operating from a 3.3 V supply and driven by the on-chip PLL that produces a coil current of 15 mA at 600 MHz. The power consumption of this NMR-on-a-chip transceiver is reported as 20 mW. In this design, the active coil provides the advantage of very short RF connections to the PA, LNA, and mixers, which enable a precise definition of the sensitive volume. Due to the amplitude and phase modulation capabilities, the on-chip electronics can use standard imaging sequences and spectroscopy techniques. The MR images with 10-μm isotropic resolutions acquired in only a few minutes of imaging time have been reported.

We will close this section with an overview of a portable NMR system by Krüger *et al.* [771], which not only enhances spectroscopy and relaxometry performance but also includes MRI. It achieves a state-of-the-art spectral resolution of <0.05 ppm (1.1 Hz), which is the highest resolution reported for portable CMOS-based systems based on permanent magnets for NMR spectroscopy. Furthermore, for MRI a spatial resolution of 67 × 67 × 83 μm^3 is achieved. The CMOS RF

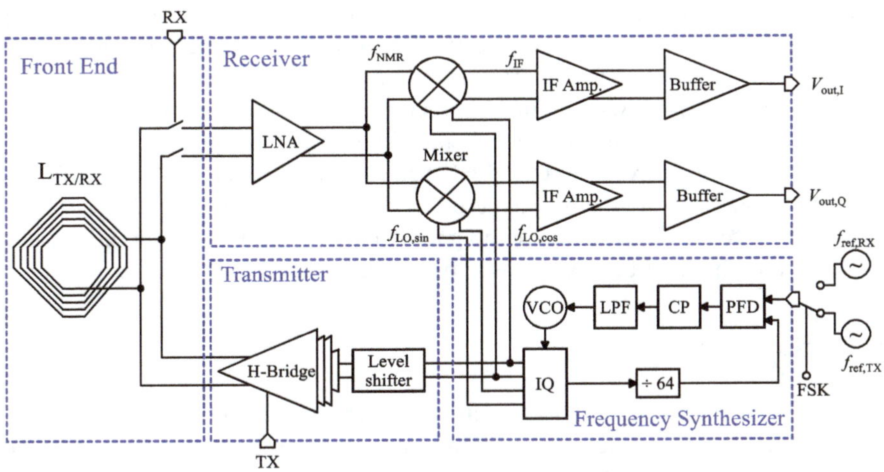

Figure 10.10 Fully integrated NMR-on-a-chip spectrometer with an on-chip planar broadband detection coil

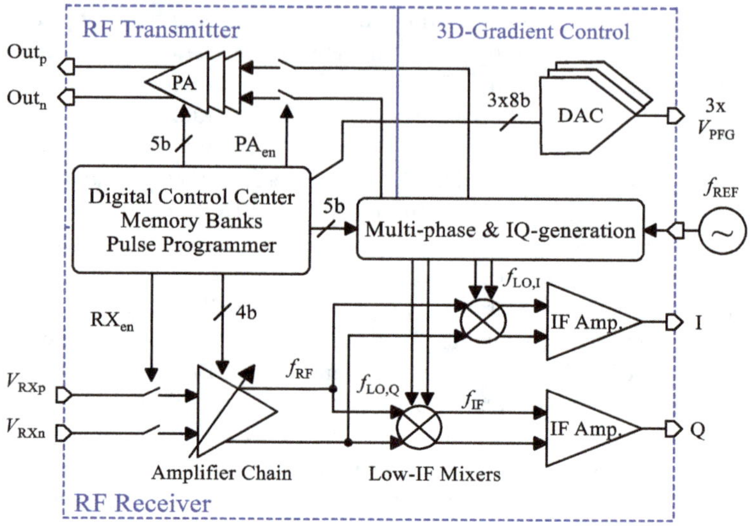

Figure 10.11 The CMOS RF transceiver IC for a portable NMR system

transceiver architecture of the NMR system is shown in Figure 10.11. The system consists of a NdFeB Halbach permanent magnet ($B_0 = 0.51$ T and $f_L \approx 21.8$ MHz), a solenoidal sample coil to generate RF magnetic fields and to pick up precession signals, and a CMOS RF transceiver IC. Two additional sets of coils are also prepared that contain a shimming coil set to make the static field more homogeneous for spectroscopy, and a gradient coil set to introduce gradients in the static field for MRI.

CMOS RF transceiver IC consists of an RF RX, TX, a 3-D-gradient control, and a digital control center that coordinates the operational timings of the building blocks for the NMR experiments. The RF RX has a heterodyne architecture that comprises an amplifier chain, quadrature mixers for frequency down-conversion, and IF-amplifiers with low-pass filter characteristics. It is notable since this system works with smaller samples for spectroscopy and larger samples for relaxometry and MRI, the RF input voltage to the RX has a broad dynamic range, and for the smallest sample, the minimum voltage of around ~0.6 μV is used. The TX has a tunable structure composed by combining a digitally controlled current mode class-D PA and a multiphase generator based on a delay-locked loop (DLL).

10.3 Summary

NMR is a robust analytical instrumentation to examine the fundamental physics of atomic nuclei, analyze molecular structures, and probe properties of materials. The NMR can also present high sensitivity together with a unique possibility of characterizing dynamic molecular interactions during the assay process. Due to its non-invasive nature and the possibility of measuring the NMR signal contactless, NMR is widely used in biomedicine and chemistry applications. It requires a biasing magnet, and its large coil size can present challenges in large array implementations. μNMR system, which utilizes a portable permanent magnet and an integrated electronic chip, has emerged to provide a practical solution featuring low cost, enhanced portability, and minimal sample consumption.

In this chapter, we review the latest developments in the miniaturization of NMR systems with an emphasis on NMR electronics. As we discussed, NMR experiments are categorized into spectroscopy, relaxometry, and MRI. For instance, Liu *et al.* [772], Sun *et al.* [759], Lei *et al.* [755,762,791], Ha *et al.* [804], Lei *et al.* [794,795], and Dreyer *et al.* [770] presented an NMR relaxometry and/or spectroscopy. Liu *et al.* [772] and Sun *et al.* [759] integrated the RF transceiver to develop a portable NMR relaxometry system for the detection of biological objects such as cancer marker proteins. Sun *et al.* [784,785] integrated the coil on chip and reduced the magnitude size, offering a compact NMR sensor solution with LoC capabilities for proton, hCG cancer, and bladder cancer detection. Sporrer *et al.* limited to MRI [775], integrated MRI RX for on-coil-operation in scanners, and its functionality was verified by imaging of a human wrist. Handwerker *et al.* [17], Lei *et al.* [776], and Krüger *et al.* [771] presented an NMR system capable of all three key NMR modalities.

Chapter 11
Thermal sensors

Temperature has an essential role in the main fields of physical, chemical, and biological. Therefore, thermal sensors are extensively employed in numerous applications, such as in medical devices [809–817], environmental thermal monitoring systems, thermal management [818–823], and radio frequency identification (RFID) tag devices [824–827]. During the last decades, complementary metal-oxide-semiconductor (CMOS)-compatible thermal sensors have become increasingly employed due to their compactness, low cost, and digital output. Such sensors can be categorized into five types based on the sensing source, including, bipolar junction transistors (BJTs) or parasitic BJT found in the CMOS process, metal oxide semiconductor field-effect transistors (MOSFETs), resistors, electro-thermal filter (ETF), and Seebeck effect. The choice of sensing source for such temperature sensors is mostly based on the application's specifications, such as accuracy, energy efficiency, power consumption, resolution, and operating temperature range. The output of such sensors is typically a temperature-dependent voltage or current signal. In many applications, it would be preferred to accomplish temperature sensors that produce a digital outcome. This sensor, which is termed a smart temperature sensor combines a sensor and interface electronics on a single chip via the CMOS process. In the following section, various CMOS-compatible sensing elements are briefly introduced.

11.1 BJT-based thermal sensors

11.1.1 Principles of operation of BJT-based thermal sensors

The strong thermal dependence characteristics of PN junction in a diode or BJT transistors make it a promising candidate for designing thermal sensors. On the other hand, in CMOS technology two types of BJT transistors including lateral and substrate (vertical) are available as parasitic devices which can be used as a thermal sensor when operated in a diode configuration. Conventionally, a basic block of BJT-based smart thermal sensor comprising several bipolar transistors followed by an analog-to-digital converter (ADC) to realize a digital representation of temperature is shown in Figure 11.1 [828]. The principle of this method is based on combining the difference between the base-emitter voltage (ΔV_{BE}) of two bipolar transistors. The difference in base-emitter voltage between two transistors biased at different current densities as one of the transistors is dimensioned N-times greater

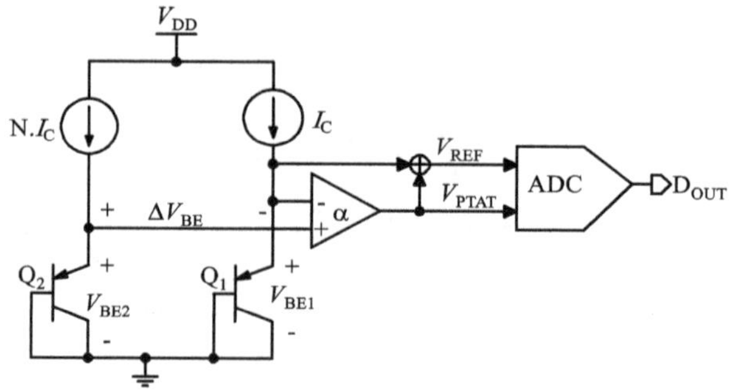

Figure 11.1 Operating principle of the BJT-based thermal sensor

than the other is used to generate a voltage proportional to absolute temperature (PTAT). ΔV_{BE} can be described by (11.1) [829].

$$\Delta V_{\text{BE}} = V_T \ln\left(\frac{NI_{\text{C}}}{I_{\text{S}}}\right) - V_{\text{T}} \ln\left(\frac{I_{\text{C}}}{I_{\text{S}}}\right) = V_{\text{T}} \ln(N) \tag{11.1}$$

where V_{T} is the thermal voltage, I_{C} is the collector current, I_{S} is the saturation current, and N is the current ratio between two bipolar transistors. In this method to implement a thermal sensor, besides the PTAT characteristic, a complementary to absolute temperature (CTAT) behavior is also required. Forward biasing of the PN junction with a constant current causes a voltage drop across the junction, which exhibits a CTAT behavior. For more clarification consider the base-emitter voltage V_{BE} of a BJT that is expressed by (11.2) [829].

$$V_{\text{BE}} = V_{\text{T}} \ln\frac{I_{\text{C}}}{I_{\text{S}}} \tag{11.2}$$

From this equation, it can be indicated that V_{BE} exhibits CTAT behavior with a slope of −2 mV/C. By scaling ΔV_{BE} with a constant α and combining PTAT ($\alpha\cdot\Delta V_{\text{BE}}$) and CTAT ($V_{\text{BE}}$) voltages, it is possible to obtain a reference voltage V_{REF} insensitive to the temperature, which is called bandgap reference voltage as described by (11.3). Finally, the ratio of the PTAT and V_{REF}, represented by K_{VT}, will be determined by ADC as shown by (11.4), to achieve a digital representation of temperature. Due to the V_{REF} being temperature-independent, the output of ADC will present a digital representation of temperature with a PTAT behavior.

$$V_{\text{REF}} = V_{\text{BE}} + \alpha\Delta V_{\text{BE}} \tag{11.3}$$

$$K_{VT} = \frac{V_{\text{PTAT}}}{V_{\text{REF}}} = \frac{\alpha\Delta V_{\text{BE}}}{V_{\text{BE}} + \alpha\Delta V_{\text{BE}}} \tag{11.4}$$

The first thermal sensor that employed the concept of ΔV_{BE} and the first integrated version of the thermal sensor were introduced by Verster [830] and Dobkin [831], respectively. After that, there have been a lot of challenges to the design of BJT-based smart thermal sensors [832–851]. One of the first efforts to implement a smart thermal sensor with an output signal of a temperature-dependent frequency was proposed by Boomkamp *et al.* [832] and fabricated in bipolar technology for biomedical applications. This sensor works with the temperature range from 30 °C to 46 °C, an accuracy of ±0.1 °C, and a resolution of 0.01 °C.

11.1.2 Bipolar transistors in CMOS technology

Forasmuch as CMOS technology became available at much lower costs than bipolar technology, from the 1990s, most BJT-based thermal sensors are based on the temperature characteristics of parasitic bipolar transistors available in CMOS technology. As mentioned earlier, in CMOS technology, two types of parasitic bipolar PNP transistors comprising lateral and vertical exist. Figure 11.2 shows the cross-section of these PNP parasitic transistors. As illustrated in Figure 11.2(a), a lateral bipolar transistor is formed between the drain region and the source region of the PMOS transistor in which the drain acts as the collector and the source region acts as the emitter, while the base is formed by the N-well [852]. By applying a positive voltage to the gate of the PMOS transistor, the conduction drain-source channel will be turned off and a depletion layer will be created to push the diffusion current of the lateral bipolar transistor beneath the surface. The structure of the vertical bipolar transistor (see Figure 11.2(b)) is similar to that of lateral PNP transistors except that the collector is formed by the substrate. Due to the silicon substrate acting as the collector, this device is also known as a substrate bipolar transistor. In this structure, the collector is not accessed directly because it is formed inside the P-substrate. Another option to perform a bipolar is the vertical NPN, which can be implemented in modern CMOS technologies with twin well or deep N-well options [839,853–856]. Figure 11.3 shows the cross-section of vertical NPN transistors in modern CMOS technology, in which the collector is directly accessible. The collector is formed by a deep N-well, the N^+ drain diffusion acts as an emitter, and the base is formed by a P-well.

It is worth mentioning that although smart thermal sensors based on lateral transistors have been developed [857–859], due to the lower sensitivity to process expansion and packaging stress and the better exponential relationship between base-emitter voltage and collector current, vertical PNP transistors for temperature measurement are preferred [833,835,860–863].

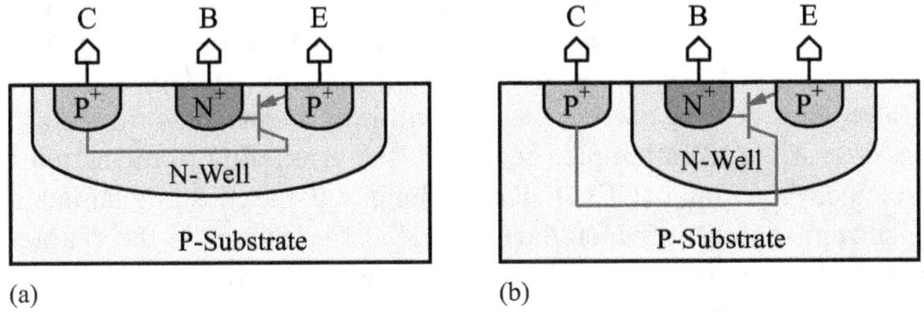

Figure 11.2 Cross-section of parasitic PNP transistors in CMOS technology, (a) lateral, (b) vertical

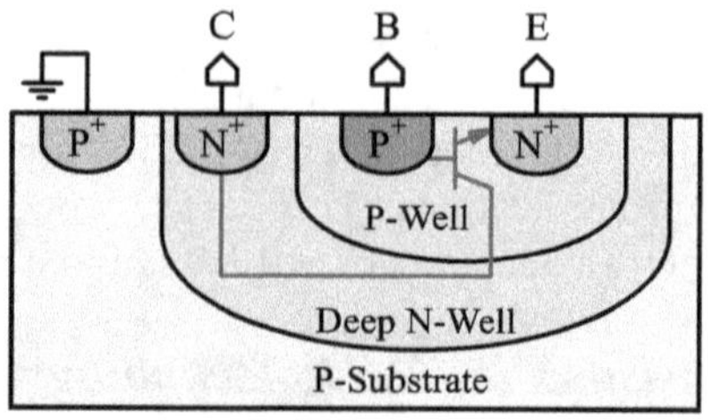

Figure 11.3 Cross-section of vertical NPN transistors in modern CMOS technology

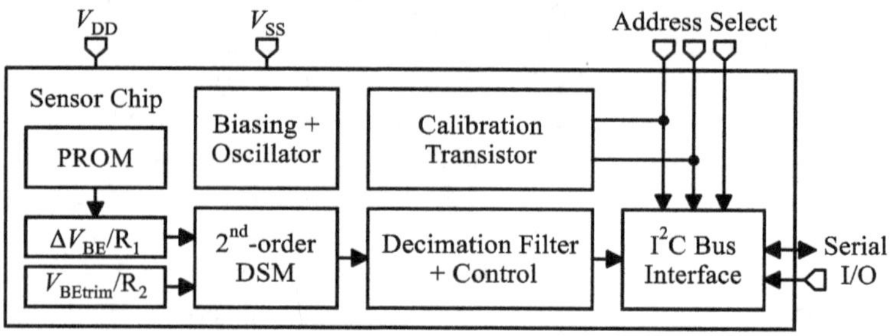

Figure 11.4 Block diagram of a smart thermal sensor with on-chip ΣΔM and digital bus interface

11.1.3 Implementation of BJT-based thermal sensors

Refereeing to (11.3), V_{BE}, ΔV_{BE}, and gain α can be affected by various non-idealities such as process spread (spread of the saturation current and spread of the current gain), curvature, finite current-gain, series resistances, transistor mismatch, etc. [828], which produce many errors and result in deviation output of ADC from the actual values. In other words, the accuracy of BJT-based thermal sensors mainly depends on the accuracy of the temperature characteristics of the bipolar transistors.

In the design of a BJT-based smart temperature sensor, two main challenges exist. First, the sensor's electronics should be designed as the spread of V_{BE} is the only major error source. Second, to determine the temperature error resulting from the spread of V_{BE}, a cost-effective calibration technique has to be developed [828]. There have been a lot of attempts that focus on eliminating the mentioned problems which will be verified briefly in the following text. For instance, the mismatch between the transistors and current sources will limit the accuracy of ΔV_{BE}. In one work, Meijer *et al.* [833] attempted to reduce the errors due to mismatch through dynamic element matching (DEM). This technique is based on dynamically interchanging current sources in which the effect of any mismatch in the characteristics of the current sources is decreased. In another effort, Pertijs *et al.* [834], with a description in detail [837], proposed a smart thermal sensor with an on-chip sigma-delta modulator (ΣΔM) and digital bus interface in a 0.5 μm CMOS process that its block diagram is shown in Figure 11.4. This sensor employed substrate PNP transistors for temperature sensing and generating the ADC's reference voltage. In this design [837], to obtain high initial accuracy, dynamic offset cancelation, and DEM in the analog front end are employed. Also, by applying second-order curvature correction, linearity over a wide temperature range is achieved. So, the spread on the base-emitter voltage will be the dominant source of errors, which is trimmed based on the results of a single-point calibration that takes place after packaging. The main task of the decimation filter is converting the bitstream output of the modulator to a digital representation of the temperature, which is transmitted to the outside world using an integrated circuit (IC) bus interface. Finally, after calibration at room temperature and trimming, the sensor shows an inaccuracy of ±0.5 °C (3σ) in the temperature range of −50 °C to 120 °C.

This group in another work [836], proposed a smart thermal sensor that extended the temperature range of −55 °C to 125 °C and reduced the temperature error inaccuracy to ±0.1 °C level in 0.7 μm CMOS. Same as in previous work in [837], substrate PNP transistors are used to extract the temperature information. In this design [836], a switched-capacitor (SC) implementation of ΣΔM instead of continuous-time implementations is used that facilitates the implementation of the various offset cancelation and DEM techniques required to achieve low inaccuracy performance. Also, the error due to the spread of the base-emitter voltage of the bipolar transistors is trimmed out based on a calibration at one temperature, by using a digital-to-analog (DAC) ΣΔ current. Although these calibrations and trimming methods are efficient, individual calibration causes a significant cost contributor. To address this problem, Aita *et al.* [864], with a description in detail [841], proposed a smart thermal sensor based on substrate PNP transistors that employed a batch calibration technique. This work achieved an inaccuracy of ±0.25 °C (3σ) from −70 °C to 130 °C, which shows a 2× improvement over the state of the art in [836]. After an individual calibration at room temperature, the inaccuracy improves to ±0.1 °C (3σ) over the military range.

Another substrate PNP-based thermal sensor, which is intended for use in RFID tags is proposed by Souri *et al.* [840] in 0.16 μm technology. The block diagram of the sensor is shown in Figure 11.5. Its structure benefits from an improved charge-balancing scheme, which reduces ADC conversion time, and a second-order zoom ADC that combines a successive-approximation (SAR) and second-order ΣΔ ADCs to realize accurate, high-resolution readout of two sensing BJTs voltages, and diminish power consumption. Moreover, a voltage calibration method [838], in which die temperature is established by measuring an on-chip ΔV_{BE}, is used, which is significantly faster compared to thermal calibration. So considering the mentioned benefits, the sensor achieves an inaccuracy of ±0.15 °C (3σ) from –55 °C to 125 °C after the thermal calibration at 30 °C and a PTAT trim [840]. It draws only 3.4 μA and operates from a 1.5 V to 2 V supply. Also, it shows a resolution of 20 mK and a supply sensitivity of 0.5 °C/V. This structure is based on batch-to-batch measurements as its average calibration parameters change significantly from one batch to another. To address this issue, Yousefzadeh *et al.* [844] (with extended and more details in [845]), presented a BJT-based temperature-to-digital converter (TDC), with sensing PNPs, which achieves state-of-the-art accuracy and preserves it from

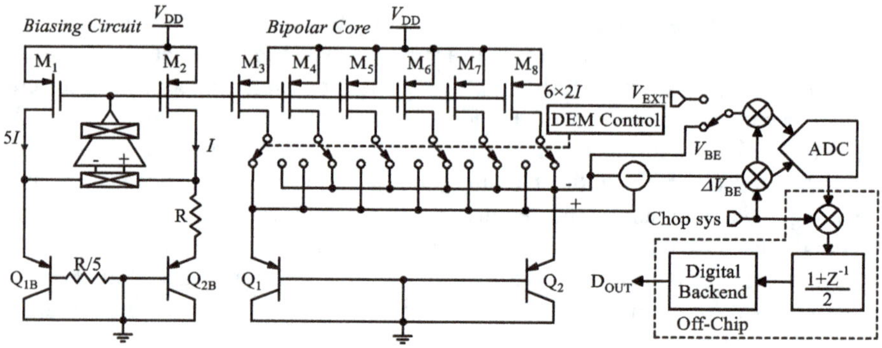

Figure 11.5 The smart thermal sensor with zoom ADC for use in RFID tags

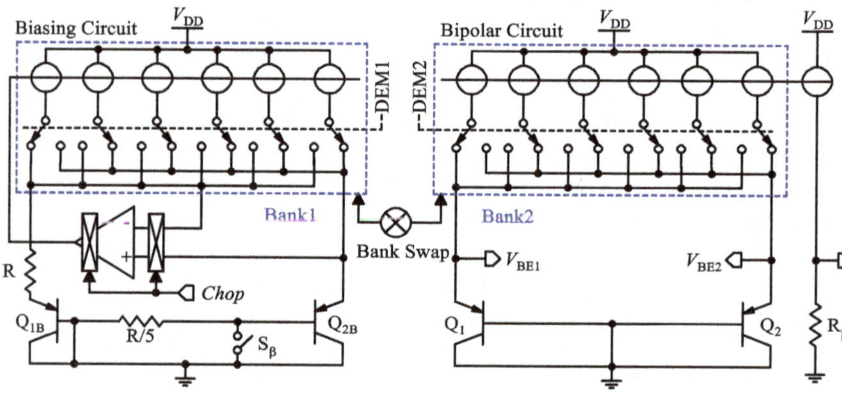

Figure 11.6 Sensing front end of the TDC with β-detection circuit

batch-to-batch. The measurements result on 20 samples from one batch show that after a single room-temperature trim, TDC obtains an inaccuracy of ±60 mK (0.06 °C) (3σ) from −55 °C to 125 °C. The sensing front end of TDC is shown in Figure 11.6. It consists of PTAT/Resistance (PTAT/R) biasing with (optional) β-compensating and a bipolar core built around four identical substrate PNPs. A PTAT/R biasing circuit provides a PTAT current with less nonlinearity compared to a constant biasing current schema. Moreover, it provides a supply-independent current, that results in an incremental accuracy of V_{BE}. On the other hand, the effect of β on V_{BE} is suppressed by using a β-compensating resistor ($R_{\beta b}$). A two-step SAR-ΣΔ architecture is used for ADC to digitize $V_{BE}/\Delta V_{BE}$. Adding an offset to $V_{BE}/\Delta V_{BE}$ effectively adds a scaled version of ΔV_{BE} to V_{BE}, and thus realizes a PTAT trim.

The time-domain thermal sensor that converted the CTAT and PTAT voltages to a duty cycle is widely used [842,843,847–849,865–867]. The basic operating principle of thermal sensors with duty-cycle-modulated output is illustrated in

Figure 11.7 [842]. Its principle operation is based on alternately charging and discharging of capacitor C as charging current (I_{Cha}) is proportional to PTAT and discharging current is proportional to CTAT (I_{Dis}). So, the duty cycle of the charging and discharging can be given by $I_{\mathrm{Cha}}/(I_{\mathrm{Cha}} + I_{\mathrm{Dis}})$, which has a linear function of temperature and is insensitive to the V_{REF}, V_1, and V_2. Heidary *et al.* [842] employed the concept of a duty cycle proportional to the thermal information in the analog front end of the sensor to implement a time-base thermal sensor, which achieves a spread of less than ± 0.15 °C (3σ) from -45 °C to 130 °C. In this architecture charging and discharging periods should be measured in a highly precise manner. Also, to generate the CTAT current from the BJT core, an extra amplifier is required. On the other hand, accurate trimming circuits are required to determine the current gains by the analog components. A capacitor-reused thermal sensor with duty-cycle-modulated output in a 0.13 μm CMOS process is reported by Tang *et al.* [868]. It has a compact area of 0.073 mm^2 and an inaccuracy of ± 0.38 °C (3σ) across the temperature of -10 °C to 100 °C. Another BJT-based thermal sensor based on a time-domain conversion scheme is introduced by Huang *et al.* [847] with an inaccuracy of ± 0.54 °C (3σ) and a wider temperature range of -40 °C to 125 °C compared to [868] in a 0.13 μm CMOS process. In this design, as illustrated in Figure 11.8, to generate the V_{PTAT} and V_{CTAT}, a PNP transistor and NPN transistor are employed. In this structure, the ramp voltage of V_{C} is created by a capacitor (C_1) and a current source. Then comparing V_{C} with the PTAT and CTAT voltages modulates the thermal information into the duty-cycle-modulated oscillation output. In this design to obtain an improvement in the accuracy of V_{PTAT}, DEM with Kelvin connection (KC-DEM) is employed. Also, a continuous-time dynamic single-threshold hysteresis comparator is proposed to enhance the robustness of the sensor. Nevertheless, this design lacks sufficiently precise measurements of phase durations to obtain accurate thermal information.

In another work, a BJT-based temperature-to-frequency converter (TFC) applicable for on-chip thermal monitoring is introduced by Park *et al.* [849] that achieves temperature information from oscillation frequency. The simplified circuit diagram of the TFC structure based on the relaxation oscillator is illustrated in Figure 11.9. It comprises an integration capacitor C_1, a PTAT current source I_{PTAT}, a CTAT reference voltage V_{CTAT} from the BJT core, and a comparator, which

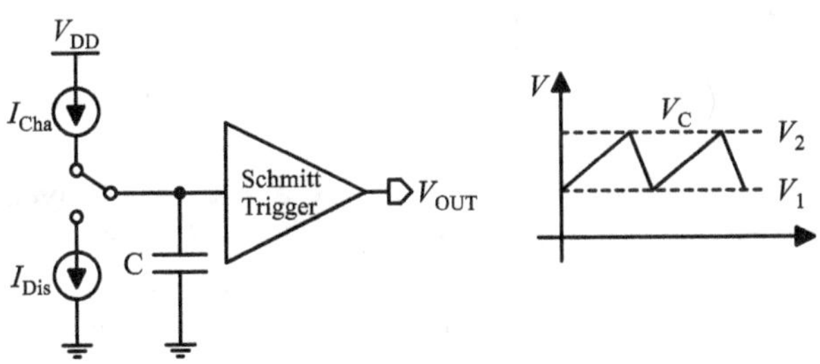

Figure 11.7 Operating principle of a thermal sensor with duty-cycle-modulated output

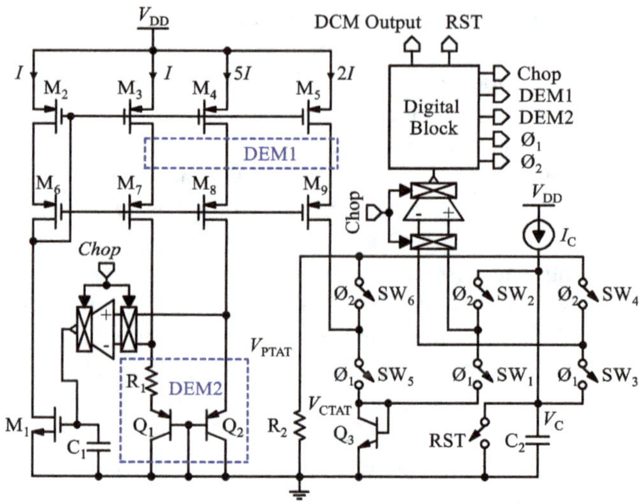

Figure 11.8 A thermal sensor with duty-cycle-modulated output employed PNP and NPN transistor

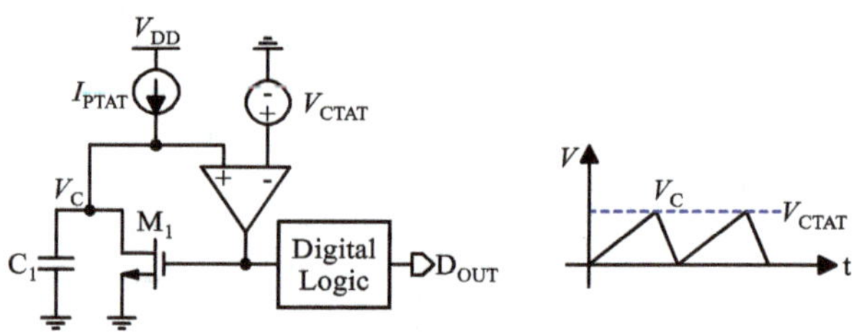

Figure 11.9 Simplified circuit diagram of a BJT-based TFC

converts the ratio of V_{CTAT} to the V_{PTAT} to frequency. The flipping technique is employed to diminish the comparator offset. The comparator inputs are swapped together by flipping the C_1 like the chopping. This approach called as capacitor flipping technique, which diminishes the comparator offset. Also, a bootstrapping circuit is used to alleviate the leakage current in the capacitor flipping. This design achieves the 3σ-inaccuracy of $\pm$1 °C from $-$40 °C to 140 °C without any trimming circuit.

The NPN transistors in CMOS technology are also employed as a sensing element for the design of thermal sensors [839,854,856,869,870]. For example, Sebastiano *et al.* [839] have used a vertical NPN transistor as the sensing element to design a thermal sensor with digital output in a 65 nm CMOS process. In this design, employing of NPN transistor enables low-temperature $-$70 °C sensing while operating from a low supply voltage of 1.2 V. The sensor achieves trimmed inaccuracy of $\pm$0.2 °C (3σ) over the temperature range from $-$70 °C to 125 °C and draws 8.3 μA from a 1.2 V supply voltage. Another work by Someya *et al.* [856] presented a low-power, NPN-based thermal sensor that achieves an inaccuracy of ±0.15 °C (3σ) after a 1-point trim from −15 °C to 85 °C. In this sensor, by use of a power-efficient dual-mode front end and a tracking ΣΔ ADC, the consumption power is significantly reduced, which is only 0.21 μW from a 1.25 V supply voltage.

BJT-based thermal sensors typically require supply voltages above 1 V, since V_{BE} will be about 0.8 V at −55 °C and some headroom is required for proper operation of the current source that biases the BJT [871]. In one effort by Tang *et al.* [872] a sub-1 V BJT-based thermal sensor is presented, which has employed the capacitive biasing principle instead of current-source biasing as shown in Figure 11.10. At first, a sampling capacitor C_S is charged to the supply voltage and then discharged in which the discharging time is set by switching the base of a PNP transistor. Finally, the voltage on C_S will be determined by the I/V characteristic of the PNP transistor [873,874]. In fact, by fixing the discharging time, the voltage on C_S will show CTAT behavior, and by subtracting the voltage on the sampling capacitor of two PNP transistors with a fixed discharging time a PTAT behavior will be achieved. So, it can be deduced that the thermal sensor based on the capacitive biasing technique can work with very little headroom and enables sub-1 V operation compared to the current-source biasing. The sensor shows a 1-point-trimmed inaccuracy of ±0.15 °C (3σ) from −55 °C to 125 °C and consumes 0.81 μW from a 0.95 V supply at room temperature.

Table 11.1 summarizes the performance of some BJT-based thermal sensors in CMOS technologies. The TDC in [845], achieved the best inaccuracy of ±0.06 °C (after 1-point trim), while the current drawn of supply is only 4.6 μA. One of the lowest power consumption BJT-based thermal sensors is reported by Someya *et al.* [856], which consumed only 0.21 μW power. The work in [836] presented a wide temperature range of −70 °C to 130 °C compared to other reported designs. The thermal sensor of Tang *et al.* [872] by employing the capacitive biasing technique enables sub-1 V operation.

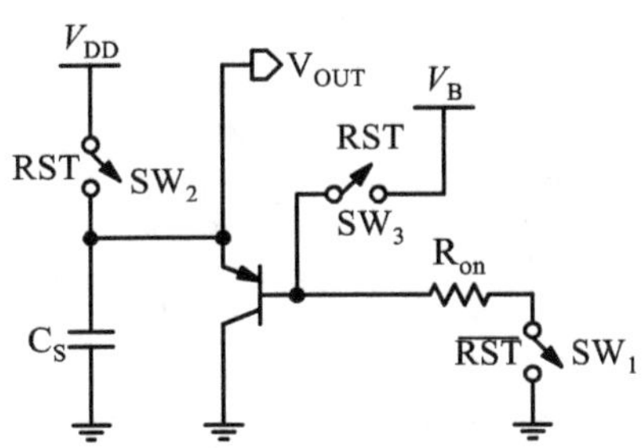

Figure 11.10 Principle operation of thermal sensor based on capacitive biasing technique

Table 11.1 Comparison of BJT-based CMOS sensor

Temperature range	Resolution	Inaccuracy (3σ)	Supply current	Power Supply sensitivity	Supply voltage	Chip area	Process	Ref.
−50 °C to 120 °C	NA	±0.3 °C @ 25 °C (1-point) ±0.5 °C @ −50 °C to 120 °C (1-point)	130 μA	0.3 °C/V	2.7–5.5 V	2.5 mm²	0.5 μm	[837]
−55 °C to 125 °C	0.01 °C	±0.03 °C @ 30 °C (1-point) ±0.1 °C @ −55 °C to 125 °C (1-point)	75 μA	0.03 °C/V	2.5–5.5 V	4.5 mm²	0.7 μm	[836]
−70 °C to 130 °C	0.025 °C	±0.25 °C @-70 °C to 130 °C ±0.1 °C@ −55 °C to 125 °C (1-point)	25 μA	0.05 °C/V	2.5–5.5 V	4.5 mm²	0.7 μm	[841]
−55 °C to 125 °C	20 mK	±0.15 °C (1-point)	3.4 μA	0.5 °C/V	1.5–2 V	0.08 mm²	0.16 μm	[840]
−55 °C to 125 °C	15 mK	±0.06 °C (1-point)	4.6 μA	0.01 °C/V	1.5–2 V	0.16 mm²	0.16 μm	[845]
−45 °C to 130 °C	3 mK	±0.15 °C (1-point)	55 μA	0.05 °C/V	2.9–5.5 V	0.8 mm²	0.7 μm	[842]
−10 °C to 100 °C	32 mK	±0.38 °C (1-point)	17 μA	0.3 °C/V	1.8–3.5 V	0.073 mm²	0.13 μm	[868]
−40 °C to 125 °C	0.06 °C	±0.54 °C (1-point)	39.7 μW**	0.21 °C/V	1.2/3.3 V	0.086 mm²	0.13 μm	[847]
−40 °C to 140 °C	144 mK	±1 °C (2-point)	3.1 μW**	0.3 °C/V	1.35–2 V	0.025 mm²	0.11 μm	[849]
−70 °C to 125 °C	0.03 °C	±0.2 °C (1-point)	8.3 μA	1.2 °C/V	1.2–1.3 V	0.1 mm²	65 nm	[839]
−15 °C to 85 °C	15 mK	±0.15 °C (1-point)	0.21 μW**	0.07 °C/V	1.25 V	0.058 mm²	0.18 μm	[856]
−55 °C to 125 °C	1.8 mK	±0.15 °C (1-point)	0.81 μW**	0.2 °C/V	0.95–1.4 V	0.25 mm²	0.18 μm	[872]

* NA: Not available ** Power consumption

11.2 MOSFET-based thermal sensors

Typically, MOSFET-based thermal sensors operate in saturation [875–878] or in subthreshold regain [879–884]. The drain current of MOSFET (I_D) in saturation regain is approximated by (11.5).

$$I_D = 0.5\mu C_{ox}\frac{W}{L}\left(V_{GS} - V_{th}\right)^2 \tag{11.5}$$

where μ is the mobility of carriers, C_{ox} is the gate-oxide capacitance, W/L is the dimension, V_{GS} is the gate-source voltage, and V_{th} is the threshold voltage of the transistor. According to (11.5), as temperature increases, both the μ and the V_{th} decrease. Reduction in μ reduces drain current while V_{th} reduction tends to increase the current. When MOSFET is biased in the subthreshold regain, the ID of a MOSFET has a temperature-dependent exponential function of the VGS and the drain-source voltage (VDS), which can be described by (11.6).

$$I_D = I_S \exp\left(\frac{V_{GS} - V_{th}}{nV_T}\right)\left(1 - \exp\left(-\frac{V_{DS}}{V_T}\right)\right) \tag{11.6}$$

$$I_S = \mu\, C_{ox}\frac{W}{L}(n-1)V_T^2 \tag{11.7}$$

where I_S is the leakage current, n is the subthreshold slope factor, V_T (k_BT/q) is the thermal voltage, k_B is Boltzmann's constant, T is the absolute temperature, and q is the electron charge [885]. The current I_D for voltage V_{DS} greater than 0.1 V can be assumed independent of V_{DS} and is expressed by (11.8).

$$I_D = I_S \exp\left(\frac{V_{GS} - V_{th}}{nV_T}\right) \tag{11.8}$$

As mentioned in the previous section, BJT-based thermal sensors typically require V_{BE} to be about 0.8 V (@−55 °C) [871], while V_{GS} of a MOSFET biased in subthreshold is smaller compared to BJT. Therefore, MOSFET-based thermal sensors biased in subthreshold can be operated with supply voltages sub-1 V and consequently consume lower power. So, the thermal sensors biasing in subthreshold regain is very desired [886–891].

11.2.1 MOSFET-based thermal sensors in subthreshold regain

Figure 11.11 shows the operating principle of MOSFET-based thermal sensors biased in subthreshold. Referring to (11.8), in this configuration, the sensing block converts temperature into PTAT current. Then this current will be turned into a digital output temperature code via various converting structures that will be examined in the following.

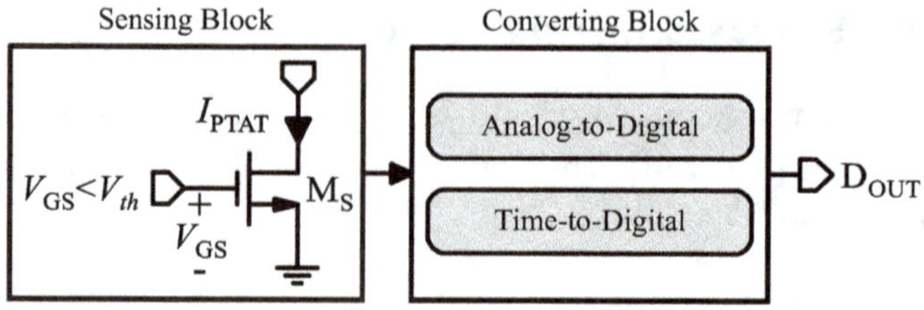

Figure 11.11 Block diagram of a MOSFET-based thermal sensor biased in subthreshold

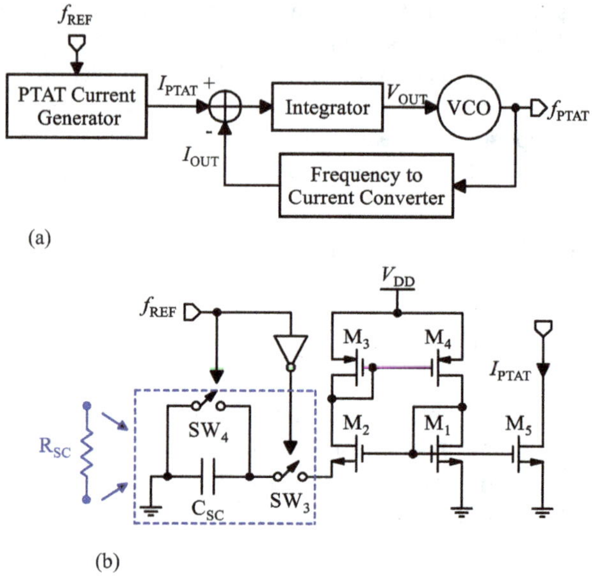

Figure 11.12 (a) Block diagram of a temperature-to-frequency converter and (b) PTAT current generator circuit

11.2.1.1 Constant threshold MOSFET-based thermal sensors

In one effort, Ueno *et al.* [881], employed the exponential current characteristic of MOSFETs operating in the subthreshold region to implement a temperature-to-frequency converter. Figure 11.12(a) shows a block diagram of the sensor, that consists of a PTAT current generator and a current-feedback loop. Figure 11.12(b) illustrates the structure of the PTAT current generator. In this circuit, MOSFETs operate in subthreshold regain, and the SC circuit operates as a resistor with average resistance R_{SC} equal to $(C_{SC}.f_{REF})^{-1}$, where f_{REF} is the external reference clock and is independent of temperature, and C_{SC} is the capacitor of SC circuit. The output current I_{PTAT} can be given by (11.9) which is proportional to temperature.

$$I_{PTAT} = \frac{nV_T}{R_{SC}} \ln\left(\frac{K_2}{K_1}\right) = f_{REF}.C_{CS}.\frac{nkT}{q}.\ln\left(\frac{K_2}{K_1}\right) \tag{11.9}$$

where K_1 and K_2 is the aspect ratio (W/L) of the transistor M_1 and M_2, respectively. As illustrated in Figure 11.12(b), in the feedback loop, the current subtractor compares I_{PTAT} with I_{OUT} to adjust V_{OUT} to produce f_{PTAT}. Finally, in the lock state, the f_{PTAT} will be proportional to the absolute temperature. It achieves inaccuracy of −1.8 °C to 1 °C over a temperature range of 10 °C to 80 °C. The power dissipation is 10 μW from a 2.2 V supply voltage. This design relies on external frequency references, which imposes an extra amount of power and area that limits the system integration and miniaturization.

In another effort, two compact frequency-output temperature sensors based on a multivibrator current-to-frequency converter are proposed by Azcona *et al.* [887] that do not require any external frequency reference. Figure 11.13(a) shows the block diagram of the sensors, which is comprised of a bidirectional current integrator driven by a voltage window comparator (VWC) and the circuits to generate the two comparator voltages V_H and V_L and the current I_{Charge}. As can be seen from the schematics of the sensor shown in Figure 11.13(b), the current I_{Charge} drives a capacitor C between the limits V_H and V_L of the VWC. The charging and discharging process of capacitor C results in a frequency of oscillation given by (11.10).

$$f = \frac{I_{Charge}}{2C(V_H - V_L)} = \frac{I_{Charge}}{2C\Delta V} \tag{11.10}$$

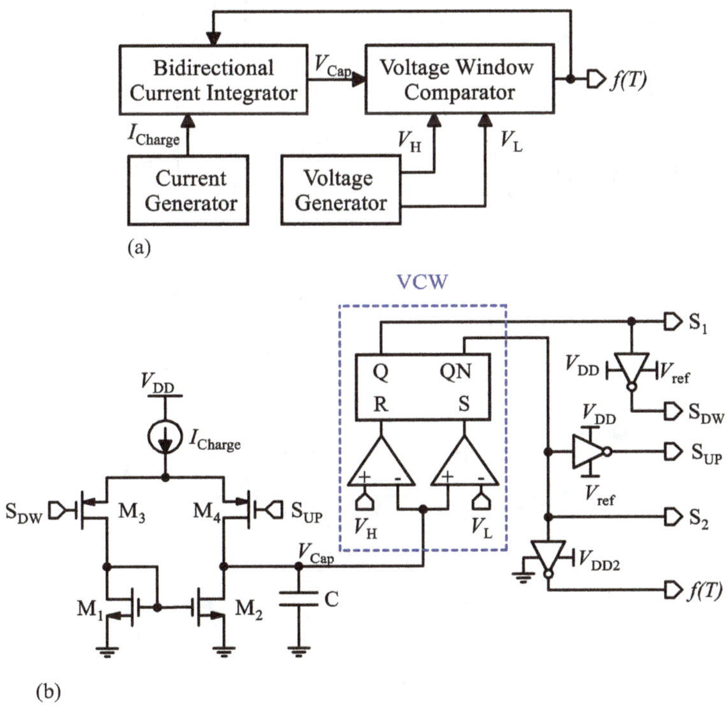

Figure 11.13 Temperature sensor with frequency-output (a) block diagram and (b) schematic

By employing this technique, the requirement for accurate frequency reference is removed. According to (11.10), two types of circuits can be used for I_{Charge} generation as shown in Figure 11.14. The structure shown in Figure 11.14(a) is designed for generating a PTAT current $I_{\text{Charge}}(T)$ and temperature independent ΔV (referring as design1) and another circuit (see Figure 11.14(b)) is used to generate a CTAT voltage of $\Delta V(T)$ and temperature independent current I_{Charge} (referring as design2). The sensor achieved an inaccuracy of ± 1 °C, over a temperature range of -40 °C to $+120$ °C with sensitivities of 335 Hz/°C and 460 Hz/°C for design1 and design2, respectively. It consumes power of below 3 μW from a 1.2 V supply voltage.

Another frequency reference-less temperature to digital converter is proposed by Zhao *et al.* [886], which works as a current/delay convertor. The temperature-sensing principle of this design is shown in Figure 11.15. Two N-type MOSFETs ($M_{1,2}$) in the subthreshold region with different gate voltages

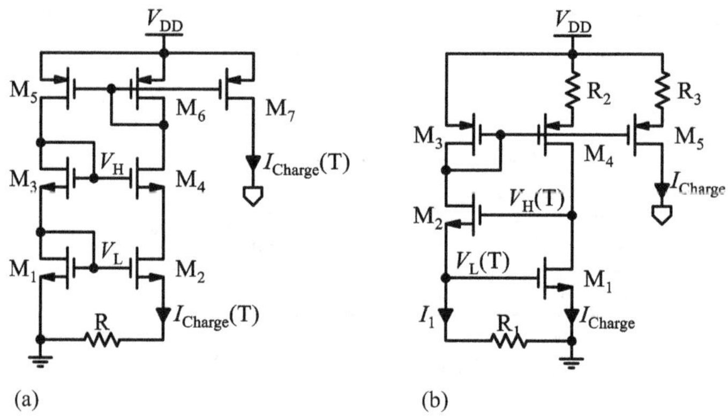

Figure 11.14 Schematics of (a) V_H, V_L, *and* $I_{Charge}(T)$ *generator, (b)* I_{Charge}, $V_H(T)$, *and* $V_L(T)$ *generator*

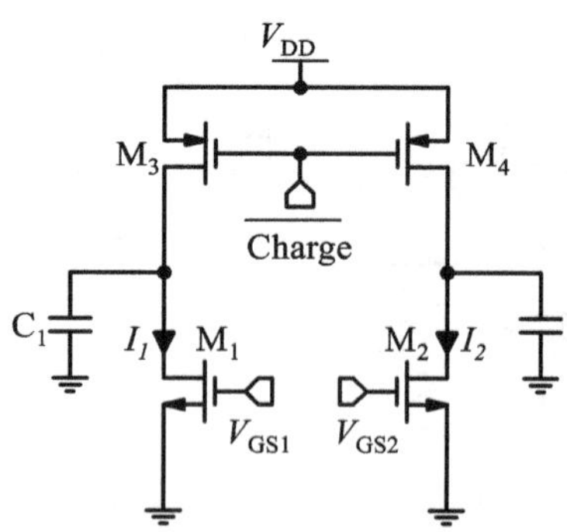

Figure 11.15 Schematic of the current/delay PTAT sensor

($\Delta V_{GS} = V_{GS2}$-V_{GS1}) were implemented to achieve a PTAT current of I_2/I_1 that is expressed by (11.11).

$$\frac{I_2}{I_1} = \exp\left(\frac{\Delta V_{GS}}{nV_T}\right) \tag{11.11}$$

So, the absolute temperature is proportional to $1/\ln(I_2/I_1)$, as described by (11.12).

$$T = \exp\left(\frac{\Delta V_{GS}}{nk_B/q}\right)\frac{1}{\ln(I_2/I_1)} \tag{11.12}$$

With a current-to-delay converter, current information is turned into delay information, which is in a digital domain. This sensor shows the inaccuracy of −1.6/1 °C over the temperature range of 0 °C to 100 °C while consuming power of 280 nW at 0.4 V supply voltage.

Another frequency reference-less temperature-to-digital converter sensor is introduced by Someya *et al.* [889], which uses a similar approach [886] but uses a different physical phenomenon to obtain the PTAT quantity. The operating principle of temperature-sensing is depicted in Figure 11.16. Two N-type MOSFETs (M_1 and M_2) operating in the subthreshold region are biased with different V_{DS} values while V_{GS} are kept at the same value. As a result, the current ratio I_2/I_1 will be a function of V_{DS1} and T as described by (11.13).

$$\frac{I_2}{I_1} = \frac{1}{1 - \exp(-V_{DS1}/V_T)} \tag{11.13}$$

In this design considering that V_{DS1} is sufficiently smaller than V_T, I_2/I_1 is approximately given by (11.14).

$$\frac{I_2}{I_1} = \frac{V_T}{V_{DS1}} + K_C \tag{11.14}$$

where K_C is a temperature-independent constant. By setting the V_{DS} for one MOSFET much larger than V_T and setting V_{DS} for the another MOSFET below V_T,

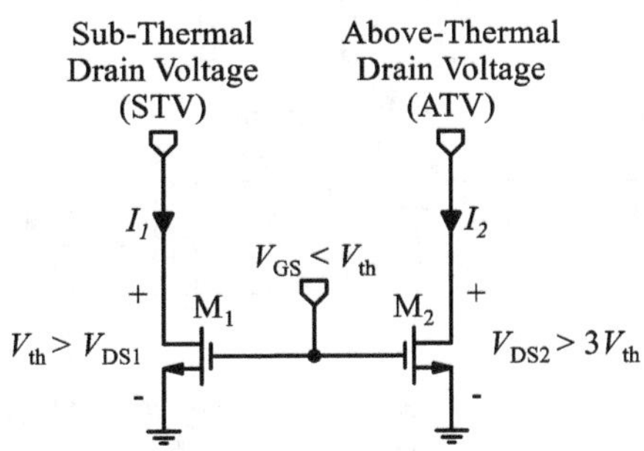

Figure 11.16 Mechanism of the temperature sensing of the temperature-to-digital converter

the I_2/I_1 provides better linearity and sensitivity than those obtained by the approaches presented in [886]. The sensor achieves −0.9/1.2 °C peak-to-peak inaccuracy over a temperature range of −20 °C to 80 °C after a two-point calibration with a resolution of 145 mK. Also, the power consumption of the sensor at a room temperature of 25 °C is only 11 nW.

11.2.1.2 Dynamic threshold MOSFET (DTMOS)-based thermal sensors

Some of the MOSFET-based thermal sensors use dynamic threshold MOSFET (DTMOST) as sensing elements [871,892–895]. A DTMOST is a standard MOSFET whose body and gate terminals are tied together [896], causing the threshold voltage to vary dynamically. The drain current I_D of a DTMOST operated in the subthreshold region can be defined by (11.15) [897].

$$I_D = \frac{W}{L}\exp\left[\frac{q}{k_B T}(V_{GS} - V_{th})\right] \tag{11.15}$$

A diode-connected DTMOST shows a near-ideal exponential relationship between I_D and V_{GS}, which is less dependent on C_{OX} and C_D [897,898]. As a result, DTMOST can be used to realize thermal sensors with a lower spread while operating at a lower supply voltage. The work by Souri *et al.* in [871], presented a thermal sensor based on DTMOST shown in Figure 11.17. As illustrated in Figure 11.17(a), the front end of the sensor of this work consists of two

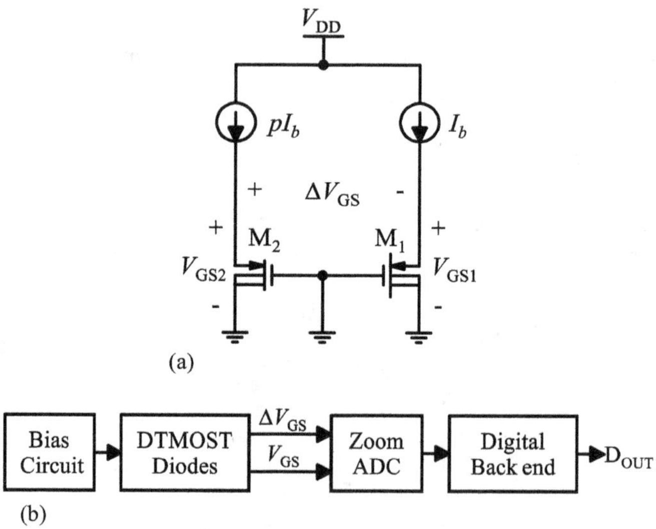

Figure 11.17 (a) Temperature dependence of V_{GS} and ΔV_{GS}, generated by two DTMOS diodes biased in sub-threshold region and (b) block diagram of the smart thermal sensor based on DTMOST

diode-connected. The V_{GS} of a DTMOST diode exhibits CTAT behavior, with a temperature coefficient near −1 mV/°C. On the other hand, the difference in the gate-source voltage ΔV_{GS} of two DTMOS diodes shows the PTAT behavior described by (11.16).

$$\Delta V_{GS} = \frac{k_B T}{q}.\ln(p) \tag{11.16}$$

where p is the bias current ratio. As shown in Figure 11.17(b), the ratio between V_{GS} and ΔV_{GS} is used to measure temperature. This sensor achieves an inaccuracy of ±0.4 °C (3σ) over the military temperature range of −55 °C to 125 °C after a single-temperature trim that improves to ±0.1 °C (3σ) after a two-temperature trim. Also, it shows a resolution of 33 mK, while dissipating 8.6 μW from a 1.8 V supply.

In another effort by Souri *et al.* [892], a thermal sensor based on DTMOST intended for RFID applications has been proposed that achieves an accuracy of ±0.4 °C over a temperature range of −40 °C to 125 °C. Also, it draws 700 nA from a 0.85 V supply. In other work by Khanpeth *et al.* [895], employed the concept of DTMOST in the design of a thermal sensor for achieving the digital representation of temperature information, a charge-balancing ΣΔ ADC is used. The sensor achieves an inaccuracy of ±0.25 °C (3σ) across all process corners and the temperature range of −20 °C to 85 °C while consuming 8.1 μW under 1 V power supply voltages.

The main performance parameters of the MOSFET-based thermal sensor are summarized in Table 11.2. MOSFETs can also be used as temperature-sensing elements with sub-1 V operation. As can be seen from Table 11.2, the work in [889], operates with 0.8 V and consumes only 11 nW. The work in [886] operates with 0.4 V supply voltage while presenting an inaccuracy of −1.6/1 °C from a temperature range of 0 °C to 100 °C. Unfortunately, although MOSFET-based thermal sensors offer sub-1 V performance, the process spread affects the V_{th} and μ MOSFETs, which results in these sensors exhibiting greater inaccuracy compared to BJT-based thermal sensors. The V_{GS}-I_D characteristic of a diode-connected DTMOS transistor, is less sensitive to V_{th} spread, compared to a diode-connected MOSFET [871,893,897,899]. The thermal sensor based on DTMOS in [871], achieves an inaccuracy of ±0.1 °C (3σ) over the temperature range of −55 °C to 125 °C after two-temperature trim, respectively.

11.3 ETF-based thermal sensors

11.3.1 Principles of operation of ETF-based thermal sensors

The operation principle of electro-thermal filter (ETF) temperature sensors is the thermal diffusivity of the silicon *D*, defined as the rate at which heat diffuses through a silicon substrate from a hot end to a cold end. *D* is strongly temperature dependent and it is approximated by a power-law [900–902] given by (11.17).

$$D = T^{-1.8} \tag{11.17}$$

Table 11.2 Comparison of different MOSFET-based sensor

Temperature range	Resolution	Inaccuracy (3σ)	Power cons.	Power supply sensitivity	Supply voltage	Chip area	Process	Ref.
10 °C to 80 °C	0.1 °C	−1.8/1 °C	10 μW @2.2 V 27 μW (@3 V	NA	2.2–3 V	0.08 mm^2	0.35 μm	[881]
−40 °C to 120 °C	460 Hz/°C @Design1 335 Hz/°C @Design1	±1 °C	1.9 μW @Design1 1.9 μW @Design1	0.45 °C/V	1.2 V	0.019 mm^2 @Design1 0.024 mm^2 @Design2	0.18 μm	[887]
0 °C to 100 °C	0.25 °C	−1.6/1 °C	280 nW	NA	0.4 V	0.022 mm^2	65 nm	[886]
−20 °C to 80 °C	145 mk	−0.9/1.2 °C	11 nW	3.8 °C/V @0.7–1.5 V	0.8 V	0.074 mm^2	0.18 μm	[889]
−40 °C to 90 °C	40 mK	−0.92/0.75 °C	25.4 μw	NA	1.8 V	0.63 mm^2	0.13 μm	[891]
−55 °C to 125 °C	33 mK	±0.4 °C (1-pint) ±0.1 °C (2-pint)	8.6 μw	0.1 °C/V	1.8 V	0.12 mm^2	0.16 μm	[871]
−40 °C to 125 °C	63 mK	±0.4 °C (1-pint)	700 nA @0.85 V	0.45 °C/V	0.85–1.2 V	0.085 mm^2	0.16 μm	[892]
−20 °C to 85 °C	NA	±0.25 °C (1-pint)	8.1 μw	NA	1 V	NA	0.18 μm	[895]

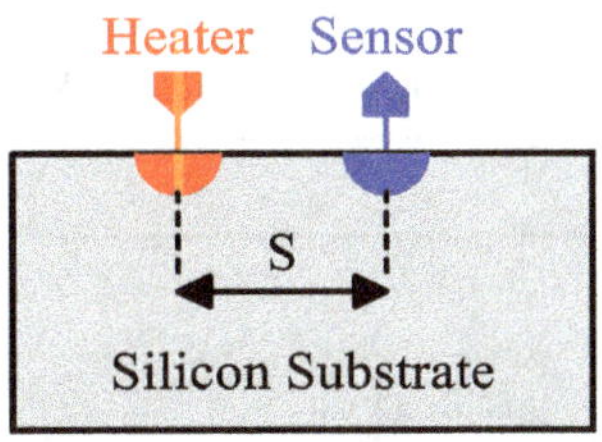

Figure 11.18 Cross-section of ETF sensing element

The thermal dependence of D is employed to develop temperature sensors. The architecture of ETF is depicted in Figure 11.18. It consists of a heating element and a temperature sensor formed by the junction of two dissimilar materials (thermocouples) fabricated at a distance of S concerning each other. Typically, instead of one thermocouple, several thermocouples connected in series are used, which is known as thermopile. The temperature sensor converts the received temperature variations induced by the thermal element into a small voltage signal. ETF due to the thermal substrate dependency behaves like a low-pass filter. Exciting the hot end of the thermal filter at a signal with a known reference frequency of f_{ref}, cause a temperature-dependent phase shift at the cold end that is given by (11.18) [903,904].

$$\varnothing_{ETF} \propto \left(S\sqrt{f_{ref}}\right)T^{n/2} \tag{11.18}$$

where, T is the absolute temperature, and n is near 1.8. The temperature-dependent phase shift can be turned to digital information by a phase domain ADC. It can be seen from the expression (11.18) that ETF is slightly nonlinear with temperature, so a linearization is required in the digital domain. The high accuracy of ETF-based sensors can be achieved without trimming, so it minimizes the calibration cost and is very suitable for low-cost applications [905].

11.3.2 Implementation of ETF-based thermal sensors

In one effort, a temperature-to-frequency converter based on the thermal diffusivity constant of silicon has been proposed by Makinwa *et al.* [903] which is shown in Figure 11.19(a). It is designed based on a frequency-locked loop in which in the lock state the output frequency will correspond to a fixed phase shift in the ETF. As a result, the output frequency will be a well-defined function of temperature. Figure 11.19(b) shows the layout of an ETF. In this design, the heater is an N $^{+}$-diffusion resistor, and the temperature sensor is a thermopile realized from P $^{+}$-diffusion/aluminum thermocouples to obtain high sensitivity. The sensor is implemented in a 0.7 μm CMOS process in which the preamp, synchronous demodulator, and two ETFs were fabricated on a 2.3 mm^2 chip area while the voltage-controlled oscillator (VCO) and the heater drive logic were implemented off-chip. Also, an off-chip instrumentation amplifier was used to convert the demodulator differential output into the single-ended control signal required by the

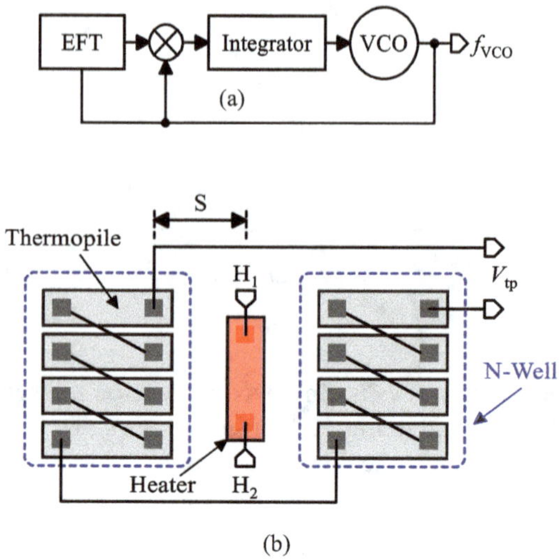

Figure 11.19 (a) Block diagram of an electrothermal sensor based on a frequency-locked loop and (b) layout of a CMOS electro-thermal filter

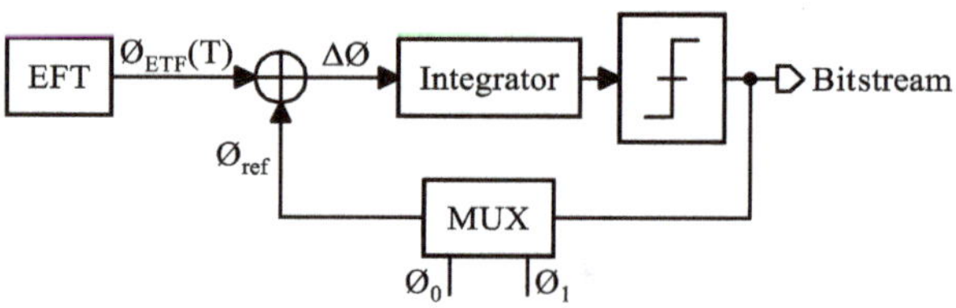

Figure 11.20 Block diagram of an ETF-based sensor utilizing PΣΔM to digitize the phase shift

VCO. The sensor achieves an accuracy of ±0.5 °C (3σ) over the temperature range of −40 °C to 105 °C. Also, it dissipates 2.5 mW from a 5 V supply voltage.

The work in [904] by Vroonhoven *et al.*, presented the temperature-dependent phase shift of an ETF digitized by a first-order single-bit phase domain sigma-delta modulator (PΣΔM) that compared to previously examined work [903], does not require any off-chip analog components, and operates over a wider temperature range. Its functional block diagram is illustrated in Figure 11.20. The average of the modulator's bitstream represents the weighted average of two reference phase shifts (φ_0, φ_1) that best approximates the ETF's phase shift. The phase difference $\Delta\varphi$ (φ_{ETF}-φ_{ref}) is obtained by multiplying the ETF's output with a reference signal, which multiplication is realized by a chopper. This design has employed the schematic layout of an ETF [903] with equal geometry. It achieves an untrimmed (but batch-calibrated) inaccuracy of ±0.5 °C (3σ) over the military temperature range −55 °C to 125 °C.

Another design by Vroonhoven *et al.* [906], used the PΣΔM same as [904] with a new schematic layout of an ETF shown in Figure 11.21 to improve inaccuracy. The spread of φ_{ETF} is mainly determined by the spread in the position of the diffusion/aluminum contacts since this defines the effective value of *S*, which is determined by the lithographic accuracy of the process [906]. In this design, by improving the interface electronics and by leveraging the more accurate lithography, greater accuracy was obtained that shows an untrimmed inaccuracy of ±0.2 °C (3σ) across −55 °C to 125 °C. The PΣΔM consumes 500 μW from a 1.8 V supply, which is 5× less than in [904] (see Table 11.3).

Sönmez *et al.* [907], presented an ETF-based sensor in 40 nm CMOS, which consists of a VCO and a multi-bit PΣΔM for ETF readout as shown in Figure 11.22. The ETF is driven with a square wave (FDRIVE), producing a phase-shifted output signal V_{ETF} that modulates the VCO's nominal frequency. Then, the output of VCO is demodulated by a PΣΔM to generate digitization of the ETF's phase shift φ_{ETF}. In this design, employing multi-bit PΣΔM compared to a single-bit modulator [908], diminishes quantization noise and the inherent cosine nonlinearity of synchronous phase demodulation [909] to negligible levels, while avoiding the complexity of a two-step conversion [910]. In this work to explore the tradeoff between resolution and accuracy, two ETFs with S of 3.3 μm and 2 μm were implemented. Each sensor dissipates 2.5 mW from the supply of 1.05 V. The sensors based on the S of 3.3 μm ETF achieve an untrimmed inaccuracy of ±1.8 °C (3σ) from −40 °C to 125 °C that improves to ±0.75 °C (3σ) after temperature trimming at 25 °C. Also, inaccuracy for S of 2 μm reduced from ±2.3 °C (3σ) to ±1.05 °C (3σ) after a single-point trimming.

Another work proposed by Pan *et al.* [911] is a hybrid thermal sensor that uses an inherently accurate, but power-hungry, ETF-based sensor [906] to achieve a self-calibrate and efficient resistor-based sensor [912]. As depicted in Figure 11.23, the Wien bridge structure is used to design the resistor-based sensor, and the ETF-based sensor employs second-order PΣΔM [906] to generate digital output. The

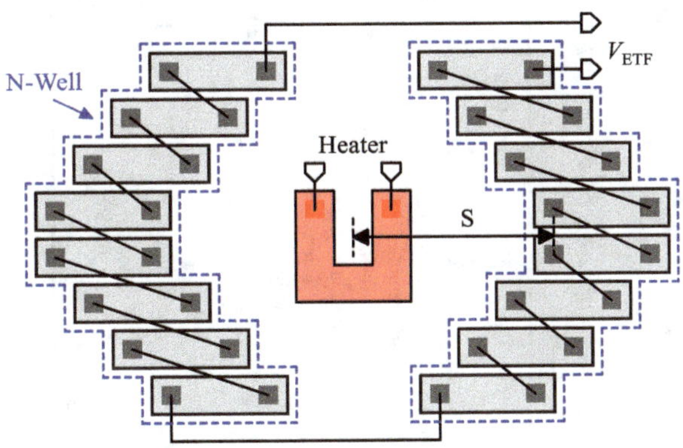

Figure 11.21 Layout of a CMOS electrothermal filter

Table 11.3 Comparison of different ETF-based sensor

Heater	Thermopile	S	Temperature range	Resolution	Inaccuracy (3σ)	Power Cons.	Supply voltage	Chip area	Process	Ref.
N+-diffused resistor	P+-diffusion/ aluminum	20 μm	−40 °C to 105 °C	NA	±0.5 °C	2.3 mW@Chip10m W@ Heater	5 V	2.3 mm^2	0.7 μm	[903]
N+-diffused resistor	P+-diffusion/ aluminum	20 μm	−55 °C to 125 °C	0.05 °C	±0.5 °C	2.5 mW@ Chip2.5 mW@ Heater	5 V	2.3 mm^2	0.7 μm	[904]
N+-diffused resistor	P+-diffusion/ aluminum	23 μm	−55 °C to 125 °C	0.02 °C (20 mK)	±0.2 °C	0.5 mW@Chip2.5 mW@ Heater	1.8 V	0.18 mm^2	0.18 μm	[906]
Resistive	NA	3.3 μm	−40 °C to 125 °C	0.36 °C	±0.75 °C	2.5 mW@Chip	0.9 to 1.2 V	1.65 mm^2	40 nm	[907]
		2 μm		0.24 °C	±1.05 °C					
N+-diffusion	P+-diffusion/ metal	24 μm	ETF:-15 °C to 105 °C	15 mK	0.2 °C	0.906 mW@Chip4.2 mW@ Heater	1.8 V	0.2 mm^2	0.18 μm	[911]
			Wien bridge: −55 °C to 125 °C	0.45 mK	0.03 °C	0.06 mW				

*NA: Not available

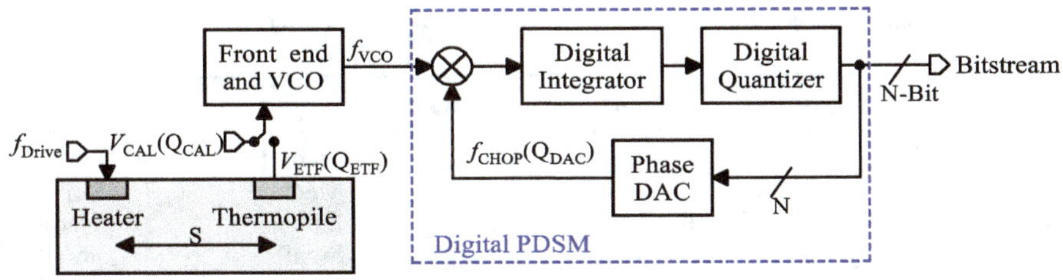

Figure 11.22 The block diagram of the ETF-based sensor comprising VCO and PΣΔM

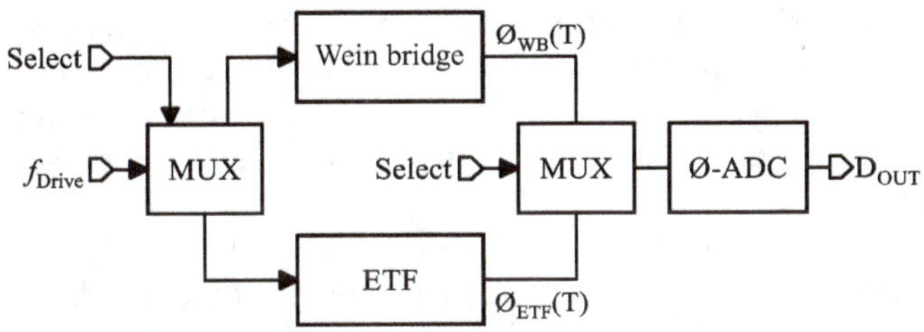

Figure 11.23 Schematic of the hybrid sensor consists of ETF and Wien bridge structures

ETF sensor achieves an inaccuracy of 0.2 °C (3σ) without trimming, while the Wien bridge sensor's no-trim inaccuracy is larger than 10 °C. After a 2-point trimming, the inaccuracy of the Wien bridge sensor improves to 0.03 °C over −35 °C and 105 °C. In ETF mode, the sensor consumes 5.1 mW (4.2 mW heater, 0.9 mW analog, and 0.6 μW digital) from a 1.8 V supply. In Wien bridge mode, its power consumption is only 66 μW. The sensor is fabricated in a standard 0.18 μm CMOS process and has an active area of 0.2 mm^2. The extended version of this design is also proposed by this research group [913].

11.4 Resistor-based thermal sensors

In CMOS technology, resistor-based thermal sensors can be realized with metal resistors, N-well resistors, diffusion resistors, poly-silicon resistors, and silicide resistors [855]. To have a comparison of different resistors' performance, their properties are illustrated in Table 11.4 [855]. The temperature dependence of a resistor can be modeled as (11.19).

$$R(T) = R(T_0)\left(1 + TC_1.\Delta T + TC_2.\Delta T^2\right) \tag{11.19}$$

where $R(T_0)$ is the resistor at a reference temperature T_0. TC_1 and TC_2 are the first-order and second-order temperature coefficients (TCs), respectively, and ΔT is the temperature difference compared to T_0 (the nominal value).

Table 11.4 Characteristics of various resistors in CMOS processes

Resistor type	Metal	Diffusion	N-Well	Poly	Silicided
First-Order TC	Large	Medium	Large	Medium or Small	Large
Second-Order TC	Medium	Medium	Large	Medium	Small
Sheet Resistance	Very Small	Large	Large	Large	Medium
Supply Dependency	Small	Medium	Large	Small	Small
1/*f* Noise	Negligible	Negligible	Negligible	Large	Small
Stress Sensitivity	Small	Large	Large	Medium	Small

To achieve a resistor-based thermal sensor with high performance, the sensing resistor should satisfy some specifications including a large temperature coefficient (high linearity) for high sensitivity, a large sheet resistance for smaller areas and low power consumption, low supply dependency, and stress sensitivity for stable operation, and low 1/*f* noise.

According to Table 11.4, metal resistors have the best performance of first-order *TC*, which causes good energy efficiency. On the other hand, its lower sheet resistance leads the sensor to occupy a large area and consume high power. The diffusion resistors have high sheet resistance, but they suffer from stress sensitivity. The N-well resistors have comparable *TC* to metal resistors and very low 1/*f* noise, but their main drawbacks are high supply dependency, high stress sensitivity, and very low doping levels, which can affect stability performance. The poly-silicon resistors have large 1/*f* noise and small temperature coefficients. Silicided resistors have a large *TC*, low 1/*f* noise, low supply dependency, and low stress sensitivity. But, they exhibit a lower sheet resistance. There are three mainstream temperature-dependent resistors in the CMOS process: Wheatstone bridge [816,914–920], Wien bridge [912,921–925], and polyphase filter (PPF) [926–930]. In the following, we will examine these structures briefly from different aspects such as principle operation, applications, readout interfaces, etc.

11.4.1 Implementation of resistor-based thermal sensors

11.4.1.1 Wheatstone bridge-based thermal sensors

The core of the Wheatstone bridge sensor consists of two different resistors ($R_N(T)$ and $R_P(T)$) types with negative and positive *TC*s. As shown in Figure 11.24, the output signal of the sensor can be implemented as a voltage mode or current mode.

To achieve high sensitivity and stability, $R_P(T)$ and $R_N(T)$ are implemented as silicided-p-poly and non-silicided n-poly resistors, respectively [816,914–918,920]. For instance, Pan *et al.* [914], proposed a Wheatstone bridge-based sensor from poly-silicon resistors. The current readout of this sensor is realized by a single-bit second-order continuous-time sigma-delta modulator (CTΣΔM) as shown in Figure 11.25. A resistor DAC is used to remove the bridge's temperature-dependent output current $I_{sig}(T)$ which drives the average value of $I_{err}(T)$ to zero by switching the DAC's resistors to the ground and thus effectively balancing the

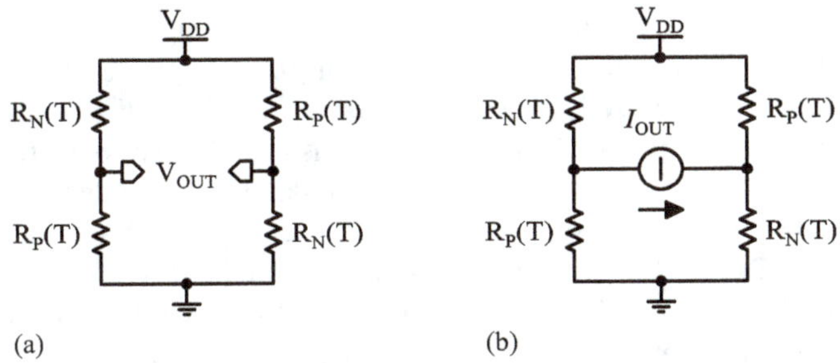

Figure 11.24 Wheatstone bridge sensor, (a) voltage readout schema, (b) current readout schema

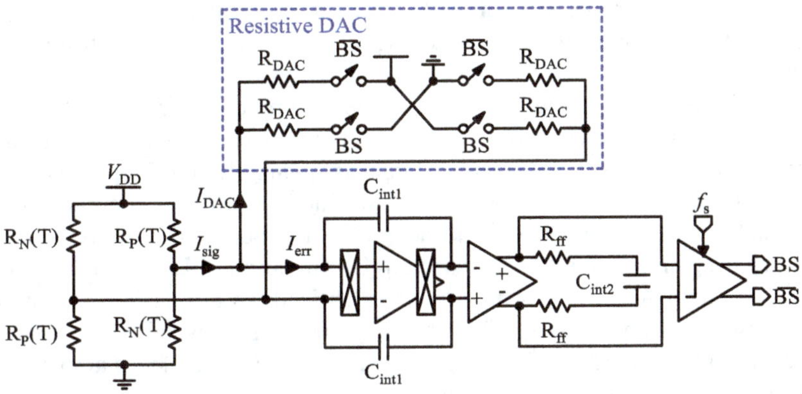

Figure 11.25 Wheatstone bridge-based sensor with CTΣΔM readout

bridge. The sensor achieves a resolution of 0.16 mK and inaccuracy ±100 mK (3σ) over the temperature range −40 °C to 85 °C. It draws 180 μW from a 1.8 V supply. In another work, Pan *et al.* [915], present a senor with multi-bit zoom CTΣΔM. The multi-bit resistor DAC reduces the magnitude of I_{err}, so it compensates the output current of the bridge more precisely. Due to the compensating most of I_{sig} by I_{DAC}, the ADC's power dissipation, and the size of the required integration capacitors, can be significantly reduced. The nonlinearity of the zoom ADC is significantly decreased by a segment averaging technique. It obtains an inaccuracy of ±0.12 °C (3σ) from −55 °C to 125 °C, power consumption of 94 μW. Also, it achieves a 3× smaller area and higher energy efficiency compared to the previous design [914]. In this design, the DAC state is determined by combining the output of a CTΣΔM with the result of an initial coarse SAR conversion. Therefore, extra logic is needed to implement the SAR conversion, as well as the segment-averaging schemes.

Pan *et al.* [918], to achieve more reduction of area and increase energy efficiency, proposed another sensor that uses a finite impulse response (FIR) DAC.

Due to the FIR-DAC not requiring over-ranging, this design results in I_{err} swing being about 2× less than that in the zoom ADC of [914]. So, the area and the power consumption will be decreased. It has an inaccuracy of ±0.14 °C (3σ) from −55 °C to 125 °C, with a resolution of 0.16 mK, and consumes 79 μW power.

In another design, Pan *et al.* [920], introduced an improved design of previous works [914,918] with two innovations of a sensitivity-preserving return-to-common-mode (RCM) DAC switching scheme and an energy-efficient operational transconductance amplifier (OTA) with a tail-resistor linearization scheme. Also, by merging the poly arms of the bridge with the unit elements of the DAC, the accuracy of the sensor is improved. The inaccuracy of the sensor is ±0.4 °C (3σ) from −55 °C to 125 °C after a ratio-based one-point trim and systematic non-linearity removal, which after a first-order fit improves to ±0.1 °C (3σ). The sensor consumes 55 μW from a 1.8 V supply. To significantly reduce the power consumption, Pan *et al.* proposed a duty cycle of a μW-power temperature sensor [816] that is dedicated to biomedical applications and draws only 6.6 μW from a 1.6 V supply.

Another work introduced by Jain *et al.* [931], implemented a very low power temperature sensor for wireless Internet of Things (IoT) devices/RFID tags that Wheatstone bridge is read out using a digital frequency-locked loop (DFLL). The system architecture is shown in Figure 11.26. The output of the thermal sensor is transmitted back to the receiver via backscatter by modulating the resonant frequency of the wireless link. In this design, by employing FLL, an on-chip clock generator, and an ADC are not required, which leads to diminishing complexity, and the power consumption of 310 nW from a 1.6 V supply is the lowest compared to other sensors reported in this section (see Table 11.6). It achieves an inaccuracy of −1.6/1.8 °C and 9.8 mK resolution.

11.4.1.2 Wien bridge-based thermal sensors

The schematic of a Wien bridge is shown in Figure 11.27. It is a second-order band-pass filter, in which the phase shift is the temperature dependence parameter. The

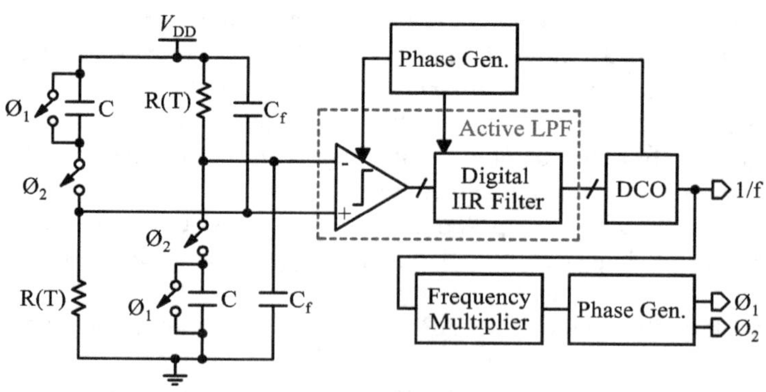

Figure 11.26 DFLL scheme for wireless IoT device

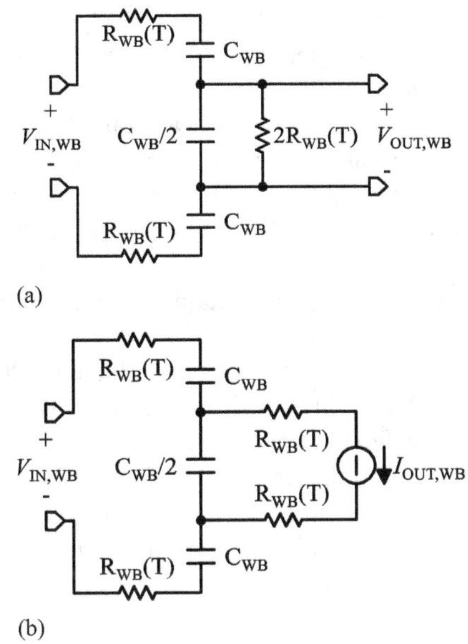

Figure 11.27 Wien bridge sensor, (a) voltage readout schema, (b) current readout schema

phase shift can be specified by reading either the voltage across the output resistor $2R_{WB}(T)$ or its current.

The phase shift can be expressed by (11.20).

$$\varnothing_{WB}(\omega) = -\tan^{-1}\left(\frac{R_{WB}(T)^2 C_{WB}{}^2\omega^2 - 1}{3R_{WB}(T)C_{WB}\omega}\right) \tag{11.20}$$

Due to the *TC* of metal-insulator-metal (MIM) capacitors being very low, from (11.20) it can be deduced that at a constant frequency, the temperature dependence of this phase shift will be mainly affected by the resistors. For the voltage and current readout circuit, the voltage and current transfer functions can be defined by (11.21) and (11.22) respectively.

$$\frac{V_{OUT,WB}(\omega)}{V_{IN,WB}(\omega)} = \frac{R_{WB}(T)C_{WB}\,j\omega}{1 - R_{WB}(T)^2 C_{WB}{}^2\omega^2 + 3R_{WB}(T)C_{WB}j\omega} \tag{11.21}$$

$$\frac{I_{OUT,WB}(\omega)}{V_{IN,WB}(\omega)} = \frac{C_{WB}.j\omega}{-R_{WB}(T)^2 C_{WB}{}^2\omega^2 + 3R_{WB}(T)C_{WB}j\omega + 1} \tag{11.22}$$

The transfer function of the current readout schema has a bandpass characteristic as the phase shift is zero at the filter's center frequency of $\omega_0 = 1/RC$ and it has a linear characteristic around this frequency. The stability of the center frequency is a very important issue for decreasing the error signal between ω_0 and the frequency of input voltage. In an effort, a Wien bridge sensor with a frequency-locked loop (FLL) readout was designed by Park *et al.* [922] so that the output frequency will be locked to ω_0. The implemented sensor structure is shown in Figure 11.28. The frequency of a VCO is divided by N to generate in-phase and quadrature-phase signals. The output current Wien bridge turns to voltage through the capacitor and adjusts the output frequency of VCO. As a result, at a steady state, the output frequency will be locked to $N.\ \omega_0$ and the phase shift of the Wien bridge will be zero. The sensor achieves an inaccuracy of less than 0.12 °C (min-max) over −40 °C to 85 °C. Also, the sensor obtains 2.8 mK resolution and consumes 31 μW.

In [921], Shahmohammadi *et al.* proposed a voltage readout schema of a Wien bridge-based sensor that phase shift is digitized by a PΣΔM. In this design to examine the properties of different kinds of resistors, four different types of resistors including, N^+ diffusion, P+ diffusion, N-poly, and N-well are used and their specifications are summarized in Table 11.5. The sensor achieves an inaccuracy of ±0.15 °C for N^+ diffusion and N-poly resistor, and ±0.25 °C for P^+ diffusion and N-well resistor over the temperature range of −40 °C to 85 °C. It obtains a resolution of 3 mK for the N-well resistor and around 6 mK for the other resistor types. According to Table 11.5, the N-poly resistor is the most accurate and the N-well shows around 2× more resolution. This sensor dissipates 36 μW from a 1.8 V supply.

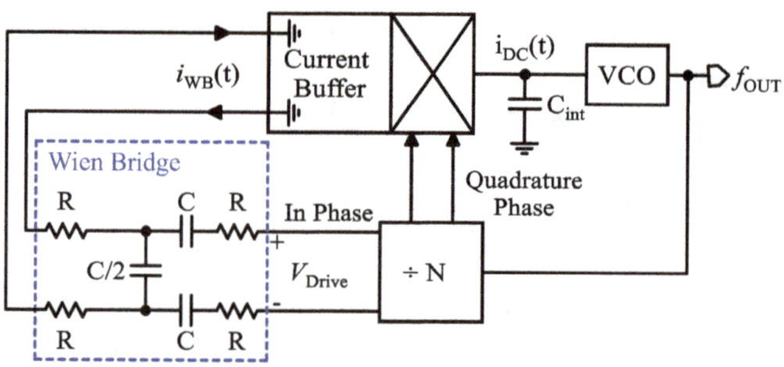

Figure 11.28 FLL-based readout method for the Wien bridge-based temperature sensor

Table 11.5 Performance of the different resistor types

Resistor type	Temperature coefficient	Inaccuracy (after second order fit)	Order of correction polynomial	Total phase shift
N+ Diffusion	0.15%	±0.15 °C	2	8 degrees
P+ Diffusion	0.15%	±0.25 °C	3	7.8 degrees
N-Poly	−0.15%	±0.15 °C	3	7.5 degrees
N-Well	0.3%	±0.25 °C	4	15 degrees

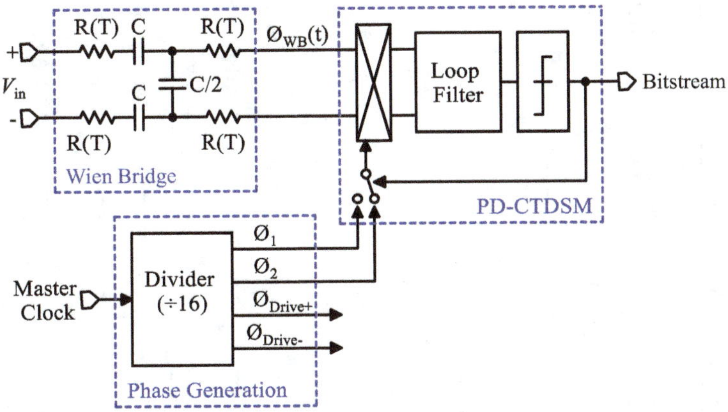

Figure 11.29 Thermal sensor based on continuous-time PΣΔM

Another Wien bridge-based sensor was presented by Pan *et al.* [912] for the temperature compensation of MEMS/quartz frequency references (see Figure 11.29). To achieve high resolution, the phase shift is digitized by an energy-efficient continuous-time PΣΔM, as the average of the output bit stream is a digital representation of the phase shift. This work to achieve sub-mK resolution, uses a second-order modulator instead of a first-order modulator that was employed in their previous design [921]. The sensor achieves an inaccuracy of ±0.2 °C (3σ) from −40 °C to 85 °C after a single-point calibration and a correction for systematic nonlinearity. The inaccuracy can be reduced to ±0.03 °C (3σ) after a first-order fit, which is one of the best inaccuracies reported by resistor-based thermal sensors (see Table 11.6). Also, it achieves a resolution of 410 μK and dissipates 160 μW from a 1.8 V supply. Another Wien bridge-based sensor was reported by Pan *et al.* in [925]. In this design, the first integrator is built around a two-stage miller compensated opamp based on current-reuse amplifiers, which allows the value of C_{int} to be reduced from the 180 pF used in [912] to 23 pF. As a result, it occupies 6× less area and achieves comparable relative accuracy over a 76% wider operating range compared to [912]. The sensor achieves 0.1 °C (3σ) inaccuracy from −40 °C to 180 °C. In this design, three pairs of identical sensors based on silicided p-poly, unsilicided n-poly, and high-resistive poly resistors were fabricated, and the silicided p-poly sensor exhibits the best resolution of 0.46 mK.

11.4.1.3 Poly-phase filter-based thermal sensors

The schematic of a poly-phase filter (PPF) as a thermal sensor is shown in Figure 11.30. It is like a first-order RC LPF that consists of two sensing resistors and two reference capacitors. The voltage transfer function and the phase shift of PPF are given by (11.23) and (11.24) respectively.

$$\frac{V_{OUT.PPF}(j\omega)}{V_{IN.PPF}(j\omega)} = \frac{1 - R_{PPF}(T)C_{PPF}j\omega}{1 + R_{PPF}(T)C_{PPF}j\omega} \tag{11.23}$$

$$\varnothing_{PPF}(\omega) = 2\tan^{-1}(-R_{PPF}(T)C_{PPF}\omega) \tag{11.24}$$

Table 11.6 Comparison of different resistor-based CMOS sensor

Sensor Structure	Resistor type	Temperature Range	Resolution	Inaccuracy (3σ)	Power cons.	Power supply sensitivity	Supply voltage	Chip area	Process	Ref.
Wheatstone Bridge	Silicided p-poly/Non-Silicided n-poly	−40 °C to 85 °C	0.16 mK	±100 mK	180 μW	16 mK/V0.016 °C/V	1.8 V	0.72 mm^2	0.18 μm	[914]
Wheatstone Bridge	Silicided p-poly/Non-silicided n-poly	−55 °C to 125 °C	0.29 mK	0.12 °C	94 μW	NA	1.8 V	0.25 mm^2	0.18 μm	[915]
Wheatstone Bridge	Silicided p-poly/Non-silicided n-poly	−55 °C to 125 °C	0.16 mK	0.14 °C	79 μW	NA	1.8 V	0.12 mm^2	0.18 μm	[918]
Wheatstone Bridge	Silicided p-poly/Non-silicided n-poly	−55 °C to 125 °C	0.16 mK	0.15 °C	55 μW	0.04 °C/V	1.8 V	0.11 mm^2	0.18 μm	[920]
	Silicided-p-diffusion/ N-poly		0.15 mK	0.1 °C						
Wheatstone Bridge	Silicided p-poly/Non-silicided n-poly	27.5 °C to 47.5 °C	0.2 mK	+0.2 °C, −0.1 °C	6.6 μW	0.004 °C/V	1.6 V	0.12 mm^2	0.18 μm	[816]
Wheatstone Bridge	Silicided p-poly	−5 °C to 95 °C	9.8 mK	+1.8 °C, −1.6 °C	0.31 μW	2.8 °C/V	0.9 V	0.47 mm^2	65 nm	[931]
Wien Bridge	N-poly	−40 °C to 85 °C	6 mK	±0.15 °C	36 μW	NA	1.8 V	0.35 mm^2	0.18 μm	[921]
Wien Bridge	Poly-silicon	−40 °C to 85 °C	2.8 mK	±0.12 °C	31 μW	-0.4°C/V	1.7 V/1 V	0.09 mm^2	0.18 μm	[922]
Wien Bridge	Silicided p-poly	−40 °C to 85 °C	0.41 mK	±0.03 °C	160 μW	-0.17 °C/V @1.6 V to 2 V	1.8 V	0.72 mm^2	0.18 μm	[912]
	Silicided n-poly					0.34 °C/V @1.6 V to 2 V				
Wien Bridge	Silicided p-poly	−40 °C to 180 °C	0.46 mK	0.1 °C	52 μW	NA	1.8 V	0.12 mm^2	0.18 μm	[925]
PPF	Silicided p-poly	−40 °C to 85 °C	0.0025 °C	±0.12 °C	68 μW	0.5 °C/V	0.85 V-1.05 V	7 mm^2	65 nm	[926]
PPF	Silicided p-poly	−50 °C to 105 °C	2.8 mK	±0.16 °C	32.5 μW	0.22 °C/V	0.85 V-1.3 V	5.8 mm^2	65 nm	[927]
PPF	Silicided poly	−55 °C to 125 °C	12.8 mK	±1.3 °C	28 μW	NA	0.9 V	2.21 mm^2	65 nm	[929]
PPF	Silicided poly	−50 °C to 125 °C	2.2 mK@1 ms C.T/ 1.05 mK@2.5 ms C.T	±0.64 °C/ ±0.12 °C	40 μW	0.24 °C/V	0.85 V-1.25 V	6.4 mm^2	65 nm	[930]

* NA: Not available

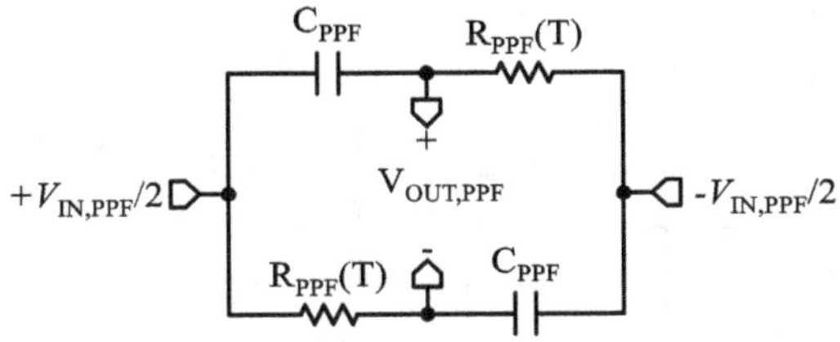

Figure 11.30 Schematic of PPF sensor

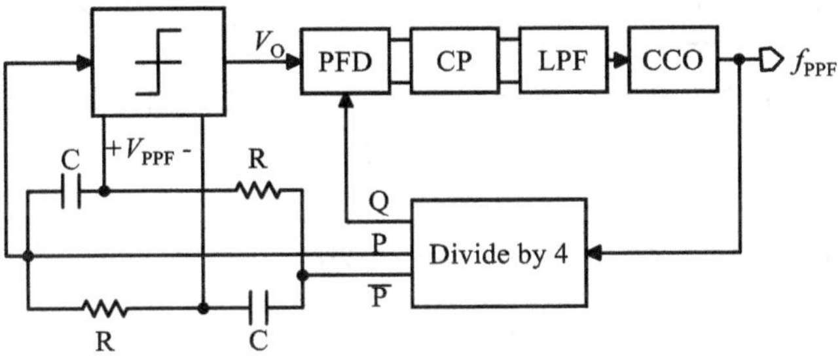

Figure 11.31 PPF-based FLL

According to (11.23) and (11.24) it can be deduced that the frequency-to-phase gain of the PPF is 1.5× larger than the Wien bridge filter's [926]. Also, in the case of adjusting the phase shift to zero, the frequency shifts for both the Wien bridge filter and the PPF are equal. As a result, by considering that the PPF is constructed via fewer elements than the Wien bridge sensor, the PPF can be implemented in a more area-efficient manner. Moreover, the output signal swing of the PPF sensor is much larger (near 7×) than that of the Wien bridge, which results in significant mitigation of error contribution of the readout circuit [926].

Figure 11.31 shows a PPF sensor reported by Martin *et al.* [926] that the phase shift is read out by a FLL. The output frequency of a current-controlled oscillator (CCO) is divided by 4 to provide an in-phase feedback signal (P) and a quadrature-phase signal (Q). Then, a zero-crossing (ZC) of the sensor output (V_{PPF}) is converted into V_O via the ZC detector, which is implemented by a comparator. So, the phase shift can be determined by a ZC detector, which allows the implementation of the rest of the FLL in an area-efficient way. The sensor achieves an inaccuracy of ±0.12 °C (3σ) from −40 °C to 85 °C and a resolution of 2.5 mK. It consumes 68 μW from supply voltages of 0.85 V and occupies an area of 7 mm^2.

In another work, Lee *et al.* [927] proposed a PPF sensor for the thermal monitoring of microprocessors and DRAMs that read out by an FLL based on a dual ZC detection scheme, which improves accuracy by reducing the impact of the ZC detector offset. Moreover, in this design, a cascaded structure of a charge pump is used, which results in less supply sensitivity as the sensor achieves a supply sensitivity of 0.22 °C/V from 0.85 V to 1.3 V, which is 2.3× better than previously

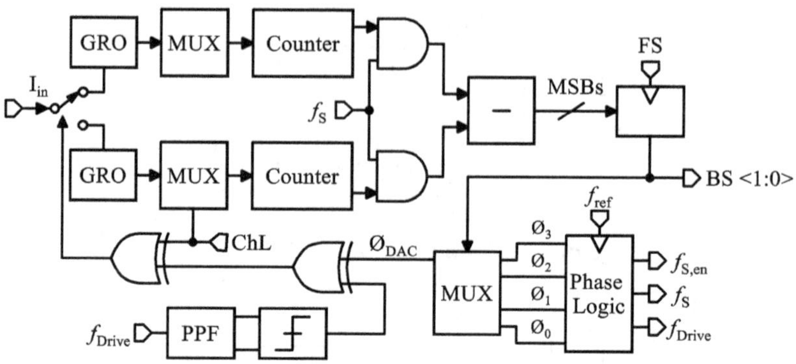

Figure 11.32 Block diagram thermal sensor based on PΣΔM with fixed frequency

examined work [926]. The sensor achieves accuracy ±1.2 °C (3σ) from a temperature range of −50 °C to 105°C, after a one-point trim. By a two-point trim, accuracy can be further improved to ±0.16 °C (3σ). Also, it achieves 2.8 mK resolution and operates from 0.85 V to 1.3 V supplying and dissipating 32.5 μW power.

In an effort, Angevare *et al.* [929], presented a PPF sensor based on a PΣΔM with two innovations (see Figure 11.32). First, to shape the counter's time-domain Q-noise, dual gated-ring oscillators (GROs) are used. Second, it operates at a fixed frequency. This design compared to previously examined FLL-based designs [927], has similar inaccuracy (±1.3 °C (3σ), 1-point trim), but over the full military temperature range of −55 °C to 125 °C. Moreover, it occupies nearly 3× less area while being highly scalable. It consumes 28 μW from a 0.9 V supply and achieves 12.8 mM resolution. In other work, Lee *et al.* [930] introduced a thermal sensor based on FLL intended for thermal monitoring in systems on chip (SoCs). In this work, the FLL uses an XOR gate and a SC loop filter (SCLF) instead of a phase-frequency detector (PFD) and a charge pump (CP), which significantly improves the sensor's inaccuracy. Furthermore, this design to achieve efficient frequency-to-digital conversion, employs a multi-phase counter at the FLL's output. The sensor achieves an inaccuracy of ±0.64 °C (3σ) from a temperature range of −50 °C to 125 °C after a one-point trimming. Also, it achieves a resolution of 2.2 mK and consumes 40 μW at a supply voltage of 0.9 V.

11.5 Seebeck effect-based thermal sensors

The last type of thermal sensor explained in this chapter is implemented based on the thermoelectric effect. The thermoelectric effect is a phenomenon by which a temperature difference is directly converted to electric voltage and vice versa. The thermoelectric effect is categorized into three separately identified phenomena, known as the Seebeck effect, the Peltier effect, and the Thomson effect [932]. Among the different thermoelectric effects, the Seebeck effect can be used to implement the thermal sensors that will be discussed in the following section.

11.5.1 Principles of operations of Seebeck effect-based thermal sensors

The Seebeck effect was discovered by Thomas Seebeck in 1821 [132]. According to this thermoelectric effect, the energy of heat converts into a voltage potential (ΔV_{SB}) between two dissimilar materials (thermocouple) that the junctions of materials maintained at different temperatures (T and $T+\Delta T_{SB}$). The Seebeck effect is shown in Figure 11.33 with thermocouple materials A and B attached, where the junction at $T+\Delta T_{SB}$ is called the hot junction, whereas at T, the cold junction. The generated voltage of ΔV_{SB}, the so-called Seebeck voltage, across the junctions, can be driven by (11.25).

$$\Delta V_{SB} = \alpha_s \, \Delta T_{SB} \tag{11.25}$$

where α_s is the Seebeck coefficient expressed in V/K. The materials for constructing the thermocouple are semiconductors and/or metals layers of CMOS technology such as polysilicon/aluminum [933,934], p-doped polysilicon/n-doped polysilicon [935,936], n-doped poly1/n-doped poly2 [937], etc. The combinations of such materials determined the Seebeck coefficients, which the semiconductor layers compared to metals show larger Seebeck coefficients [938–940]. Typical values reported for the Seebeck coefficient for CMOS process materials are 500–700 μV/K [941].

The Seebeck effect can be used to implement the thermal sensors if the hot junction is placed on a thermally isolated structure like a membrane, micro-bridge, etc., and the cold junction is located on the bulk chip with the thermally highly conducting silicon underneath [20]. The arrangement of these layers is shown in Figure 11.34. The thermocouples as a sensing element, are placed on top of the suspended dielectric layer where the silicon underneath is removed. Because the output voltage of a single thermocouple is usually not sufficient, several thermocouples are connected in series to form a so-called thermopile. On the other hand, by performing high thermal isolation between hot and cold junctions through a thermally isolated stage, which here is a membrane, the temperature difference will be increased. Also, the membrane keeps the sensing element at a high temperature, while protecting the other components from heating. A noteworthy feature of the high-temperature operation sensors, which demand superb thermal insulation, is the utilization of a micro-hotplate structure. For example, the operating temperatures of the metal-oxide gas sensor are typically up to 350 °C which can be implemented

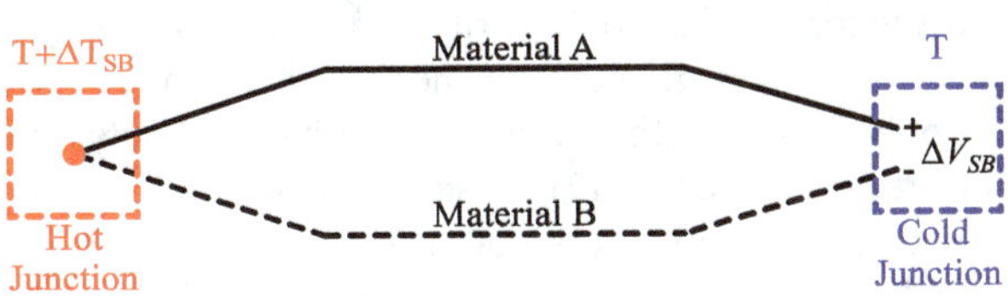

Figure 11.33 Thermocouples convert temperature difference to electrical voltage according to the Seebeck

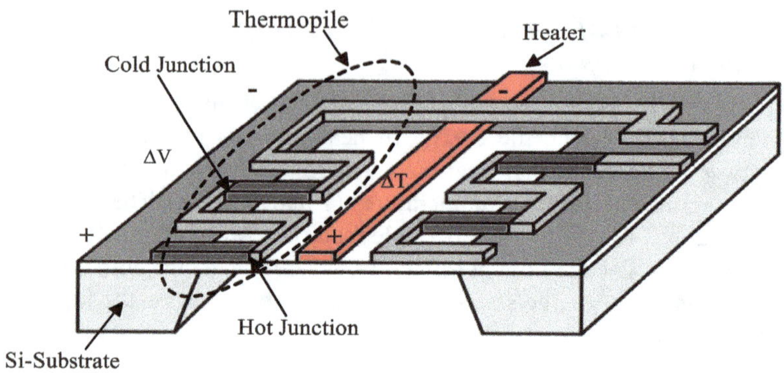

Figure 11.34 Layers arrangement to implement thermal sensor based on Seebeck effect using CMOS technology

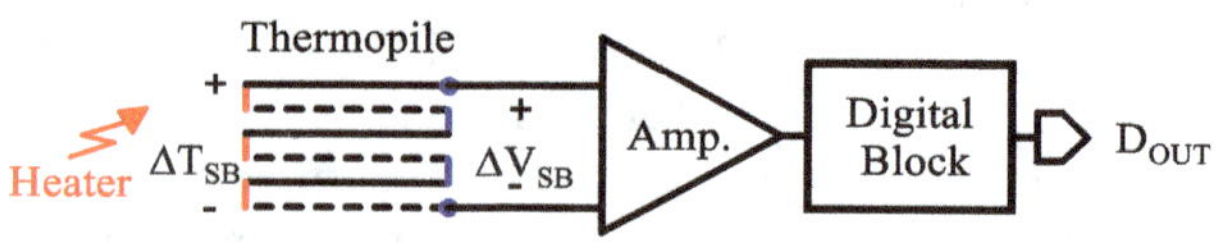

Figure 11.35 The schematic of the thermal sensor based on the Seebeck effect

with a micro-hotplate. More details about micro-hotplate have been given in Appendix D.

Generally, this structure that directly converts heat to voltage is commonly referred to as a thermoelectric sensor, and its main building block is shown in Figure 11.35. The small thermo-voltage produced by the thermopile is amplified using devices like chopper-based amplifiers [942], after which it is converted into digital data using a digital processing block.

11.5.2 Implementation of Seebeck effect-based thermal sensors

There have been reported thermal sensors based on the Seebeck effect [934,937,941,943–948]. For instance, a microsystem consisting of two CMOS thermal converters on one single oxide/nitride membrane in a 2 μm CMOS process is introduced by Jaeggi *et al.* [946]. In this design, the configuration of two sensors has been employed to overcome the nonlinearity introduced at high power levels in which one sensor acts as a measuring device and the other as a reference. Each sensor consists of a 1 kΩ polysilicon heating resistor and a bidoped polysilicon thermopile.

In another effort, Koll *et al.* [947] presented a sensor for recording the temperature variations caused by analyte sorption in the polymer that was fabricated using a 1 μm CMOS process. In this design, the membranes have a sensor/reference configuration, in which the sensor is coated with the gas-sensitive polymer, while the reference is uncoated to provide compensation for gas-induced temperature fluctuations. To form dielectric membranes, dielectric CMOS layers on top of the silicon substrate are released using a post-processing anisotropic KOH etching step. Then, the wafers are etched for 4 h in a 6 molar KOH solution at 95 °C with the field oxide layer serving as an etch-stop layer [947]. Each of the thermopiles integrated on the membrane consists of 44 thermocouples fabricated with the N $^{+}$-doped polysilicon and aluminum layers of the CMOS process and the reported Seebeck coefficient for this junction is about 109 μV/K [949]. In this work, to achieve the temperature differences between the two membranes, low-noise chopper-stabilized instrumentation amplifiers are fabricated in the same chip of the sensor/reference structure with a membrane area of 700 × 1,500 μm^2. A polysilicon heating resistor is located on both membranes center for calibration, which results in a sensitivity (the voltage versus heating power) of 51.1 mV/μW and 51.3 mV/μW for the polymer-coated sensor and uncoated reference, respectively.

The improved version of the previously reviewed design of Koll *et al.* [947], is also reported by the same group [934] that is based on an N-well island structure and fabricated using a 0.8 μm CMOS process. Within this sensor, the N-well islands are suspended by a dielectric membrane. The thermocouple's hot junction is situated on the rim of the island structure, while the cold junction is positioned on the bulk silicon, extending beyond the membrane's edge. During the post-processing KOH etching step with an electrochemical etch stop, the island structures were released. Employing the island structure in the membrane provides an improvement in mechanical stability and a more homogenous temperature distribution in the sensing area [950]. Also, this design benefits from covering the heating resistor in the same area as the polymer layer, which results in providing a more precise calibration compared to the previous design [949]. On the other hand, diodes are implemented on and off the membrane, enabling a precise temperature measurement. In this work, two sensors are reported with square and rectangular-shaped membranes. In the case of the square membrane, the chip occupies an area of 1,168 × 1,168 μm^2 and incorporates 274 thermocouples, a polysilicon heater, and an on-membrane diode. Meanwhile, for the rectangular structure, the chip area measures 595 × 2,162 μm^2, which contains 300 thermocouples, a metal heater, and an on-membrane diode. The sensor achieves a physical sensitivity of 34 mV/μW and 26.5 mV/μW for the square and rectangular-shaped membrane devices, respectively. More detail about this sensor is summarized in Table 11.7.

In a separate study, Voiculescu *et al.* [948] reported the development of a thermal sensor that utilizes the Seebeck effect. They explored various configurations, lengths, and quantities of N-polysilicon/aluminum thermocouples to enhance temperature measurement accuracy. This sensor was fabricated using a double-poly CMOS AMIS 1.5 μm process. Figure 11.36 depicts the configuration of thermopiles, each measuring 200 μm in length, with the inclusion of the 11th thermocouple on one side of the heater. The presented results indicate that the arrangements of thermopiles based on the same membrane as the heater achieve the best thermopile sensitivity characteristics of 6.4 V/W. In this work [948], during the CMOS fabrication process, some areas of the chip are designated to be "open"

Table 11.7 Comparison of Seebeck effect-based CMOS sensors

Heater	Membrane type /Shape	Thermocouple type/ Number	Sensitivity	Seebeck coefficient	Heating power	Readout	Membrane area	Process	Ref.
Resistive polysilicon	oxide/nitride, NA	Bidoped polysilicon	NA	NA	NA	Two-point regulator with an adaptive bitstream level.	NA	2 μm	[946]
Resistive polysilicon	Dielectric CMOS layers, NA	N^+-doped polysilicon/ Aluminum, 44	51.1 mV/ μW	109 μV/K	0-12 μW	Low noise chopper amplifier	700 × 1.500 μm^2	1 μm	[947]
Resistive polysilicon	N-well island, square	N-doped polysilicon/ Aluminum, 274	34 mV/μW	111.3 μV/K	0-30 μW	Low noise chopper amplifier	1,168 × 1,168 μm^2	0.8 μm	[934]
Resistive Metal	N-well island, rectangular	N-doped polysilicon/ Aluminum, 300	26.5 mV/ μW				595 × 2,162 μm^2		
Resistive polysilicon	NA, NA	N-doped polysilicon/ Aluminum , 11	6.4 V/W	60 μV/K	0–35 mW	Kethley source meter	NA	1.5 μm	[948]
Diode connected N-MOSFET	NA	N-doped poly1/N-doped poly2 N-doped poly1/P $^+$-diffusion	101.55 mV/ W 267.22mV/ W	NA	0-48 mW	Agilent 34401 A Digital Multimeter	NA	0.35 μm	[937]

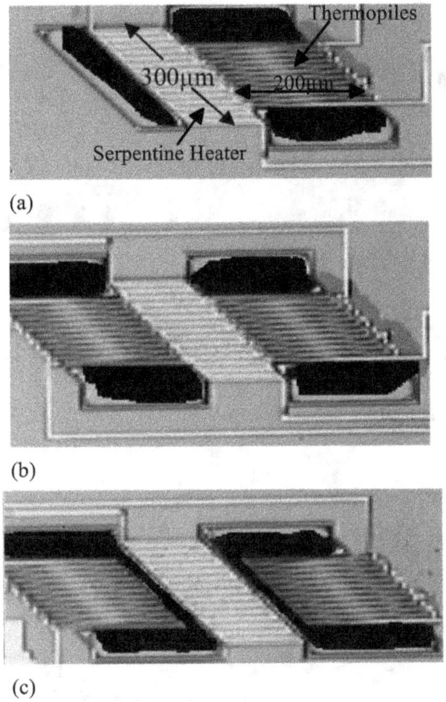

Figure 11.36 Photomicrograph of the thermopile arrangements (a) one side of the heater on the same membrane, (b) both sides of the heater on the same membrane, and, (c) in separate membranes [948]

for post-processing as shown in Figure 11.37. In "open" regions, all the thin films that cover the substrate are absent, so the substrate can be etched in the post-processing step to release the membrane support structures. In this work, the post-etching is done from the backside of the wafer with a backside patterning step and deep reactive-ion etching (DRIE). During this procedure, the back side of the chip is spin-coated with a negative photoresist. Then, the resist is exposed using a mask designed for the die as needed, and each die is DRIE etched until the bottom oxide layer is reached.

In another effort, a differential thermal sensor based on the Seebeck effect with two implemented thermopiles consisting of the same structure but different materials fabricated in a standard 0.35 μm CMOS process is proposed by Aldrete-Vidrio *et al.* [937]. Figure 11.38 shows the photograph and schematic diagram of two implemented thermopiles that consist of eight thermocouples serially connected (16 stripes) which are made of poly1 (N-doped) and poly2 (N-doped) in the first thermopile and of poly1 and P^+ diffusion in the second one. The distance between the hot junction and the cold junction is assumed 500 μm. An NMOS transistor in a diode connection structure acted as a heat source and was placed in the same silicon die of thermopiles at 5 μm from the hot junction (see Figure 11.38(b)). When the

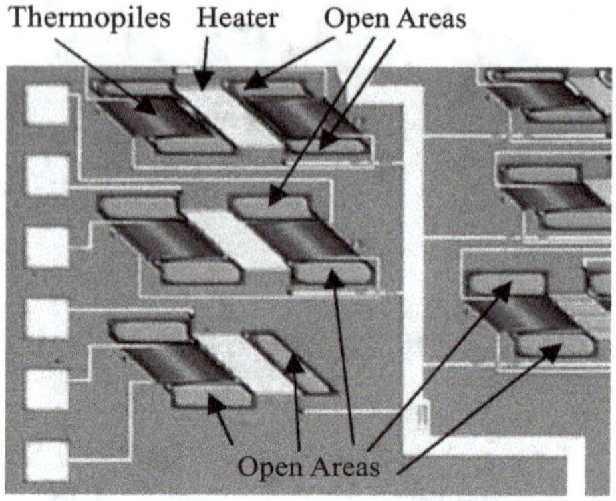

Figure 11.37 Photomicrograph of a portion of the chip before post-processing [948]

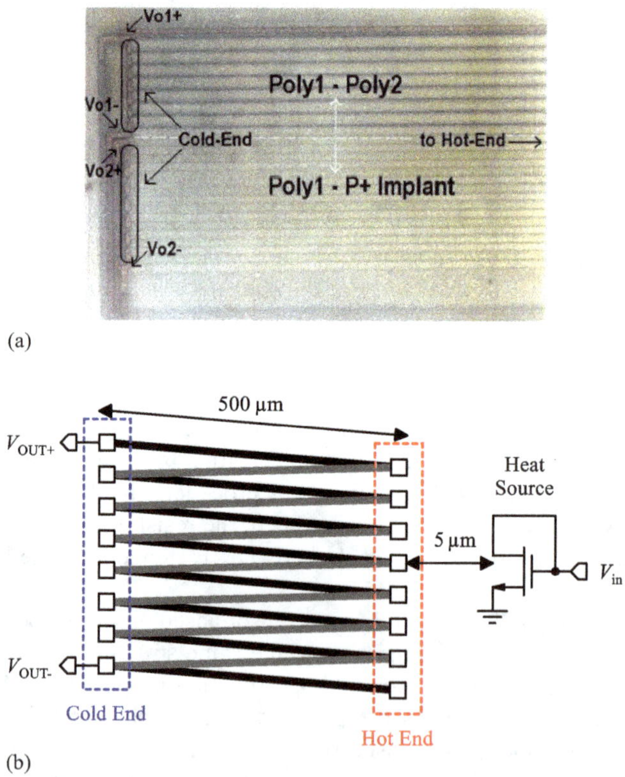

Figure 11.38 (a) Photograph of integrated thermopiles [937], (b) schematic of a thermopile and the dissipating transistor acting as a heat source

diode-connected transistor is activated, the temperature at the silicon surface around this device increases and will be recorded by the differential temperature sensors. The output voltage of the sensor is monitored by an Agilent 34401 A Digital Multi-meter. Experimental results show that the sensitivity of poly1-poly2 and poly1-P^+ diffusion thermopiles are 0.10155 V/W, and 0.26722 V/W, respectively.

11.6 Summary

In this chapter, CMOS-based thermal sensors were discussed and based on the sensing element categorized into five general groups including BJT-based, MOSFET-based, ETF-based, resistor-based, and Seebeck effect-based are examined. Depending on the application requirements, the sensing element of thermal sensors should be selected by considering some design parameters such as the accuracy, resolution, operating temperature range, operating supply voltage, power consumption, power supply sensitivity, area, application, etc. For example, BJT-based thermal sensors show high accuracy over wide temperature ranges but are not suited well for operating at low supply voltages. MOSFET-based thermal sensors, especially DTMOSTs enable sub-1 V operation with appropriate accuracy over a wide temperature range. Also, providing a temperature measurement in a digital format is another important issue that should be considered in the design of thermal sensors. So, various readout methods for digitization temperature information are reported by researchers such as analog-to-digital converting and time-to-digital converting, which can be implemented with different circuits such as zoom-ADC, ΣΔ ADC, PLL, FLL, duty-cycled modulator, etc.

Chapter 12
CMOS-based actuators

Manipulation and characterization of biological particles are crucial for diagnostic and clinical applications [951–953]. Separation of live and dead cells to diagnose early-stage diseases [954,955], isolation and trapping of certain cells for disease detection [956–958], purification, filtering of micro- or nano-sized particles, viruses, cells, and proteins are crucial instances in a variety of biological applications [959–962].

Miniaturized lab-on-chip (LoC) devices have been widely used to control and manipulate biological targets such as engineering particles, droplets, and cells within a microfluidic environment. There are many ways to manipulate and detect bioparticles within a microfluidic LoC device, such as optical tweezers, magnetic, dielectrophoresis (DEP), and hydrodynamic technique.

A brief description and comparison between the mentioned techniques are as follows. The optical tweezers can induce strong trapping force via photon refraction entering the cell with sub-nanometric precision [963]. However, utilizing a highly focused laser beam requires high power, expensive, and bulky tools, making them non-ideal for commercialization applications [964–968]. Magnetic manipulation, despite their good precision, has the same problem as optical ones [969–973]. Furthermore, magnetic manipulation requires cell labeling with nanoparticles. Hydrodynamic flow is the most straightforward method, but it has a low trapping resolution [974–977]. DEP offers excellent benefits in terms of cost, operational power, and requiring label-free sample preparation [18,958,978,979]. Also, in addition, DEP benefits from easier microsystem set-ups in comparison to other techniques such as magnetic manipulation which has more constraints, or optical manipulations. Magnetic or optical manipulation systems, when integrated on a complementary metal-oxide-semiconductor (CMOS) chip, need a conversion of electrical signals into a magnetic field or converting an optical effect into an electrical signal. Compared with DEP and optical tweezers, magnetically actuated technology has its advantages in biological compatibility and micro-flexibility [980], as DEP may damage biological cells while magnetic fields are transparent to the cells [969]. Among different kinds of manipulation techniques, the focus of this chapter is to examine manipulation targets using magnetic and DEP methods with an emphasis on implementing them by CMOS technology and the comparisons of them are summarized in Table 12.1.

Table 12.1 Comparison of various manipulation methods

Method	Current/ Diameter (Coil)	Voltage/ Frequency	Bead diameter (μm)	Array	μ fluidic	Manipulated target	Function	Chip Area (mm^2)	Proc.	Ref.
Magnetic	20 mA / 20 μm	NA	8.5	8 × 8 (Coil)	PDMS	Yeast Cell, Bovine Capillary Endothelial (BCE) Cell	Manipulation	5 × 2	0.18 μm	[970]
Magnetic	40 mA / 80 μm	NA	2.8	8 × 8 (Coil)	Open-Cavity	Biomedical Magnetic Bead	Manipulation	1.5 × 1.5	0.5 μm	[972]
Magnetic	10–30 mA /1,10 μm	NA	1	8 × 8 (Coil)	NA	Magnetic Microbead	Manipulation/ Capacitive Sensor	NA	0.35 μm	[104]
DEP	NA	3.3 V1 MHz, 800 kHz,500 kHz	NA	320 × 320 (Microsite)		Clusters of Saccharomyces Cerevisiae, Polystyrene Beads, K562 Tumor Cells	Manipulation/ Optical Sensor	8 × 8	0.35 μm	[18]
DEP	NA	3-5 V0-1.8 MHz	NA	128 × 256 (Pixel)	PDMS	Yeast Cells, Rat Alveolar, Macrophages	Manipulation	2.32 × 3.27	65 nm	[1001]
DEP	NA	≥ −1.2 V and ≤ 1.3 V0-1 MHz	NA	NA		PC05N, PA04N, PS03N	Manipulation/ Capacitive Sensor	1.5 × 1.2	0.18 μm	[109]
twDEP	NA	1.26 V100 Hz-20 MHz	NA	30 × 60 (Microelectrode dot array) 8 × 8 (Sub-array)		MDA-MB-231 cells, HELA Cells	Manipulation	3.2 × 2.2	0.18 μm	[978]

NA: Not available

12.1 Magnetic manipulation

Manipulation of magnetic particles using magnetic fields within a microfluidic LoC device plays an important role in biomedical applications such as biochemical assays [973], medical diagnostics [971,981], and drug development [982]. Magnetic manipulation has some inherent outstanding features including: (1) magnetic fields are biocompatible [980,983], (2) the fields are not influenced by the medium, ensuring high selectivity of magnetically-tagged cells, (3) magnetic particles can be manipulated in numerous ways with the aid of controllable magnetic forces through designing magnetic fields [973,984], and (4) magnetic particles can be coated with biological molecules to modify surface chemistry and physical properties [985,986].

12.1.1 Implementation of magnetic manipulation

The manipulation of particles using magnetic fields has received growing interest from the field of biotechnology in recent years. As reported in Table 12.1, in the magnetic method to produce magnetic fields, the coil arrays with current control are used. For instance, Lee *et al.* [969,970] presented an integrated circuit (IC)/microfluidic hybrid system for magnetic manipulation of biological cells using IC chips as an active, versatile actuator. As shown in Figure 12.1(a), the IC contains a microcoil array circuit that produces spatially patterned magnetic fields utilizing an integrated micro-electromagnet array, which is controlled by integrated electronics. Figure 12.1(b) illustrates the architecture of the CMOS IC that contains five 8×8-microcoil arrays. For independent magnetic field control, each microcoil is connected to its on-chip current source. The operating principle of the microcoil array for cell manipulation is to create and move magnetic field peaks by modulating currents in the microcoil. For instance, by activating only one microcoil in the array, a magnetic bead suspended in the fluid will be attracted to the field peak at the center of the microcoil on the surface of the IC. Programmable and rapid reconfiguration of the field pattern allows an efficient simultaneous manipulation of multiple individual bead-bound cells with precise position control using the IC. Two prototypes, SiGe/microfluidic and CMOS/microfluidic hybrid systems validated the proposed approach.

In another effort, Lee *et al.* [971] extended the hybrid system concept to biological and biomedical applications. In this work, a new set of advanced manipulations of biological cells is reported which uses a CMOS/microfluidic hybrid system, with an emphasis on sustaining biocompatible environments during the operation. This design employs bovine capillary endothelial (BCE) cells as a representative manipulation target. In this work, various magnetic field patterns can be generated by controlling currents in the microcoils to provide more flexibility in cell manipulation.

In another work, Abu-Nimeh and Salem [972] presented a low-cost integrated system designed in bulk 0.5 μm CMOS to manipulate and separate biomedical magnetic beads. This system is designed to operate as an "open-cavity" device where the entire surface of the chip is exposed to the sample or biomaterial of interest; so, it does not require specialized post-processing for

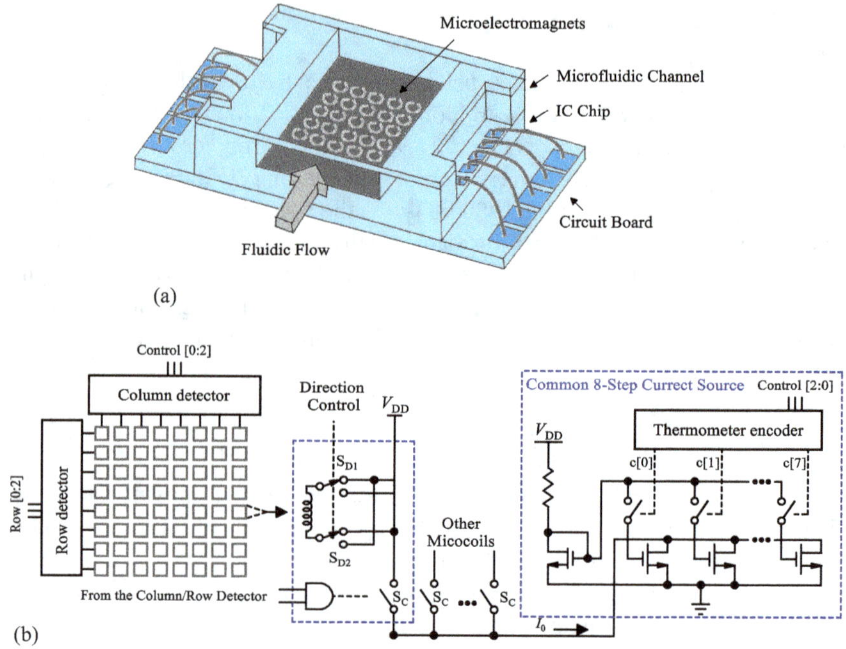

Figure 12.1 (a) Schematic of an IC/microfluidic hybrid microsystem and (b) architecture of the CMOS IC for magnetic manipulation of biological cells

microfluidic structures. The system-level block diagram depicted in Figure 12.2 (a) consists of four major components: the 8 × 8-coil array, 3-to-8 row and column decoders, a variable power bidirectional current source, and the global digital logic control. The coil array can be controlled individually for bead manipulation or collaboratively in a pseudo-parallel fashion. The photograph of the 1.5 mm × 1.5 mm CMOS die as an open-cavity device is shown in Figure 12.2(b). A set of eight parallel NMOS transistors is cascaded to create a binary controllable current sink, whose simplified configuration is shown in Figure 12.2(c). Figure 12.2(d) depicts multiple frames taken from a video sequence while a bead is being manipulated to the right. In frames 4 and 5, the bead is attracted closer to the surface by switching on the coil on the left. In the remaining frames, the bead is transported to the right by switching off the coil on the left and switching on the coil on the right.

12.1.2 Implementation of magnetic manipulation and sensing

An ideal LoC is capable of both manipulating and sensing biological samples. There have been reported many LoC structures for the manipulation and detection of magnetic microparticles [987–989], most of which require external

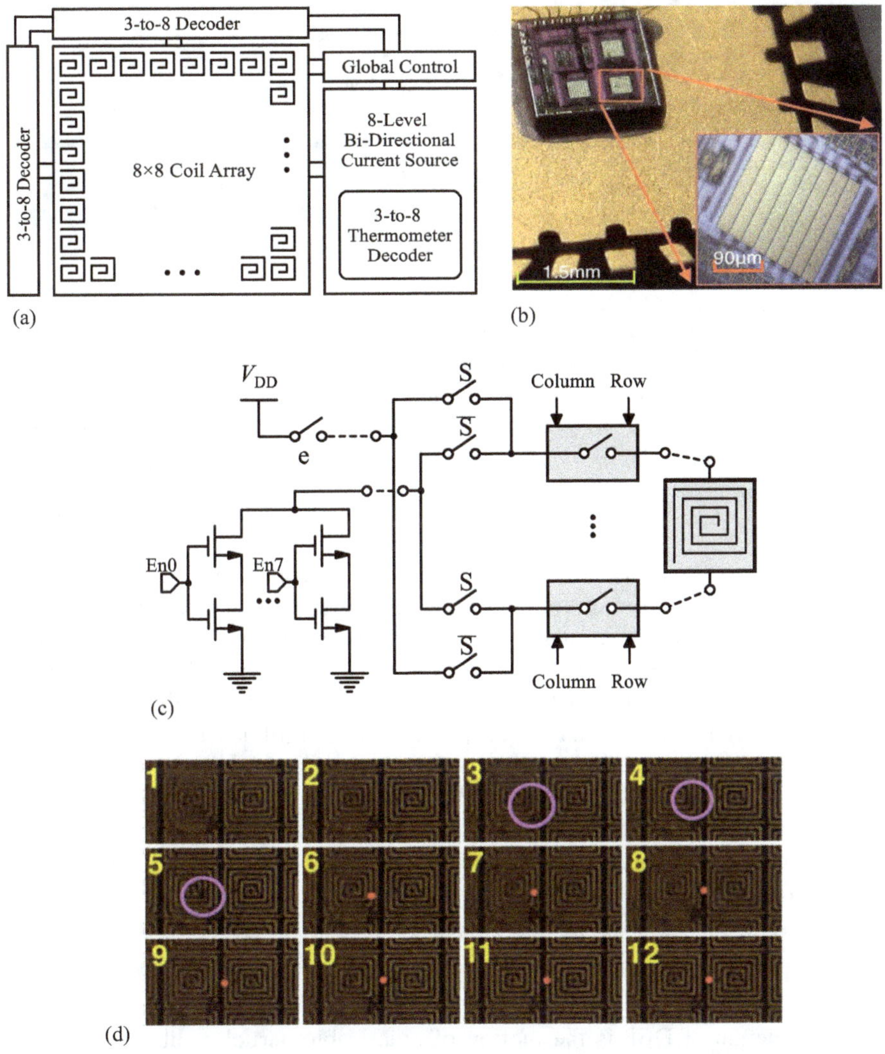

Figure 12.2 (a) System-level block diagram, (b) photograph of CMOS die, (c) variable power bidirectional current sink with a simplified coil path, and (d) multiple frames taken from a video sequence while a bead is being manipulated [972]

instruments to perform one of the functions, leading to increased size, complexity, and cost of the overall system [40]. In the work reported by Chang and Lu [104,990], an integrated platform was developed to perform these two functions by implementing electromagnetic microcoils and capacitive biosensors on a 0.35 µm CMOS, which does not require any externally applied magnetic fields. The 8 × 8-micro-manipulation and sensor array is shown in Figure 12.3(a). Each

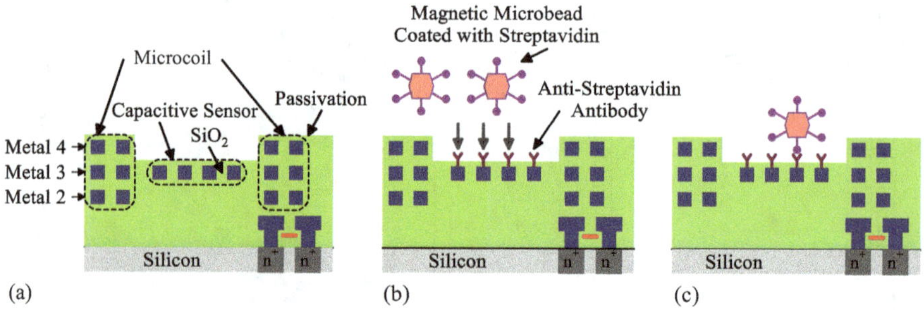

Figure 12.3 Operating principle of magnetic micro-bead-based manipulation and capacitive detection: (a) Cross-sectional view of the fabricated CMOS chip, (b) magnetic actuation of streptavidin-coated magnetic micro-beads toward the capacitive sensor with functionalized anti-streptavidin antibody, and (c) capacitive detection of the remaining micro-bead after specific binding

basic element in the array consists of a microcoil for electromagnetic actuation and a set of interdigitated microelectrodes in its center for capacitive detection. The microcoil is constructed using multiple metal layers provided by the CMOS process to increase the number of turns, and thus, the produced electromagnetic force. The thin inter-metal silicon dioxide is used for immobilizing biorecognition elements (BREs), such as the anti-streptavidin antibody used in this work. As shown in Figure 12.3(b), magnetic microbeads functionalized with streptavidin electromagnetically move toward the sensor in the microcoil center. The electrode is slightly larger than a micro-bead, thereby it allows only one micro-bead to remain on its surface for specific binding and the following capacitive detection (Figure 12.3(c)).

12.2 Dielectrophoresis (DEP)

The phenomenon of DEP is the motion of polarizable particles subjected to non-uniform electric fields [991]. The transportation, accumulation, and separation of the target cells can be achieved according to their dielectric properties [978]. Generally, the DEP force is divided into two parts: (1) classical DEP and (2) traveling-wave DEP (twDEP). For classical DEP, the particles according to their physical properties and medium experience a net force directed toward locations with increasing or decreasing field intensity, which is called positive DEP and negative DEP, respectively. DEP has a wide range of applications and has been used to manipulate nanoparticles [992], viruses [951], single molecules [993], and deoxyribonucleic acid (DNA) separation [994]. The twDEP is generated by multi-phase-quadrature signals [978,991,995–998]. The motion speed and direction by twDEP force are unique for each kind of nanoparticle, which means that the twDEP has better selectivity than the classical DEP in the digital microfluidic system [978].

Several approaches for target manipulation have been developed based on DEP and twDEP, and their overview in terms such as voltage, frequency, array, and chip area are presented in Table 12.1.

12.2.1 Implementation of classical DEP

In one reported work, Manaresi *et al.* [18] presented a microsystem implemented in 0.35 μm CMOS technology that individually detects and manipulates more than 10,000 cells in a parallel fashion and allows programmable manipulation of individual particles with no need for fluid flow or micromachining, combined with embedded optical detection. The general architecture of the chip with the dimension of 8 × 8 mm^2 is shown in Figure 12.4(a) where the main blocks include the array of microsites composed of 320 × 320 elements of 20 μm pitch, 9-bit static column/row decoders for random access, the bias generator block, and the readout circuit block. The LoC architecture is based on a two-dimensional (2D) array of microsites to generate the electric field necessary to create dielectrophoretic cages and detect the presence of single particles or clusters trapped in cages by using optical sensing. A closed DEP cage in the spatial region above a microsite can be created by connecting the associated electrode and the microchamber lid to a counter-phase (V_{phim}) sinusoidal voltage, while the electrode of the neighboring microsites is connected to an in-phase sinusoidal voltage (V_{phip}). Each microsite consists of an actuation electrode, implemented with a top metal plate, and underlying embedded circuitry for programming and detection. Microsites can be addressed in a random-access mode utilizing row and column decoders, for both actuation and sensing. The presence of particles is detected by photodiodes, embedded in the substrate, that measure signal variations from uniform light impinging on the chip surface. Figure 12.4(b) shows the chip in the microfluidic

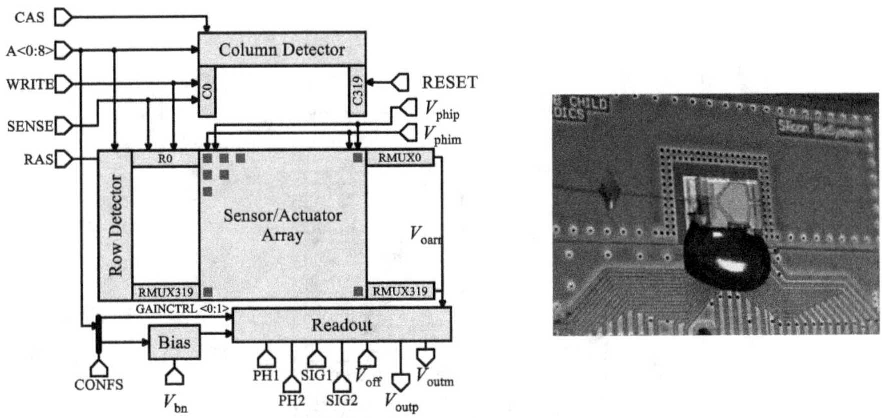

(a) (b)

Figure 12.4 (a) Chip architecture and (b) chip in microfluidic packaging [18]

package. After a conductive-glass lid with SU8 walls is glued to the chip to define the microchamber, the die is mounted directly on the printed circuit board (PCB) with a chip-on-board technique. Figure 12.5 demonstrates how, after applying a pattern implementing an array of DEP cages, the micro-beads are arranged correspondingly, and how they can be detected with the embedded optical sensors. The mentioned technique involves the generation of an electrical field by applying two 180-phase-shifted signals. Another technique consists of applying three 90-phase-shifted signal voltages on planar electrodes to move cells in a given direction [995,999,1000].

In another design, Hunt *et al.* [1001] presented a hybrid IC/microfluidic system containing a 128 × 256-array of 11 × 11 mm^2-pixels that can trap and move cells or liquid droplets in a microfluidic chamber using DEP (Figure 12.6). The dimension of the chip implemented in 0.35 μm technology is 2.32 × 3.27 mm^2. The row control circuits that are shown in Figure 12.6 select a row of pixels and the bit control circuits select a column, allowing individual pixels to be addressed. An

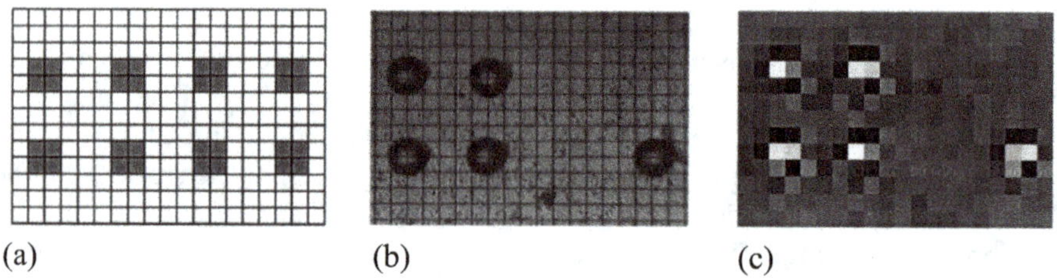

Figure 12.5 Manipulation and detection of 50 μm polystyrene beads: (a) Actuation pattern, (b) microscope image, and (c) embedded optical sensors image [18]

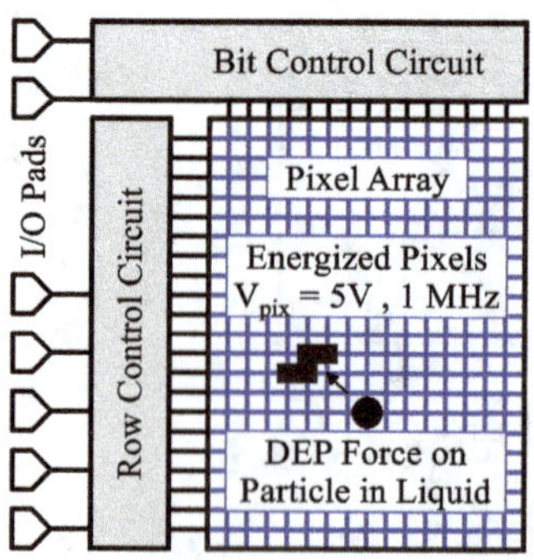

Figure 12.6 IC block diagram of DEP manipulator

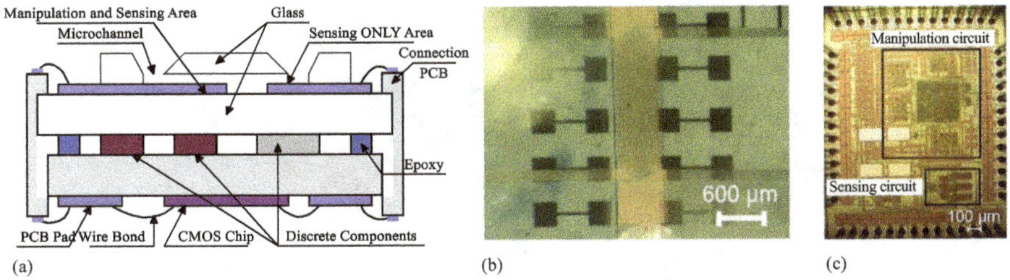

Figure 12.7 (a) DEP-based LoC microsystem architecture using planar electrodes, (b) microfluidic architecture [109], and (c) CMOS chip [109]

image of the energized pixels is stored in static random-access memory (SRAM) immediately below the pixel array. In this design, each pixel is individually driven with a radio frequency (RF) voltage with an amplitude of 5 V_{pp} at frequencies from DC (Direct-Current) to 1.8 MHz, creating a local electric field that exerts a force via DEP on a cell or fluid droplet in the microfluidic chamber above the IC's surface.

As another implementation, Miled and Sawan [109] proposed a LoC architecture for DEP-based cell manipulation, detection, and capacitive measurement that is fabricated in a 0.18 μm CMOS process. A LoC consists of two main components, including the microfluidic and the microelectronics parts. Figure 12.7(a) summarizes the different components of the proposed LoC. The discrete components are placed on the bottom side of the PCB to minimize the occupied area. The microfluidic platform is stuck on the top of the PCB and connected by wire bonding to the latter. The CMOS chip includes digital and analog building blocks [1002]. The digital module monitors the parameters of generated signals, such as frequency and phase. The analog part converts digital data into analog signals while adapting them to the microfluidic architecture.

The proposed microfluidic architecture includes 64 electrodes to perform various manipulation operations such as cell motion, separation, and rotation. The 64 in-channel electrodes are implemented as follows: 32 L-shaped electrodes to control the spread of the horizontal electric field and 32 electrodes to control the vertical spread for horizontal cell manipulation and vertical cell manipulation respectively. In addition, a 3-level mixing architecture is proposed for efficient mixing cell manipulation. The microfluidic design includes one large, six medium, and 16 small mixing areas whose sizes are 600 μm × 600 μm, 100 μm × 100 μm, and 50 μm × 50 μm, respectively, which, in turn, represent the first, second, and third mixing level. In addition, an interdigitated electrode (IDE) (Sens) is added to detect the variation of capacitance in the microchannel shown in Figure 12.8. The obtained results show an efficient nano and microparticles (PC05N, PA04N, and PS03N) separation based

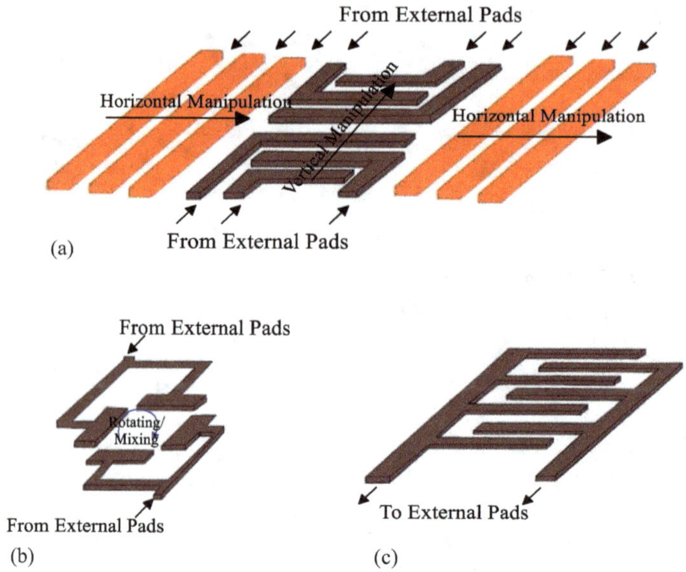

Figure 12.8 Manipulation electrodes: (a) Horizontal and vertical motion and separation electrode, (b) mixing electrodes, and (c) sensing IDEs

on frequency segregation with low voltages of less than 1.7 V and a fully integrated and reconfigurable system.

12.2.2 Implementation of twDEP

Chan *et al.* [978] used twDEP to implement a digital microfluidic biochip for cell manipulation and separation in a standard 0.18 μm CMOS process. The programmable microelectrode dot array with 115,200 sub-microelectrodes is used to generate a traveling wave electric field on the proposed chip. The microelectrode dot array is composed of 1,800 actuated microelectrodes with an area size of 52 μm × 52 μm. As illustrated in Figure 12.9, multi-tasks can be achieved on the proposed biochip. In Figure 12.9(a), two tasks of separation are processed. One task of separation of blue particles and red particles is processed in droplet A. The other task of separation of green particles and red particles is processed in droplet B. The traveling field can travel in two directions on the chip, horizontal and vertical. The target cells can be manipulated along the traveling field. Each microelectrode on the proposed biochip contains a subarray, 16 control units, the drivers for phase-quadrature signals, and a 1-bit storage element. The detailed schematic of the microelectrode is shown in Figure 12.9(b). The chip photo is illustrated in Figure 12.10(a,b) whose area is 3.2 mm × 2.2 mm. The die is mounted on a PCB and the bonding wire is glued to achieve a waterproof system. Experimental results show that MDA-MB-231 cells and HELA cells can be manipulated at the speed of 1.08 μm/s and 1.5 μm/s,

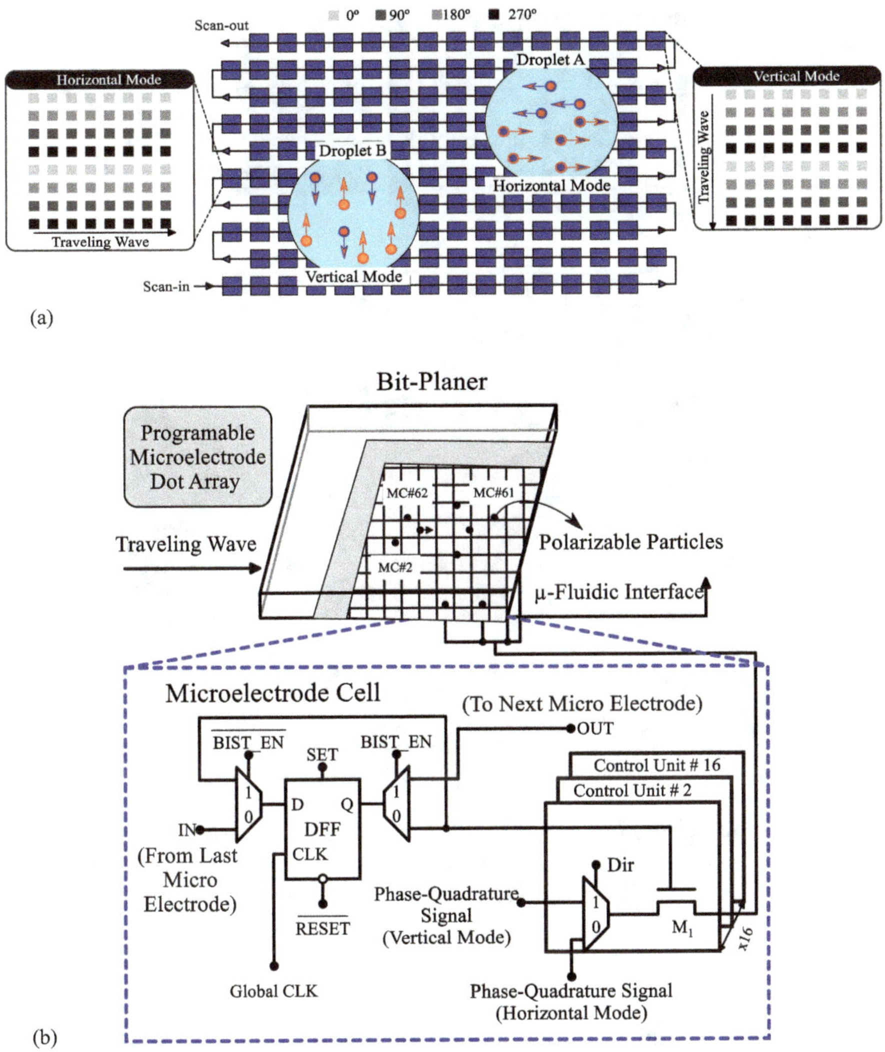

Figure 12.9 *(a) The illustration of multitasking on the proposed microelectrode dot array with two moving directions is supported by each microelectrode of an array of 8 × 8 sub-microelectrodes and (b) the schematic of each proposed microelectrode and the sub-array for cell manipulation*

respectively, under the supply voltage of 1.8 V. The results for the separation of polystyrene particles and HELA cells at 100 kHz under vertical mode are shown in Figure 12.10(c), where red circles and white circles represent polystyrene particles and HELA cells, respectively.

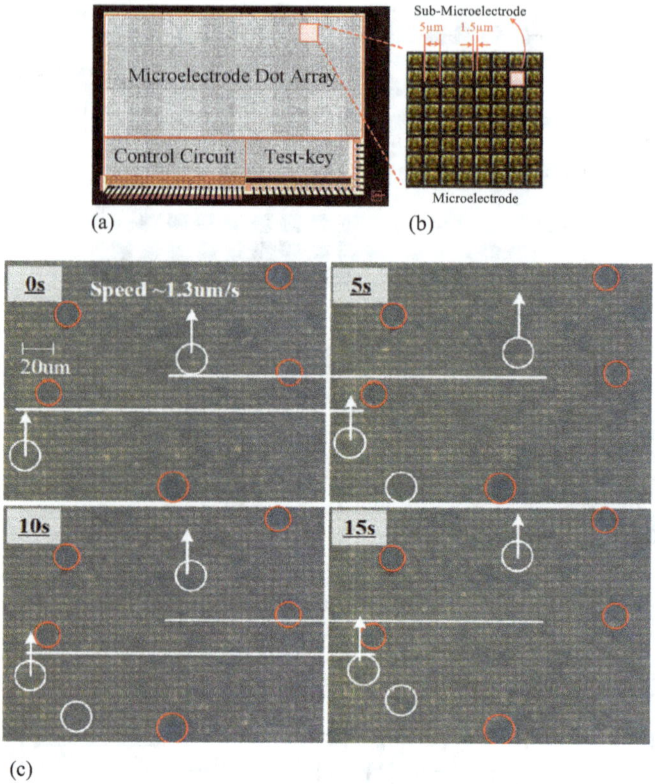

Figure 12.10 (a) Chip photo, (b) the microelectrode of an array of 8 × 8 sub-microelectrodes, and (c) the separation of polystyrene particles and HELA cells at 100 kHz under vertical mode [978]

12.3 Summary

Manipulation of biological cells using magnetic beads is a common method in biological and biomedical studies. By functionalizing the bead's surface with antibodies, peptides, or lectins, magnetic beads can be attached to or engulfed by cells with high selectivity, making it possible to manipulate cells with external magnetic fields. Various magnetic manipulation tools, which use macroscopic external magnets or lithographically patterned electromagnets as the magnetic field source, have been developed for specific experimental purposes.

DEP is the movement of a particle in a non-uniform electric field due to the induced dipole moment of the particle relative to the surrounding medium. An advantage of DEP is the ability to manipulate and sort cells without the need for biochemical labels or other bioengineered tags, and without contact with any surfaces. Micro-fabricated LoC devices for cell manipulation and detection using electrical approaches are more promising than magnetic approaches due to low cost, smaller form factor, and more functionality resulting from integration with electronics. Specifically, DEP-based manipulation techniques and detection of biological entities in miniaturized LoC format have been proven to be highly selective and versatile for smaller biological particles such as cells, viruses, and bacteria. Another advantage of DEP is the ability to manipulate and detect cells without the need for biochemical labels or bioengineered tags and without the need to maintain contact with any surfaces.

Chapter 13

Microfluidic packaging techniques

Microfluidic platforms encompass two broad categories: those that simulate chemical and physical injury, and those that simulate the regenerative environment [1003]. While they have been utilized for *in vivo* purposes, such as precise drug delivery from on-chip reservoirs in neural implants [1004], their primary use has been for *in vitro* investigations involving the delivery and manipulation of biochemical molecules necessary for tasks like deoxyribonucleic acid (DNA) sequencing [1005] and protein analysis [1006], as well as various cell-based applications. Specifically, microfluidics have been employed for flow cytometric cell counting, sorting, and analysis [1007]; controlled sequential delivery of cell adhesives and target cells to facilitate micropatterned cell growth on substrates [1008–1011]; generation of simulated *in vivo* tissues through gradual deposition of stromal support molecules within cell culture chambers, enabling the examination of the impact of diverse three-dimensional (3D) structures on cell function [1009,1010]; exploration of the effects of forces such as laminar flow on endothelial and other cell functions [1012,1013]; simulation of circulation between organ-simulating cell culture chambers to evaluate the pharmacokinetics of various agents [1014]; delivery of test specimens to chambers housing cell-based biosensors that employ cells as transducers to detect potential toxins and agents [289,1015]; and investigation of fundamental cell physiology and cell-cell interactions at a single or near-single cell level within controlled environmental chambers [1016,1017]. In recent times, advanced microfluidic cell culture systems have surfaced, demonstrating the ability to mimic complex 3D structures of tissues and organs *in vitro*. These systems serve as valuable biological models for studying the mechanisms and functions of human tissue structures, as well as investigating the onset and progression of diseases like cancer. Microfluidic platforms and organ-on-a-chip systems are employed as advanced tools to enhance the physiological relevance of coculture cell models and investigate cell-cell interactions. The application of micromachining technologies in biomedical research over the past two decades has given rise to miniaturized assays for advanced *in vitro* cell analysis, referred to as cell-based microfluidic platforms [1018].

In this chapter, we focus on microfluidic packaging methods that are designed to be compatible with both integrated circuit (IC) technology and aqueous environments.

13.1 Materials and challenges

Upon initial examination, it may seem feasible to simply drop a liquid sample onto the sensing region of a complementary metal-oxide-semiconductor (CMOS) sensor while passivating the bonding wires. However, depending on the application and the properties of the surface material, droplet dispersion and/or evaporation may occur. This necessitates the use of fluidic devices to contain and direct the sample through the CMOS sensor. Generally, three approaches have been reported to achieve this, as illustrated in Figure 13.1(a–c): (1) a large chamber with a small hole to access an open-top CMOS sensor, (2) single-channel microfluidics directing the liquid through one channel towards the sensor, and (3) multi-channel microfluidics directing the liquid through multiple channels towards different sensing regions arranged in an array structure. The latter is useful for high-throughput experiments but is more complex than the other two configurations.

Initially, microfluidics was fabricated using thermally and chemically stable materials such as Si and glass, which require complicated and costly techniques. However, the development of polymers such as SU8, poly(methyl-methacrylate) (PMMA), and poly(dimethylsiloxane) (PDMS) has garnered widespread interest due to their biocompatibility. Among these polymers, PDMS is the most commonly used due to its low cost, transparency, compatibility with rapid prototyping, and biocompatibility [1019].

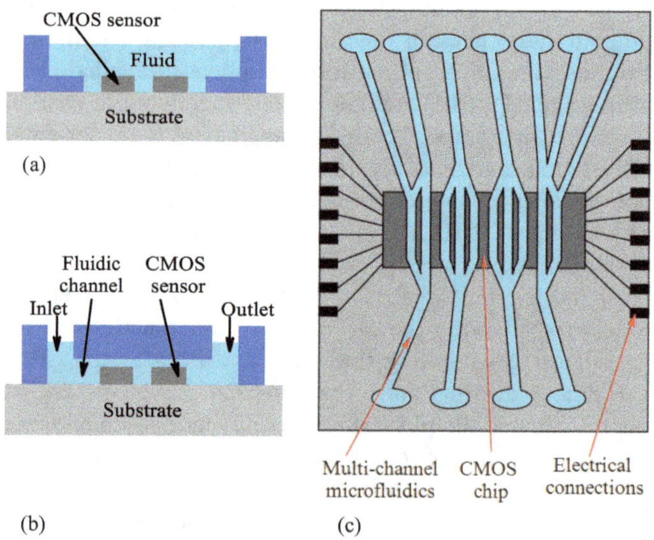

Figure 13.1 (a) The cross-section of the internal structure of an open-top microfluidic packaging. (b) The internal structure of a single-channel microfluidic packaging. (c) An overhead view displaying the layout of a multi-channel microfluidic packaging

The development of hybrid CMOS-microfluidic devices presents several challenges that need to be addressed to create functional and effective devices. One major challenge is the selection of materials that are biocompatible with the liquid sample and can protect it without causing any toxicity. Another challenge is the susceptibility of the IC to interfering signals caused by changes in the dielectric constant of the medium surrounding the chip, which can result in inaccurate readings. The heterogeneous integration of fluidics and electronics also poses challenges due to uncertain and high-priced implementation processes. Developing custom packages may be necessary, but this can increase production costs. Sealing microfluidics with IC also presents a challenge, as it is important to find simple and reliable approaches that prevent leaks. Additionally, the package should be reusable and able to be cleaned easily after each experiment. The lack of mature and reliable simulation and modeling software is also a challenge that needs to be addressed. Topographical mismatches between the footprints of electrical connections and microfluidic channels can cause issues, and precise matching is required for optimal performance. The packaging method should also isolate the electrical part from fluids without any contradictions in the required conditions. Lastly, the package should be usable in specific conditions such as incubators used for cell culture. These challenges must be overcome to develop hybrid CMOS-microfluidic devices that can perform their intended functions effectively.

Numerous techniques have been suggested in academic literature to tackle the aforementioned obstacles, which are explored and analyzed in detail in this chapter.

13.2 IC-microfluidic packaging techniques

The packaging techniques for IC-microfluidic devices that are discussed in this section can be classified into four main categories: rapid prototyping, soft-lithography-based packaging, direct-write fabrication process (DWFP), and other microfabrication methods.

13.2.1 Rapid prototyping methods

Numerous methods have been proposed to package CMOS chips with microfluidic devices in the literature. One of the most popular methods is rapid prototyping. This technique utilizes an open-top standard cell-culture well, in which an insulating layer covers non-sensitive areas of the chip [3,29,41,1020,1021].

Parylene is frequently used as the insulating material, as it is bio-friendly and adheres well to Si, silicon oxides, and epoxy [1022]. In one study, Prodromakis *et al.* [1023] used parylene to encapsulate their CMOS chip and employed an automated laser micro-machining system to create openings for sensing areas. The parylene-membrane was then lifted off using ultrasonics. Prakash *et al.* [29] utilized a CMOS capacitive sensor combined with a cell-culture well that was affixed on top of the packaged chip to monitor cancer cell growth. Similarly, Lopez *et al.* [73] secured a glass ring on top of the carrier printed circuit board (PCB) to hold the cell medium and the cells. In separate experiments, Ballini *et al.* [43] and Viswam *et al.* [44] implemented a polycarbonate ring and a biocompatible plastic epoxy well, respectively, to confine the medium and protect the wire bonds by utilizing epoxy.

In a design [673], a PDMS reservoir is constructed on top of the CMOS sensor chip to house the sensing cells and test chemicals. As described in [673], the reservoir is formed by mixing the PDMS base material and its curing agent in a 10:1 ratio to create a sealing material. A cylindrical reservoir is then created using a short piece of plastic tubing. This plastic reservoir is placed over the sensor chip, aligned with the sensing area, and secured with PDMS adhesive. The entire assembly is thermally cured for 1 h at 50 °C.

In another design [648], a CMOS biochip is designed for the quantification of prostate-specific antigen (PSA). The PSA immunoassay protocol necessitates a simple droplet-based liquid handling method using pipettes. To fulfill this requirement, an open cavity approach without integrated microfluidics is employed to deliver liquids. Polyurethane (PUR) is used as a chemically inert material to construct the cavity, and the bond wires are sealed with a biocompatible and electrically isolating compound. Several experiments are conducted to identify a bio-compatible material suitable for all process steps and chemicals used for probe immobilization and the PSA immunoassay. Eventually, Structalit 5891 T, a two-component black epoxy resin glue, was chosen for its superior stability and compatibility.

Datta-Chaudhuri *et al.* [1024] implemented a photolithographic post-processing and microfluidic integration by embedding their chip in an epoxy handle wafer. The off-chip electrical connection was made through thin-film metal traces passivated with parylene-C. The packaging procedure is illustrated in Figure 13.2(a) to (e). First, the chip was placed face-down in a glass petri dish coated with PDMS to act as a mold. Then, epoxy was poured around the chip and cured to form the handle wafer. Next, Cr/Au was sputter-deposited and patterned onto the front side of the surface. Poly(3,4-ethylenedioxythiophene) doped with poly(styrenesulfonate) (PEDOT:PSS) was then electrodeposited on the Au electrodes. Parylene was deposited and patterned to cover the input/output (I/O) region. The metal and parylene were patterned using a two-step process of dry etching and lift-off. The same group also reported a similar approach for developing a bioelectronics nose, including well and microfluidic channels [1022].

Musayev *et al.* [1025] utilized epoxy to insulate the CMOS chip wire-bonded to a ceramic package and to create a reservoir for holding the solution above the chip (refer to Figure 13.3).

Halonen *et al.* [1026] developed a durable package compatible with cell culture by combining a second generation of low-temperature co-fired ceramic (LTCC) technology with flip-chip bonding. The conductor lines were screen printed on commercial Dupont 951 Green Tape™ LTCC. The package was manufactured using a standard LTCC process, while a special process was developed for the microfluidic channels. Isotropic conductive adhesive (ICA) was applied to the LTCC contact pads through a stamping process using a laser-processed alumina stamp. The chip was aligned using a flip-chip bonder and glued onto the bumps

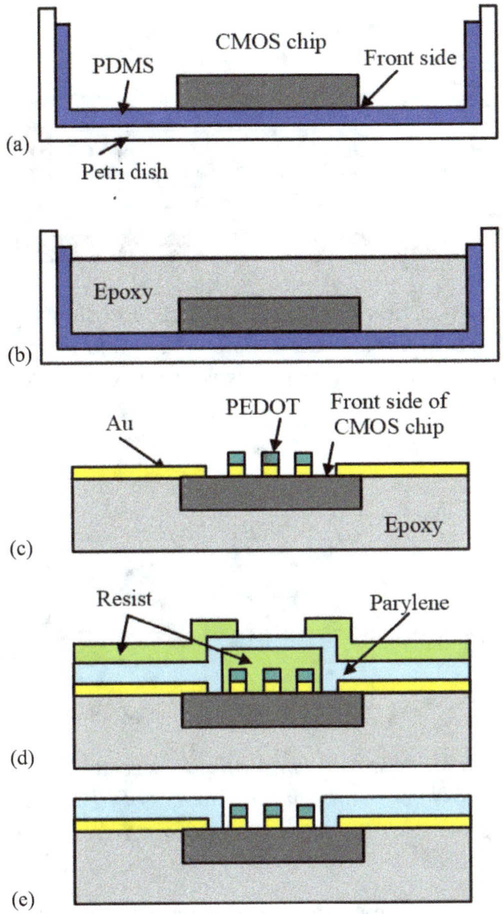

Figure 13.2 (a) The chip is positioned with its face down in the middle of a PDMS-coated glass petri dish. (b) Epoxy is poured around the chip and allowed to cure, creating a handle wafer. (c) The wafer is taken out of the petri dish, leaving the chip with its face up. A layer of Cr/Au is sputter deposited onto the chip's surface, and PEDOT:PSS is electrodeposited onto the Au electrodes located at the center of the chip. (d) and (e) The parylene passivation layer is deposited and patterned onto the chip

with epoxy, which was cured on a hot plate. To seal the chip and the LTCC against liquid and provide additional attachment strength, an epoxy underfill was applied around the bumps and cured. The same underfill material was used to glue the well onto the package. A special lamination technique involving sacrificial carbon tape and lower pressure was used to create fluidic channels.

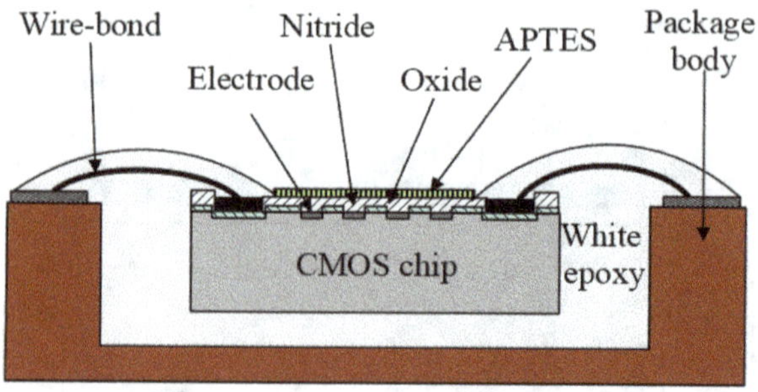

Figure 13.3 The cross-section of a packaging method showing the wire bonding of a CMOS chip to a ceramic package and the usage of white epoxy to insulate the wires and construct a reservoir to hold the sample solution

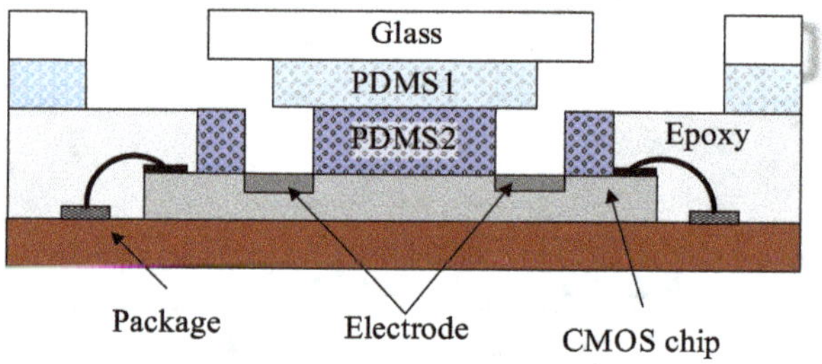

Figure 13.4 A depiction of the rapid microfluidic packaging technique reveals the creation of microwells on three layers of PDMS2, and PDMS1, which are aligned with the open-top packaged die using a flip-chip bonder

PDMS is another popular material for making reservoirs on top of chips, as it is flexible and easy to work with [673]. Nabovati *et al.* [2] developed a rapid microfluidic packaging process using two layers of PDMS and a glass layer, as shown in Figure 13.4. First, PDMS is spin-coated onto a petri dish and cured. The PDMS is then patterned and cut using a charge pump laser machine. The top glass layer is also patterned and cut using a mask-less laser method to create microwells. Next, the layers are thoroughly cleaned and dried. The dies are packaged on open-top packages and located and fixed with tape on a vacuum plate. A flip-chip bonder is used to stack and align the three layers of PDMS2, PDMS1, and glass on the open-top packaged die, respectively. Two cameras mounted on the device help achieve precise alignment. The layers are subsequently compressed onto the die by employing a femto pick and a placement nozzle to eliminate any excess air trapped between them. As the pressing occurs, the hot plate is activated, and epoxy resin is poured to secure the layers, bridge the gap between PDMS1 and the chip, and encapsulate the wire bonds. The temporary hermetic bond created by the soft PDMS1 layer prevents leakage of the epoxy resin underneath. The layers are constantly pressed by the nozzle while the epoxy cures, and after curing, the chip is removed from the hot plate.

Li *et al.* [1020] employed a different approach to insulating the package contact pads, wire bonds, and electrode array chips from each other (Figure 13.5(a–d)). They used a layer of parylene to coat all surfaces within the package and then selectively removed parylene from only the electrode array area using reactive ion etching (RIE) with oxygen gas. However, traditional masking materials like metal and photoresist could not be used due to the complex 3D structure. Instead, they used a PDMS cylinder of the same size as the uncovered area of the chip's surface and attached it to a slightly smaller silicon chip using oxygen plasma. The glass slide was attached to the other side of the PDMS cylinder, and the resulting assembly was clamped over the center of the chip, which was coated with parylene, as illustrated in Figure 13.5(b). To fill the package cavity, crystal adhesive was melted at the point where the

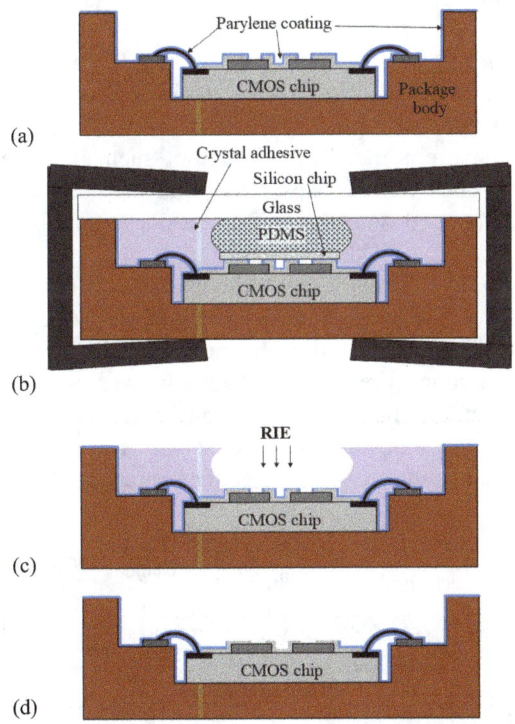

Figure 13.5 The sequence of steps involved in sealing the chip-in-package is as follows: (a) The chip is wire-bonded to the package and coated with a layer of parylene, (b) a PDMS cylinder and silicon chip are affixed to a glass slide and securely fastened to the parylene-coated package using clamps, (c) the glass slide is detached, and the parylene is selectively etched, (d) the crystal adhesive is removed, resulting in the formation of the final package

PDMS/silicon cylinder was held. Afterward, the assembly of the slide/PDMS/silicon chip was taken apart, and the parylene layer that covered the on-chip electrodes was removed using RIE. Finally, the packaging was completed by removing the crystal adhesive with acetone (ACE).

While low-cost rapid prototyping techniques [2,3,29,259,264,1020,1025, 1027,1028] are available, they may not be suitable for high-precision applications that require numerous wells and microchannels.

13.2.2 Soft lithography

Soft lithography is a commonly used technique for creating integrated microfluidic systems, with PDMS being a popular material for bonding microfluidics to a CMOS chip [1029,1030]. In one example, Welch *et al.* [1031] developed a PDMS microfluidic device that was bonded to a flexible PCB using oxygen plasma, as depicted in Figure 13.6. To improve the bonding process, the PCB surface was coated with polyimide. Additionally, a CMOS chip was flip-chip bonded to the same PCB.

Li and Yin *et al.* [1030,1032] attempted to develop a chip-in-carrier process that is compatible with CMOS. The process involved fixing the chip onto a glass using epoxy, as illustrated in Figure 13.7. Other microfabrication processes were used to deposit and pattern metallic conductors such as screening silver (Ag) electrodes. The electrodes were encapsulated with parylene and a PDMS microfluidic structure was bonded onto the chip-in-carrier assembly. The PDMS structure was created by using an SU8 mold [1032].

Zhang *et al.* [1033] also explored a flexible microfluidic-CMOS packaging technique that involves bonding a PDMS substrate, a CMOS chip, and a PDMS microfluidic device. The microfluidic channels are filled with liquid metals to create conductive interconnections. Soft-lithography was used to effectively create the microfluidic channels and electrical connections. Figure 13.8 depicts the packaged system. The packaged system consists of two PDMS layers: a top microfluidic layer and a bottom CMOS layer. The bare CMOS die is located upside

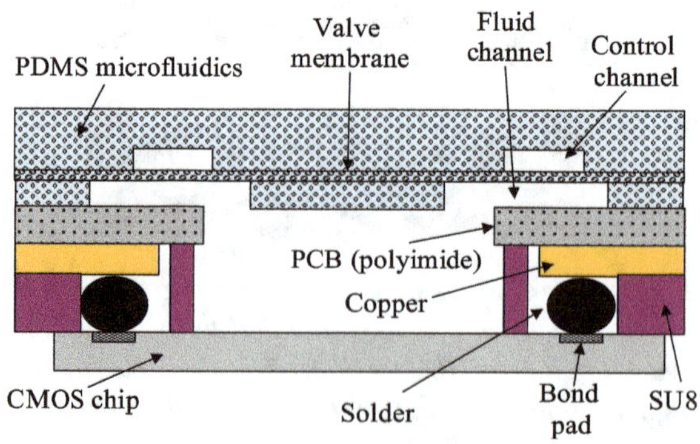

Figure 13.6 The cross-sectional view of a PDMS microfluidic device bonded to a flexible PCB using oxygen plasma

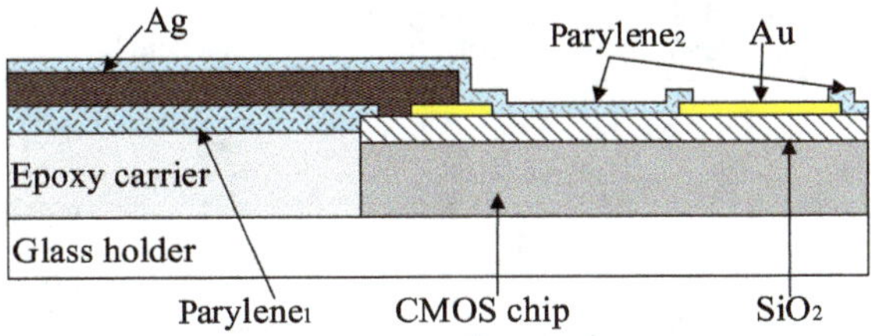

Figure 13.7 The cross-sectional representation of a chip-in-carrier device accompanied by screen-printed planar interconnects

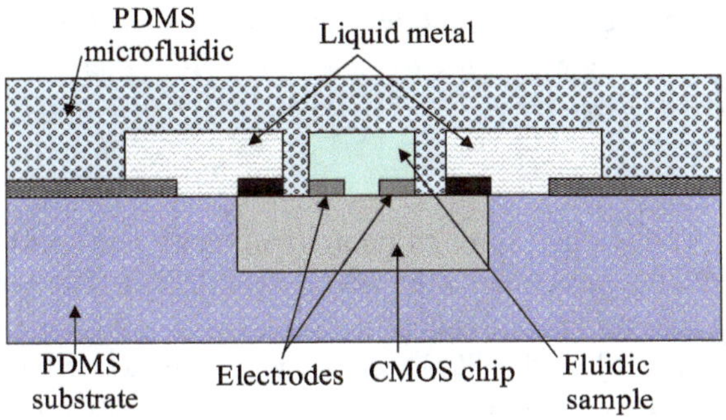

Figure 13.8 A cross-sectional view of a packaging method, illustrating the utilization of liquid metal for establishing electrical connections between the IC and the substrate

down on a flat silicon wafer and held in contact with the wafer by applying a small pressure to its backside. Degassed PDMS prepolymer is then poured onto the CMOS die and baked, after which the CMOS die is peeled off from the silicon wafer and embedded in a piece of flat PDMS while its active area is still exposed. Microfluidic layers were fabricated using conventional soft lithography, with a patterned SU8 photo-resist master mold fabricated on a silicon wafer utilizing photolithography. After pouring PDMS prepolymer onto the mold and degassing it in a vacuum chamber and baking it, the partially cured PDMS is peeled off from the mold to form the microfluidic layer. Liquid inlet/outlet ports are formed by punching through the whole layer. A strong bond between microfluidic and CMOS layers was achieved through air or oxygen plasma treatment. After alignment of the microfluidic layer on top of the CMOS layer, the package is baked. Finally, the liquid metal Galinstan was used to fill some of the microfluidic channels as the electrical interconnect to the CMOS chip.

In another effort, Wu *et al.* [1034] have developed a hybrid system that integrates a microfluidic structure onto a flexible PCB. This technique is illustrated in Figure 13.9. The microfluidic layers are composed of a glass pneumatic valve actuation layer, a PDMS valve membrane, and a glass fluidic layer with channels.

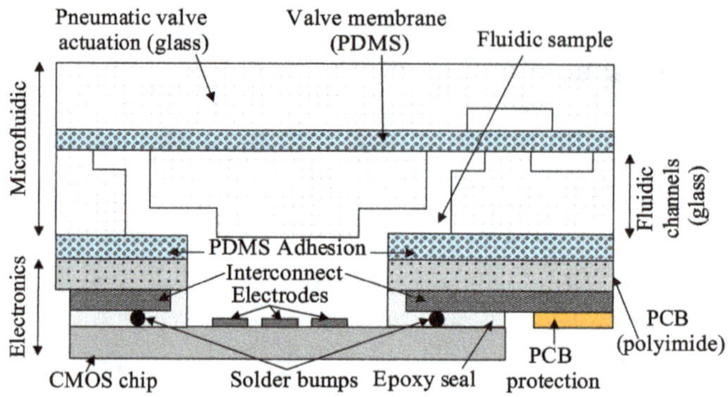

Figure 13.9 A cross-sectional depiction of a hybrid system featuring microfluidic layers comprised of a glass pneumatic valve actuation layer, a PDMS valve membrane, and a glass fluidic layer containing channels

The glass wafers were etched through photolithography and wet etching, and the holes for inlets, outlets, and pneumatic ports were drilled. The wafers were then bonded together using a PDMS membrane.

Burdallo *et al.* [1035] used a PCB with a hole as the carrier for a PDMS microfluidic structure, which was bonded using adhesive materials coated on the PCB. Berdat *et al.* [92] created a molded PDMS microfluidic device that was clamped between two PMMA micro-machined slides and used in conjunction with a biochip.

A lab-on-CMOS integration process using chip-in-carrier assembly was described in [1036] where, to expand the surface area beyond a CMOS chip, a silicon substrate carrier was used. A photoresist was spin-coated on a silicon wafer, patterned with a mask, and the through-wafer cavities within each carrier were opened using deep reactive ion etching (DRIE). The silicon carrier wafer was insulated with thermal oxide after patterning and etching the dicing streets. The carriers were singulated along with them or alternatively, the combination of all carriers and CMOS chips at the wafer scale could be singulated. The lab-on-CMOS integration process flow was then performed as shown in Figure 13.10:

1. To create a chip-in-carrier assembly, a Si substrate carrier was mounted onto a wax-coated glass handling wafer and a CMOS chip was placed in the cavity with its active surface facing down. Heating the assembly and cooling allowed the carrier and chip to attach to the wax while expelling air bubbles.
2. Epoxy was used to fill the gap between the chip and carrier, and a thin glass was added to the top to provide additional mechanical strength and create a vacuum seal. The glass was then separated by melting the wax, and any excess epoxy was removed with oxygen plasma.
3. A layer of polyimide was applied to insulate the exposed silicon and smooth out any sharp edges, and a Ti/Au thin film was deposited and patterned using thermal vapor deposition and lift-off process, respectively.

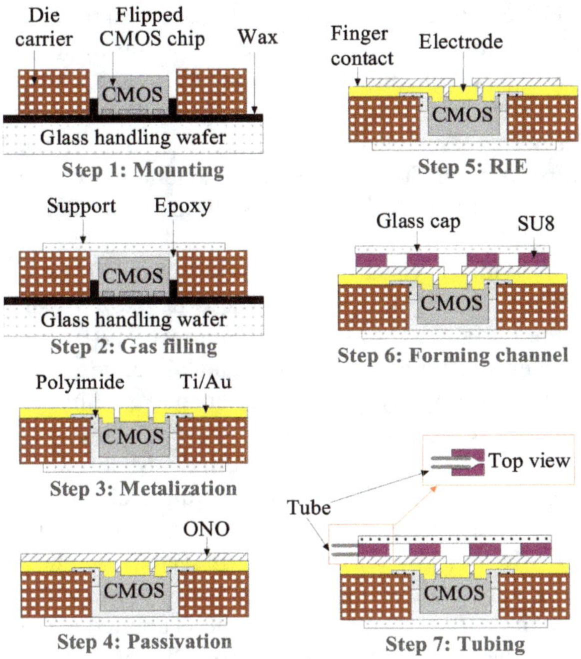

Figure 13.10 A depiction of the integration process flow for the lab-on-CMOS system using chip-in-carrier assembly

4. The interconnects were insulated with an oxide/nitride/oxide (ONO) passivation layer using a plasma-enhanced chemical vapor deposition (PECVD) system at a low temperature to ensure the stability of the polyimide and epoxy within the assembly.
5. Following metallization, the ONO layer was selectively etched using RIE to create open microfluidic channels while protecting the CMOS chip surface with the ONO layer and over-glass layer. The Au metallization and ONO layer were effective in sealing the CMOS bonding pads and sensor electrodes.
6. To form the open microfluidic channels, SU8 was spin-coated on the chip-carrier assembly, exposed to light, and developed. A glass slide was used to flatten the top surface of the SU8 channels, and an oxygen plasma treatment was applied to remove any SU8 residues from the Au electrodes and convert the surface from hydrophobic to hydrophilic. A thin layer of SU8 glue was added to the top of the open channels, and a glass cap was attached to cover them.

The final device was baked, and inlet and outlet tubes were inserted laterally into the SU8 layer through taper joints.

In the study by Norian *et al.* [1037], the active microfluidic region was defined and encapsulated using SU8. The wire bonds were also encapsulated, and an indium-tin-oxide-coated polyethylene naphthalate coverslip was added.

Huang *et al.* [1038] utilized a shifted Pt electrode design for their device, and SU8 was spin-coated and developed to create the insulation layer and cell-trapping microwells. Oxygen plasma was applied to the chip to increase the hydrophilicity of the surface and remove any residual SU8 from the electrode. The Pt electrodes were designed to cover only a portion of the bottom of the SU8 microwells, preventing the occurrence of conducting connections between adjacent electrodes after the Pt lift-off process. Figure 13.11 depicts the shifted Pt electrode strategy.

Figure 13.12 provides a visual representation of the fabrication process for integrating the CMOS chip with microfluidics to develop a magnetic flow cytometer [669]. In the initial step, as shown in Figure 13.12(b), a standard PCB is employed as a holder substrate to offer mechanical support and facilitate electrical interconnects. Additionally, a cavity is created within the PDMS microchannel to accommodate wire bonds in subsequent stages. Subsequently, an adhesive tape is utilized to temporarily secure the CMOS chip within the PCB cavity, preventing disruptions in the final microfluidic channel that may lead to undesired turbulence. The cavity is then filled with epoxy and allowed to cure (Figure 13.12(b)). Moving on, the tape is removed from the PCB, and wire bonding is performed to establish connections between the surface chip and the PCB. To safeguard the wire bonds, a crystal band is applied (Figure 13.12(c)). In this specific design, the PDMS microchannel is fabricated using an SU8 mold. The PDMS is prepared by combining the polymerizing agent in a 10:1 ratio, followed by degassing for 30 min. The mixture is poured onto the mold and cured at 60 °C for 3 h. Finally, as depicted in Figure 13.12(d), the PDMS microchannel is aligned with the chip's detection coil using a microscope for visual alignment. A glass slide applies uniform pressure, and compression is employed to seal the channel against the PCB with the assistance of acrylic supports. The resulting integration of the CMOS chip and microfluidics is displayed in Figure 13.12(e) [670]. It is important to note that integrating the CMOS chip with the microfluidic channel presents additional challenges in layout design. For example, as illustrated in [15], the pads are positioned exclusively on the two opposite sides of the chip to ensure that wire bonds do not interfere with fluid flow. Additionally, the sensor is placed near the center of the chip to prevent leakage near the

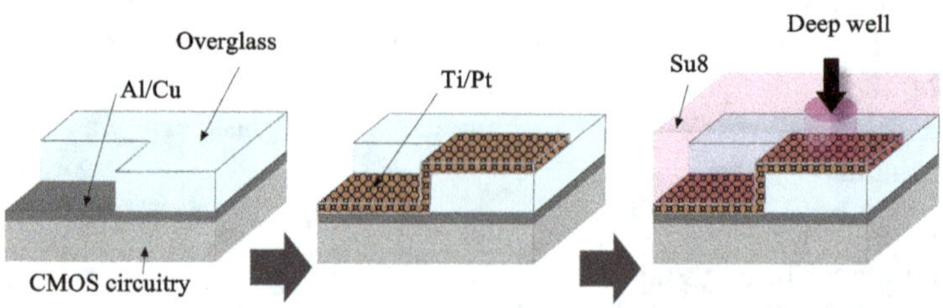

Figure 13.11 The post-fabrication process of the shifted electrode design featuring SU8 microwells

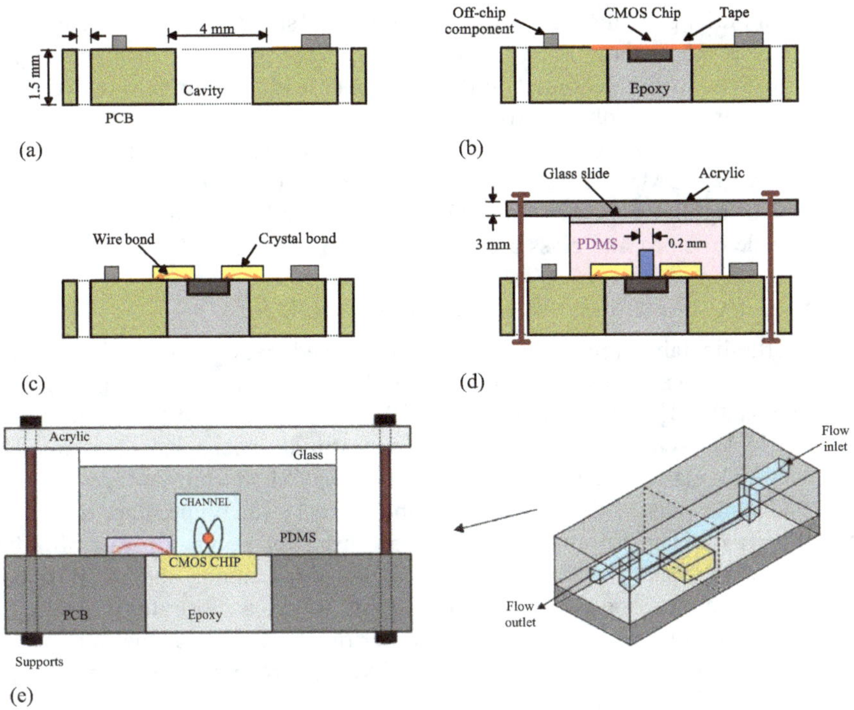

Figure 13.12 Fabrication process to integrate the CMOS chip with PDMS microfluidic. (a) PCB uses a holder substrate and a cavity is made to support the chip. (b) The chip is placed on the tap, the tap is stuck to the PCB, and the cavity is filled with epoxy. (c) Chip is wire-bonded to the PCB. (d) PDMS microchannel is placed on the PCB. (e) Microfluidics integration of the CMOS chip

interface between the chip edge and the microfluidic channel. In the context of [15], a 3D printed mold with wire bond cavities is developed to fabricate the PDMS microchannel. O_2 plasma treatment is subsequently applied to enhance hydrophilicity. The microchannel is precisely aligned over the chip's sensor, and an acrylic piece is employed to securely fit the microfluidic channel against the chip, thereby preventing leakage.

In a design mentioned in [730], a cost-effective PDMS microfluidic structure comprising fluidic channels and control valves is fabricated and bonded to an 8-parallel sensor cell frequency-shift CMOS magnetic biosensor chip. This microfluidic structure supports eight pairs of independent differential-sensing chambers, each with a volume of less than 0.2 nL, enabling simultaneous sensing of multiple samples for high throughput.

In [555], an optically transparent fluidic cap is utilized to confine the aqueous sample on top of the chip and facilitate easy insertion and removal through an inlet port and an outlet port, respectively. The fluidic cap structure is constructed using multilayer PMMA material and bonded with a silicone pressure-sensitive adhesive to create the resulting inlet/outlet fluidic channels. It is then further bonded to the FR4 PCB using epoxy adhesive [609].

Polymeric microfluidics are commonly fabricated using soft lithography methods, with PDMS being a popular choice [1031–1033,1035,1039,1040]. However, for applications that require high thermal and chemical stability, fluidic devices made of silica and glass are preferred [1034].

13.2.3 Direct-write fabrication process (DWFP)

3D microfluidic fabrication techniques have gained popularity in recent years, including jetting, Stereolithography (SLA), laser melting, laser sintering, and DWFP [1041–1043]. One approach using DWFP was proposed by Ghafar-Zadeh *et al.* [21,35,174,1044] for hybrid CMOS-microfluidic packaging. This involves encapsulating a paste-like sacrificial layer on the CMOS chip surface to create microfluidic channels, which is illustrated in Figure 13.13. While other techniques such as SLA or resin printing are commonly used for polymeric microfluidics, DWFP offers advantages for hybrid packaging applications that require both microfluidic channels and electronic circuitry on the same substrate.

The DWFP presented in [21,46] uses a sacrificial layer consisting of a mixture of microcrystalline wax ink and petroleum jelly. The process for packaging a chip in a DWFP involves six steps:

1. The first step is encapsulating the bonding wires and pads with a partially cured epoxy resin that flows around the wires but stops near the pads due to its high viscosity and surface tension. The deposition of ink on the sensing electrodes follows a precise trajectory path, which is measured using an optical profiler and programmed into a robot-driven dispensing system.
2. Next, a paste-like ink is extruded through a micronozzle and deposited on the substrate. The parameters that require control in this process include the velocity of the moving micronozzle, the air pressure applied to extrude the ink

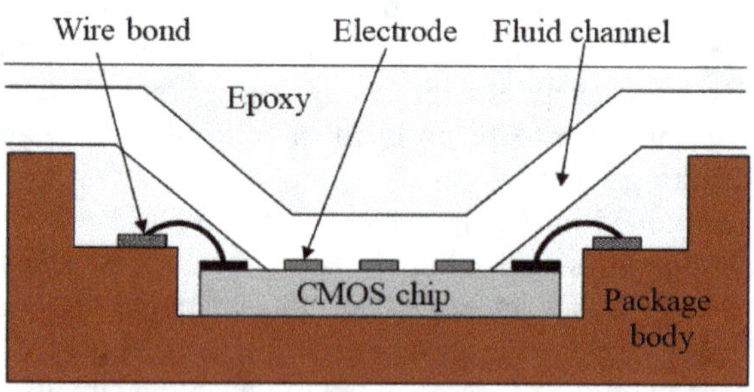

Figure 13.13 A cross-sectional view of the hybrid microfluidic-CMOS capacitive sensor, where the microfluidic component is fabricated using DWFP

through the nozzle, the microcrystalline fraction of the organic ink mixture, and the relative height between the nozzle and substrate.

3. The microscale fluidic fittings are fixed near the deposited ink using hot glue, and additional fugitive ink is deposited to fill the space between the fluidic connection and the ink filament. This step prevents the infiltration of epoxy into the fitting during the encapsulation process.
4. Another ink deposition is done in the boundary of the epoxy encapsulation, forming a fugitive dam.
5. A low-viscosity epoxy resin is dispensed and cured on the deposited ink within the encapsulation boundary to encapsulate the fugitive ink.
6. Finally, the fugitive ink is melted and expelled under air pressure or a light vacuum, and the remnants of the ink are removed by injecting hot water through the channel for analyte injection.

The DWFP technique [21,35,174] provides a distinct advantage over soft lithography in that it allows for direct fabrication on ICs, without requiring the use of molds or masks. This makes it well-suited for creating complex 3D microfluidic structures using polymer-based materials.

13.2.4 Other microfabrication methods

There exist other approaches for manufacturing IC-microfluidic packaging. Lee *et al.* [970] have proposed a microfluidic structure that can be post-fabricated on a SiGe IC, as shown in Figure 13.14. The use of magnetic microcoils on a CMOS chip for cell manipulation is reported in this system. The proposed packaging technique has the potential to be utilized for packaging microfluidics and other ICs such as CMOS capacitive sensors. The process involves dicing the IC from the wafer with an area larger than the required active area to be used as the carrier for microfluidics. The sidewalls of microfluidic channels are defined using spin-coating and a layer of polyimide is patterned. A glass coverslip is then sealed on top of the structure, and plastic fluidic tubes are attached to its inlet and outlet. The SiGe IC-microfluidic system is positioned above a Si substrate with electrical leads as electrical connections.

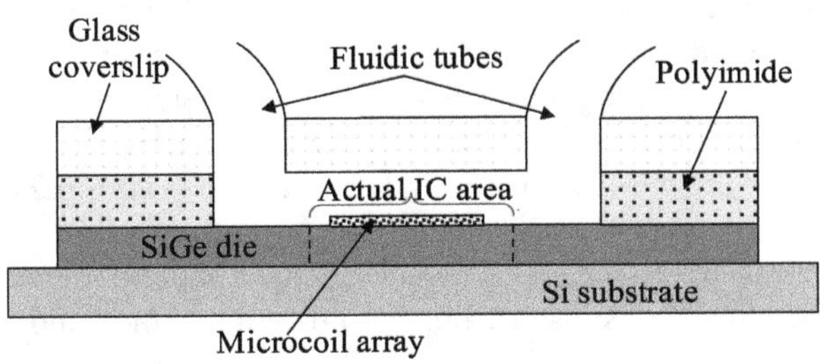

Figure 13.14 A cross-sectional representation of a microfluidic structure, which is directly fabricated on the surface of a SiGe IC for magnetic manipulation

A scalable heterogeneous integration of Si ICs and microfluidics was demonstrated by Lindsay *et al.* [1045] using a fan-out wafer-level packaging technique. They employed a stacked fluidic assembly, which was laser-cut into a two-sided pressure-sensitive adhesive sheet, with a simple Y-channel for fluid flow. This assembly was then sandwiched between glass/acrylic layers.

Karl *et al.* [1046] proposed a packaging technique using laminated photosensitive dry-film resist (DFR) for the realization of passive microfluidic channels and chambers, as well as adhesive wafer bonding to achieve heterogeneous integration of different substrates on the wafer level. Their packaging approach involves a three-layer bonded sandwich consisting of a CMOS substrate, patterned DFR, and a glass cap.

DFR laminates have the potential for the realization of microfluidic channels and adhesive wafer bonding for heterogeneous integration on the wafer level. Karl *et al.* [1046] proposed a packaging technique that involves bonding a three-layer structure consisting of a CMOS substrate, a patterned DFR layer, and a glass cap.

Mujika *et al.* [1047,1048] employed a technique for creating 3D multilayer microstructures by bonding two SU8 layers (SU8 50 and SU8 5) on separate wafers using adhesive bonding. The process relied on the CMOS-compatible adhesive bonding and releasing steps proposed by Agirregabiria *et al.* [1049]. Pyrex wafers were used as the substrate, and propylene glycol monoether acetate (PGMEA) was utilized as a developer. The process involved the following four main steps:

1. Substrate wafer preparation: Two substrates were prepared, a bottom substrate and a temporary top substrate. A highly polymerized thin SU8 2 layer was used for the bottom substrate, and an optimized photolithographic process of SU8 5 was used to enhance the adhesion between Pyrex glass and the thick SU8. Kapton films bonded to a Pyrex wafer were employed as the top SU8 substrate. The positive photoresist (S1818) was spun on the Pyrex wafer, and both substrates were brought into contact under a vacuum in a substrate bonder, pressed together, and heated.
2. Photolithography of SU8 50 layers: SU8 50 was photo-patterned on both substrates using two photolithographic processes. To achieve uniform SU8 films with good adhesive properties, a five-step approach including dynamic spin coating, soft baking with progressive temperature ramps, ultraviolet (UV) exposure, post-baking using progressive temperature ramps, and developing in a PGMEA bath and rinsing in isopropanol (IPA) was employed.
3. Bonding the two photo-patterned layers: The two photo-patterned wafers were put into contact using a substrate bonder, pressed together, and heated to bond them together.
4. Release of the temporary film: The SU8 stack and bottom substrate were separated from the top Pyrex wafer and the Kapton film. After rinsing the wafers in an IPA ultrasonic bath, the top Pyrex wafer and the Kapton film were

detached by the insertion of a razor blade. The Kapton film was then peeled off from the SU8 stack.

After completing these four steps, multilayered SU8 stacks can be obtained by preparing, bonding, and releasing more Kapton substrates. Figure 13.15 illustrates the four main steps involved in this approach.

In Figure 13.16, Inac *et al.* [1050] presented a packaging process for a three-wafer stack. The first wafer contained both sensors and active circuitry, while the inlet and outlet of microfluidics were etched from the backside using localized backside etching (LBE). The second wafer was utilized for patterning and etching

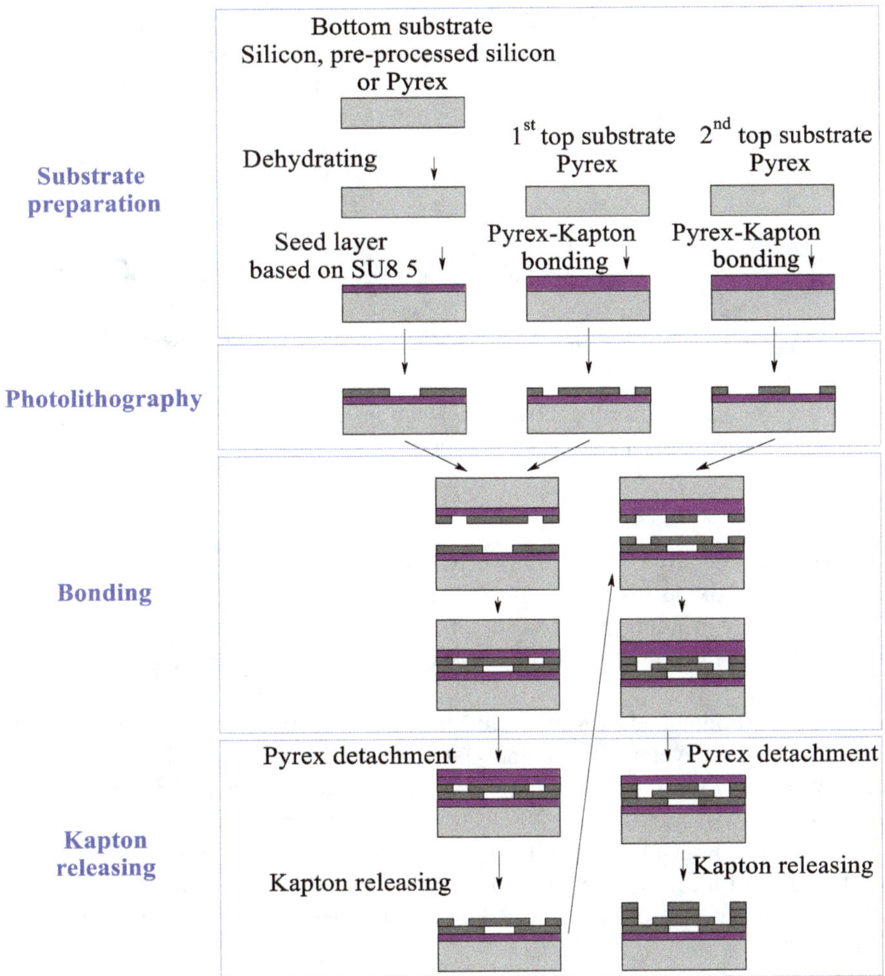

Figure 13.15 *The procedure for creating SU8 multilayer microstructures using a sequence of adhesive bonding and releasing steps that are compatible with CMOS technology*

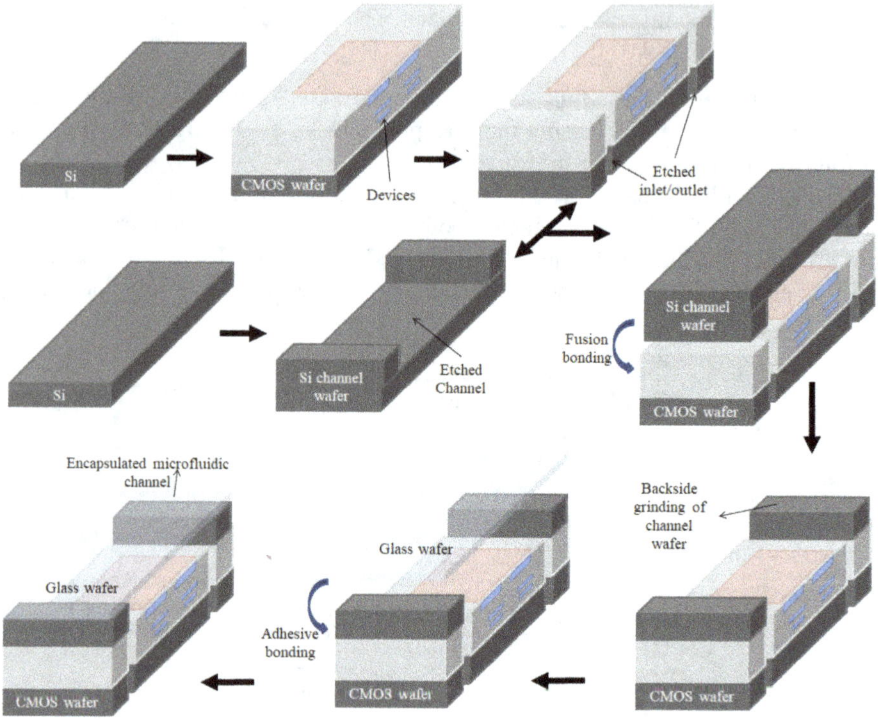

Figure 13.16 Fabrication of CMOS chip and microfluidic channel and three-wafer-stack bonding process

the channels. The front sides of these two wafers were bonded through plasma-activated oxide-oxide fusion bonding. The microfluidic channel's desired height was achieved by grinding the backside of the channel, followed by the adhesion of a glass wafer (the third wafer) to the top of the microfluidic channel, completing the sealing process.

In a study conducted by Matbaechi Ettehad *et al.* [1051], a microfluidic lab-on-chip (LoC) was developed to manipulate and analyze yeast cells using dielectrophoresis (DEP) forces. The chip was constructed using a three-wafer-stack approach, which integrated CMOS electronics, Si channels, and a glass wafer. To control fluid flow and sample injection through the microfluidic channel, an external macrofluidic technology was employed. Additionally, a PMMA fluidic manifold with a cavity of the same size as the chip was created using a commercial 3D printer.

In their study, Rasmussen *et al.* [1052] discussed surface micro-machining as an alternative technique for creating microchannels. This involves etching a sacrificial layer below a deposited material, as depicted in Figure 13.17(a) and (b). The use of a standard metal layer inside the CMOS chip that is patterned

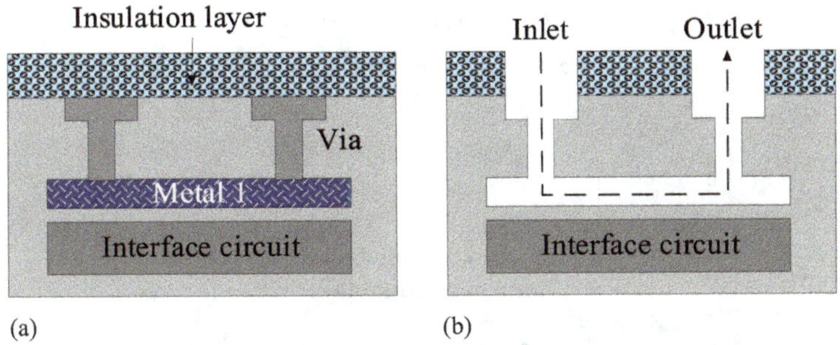

Figure 13.17 The implementation of a microchannel using surface micromachining is illustrated in two stages: (a) before etching, and (b) after etching

using traditional computer-aided design (CAD) tools creates a shallow microchannel. The conductors and vias that are etched play a vital role as the sacrificial layer.

Man *et al.* [1040] used a batch lithographic process to create large-volume plastic capillaries on planar substrates for deoxyribonucleic acid (DNA) analysis. Two types of capillaries were proposed using a simplified 3-polymer, 3-mask, low-temperature process, as illustrated in Figure 13.18. Initially, a low-tensile stress SiN or equivalent SiO_2 layer was deposited on p-type silicon substrates. The SiN (or SiO_2) layer was then etched on the backside at the outlets and inlets of the capillaries. Next, the pyramidal outlet and inlet access holes were formed by anisotropical etching of the sample in ethylenediamine pyrocatechol (EDP) (or tetramethyl ammonium hydroxide and water (TMAHW)). A thin layer of poly-p-xylylene (parylene C) was deposited on the samples, followed by the spin coating and lithographic patterning of a sacrificial layer of photoresist on the front of the wafers to define the height of the capillaries, which was then hard-baked. A second layer of poly-p-xylylene was deposited and sealed with resist. The p-xylylene layer played a crucial role as a barrier to prevent the soft sacrificial resist from being attacked by the polyimide solvent. Finally, a layer of photosensitive polyimide was spin-coated and soft-baked, and after a short plasma treatment to roughen the pxylylene surface, the capillaries were ready for integration with microfluidic and circuit elements for DNA analysis. Next, the sacrificial resist is removed, and a layer of evaporated Al is used as a mask to etch the top p-xylylene layer using oxygen plasma. This step is done until the sacrificial resist is exposed on the capillaries with front inlets. Then, the Al mask is removed using a wet Al etchant, and a resist layer is applied to protect the front side of the wafer. The backside outlets and inlets are created using a combination of oxygen, CF_4, and oxygen plasma. The protective resist and sacrificial layer are then eliminated using warm ACE, followed by rinsing, drying, and dicing of the samples.

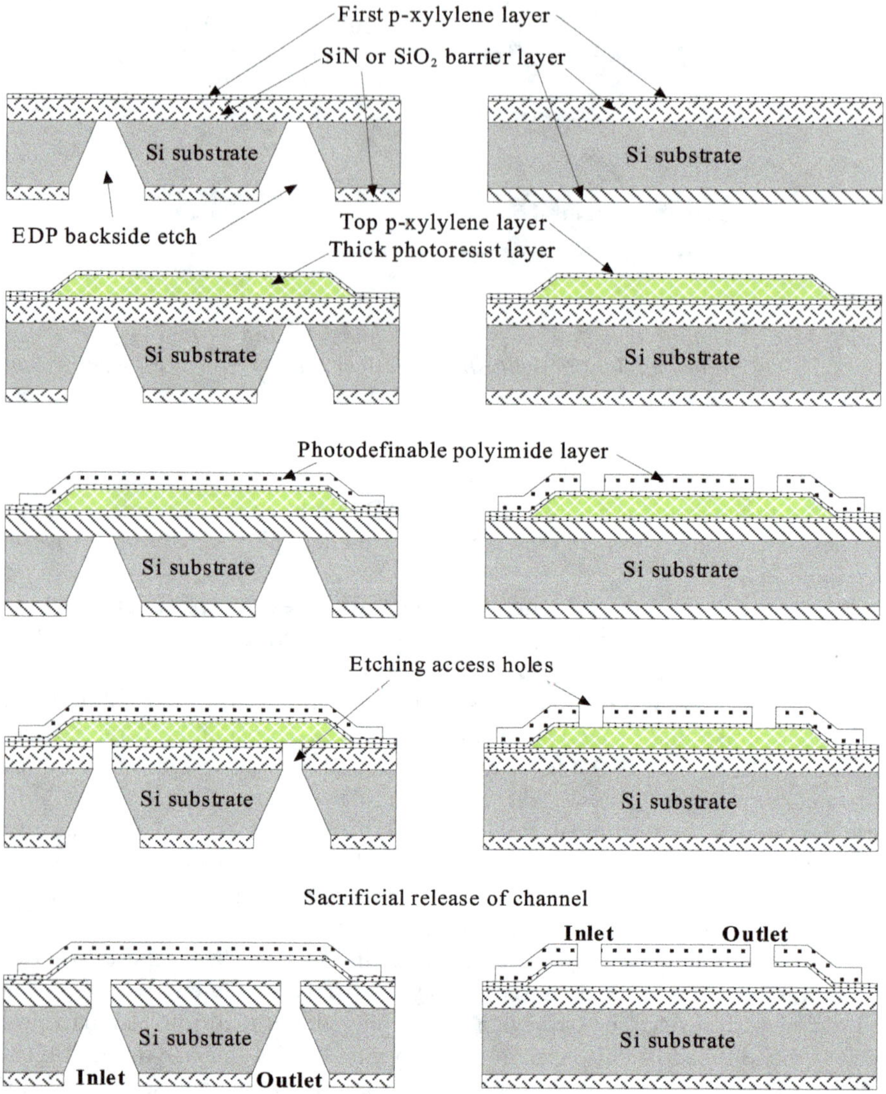

Figure 13.18 The process of fabricating front and back access capillaries

13.3 Discussion on IC-microfluidic packaging

Table 13.1 provides a comparison of various fluidic packaging methods, which is intriguing for several reasons. First, it is essential to securely fix the entire microsystems onto a carrier to enable easy handling.

Table 13.1 shows several options suitable for a career, including a standard chip package [1020], a Si wafer [1030,1032,1055], or PCB [1035]. However, for

Table 13.1 Summary of different IC-microfluidic packaging approaches

Application	**Fluidic device**	**Critical isolation material from an aqueous environment**	**Carrier**	**Metallic connections**	**Chip-in-carrier?**	**Integrated valves?**	**Ref.**
Tracking cancer cell proliferation	Well	Loctite 3340 (A polymer)	DIP40 ceramic carrier	Wire bonding	No	No	[29]
Real-time measurements of cell proliferation	Well (Cut Eppendorf Flex Tube, 1.5 mL)	Epoxy	DIP40 ceramic carrier	Wire bonding	Yes	No	[3]
Impedance spectroscopy and electrophysiological imaging of cells	A plastic epoxy well	Epoxy	PCB	Wire bonding	No	No	[44]
Cell viability monitoring	Well	Epoxy	LTCC package	Wire bonding	No	No	[1026]
Recording and stimulation of electrogenic cells *in vitro*	A polycarbonate ring	Epoxy	PCB	Wire bonding	No	No	[43]
Intracellular AP measurements and impedance spectroscopy in drug-screening applications	A glass ring	NA	PCB	Wire bonding	No	No	[73]
Single-cell recording	A glass ring	NA	Ceramic substrate	Flip-chip	No	No	[264]
Monitoring of electrogenic cells	A glass O-ring	Epoxy and PDMS	Package	Wire bonding	No	No	[259]
CV of potassium ferrocyanide	PDMS reservoir	Parylene	Ceramic package	Wire bonding	No	No	[1020]

(Continues)

Table 13.1 (Continued)

Application	Fluidic device	Critical isolation material from an aqueous environment	Carrier	Metallic connections	Chip-in-carrier?	Integrated valves?	Ref.
Detection of cell- or tissue-specific responses	A PDMS reservoir	PDMS/plastic	Brass substrate	Wire bonding	No	No	[673]
Cell culture monitoring and drug test	Microwells created on three layers of PDMS and glass	Epoxy	TQFP-44 package	Wire bonding	No	No	[2]
DNA detection	A reservoir made out of white epoxy	White epoxy	Ceramic package	Wire bonding	No	No	[1025]
Neuronal recordings and as an electric cell-substrate impedance sensor (EIS)	Glass cylinder culture chamber	Silastic 9161 RTV	Ceramic package	Wire bonding	No	No	[1027]
Neuronal recordings and as an electric cell-substrate impedance sensor (EIS)	Epoxy window frames	Hysol CB064	Ceramic package	Wire bonding	No	No	
pH sensing	Parylene encapsulated gate	Parylene	PCB	Wire bonding	No	No	[1023]
Real-time holistic cellular characterization and cell-based drug screening	Cell-culture plate with a laser-cut bottom	Epoxy, PDMS	PCB	Wire bonding	No	No	[1021]
LoC devices	Laser-cut pressure-sensitive adhesive sheet sandwiched between glass/acrylic layers	Pressure-sensitive adhesive, glass/acrylic	NA	Planar thin-film metallization	No	No	[1045]
Bacteria growth monitoring	Epoxy microfluidics (DWFP)	Epoxy	Package	Wire bonding	No	No	[35]
LoC devices	PDMS microfluidics	Polyimide/SiO_2	Si	Planar thin-film metallization	Yes (Chip-in-hole)	NA	[1039]

LoC devices	PDMS microfluidics	Parylene	Epoxy	Screen printed metallization	Yes	No	[1032]
pH sensing	PDMS microfluidics	DABA (Ebecryl 600)	PCB	Wire bonding (protected by Si rubber)	Yes (chip-in-hole)	No	[1035]
Detection of DNA	PDMS microfluidics	PDMS, PMMA	PCB	NA	No	No	[92]
Bioelectronic nose	PDMS microfluidics/ well (cut 15 mL centrifuge tube)	Parylene	Epoxy	Planar thin-film metallization	Yes	No	[1022]
Wearable health monitoring and environmental sensing	PDMS microfluidics	PDMS	PDMS substrate	Galinstan (liquid metal interconnection)	Yes	No	[1033]
LoC devices	PDMS microfluidics	SU8	Polyimide flexible PCB	Flip-chip	No	Yes	[1031]
Automated detection of magnetically labeled serum protein–PAH adducts	A glass microfluidic layer	Epoxy	Polyimide	Flip-chip	No	Yes	[1034]
Multi-wavelength chemical detectionsystems-on-a-chip	PDMS microfluidics	SiO_2	Si substrate	Wire bonding	No	Yes	[1053]
Magneticmanipulation of biological cells	Polyimide channel sidewalls covered with a glass coverslip	Polyimide	Si substrate	Planar thin-film metallization	No	No	[970]
	SU8 open microfluidics sealed with a cured PDMS layer	SU8	Si/SiO_2 substrate	Planar thin-film metallization	No	No	
Amperometry experiments	SU8 open microchannels covered with a glass cap	Oxide/nitride	Si substrate	Planar thin-film metallization	Yes (chip-in-hole)	No	[1036]
Amperometry measurements	SU8 microwell	SU8	PCB	Wire bonding	No	No	[1038]

(Continues)

Table 13.1 (Continued)

Application	**Fluidic device**	**Critical isolation material from an aqueous environment**	**Carrier**	**Metallic connections**	**Chip-in-carrier?**	**Integrated valves?**	**Ref.**
Quantitative-polymerase chain-reaction LoC	SU8 open microchannels covered with a glass cap	SU8, Parylene-C	Package	Wire bonding	No	No	[1037]
DEP manipulation of cells	Polycarbonate fluidic cap	Polycarbonate and soft glue	PCB	Wire bonding	No	No	[1054]
DNA analysis systems	Polymeric capillaries with etched vias as inlet and outlet	Polyimide / p−xylylene	Si substrate	Wire bonding (protected by a lid)	No	No	[1040]
Biomedical applications	Acrylic microfluidic	Spin-on-glass	Si substrate	Planar thin-film metallization	Yes	No	[1055]
Cytometry on a chip	Glass microfluidic	PDMS	Glass substrate	Flip-chip	No	No	[1056]
DNA analysis	Glass microfluidic	SiO_2	Si substrate	Wire bonding	No	No	[1057]
pH sensing	Pre-cut gate windows encapsulated by an adhesive like B-staged epoxies, pressure-sensitive thermosetting Si or acrylics	Polyimide	NA	TAB	No	No	[1058]
Detection of *E. coli* O157: H7	Two photolithographically patterned SU8 layers bonded on separate wafers	SU8	NA	NA	No	No	[1048]
DEP immobilization of yeast cells	Si microfluidic channel covered by glass cap	Si	Si	NA	No	No	[1051]

NA: Not available, D: Diameter

wearable health monitoring systems, the substrate needs to be flexible and skin-mountable, which can be achieved using flexible substrates made of PDMS [1033], polyimide [1034], or other flexible materials. The chip-in-carrier structure [1032,1039] is a suitable structure for high throughput analysis because of the extended surface area of the carrier which is useful for microfluidics. Typically, such carriers are made of Si or epoxy. It is worth mentioning that the chip-in-hole process is one of the techniques used in chip-in-carrier systems, like [1035], to integrate different chips heterogeneously into an active substrate [1022]. In this process, a pre-formed hole in the carrier is allocated to the IC. However, the size mismatch between the cavity and the IC might result in an uneven surface, which can be fixed by ensuring an even surface and filling the gap [1036].

Table 13.1 outlines the prevalent methods employed for interconnecting ICs and carriers. These methods include bump-bonding such as flip-chip [1031,1034,1056] and tape-automated bonding (TAB) [1058], as well as wire bonding [1023,1035,1040,1053,1054,1057]. However, wire bonds are vulnerable to corrosion and are unsuitable for integration with CMOS microfluidics [1034]. In situations where the chip is connected to a substrate with a fluid-containing window, flip-chip is the more appropriate option. However, the process is complicated and costly due to the pre-processing steps and additional materials required. Although TAB is similar to flip-chip, it has temperature limitations and is not suitable for CMOS-microfluidic integration [1032]. Screen printing [1032] and planar thin-film metallization [1022,1036,1039] are also useful techniques for making electrical connections, particularly for routing signals off the CMOS IC and onto the carrier in chip-in-carrier systems. These techniques are suitable for creating complex microfluidics on top of the chip and surrounding carrier and do not suffer from interference of interconnects. Photolithography facilitates well-defined planar thin-film deposition, while screen printing uses a predefined mask or screen to deposit conductive inks. Although thicker interconnections in screen printing are more dependable, they are not as precisely patterned as planar thin films [1032]. In [1033], the carrier is formed around the die, and Galinstan is employed to create liquid-metal-filled microchannels for electrical connections. This technique does not require a perfectly flat surface and is different from the chip-in-hole process.

An insulating layer, such as polymer, nitride, or oxide, should cover the CMOS circuits of a biosensor to prevent corrosion when in contact with fluid [1022]. Chemical-inert covers and sealants, such as epoxy, PDMS, and parylene, should also be added around electrical connections to protect them from corrosion [1019,1059]. Epoxy can be selectively applied or exposed to UV radiation, but it requires high temperatures to solidify (100 °C), which may not be suitable for some applications. Alternatively, UV-curable polymers, such as UV-curable epoxy, can be used as a patternable passivation material to protect wire bonds by easier procedures [1035,1060]. Waterproof adhesives or passivation materials are used to underfill flip-chips and protect electrical connections [1034]. Wires are usually placed in a thick polymeric well [1061] or covered by passivation materials [1035]. Some structures use etched vias to make fluidic connections from the back of the chip, with the wire located in a cut-out area in front of the chip, separated from the aqueous environment [1040].

In several systems [1032–1035,1039,1054,1055], microfluidics are created separately using materials like PDMS, SU8, polyimide, laminated photosensitive DFR, diacrylate bisphenol A (DABA) photocurable polymer, and Benzocyclobutene (BCB), which are then adhesively bonded on top of the CMOS chip using glue or plasma-activated bonding. Polymeric microchannels are usually replicated by molding [1032,1035,1039], glass microfluidics [1034], or hot embossing [1054] techniques. However, it is important to note that the adhesive may not offer a complete hermetic bonding for different fluid samples [56], and it may also take up a significant amount of space.

13.4 Summary

In this chapter, various IC-microfluidic packaging techniques were evaluated for their potential use in CMOS-based LoC applications. These techniques included rapid prototyping, soft lithography, direct write, and soft lithography, as well as other microfabrication approaches like micromachining. Rapid prototyping, while low-cost, was deemed unsuitable for large arrays of microchannels and wells. Soft lithography was found to be a popular technique but the issue of thermal and chemical stability must be taken into account. Direct write and soft lithography are capable of fabricating complex 3D microfluidic structures directly on top of the IC without the need for masks or molds. Despite reviewing these techniques, various materials used, and their associated challenges were also discussed.

While conventional methods like wire bonding and flip-chip are used for the electrical packaging of CMOS chips, non-conventional methods are required for microfluidic packaging. There is a need for further research to standardize low-cost and reliable packaging techniques for such hybrid systems. Although CMOS technology is efficient for the affordable mass production of microelectronic devices, there is growing interest in developing generic and affordable microfluidic packaging methods incorporated with reusable CMOS biosensors for mass-production purposes. As a result, one of the main goals is to develop novel contact-free hybrid microfluidic/CMOS techniques for the development of affordable and portable biosensing platforms, especially for point-of-care (PoC) applications.

Chapter 14
Life science protocols and applications

Complementary metal-oxide-semiconductor (CMOS) biosensors have emerged as versatile tools for the detection and monitoring of various chemical and biological entities, including cancer cells [29,641], cardiomyocytes [1062], neural cells [1062], bacteria [4,35], viruses [1063], nucleic acids [1064,1065], proteins [1066–1073], small organic molecules [1072,1074,1075], saccharides [1076], and similar species. As aforementioned, certain biosensing methods, such as optical or magnetic approaches, usually necessitate labeling for the detection of these targets. However, alternative techniques like capacitive, impedimetric, voltammetric, and biological/chemical field-effect transistor (Bio/ChemFET)-based sensors eliminate the need for labeling. The key to generating target-specific signals lies in the collision, interaction, and binding of the target species with the capturing layer, resulting in measurable signals.

The interface between the transducer and the target species is a critical component of biosensors, enabling the detection of ultra-low concentrations of analytes. In certain applications, such as cell growth monitoring, the bio-interface layer naturally forms on the sensing transducer, as exemplified by the biofilm in this case, or in organic solvent sensing. However, in other applications, the sensing surface must be functionalized with recognition elements that possess specific selectivity to interact with the target species [56]. This group employs a recognition layer as an interface, where biological capturing probes are immobilized on the surface to bind with the target species in the test solution.

This chapter categorizes life science applications of CMOS biosensors into two main groups: (1) chemical assays and (2) biological assays. After a concise introduction to recognition layers necessary for biological assays, we will present an overview of the state-of-the-art CMOS biosensors specifically designed for these applications.

14.1 Biorecognition layers

Biorecognition elements (BRE) used in biosensors (which were previously introduced in Chapter 1) can be classified into two types: classical and artificial. Traditional BREs, including nucleic acids, antibodies, enzymes, and whole cells, have long been employed in electrochemical biosensors (Figure 14.1).

Recent advancements in biochemical and nano-sensors have also introduced artificial BREs such as phages, aptamers, affibodies, and molecularly imprinted polymers (MIPs) [1077].

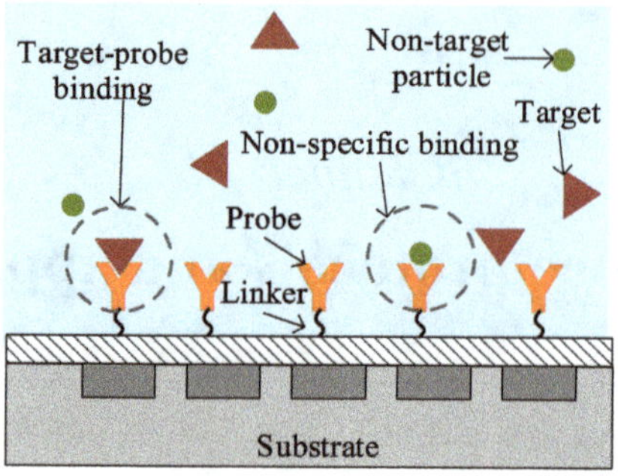

Figure 14.1 A selective sensor designed for detecting specific biospecies such as DNA, antibodies, antigens, enzymes, and more

To address the limitations associated with antibody BREs, alternative non-antibody BREs have been developed. These include aptamers [1066,1078], affibodies [1079,1080], and affimers [1068,1081]. Aptamers are short oligonucleotide (single-stranded deoxyribonucleic acid (ssDNA) or ribonucleic acid (RNA)) or peptide molecules that exhibit specific binding to target molecules. They offer advantages such as reusability, ease of modification, versatility, and thermal stability [1077]. Affibodies are newly engineered proteins that demonstrate high specificity and affinity in binding to target proteins or peptides. On the other hand, affimers represent a peptide aptamer variant that features peptide loops, and their size is twice that of affibodies. Notable characteristics of affimers include small size, high specificity, and the ability for flexible functionalization [1082].

Bioimprinting is a technique used to create highly selective recognition layers [1072,1083,1084] for the development of biomimetic sensors with superior performance. Various strategies, such as soft lithography [1085], template immobilization [1086], emulsion polymerization [1074], surface grafting [1087], and nano-imprint lithography (NIL) [1088], are employed to generate recognition cavities that accurately match the shape, position, size, and orientation of the target (template) compounds [1089]. MIPs, also known as artificial antibodies, are preferred over unstable antibodies and enzymes due to their selective binding capabilities to target (template) molecules [1072]. Moreover, MIPs offer additional advantages such as high thermostability, rapid synthesis, and cost-effectiveness [1077].

Furthermore, nano-sized materials like nanoparticles (NPs) [1090], carbon nanotubes (CNTs) [1091], and nanorods (NRs) [1092] have garnered significant interest in enhancing analytical assays when combined with BREs such as enzymes and antibodies [1077].

Affinity-based biosensors heavily rely on the electrical material and the method employed to form the intermediate layer for capturing target biospecies. Various immobilization techniques for receptors have been reported [1093], including covalent attachment using crosslinking agents like silanes [92,1094–1096], physical adsorption through non-covalent interactions [1094,1097], and incorporation into membranes [1098,1099] or polymers [1100,1101]. Receptors have been immobilized on Si/SiO_2 surfaces through covalent attachment in studies [1063,1102–1106], or incorporated into Langmuir–Blodgett films on Si/SiO_2 structures [1107]. Polymeric membranes have also been utilized to immobilize receptors on Si/SiO_2 structures [1106,1108,1109]. In another study mentioned in [1110], receptors were immobilized on the surface of Si/SiO_2, and a poly(ethylene glycol) (PEG) layer was utilized to cover the receptor layer and prevent ion interference. Covalent techniques employing functional amino acid moieties or Au-thiol bonds are typically employed for immobilizing protein receptors on Au electrodes, allowing for the formation of self-assembled monolayers (SAMs) [1090,1093, 1111–1114]. Tantalum oxide emerged as a viable material for the electrode surface, offering potential for receptor immobilization [1115]. Additionally, Pt electrodes with polymeric membranes demonstrated utility in this context [1116,1117].

Preparing the silicon chip surface for probe attachment requires different functionalization approaches. For microarray applications, attachment chemistries such as hydroxyl, poly-L-lysine, amine, aldehyde, and epoxy coatings are commonly used for antibody and nucleic acid attachment. Poly-L-lysine surfaces are typically employed for DNA attachment to facilitate electrostatic adsorption, while epoxy surfaces are prepared for covalent attachment of capture antibodies. Literature suggests that epoxy functionalization yields higher binding densities compared to amine, aldehyde, or poly-L-lysine [1118]. Surface functionalization aims to make the sensor surface reactive and uniform, maximizing bioreceptor binding and ensuring sensor-to-sensor reproducibility by eliminating non-uniformities such as microbubbles or particulate debris [609].

14.2 Chemical assays

Chemical tests, which encompass evaluations involving organic solvents, gas detection, and pH determination, play a pivotal role in numerous applications, such as the assessment of food, water, drug, and soil toxicity. Table 14.1 summarizes the CMOS biosensors reported for chemical assays.

14.2.1 Chemical solvent detection

Capacitive sensors are extensively used for chemical solvent sensing, particularly for polar solvents due to their low electrical conductivity. By incorporating passivation or insulation layers above the sensing electrodes, leakage current can be reduced, leading to improved capacitive sensing properties [56].

In gaseous solvent sensing applications, polymer films such as polyether urethane (PEUT), poly(cyanopropylmethylsiloxane) (PCM), ethylcellulose (EC), and poly(epichlorohydrin) (PECH) are commonly used to coat the sensing electrode. These films absorb analytes, altering their mass and dielectric properties, enabling the detection of solvent concentrations based on changes in the polymeric dielectric constant (refer to Figure 14.2) [52]. For example, Hagleitner and colleagues [50] presented a capacitive sensor designed for the detection of toluene (TOL) and ethanol (EtOH), utilizing a polymeric sensing layer called PEUT. The physical

Table 14.1 Comparison of different CMOS biosensors used for chemical sensing

Chemical solvent	Transducer's surface material	Sensor type	Label (Y/N?)	CMOS technology (nm)	Ref.
TOL, EtOH	Al/PEUT	Capacitive	N	800	[50]
ACE, MeOH, EtOH, DCM	Al/SiO_2	Capacitive	N	350	[54]
DI water, ACE, MeOH, IPA	$Al/SiO_2/Si_3N_4$	Capacitive	N	350	[3]
DI-water, MeOH, Propanol	Al/Al_2O_3	Capacitive	N	350	[115, 148]
(EtOH, MeOH, IPA and 2-butyl-alcohol); (DMSO), (ethylene glycol); (ACE); (ethyl-acetate); (xylene)	Al/Al_2O_3	Capacitive	N	90	[120]
Air, EtOH, IPA, DI water	$Al/SiO_2/Si_3N_4$	Capacitive (CMOS/MEMS)	N	350	[1119]
H^+	Si_3N_4	ISFET	N	350	[473]
K^+, Na^+, Ca^{2+}, H^+	Si_3N_4 and polymeric analyte-specific ionophore membranes	ISFET	N	350	[1120]
H^+	Ta_2O_5	ISFET	N	150	[462]
H^+	Si_3N_4	ISFET	N	350	[469]
H^+	Si_3N_4	ISFET	N	350	[477]
H^+	Si_3N_4	ISFET	N	180	[479]
H^+ and KCl	SiNW (Si/SiO_2)	ISFET	N	350	[483]
Pb	Au	Amperometric	N	500	[211]
$[Fe(CN)_6]^{4-}$	Au	CA	N	2500	[201]
$K_3[Fe(CN)_6]$	Au	Linear sweep, CA, CV, PV	N	500	[199]
$K_3[Fe(CN)_6]$	Au	CV	N	130	[192]
2-propanol	Au	Amperometric	N	500	[235]
O_2 and H_2O_2	Pt	Amperometric	N	180	[236]
Luminol	NA	Chemiluminescent	N		[1121]

NA: Not available, PV: Pulse voltammeter, Y: Yes, N: No

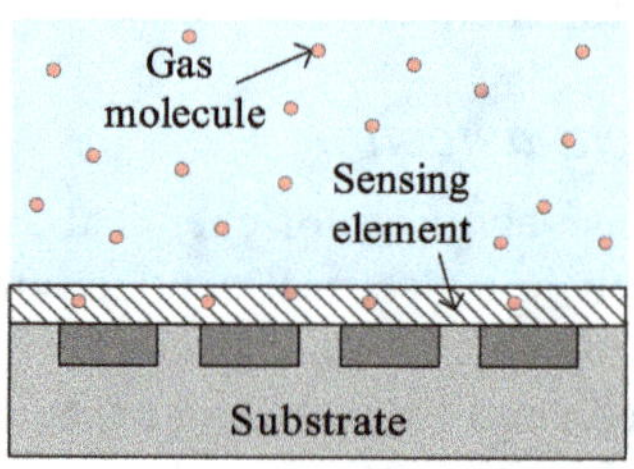

Figure 14.2 The conceptual perspective of gas sensing

characteristics of the polymeric film, including mass and dielectric properties, can be altered by the absorption of analytes [50]. As a result, changes in the density of diffused molecules and the corresponding variations in the polymer's dielectric constants are directly proportional to the solvent concentration. Consequently, monitoring the fluctuations in the polymeric dielectric constant provides valuable insights into the dielectric properties of the analyte.

For sensing liquid-phase chemical solvents using capacitive sensors, the sensing capacitance varies according to the dielectric constant of the liquid solvent. Several studies have explored the removal of the passivation layer above the electrode to enhance the sensitivity of solvent detection, and distinct patterns corresponding to solutions with varying conductivities have been observed [47,54,174]. For example, Nabovati *et al.* [54] suggested the removal of the passivation layer on the topmost metal layer of CMOS electrodes and evaluated their capacitive sensor's performance by testing it with polar chemical solvents, including EtOH, methanol (MeOH), acetone (ACE), and dichloromethane (DCM), which possess varying dielectric constants. Osouli *et al.* [115,148], in a separate study, also removed the passivation layer above the electrode using pad masks. They investigated the behavior of their capacitive sensor with different chemical solvents, such as deionized (DI)-water, MeOH, and propanol, observing distinct patterns corresponding to solutions with varying conductivities, ranging from low to high.

Helmy *et al.* [120] introduced a self-sustained CMOS microwave chemical sensor capable of detecting the dielectric constants of organic chemicals, and binary mixtures, and estimating the fractional volume of constituent materials. They exposed the sensor's electrode by removing the passivation layer and tested it with various substances, including alcohols (EtOH, MeOH, isopropanol, and 2-butyl-alcohol), organosulfur compounds (di-methyl-sulphoxide (DMSO)), glycols (ethylene glycol), ketones (ACE), esters (ethyl-acetate), and alkyl benzenes (xylene). In another work, Senevirathna *et al.* [3] employed their CMOS capacitive sensor to detect MeOH, deionized (DI) water, ACE, and isopropyl alcohol (IPA).

Furthermore, Chiang *et al.* [1119] presented a CMOS MEMS capacitive oscillator designed for resonant sensing. They validated the sensor's functionality by testing it with EtOH, IPA, and deionized water (DI water).

In another investigation, Yang *et al.* [235] proposed an amperometric sensor for measuring different concentrations of 2-propanol. They employed gold working electrodes (WEs), which were modified with 3-mercaptopropionic acid (MPA), β-nicotinamide adenine dinucleotide phosphate ($NADP^+$), toluidine blue O (TBO), and 2°ADH. The modified electrodes were exposed to varying concentrations of 2-propanol in phosphate-buffered saline (PBS).

14.2.2 Ion sensing

Ion-selective field-effect transistors (ISFETs) fabricated through the CMOS process utilize an extended gate that reaches the top metal layer and usually the CMOS passivation layer (Si_3N_4). The passivation layer serves as the sensing interface in contact with the liquid sample. Si_3N_4, being sensitive to hydrogen ions (H^+), which are prevalent in solutions, can effectively measure the hydrogen-ion concentration, also known as pH. For instance, the functionality of unmodified CMOS-fabricated ISFET arrays for pH measurement was demonstrated by Miscourides *et al.* [469]. They utilized in-house pH buffers of pH 4, 7, and 10, prepared by mixing DI water,

KCl, and Tricine as the main buffer solution. The pH was adjusted by adding 1 M hydrochloride (HCl) (acid) or 1 M sodium hydroxide (NaOH) (base) to decrease or increase it, respectively. Similarly, Cacho-Soblechero *et al.* [477] tested their ISFET-based sensor chip by measuring the pH of known-value pH buffers (pH 4, 7, and 10) using the Si_3N_4 layer of standard CMOS. In another study, Liu *et al.* [479] employed the CMOS passivation layers again as the sensing membrane and examined the short-term and long-term pH response and diffusion effects of the sensors. They also accomplished this by introducing various amounts of HCl and NaOH into a pH 10 buffer solution to alter the pH. Zeng *et al.* [473] proposed an ion-sensing lab-on-chip (LoC) platform utilizing a 128 × 128 CMOS ISFET array for high-speed visualization of NaOH powder diffusion in water. The passivation layer of the CMOS technology (Si_3N_4) served as the sensing membrane. Figure 14.3 demonstrates ion imaging with a frame rate of approximately 1,000 fps. Additionally, Lee *et al.* [462] suggested the use of a Ta_2O_5 membrane in the CMOS process, achieving a pH sensitivity of 55.1 mV/pH.

Moreover, to achieve specificity towards various ions, it is feasible to apply ion-selective membranes onto the passivated surface. Several reports have documented ISFET-based sensors designed for the detection of [K^+], [Na^+], [Ca^{2+}], and [Mg^{2+}] ions [1122–1126]. For instance, Moser *et al.* [1120] proposed an array consisting of 1,024 ISFETs to enable multi-ion imaging, enabling simultaneous monitoring of potassium ([K^+]), calcium ([Ca^{2+}]), sodium ([Na^+]), and hydrogen ([H^+]) ions. To attain the desired quasi-Nernstian sensitivity to K^+, Ca^{2+}, or Na^+, analyte-specific ionophore membranes were deposited on the ISFET surface, while the standard Si_3N_4 passivation layer was used for pH sensing on uncoated pixels.

Another variant of the chemical field-effect transistor (ChemFET) utilized for chemical sensing is the silicon nanowire (SiNW) FETs. These devices consist of a single SiNW that is electrically connected between the source and drain electrodes, offering rapid response and a high surface-to-volume ratio. Furthermore, it is possible to integrate NW FETs into the CMOS process. For example, Livi *et al.* [483]

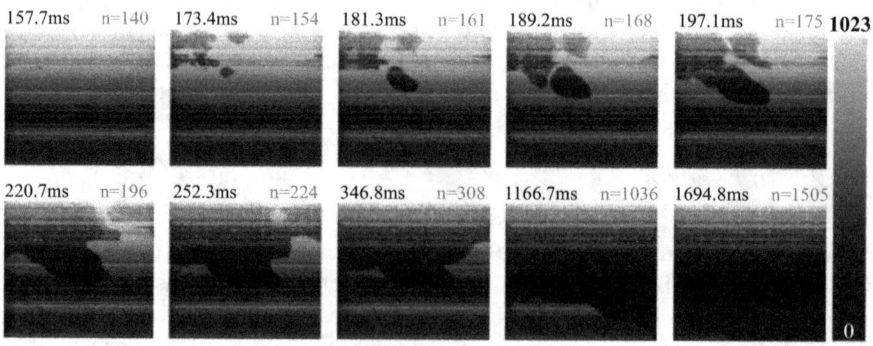

Figure 14.3 Real-time ion imaging capturing the diffusion of NaOH powder (frame rate approximately 1,000 fps, timestamp of each frame in the top left corner, n representing the frame number) [473]

proposed a micro/nanosensor that incorporated an array of SiNW FETs, combined with a CMOS readout chip, demonstrating its potential for conducting parallel, high-temporal-resolution experiments involving pH and KCl measurements.

14.2.3 Metal sensing

CMOS voltammetric systems are among the sensors that have been used for the detection of heavy metals by distinguishing between metals with different oxidation states. Researchers have evaluated the system's performance using solutions containing specific metals, such as $K_3[Fe(CN)_6]$. A study by Martin *et al.* [211] introduced a CMOS voltammetric system capable of detecting heavy metals such as lead (Pb) by distinguishing between metals with different oxidation states. The system employed a three-electrode configuration, integrating a gold electrode (working electrode (WE)), a platinum electrode (auxiliary or counter electrode (CE)), and a silver/silver chloride electrode (reference electrode (RE)) for accurate measurements. The researchers initially evaluated the system's performance using a solution containing 1 mM $K_3[Fe(CN)_6]$ and 100 mM KCl. Similarly, other researchers such as Li *et al.* [199] and Jafari *et al.* [192] assessed their voltammetric sensors by subjecting them to cyclic voltammetry (CV) with various concentrations of potassium ferricyanide ($K_3[Fe(CN)_6]$). In a different study, Sutula *et al.* [201] utilized ferrocyanide ion $[Fe(CN)_6]^{4-}$ as a characterization compound for their sensor. Notably, ferrocyanide ion is commonly used as a standard compound due to its distinct properties, including high reversibility.

14.2.4 Sensing other chemical materials

Inoue *et al.* [236] utilized a CMOS amperometric sensor with a mode-selectable function that allowed the independent operation of 400 electrodes in off, electrometer, V_1, and V_2 modes. The researchers demonstrated a dot picture based on the redox reaction of ferrocenemethanol by applying different voltages using the V1 and V_2 modes. They also showcased the sensor's capability for simultaneously detecting O_2 and H_2O_2 using the V_1 and V_2 modes. By leveraging the mode-selectable function, selected electrodes were modified with an osmium-polyvinylpyridine gel polymer containing horseradish peroxidase (Os-HRP) using the off and V_1 modes. The Os-HRP-deposited electrodes exhibited noticeable current responses to H_2O_2, whereas the non-deposited electrodes did not exhibit significant responses.

Optical sensors also offer an alternative approach to chemical sensing. For instance, Singh *et al.* [1121] introduced a novel optical sensor designed for the detection of luminol through chemiluminescence and electrochemiluminescence assays. The sensor utilizes a CMOS-based microsystem, where luminol and hydrogen peroxide are introduced and mixed via a diffusion-based approach. When luminol reacts with hydrogen peroxide, it enters an excited state known as aminophthalate, emitting a vivid blue light. The chemiluminescent reaction, characterized by its transient nature lasting mere seconds, is captured by an imager for detection and analysis.

14.3 Biological assays

Biological assays can be divided into six groups: (1) antibody/antigen-based assays, (2) nucleic acid-based assays, (3) cellular assays, (4) neural recording and neurotransmitter-based assays, (5) saccharide-based assays, and (6) other applications. Each group's sample preparation protocols, sensing mechanisms, and concepts reported in the literature will be discussed in detail in this section. Table 14.2 compares different CMOS biosensors reported for biological assays.

14.3.1 Antibody/antigen-based assays

Antibodies serve as widely utilized BREs [1063,1092,1097,1099,1127], capable of specifically binding to distinct fragments of antigens known as epitopes (refer to Figure 14.4). These antibodies can be either monoclonal or polyclonal, depending on whether they are derived from a single-cell clone or multiple immune cells, respectively. While polyclonal antibodies can bind to multiple epitopes on the same antigen, monoclonal antibodies exhibit binding specificity towards a singular epitope on the antigens [1077]. Monoclonal antibodies possess greater selectivity compared to polyclonal ones, albeit with a higher production cost. Numerous electrochemical biosensors operate on the principle of antibody-antigen interaction [1108,1111,1112,1117].

The work of Gambini *et al.* [728] introduced a CMOS microsystem designed for detecting microparticles in magnetic immunoassays. In the field of detector design, the utilization of the super-paramagnetic properties of labels is an option that allows for accurate measurement of magnetic label concentration, eliminating the need for baselines and significantly enhancing system reliability. By exploiting the inherent magnetization dynamics of super-paramagnetic materials, their approach achieved a remarkable reduction in measurement baselines, surpassing conventional methods by over 300 times. Moreover, their system demonstrated a signal-to-baseline ratio exceeding 1, presenting an improvement of over 50 compared to systems that overlook bead dynamics. To authenticate the sensor's performance within an assay setting, an inverse-phase direct assay focusing on human serum albumin (HSA) was executed. The assay employed 2.8 μm Dynal beads as labels, and the sensor surface was initially functionalized with a known concentration of HSA via physical adsorption. Following surface blocking with 1% bovine serum albumin (BSA), magnetic beads coated with anti-HSA antibodies were introduced and incubated for 2 min. Subsequently, non-specifically bound beads were eliminated using external magnets.

A distinctive study by Kuo *et al.* [1133] resulted in the development of a CMOS-based LoC immunoassay system that is controlled by a microcontroller unit (MCU). This innovative technology facilitated the automation of various detection processes, encompassing blood filtration through a nonporous aluminum oxide membrane, conjugation of biomolecules with magnetic beads, electrolytic pumping, magnetic flushing, and threshold detection utilizing a Hall sensor array for readout analysis. The functionality of this chip was verified through *in vitro*

Table 14.2 Comparison of different CMOS biosensors used for biological sensing

Target	Transducer's surface material	Sensor type (Transducer)	Label (Y/N?)	Linker	BRE	Sample/ Medium	Application	CMOS technology (nm)	Ref.
E. coli	Al/ CMOS passivation layers	Capacitive	N	None	None	Culture medium	Bacteria growth monitoring	180	[35]
S. epidermidis	Al/Al_2O_3	Capacitive	N	None	None	Buffer	Bacteria detection	250	[33]
S. epidermidis	Al/Al_2O_3	Capacitive	N	None	None	Buffer	Detection of single bacterial cell	250	[4]
MDA-MB-231	Al/One of the CMOS passivation layers	Capacitive	N	None	None	Culture medium	Tracking breast cancer cell proliferation	500	[29]
H1299	Al/PDMS	Capacitive	N	None	None	Culture medium	Human lung carcinoma cell growth screening	350	[37]
H1299 and Hek293 cells	Al/PEM films	Capacitive	N	None	None	Culture medium	Human lung carcinoma cell culture monitoring and drug test	350	[2]
CP70 and A2780	Al/CMOS passivation layers	Capacitive	N	None	None	Culture medium	Measurements of Ovarian cancer cell proliferation	350	[3]
Neutrophil granulocyte	Al/SiO_2/ Si_3N_4 passivation layer	Capacitive	N	None	None	Buffer	NP activated neutrophils-on-a-chip	350	[1128]
Oral neutrophils	Al/Al_2O_3,	Capacitive	N	None	None	Saliva	Oral cell-on-chip for the detection of gum diseases	350	[1129]
A2780 and CP70	Al/SiO_2/ Si_3N_4	Capacitive	N	None	None	Culture medium	Monitoring of chemotherapeutic agent potency in cancer cells	350	[1130]
MCF-10A and MDA-MB-231, K-562 and THP-1	Al	Microwave-frequency dielectric spectroscopy	N	None	None	Culture medium	Single-cell spectroscopy of breast cell lines and, two leukocytes cell lines	65	[129]

(Continues)

Table 14.2 (Continued)

Target	Transducer's surface material	Sensor type (Transducer)	Label (Y/N?)	Linker	BRE	Sample/ Medium	Application	CMOS technology (nm)	Ref.
E. coli	Si_3N_4	BioFET	N	None	None	Culture medium	Bacterial metabolism monitoring	180	[470]
E. coli	Si_3N_4	BioFET	N	None	None	Culture medium	Food safety testing	65	[7]
Cosmarium turpinii cell	TiN	Impedimetric	N	None	None	NA	Measurements of sub-cellular permittivity distributions within single algae cells	180	[1131]
E. coli	Al/ CMOS passivation layers	Magnetic	N	None	None	Culture medium	Monitoring bacterial colony growth	65	[124]
G3.5 and G4 on Caco-2 cells	p+/ n-well photodiode	Optical	Y	None	None	Buffer	Determining the cytotoxicity of two types of PAMAM dendrimers on human intestinal adenocarcinoma cells	500	[1132]
NADH						Culture medium	Quantifying the levels of NADH in yeast cells		
BAOSMC						Culture medium	Assessing the cell viability and metabolic activity of BASMCs		
MCF-7	Au	Optical	N	Thiol	Anti-P-glycoprotein Ab	Buffer	Detection of breast cancer cells	350	[641]
HSA	NA	Magnetic	Y	NA	Anti-HSA Ab	NA	Immunoassay	180	[728]
TNF-α NT-proBNP	NA	Magnetic	Y	Biotin	TNF-α Ab NT-proBNP Ab	Whole blood and filtered serum	Rapid blood screening test for risk prediction	350	[1133]
CD29 antigen of mEFs cells	Al/ CMOS passivation	Magnetic	Y	Streptavidin-biotin	CD29 Ab	Buffer	Flow cytometer for detecting cancer and fibroblast cells	65	[15]

EpCAM antigen (CD326) of SKBR3 cells	layers				Anti-EpCAM (CD326) Ab				
M13KO7	Al/SiO_2	Capacitive	N	NA	Anti-M13/ HRP Ab	Buffer	Bacterial virus detection	1500	[1063]
42-base DNA	p-sub/n-well photodiode	Optical	Y	Streptavidin-biotin-PEG	DNA probe	NA	DNA hybridization assay	65	[632]
IL-6 cytokines				Silane chemistry	Human IL-6 Ab		Sandwich immunoassay		
FluA pathogen DNA	SiO_2	Optical	Y	GOPS	26-base probe DNA 1	NA	Detection of human respiratory viruses	250	[555]
FluB pathogen DNA					37-bases probe DNA 2				
RSV pathogen DNA					39-base probe DNA 3				
HPIV pathogen DNA					36-base probe DNA 4				
AdvC2 pathogen DNA					45-base probe DNA 5				
AdvE pathogen DNA					37-base probe DNA 6				
31-base DNA 7	Si_3N_4/ Poly-L-lysine	Magnetic	Y	Streptavidin-biotin	31-base capture DNA 7, 15-base probe DNA 7	Buffer	DNA hybridization sandwich assay	45	[735]
IFN-γ protein				Glu	IFN-g capture Ab		Sandwich assay for antigen detection		
90-base DNA	CMOS passivation layers	Magnetic	Y	Streptavidin-biotin	42-base probe and capture DNAs	Buffer	DNA hybridization detection	65	[738]
E. faecalis-derived DNA	NA	NMR relaxometry	Quasi label-free *	None	Probe-decorated MNPs	NA	Selective biological target pinpointing for healthcare applications	180	[774]
Human IgG, β-LG					MNPs functionalized with Protein A	NA	Protein state analysis for food applications		

(Continues)

Table 14.2 (Continued)

Target	Transducer's surface material	Sensor type (Transducer)	Label (Y/N?)	Linker	BRE	Sample/ Medium	Application	CMOS technology (nm)	Ref.
PNIPAM					None	Water	Solvent-polymer dynamics for colloidal applications		
19-base DNA	Al/Al_2O_3	BioFET	N	None	None	Buffer	Immobilization-free monitoring of DNA concentrations	1000	[467]
22-base DNA 8	$Si/SiO_2/Si_3N_4$	BioFET	N	Glu	22-base probe DNA 8	Buffer	DNA hybridization detection	NA	[463]
NDM carbapenemase	$Si/SiO_2/Si_3N_4$	BioFET	N	NA	LAMP primers	NA	Detection of the NDM carbapenemase gene in *E. coli*	350	[503]
19 base DNA 9	Extended electrode (M3)/ SiO_2	BioFET	N	Glu	19-base probe DNA 9	Buffer	Detection of HBV DNA hybridization	350	[480, 482],
The gene kelch 13 (PfK13) in *P. falciparum*	SiO_2/ Si_3N_4	BioFET	N	NA	LAMP-PfK13 primer	Buffer	*P. falciparum* malaria diagnosis and artemisinin-resistance detection	350	[442]
Prostate cancer DNA	Au	Impedimetric	N	NA	Prostate cancer ssDNA probe	Buffer	Prostate cancer DNA hybridization detection	130	[242]
BSA Herring sperm derive ssDNA (148, 309, 964, 1,389, and 2,129 bases)	Al/ Al_2O_3	Impedimetric	N	None	None	DI water	Characterization of BSA Characterization of DNA	350	[1134]
ssDNA	Al/Ni/Au	Impedimetric	N	Thiol	30 base probe DNA 10	Buffer	Detection of DNA hybridization	350	[39]
Protein-G				11-MUA	Protein-G	Buffer	Protein detection		
18-base DNA 11	Al/Cr/Au	Capacitive	N	Amine-modified PEG	18-base PNA probe 11	Buffer	Detection of DNA hybridization	350	[112]

Dry DNA mass	Al/Al_2O_3	Capacitive	N	None	None	Molecular level water	Dry DNA storage monitoring	350	[1135]
18-base DNA	Au	CV	N	Thiol	20-base poly-TC DNA probe	Buffer	DNA hybridization detection	250	[231]
5′-biotin modified 24-base DNA	Au	CV	N	Thiol	5′-thiol modified 24-base probe	Buffer	DNA hybridization detection	500	[232]
20-base DNA 12	Au	CV	N	Thiol	20-base probe DNA 12	Buffer	DNA hybridization detection	250	[233]
20-base DNA 13					20-base probe DNA 13				
18-base H5N1 derived DNA 14	Al/Cr/Au	BioFET	N	Thiol	18-base probe DNA	Buffer	AIV diagnosis	350	[443]
Glucose	Au	Capacitive	N	SAM of octanedithiol attached with melemide-NTA-Ni^{2+}	His-tagged GLK	Buffer	Glucose biosensing	CMOS compatible	[1136]
Glucose	Si_3N_4	BioFET	N	NA **	GO_x	NA	Diabetes management	250	[501]
Glucose	Pt-nanoS with MWCNTs	CA	N	NA	GO_x	Buffer	Glucose measurements	180	[198]
Lactate		CA			LO_x	Buffer	Lactate measurements		
Etoposide		CV			NA	DMSO	Anti-cancer drug Etoposide measurements		
Glucose	Pt-nanoS	CV, and CA	N	Glu	GO_x	Buffer	Detection of glucose and other metabolites	350	[237]
Neurotransmitter DA	Al/ SiO_2	Capacitive	N	APTES	CPBA	NA	Detection of neurotransmitter DA	350	[27, 32]
Norepinephrine dissolved oxygen	Pt	Amperometric	N	NA	None	NA	Multiple target detection	600	[230]
Norepinephrine and dissolved uric acid		CV							
Neurotransmitter DA	Au/ Nafion film	Voltammetric	N	NA	NA	Buffer	Detection of neurotransmitter DA	350	[226]

(Continues)

Table 14.2 (Continued)

Target	Transducer's surface material	Sensor type (Transducer)	Label (Y/N?)	Linker	BRE	Sample/ Medium	Application	CMOS technology (nm)	Ref.
DA	Au	Amperometry	N	NA	NA	Buffer	Parallel recording of neurotransmitters release from chromaffin cells	NA	[193]
Lymnaea stagnalis neuron	Ti/Pt/ Poly-l-lysine	OSFET-MEA (Extracellular voltage recording)	N	None	None	Culture medium	Extra-cellular recording of neuralactivity	500	[258]
Cardiac cells, mESC Purkinje neurons	Pt	Impedimetric + EAP recording	N	None	None	Buffer An acute cerebellar slice of a wild-type C57b/6 mouse	Impedance spectroscopy and electrophysiological imaging of cells	180	[44]
DA release from SH-SY5Y cells	Au	Amperometry	N	None	None	Culture medium	Investigating Parkinson's disease	350	[224]
K^+	$Al/SiO_2/Si_3N_4/$ PVC-based potassium sensitive membrane	BioFET	N	None	λcI857 and T_6 phage	Buffer	*E. coli* diagnosis	180	[416]

NA: Not available, Y: Yes, N: No, Ab: antibody, GOPS: (3-glycidoxypropyl) trimethoxysilane, GM-CSF: Granulocyte-macrophage colony-stimulating factor, LO_x: Lactate oxidase, BioFET: Biological field-effect transistor, *: It offers label-free-like characteristics while still employing MNPs as the detection element, ** Covalent binding, 26-base DNA probe DNA 1: GAC CTA GTT CTC GCC AGT GGA GGT CC, 37-base probe DNA 2: CAC CGC AGT TTC AGC TGC TCG AAT TGG GAG TTG AGG A, 39-base probe DNA 3: CTG TGT ATG TGG AGC CTT CGT GAA GCT TCC ATA TTT GCC , 36-base probe DNA 4: GCG ATT GAT TCC ATC ACT TAG GTA AAT TGG AAA TGC, 45-base probe DNA 5: GCT CCA CAT AAT CTA ACA CAA ACT CCT CAC CCC AGG TAA GAT CGA , 37-base probe DNA 6: TTT AAT CAT GGT TCC TGT TCT TCC CTC CCA AAT TGC A, 31-base DNA 7: 5′-ATC CAA TCA ACC CAA CAA TAT TGA TAA GGA T-3′, 31-base capture DNA 7: 5′-TTT TTC TGG TTG GGT TGA TTG GAT TTA GCT TGG C-3′, 15-base probe DNA 7: 5′-Biotin-ATC CTT ATC AAT ATT-3′ , 22-base DNA 8: 5′-GAT CAG ACA CTT CAA GGT CTA G-3′, 22-base probe DNA 8: 5′-CTA GAC CTT GAA GTG TCT GAT C-3′ , 19 base DNA 9: 5′-GTT CCG CAG TAT GGA TCG G-3′, 19-base probe DNA 9: 5′-CCG ATC CAT ACT GCG GAA C-3′, 30 base probe DNA 10: 5′-TGA TAG CCC TGT ACA ATG CTG CTA AAA AAA-Thiol-3′ , 18-base DNA 11: 5′- TGA GCA GAA TAA ACC ATT-3′, 18-base PNA probe 11: NH_2-O-AAT GGT TTA TTC TGC TCA, 20-base DNA 12: T1 5′-FcCAC TTG CAG AAT TTA AAA-3′, 20-base probe DNA 12: P1 5′-TTT TAA ATT CTG CAA GTG AT-thiol-3′, 20-base DNA 13: T2 5′-FcAAA AAG GAA GGA AAA AAA-3′, 20-base probe DNA 13: P2 5′-TTT TTT TCC TTC CTT TTT TT-thiol-3′, 18-base H5N1 derived DNA 14: CCA AGC AAC AGA CTC AAA

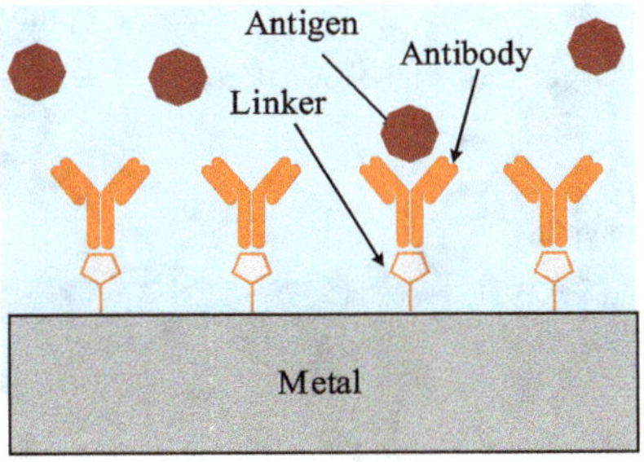

Figure 14.4 An immunosensor, demonstrating the interaction between antibodies and antigens

experiments targeting tumor necrosis factor-α (TNF-α) and N-terminal pro-brain natriuretic peptide (NT-proBNP) concentrations, which serve as indicators of anti-tumor treatment response and heart failure, respectively. Sensitivity assessments were performed using a wide range of TNF-α and NT-proBNP solutions, spanning from 0.8 pg/mL to 800 ng/mL. The linear dynamic range for biomarker concentration detection was established up to 10 ng/mL. When the biomarker concentration exceeded this range, the sensors detected saturated outcomes. To expand the linear dynamic range for higher biomarker concentrations, modifications such as reducing the anodic aluminum oxide filtration period, optimizing the first and second conjugation times, and enhancing the comparator were implemented to meet the requirements of high-concentration detection.

In [15], the functionality of a CMOS microflow cytometer has been verified for detecting magnetic labels in diverse cell types, including mouse embryonic fibroblast cells (mEFs) and cancer cells (SKBR3), employing specific implementations utilizing antibodies as BREs. The magnetizing coil is created using a three-layer structure, with each layer forming a single-turn spiral inductor. These inductors are fabricated on the top three metal layers of the CMOS process. Figure 14.5(a) displays an image where an mEFs, measuring between 10 and 15 μm in diameter, is marked with biotinylated anti-mouse/rat CD29 antibody linked to streptavidin-coated superparamagnetic Dynabeads, with a diameter of 4.5 μm [21]. To showcase the ability to distinguish between multiple magnetic labels, Dynabeads containing iron oxide NPs and silica beads containing cobalt NPs are utilized, each having a diameter of 8 to 10 μm, and their phase is measured. The scatter plot of the phase values obtained for both label types provides clear differentiation based on material [21]. Furthermore, in [15], SKBR3 cancer cells, known for their EpCAM antigen overexpression and measuring approximately 15 μm in diameter, are labeled using a two-step process (Figure 14.5(b)). First, the EpCAM antigen on the cell surface is linked with a biotinylated anti-EpCAM (CD326) antibody, and second, the biotin is coupled with magnetic labels coated in streptavidin. The flow cytometer is capable of detecting three label classes, namely Fe_3O_4, $MnFe_2O_4$, and $CoFe_2O_4$, as demonstrated in Figure 14.5(b), employing phase data at frequencies of 0.9 GHz and 2.6 GHz to accurately classify the labels [15].

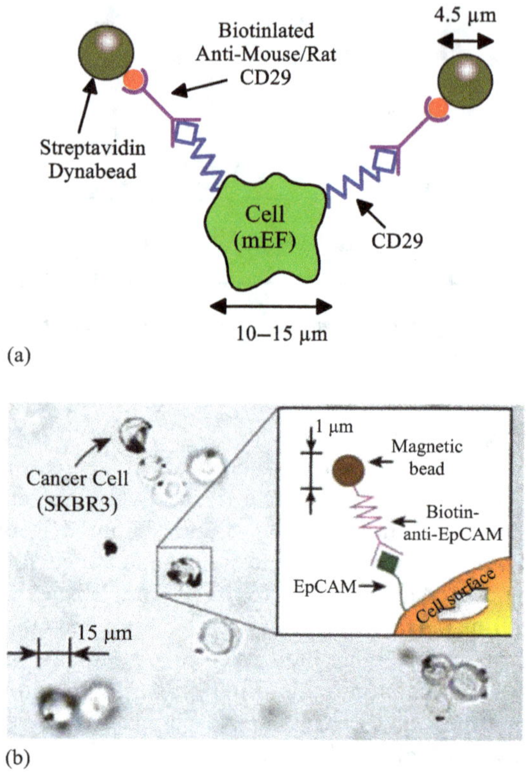

Figure 14.5 Magnetic labeling for cells: (a) mEf cells and (b) SKBR3 cells [15]

Musayev *et al.* [641] employed antibodies on a CMOS image sensor featuring a 32 × 32-pixel array, purpose-built for cell capture, detection, and quantification. The sensor incorporated a specialized light mask on its pixels, allowing it to detect cells smaller than the pixel size. Cells as small as 3 μm in diameter could be successfully detected using this pixel structure. To demonstrate the proof of concept, MCF-7 breast cancer cells were captured and imaged on the modified sensor surface. The gold-coated CMOS sensor pixels were activated using thiol-modified antibodies capable of capturing MCF-7 cells. Optical detection of the cells was performed under green light-emitting diode (LED) illumination. The necessary materials for the experiment included the cross-linker sulfosuccinimidyl 6-[3′(2-pyridyldithio)-propionamido] hexanoate, TCEP·HCl, the anti-P-glycoprotein antibody, and PBS. The activation protocol involved incubation of the sensor surface with the antibody-cross-linker solution, followed by rinsing and drying. MCF-7 cells suspended in 1× PBS were then added to the sensor surface and bound to the antibodies. The sensor images were obtained as the liquid on the surface neared complete evaporation, and a threshold was applied to extract cell position data from the acquired images. The cell detection performance was evaluated visually based on the outcomes of the image processing results.

In another effort, Balasubramanian *et al.* [1063] outlined the creation of a CMOS capacitive sensor for virus detection. The attachment of antibodies to Si-based materials was demonstrated using an enzyme-linked immunosorbent assay (ELISA), which involved the immobilization of anti-M13KO7 antibodies on a silicon wafer as a capture antibody. The capture antibody-coated surface was exposed to M13KO7 bacterial virus in solution. Detection of captured M13KO7 was performed using horseradish peroxidase (HRP)-labeled anti-M13 antibodies. The experimental setup involved diluting the capture antibody in PBS and spotting it onto the silicon wafer surfaces. Incubation, washing, and blocking steps were performed to prevent nonspecific binding. One surface was exposed to diluted M13KO7 phage, while the other surface remained in PBS-T as a control. After washing, both surfaces were incubated with HRP-labeled anti-M13 antibodies, washed again, and subjected to enhanced chemiluminescence substrate to visualize the presence of anti-M13/HRP bound to M13KO7. The results showed a signal intensity dependent on the concentration of capture anti-M13 antibodies, confirming antibody attachment to the Si surface. Additionally, the attachment and detection of viruses were demonstrated using a sub-nanoliter drop of anti-M13 deposited onto a sense capacitor element. After processing and washing, the chip was exposed to an M13K07 virus solution. Fluorescent detection using anti-M13/Fluorescein was also performed, and the presence of viruses was visualized on the chip. Output voltages of the detection circuit before and after exposure to the virus were measured, showing a change in voltage indicating successful virus detection.

14.3.2 Nucleic acid-based assays

Nucleic acids serve as both target molecules and BREs within biosensors. For applications like DNA hybridization detection and sequencing, immobilizing probes or BREs on the sensor surface is necessary. However, in situations like DNA concentration or mass detection, probe immobilization is not required.

14.3.2.1 Immobilization-based applications

Within genosensors [38,63], the immobilization of nucleic acids occurs on electrode surfaces, utilizing either the BRE or RNA/DNA probe (refer to Figure 14.6). The aim is to facilitate the detection of RNA/DNA through the binding of target

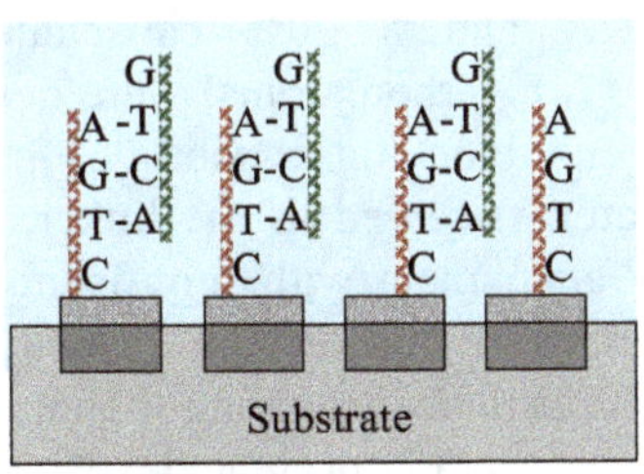

Figure 14.6 The DNA detection method using hybridization-based techniques

RNA or ssDNA molecules with their complementary RNA or ssDNA counterparts within the recognition layer. Consequently, this interaction impacts the measured capacitance signal, thereby enabling the purposeful detection of RNA/DNA.

In a study, Manickam *et al.* [39] developed a fully integrated electrochemical impedance spectroscopy (EIS) IC-based system with gold plating as the only additional postprocessing step. In their study, they directly immobilized thiolated ssDNA molecules onto the Au electrode surface as capture probes. The specific probe sequence used was 5′-TGATAGCCCTGTACAATGCTGCTAAAAAAA-Thiol-3′, with an approximate probe concentration of 7×10^6 per electrode. The admittance spectrum of the immobilized probes was measured in 40 μL of 4xSSC (saline sodium citrate) buffer. Subsequently, a complementary ssDNA strand was added to saturate the electrode surfaces, and the system's admittance spectra were measured at equilibrium. The frequency range chosen emphasized the interfacial impedance's contribution to the response, as the bulk resistance of the solution dominated at higher frequencies. The system's limit of detection (LoD) was approximately 10^5 molecules (6.25×10^9 molecules/cm^2), and preliminary measurements demonstrated label-free detection of DNA hybridization and protein attachment. In a work by the same group [132], various design and optimization aspects of integrated EIS systems were investigated, addressing challenges such as limited available materials and surface options, semiconductor-compatible biofunctionalization, and hybrid fluidic-electronic packaging and handling. Immobilization of biomolecules on gold surfaces involved the direct attachment of thiol compounds, such as thiolated DNA strands, via sulfhydryl (SH) group. Protein molecules could attach directly through hydrophobic and electrostatic interactions. Linker molecules like functional alkanethiols (e.g., 11-MUA, 3-MPA, thioctic acid) improved stability and surface coverage by forming SAM on the gold surface. N-Hydroxysuccinimide (NHS) esters were formed by modifying COOH groups in the linker molecules, facilitating strong covalent binding with amine groups in proteins. EIS experiments on gold surfaces with different thiol-based monolayers revealed that thicker monolayers exhibited increased surface coverage, higher leakage resistance, and impedance closer to an ideal capacitor. Thicker monolayers also effectively block non-specific molecules. However, longer monolayers (>10 carbon atoms) showed lower transduction gain and smaller capacitance changes upon molecular binding. Impedance measurements before and after protein attachment on 11-MUA and 3-MPA monolayers indicated minimal surface impedance change for 11-MUA and a 20% capacitance change for 3-MPA. Intermediate-length linkers (7–8 carbon atoms) were deemed important for balancing sensitivity, stability, and blocking capability. In antigen-antibody assays, thioctic acid (7–8 carbon atoms) served as the linker, enabling the detection of multiple levels of antibody attachment with significant capacitance changes. EIS also facilitated quality control by identifying defects or holes in monolayers through changes in surface impedance, ensuring monolayer integrity.

In another effort, Lee *et al.* [112] presented a label-free and fully electronic method for DNA hybridization detection using a 16×8 microarray sensor and CMOS technology-based impedance spectroscopy. The technique is dependent on gauging alterations in responsive capacitance and charge-transfer resistance when interacting with complementary DNA targets during hybridization. The authors introduce an impedance extraction method employing triangular wave voltage excitation, enabling reliable measurement of both capacitance (C) and resistance (R) and ensuring stable sensor operation regardless of external factors. The detection process involves four conditions: (1) bare gold electrode without any treatments, (2) electrode immobilized with probe peptide nucleic acid (PNA), (3) electrode hybridized with complementary DNA strands, and (4) electrode bound with non-complementary DNA strands as a negative control. The output signal of the sensor is measured under each condition by immersing the electrode surface in a buffer solution. To ensure accurate and reliable measurements, a constant triangular waveform voltage is applied during the measurement, while other bias conditions remain constant to avoid ion scattering and impedance changes. The measured waveform of the sensor chip demonstrates decreased electrode currents upon hybridization compared to the bare gold electrode. The low-frequency triangular waveform voltage, with an amplitude of 400 mV and a frequency of 350 Hz, effectively captures conformation changes in DNA resulting from hybridization. Sensor outputs of capacitance and resistance show significant changes upon hybridization with complementary samples compared to non-complementary samples. The sensor has an LoD of 10 nM, and both capacitance and resistance change linearly with the target concentration. The maximum relative variation in capacitance is 31.5% with 1 M complementary target DNA, while the resistance variation is 68.6%. Both capacitance and resistance serve as indicators of DNA hybridization, allowing the prediction of solution material properties, insulating layer characteristics, and the physical and chemical features of the immobilized bio-layer on the surface.

CMOS capacitive sensors have also been reported for DNA analyses. For instance, Stagni *et al.* [38] developed a CMOS capacitive biosensor that utilizes a nucleic acid-based approach for DNA detection. The biosensor integrates Au electrodes that are functionalized with DNA probes. Another study by the same group [156] introduced another CMOS DNA chip for fully electronic, label-free DNA recognition through capacitance measurements. The immobilization process involves cleaning the gold electrodes using plasma oxygen exposure and subsequently immobilizing alkanethiol-modified ssDNA molecules onto the electrodes through covalent S-Au bonds. The chip is electrically characterized and tested using various DNA samples, successfully detecting significant capacitance variations resulting from DNA hybridization. The gold surfaces are rinsed extensively with ultra-pure water to remove non-covalently bound molecules. The target DNA solution is then applied to the electrodes, heated, and subsequently rinsed to remove unbound DNA. The detection of DNA is demonstrated by comparing measurements on electrode couples with different DNA strands bound to the surface, one complementary to the target and the other non-complementary as a negative control. All measurements are performed in the same saline solution used during the hybridization step. After functionalization, capacitance measurements are performed, and the values serve as a reference for comparison with the results obtained after hybridization. The absolute capacitance values after functionalization show a decrease of approximately 25% compared to clean gold for each electrode couple. Results from the experiments demonstrate significant capacitance decreases resulting from hybridization, while smaller or opposite variations are observed in the case of non-complementary binding. Following the measurements, the sensing layers are regenerated by rinsing with hot ultrapure water to break the hydrogen bonds between the two DNA strands, leaving only the probe oligonucleotide chains covalently bound to the surface. New capacitive measurements are then conducted. As anticipated, an increase in interface capacitance is observed at sites where hybridization occurred, while electrodes with non-complementary probes exhibit negligible variations. It is noteworthy that the absolute capacitance variations in dehybridized chips are equivalent to those observed during hybridization.

In their work, Musayev *et al.* [1025] presented a label-free DNA detection approach utilizing a CMOS charge-sensitive microarray sensor comprising 1,024 detector elements. The sensor exhibited low noise levels, enabling the detection of DNA at a concentration as low as 1 pM. This sensitivity surpassed that of similar chips and even rivaled conventional fluorescence or gravimetric detection techniques. The experiments involved testing 10 base pairs of DNA solutions containing 13 μM probe (5′-TCTCACCTTC-3′) and 1 pM target (3′-AGAGTGGAAG-5′) oligomers. The charges associated with the DNA molecules interacting with the sensor surface were effectively detected.

In the realm of DNA analyses, voltammetric sensors have emerged as a powerful class of biosensors, offering valuable insights into DNA detection and hybridization processes. One remarkable example of voltammetric DNA analysis is the work of Kruppa *et al.* [234], who introduced a CMOS DNA chip that employs chronocoulometric measurement for DNA analysis. The electrode surface is covered with a mixed film consisting of OH-terminated alkanethiol-SAM and thiolated DNA probes hybridized with signaling oligonucleotides. In another work, Levine *et al.* [231] developed a 4 × 4 active sensor array in CMOS for electrochemical biomolecular detection. They successfully operated the array for biomolecular detection, including CV measurements of a reversible redox species, characterization of DNA probe density, and specific DNA hybridization detection. They performed CV measurements using potassium ferricyanide as a redox species, achieving matching results to a commercial potentiostat. The reduction peak current correlated with the electrode area. The team determined DNA probe surface density by functionalizing the gold electrodes with ssDNA probes and conducting CV measurements with redox species. The CV curves exhibited different shapes based on probe coverage. They used the active CMOS sensor array to detect DNA probe-target hybridization, measuring charge transfer to quantify bound target DNA. The experiments successfully detected DNA hybridization and distinguished between fully complementary targets and targets with single-base mismatches.

Furthermore, Schienle *et al.* [232] presented a chip featuring a 16 × 8 sensor array for fully electronic DNA detection based on an electrochemical redox cycling process. The chip was fabricated using an extended CMOS process. To demonstrate its biological functionality, DNA probe sequences were functionalized on two rows of the sensor array using a contact printer. Two different probe sequences were immobilized on specific positions of the array. Immobilization was carried out at room temperature for 60 min, followed by surface washing and blocking with BSA to minimize background signal from non-specific bindings. Hybridization was performed with a biotin-modified target sequence that is complementary to one of the immobilized probes. Hybridization took place at room temperature for 30 min, followed by another washing step. An enzyme complex (alkaline phosphatase conjugate) was added, which binds to the biotin-binding site of the hybridized target molecules. After the final washing step, the chip was placed into a measurement flow cell that covered a portion of the sensor field. The chip was flooded with para-aminophenylphosphate, and the sensor signal was measured after the flow was stopped. The electrical response of the sensor array exhibited good homogeneity, with a 3σ-homogeneity better than 6% (ranging from 10^{-11} A to 10^{-7} A) and better than 20% (ranging from 10^{-12} A to 10^{-7} A).

In a study, Maruyama *et al.* [463,464] presented a CMOS DNA image sensor consisting of a 32 × 32 array (1,024 pixels) that utilized the charge transfer technique. Originally designed as a pH sensor, the sensor exhibited sensitivity comparable to FET-type DNA sensors. However, by implementing the charge accumulation method, also known as the charge transfer procedure, the sensor achieved enhanced sensitivity and signal-to-noise ratio (SNR). To immobilize the probe DNA, the sensor surface was silanized using 3-aminopropyltriethoxysilane, introducing an amino group. Subsequently, the amino-silanized sensor surface and amino-linked probe DNA were connected through Glu crosslinking. The probe DNA had a base sequence of 5′-CTAGACCTTGAAGTGTCTGATC-3′ (22 bases), with an amino base attached at the 3′-end. The target DNA sequences for full mismatch and full match were 5′-CTAGACCTTGAAGTGTCTGATC-3′ (22 bases) and 5′-GATCAGACACTTCAAGGTCTAG-3′ (22 bases), respectively. The immobilization and hybridization processes involved several steps: (1) soaking the CMOS DNA image sensor in 1 M NaOH solution, rinsing it in deionized water, and drying it in a vacuum chamber; (2) dipping the sensor in a 2 wt% 3-aminopropyltriethoxysilane solution to silanize the surface of the Si_3N_4 layer, followed by rinsing and drying; (3) dipping the sensor in a 25 wt% Glu solution containing 10 g/L of sodium cyanoborohydride, followed by washing with deionized water and drying in a vacuum chamber; (4) immobilizing the probe DNA by dropping a 25 M DNA solution containing 10 g/L of sodium cyanoborohydride onto the sensor and storing it in a chamber at 50 °C for 15 h; (5) rinsing and drying the sensor, then dipping it in a 1 M glycine solution to block non-DNA immobilized surfaces, followed by rinsing in PBS and drying in a vacuum chamber; (6) hybridizing the target DNA by dropping a 25 M DNA solution diluted with SSC buffer containing 0.1 wt% SDS onto the same spot as the probe DNA for 15 h at room temperature, followed by washing with diluted SSC buffer and drying in a vacuum chamber. After immobilization and hybridization, significant output voltages of 76.4 ± 16.5 mV and 64.5 ± 15.7 mV were measured for DNA molecules containing 22 bases. Based on these results, the densities of immobilized and hybridized DNA were estimated to be $6.3 \pm 1.4 \times 10^8$ cm^{-2} and $5.3 \pm 1.3 \times 10^8$ cm^{-2}, respectively.

The DNA LoD was calculated to be approximately 2.7×10^7 cm^{-2} molecules (22 bases).

The arrangement of a CMOS biosensor for DNA analysis described in reference [553] involves placing the capture probes on the chip's surface and establishing an optical connection between them and the CMOS detector using a fiber-optic face-plate (FOF). This integration enables the seamless incorporation of the biochemical components of the assay with the fluorescent detector. Figure 14.7 presents a cross-sectional view of the biosensor, illustrating the integration of an FOF to the top surface of the biosensor chip through post-fabrication procedures. The FOF, with a thickness of 3 mm, acts as a thermal barrier between the microarray assay (operating at temperatures of 40 °C to 60 °C) and the CMOS chip. It also maintains a sufficient distance between the solution and the chip while minimizing signal loss. The exposed surface of the FOF is composed of polished glass (SiO_2), providing an optimal platform for attaching DNA-capture probes using standard aldehyde-modified surfaces [1137]. The photodiode dimensions are 50 μm × 50 μm, with a pitch of 250 μm × 250 μm between each array, ensuring compatibility with microarray probe spotting instruments [1138].

In Reference [555], the DNA capture probes are immobilized on top of a SiO_2 filter surface. Typically, the direct attachment method (spotting) [553,611,645,

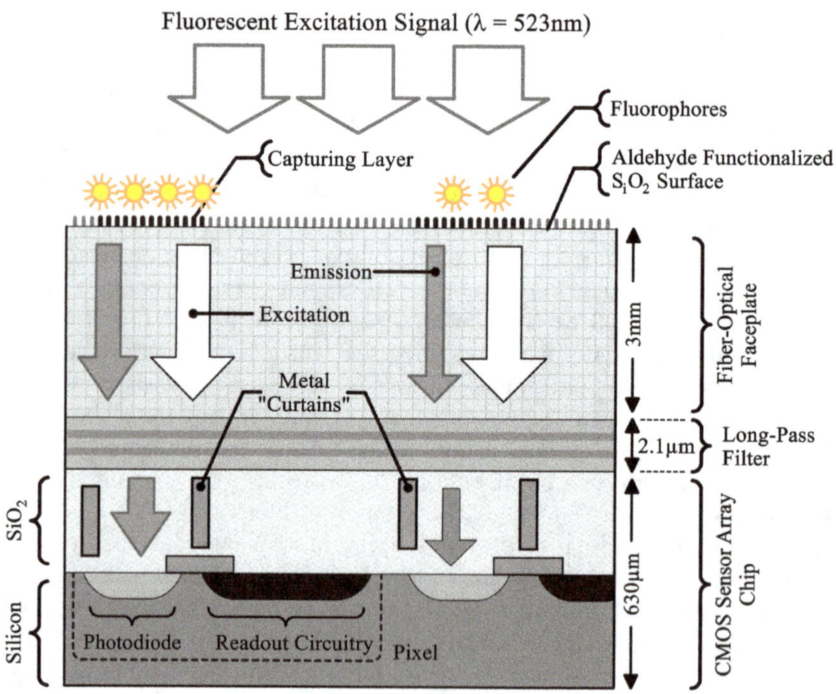

Figure 14.7 A cross-sectional view of the microarray biosensor based on CMOS technology, utilizing fluorescence as the detection method [553]

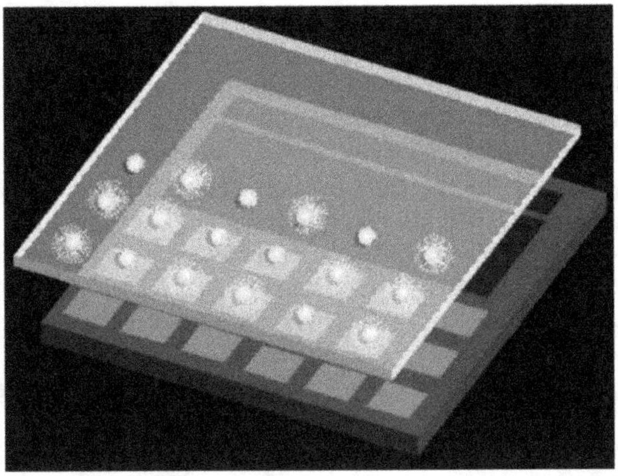

Figure 14.8 The detection SoC coupled to a fiber-optic faceplate featuring luminescent assays [586]

1139,1140] is widely employed for functionalizing the probe in each array. This involves depositing small amounts of sample probes onto a surface using a microarray spotter element and chemically attaching them to the surface through linker molecules. In contrast, in Reference [581], the probes are directly immobilized on the chip surface, eliminating losses associated with large, complex optics and enabling the efficient collection of fluorescent molecules placed directly on the photodiode. Another example is the CMOS bioluminescence detection LoC biosensor depicted in Figure 14.8, where a simple FOF is utilized. In this design, luminescent reporters/probes are immobilized on the FOF and directly coupled to the CMOS chip [586]. The photodiode size in this configuration is 150 μm × 150 μm, ensuring compatibility with existing microarray spotting equipment, microfabricated microfluidic systems for reagent delivery and recycling, as well as manual spotting.

In the realm of healthcare, sensing, imaging, and diagnostics, optical spectrometry in the visible and near-infrared (near-IR) range finds extensive utility. A pioneering investigation conducted by Hong *et al.* [632] introduced the initial completely integrated optical spectrometer using a conventional bulk CMOS process, eliminating the necessity for bespoke manufacturing, subsequent treatments, or external passive optical components such as lenses, gratings, collimators, or mirrors. This groundbreaking work showcased the potential of metal-optic nanostructures to downsize intricate optical instrumentation, leading to the emergence of optics-free CMOS-based systems-on-chip for diverse sensing and imaging applications in the visible and near-IR domains. To exemplify the spectrometer's applicability in biosensing scenarios, the researchers focused on fluorescence spectrometry. Specifically, they employed quantum dots (Qdots) as fluorescent tags due to their exceptional properties such as photo-stability, brightness, and large Stokes shift, surpassing traditional organic fluorescent dyes. Moreover, the relatively small size of Qdots eliminates significant steric effects in assay experiments, making them widely accepted as biological labels for various applications, including nucleic acid and protein detections. The CMOS spectrometer chip developed by Hong *et al.* successfully measured two types of Qdots with distinct wavelengths (approximately 705 nm and 800 nm) and compared them with their reference spectra obtained using bench-top spectrometers. The researchers initially employed the Qdots as fluorescence tags in sandwich immunoassays and DNA hybridization assays. Specifically, they conducted assays targeting a 42-base target DNA and IL-6 cytokines, which play a crucial role in regulating biological processes and auto-immune responses in numerous diseases. Biotinylated capture DNA and human IL-6 antibody probes, along with biotin-PEG linkers for nucleic acid assays and silane chemistry for immunoassays, were utilized. Streptavidin-conjugated Qdots served as the fluorescence labels, and the assays were tested with varying target concentrations ranging from 1–100 pM to 5–125 pM for DNA hybridization and IL-6 cytokines, respectively. The intensity exhibited a linear correlation with a concentration in both cases, confirming the efficacy of Qdots as tags for affinity-based assays. To spectrally characterize the fluorescence emission of the two types of Qdot tags, a fiber fan-out cable was employed to collect the emitted light. One branch of the fiber was positioned vertically, and the Qdots were dropped onto it using a pipette in small volumes (0.2 μL per drop). Excitation of both Qdot types was achieved using a diode laser operating at a wavelength of 405 nm. The fluorescence spectra measured by the CMOS spectrometer chip for the 705 nm and 800 nm Qdots were then compared to their reference spectra obtained using a bench-top spectrometer. The results demonstrated a reasonably close agreement, even accurately predicting the shift in the fluorescence emission peak. Enhanced optical shielding and packaging could further enhance the resolution by minimizing external variable scattering.

Advancements in CMOS technology have also unlocked the possibility of the development of magnetic biosensors tailored for DNA analyses. One such study [730] presents an experimental arrangement and provides the DNA strand sequences (Figure 14.9(a)). To conduct a sandwich assay for DNA measurement,

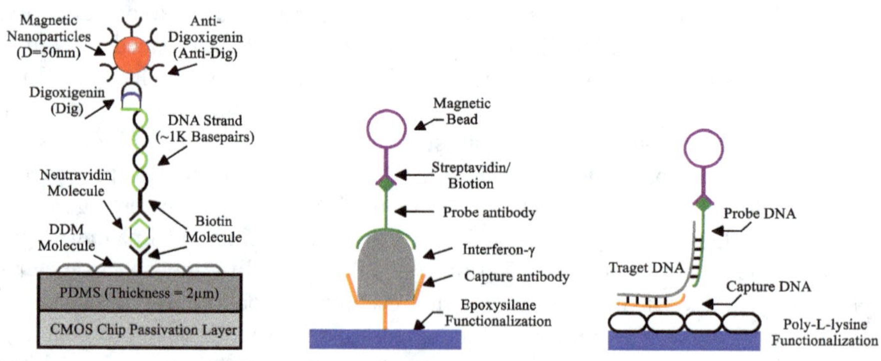

Figure 14.9 Sandwich assay. (a) DNA sample, (b) IFN-g protein, and (c) DNA oligomer

biotin-labeled DNA probes are initially immobilized on the biotin-modified PDMS bottom surface using neutravidin molecules, while β-D-dodecyl-N-maltoside (DDM) molecules prevent nonspecific binding between neutravidin and PDMS surface [1141]. Subsequently, in the presence of digoxigenin (dig)-modified target complementary DNA strands, antidig-labeled magnetic nanoparticles (MNPs) are captured on the sensor surface through the dig-antidig link. Finally, a sensor frequency-shift response ($\Delta f/f_0$) is achieved for 1 nM DNA samples (1 k base pairs in length) [730].

In a different configuration presented in [735], a sandwich assay is employed for the detection of both interferon-g (IFN-g) protein and DNA oligomers. Figure 14.9(b) illustrates the surface functionalization process for IFN-g, involving the covalent attachment of the capture antibody through an epoxy silane group. The probe antibody binds to a distinct epitope of IFN-g, while the magnetic bead is linked to the probe antibody via a streptavidin/biotin interaction. Additionally, Figure 14.9(c) demonstrates the functionalization of magnetic beads with the desired DNA molecules. A capture DNA sequence, which complements a distinct section of the oligostrand, adheres to the poly-L-lysine surface through electrostatic attraction. The probe DNA sequence, corresponding to another segment of the target strand, connects to the bead via a streptavidin/biotin linkage. Furthermore, in [738], streptavidin-biotin chemistry is employed to attach the DNA target to 1-μm Dynabeads [1142].

In their study, Lee *et al.* [443] presented a CMOS ISFET biosensor designed for the detection of DNA hybridization. The sensor successfully demonstrated specific detection of oligonucleotide sequences derived from the H5N1 avian influenza virus (AIV). The experiment utilized a specific oligonucleotide derived from H5N1, along with thiol-functionalized probes that were modified with a six-carbon atom spacer and were complementary to the H5N1 oligonucleotide sequence. A non-specific target sequence was also used as a negative control to verify the hybridization between the probe and complementary target strands. Before conducting the DNA experiment, the sensor electrodes underwent a thorough cleaning process using ACE, MeOH, and distilled water. The electrode surfaces were then activated through O_2 plasma treatment. The chip's surface was incubated with a layer of 10 M DNA probes for 6 h at room temperature to immobilize the probes. Subsequently, the chip's surface was immersed in a solution of 10 mM mercaptohexanol in EtOH for 5 min at room temperature to passivate the unreacted electrode surface and reduce non-specific binding. The hybridization of target DNA strands with the probes was carried out in a 5× SSC buffer containing 0.1% sodium dodecyl sulfate (SDS) to minimize electrostatic repulsion between the probe strands. Impedance measurements were conducted in 0.1 mM PBS (pH 7.4) at room temperature, utilizing low ionic strength to ensure the occurrence of biomolecular reactions within the Debye length. The experiments yielded a LoD of 100 pM and included a test for single-base mismatches in 18-mer oligonucleotides.

The label-free detection of biomolecules using ISFETs in physiological solutions faces challenges due to the Debye charge-screening effect of the electrical double layer. To overcome this limitation, the CMOS ISFETs reported in [480] incorporate a capacitor structure in the oscillator readout, allowing modulation of the surface electric field to reduce charge screening and nonideal signal drift. Frequency modulation of signals is employed at frequencies exceeding the cut-off frequency influenced by ion strength, with the highest utilized frequency reaching around 100 MHz among the four variations. The measured results demonstrate a pH sensitivity ranging from 52.3 to 54.4 mV/pH in terms of the potential change of the floating gate. The study successfully performs direct DNA hybridization in a high ionic-strength solution with near picomolar-level resolution. Hepatitis B virus (HBV) DNA strands are employed to test DNA hybridization, with specific probes and target sequences. The CMOS chips undergo thorough cleaning with EtOH and ACE solutions to eliminate contaminants, followed by treatment in an ultraviolet/ozone dry stripper to introduce hydroxyl groups (–OH). Subsequently, the sensor surface is immersed in a 2% solution of 3-aminopropyltriethoxysilane for 30 min to introduce amino groups ($-NH_2$). The next step involves producing aldehyde groups (–CHO) by immersing the sensors in 10 mM PBS containing 2.5% Glu for 2 h. The 5′-aminomodified HBV probe DNAs (1 μM) are immobilized on the sensor surface by reacting with the aldehyde groups over 20 h. Unreacted aldehyde groups are blocked by adding 50 mM ethanolamine mixed in 1× PBS for 1 h. The immobilization and hybridization processes are verified on a CMOS chip using a fluorescence-tagged probe and target DNA strands. The study measures DNA hybridization by increasing the target DNA concentration from 1 fM to 10 μM, with the target DNA solution heated to 44 °C before being added for hybridization for 20 min. Finally, the chips are rinsed with deionized water to remove unbound DNA strands before measurement in 1× PBS.

In their investigation, Chang *et al.* [482] conducted a study utilizing CMOS ISFETs with three different designs featuring varying electrode and transistor sizes. The pH sensitivities were determined to range between 27 and 32 mV/pH, based on the shift in threshold voltage. The detection of DNA hybridization was evaluated using HBV DNA strands with specific probes and target sequences. The study successfully detected the increase in current in the p-type ISFETs caused by the hybridization of target DNA molecules, achieving detection at near femtomolar concentrations.

In a separate study, Malpartida-Cardenas *et al.* [442] developed an in-house LoC platform using BioFETs for specific, quantitative, and fully electronic detection of *Plasmodium falciparum*, the major malaria-causing parasite globally. The platform also enabled on-chip detection of the C580Y single-nucleotide polymorphism (SNP) associated with artemisinin-resistant malaria. The detection was achieved using a loop-mediated isothermal amplification (LAMP) primer set, specifically designed for the Kelch 13 gene of *P. falciparum*. The primer set was designed based on consensus reference genomic sequences obtained from various *Plasmodium* species and optimized for specificity. The amplification process involved the use of an isothermal buffer, deoxynucleotide triphosphates (dNTPs), BSA, Betaine, DNA polymerase, synthetic DNA or genomic DNA (gDNA) samples, and the LAMP primer mixture. The reactions were performed at 63 °C for 35 min, followed by a melting cycle to validate specificity. The experiments were conducted using a polymerase chain reaction (PCR) system. To ensure compatibility with ISFET sensors, the LAMP reaction mix was modified to pH-LAMP, including NaOH and a customized buffer for pH measurements. Analytical sensitivity evaluations were performed using synthetic DNA and gDNA samples, and cross-reactivity tests were conducted using clinical isolates. SNP discrimination was achieved through modified SNP-based LAMP (sbLAMP) primers. Gel electrophoresis was employed for visualizing LAMP products. Statistical analysis was carried out, including the calculation of correlation coefficients. The study also discussed the advantages and challenges associated with ISFET-based sensing, highlighting the pH dependence observed in ISFETs and compensation methods for sensor offset and drift.

In another work, Miscourides *et al.* [503] introduced a large-scale CMOS array consisting of 12.8 k BioFET sensors arranged in a 2T per pixel topology. The array was designed for real-time ion imaging and detection of DNA amplification. Genomic DNA was extracted from carbapenem-resistant *Escherichia coli* strains carrying the New Delhi metallo-β-lactamase (NDM) gene. The purified DNA samples were either immediately analyzed or stored at −80 °C for future experiments. Specific LAMP primers were utilized to confirm the presence of the target NDM gene in the samples. Real-time LAMP experiments, including negative and positive controls with 10^5 genomic copies per reaction, were conducted in parallel using both a conventional real-time instrument and the on-chip setup. The LAMP assay was optimized for pH-based detection on the chip. A 20-μL solution was loaded into either 0.2 mL PCR tubes or the microfluidic chamber on the chip and heated at 63 °C for 20 min. The system achieved a sensitivity of 1.03 μA/pH with a pH resolution of 0.101 pH units, making it suitable for the real-time detection of the NDM carbapenemase gene in *E. coli* using LAMP.

A fully integrated CMOS fluorescence biochip for DNA/RNA testing in molecular diagnostics was presented by Manickam *et al.* [555]. They successfully utilized the array and its thermal cycling capability to adopt a 7-plex panel for the detection of six human upper respiratory viruses, including Influenza A (FluA), Influenza B (FluB), Respiratory syncytial virus (RSV), Human parainfluenza virus (HPIV), Human adenovirus C (AdvC2), and Human adenovirus E (AdvE). This board utilizes a nucleic acid amplification testing (NAAT) approach that incorporates PCR. The workflow of the chip involves a reverse transcription PCR enzymatic process in the initial phase to amplify specific DNA/RNA sequences unique to pathogens in the sample. This amplification is achieved through thermal cycling steps and the addition of primers. An inverse fluorescence detection assay was employed to enable the detection of amplicons. In this approach, PCR primers labeled with quencher molecules are incorporated into the amplicons. Fluorescent labeled DNA capture probes are placed at fixed coordinates on the array, and during the capturing step, the generated amplicons hybridize with the corresponding probes, leading to a temporary reduction in fluorescence at specific locations. Positive and negative controls, quality control, and calibration probes are included in the array to verify system functionality, determine repeatability, and correct fluorophore properties. High-resolution melt analysis was performed to confirm the correct target reducing the signal at corresponding probe coordinates. The entire experiment takes approximately 2 h, including PCR cycling, hybridization/capturing, and melt stages.

In another effort, Ma *et al.* [1134] developed an integrated biosensor capable of performing dielectric spectroscopy for the analysis of biological samples. The biosensor demonstrated a strong correlation between the impedance of the sample solution and the length of DNA fragments, enabling the validation of PCR product sizes using a chip-based system that is compatible with biological samples. The scientists utilized concentrated ssDNA fragments sourced from herring sperm, displaying a consistent length of around 700 base pairs. These DNA fragments were dissolved in DI water and diluted to create six different concentrations. A biochip-based system was employed to measure the impedance of the DNA solutions at various concentrations. The results revealed that higher concentrations of ssDNA exhibited lower impedance due to a higher charge carrier density, resulting in decreased conductivity. Additionally, the system was tested with solutions of BSA at different concentrations, yielding impedance trends similar to those observed with DNA samples. The impedance data for both DNA and BSA samples were plotted on a logarithmic scale, showing a linear relationship between impedance and concentration. Furthermore, the system was capable of analyzing digested ssDNA samples, where the impedance values correlated with the reaction time, indicating the detection of changes in DNA fragment length. The system was further validated using DNA fragments of known lengths generated through PCR, successfully distinguishing between different fragment sizes based on impedance measurements. Notably, the impedance values decreased with increasing stimulus frequency, and a linear relationship between DNA fragment length and impedance was observed at a specific frequency.

Impedimetric sensors have proven to be highly valuable tools for DNA analyses, offering a label-free and real-time approach to detecting and quantifying DNA molecules. By monitoring impedance in response to DNA binding events, impedimetric sensors enable researchers to precisely and rapidly analyze DNA sequences and identify genetic variations. For example, Jafari *et al.* [242] introduced a 16-channel, mixed-signal CMOS DNA analyzer that utilizes frequency response analysis (FRA) to analyze the impedance components of the biosensor. The system was extensively validated for DNA analysis, focusing on the detection of prostate cancer. On-chip Au electrodes were employed, and they were hybridized with a 5 M ssDNA probe specific to prostate cancer. Introducing a 5 M noncomplementary target DNA did not cause significant changes in the impedance components, indicating minimal non-specific adsorption. Nonetheless, the introduction of a 5 M complementary target DNA that is specific to prostate cancer led to a significant decrease in the redox current of potassium ferrocyanide, causing an elevation in both the actual and imaginary parts of the electrode impedance. This change was attributed to the additional negative surface charge arising from the formation of double-stranded DNA on the electrode surface.

Nuclear magnetic resonance (NMR) is also a valuable technique for investigating molecular-level information in samples. In their study, Lei *et al.* [774]

developed a handheld μNMR system suitable for multi-type biological/chemical LoC assays. The compact and lightweight system (14 × 6 × 11 cm^3, 1.4 kg) enables the detection of *Enterococcus faecalis*-derived DNA at concentrations below 100 pM using only a small 2.5 μL sample. This system offers selective identification of biological targets, analysis of protein states, and examination of solvent-polymer dynamics. By employing Protein A-functionalized water-soluble MNPs, the presence of human immunoglobulin G (IgG) can be detected based on changes in relaxation time. Moreover, the platform allows for the analysis of β-lactoglobulin (β-LG) protein in food and the investigation of solvent-polymer dynamics using poly(N-isopropylacrylamide) (PNIPAM) as an example. The versatility of this platform is demonstrated through its ability to probe molecular structures and perform thermal profiling.

14.3.2.2 Immobilization-free applications

In certain applications, immobilization is not required. For example, a label-free CMOS field-effect transistor (FET) sensing array was developed by Seong-Jin *et al.* [467] to detect changes in surface potential caused by negative charges in DNA molecules, enabling real-time monitoring and quantification. The sensor array integrated detection does not rely on surface binding, enabling the ongoing tracking of DNA concentrations without needing to immobilize them on the sensing surface. The prototype device fabricated in the CMOS process successfully discriminated various concentrations of 19-bp oligonucleotide solution. The LoD was 1.1 ng/μL with a dynamic range of 40 dB, and the transient response time was less than 20 s. The study focused on detecting and quantifying DNA molecules in a buffer solution by measuring the change in the surface potential of the sensing electrodes. The experimental results demonstrated the capability of the sensor to discriminate 19-base pair single-stranded oligonucleotides at different concentrations. The use of a low-concentration PBS helped increase the detectable range. The sensor array was equilibrated in the buffer solution before each experiment, and the measured surface potential difference between the sample and reference chambers showed the potential for real-time quantification. Despite some drift and noise, the signals obtained from various DNA concentrations exhibited an increasing difference between the bio-sample signals and the reference signal. The study identified a minimum voltage level for DNA quantification and highlighted the challenges associated with drift and noise in the detection process.

DNA-based data storage for emerging DNA memories has emerged as a promising solution with an impressive data density capability of gigabytes per nanogram of DNA. However, achieving widespread adoption of this technology poses several challenges. Existing research has primarily focused on key aspects such as converting digital data into synthesized nucleotides, physically storing DNA, reading out the data through sequencing, and performing computational processes. The idea of utilizing DNA-based data storage in the development of DNA memories has demonstrated great potential as a viable solution, offering an exceptionally high data storage capacity of gigabytes per nanogram of DNA mass. Numerous challenges exist in realizing a widespread technology, and current investigations have primarily focused on converting digital data into synthesized nucleotides, physically storing DNA, extracting data through sequencing, and managing computational procedures. Osouli *et al.* [1135] have addressed the challenge of monitoring the physical storage of dried DNA material by developing a CMOS capacitive sensor that detects the mass of dried DNA. The study involved two independent protocols for investigation. In the first protocol, a droplet of the DNA sample was pipetted onto the sensor chamber and allowed to dry. After complete evaporation, a water droplet was placed on the sensor to wash the previous DNA sample. The water droplet remained for 30 s before being removed without any additional pipetting or brushing. This test was performed for various concentrations, and three repeats were conducted for each concentration. The results initially showed some variation and insignificant repeatability. To improve accuracy, modifications were made, including using a 0.7 μL droplet of DNA sample that remained on the sensor until complete evaporation. Subsequently, 1.5 μL of molecular level water was added, and the sensor was pipetted with water at least 10 times and brushed to ensure the complete removal of the DNA sample. The second protocol aimed to investigate the possibility of sensing the accumulation of DNA by continuously adding droplets of DNA. A 0.7 μL droplet of DNA with a concentration of 117 ng/μL was initially placed on the sensor to evaporate. Without any washing process, a second droplet was added to increase the DNA concentration on the sensor. This process was repeated with five droplets added to the sensor surface in a similar pattern, and the resulting data were analyzed. These two protocols provide valuable insights into monitoring the physical storage of dried DNA using the developed CMOS capacitive sensor. This sensor utilizes the highly integrable nature of CMOS chips and demonstrates successful detection down to a mass of 45 ng/μL of dried DNA.

14.3.3 Neural investigations

The realm of neural investigations encompasses a diverse array of applications that harness the capabilities of CMOS sensors to unravel the intricacies of the nervous system. These innovative techniques enable researchers to delve into the realms of cell monitoring, neurotransmitter assays, and brain slice analysis, unlocking deeper insights into neural functionality and paving the way for breakthroughs in neuroscience. By seamlessly integrating cutting-edge CMOS technology with the intricacies of neural research, these applications provide invaluable tools for comprehending the complex dynamics that govern neuronal activity and communication.

14.3.3.1 Neuronal cell analyses

Neural tissue is composed of a variety of cell types, including neurons, astrocytes, oligodendrocytes, and microglia. This complexity poses challenges for neuroscientists in determining the specific cells affected by drugs or manipulations and discerning direct or indirect effects. Cultured neuronal networks on microelectrode arrays (MEAs) are typically studied using dissociated neurons from embryonic or neonatal animals like rats or mice. Different types of neurons at specific developmental stages have been employed in various cell culture systems and techniques, such as E18 rat hippocampal neurons [1143–1148], E14-15 mouse spinal cord or cortical neurons [1148–1150], E17-18 rat cortical neurons [282,1149,1151], P0-P3 rat cortical neurons [1152,1153], P3 rat hippocampal neurons [1154], and embryonic stem cell-derived neurons [1155,1156].

In many cases, cultures are maintained under high-density conditions in a serum-containing medium to promote the growth of glial cells critical for long-term culture viability. However, alternative approaches like glia-conditioned medium [1146,1157] or serum-free conditions [303,1144,1145,1158,1159] have also been explored. These systems have specific requirements and necessitate skilled personnel with expertise in the field. It is crucial to note that no *in vitro* condition fully replicates the complex mechanisms observed in living organisms. Cell culture systems can generally be categorized into primary, secondary, and continuous cell cultures [1160,1161].

Primary culture involves directly deriving cells from tissues through enzymatic and mechanical dissociation. This results in a suspension of individual cells plated onto a suitable culture surface, such as a dish (Figure 14.10(b)). Despite challenges in obtaining a heterogeneous population of cells and a relatively low proliferation rate, primary cultures closely resemble the cells' original form in the tissues they were derived from [1162]. The process can be visualized through tissue explant culture.

In tissue organization, cells are interconnected through intercellular and cell matrix connections. Culturing these cells requires disrupting these connections to obtain a single-cell suspension. This can be achieved through enzymatic or chemical digestion using proteolytic enzymes like trypsin or collagenase, or mechanical separation methods such as tissue mincing with surgical knives. The resulting cell suspensions are purified through serial dilution and centrifugation before transfer to culture vessels. During this process, many cells adhere to the culture vessel while being freed from cellular connections. The primary culture is considered complete once non-adherent cells and their residues are removed from the culture environment. However, it is common for other cell types, especially fibroblasts, to also attach to the culture vessel, alongside the desired cells initially targeted for the primary culture [1162].

In certain instances, cells derived from the embryonic brain or astroglia may undergo a phase of proliferation, followed by differentiation upon reaching confluence and establishing contact with neighboring cells [1163]. Primary cultures are

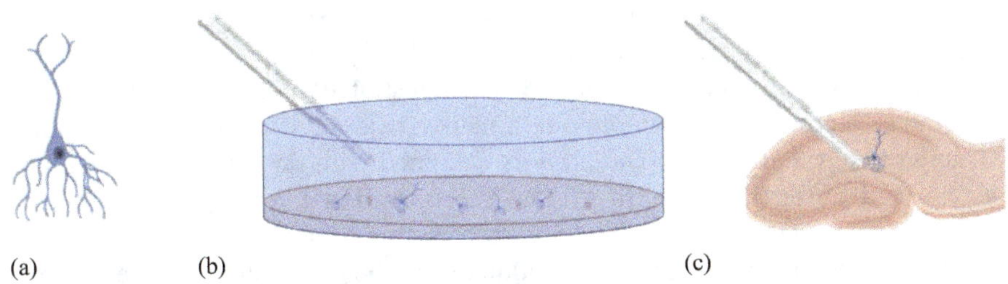

Figure 14.10 (a) Neuron, (b) neuron culture, and (c) acute slice

highly valuable as they closely mimic the characteristics of cells in normal tissues. Nevertheless, they do have certain limitations, such as a finite lifespan. As a result, these cells need to be reproduced through successive passages and preserved. In certain cases, reproducing specific cell types of interest can be challenging due to their lower abundance compared to other cell types, necessitating intervention with specialized media or growth factors. Another significant hurdle in primary cultures is the potential loss of structural and functional properties when transitioning cells from *in vivo* conditions to the culture environment. Consequently, cells that exhibit distinct morphologies in living tissue may demonstrate similar morphologies in the culture environment [1162].

Additionally, cells in primary culture have a limited lifespan over time due to the rapid depletion of essential substances and changes in metabolite levels as the proliferating areas decrease. However, by transferring cells obtained from the primary culture to a new culture environment under appropriate conditions, they can continue to proliferate, resulting in secondary cultures. In secondary cultures derived from passaging, the addition of growth factors specific to the desired cell type can promote the proliferation of the intended cells while suppressing others [1164,1165].

Although primary neural cell cultures have some disadvantages, such as the loss of normal synaptic pathways, disruption of normal anatomical relations, and changes in gene expression, studies have shown that most neural cell types express similar properties both *in vitro* and *in vivo*. For instance, astroglia produces vimentin and glial fibrillary acidic protein as typical intermediate filament proteins, along with neurotrophic factors. Oligodendroglia and Schwann cells are capable of producing myelin, while cultured neurons can form synapses, express appropriate neurotransmitter receptors, conduct action potentials (APs), and exhibit neurofilaments and growth cones [1162].

The transfer of cells from living tissue to a culture environment induces cellular stress, necessitating the establishment of a suitable culture environment tailored to the specific requirements of the cells. The cell culture comprises two essential components: the physiological environment and the physicochemical environment. The physicochemical environment encompasses factors such as temperature, atmospheric conditions (including oxygen and carbon dioxide concentrations), and pH. Conversely, the physiological environment includes growth factors and nutrients that are vital for the cells' growth and reproduction, primarily provided through culture media. The surface characteristics of the culture vessel contribute to the physicochemical environment [1162].

As an example, Eversmann *et al.* [258] demonstrated a FET-based, 128 $\times$ 128-MEA (microelectrode array) employing a matrix readout scheme for extracellular recording of neural activities. The array is equipped with on-chip amplifiers, multiplexers, and buffers (Figure 14.11(a)). The operational principle of this array relies on the oxide-semiconductor field-effect transistor (OSFET) mechanism, where the transistor's drain current is modulated by the potential of the cleft formed between the cell and the sensor surface. Extracellular recording capabilities were tested using neurons from the pond snail, *Lymnaea stagnalis*. Before cultivation,

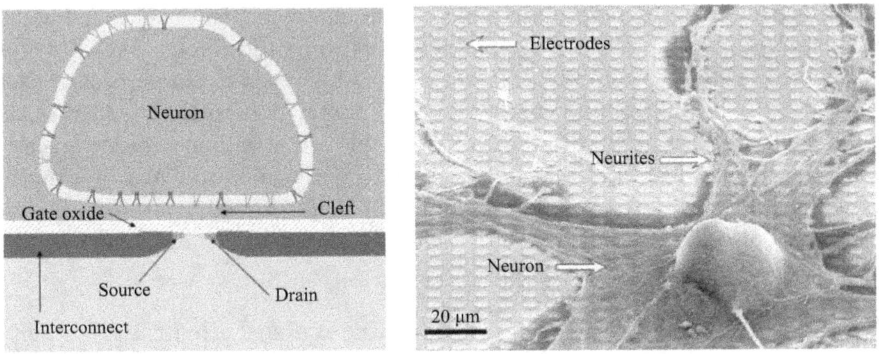

(a) (b)

Figure 14.11 (a) OSFET and (b) snail neuron on sensor chip in culture [258]

the chips underwent cleaning, sterilization, and coating with a cell-adhesive layer of poly-l-lysine. Extracted neurons were placed on the sensor chip and cultured for 1 day, demonstrating biocompatibility with neurite growth. Figure 14.11(b) illustrates the Snail neuron cultured on the sensor chip, confirming the sensor surface's biocompatibility for neurite growth.

Human neuroblastoma cells, SH-SY5Y, are commonly employed as a neuronal model to investigate Parkinson's disease and dopamine (DA) release in the substantia nigra. By utilizing amperometric single-cell recordings of single vesicle release events, researchers can examine the effects of molecular manipulations on DA release and gain insights into the underlying mechanisms of neurological disorders. However, traditional techniques for single-cell analysis of neurotransmitter release suffer from low throughput. To address this limitation, White *et al.* [224] discussed the development of a monolithically-integrated CMOS sensor array capable of high-throughput observation of DA release from multiple single cells at single-vesicle resolution. The CMOS device integrates 1,024 on-chip gold electrodes, each with a size of 15 μm × 15 μm, directly on 1,024 transimpedance amplifiers through post-CMOS processing. SU8 traps are fabricated on individual electrodes to enable the interrogation of single cells while rejecting multicellular clumps. This study demonstrates the capability of the CMOS device to record vesicle secretion at a single-cell level, utilizing 1,024 parallel channels to provide detailed information on the dynamics of DA release with single-vesicle resolution. The SH-SY5Y neuroblastoma cells are cultured in a growth media composed of a 1:1 ratio of MEM with L-Glutamate and F12 media, supplemented with fetal bovine serum (FBS) and antibiotic/antimycotic. The cells are cultured under specific conditions and passaged when reaching 80% confluence. For the recording of neurotransmitter releases, the CMOS device is loaded with suspended SH-SY5Y cells in culture media and placed in an incubator to allow cell capture and settling onto the chip surface. The deposited cells are distributed in a random manner, with certain cells adhering to the SU8 surface while others find their way into the SU8 traps. Once the cells are trapped and adhered, the culture media is exchanged with a recording buffer containing specific concentrations of various ions. The chip with cells is then ready for recording, and an Ag/AgCl reference electrode is connected to the chip. To stimulate the cells, a high KCl stimulation buffer is added to the recording buffer, inducing depolarization of the cell membrane and triggering vesicle secretions. The functionality of the CMOS device is validated by simultaneously monitoring DA secretions from 76 SH-SY5Y cells, resulting in the observation of 7,147 single vesicle release events during a 42-s measurement.

14.3.3.2 Brain slice analyses

MEAs provide a valuable approach for monitoring neuronal electrical activity in a three-dimensional (3D) setting by placing acute brain slices onto them (Figure 14.10(c)). This technique allows imaging and mapping of deep-seated neurons within the brain [311]. The *in vitro* brain slice preparation, pioneered by Henry McIlwain in the 1950s and 1960s, has become a powerful tool for neuroscience research [1166,1167]. Culturing cells as 3D models offers a more physiologically and biochemically relevant system that closely mimics *in vivo* conditions [1168]. This approach has demonstrated its realism and applicability in translating research findings to *in vivo* situations, with over 380 cell lines being studied using this method [1169].

High-density microelectrode arrays (HD-MEAs) have been effectively used to record electrical images of neuronal network activity in various brain regions, such as the hippocampus and suprachiasmatic nucleus, with high spatial resolution [1170]. Neurons in brain slices can remain viable for up to 10 h when perfused with fresh oxygenated artificial cerebrospinal fluid, maintaining their network structure akin to *in vivo* conditions. Brain slice preparations are widely employed in neuroscience experiments, enabling diverse investigations ranging from immunohistochemistry, brain anatomy, and biomolecular studies to pharmacological research, channelopathies, electrophysiology of individual neurons and synapses, as well as characterizing glial and neuronal networks [1171].

The *in vitro* brain slice preparation offers a unique opportunity to study metabolic and electrophysiological properties without interference from anesthetics, muscle relaxants, or intrinsic regulatory substances [1172]. It also provides new avenues for studying physiological and pharmacological mechanisms within the central nervous system (CNS) [1173–1175]. Moreover, obtaining multiple slices from the same animal reduces the need for using a large number of animals for each experiment. Brain slices and slice cultures are increasingly used as models for investigating various neurological conditions, including synaptic plasticity, epilepsy, ischemia, regeneration, demyelination, apoptosis, and degenerative diseases [1176–1201].

Brain slice preparations support various methodological approaches, with electrophysiology, including extracellular or intracellular patch clamp techniques, being widely utilized for functional studies [1179,1202–1205]. Additionally, researchers have successfully applied diverse analyses, such as live confocal imaging, transfection methodologies, and genomic analyses, to slices or slice cultures [1205–1216].

The rapid preparation time of brain slices minimizes the use of anesthetics, and their structural integrity allows the study of specific circuits and brain networks in isolation, like the thalamocortical pathway [1217–1220]. Electrophysiological recordings in acute slices offer greater stability compared to *in vivo* recordings as they eliminate the influence of the experimental animal's heartbeat and respiration, enabling longer cellular recording periods [1172]. Furthermore, the direct visualization of the slice facilitates cell identification, access for investigation, and local drug application, which is restricted by the blood–brain barrier.

The use of brain slices has significantly advanced our understanding of the mammalian CNS and is anticipated to remain a valuable experimental method in the future. However, it does have limitations, with the lifespan of brain slices from rats or mice usually limited to 6–12 h, and in thin hippocampal and neocortical slices, most cells can only be maintained in a healthy state for approximately 4 h [1173].

In a study, Viswam *et al.* [44] developed a monolithic multi-functional CMOS MEA system that enables electrical impedance imaging and electrophysiology recordings conducted on cardiac cells and brain slices. Two bioexperiments were conducted to validate the system's performance. The first experiment focused on embryoid bodies (EBs), while the second experiment examined acute brain slices. Impedance and electrophysiological measurements were taken using $3 \times 7.5\ \mu m^2$ bright Pt electrodes with impedance magnitude values of $2.3 \pm 0.04\ M\Omega$ and phase values of 0.7 ± 0.5 degrees at 10 kHz. Pt-black electrodeposition was utilized to reduce impedance and thermal noise without altering electrode geometry, resulting in a 30–40-fold impedance reduction compared to bright Pt electrodes. Before each bioexperiment, Pt-black was redeposited, and electrode impedances were measured to ensure uniform impedance across the array and enable assessment of spatial impedance features. In the cardiac cell experiment, mouse embryonic stem cells (mESC) formed 3D multicellular aggregates (EBs) and were monitored for cardiac differentiation using impedance measurements and electrophysiological sensing. After differentiation, EBs were plated on an HD-MEA electrode surface coated with Fibronectin. Impedance measurements captured the position, adhesion, spreading, and growth of the cells over 5 days, with cells at the tissue circumference forming a tightly attached thin layer to the electrode array. Electrophysiological activity of differentiated EBs, exhibiting an average beating rate of 36 per min, was recorded using extracellular action potential (EAP) recording channels. In the impedance imaging experiment with acute brain slices, an acute cerebellar slice from a wild-type mouse was placed on the HD-MEA. Impedance measurements at 10 kHz were performed using 110 different electrode configurations, acquiring impedance information from 3,520 uniformly distributed electrodes on the array. Impedance magnitude and phase changes enabled the differentiation of different cell layers, including white matter, the granular cell layer, the Purkinje cell layer, and the molecular layer. Spontaneous electrical activity of Purkinje neurons was recorded, demonstrating good SNRs. These bioexperiments demonstrated the efficacy of the EIS system for monitoring cardiac differentiation in EBs and imaging distinct cell layers in brain tissue using impedance measurements and electrophysiological sensing.

14.3.3.3 Neurotransmitter assays

In the field of biosensing, capacitive sensors have proven to be valuable tools for detecting neurotransmitters. For example, Lu *et al.* [27] developed a remarkable CMOS capacitive biosensor where they immobilized 4-carboxyphenylboronic acid (CPBA). The positive charges resulting from the protonation of boronate and amino groups enhanced the electrode-solution capacitance, while the binding of the neurotransmitter DA and CPBA molecules generated negative charges, leading to a reduction in capacitance (refer to Figure 14.12). The surface of the interdigitated electrodes (IDEs), covered by a CMOS SiO_2 thin film, underwent treatment with a UV-ozone dry stripper, followed by immersion in 3-Aminopropyltriethoxysilane (APTES). The UV light decomposed ozone into oxygen molecules and atoms, which effectively cleaned and activated the surface by introducing –OH to the oxide surface. The amino group introduced by APTES was protonated to –NH3+, enabling its coupling to CPBA using 1-ethyl-3-(3-dimethylaminopropyl) carbodiimide (EDC) activation in the 2-(N-morpholino)ethanesulfonic acid (MES) buffer.

In the realm of voltammetric sensors for neurotransmitter-based applications, an interesting study by Tedjo *et al.* [230] sheds light on an electrochemical imaging system integrated with a custom CMOS microchip. The microchip encompassed a sensing area of 3.6 mm × 3.6 mm, containing 16,064 Pt MEA and their corresponding integrated read channels and digital control circuits. The system employed a unique three-electrode geometry with a spatial pitch of 27.5 μm, enabling high-resolution chemical gradient imaging at the cellular level. The on-chip read channel array exhibited low noise levels, enabling sensitive amperometric detection of neurotransmitters like norepinephrine within a concentration range of 4 μM to 512 μM, with a sensitivity of 4.7 pA/μM (R^2 = 0.98). Furthermore, the system showcased its capability to detect dissolved oxygen concentration by characterizing the electrochemical response using deoxygenated deionized water with oxygen concentrations ranging from 5% to 80% of the ambient levels,

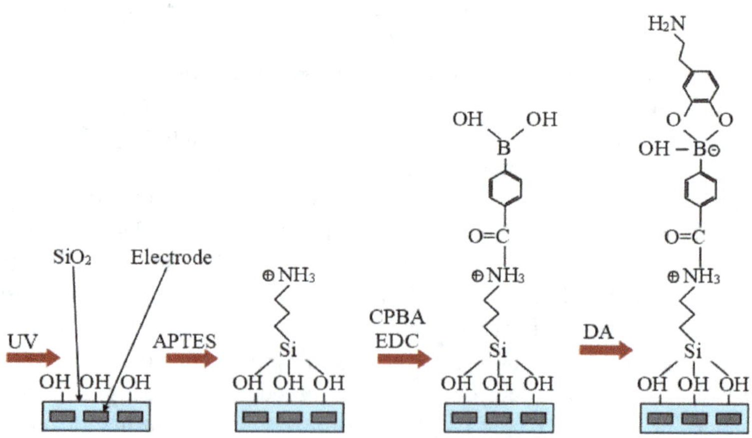

Figure 14.12 Detection of the neurotransmitter DA using capacitive sensing

achieving a sensitivity of 86 pA/mg/L ($R^2 = 0.89$). The system also demonstrated its selectivity to different target analytes by simultaneously detecting norepinephrine and uric acid using the CV method. To ensure the viability of biosamples during experiments, a custom indium tin oxide electrode with deposited gold (Au) glass was integrated into the microfluidic system for pH measurement. Electrochemical images captured by the system showcased both spatial and temporal performance at a rate of four frames per second while maintaining sensitivity to the target analytes.

In the quest for enhanced DA detection, Huang *et al.* [226] conducted a study focusing on electrode separation reduction. They successfully fabricated sub-μm microelectrodes using nanoimprint lithography, which eliminated the need for expensive electron-beam lithography. The reduction and oxidation potentials for DA were determined through CV at –0.2 V and 0.6 V, respectively. The measured currents aligned well with the expected values. Sensitivities of 2.89 nA/μM and 1.73 nA/μM were achieved for electrode separations of 300 nm and 600 nm, respectively, with average collection efficiency values of 97.9% and 95.4%. The study also demonstrated signal transduction using a CMOS sensing circuit, with a detectable resolution of approximately 0.1 μM. To ensure selectivity, Nafion–coated electrodes were employed, exhibiting a high selectivity ratio of 41:1 for DA over ascorbic acid. The microelectrode chips were securely connected to printed circuit boards (PCBs), and each chip was enclosed within a glass ring containing the solution. The electrochemical experiments were conducted at room temperature and atmospheric pressure, and CV was utilized to determine the oxidation and reduction potentials of DA in PBS. The sequential addition of DA solutions of varying concentrations to the electrodes resulted in increased steady-state currents, confirming the enhanced collection efficiency with smaller electrode separations. The CMOS sensing circuit generated output waveforms proportional to DA concentrations, and the minimum detectable concentration was estimated at approximately 0.1 μM. This study demonstrated the effectiveness of Nafion-coated electrodes and the potential of sub-μm microelectrodes for improved DA detection.

The modulation of neurotransmitter release is influenced by various pharmaceutical agents and molecular manipulations. Kim *et al.* [193] introduced an active CMOS-based electrochemical biosensor array capable of high-throughput measurement (100 electrodes) for on-chip detection of neurotransmitter release. The biosensor array, fabricated using CMOS IC technology and photolithography, integrated platinum on-chip WEs through conformal deposition and lift-off photolithography. The biosensor's performance was validated by simultaneously measuring amperometric currents from 100 electrodes in response to DA injection, providing insights into DA diffusion along the biosensor array surface. The biosensor successfully recorded neurotransmitter release from multiple individual living chromaffin cells, capturing small and fast amperometric spikes corresponding to single vesicle secretions. The biosensor exhibited stable baseline currents and nominal electrode resistance even after several days of cell culture. Amperometric events were observed from most cell-covered electrodes after high K^+ stimulation, representing quantal release events from individual vesicles. The biosensor demonstrated its capability to detect release events with a sensitivity of 6,000 molecules, offering pico-ampere current resolution and sub-millisecond time resolution, enabling the analysis of single vesicle secretion events from chromaffin cells.

14.3.4 Cellular assays

Beyond neural cells, CMOS biosensors provide a versatile and robust platform for a wide range of cellular-based assays, including monitoring cancer cells [29], bacteria [33], cardiomyocytes [1062], and alike. CMOS biosensors hold significant promise in advancing our understanding of cellular processes and their implications in various fields, including biomedical research, drug discovery, and personalized medicine.

CMOS capacitive biosensors have proven to be valuable tools for cellular assays, enabling the detection and monitoring of various living organisms, including bacteria and different cell types [4,29,35,1221] (Figure 14.13). These biosensors utilize the deposition of cells onto the electrodes to create a biointerface layer. As the cells interact and undergo dynamic activities, such as viability, growth inhibition, adherence, proliferation, and migration, they influence the local microenvironments' electrical characteristics, leading to measurable changes in the detected signal.

14.3.4.1 Bacteria analyses

CMOS sensors find a valuable application in the realm of bacteria detection and growth monitoring. These sensors play a pivotal role in enabling the real-time identification and tracking of bacterial populations, facilitating a deeper understanding of bacterial behavior, and aiding in the development of strategies for disease control and antimicrobial research. By harnessing the capabilities of CMOS technology, researchers can create highly sensitive and efficient platforms that contribute to advancing our knowledge of bacterial dynamics and enhancing our ability to combat bacterial infections.

As an example, Couniot *et al.* [33] put forth a capacitive sensor in CMOS technology designed to detect *S. epidermidis* in a solution of PBS. To prepare the sample, *S. epidermidis* was cultivated on agar plates and subjected to incubation at 37 °C. Following this, a single colony was suspended in Tryptic Soy Broth (TSB) and allowed to incubate until the bacterial culture reached a stationary phase. The culture was then centrifuged and the cells were resuspended in PBS 1×, with this

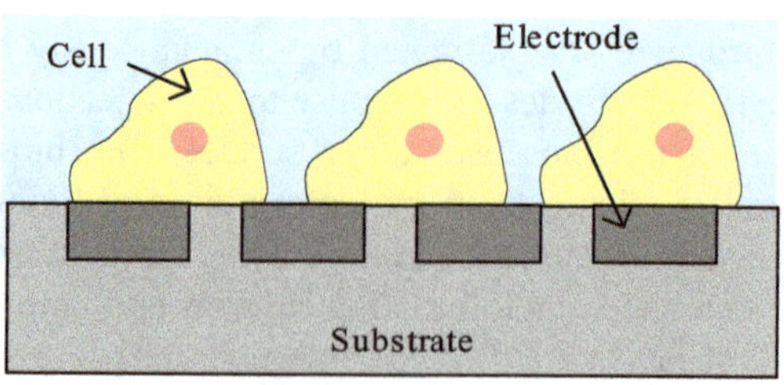

Figure 14.13 Cellular biosensor

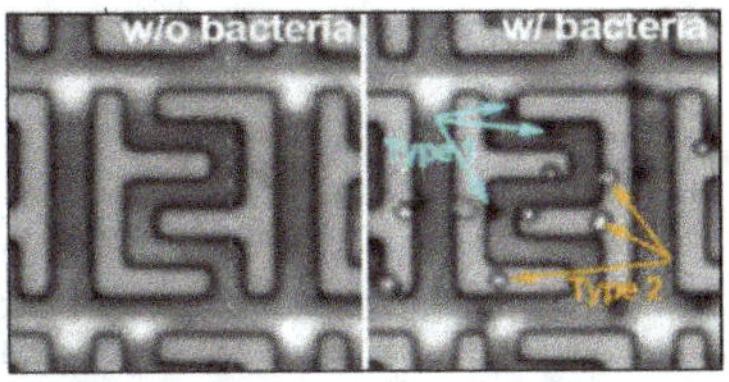

Figure 14.14 Pre- and post-binding of bacterial cells to the electrodes employed in [4] illustrating two distinct modes of bacterial adhesion (type 1: bacteria between electrodes, type 2: bacteria on top of electrodes)

step repeated twice. By utilizing Al/Al_2O_3-electrodes, they managed to achieve an LoD of 10^7 cfu.mL^{-1}. The same research group [4] developed an alternative CMOS capacitive biosensor specifically designed to detect the suspension of the aforementioned bacteria in a diluted PBS solution (1:1,000 dilution). This biosensor demonstrated an LoD of approximately seven bacteria. Figure 14.14 visually illustrates two potential scenarios for bacteria binding: either between the electrode surfaces (type 1) or on top of the electrodes (type 2). In another effort, Ghafar-Zadeh and colleagues [35] introduced a capacitive biosensor based on CMOS technology to monitor bacterial growth. Their study focused on the examination of *E. coli* suspended in Lysogeny broth (LB) medium.

In another study, Duan *et al.* [470] introduced a CMOS ISFET array featuring a floating gate design, which demonstrated improved tolerance to offset distribution, resolution, and linearity for long-term monitoring of bacterial metabolism. The system effectively tracked pH changes during an 11-h growth period of *E. coli* and its response to antibiotics, showcasing its ability for long-term monitoring of bacterial metabolism. pH solutions were prepared using 1 M KOH and HCl reagents, creating a range of solutions from pH 4 to 10.26 by adjusting the ratios of KOH and HCl. Fresh solutions were prepared for each measurement. To compensate for offsets and calibrate the gain, the chip was initially immersed in a pH 7.01 solution. Between testing different buffer solutions, the chip and flow chamber underwent thorough rinsing with DI water to prevent contamination. The ISFET sensor array demonstrated an average passivation sensitivity of 30 mV/pH and achieved high linearity with input-output relations of $R^2 = 99.89\%$ and 99.94%. In the bacterial growth monitoring experiment, *E. coli B* was cultured in LB at 37 °C on an incubator shaker, with 5 mL of bacterial solution used for each measurement. To ensure stability, the chip was immersed in LB for 14 h before measurements. The growth curve, measured using optical density (OD), displayed an exponential pattern, reaching a stationary phase within 6 h. The ISFET array and reference pH measurements confirmed the alkalization of the culture medium, which was reproduced by adding fresh LB. The pH monitoring range of the ISFET array was wider compared to that of the commercial optical instrument. The addition of antibiotics (penicillin and streptomycin) resulted in a decrease in bacterial OD, indicating their efficacy. However, the pH did not stabilize, suggesting the viability of the bacteria. The microscopic examination further confirmed the presence and movement of bacteria despite the antibiotic addition, aligning with the conclusions drawn from the optical and pH measurements.

In their research, Jiang *et al.* [7] developed an ISFET sensor for efficient detection of *E. coli* bacteria, providing results in a shorter screening time of 4 h compared to the conventional 24-h direct plate counting method. Bacterial samples with initial concentrations of 5×, 10×, and 50× cfu/mL were incubated for 7 h, with pH measurements recorded at hourly intervals. As the bacteria proliferated and metabolized, the culture media underwent depletion, resulting in the production of acidic substances and a subsequent decrease in pH values. Initially, the bacterial growth rate was slow, but it accelerated over time, eventually reaching a plateau. Due to the low bacterial density (14, 28, and 140 cfu/mL), it took several hours for noticeable pH changes to occur. The experiment commenced after a 3-h culture period, and pH measurements were taken accordingly. Minimal pH changes were observed during the initial 3 h of culture. Subsequently, the absolute pH values of the samples decreased within the first 5 h of culture, consistent with measurements obtained using a commercial pH meter. The subsequent increase in pH after 5 h could be attributed to a limited dissolved oxygen supply, restricting bacterial respiratory activity and leading to the production of alkaline substances. Another possibility was the depletion of glucose or specific nutrients in the culture medium, prompting bacteria to seek alternative sources for metabolism, resulting in a diauxic-like pH change. In terms of *E. coli* concentration prediction, after 4 h of incubation, the pH difference between samples with initial concentrations of 14 and 28 CFU/mL was 0.01 pH, while it increased to 0.045 pH between samples with concentrations of 14 and 140 cfu/mL. A clear trend was observed, where higher bacterial density corresponded to increased pH difference. This allowed the determination of an unknown sample's *E. coli* level by comparing its pH difference to a reference sample. The ISFET sensing system demonstrated improved time efficiency and a satisfactory low LoD compared to the traditional direct plate counting method.

Mitsunaka *et al.* [124] presented a CMOS biosensor that leverages 120 and 60 GHz oscillator arrays to detect dielectric relaxations of biological water. The study focused on monitoring the growth of bacterial colonies. The experimental procedure involved culturing *E. coli* bacterial seeds in Trytic Soy Broth agar, followed by spraying the bacterial suspension onto a nutrient agar plate positioned on the IC surface. Cultivation took place in a thermostatic chamber at 30 °C under high humidity conditions. Remarkably, the biosensor IC demonstrated remarkable capability in detecting bacterial colonies by analyzing the frequency shift distribution, even in scenarios where microscopic images had limited visibility. The biosensor's sensitivity to dielectric loss disparities between bulk water and bacterial colonies showcased its potential for automated bacterial detection. Moreover, the study explored the potential of the biosensor in diverse applications, such as label-free detection of target proteins and microorganism detection using specialized interfaces. With its utilization of higher frequencies and smaller sensing elements, the developed biosensor IC showcased considerable promise in bacterial colony detection and the visualization of target distributions.

14.3.4.2 Detecting cancer cells and their proliferation

An impactful application of CMOS sensors lies in their capability to detect cancer cells and monitor their proliferation dynamics. This pioneering use of CMOS technology entails harnessing the sensor's unique attributes to identify and track cancerous cells, shedding light on their behavior and aiding in understanding the progression of cancer diseases. For example, a research effort by Prakash *et al.* [29] involved the development of a CMOS capacitive biosensor to track the adhesion and proliferation of MDA-MB-231 breast cancer cells. The experimental setup included culturing MDA-MB-231 cells in a flask containing a growth medium comprising an improved minimum essential medium, FBS, and penicillin-streptomycin (100×). The cells took approximately 48 h to grow in the flask, after which they were detached using Trypsin/Ethylenediaminetetraacetic acid (EDTA). A cell suspension with the desired density was prepared in the growth medium, and the cells were loaded onto the chip without any additional surface modification or functionalization.

In another study, Senevirathna *et al.* [3] presented a capacitive biosensor based on CMOS technology for real-time measurements of ovarian cancer cell proliferation. The experimental procedure involved adding a prepared cell suspension to the device and allowing adherent cells to grow on the chip array. In a study conducted by the same research group [1130], a chemotherapeutic assay was carried out to investigate the cytotoxic effects of drug-resistant (CP70) and drug-sensitive (A2780) cancer cell lines. Human ovarian cancer cells, CP70 and A2780, were cultured in T25 flasks using growth media. The media consisted of RPMI 1640 serum supplemented with 10% FBS, and 1% penicillin and streptomycin. The cells were grown in a standard T-cell flask within an incubator set at 37 °C and 5% CO_2 until they reached the logarithmic growth phase. To detach the cells from the flasks, 0.25% trypsin/EDTA solution was used, followed by pelleting and resuspension in 12.5 mL of fresh cell media. Cisplatin, a commonly used platinum-based chemotherapeutic drug, was employed as the anti-cancer agent. It was obtained as a dry powder and prepared as a 1 mM stock solution with DI water, stored at room temperature away from light. For treatment, the cells were plated in the device and allowed to grow for 48 h before the application of the chemotherapeutic agent. Cell counting was performed using a hemocytometer and optical microscope, with counts repeated three times and averaged. The plating area of the device maintained a cell density of 105 cells/ml for each experiment. Figure 14.15 showcases time-lapse images captured by the in-incubator imaging system, along with corresponding capacitance measurements from four sensors in an array. The researchers then examined the dose-dependent effects of cisplatin on the two cell lines. Both cell types were allowed to grow for a minimum of 48 h, after which varying concentrations of cisplatin were applied. Control experiments were conducted, where both cell lines were plated and allowed to proliferate without treatment during the experimental period. According to the findings, concentrations higher than 5 μM demonstrated cytotoxic effects on both cell lines, while a dose of 1 μM enabled differentiation between the two cell types.

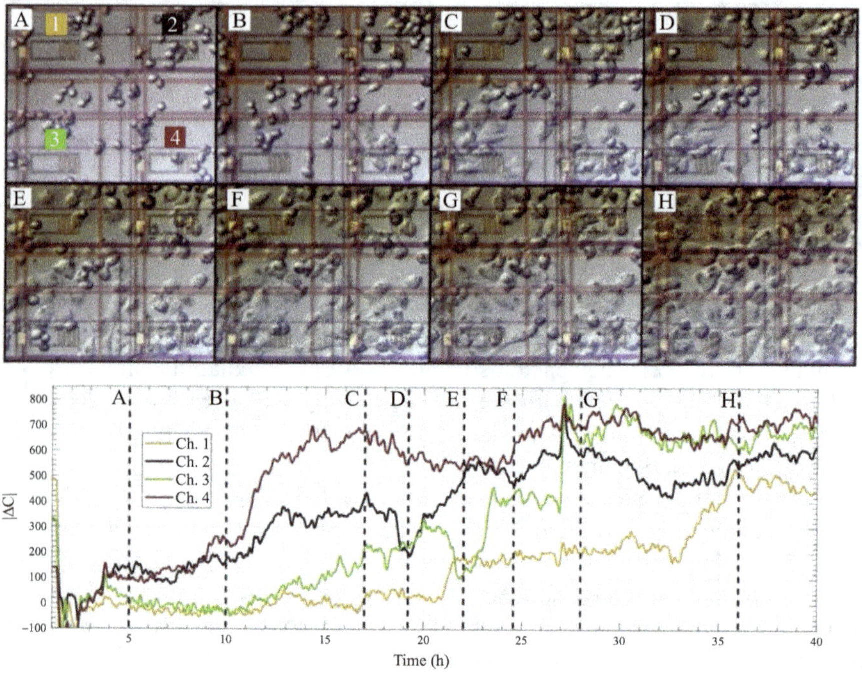

Figure 14.15 The relationship between capacitance measurements and time-lapse imaging of cell-chip interactions during cell-substrate binding [1130]: The top section presents images captured at different time points for four sensors. The bottom section illustrates the corresponding responses of the sensors, indicating the measured changes in capacitance over time. Letter markers are used to indicate the time points corresponding to each frame

Nabovati and colleagues [37] employed CMOS-based capacitive biosensors to facilitate efficient monitoring of cell growth rates. The study focused on utilizing the H1299 cell line, derived from human lung carcinoma. To enhance the biocompatibility of the CMOS technology's topmost metal electrodes, a thin layer of PDMS was spin-coated onto them. The researchers prepared cell cultures in a medium and then appropriately diluted aliquots of the cell suspension based on the size of the wells on the CMOS chip. In Figure 14.16(a), the response to 200 μL of cell suspension with a concentration of 100 kCells.mL^{-1} in the media is demonstrated, spanning 12 h and encompassing phases such as cell seeding, adherence, and proliferation. Another investigation by the same group [2] examined the CMOS chip's capabilities for monitoring the growth of H1299 and Hek293 cell lines, as well as assessing drug cytotoxicity (refer to Figure 14.16(b)). A conventional AlamarBlue cell-based assay and the selective antibiotic Geneticin (G418) were employed as controls, while cytotoxic drugs were administered to non-resistant

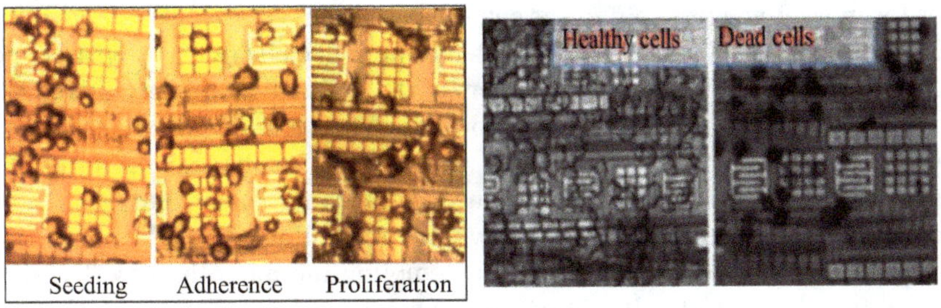

(a) (b)

Figure 14.16 *(a) The 12-h response of the CMOS chip, as described in [37], to a concentration of 100 k Cells.mL*$^{-1}$*. On the other hand, (b) showcases the cell viability before and after the introduction of antibiotics to the cell growth medium on the intelligent Petri dish, as reported in [2].*

H1299 and resistant Hek293 cell lines, respectively. To enhance biocompatibility, the researchers employed a layer-by-layer (LBL) polyelectrolyte deposition technique to build polyelectrolyte multilayer (PEM) films on the CMOS chips, generating robust electrostatic interactions between the PEM layers. These layers consisted of a precursor base layer of positively charged poly(ethyleneimine) (PEI), followed by sequential adsorption of polyanion (poly(sodium styrene sulfonate) (PSS)) and polycation (poly(allylamine) hydrochloride (PAH)) layers, forming five polyelectrolyte bi-layers [PSS/PAH].

In another study [129], a microfluidics-integrated label-free flow cytometry-on-a-CMOS platform is presented for assessing the cytoplasmic dielectric properties at microwave frequencies. Unlike MHz impedance cytometers, operating at GHz frequencies enables direct probing of intracellular permittivity due to the penetration of electric fields through the cellular membrane. To address detection challenges at high frequencies, the spectrometer utilizes on-chip oscillator-based sensors that integrate simultaneous frequency generation, electrode excitation, and signal detection capabilities. By employing 3D hydrodynamic focusing microfluidics, characterization is performed on four distinct cell lines, including two breast cell lines (MCF-10A and MDA-MB231) and two leukocyte cell lines (K-562 and THP-1). The mammalian cells are cultured in 10 cm tissue culture-treated plastic dishes at 37 °C with 5% CO_2. MCF-10A human mammary epithelial cells are maintained in DMEM-F12 supplemented with 5% horse serum, epidermal growth factor, hydrocortisone, cholera toxin, and insulin. MDA-MB-231 human mammary epithelial carcinoma cells are maintained in DMEM GlutaMAX with FBS. THP-1 human monocyte cells are maintained in RPMI-1640 supplemented with FBS, sodium pyruvate, 4-(2-hydroxyethyl)-1-piperazineethanesulfonic acid (HEPES), glucose, and beta-mercaptoethanol (BME). K562 leukemic cells are maintained in RPMI-1640, FBS, sodium pyruvate, and HEPES. Before the experiments, adherent cells are detached using trypsin–EDTA and inactivated with serum-containing media, while suspension cells are directly diluted into an isotonic buffer for measurements. To prevent channel blockage, both cells and polystyrene beads larger than 6 μm in diameter are filtered through CellTrics® filters to remove larger debris. Smaller beads with diameters of 0.5 μm and 1.0 μm are filtered using syringe filters with 0.7 μm and 1.2 μm pore sizes, respectively. Syringe pumps are employed to deliver both samples and sheath flows to the microfluidics. Each cellular experiment is conducted within 1 h at room temperature. To ensure cleanliness, the fluidic channel is rinsed with bleach followed by DI water between each batch of measurements without disassembling the microfluidics. The use of bleach is found to effectively remove residues and debris on the CMOS surface without compromising electrode integrity and sensitivity. Subsequently, four different cell lines are characterized to determine whether they exhibit distinguishable permittivity dispersions at GHz frequencies. The findings from the experimentation reveal that upon standardizing higher frequency signals with the 6.5 GHz signals, the size-agnostic dielectric opacity exhibits a noticeable disparity at 17.5 GHz when comparing normal breast cells to those with a high level of metastasis.

14.3.4.3 Neutrophil analyses

Bunnfors *et al.* [1128] utilized LoC-based capacitive biosensors to monitor the activity of neutrophil granulocytes and their oxidative stress when stimulated by cerium-oxide-based NPs. Neutrophils were isolated from venous whole blood, which was anticoagulated with heparin, brought to room temperature, and layered on a density gradient before being centrifuged. The separated neutrophils were then resuspended in NaCl and PBS and subjected to further centrifugation. To remove any remaining erythrocytes, the cells underwent two brief hypotonic treatments in ice-cold distilled water, followed by washing in a PBS solution containing NaCl and HEPES. The isolated neutrophils were subsequently resuspended in a HEPES solution and their cell concentration was adjusted as desired, after which they were kept on ice until use. During the experiment, the neutrophils were incubated at 37 °C without CO_2. The researchers prepared cerium-oxide-based NPs, specifically Ce_2O_3 and gadolinium (Gd) alloyed Ce_2O_3, using a room-temperature wet-chemical synthesis procedure described by Eriksson *et al.* [1222]. The concentrations of cerium (Ce) and Gd in the NPs were measured. These NPs served as triggers for initiating cell activation. In Figure 14.17, the neutrophils on the insulator surface are depicted in their initial activation stage and subsequently. The experimental findings revealed that while the sensor was capable of detecting the neutrophil response, it was unable to detect the stimuli.

Osouli *et al.* [1129] conducted a study introducing a CMOS capacitive sensing platform as an alternative method for analyzing salivatory cells, specifically oral neutrophils. Oral cells, unlike other types, exhibit a limited ability to thrive in culture conditions. Consequently, they display less adherence to the surface, even when in a culture medium [1223,1224]. To obtain neutrophil samples, participants were instructed to consume only water for 30 min before collection. Following this, they rinsed their oral cavities with 10 mL of tap water for 15 s and discarded the rinse. A 2-min interval was observed before rinsing the oral cavity again, this time with 10 mL of a 0.9% w/v salt water solution (normal saline), measured using a

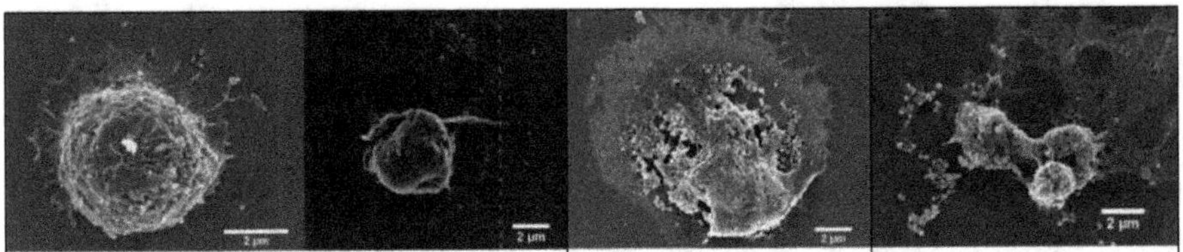

Initial activation stage	Strongly activated neutrophils showing irregular reticular structure	Globular domains and neutrophil extracellular traps

Figure 14.17 The sequential activation stages of neutrophil granulocytes on the chip as documented in the study by Bunnfors et al. [1128]

15 mL falcon tube, and ejecting the sample into a 50 mL falcon tube. This process was repeated five times, with a 2-min waiting period between each wash. Subsequently, the final sample was gently filtered through 40 μm filters to isolate cells from debris larger than 40 μm. The falcon tubes were then centrifuged at 4 °C for 5 min at 2,600 RPM (revolutions per minute), and the supernatant was discarded until 1 mL of the sample remained. To this remaining volume, 1 mL of ultrapure water was added and mixed with the cells. The mixture was then dried after adding EtOH and placed on the hemocytometer counting chamber. To establish a baseline, the output of the chip was evaluated without introducing any samples, to assess the inherent offset capacitance of the IDE. In wet mode, the presence of water on non-passivated chips could cause a saturation level increase in capacitance. Hence, data obtained before sample introduction and water evaporation were considered to determine the cell coverage on the electrode. Figure 14.18 (a–f) presents microscopic images of the chip surface at various stages: before sample introduction, after the first, second, third, and fourth samples, and after electrode washing. The smaller spherical cells represent neutrophils, while the larger cells with nuclei correspond to epithelial cells. The researchers measured the magnitude of capacitance changes after each run relative to the baseline before the first run and established a correlation between the number of cells covering the electrode surface and the capacitance variation.

14.3.4.4 Analyses of other types of cells

CMOS biosensors have been developed for various types of cells, including optical biosensors, which offer non-invasive and real-time monitoring of diverse cellular processes. Among these optical biosensors, handheld fluorometers stand out as versatile devices capable of measuring the metabolic activity and viability of cells. Nelson *et al.* [1132] developed a series of simple handheld fluorometers designed for bioassays, enabling the measurement of cellular metabolic activity and viability. These fluorometers utilize distinct chromophores with long-pass wavelengths of 400 nm and 540 nm, which allow for the detection of inherent fluorescence within cells. Of particular interest is the autofluorescent compound nicotinamide adenine

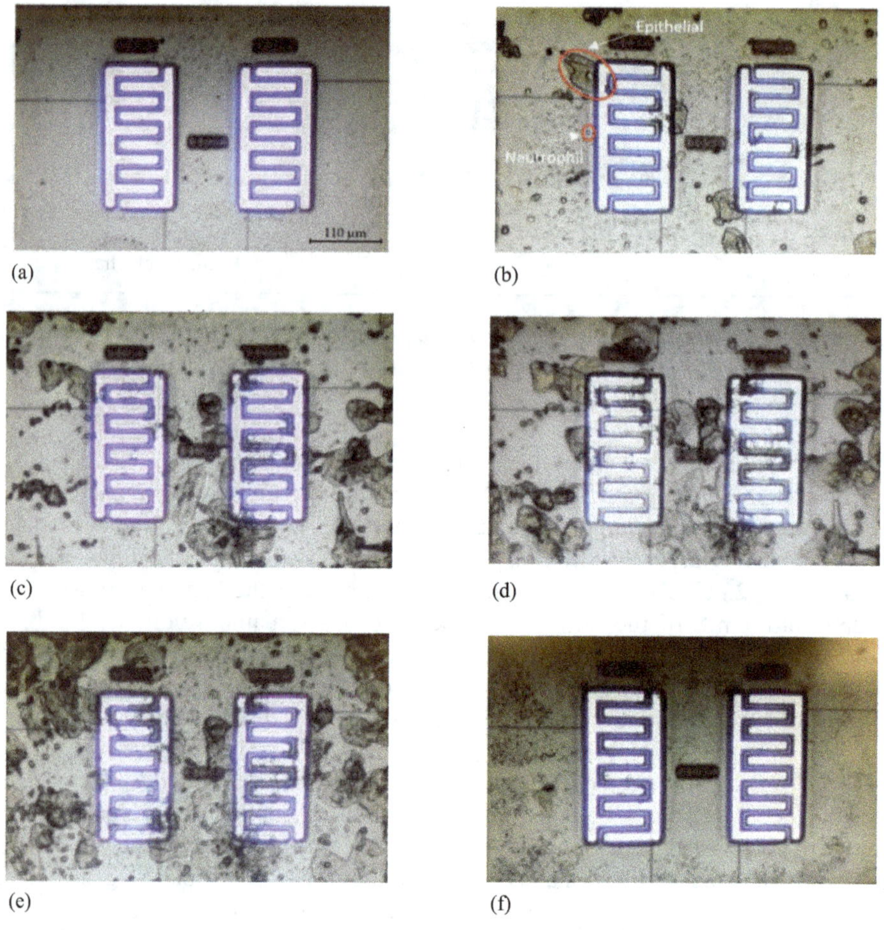

Figure 14.18 The surface of the chip under microscopic observation: (a) before sample introduction, (b) after the first sample, (c) after the second sample, (d) after the third sample, (e) after the fourth sample, and (f) after electrode washing [1129]

dinucleotide (NADH), which serves as an indicator of potential pathogens during cellular respiration. The fluorometers provide both end-point and online measurements, facilitating real-time analysis of samples. The handheld fluorometers are equipped with a differential active pixel sensor. Additionally, the devices feature an integrated optical filter containing a high-rejection chromophore embedded in a polymer film, which is cast onto the chip. The bioassays conducted with these handheld fluorometers involve the use of two different single-color fluorometers based on the chromophores 2-(2'-hydroxy 5′-methylphenyl) benzotriazole and Sudan II, with long-pass wavelengths of 400 nm and 540 nm, respectively. In one experiment, the study focuses on evaluating the cytotoxicity of two types of Poly (amidoamine) (PAMAM) dendrimers using the fluorometer. These dendrimers have potential applications in oral drug delivery, but their cytotoxicity needs to be well-characterized. The fluorometer utilizes an AQUA live/dead stain that reacts with amine groups inside cells, producing fluorescence intensity that indicates cell viability. Human intestinal adenocarcinoma cells (Caco-2 cell line) were exposed to two types of PAMAM dendrimers (G3.5 and G4), and their viability was measured using the handheld fluorometer. The results demonstrate that the G4 dendrimer family is cytotoxic, while the G3.5 family is not, consistent with previous assays. The handheld fluorometer can discriminate between viable and compromised cells based on fluorescence intensity, but it is unable to distinguish between different dosages of dendrimers. The experiment closely matches traditional cell viability assays' results, confirming the fluorometer's reliability. Additionally, a calibration curve of the sensor's response to the AQUA stain shows that it can reliably detect less than 12% of the recommended dosage. The second experiment explores the use of the fluorometer to measure the metabolic activity of cells based on the fluorescence of NADH. The fluorometer is tested using yeast cells as a test pathogen. A calibration curve is obtained for NADH measurement, and the sensor demonstrates the ability to detect low concentrations of NADH, though it exhibits self-quenching at very high concentrations. The NADH levels in a yeast culture are monitored over time, showing variations influenced by glucose concentration, dissolved oxygen, and pH. An online assay is conducted to monitor changes in intracellular NADH concentration during yeast spore germination. The fluorescence intensity reflects changes in NADH concentration as the cells switch between aerobic and anaerobic respiration. The availability of glucose also influences NADH production. However, in this experiment, the sensor cannot distinguish between extracellular and intracellular NADH. The third experiment involves using AlamarBlue dye as a cell viability indicator to assess the metabolic activity of cells through fluorometric analysis. The experiment is conducted with bovine aortic smooth muscle cells (BAOSMCs) in a custom sample well. The AlamarBlue dye exhibits increased fluorescence when metabolically active cells reduce it from a nonfluorescent to a fluorescent form. The handheld fluorometer, equipped with an emission filter suitable for AlamarBlue's excitation range, is placed in a humidified incubator at 37 °C and 5% humidity for 10 h. Fluorescence readings are taken at 10-min intervals. The dye is observed to be reduced over time, indicating metabolic activity in the cells. At the end of the 10 h, the dye is fully reduced, and further growth medium replenishment is required for continued cell viability assessment. The dye's reduction process can be visually confirmed by observing the change in color from dark blue to pink. The presence of viable BAOSMC cells as the source of dye reduction is confirmed through microscopic observation.

Among the diverse subjects studied using CMOS sensors, algae cells hold particular significance due to their wide range of shapes and sizes. Hu *et al.* [1225] conducted a study utilizing a high-frequency EIS impedance sensor composed of a grid of electrodes with a 10 μm pitch and successfully imaged individual *Cosmarium* cells, but sub-cellular details remained unresolved. To achieve simultaneous optical and electrochemical visualization, they enclosed the sensor in a small open fluid chamber and placed it under an inspection microscope. Their findings revealed a strong correlation between optical and impedance images, with improved impedance image quality at higher frequencies. They were able to detect different algae species, such as *Cosmarium turpinii* and *Closterium acerosum*, although subcellular structures were only occasionally visible. Furthermore, the study investigated the challenges associated with detecting and classifying smaller microalgae like *Cyclotella* sp., which appeared as single points in impedance images. The temporal analysis provided insights into the movement behavior and cellular growth of *Closterium* cells. Shape metrics and a classifier were employed to accurately label cells, and the potential for machine learning in analyzing electrochemical images shows promise for automated phenotype classification. In a separate endeavor, the same group [1131] specifically selected *Cosmarium turpinii* for its moderate cell size (around 50 μm in diameter) and distinctive bilobal shape, enabling better visualization of subcellular structures. To enhance impedance image resolution, they implemented a technique inspired by high-resolution optical image assembly methods. Multiple realigned impedance images were merged computationally, taking advantage of the varied signal provided by different kernel offsets. The approach involved upsampling the original images, compensating for offset-dependent image skew through shifts and shear mapping, and ultimately summing the stack of realigned images. The resulting composite super-resolution impedance image showcased the potential of this concept, although further algorithm development for image alignment and merging is necessary. Different kernel offsets emphasized specific features within the algae cells, suggesting the potential integration of these measurements with electrical impedance tomography algorithms to reconstruct a three-dimensional representation of the sample's permittivity.

14.3.5 Saccharide-based assays

Saccharide-based assays play a crucial role in various fields, ranging from medical diagnostics to food safety and environmental monitoring. These assays involve the detection and quantification of sugars and carbohydrates, which are essential biomolecules with numerous biological and industrial applications. The development of glucose biosensors has gone through several generations (see Figure 14.19(a–c)).

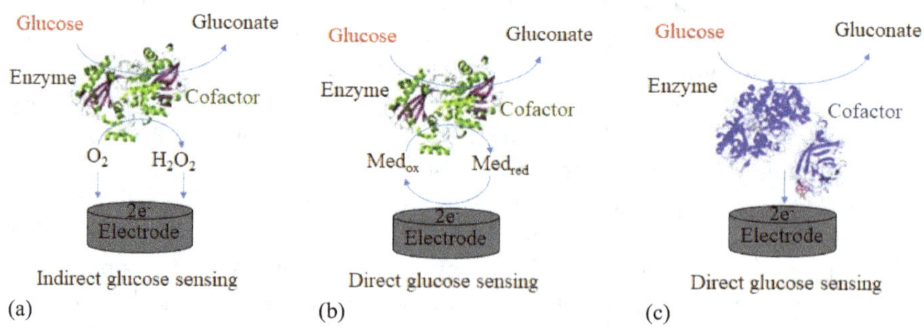

Figure 14.19 Different generations of glucose biosensors: (a) First generation, (b) second generation, and (c) third generation

The first-generation glucose biosensors was invented in 1962 by Clark and Lyons [1226]. The first glucose biosensor used an oxygen electrode sensor and glucose oxidase (GO_x) to catalyze the oxidation of glucose into gluconic acid. The resulting hydrogen peroxide production was measured using amperometric detection. The second-generation glucose biosensors were developed during the 1980s [1227]. Second-generation glucose biosensors introduced redox mediators to enhance electron transfer between the enzyme (GO_x) and the WE, thereby improving sensor performance. Third-generation glucose biosensors emerged in the 1990s [1228]. Third-generation glucose biosensors were reagentless and utilized direct electron transfer between the enzyme and the electrode without the need for mediators. These biosensors often used organic conducting materials based on charge-transfer complexes, allowing for continuous *in vivo* monitoring of blood glucose with superior selectivity [1229].

To meet the growing demand for sensitive and rapid saccharide detection, researchers have explored the potential of different CMOS biosensors, each offering unique advantages for this application.

One prominent CMOS biosensor used in saccharide-based assays is the capacitive sensor. Capacitive biosensors utilize changes in electrical capacitance caused by interactions between target saccharides and specific BREs, such as enzymes or receptors. Among the notable examples is the work of Ghafar-Zadeh and colleagues [1136], who developed a CMOS-compatible protein-based biosensor for glucose detection using genetically engineered glucokinase (GLK) receptor proteins. The biosensing system utilized a gold electrode surface modified with an octanedithiol SAM, coupled with a nitrilotriacetic acid (NTA) Ni^{2+} linker to bind histidine-tagged recombinant human GLK enzyme (refer to Figure 14.20). The surface preparation process involved the following steps: (1) The Au electrodes underwent thorough cleaning with H_2SO_4/H_2O_2, DI water, and absolute EtOH, followed by drying under nitrogen gas. (2) SAMs of 1,8-octanedithiol were formed on the electrodes through immersion in an EtOH solution. (3) Physioabsorbed materials were eliminated by rinsing the modified electrodes with EtOH. (4) The modified electrodes were immersed in a maleimide-C3-NTA solution to facilitate the coupling reaction between maleimide-C3-NTA and the sulfhydryl surface. (5) To remove any remaining physio-absorbed material, the electrodes were rinsed with DI water and MeOH. (6) Electrodes with NTA ligands were subjected to immersion in $NiCl_2$ to load Ni cations (Ni^{++}). (7) The electrodes underwent rinsing with DI water to eliminate residual metal and ion solution. (8) His-tagged GLK proteins were coordinated with Ni^{++}. (9) The electrodes were cleaned with DI water and buffer. (10) The biosensor was stored in a fresh buffer solution for preservation. The research team achieved remarkable sensitivity with a detection limit of 0.5 mM within the concentration range of 0.5 mM to 7.5 mM.

Bio/ChemFET-based sensors are another promising category of CMOS sensors extensively explored for saccharide-based applications. For example, Georgiou *et al.* [501] explore the utilization of silicon devices, specifically ISFETs, for modeling metabolic functions in the field of biology. The researchers present a biologically inspired silicon beta cell as an essential component of an artificial

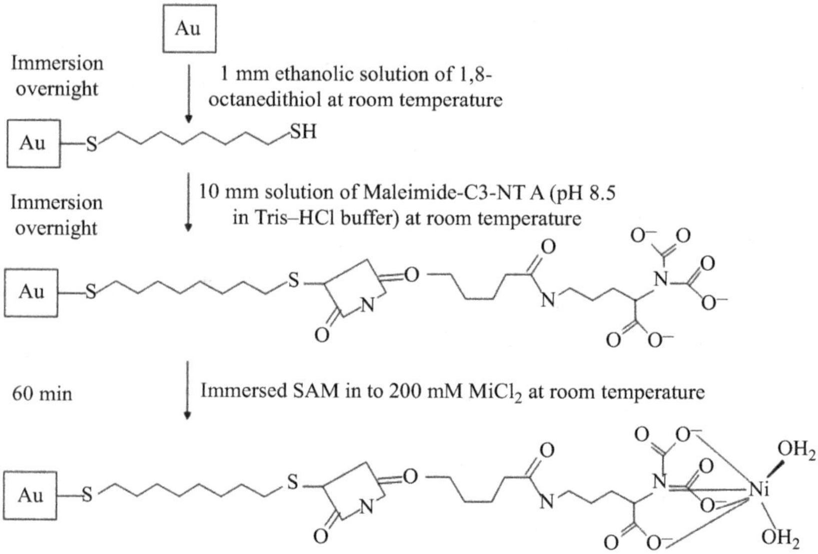

Figure 14.20 The surface reaction mechanism of a protein-based biosensor, illustrating the binding of NTA functional groups to the available sulfhydryl sites of a SAM formed by 1,8 octanedithiol on an Au electrode

pancreas, aiming to enable real-time glucose sensing and insulin release for individuals with diabetes. This study introduces the first silicon-based implementation of a metabolic cell that exhibits variable bursting behavior when stimulated by glucose. By leveraging the Hodgkin and Huxley formalism, the researchers achieve dynamics similar to natural beta cells by operating the devices in the subthreshold regime. The circuit design considers the glucose input as a current, which maps the physiological glucose range of 5 to 20 mM to an equivalent current of 5 to 20 nA. One possible approach to generate this current, allowing for monolithic integration, involves utilizing an ISFET based on glucose.

Another approach for saccharide-based assays involves voltammetric techniques. For instance, Ghoreishizadeh *et al.* [149,198] developed an integrated biosensor system that combines CV and chronoamperometry (CA) for detecting multiple molecular targets. Their custom IC allows for voltage control and current readout of the biosensors. The biosensors were fabricated using nanostructured WEs with multi-walled carbon nanotubes (MWCNTs) and enzyme drop-casting. CA was employed for glucose and lactate sensing, while CV was used to identify etoposide oxidation peaks. The IC measurements exhibited slightly higher LoD due to accuracy limitations and increased noise. However, the biosensing platform met the accuracy requirements for glucose monitoring systems.

In another effort, the same group [237] introduced a CMOS miniaturized amperometric electrochemical sensor with a fully integrated current readout circuit

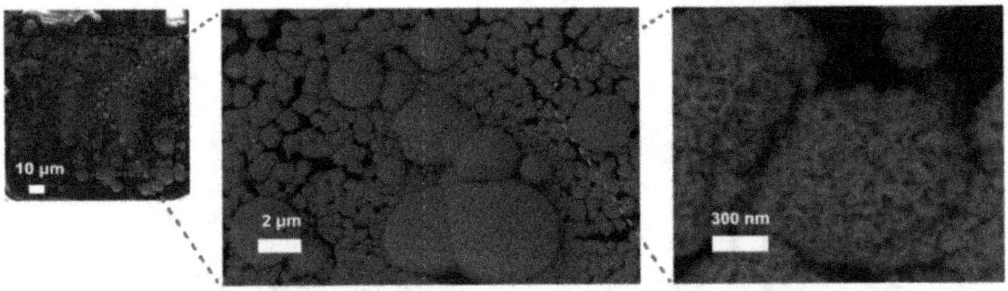

Figure 14.21 SEM images obtained at various magnifications to showcase the Pt-nanoS grown on the top metal layer of a CMOS substrate [237]

for differential current measurement. The sensor utilized platinum nanostructures (Pt-nanoS) grown on top of electrodes implemented with the top metal layer (shown in Figure 14.21). The incorporation of Pt-nanoS enabled highly sensitive and offset-free detection of metabolites. The system was tested for glucose measurement in submillimolar concentrations using CA. The results demonstrated that the introduction of Pt-nanoS significantly enhanced the sensor's sensitivity. Additionally, CV was conducted to detect H_2O_2, showcasing the successful detection capability of the system. SEM images confirmed the presence of Pt-nanoS on the electrodes. The on-die Pt-nanoS sensors exhibited variability, which could be addressed using differential measurements and a two-step calibration scheme. This work demonstrated the integration of miniaturized and highly sensitive sensors with CMOS readout systems, improving the detection of metabolites with enhanced SNR and LoD. The system achieved a significant cross-scale dynamic range, making it a pioneering achievement in the field of integrated readout circuits. The heterogeneous integration of Pt-nanoS sensors on CMOS technology was also a notable advancement showcased in this research.

14.3.6 Other applications

Bacteriophages (phages), which are viruses targeting bacteria, play a crucial role in this system. Phage libraries with specificities for various bacteria are commercially available, offering cost-effective and concentrated sources that can be stored for extended periods. The structure of a typical phage is depicted in Figure 14.22(a). Upon infecting their target bacteria, phages inject their DNA into the host cells, leading to the release of new phage entities.

In an effort, Nikkhoo *et al.* [416] introduced an innovative integrated system capable of detecting live bacteria in under 10 min. The setup merges the precision of bacteriophages as biological detection agents with the responsiveness of integrated ISFETs, realized using standard CMOS technology. To enhance its capabilities, a polyvinylchloride (PVC)-based potassium-sensitive membrane is incorporated as an additional post-process. Instead of relying on cell lysis, this system capitalizes on the chemical changes occurring during the initial infection process when the phage introduces its DNA into the cell. The output of the biological transducer is the

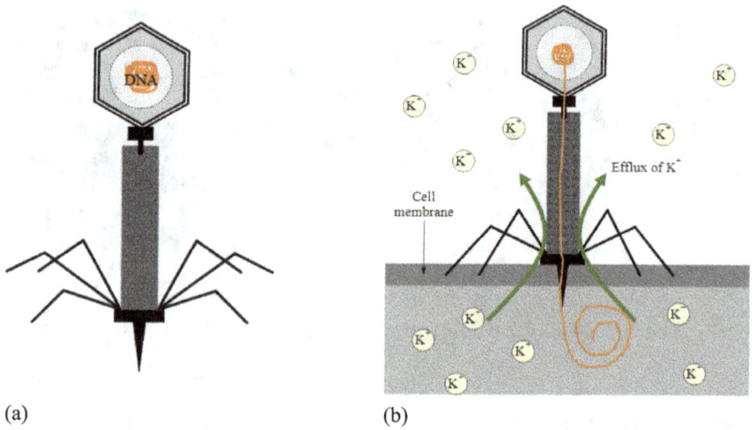

Figure 14.22 (a) Structure of a typical phage and (b) structure of phage attachment to the bacterial cell membrane, DNA injection, and efflux of potassium ions.

increased concentration of potassium ions in the sample after infection (as shown in Figure 14.22(b)). The experiments involved different strains of *E. coli*. Before each experiment, overnight cultures of *E. coli* were prepared by adding frozen stock aliquots of bacterial cells to LB-Lennox Broth (LB) and allowing them to grow overnight in a shaking incubator at 37 °C. On the day of the experiment, the selected overnight cultures were added to LB and grown further. The cell density was monitored using a spectrophotometer at a wavelength of 600 nm until the desired Optical Density (OD) was achieved. Independent experiments were conducted to establish the relationship between the measured OD values and the actual concentration in colony-forming units per mL (cfu/mL) for each cell type. Once the desired OD was reached, the cultures were collected by centrifugation, washed, and suspended in a 0.5 ml suspension media. The suspension media used for lambda phage experiments was SM (100 mM NaCl, 10 mM $MgSO_4$, and 50 mM Tris at pH 7.5), while 10% L-tryptophan was added to the SM for T_6 phage experiments. The prepared cells were immediately utilized for the measurements. Two specific phages, phage λcI857, and T_6 phage, were employed in the experiments. The phage stocks were prepared at a titer of 1,011 pfu/ml from low-concentration phage lysate stocks. Each phage stock was then dialyzed in clean SM media, stored at 4 °C, and used as needed for several months without a significant drop in titer. The experimental results are divided into three groups, involving four *E. coli* strains, two bacteriophages, and two different temperatures. The results include both positive and negative control experiments, demonstrating the sensor's specificity and detection capabilities. In the positive control experiment, the sample contains bacterial cells that are susceptible to the designated detector phage. As a result of phage infection in these sensitive cells, an expected release of potassium (K^+) ions occurs from the interior of the bacterial cells to the surrounding sample, where ISFET measurements are conducted. Conversely, the negative control experiment employs the same detector (phage), but the sample lacks sensitive cells. Instead, the sample contains bacterial cells at the same concentration as the positive control experiment, but they are resistant to the chosen phage. Since no infection takes place upon phage addition, no changes in potassium concentration are anticipated in the sample.

14.4 Summary

This chapter comprehensively categorizes the diverse life science applications of CMOS biosensors into six distinct groups, elucidating sample preparation protocols, sensing mechanisms, and conceptual frameworks detailed in the literature for each category. The versatility of CMOS biosensors has enabled the detection and monitoring of a wide spectrum of biological entities, encompassing cancer cells, cardiomyocytes, neural cells, bacteria, viruses, nucleic acids, proteins, small organic molecules, and saccharides. These label-free biosensors, employing techniques such as capacitive, impedimetric, voltammetric, and Bio/ChemFET-based sensors, offer a powerful alternative to labeling-dependent methods. Central to the effectiveness of these biosensors is the transducer-target interface, enabling the precise detection of ultra-low analyte concentrations. While some applications naturally form biointerface layers on the sensing surface, others necessitate functionalization with specific recognition elements for target species binding. Traditional BREs such as nucleic acids, antibodies, enzymes, and whole cells have found widespread use, while novel artificial recognition elements like phages, aptamers, affibodies, and molecularly imprinted polymers have emerged. The immobilization of receptors on biosensor surfaces employs diverse techniques, ranging from covalent attachment to physical adsorption and incorporation into membranes or polymers. Surface functionalization strategies encompass various approaches, such as hydroxyl, poly-L-lysine, amine, aldehyde, and epoxy coatings, tailored to specific applications. As CMOS biosensors continue to advance, their versatile and adaptable nature holds significant promise for advancing our understanding of complex biological processes and driving innovations in the life sciences.

Chapter 15
State-of-the-art and future works

This book explores the design and implementation techniques of complementary metal-oxide-semiconductor (CMOS)-based biosensors and actuators, highlighting their significant contributions to various life science applications. The research and development efforts in this field have led to the successful integration of these technologies into the market, as described in the following section. The wide range of applications, including disease diagnostics and automation in biological and chemical laboratories, necessitates the continuous development and customization of new devices to meet these diverse needs. In this direction, several challenges are discussed briefly in this chapter that need to be effectively addressed.

15.1 CMOS biosensors in the market

CMOS biosensors have emerged as a highly promising field of research, attracting significant attention for both scientific exploration and commercialization purposes. Over the years, thousands of papers have been published, highlighting the advancements and potential applications of CMOS biosensors. The appeal of CMOS biosensors stems from their utilization of CMOS platform technology, which offers numerous advantages, including low cost and compatibility with established manufacturing processes. Furthermore, the prospect of commercialization has significantly contributed to the popularity of CMOS biosensors. The utilization of CMOS platform technology offers cost advantages, making the production of CMOS biosensors more affordable compared to traditional biosensing technologies. This affordability, coupled with the scalability and mass-production capabilities of CMOS technology, has piqued the interest of industry stakeholders and entrepreneurs, fostering the rapid translation of CMOS biosensor research into commercial products. Some of the examples are described below.

15.1.1 CMOS deoxyribonucleic acid (DNA) sequencing

Ion Torrent, a company specializing in DNA sequencing, has developed CMOS-based DNA sequencing platforms. These platforms utilize CMOS biosensors to enable fast and cost-effective DNA sequencing by detecting pH changes during DNA synthesis. The Ion Torrent CMOS chip, with its integrated circuitry and millions of pH sensors, offers high-throughput DNA sequencing capabilities (Figure 15.1(a)). This commercialized CMOS DNA sequencing technology has revolutionized the field of genomics and has been widely adopted in research and clinical applications [1230].

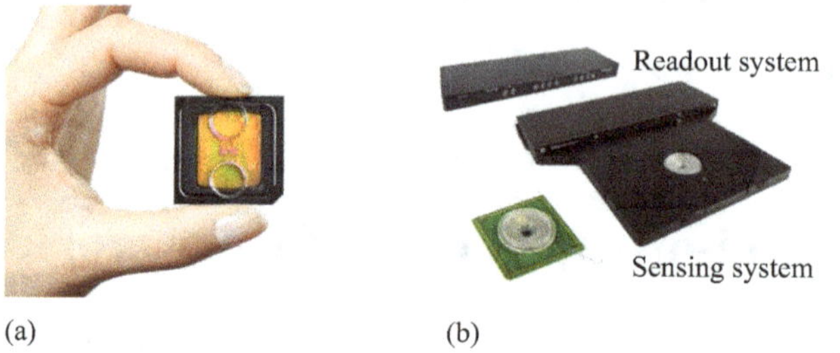

Figure 15.1 *Commercially available CMOS-based biosensors. A photograph of CMOS-based (a) DNA sequencing chip (Ion Torrent) and (b) electrophysiological recording chip (MEA5000 System, Multichannel system).*

15.1.2 CMOS-MEA (multielectrode array)

15.1.2.1 Recording system

Multi-channel systems have introduced a CMOS-based MEA recording system for extracellular recordings and stimulation. This system features a chip with more than 4,000 recording sites, each sampled at a high rate of 25 kHz. With on-chip amplification and noise reduction, the CMOS-MEA system ensures high-quality and high-resolution spatiotemporal recordings, allowing for advanced neuroscience research and neural interface applications (Figure 15.1(b)) [1231].

15.1.3 CMOS imagers

CMOS-based imagers have been commercialized and widely used in various applications, including microscopy and near-field imaging. These images can be attached directly to biological materials or integrated with microscopes, providing high-resolution imaging capabilities. The integration of CMOS imagers with biological materials enables label-free imaging and real-time monitoring of biological processes at the cellular and subcellular levels [1232].

These commercialized CMOS biosensors represent the growing trend of bringing advanced technologies to the market. They offer new avenues for research, diagnostics, and applications in fields such as genomics, neuroscience, and imaging. The integration of CMOS technology with biosensors has opened up possibilities for miniaturization, increased sensitivity, and higher throughput, presenting significant opportunities for future developments and advancements in the field. The potential market for wearable sensors and handheld point-of-care (PoC) devices is highly promising.

15.2 Biological CMOS (BioCMOS) challenges

Despite significant advances in BioCMOS systems, several challenges remain unaddressed. These challenges encompass various stages of the system, ranging from design and simulation to electronic circuit implementation, post-CMOS processes, microfluidics, and biorecognition elements (BREs). Each of these stages presents unique hurdles that need to be overcome to fully unlock the potential of BioCMOS technology.

15.2.1 Integrated sensor design and simulation

The field of integrated circuit design has a long history of extensive research and documentation. However, the application of integrated sensors using CMOS processes represents a relatively new area of exploration. As a result, dedicated software tools like Multiphysics COMSOL have not yet fully integrated with standard circuit design and simulation computer-aided design (CAD) tools. This lack of integration poses challenges in accurately estimating the sensor output, such as capacitance changes, based on the presence of specific target molecules or cells. Consequently, precise estimation of the sensor output becomes crucial for the successful development of effective readout circuitry.

15.2.2 Microfluidic packaging

The integration of CMOS and microfluidic chips poses a significant challenge, primarily arising from the disparity in dimensions between the silicon chip, which is typically a few millimeters in size, and the microfluidic chip, which can be a few centimeters in size. The limited space available on the small silicon chip makes it difficult to establish proper bonding with the microfluidic chip. Additionally, the roughness of the chip's surface further complicates achieving a hermetic bonding between the CMOS and microfluidic chips.

15.2.3 Selectivity

The immobilization of BREs on the chip's surface presents another challenging aspect, necessitating the use of a suitable linker to bind the BRE with CMOS process-compatible materials like silicon oxide or Aluminum. The selection of the appropriate BRE for each target molecule/cell requires careful consideration, along with the synthesis of the corresponding linker. In some cases, the process might involve the deposition of a layer of materials, such as a thin layer of gold, onto the CMOS chip. However, the deposition of materials on the CMOS chip introduces additional complexities, especially when adapting photolithography techniques to work with small die sizes. To mitigate these challenges, it may be advantageous to apply the required post-CMOS processes at the wafer level before dicing the chip into individual dies. This approach could streamline the immobilization process and enhance the integration of BREs onto the CMOS chips.

15.2.4 Reusability

The immobilization of BREs on CMOS chips or the deposition of bio-friendly materials presents a significant challenge for chip reusability. Ensuring an effective washing process is a key concern in this regard. Additionally, the exposure of

CMOS chips to biological agents raises considerations about treating them as single-use devices due to FDA restrictions. Despite numerous efforts to develop interfaces between CMOS chips and biomaterials to enable chip reusability, this challenge remains largely unaddressed.

It is important to note that the batch production of CMOS chips allows for the development of very low-cost small CMOS biosensors. However, the challenge of achieving reusability for larger CMOS sensors has not been adequately met. Finding solutions to enable the reusability of large CMOS sensors is an ongoing area of research and development in the field of bioelectronics.

15.3 Summary

In this chapter, a brief introduction to the current and potential market for CMOS-based biosensors and actuators was provided. Additionally, the challenges associated with these technologies were discussed briefly. By fostering collaboration between the integrated circuit research community and other scientists, these challenges can be effectively addressed, bringing this technology one step closer to the market.

Appendix A

Deposition of gold on aluminum electrodes

This appendix is dedicated to explaining the techniques for depositing gold (Au) on aluminum (Al) electrodes in CMOS technology.

A.1 Photolithographic deposition of Au on Al

The photolithographic deposition method involves using standard photolithography steps to deposit Au, as shown in Figure A.1. This technique utilizes a photomask to pattern the positions of Au electrodes on a sacrificial layer of photoresist. An interlayer of nickel (Ni) or chromium (Cr) is then deposited using e-beam or thermal evaporation techniques [51,60]. The next step involves depositing Au using either sputtering or e-beam techniques. The excess Au on unrequired regions is removed using the lift-off process [60].

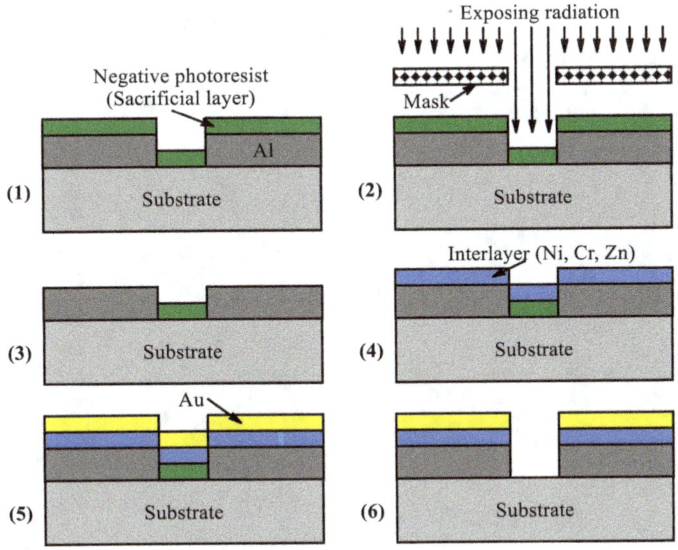

Figure A.1 The procedure of photolithographic deposition of Au

A.2 Electroless nickel immersion gold (ENIG)

The problem of photolithography can be addressed by using electroless deposition, which is also known as ENIG. This method involves a three-step process, as illustrated in Figure A.2. First, the chip is cleaned, rinsed, and dried, and then placed in a zinc (Zn) plating solution to replace Al ions with Zn ions. In the second step, the chip is immersed in a Ni-plating solution to replace the Zn layer with Ni. Finally, Au atoms are deposited on the Ni-plated regions to create an adhesion of Au to Al [60]. Another two-step electroless deposition process was presented in [66], which uses a small droplet of electroless solution without an interlayer. After treating the Al surface with Atomex solution, Au is deposited by galvanic displacement. Then, the Catagold 2A solution is used for electroless deposition. This method is cost-effective and allows for the creation of trench structures, making it suitable for the fabrication of bacteria-sized Au microelectrodes. The trench structure improves sensitivity by reducing the current flow from the peripheral part of the microelectrode, which is unaffected by the target. This structure also allows for decreased pitches by mitigating the expansion of the diffusion layer [1233,1234].

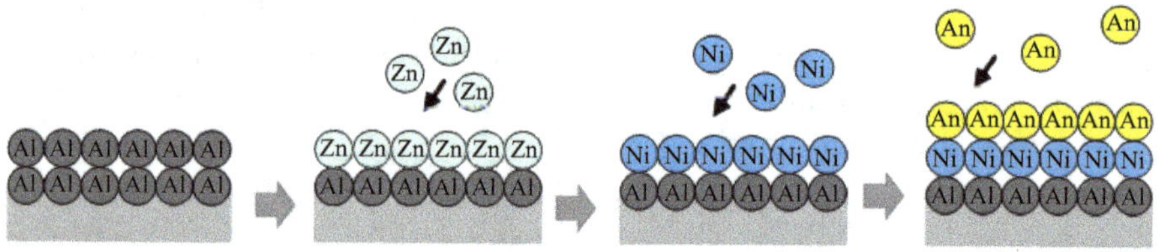

Figure A.2 The procedure of ENIG

Appendix B
Test bench details for a two-electrode capacitive sensor

This appendix is devoted to outlining the test bench design employed to test the capacitive sensor documented in [115].

B.1 Chip pin configuration and description

Table B.1 presents the pin configuration and corresponding explanations for each pin of the chip. The table pertains to the chip's pin arrangement within an 85-pin surface-mounted device (SMD) grid array package, manufactured by KYOCERA Corporation.

Figure B.1 illustrates the printed circuit board (PCB) that hosts the chip, shown as U0, four 3.3 low dropout (LDO) voltage regulators for analog and digital supply voltages (AVDD and DVDD, respectively) of right and left blocks shown as RAVDD, RDVDD, LAVDD, and LDVDD, a 1.85 V voltage regulator, shown as Vref, serving as a reference for the current-controlled oscillator (CCO), and a rail-to-rail multiplexer to switch between the right and left blocks shown as U1.

Table B.1 Pin configuration and explanation for each of the pins

Pin number on the package	Description	The physical location of the Pin on the Package
16	AVDD-R	H1
17	AGND	H2
18	Vref	J1
19	Nout-R	K1
20	DVDD-R	J2
21	DGND	L1
24	Phi3	L2
25	Phi2	L3
26	Phi1	K4
27	G3	L4
28	G2	J5
29	G1	K5
30	Sclk	L5
31	Sin	K6

(Continues)

Table B.1 (Continued)

Pin number on the package	Description	The physical location of the Pin on the Package
32	DVDD-Pad	J6
33	Ca7-R	J7
34	Sin	L7
35	Sclk	K7
36	G1	L6
37	G2	L8
38	G3	K8
39	Phi1	L9
40	Phi2	L10
41	Phi3	K9
43	DGND	K10
44	DVDD-L	J10
45	Nout-L	K11
46	Vref	J11
47	AGND	H10
48	AVDD-L	H11

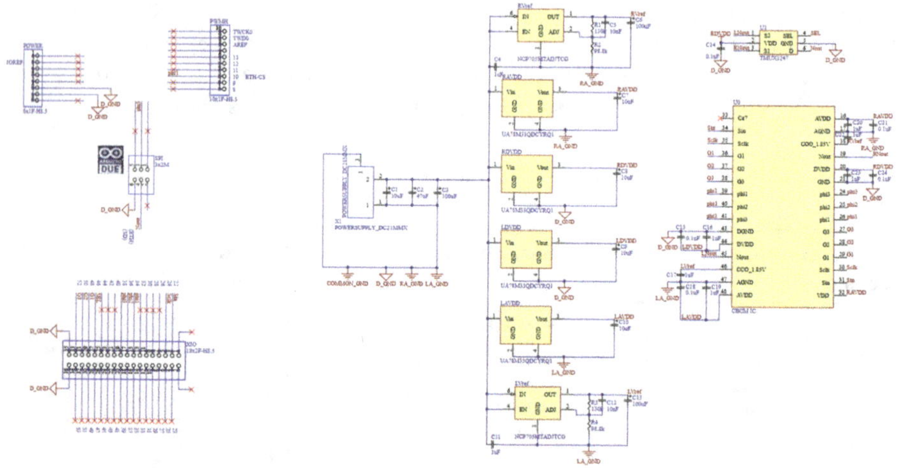

Figure B.1 Schematic block diagram of the implemented PCB

B.2 Test bench circuit and system

The board incorporates five distinct voltage regulators and a single multiplexer, as illustrated in the circuit schematic depicted in Figure B.1. Among these, four voltage regulators, namely TLV75533PDBVR, are employed. These regulators offer a 500-mA capacity, high power supply rejection ratio (PSRR), low quiescent current (IQ), and low dropout voltage characteristics, with a 5 V enable voltage requirement. Housed in small outline transistor (SOT)-23 packages and operating within a temperature range of −40 to 125 °C, two of these regulators are designated for the right side and the remaining two for the left side of the board, supplying 3.3 V to both analog and digital voltages.

To establish the necessary 1.85 V reference voltage for the oscillator's Vref pin, the TLV713185PDBVT regulator is employed. This regulator shares similar specifications, offering 500-mA capacity, high PSRR, low IQ, and low dropout voltage, with a 5 V enable voltage requirement. Housed in a SOT-23 package and operational within a temperature range of −40 to 125 °C.

Incorporated into the design is the TMUX1247DCKR, a precision analog multiplexer featuring a 30 Ohm ON resistance and rail-to-rail capability. To mitigate noise, decoupling capacitors are strategically positioned across all component input and output power supplies.

For a comprehensive overview of the chip's connectivity, Table B.1 provides an exhaustive breakdown of all the chip's pins. Refer to Figure B.1 for a detailed schematic representation.

B.2.1 Required timing scheme

The current output from enhanced charge-based capacitance measurement (ECBCM) (i'_{out} in Figure 4.46) is directed into a current-controlled relaxation oscillator, employing a similar configuration as proposed in [1235]. The arrangement is illustrated in Figure B.2(a). Adjustments have been made to the transistor's width/length (W/L) ratio (refer to Table B.1 for dimension details) within this oscillator to enhance the CCO's input dynamic range (IDR) to approximately 280 μA, thereby allowing the oscillation frequency to extend up to 300 MHz. Incorporating C_{O1}, C_{O2}, and C_{O3}, along with S_{O1}, S_{O2}, and S_{O3} switches offers a mechanism to flexibly adjust the oscillation frequency for testing objectives. The integration of the P_8 and N_6 pathway establishes a connection between ECBCM and the CCO. For deriving a digital output from the oscillation frequency, a counter mechanism is incorporated. Accompanying this is the design and integration of a serializer, responsible for loading the counter value in each measurement cycle and subsequently producing the value in the following cycle (refer to Figure B.2(b) for the counter and serializer configuration, Figure B.2(c) for the timing signal generation circuit, and Figure B.2(d) for the timing diagram). During the low phase of Φ_1, charging takes place for C_R and C_{IDE}, necessitating a pulse count while Φ_1 remains low. This criterion is met by the NOR gate at the counter's input. The serializer, synchronized with the clock

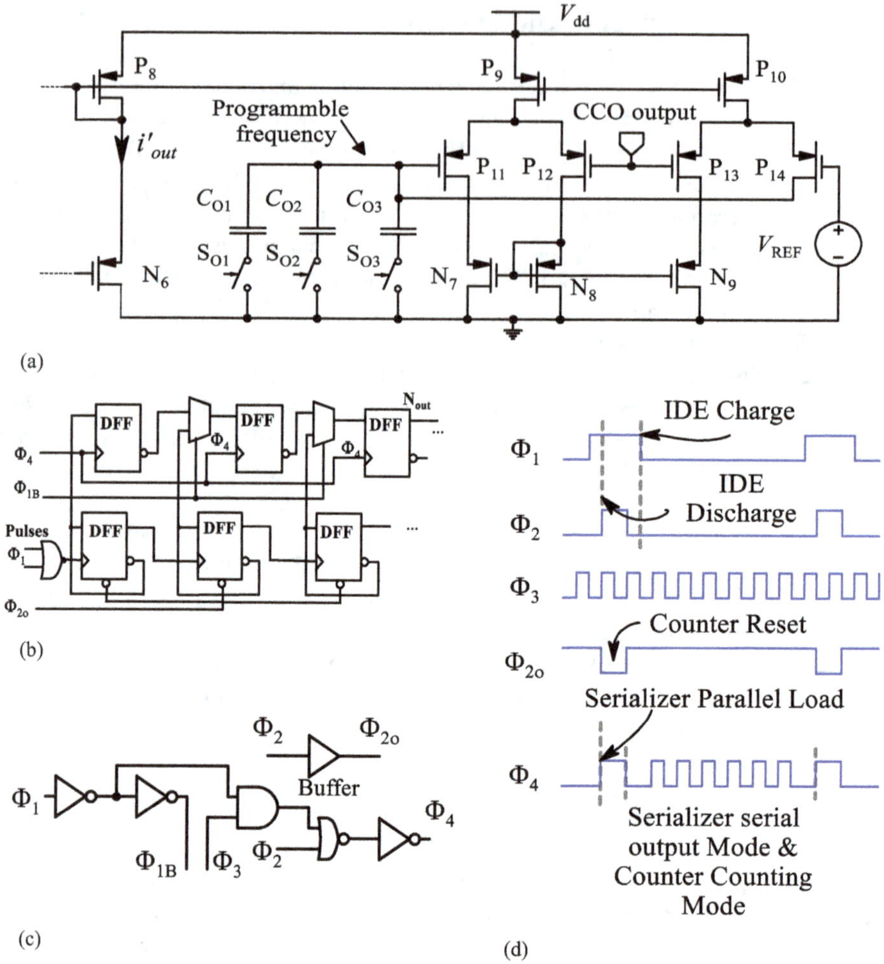

Figure B.2 Utilized circuit configurations. (a) The relaxation oscillator-based linear CCO as presented in [1235]. (b) Utilization of digital flip-flops in the counter-serializer configuration. (c) The circuit responsible for producing timing signals. (d) Timing diagram for the counter-serializer configuration

pulse Φ_3, outputs the previously loaded value from the preceding cycle. Within each cycle, as Φ_2 transitions to a high state, the generated number is parallel-loaded into the serializer. To maintain the counter value at zero while capacitors discharge, the counter's reset signal is the inverse of Φ_2. Notably, although the actual counter-serializer encompasses 12 bits, for simplicity, only 3 bits are displayed in Figure B.2(b). It is imperative to select a suitably high frequency for Φ_3, ensuring the output of all 12 bits when Φ_1 is low.

B.2.2 Serial peripheral interface (SPI) port communication

The serial peripheral interface (SPI) protocol employed here is a synchronous 4-wire, full-duplex communication scheme. Alternatively, half-duplex mode can be utilized when unidirectional communication suffices. In our specific scenario, the microcontroller unit (MCU) has been configured as the master, while the chip serves as the slave. Data transmission transpires from the chip to the MCU via a shared clock line, denoted as Φ_3.

Data reception is triggered by a falling edge on the slave select (ss) pin. In our system, Φ_3, serving as the pulse width modulation (PWM) signal output from the MCU, functions as the common clock signal. The Multiple Input/Single Output (MISO) pin on the chip connects to the MISO pin on the microcontroller board (Arduino DUE), establishing the output pathway.

Additionally, Φ_1 assumes the role of the ss pin. It facilitates SPI data reception when brought low, as illustrated in the timing diagram shown in Figure B.2(d) of the timing diagram section. The interconnections between these components are visually outlined in Figure B.1.

B.3 Printed circuit board and assembly

Figure B.3 illustrates the PCB board of the circuit schematic shown in Figure B.1. The list of the used components is given in Table B.2.

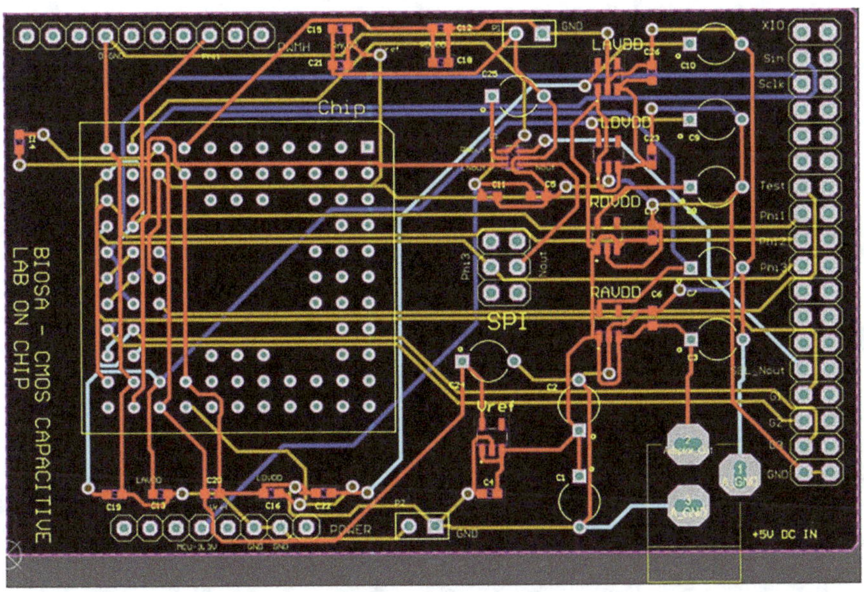

Figure B.3 The implemented PCB for the circuit shown in Figure B.1

Table B.2 List and detail of the components on the PCB

Designator	Comment	Link
C1, C2, C3, C7, C8, C9, C10, C24, C25	Cap	–
C4, C5, C6, C12, C14, C15, C17, C19, C20, C22, C23, C26	CAP	https://www.digikey.ca/en/products/detail/taiyo-yuden/EMK107BJ105KA-T/931028
C11, C13, C16, C18, C21	Cap	https://www.digikey.ca/en/products/detail/samsung-electro-mechanics/CL10B104KB8NNNL/10320686
LAVDD, LDVDD, RAVDD, RDVDD	TLV75533PDBVR	https://www.digikey.ca/en/products/detail/texas-instruments/TLV75533PDBVR/9685550
P1, P2	DGND and AGND ground Header 2	–
POWER	8x1M-H8.5	–
PWMH	10x1M-H8.5	–
SPI	3x2F-H8.5	–
U0	Chip Socket	–
U1	TMUX1247DCKR	https://www.digikey.ca/en/products/detail/texas-instruments/TMUX1247DCKR/10445173
Vref	TLV71385PDBVT	https://www.digikey.ca/en/products/detail/texas-instruments/TLV713185PDBVT/10665155
XIO	18x2F-H8.5	–

B.4 Microcontroller design

The test bench employs an Arduino DUE development board as the microcontroller, which is assigned various specific tasks outlined as follows. Each section provides a concise task description along with the corresponding disclosed firmware.

B.4.1 Generation of Φ_1, Φ_2, Φ_3 clock pulses

Three PWM channels of the Arduino Due board have been used to generate Φ_1 (15 μs period, 12 μs duty cycle), Φ_2 (15 μs period, 8 μs duty cycle, and 2 μs delay), and Φ_3 (1 μs period, 0.5 μs duty cycle). The corresponding code for the clock pulse generation is given below with comments on each line of the code.

```
REG_PMC_PCER1 |= PMC_PCER1_PID36;// Power up PWM clock
 REG_PWM_WPCR = PWM_WPCR_WPCMD(0);//Unlock user interface for PWM
 REG_PWM_CLK = PWM_CLK_PREA(0) | PWM_CLK_DIVA(21); //config the PWM clock generator: clock A,
84/21 = 4MHz
 REG_PWM_SCM = PWM_SCM_SYNC0 | PWM_SCM_SYNC1; //Synchronize phi1 and phi2 for they have the
same frequency
 //config phi1, Output pin PC2, Digital pin 34
 REG_PIOC_ABSR |= PIO_ABSR_P2; // Transfer Pin control from PIO to PWM
 REG_PIOC_PDR |= PIO_PDR_P2;   // Set PWM pin to an output
 REG_PWM_CMR0 = PWM_CMR_CPRE_CLKA; //mode configuration for channel 0: clock A, left aligned,
waveform start at high
 REG_PWM_CPRD0 = PWM_CPRD_CPRD(60);//config period for pwm 60/(4MHz)
 REG_PWM_CDTY0 = PWM_CDTY_CDTY(16);// 80/3% duty cycle(16)
 REG_PWM_ENA = PWM_ENA_CHID0;// enable PWM channel0

 //config phi2, Output pin PC4, Digital pin 36
 REG_PIOC_ABSR |= PIO_ABSR_P4; // Transfer Pin control from PIO to PWM
 REG_PIOC_PDR |= PIO_PDR_P4;   // Set PWM pin to an output
 REG_PWM_CMR1 = PWM_CMR_CPRE_CLKA | PWM_CMR_DTE; //mode configuration for channel 0: clock A,
left aligned, waveform start at high
 REG_PWM_CPRD1 = PWM_CPRD_CPRD(60);//config period for pwm 60/(4MHz)
 REG_PWM_DT1 = PWM_DT_DTL(4); //create the 1us(delay)
 REG_PWM_CDTY1 = PWM_CDTY_CDTY(12);// 80/6% duty cycle(12)
 REG_PWM_ENA = PWM_ENA_CHID1;// enable PWM channel1

 //config phi3, Output pin PC6, Digital pin 38
 REG_PIOC_ABSR |= PIO_ABSR_P6; // Transfer Pin control from PIO to PWM
 REG_PIOC_PDR |= PIO_PDR_P6;   // Set PWM pin to an output
 REG_PWM_CMR2 = PWM_CMR_CPRE_CLKA;//mode configuration for channel 0: clock A, left aligned,
waveform start at low |PWM_CMR_CALG
 REG_PWM_CPRD2 = PWM_CPRD_CPRD(4);//config period for pwm 2/(4MHz)
 REG_PWM_CDTY2 = PWM_CDTY_CDTY(2);//50% duty cycle(1)
 REG_PWM_ENA = PWM_ENA_CHID2;// enable PWM channel2
```

B.4.2 SPI communication

SPI communication occurs utilizing Φ_3 as the SPI clock and Φ_1 as the slave select, as elaborated in the SPI communication section. The associated code for generating clock pulses is provided below, complete with explanatory comments for each line of the code.

```
//SPI serial receive
 REG_PMC_PCER0 |= PMC_PCER0_PID24;// Power up SPI clock
 REG_SPI0_WPMR = 0 << SPI_WPMR_WPEN; //Unlock user interface for SPI
 //Instance SPI0, MISO: PA25, (MISO), MOSI: PA26, (MOSI), SCLK: PA27, (SCLK), NSS: PA28, Digital pin 10
 REG_PIOA_ABSR |= PIO_ABSR_P25; // Transfer Pin control from PIO to SPI
 REG_PIOA_PDR |= PIO_PDR_P25;  // Set MISO pin to an output
 REG_PIOA_ABSR |= PIO_ABSR_P26; // Transfer Pin control from PIO to SPI
 REG_PIOA_PDR |= 0 << PIO_PDR_P26; // Set MOSI pin to an input
 REG_PIOA_ABSR |= PIO_ABSR_P27; // Transfer Pin control from PIO to SPI
 REG_PIOA_PDR |= 0 << PIO_PDR_P27; // Set SCLK pin to an input
 REG_PIOA_ABSR |= PIO_ABSR_P28; // Transfer Pin control from PIO to SPI
 REG_PIOA_PDR |= 0 << PIO_PDR_P28; // Set NSS pin to an input
 //REG_ISER0 = 1<<24; //Enable interrupt controller
 REG_SPI0_CR = 1; // Enable SPI
 REG_SPI0_MR = 0; // Slave mode
 REG_SPI0_IER = 1 << SPI_IER_RDRF | 1 << SPI_IER_OVRES | 1 << SPI_IER_NSSR; //Enable interrupts
 SPI0->SPI_CSR[0] = SPI_CSR_BITS_12_BIT; // Capture on the falling edge and transfer 12 bits.
```

B.4.3 Setting the value of the reference capacitor

When the right circuit is chosen, the microcontroller employs General Purpose Input/Output (GPIO) pins to load 8 bits into the shift register. Similarly, when the left circuit is selected, 7 bits are loaded into the shift register. The subsequent code exemplifies this process and is annotated for clarity.

```
void inputSig(char testSeq[])
{
 delayMicroseconds(5);
 //fully reset the input registers
 for (int i = 13; i > 0 ; i--)
 {
  digitalWrite(22, LOW);
  digitalWrite(24, LOW);
  delayMicroseconds(5);
  digitalWrite(24, HIGH);
  delayMicroseconds(5);
 }
 //apply input sequences
 for (int i = 0; i <=7 ; i++)
 {
  if (testSeq[i] == '0')
  {
  digitalWrite(22, LOW);
  digitalWrite(24, LOW);

  delayMicroseconds(5);
  digitalWrite(24, HIGH);
  delayMicroseconds(5);
 }
 else if (testSeq[i] == '1')
 {
  digitalWrite(22, HIGH);
  digitalWrite(24, LOW);
  delayMicroseconds(5);
  digitalWrite(24, HIGH);
  delayMicroseconds(5);
 }
 else
 {
  digitalWrite(22, LOW);
  digitalWrite(24, LOW);
  delayMicroseconds(10);
 }
}
digitalWrite(22, LOW);
digitalWrite(24, LOW);
delayMicroseconds(5);
```

B.5 Graphical user interface (GUI)

B.5.1 Basic communication using MCU

The microcontroller and GUI establish communication through four simple commands transmitted via the serial port: 'S', 'G', 'W', and 'R'. These commands serve to set the number of samples, configure the CCO, assign the value of C_R, and receive data from the SPI port.

Upon the 'S' command, the GUI transmits the count of samples, which informs the microcontroller about the required repetitions for improved precision and noise reduction. Following the 'G' command, the GUI sends a user-defined string, such as 000, 001, 011, or 111. The oscillator's inherent frequency is highest when the input is 000 and lowest when it's 111. The order of bits in these strings holds no significance. Subsequently, upon the 'W' command, a string of ones and zeros (7 or 8 bits) is transmitted to set values within the shift register. The specific bits are assigned to either the left or right block of capacitors.

Upon the 'R' command, the microcontroller retrieves data from the SPI port receive buffer. The following code section provides an annotated explanation for this functionality.

```
void loop() {
 char calin_i[8];
 char calin_g[3];
 if (Serial.available() > 0) // check if serial is available
 {
  //Serial.println(2);
  char cmd = Serial.read(); // read the serial and put in cmd
  switch (cmd)
  {
   case 'S': // if the command was 'S'
   {
    String N_sample_s = Serial.readStringUntil('\0'); // read the serial again in the form of nonstop until reaching a '\0'
and put in N_sample_s
    N_sample = N_sample_s.toInt(); // Convert N_sample_s to integer and put in N_sample

    Serial.print(N_sample);
    Serial.print(TERMINATION_CHAR);
    break;
   }
   case 'G': // if the command was 'G', the oscillation frequency will be set
   {
    for (int i = 0; i < 3; )
    {
     if (Serial.available() > 0)
     {
      calin_g[i] = Serial.read(); // read each value in the CCO box (3 digits must be entered) and set G1(48+0),
G2(48+2) and G3(48+4)
      if (calin_g[i] == '1')
       digitalWrite(48 + 2*i, HIGH);
      else
       digitalWrite(48 + 2*i, LOW);
      i++;
     }
    }
    calin_g[3] = '\0'; // write a \0 at the end of the calin_g[i] string to make it ready for serial print to show the
message on the GUI
    Serial.print(calin_g); // print out calin_g
    Serial.print(TERMINATION_CHAR);
    break;
   }
```

```
        case 'W': // if the command was 'W' start setting the CR value
        {
         for (int i = 0; i <= 7; )
         {
          if (Serial.available() > 0)
          {
           calin_i[i] = Serial.read(); // read the string of size 8 and put in calin_i
           i++;
          }
         }
         if(calin_i[7] == ' ') // if '0' -> ch=Left, if '1' -> ch=Right
         {
          digitalWrite(46, LOW);
         }
         else
         {
          digitalWrite(46, HIGH);
         }

         inputSig(calin_i); //Now that the calin_i has been filled with the values sent from the GUI, call the inputSig
    function to apply it
         calin_i[8] = '\0'; // put a \0 at the end of the string and send out the whole string
         Serial.print(calin_i);
         Serial.print(TERMINATION_CHAR);
         break;
        }
        case 'R': //If the command was R, then it is time to read the data from the SPI port
        {
         int buf[N_sample];
         for (int i = 0; i < N_sample;)
         {
          if (REG_SPI0_SR & 1)
          {
//           buf[i] = i;
           buf[i] = REG_SPI0_RDR;
           i++;
          }
         }

         char bufferr[4*N_sample+N_sample];
         int n = 0;
         for (int i = 0; i < N_sample; i++) {
         n += sprintf (&bufferr[n], "%04d,", buf[i]);
        }
        bufferr[4*N_sample+N_sample-1] = '\0';

        Serial.print(bufferr);
        Serial.print(TERMINATION_CHAR);
        break;
       }
       default:
       {
        Serial.print("Meow");
        Serial.print(TERMINATION_CHAR);
        break;
       }
      }
     }
    }
```

B.5.2 GUI main configuration window

The commands are generated based on various input options integrated into the graphical interface. Figure B.4 illustrates the designed GUI. Upon executing the GUI (by opening the Windows command prompt through a search for "cmd," navigating to the GUI's directory address, and inputting "main.exe"), the GUI will automatically detect a connected microcontroller board via the universal serial bus (USB) port on the personal computer (PC). Clicking the "Connect" button establishes serial communication with the microcontroller, enabling user interaction. Once connected, input fields become active, allowing users to configure the measurement process.

As this GUI caters to both the 2-electrode complementary metal-oxide-semiconductor (CMOS) chip and the Array chip, an initial selection between "Cmos" or "Array" must be made based on the utilized chip. Subsequently, the channel can be specified as "Left," "Right," "Internal Capacitance," or "Left&Right." Opting for the "Internal Capacitance" mode opens the transmission gate on the right, effectively disconnecting interdigitated electrode (IDE) from the circuit.

Measurement mode offers two choices: "Capacitance Sweep" or "Fixed Point." In the "sweep" mode, the programmable reference capacitance undergoes a complete range sweep, while the "fixed point" mode necessitates user input for the reference capacitance value, ranging from 0 to 127 (representing 0 fF to 1270 fF, respectively).

CCO resolution, denoted by values such as 000, 001, 011, or 111, can be chosen. Opting for 000 offers the highest resolution. Users input the desired number of measurement repetitions in the "Number of Samples" field. Raw measurement values are sequentially recorded in the Excel sheet, while plot creation is based on their averages. A higher sample count results in smoother curves but extends measurement time due to repetition.

The total number of experimental repeats is determined by the value entered in the "Repeats" field. For instance, entering 20 will yield 20 distinct curves and corresponding data rows in the Excel sheet. The GUI accommodates both time-based and repeat-based measurements. If nonzero values are provided for "Repeat Interval" and "Stop Time," the GUI calculates the number of repeats by dividing stop time by the interval, suitable for uncertain experiment durations. If the user-input interval is shorter than the measurement time, a warning appears in the measurement log. While this informs the user, it does not interrupt the GUI's measurement process. In cases where either repeat interval or stop time is set to zero, the "Repeats" field determines the repetition count for measurements.

B.5.3 Plot types

As depicted in Figure B.4, the interface features a field for selecting either a two-dimensional (2D) or three-dimensional (3D) plot type. The primary distinction between these modes lies in the inclusion of a time axis within the 3D view, while the 2D view consolidates all time-resolved curves onto a single 2D plot. Notably, when the right electrode is chosen, an additional persistent 2D curve is displayed, showcasing the extracted capacitance value for the right side. This technique involves utilizing the pulse count acquired from the internal curve for $C_R = 400$ fF and subsequently identifying capacitance values within the right block that correspond to the obtained pulse count. This process aids in deriving the accurate capacitance measurement for the right side.

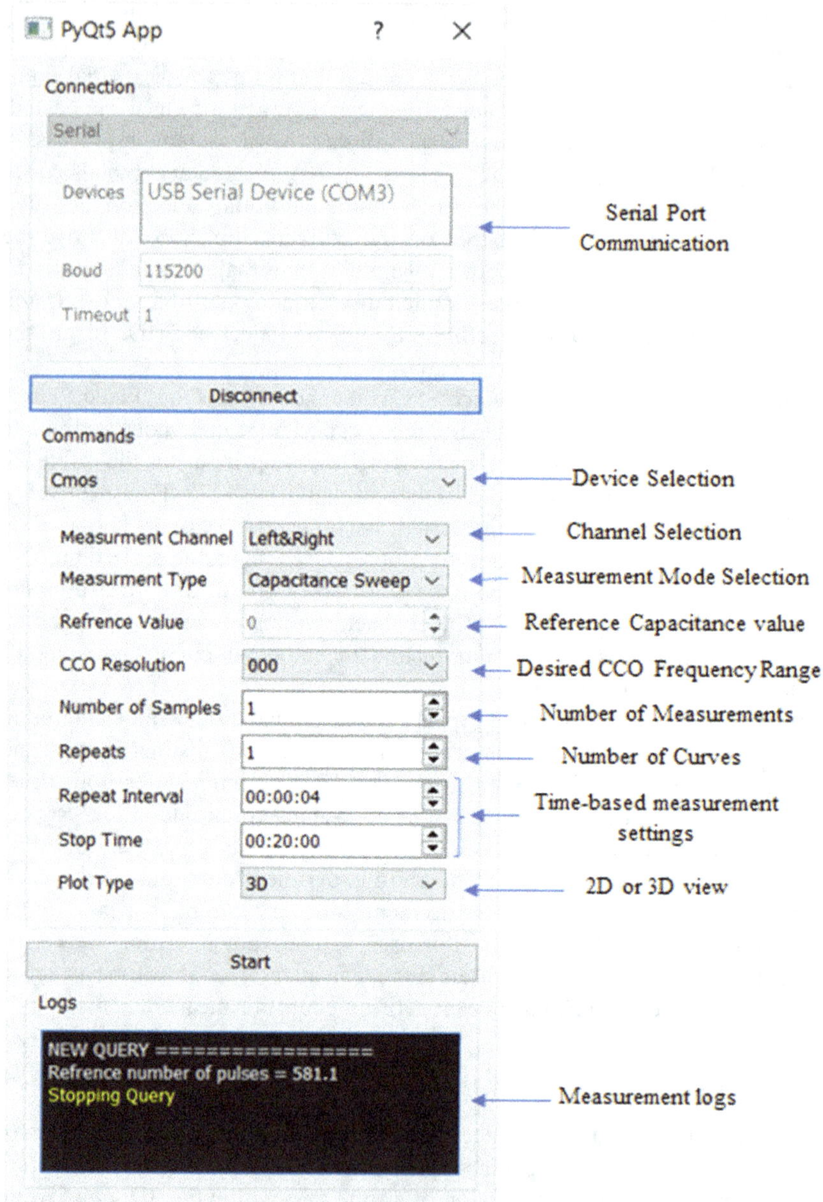

Figure B.4 A snapshot of the GUI

B.5.4 Saved Excel sheets

The data is written in real-time in Excel sheets. As a result, the Excel sheets must not be kept open during the experiments. The Excel sheets can be found in the

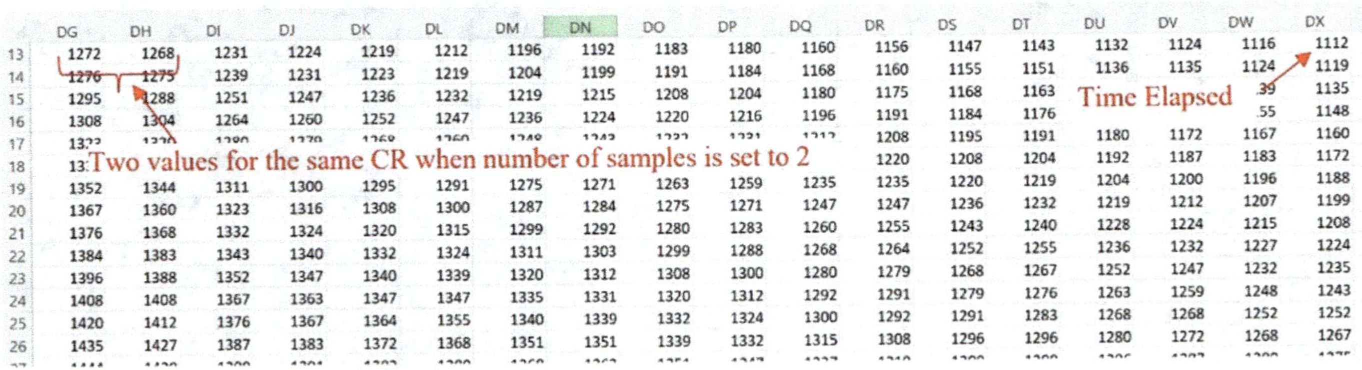

	DG	DH	DI	DJ	DK	DL	DM	DN	DO	DP	DQ	DR	DS	DT	DU	DV	DW	DX
13	1272	1268	1231	1224	1219	1212	1196	1192	1183	1180	1160	1156	1147	1143	1132	1124	1116	1112
14	1276	1275	1239	1231	1223	1219	1204	1199	1191	1184	1168	1160	1155	1151	1136	1135	1124	1119
15	1295	1288	1251	1247	1236	1232	1219	1215	1208	1204	1180	1175	1168	1163	[illegible]	[illegible]	[illegible]39	1135
16	1308	1304	1264	1260	1252	1247	1236	1224	1220	1216	1196	1191	1184	1176	[illegible]	[illegible]	[illegible]55	1148
17	[illegible]	[illegible]	[illegible]	[illegible]	[illegible]	[illegible]	[illegible]	[illegible]	[illegible]	[illegible]	[illegible]	1208	1195	1191	1180	1172	1167	1160
18	[illegible]	[illegible]	[illegible]	[illegible]	[illegible]	[illegible]	[illegible]	[illegible]	[illegible]	[illegible]	[illegible]	1220	1208	1204	1192	1187	1183	1172
19	1352	1344	1311	1300	1295	1291	1275	1271	1263	1259	1235	1235	1220	1219	1204	1200	1196	1188
20	1367	1360	1323	1316	1308	1300	1287	1284	1275	1271	1247	1247	1236	1232	1219	1212	1207	1199
21	1376	1368	1332	1324	1320	1315	1299	1292	1280	1283	1260	1255	1243	1240	1228	1224	1215	1208
22	1384	1383	1343	1340	1332	1324	1311	1303	1299	1288	1268	1264	1252	1255	1236	1232	1227	1224
23	1396	1388	1352	1347	1340	1339	1320	1312	1308	1300	1280	1279	1268	1267	1252	1247	1232	1235
24	1408	1408	1367	1363	1347	1347	1335	1331	1320	1312	1292	1291	1279	1276	1263	1259	1248	1243
25	1420	1412	1376	1367	1364	1355	1340	1339	1332	1324	1300	1292	1291	1283	1268	1268	1252	1252
26	1435	1427	1387	1383	1372	1368	1351	1351	1339	1332	1315	1308	1296	1296	1280	1272	1268	1267

Figure B.5 A snapshot of the saved data in the Excel sheet

GUI's directory in a folder named output. The name of each Excel sheet is very meaningful. It starts with *Right*, for the right channel, *Left*, for the left channel, and C_{IDE} for the values of the extracted capacitance. Then, the type of measurement is shown with the name of *SM* or *FP* representing *sweep* mode or *fixed-point* experiment. Next, the data followed by the time of measurement appears. Finally, the number after *Cmos* represents the number of samples that the user has inputted. Inside the Excel sheet, each row corresponds to one measurement curve. In the case that the number of samples is more than one, the repeated numbers show the obtained values for a given C_R beside each other. At the end of the row, the total time spent on data acquisition of the row is shown. This is valuable data showing the user the amount of real-time passed for obtaining each curve. The C_{IDE} is the extracted capacitance from each curve. As a result, it is only one point value at each row. See Figure B.5 for details.

B.6 Steps for performing the experiments

1. Begin by placing the chip in position before proceeding to connect the power and USB.
2. Initiate the process by connecting the USB cable as illustrated in Figure B.6. Utilize the programming port of the Arduino board, specifically the micro USB port located near the power jack.
3. Establish a connection between the PCB (not the Arduino board) and the AC power adapter, following the guidance provided in Figure B.6.

Figure B.6 A snapshot of the Test bench

4. Launch the GUI. To initiate the GUI, familiarize yourself with the directory containing all GUI files. Open the Windows command prompt by typing "cmd" into the Windows search bar. Set the directory within the command prompt window, which can be accomplished by copying the folder's address and pasting it into the window after typing "cd". Ensure there is a space after "cd" before pasting the address. Type "main.exe" (refer to Figure B.7). This action will execute the GUI.
5. Access the Excel sheets by navigating to the "outputs" folder within the same directory where the GUI is located.
6. Before chip removal, always disconnect the power source and USB connection to ensure safe handling.

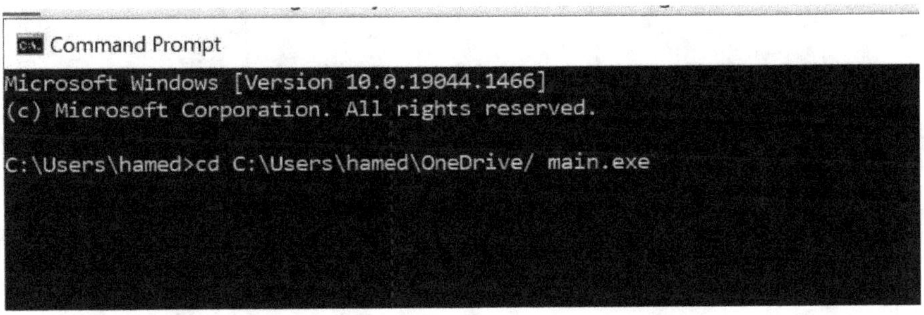

Figure B.7 A snapshot of the command prompt window to run the GUI

Appendix C
Test bench details for an array of core-CBCM capacitive sensors

This appendix is devoted to outlining the test bench design employed to test the array capacitive sensor as documented in [114].

C.1 Chip pin configuration and description

Table C.1 provides details regarding the pin configuration and corresponding explanations for each pin of the chip. The table is dedicated to the pin

Table C.1 Pin configuration and explanation for each of the pins

Up/Down circuit	Pin Number	Pin Code On the package	Signal	On The PCB
Up circuit	18	J1	VA_U	AVDD_U
	19	K1	GND	GND
	20	J2	Vdd_U	DVDD_U
	21	L1	GND	GND
	28	J5	A_U	A_U
	29	K5	B_U	B_U
	30	L5	C_U	C_U
	31	K6	D_U	D_U
	32	J6	Vdd_U	DVDD_U
	33	J7	Cal0_U	Cal0_U
	34	L7	Sin_U	Sin_U
	35	K7	Clk_U	Sclk_U
	36	L6	E_U	E_U
	37	L8	F_U	F_U
	38	K8	G_U	G_U
	39	L9	Phi1_U	Phi1_U
	40	L10	Phi2_U	Phi2_U
	41	K9	Phi3	Phi3
	43	K10	GND	GND
	44	J10	Vdd_U	DVDD_U
	45	K11	Nout_U	Nout_U

(Continues)

Table C.1 (*Continued*)

Up/Down circuit	Pin Number	Pin Code On the package	Signal	On The PCB
	46	J11	Vref	Vref
	47	H10	GND	GND
	48	H11	VA_U	AVDD_U
Down circuit	1	B2	GND	GND
	2	C2	Vdd_D	DVDD_D
	3	B1	GND	GND
	4	C1	VA_D	AVDD_D
	58	D11	VA_D	AVDD_D
	59	D10	GND	GND
	60	C11	Vref	Vref
	61	B11	Nout_D	Nout_D
	62	C10	Vdd_D	DVDD_D
	63	A11	GND	GND
	65	B9	Phi3	Phi3
	66	A10	Phi2_D	Phi2_D
	67	A9	Phi1_D	Phi1_D
	68	B8	G_D	G_D
	69	A8	F_D	F_D
	70	B6	E_D	E_D
	71	B7	Clk_D	Sclk_D
	72	A7	Sin_D	Sin_D
	73	C7	Ca10_D	Ca10_D
	74	C6	Vdd_D	DVDD_D
	75	A6	D_D	D_D
	76	A5	C_D	C_D
	77	B5	B_D	B_D
	78	C5	A_D	A_D

arrangement of the chip, enclosed within an 85-pin surface-mounted device (SMD) grid array package from KYOCERA Corporation. The chip is symbolized by "chip" and is an integral part of the circuit schematic depicted in Figure C.3.

C.2 The test bench circuit and system

The circuit board features the incorporation of five distinct voltage regulators and one multiplexer, as illustrated in the circuit schematic presented in Figure C.1. Four of these regulators are TLV75533PDBVR units, each capable of delivering 500 mA, boasting a high power supply rejection ratio (PSRR), low quiescent current (IQ), and low dropout characteristics. Operating within the temperature range of −40 to 125 °C, they require a 5 V enable voltage and are packaged in small outline transistor (SOT)-23 casings. These regulators are judiciously allocated to

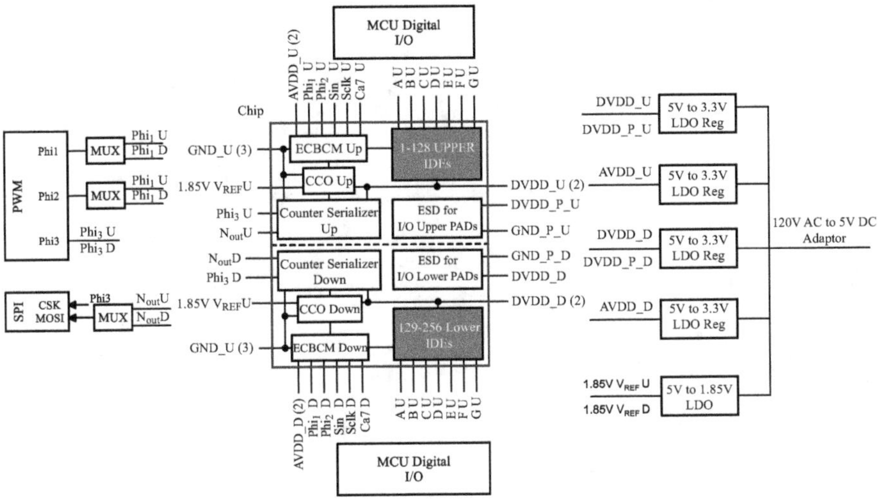

Figure C.1 Circuit schematic of the test bench for the array chip

supply 3.3 V to both analog and digital voltages, with two catering to the right side and two to the left side.

To establish the essential 1.85 V reference voltage for the oscillator's Vref pin, the TLV713185PDBVT regulator is enlisted. This regulator, similar in specifications to the aforementioned units, offers 500 mA capacity, high PSRR, low IQ, and low dropout characteristics. Operating within a temperature range of −40 to 125 °C, it also demands a 5 V enable voltage and resides within a SOT-23 package. All enable pins of the low-dropout regulators (LDOs) are tied to the input voltage.

The TMUX1247DCKR, characterized by its precision analog multiplexing capability, 30 Ohm ON resistance, and rail-to-rail functionality, has been integrated. In addition, decoupling capacitors are strategically placed to mitigate noise across the input power and output power supplies of the components.

For an intricate insight into the circuit's interconnections, refer to Figure C.2. A comprehensive breakdown of all chip pins is available in Table C.1. Furthermore, the actual fabricated printed circuit board (PCB) is showcased in Figure C.3.

C.3 Microcontroller design

C.3.1 Microcontroller commands

The test bench employs an Arduino DUE development board as the microcontroller. In this design, we have consolidated all commands within an integer

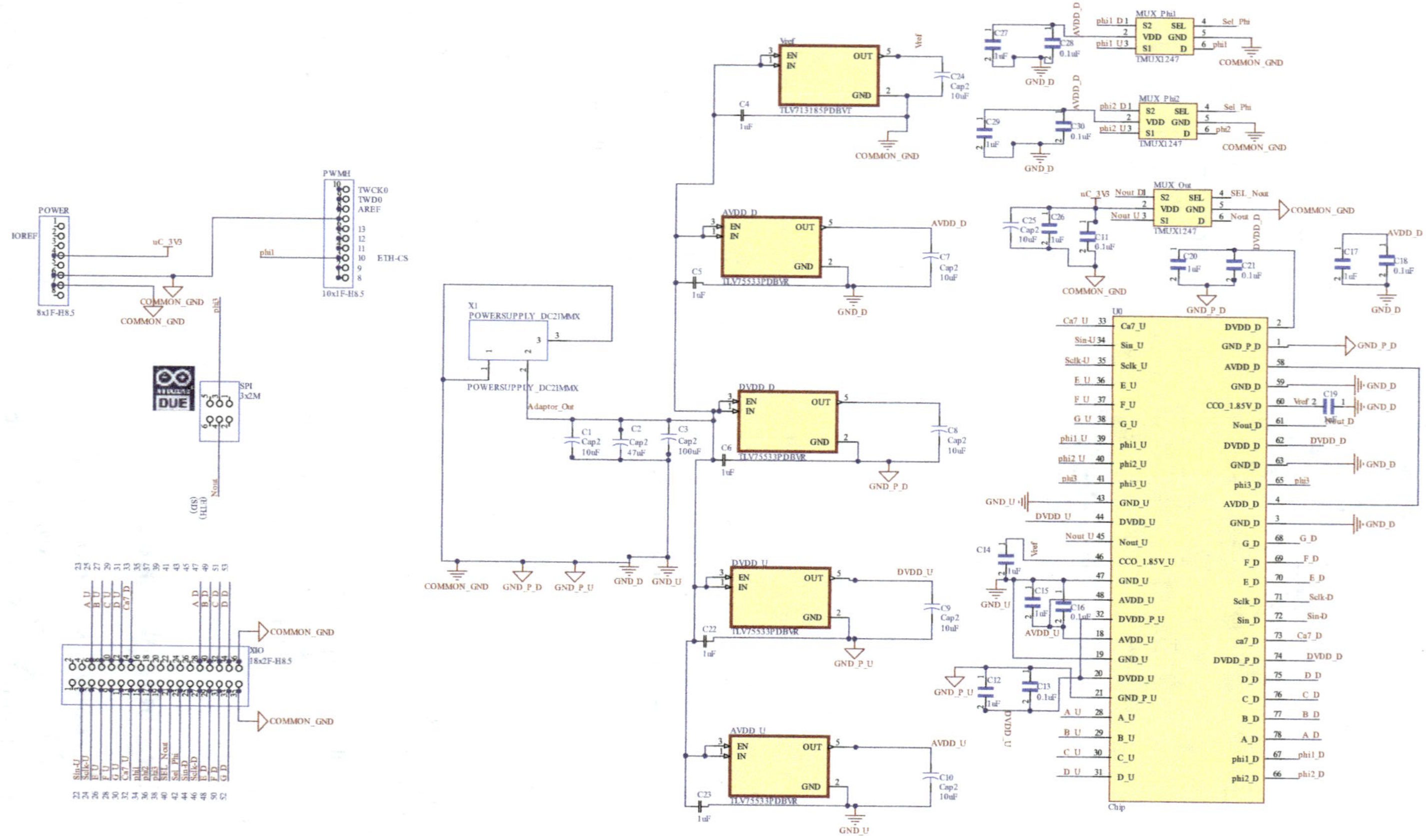

Figure C.2 PCB circuit schematic of the test bench for the array chip

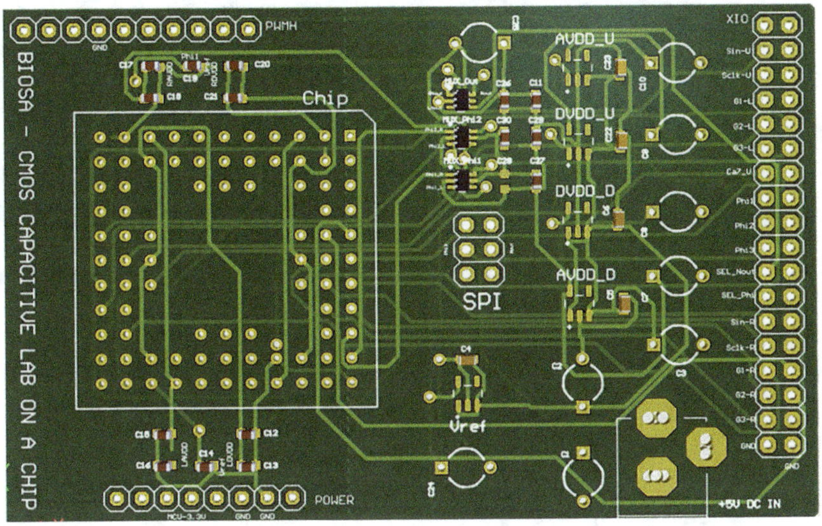

Figure C.3 PCB schematic of the test bench for the array chip

array. This array consists of 23 bits, with the specifics of each element elucidated below.

```
String command;
//{[0]UorD,
[1]NS{0}, [2]NS{1}, [3]NS{2},
[4]G, [5]F, [6]E, [7]D, [8]C, [9]B, [10]A,
[11]ca10,[12]ca9,[13]ca8,[14]ca7,[15]ca6,[16]ca5,[17]ca4,[18]ca3,[19]ca2,[20]ca1,
[21]Cs,
[22] Sweep(S) or Known Reference(R)]
```

The initial integer signifies the specific circuit being measured, whether it is the upper circuit (0) or the lower circuit (1). After this, the following three integers represent values ranging from 0 to 999, indicating the desired number of samples for measurement. The ensuing seven integers dictate the multiplexer setting for interdigitated electrode (IDE) selection. Proceeding, the subsequent ten integers configure values for the 10-bit bank of capacitors, spanning from 11 to 20. The 21st integer governs the activation of the two transmission gates integrated within the R1C1 and R16C1 IDEs. Finally, the concluding element in the array corresponds to the ASCII (American Standard Code for Information Interchange) code of either 'S' or 'R', determining the mode of measurement as either a capacitance sweep (S) or a fixed reference (R).

C.3.2 *Microcontroller (MCU) flowchart and the code snippet*

The operational sequence guiding the microcontroller's function is illustrated in Figure C.4. Following the establishment of a serial communication link between the

MCU and the personal computer (PC), the command outlined in Section C.3.1 is decoded upon reception from the graphical user interface (GUI). Subsequently, the decoded command is utilized to configure parameters including the number of samples, measurement mode, and multiplexer inputs. Once the measurement is executed, the gathered data is transmitted to the GUI via the serial peripheral interface (SPI) port.

The code modules responsible for configuring the bank of capacitors and specifying the number of samples, along with the SPI communication, remain consistent with the code modules used in the non-array test bench. Below is the code designed for execution on the Arduino DUE board, accompanied by explanatory comments.

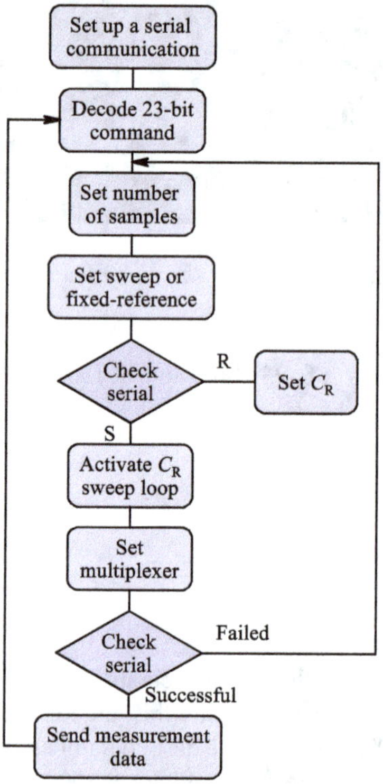

Figure C.4 Flowchart of the MCU code for the array test bench

```
const char TERMINATION_CHAR = '#';
const char Operation_Done= '-';
const int Mask=0x800;
int Data;
int Sat_Counter=0; // counts the number of times that the output comes saturated by mistake
String command;
//{[0]UorD,[1]NS{0},[2]NS{1},[3]NS{2},[4]G,[5]F,[6]E,[7]D,[8]C,[9]B,[10]A,[11]ca10,[12]ca9,[13]ca8,[14]ca7,[15]
ca6,[16]ca5,[17]ca4,[18]ca3,[19]ca2,[20]ca1,[21]Cs,[22]Sweep(S) or Known Reference(R)]
short result = 0;
int N_sample = 10;
 // Pin Definition
  #define Sin_U 22
  #define Sclk_U 24
```

```
#define E_U 26
#define F_U 28
#define G_U 30
#define Cal0_U 32
#define SEL_Nout 40
#define Sel_Phi 42
#define Sin_D 44
#define Sclk_D 46
#define E_D 48
#define F_D 50
#define G_D 52
#define A_U 25
#define B_U 27
#define C_U 29
#define D_U 31
#define Cal0_D 33
#define A_D 47
#define B_D 49
#define C_D 51
#define D_D 53
void setup() {
 Serial.begin(115200);
 pinMode(Sin_U, OUTPUT);
 pinMode(Sclk_U, OUTPUT);
 digitalWrite(Sclk_U,LOW);
 pinMode(E_U, OUTPUT);
 pinMode(F_U, OUTPUT);
 pinMode(G_U, OUTPUT);
 pinMode(Cal0_U, INPUT);
 pinMode(SEL_Nout, OUTPUT);
 pinMode(Sel_Phi, OUTPUT);
 pinMode(Sin_D, OUTPUT);
 pinMode(Sclk_D, OUTPUT);
 digitalWrite(Sclk_D,LOW);
 pinMode(E_D, OUTPUT);
 pinMode(F_D, OUTPUT);
 pinMode(G_D, OUTPUT);
 pinMode(A_U, OUTPUT);
 pinMode(B_U, OUTPUT);
 pinMode(C_U, OUTPUT);
 pinMode(D_U, OUTPUT);
 pinMode(Cal0_D, INPUT);
 pinMode(A_D, OUTPUT);
 pinMode(B_D, OUTPUT);
 pinMode(C_D, OUTPUT);
 pinMode(D_D, OUTPUT);
 delay(10);
 //Put your setup code here, to run once:
 REG_PMC_PCER1 |= PMC_PCER1_PID36;// Power up PWM clock
 REG_PWM_WPCR = PWM_WPCR_WPCMD(0);//Unlock user interface for PWM
 REG_PWM_CLK = PWM_CLK_PREA(0) | PWM_CLK_DIVA(21); //config the PWM clock generator: clock A,
84/21 = 4MHz
 REG_PWM_SCM = PWM_SCM_SYNC0 | PWM_SCM_SYNC1 ; //Synchronize phi1 and phi2 in the new version
for they have the same frequency
 //config phi1, Output pin PC2, Digital pin 34, has PWML0
 REG_PIOC_ABSR |= PIO_ABSR_P2; // Transfer Pin control from PIO to PWM
 REG_PIOC_PDR |= PIO_PDR_P2;  // Set PWM pin to an output
 REG_PWM_CMR0 = PWM_CMR_CPRE_CLKA | PWM_CMR_CPOL; //mode configuration for channel 0: clock
A, left aligned, waveform start at high
 REG_PWM_CPRD0 = PWM_CPRD_CPRD(60);//config period for PWM 60/(4MHz)
 REG_PWM_CDTY0 = PWM_CDTY_CDTY(16);// 80/3% duty cycle(16)
 REG_PWM_ENA = PWM_ENA_CHID0;// enable PWM channel0
 //config phi1 inverted to be used as NSS in the SPI, Output pin PC3, Digital pin 35, has PWMH0 which is the
complementary pin of PWML0 (Phi1 )
 REG_PIOC_ABSR |= PIO_ABSR_P3; // Transfer Pin control from PIO to PWM
 REG_PIOC_PDR |= PIO_PDR_P3;  // Set PWM pin to an output
 //config phi2, Output pin PC4, Digital pin 36
 REG_PIOC_ABSR |= PIO_ABSR_P4; // Transfer Pin control from PIO to PWM
 REG_PIOC_PDR |= PIO_PDR_P4;  // Set PWM pin to an output
 REG_PWM_CMR1 = PWM_CMR_CPRE_CLKA | PWM_CMR_DTE| PWM_CMR_CPOL; //mode configuration
for channel 0: clock A, left aligned, waveform starts at high, with deadtime generator enabled because we want some
delay in this channel
```

```
REG_PWM_CPRD1 = PWM_CPRD_CPRD(60);//config period for PWM 60/(4MHz)
REG_PWM_DT1 = PWM_DT_DTL(4); //create the 1us(delay)
REG_PWM_CDTY1 = PWM_CDTY_CDTY(12);// 80/6% duty cycle(12)
REG_PWM_ENA = PWM_ENA_CHID1;// enable PWM channel1
//config phi3, Output pin PC6, Digital pin 38
REG_PIOC_ABSR |= PIO_ABSR_P6; // Transfer Pin control from PIO to PWM
REG_PIOC_PDR |= PIO_PDR_P6;  // Set PWM pin to an output
REG_PWM_CMR2 = PWM_CMR_CPRE_CLKA | PWM_CMR_CPOL ;//mode configuration for channel 0: clock
A, left aligned, waveform start at low |PWM_CMR_CALG
REG_PWM_CPRD2 = PWM_CPRD_CPRD(3);//config period for PWM 2/(4MHz)
//REG_PWM_DT2 = PWM_DT_DTL(1); //create the 1us(delay)
REG_PWM_CDTY2 = PWM_CDTY_CDTY(2);//50% duty cycle(1)
REG_PWM_ENA = PWM_ENA_CHID2;// enable PWM channel2
//SPI serial receive
REG_PMC_PCER0 |= PMC_PCER0_PID24;// Power up SPI clock
REG_SPI0_WPMR = 0 << SPI_WPMR_WPEN; //Unlock user interface for SPI
//Instance SPI0, MISO: PA25, (MISO), MOSI: PA26, (MOSI), SCLK: PA27, (SCLK), NSS: PA28, Digital pin 10
REG_PIOA_ABSR |= PIO_ABSR_P25; // Transfer Pin control from PIO to SPI
REG_PIOA_PDR |= PIO_PDR_P25;  // Set MISO pin to an output
REG_PIOA_ABSR |= PIO_ABSR_P26; // Transfer Pin control from PIO to SPI
REG_PIOA_PDR |= 0 << PIO_PDR_P26; // Set MOSI pin to an input
REG_PIOA_ABSR |= PIO_ABSR_P27; // Transfer Pin control from PIO to SPI
REG_PIOA_PDR |= 0 << PIO_PDR_P27; // Set SCLK pin to an input
REG_PIOA_ABSR |= PIO_ABSR_P28; // Transfer Pin control from PIO to SPI
REG_PIOA_PDR |= 0 << PIO_PDR_P28; // Set NSS pin to an input
//REG_ISER0 = 1<<24; //Enable interrupt controller
//REG_SPI0_CR = 1; // Enable SPI it should move to where we want to read the SPI port
REG_SPI0_MR = 0; // Slave mode
REG_SPI0_IER = 1 << SPI_IER_RDRF | 1 << SPI_IER_OVRES | 1 << SPI_IER_NSSR; //Enable interrupts
SPI0->SPI_CSR[0] = SPI_CSR_BITS_12_BIT; // Capture on the falling edge and transfer 12 bits.
}
void loop() {
 int i=0;
 REG_SPI0_CR = 0; //First disable SPI reading to config the hardware
 if (Serial.available() > 0) // check if serial is available
 {
  command = Serial.readStringUntil('\0'); // read the serial and put in cmd
   if(command[22]=='R'){
    here_R:
   if (command[0]=='U')
   {digitalWrite(SEL_Nout,HIGH); digitalWrite(Sel_Phi,HIGH);
   if ( command[1]<'0' || command[1]>'9'){Serial.print("E1");return;}
   if ( command[2]<'0' || command[2]>'9'){Serial.print("E2");return;}
   if ( command[3]<'0' || command[3]>'9'){Serial.print("E3");return;}
   N_sample = 100*(command[1]-'0')+10*(command[2]-'0')+command[3]-'0'; // Convert N_sample_s to integer and
put in N_sample
   if (command[4]=='0') {digitalWrite(G_U,LOW);} else if (command[4]=='1') {digitalWrite(G_U,HIGH);} else
{Serial.print("E4");return;}
   if (command[5]=='0') {digitalWrite(F_U,LOW);} else if (command[5]=='1') {digitalWrite(F_U,HIGH);} else
{Serial.print("E5");return;}
   if (command[6]=='0') {digitalWrite(E_U,LOW);} else if (command[6]=='1') {digitalWrite(E_U,HIGH);} else
{Serial.print("E6");return;}
   if (command[7]=='0') {digitalWrite(D_U,LOW);} else if (command[7]=='1') {digitalWrite(D_U,HIGH);} else
{Serial.print("E7");return;}
   if (command[8]=='0') {digitalWrite(C_U,LOW);} else if (command[8]=='1') {digitalWrite(C_U,HIGH);} else
{Serial.print("E8");return;}
   if (command[9]=='0') {digitalWrite(B_U,LOW);} else if (command[9]=='1') {digitalWrite(B_U,HIGH);} else
{Serial.print("E9");return;}
   if (command[10]=='0') {digitalWrite(A_U,LOW);} else if (command[10]=='1') {digitalWrite(A_U,HIGH);} else
{Serial.print("E10");return;}
// Configuring the CR for the Upper circuit
 //turn off the phi clock pulses
 REG_PWM_DIS = PWM_DIS_CHID0;
 REG_PWM_DIS = PWM_DIS_CHID1;
 REG_PWM_DIS = PWM_DIS_CHID2;
 delayMicroseconds(5);
 //fully reset the input registers
 for ( i = 13; i > 0 ; i--)
 {
  digitalWrite(Sin_U, LOW);
  digitalWrite(Sclk_U, LOW);
  delayMicroseconds(5); // Originally 5, I changed to 1
```

```
  digitalWrite(Sclk_U, HIGH);
  delayMicroseconds(5); // Originally 5, I changed to 1
 }
 //apply input sequences
 for ( i = 11; i <=21 ; i++)
 {
  if (command[i] == '0')
  {
   digitalWrite(Sin_U, LOW);
   digitalWrite(Sclk_U, LOW);
   delayMicroseconds(5); // Originally 5, I changed to 1
   digitalWrite(Sclk_U, HIGH);
   delayMicroseconds(5); // Originally 5, I changed to 1
  }
  else if (command[i] == '1')
  {
   digitalWrite(Sin_U, HIGH);
   digitalWrite(Sclk_U, LOW);
   delayMicroseconds(5); // Originally 5, I changed to 1
   digitalWrite(Sclk_U, HIGH);
   delayMicroseconds(5); // Originally 5, I changed to 1
  }
  else {Serial.print("E"); Serial.print(i);return;}
 }
 digitalWrite(Sin_U, LOW);
 digitalWrite(Sclk_U, LOW);
 delayMicroseconds(5);
//turn the phi pulses back on
 REG_PWM_ENA = PWM_ENA_CHID0;
 REG_PWM_ENA = PWM_ENA_CHID1;
 REG_PWM_CMR2 = 0;
 REG_PWM_CMR2 = PWM_CMR_CPRE_CLKA;
 REG_PWM_CMR2 &= ~PWM_CMR_CPOL;
 REG_PWM_ENA = PWM_ENA_CHID2;
 delayMicroseconds(20);
     delayMicroseconds(50);
    //if (!((digitalRead(Ca10_U)==LOW && command[11]=='0')||(digitalRead(Ca10_U)==HIGH &&
command[11]=='1'))){Serial.print("EM");}
    //Serial.println(TERMINATION_CHAR);
    }
    else if (command[0]=='D') {digitalWrite(SEL_Nout,LOW); digitalWrite(Sel_Phi,LOW);
    if ( command[1]<'0' || command[1]>'9'){Serial.print("E1");return;}
    if ( command[2]<'0' || command[2]>'9'){Serial.print("E2");return;}
    if ( command[3]<'0' || command[3]>'9'){Serial.print("E3");return;}
    N_sample = 100*(command[1]-'0')+10*(command[2]-'0')+command[3]-'0'; // Convert N_sample_s to integer and
put in N_sample
    if (command[4]=='0') {digitalWrite(G_D,LOW);} else if (command[4]=='1') {digitalWrite(G_D,HIGH);} else
{Serial.print("E4");return;}
    if (command[5]=='0') {digitalWrite(F_D,LOW);} else if (command[5]=='1') {digitalWrite(F_D,HIGH);} else
{Serial.print("E5");return;}
    if (command[6]=='0') {digitalWrite(E_D,LOW);} else if (command[6]=='1') {digitalWrite(E_D,HIGH);} else
{Serial.print("E6");return;}
    if (command[7]=='0') {digitalWrite(D_D,LOW);} else if (command[7]=='1') {digitalWrite(D_D,HIGH);} else
{Serial.print("E7");return;}
    if (command[8]=='0') {digitalWrite(C_D,LOW);} else if (command[8]=='1') {digitalWrite(C_D,HIGH);} else
{Serial.print("E8");return;}
    if (command[9]=='0') {digitalWrite(B_D,LOW);} else if (command[9]=='1') {digitalWrite(B_D,HIGH);} else
{Serial.print("E9");return;}
    if (command[10]=='0') {digitalWrite(A_D,LOW);} else if (command[10]=='1') {digitalWrite(A_D,HIGH);} else
{Serial.print("E10");return;}
    // Configuring the CR for the Upper circuit
  //turn off the phi clock pulses
  REG_PWM_DIS = PWM_DIS_CHID0;
  REG_PWM_DIS = PWM_DIS_CHID1;
  REG_PWM_DIS = PWM_DIS_CHID2;
  delayMicroseconds(5);
  //fully reset the input registers
  for ( i = 13; i > 0 ; i--)
  {
   digitalWrite(Sin_D, LOW);
   digitalWrite(Sclk_D, LOW);
   delayMicroseconds(5); // Originally 5, I changed to 1
```

```
  digitalWrite(Sclk_D, HIGH);
  delayMicroseconds(5); // Originally 5, I changed to 1
 }
 //apply input sequences
 for ( i = 11; i <=21 ; i++)
 {
  if (command[i] == '0')
  {
   digitalWrite(Sin_D, LOW);
   digitalWrite(Sclk_D, LOW);
   delayMicroseconds(5); // Originally 5, I changed to 1
   digitalWrite(Sclk_D, HIGH);
   delayMicroseconds(5); // Originally 5, I changed to 1
  }
  else if (command[i] == '1')
  {
   digitalWrite(Sin_D, HIGH);
   digitalWrite(Sclk_D, LOW);
   delayMicroseconds(5); // Originally 5, I changed to 1
   digitalWrite(Sclk_D, HIGH);
   delayMicroseconds(5); // Originally 5, I changed to 1
  }
  else {Serial.print("E"); Serial.print(i);return;}
 }
 digitalWrite(Sin_D, LOW);
 digitalWrite(Sclk_D, LOW);
 delayMicroseconds(5);
 //turn the phi pulses back on
 REG_PWM_ENA = PWM_ENA_CHID0;
 REG_PWM_ENA = PWM_ENA_CHID1;
 REG_PWM_CMR2 = 0;
 REG_PWM_CMR2 = PWM_CMR_CPRE_CLKA;
 REG_PWM_CMR2 &= ~PWM_CMR_CPOL;
 REG_PWM_ENA = PWM_ENA_CHID2;
 delayMicroseconds(20);
    delayMicroseconds(50);
    //if (!((digitalRead(Ca10_D)==LOW && command[11]=='0')||(digitalRead(Ca10_D)==HIGH &&
command[11]=='1'))) {Serial.print("EM");}
    //Serial.println(TERMINATION_CHAR);

    }
    else {Serial.print("E0"); return;}
     /// Start Reading the data
     int buf[N_sample];
     REG_SPI0_CR = 1; //Enable SPI now to read the chip output
     for (int i = 0; i < N_sample;)
     {
      if (REG_SPI0_SR & 1)
      {
       Data = REG_SPI0_RDR | Mask;
       if (Data==0xFFF && Sat_Counter<2 ) {Sat_Counter++; goto here_R;}  // Checks if the output is
4097(Saturated). If it happens it sets the signals again and captures new data
       //and increments Sat_Counter. If it happens again, again ignores and sets the signals and captures new data. If it
happens for the third time it means the output has a huge capacitance so it just records it in the buffer.
       else if (Data==0xFFF && Sat_Counter>=2){for (i=0; i<N_sample; i++) {buf[i]=0xFFF;} Sat_Counter=0; goto
Print_Data_R;}
       Data = Data - 2047;
       buf[i] = Data;
       //REG_SPI0_RDR=0;
       i++;
      }
     }
     Print_Data_R:
     char bufferr[4*N_sample+N_sample];
     int n = 0;
     for (int i = 0; i < N_sample; i++) {
        n += sprintf (&bufferr[n], "%04d,", buf[i]);
     }
     bufferr[4*N_sample+N_sample-1] = '\0';

     Serial.print(bufferr);
     Serial.print(TERMINATION_CHAR);
```

```
    //Serial.print("\n");
  }
///////////\\\\\\\\\\\\\\\\\\\\\\\\\\\\\\\\\\\\\\\\\\\\\\\\\\\\\\\\\\\\\\\\\\\\\\\\\\\\\\\\\\\\\\\\\\\\\\\\\\\\\\\\\\\\\\\\\\\\\\\\\\\\//For R
  else if (command[22]=='S')
  {
   for (int j=0 ; j<1024 ; j++)
   {
    int k=j;
    if ((k&1)==0){command[20]='0';} else if ((k&1)==1) {command[20]='1';} else {Serial.print("error 20");}
    k=k>>1;
    if ((k&1)==0){command[19]='0';} else if ((k&1)==1) {command[19]='1';} else {Serial.print("error 19");}
    k=k>>1;
    if ((k&1)==0){command[18]='0';} else if ((k&1)==1) {command[18]='1';} else {Serial.print("error 18");}
    k=k>>1;
    if ((k&1)==0){command[17]='0';} else if ((k&1)==1) {command[17]='1';} else {Serial.print("error 17");}
    k=k>>1;
    if ((k&1)==0){command[16]='0';} else if ((k&1)==1) {command[16]='1';} else {Serial.print("error 16");}
    k=k>>1;
    if ((k&1)==0){command[15]='0';} else if ((k&1)==1) {command[15]='1';} else {Serial.print("error 15");}
    k=k>>1;
    if ((k&1)==0){command[14]='0';} else if ((k&1)==1) {command[14]='1';} else {Serial.print("error 14");}
    k=k>>1;
    if ((k&1)==0){command[13]='0';} else if ((k&1)==1) {command[13]='1';} else {Serial.print("error 13");}
    k=k>>1;
    if ((k&1)==0){command[12]='0';} else if ((k&1)==1) {command[12]='1';} else {Serial.print("error 12");}
    k=k>>1;
    if ((k&1)==0){command[11]='0';} else if ((k&1)==1) {command[11]='1';} else {Serial.print("error 11");}
    k=k>>1;

        here_S:
    if (command[0]=='U')
    {digitalWrite(SEL_Nout,HIGH); digitalWrite(Sel_Phi,HIGH);
    if ( command[1]<'0' || command[1]>'9'){Serial.print("E1");return;}
    if ( command[2]<'0' || command[2]>'9'){Serial.print("E2");return;}
    if ( command[3]<'0' || command[3]>'9'){Serial.print("E3");return;}
    N_sample = 100*(command[1]-'0')+10*(command[2]-'0')+command[3]-'0'; // Convert N_sample_s to integer and
put in N_sample
    if (command[4]=='0') {digitalWrite(G_U,LOW);} else if (command[4]=='1') {digitalWrite(G_U,HIGH);} else
{Serial.print("E4");return;}
    if (command[5]=='0') {digitalWrite(F_U,LOW);} else if (command[5]=='1') {digitalWrite(F_U,HIGH);} else
{Serial.print("E5");return;}
    if (command[6]=='0') {digitalWrite(E_U,LOW);} else if (command[6]=='1') {digitalWrite(E_U,HIGH);} else
{Serial.print("E6");return;}
    if (command[7]=='0') {digitalWrite(D_U,LOW);} else if (command[7]=='1') {digitalWrite(D_U,HIGH);} else
{Serial.print("E7");return;}
    if (command[8]=='0') {digitalWrite(C_U,LOW);} else if (command[8]=='1') {digitalWrite(C_U,HIGH);} else
{Serial.print("E8");return;}
    if (command[9]=='0') {digitalWrite(B_U,LOW);} else if (command[9]=='1') {digitalWrite(B_U,HIGH);} else
{Serial.print("E9");return;}
    if (command[10]=='0') {digitalWrite(A_U,LOW);} else if (command[10]=='1') {digitalWrite(A_U,HIGH);} else
{Serial.print("E10");return;}

// Configuring the CR for the Upper circuit
  //turn off the phi clock pulses
  REG_PWM_DIS = PWM_DIS_CHID0;
  REG_PWM_DIS = PWM_DIS_CHID1;
  REG_PWM_DIS = PWM_DIS_CHID2;
  delayMicroseconds(5);
  //fully reset the input registers

  for ( i = 13; i > 0 ; i--)
  {
   digitalWrite(Sin_U, LOW);
   digitalWrite(Sclk_U, LOW);
   delayMicroseconds(5); // Originally 5, I changed to 1
   digitalWrite(Sclk_U, HIGH);
   delayMicroseconds(5); // Originally 5, I changed to 1
  }
  //apply input sequences
  for ( i = 11; i <=21 ; i++)
  {
   if (command[i] == '0')
```

```
    {
     digitalWrite(Sin_U, LOW);
     digitalWrite(Sclk_U, LOW);
     delayMicroseconds(5); // Originally 5, I changed to 1
     digitalWrite(Sclk_U, HIGH);
     delayMicroseconds(5); // Originally 5, I changed to 1
    }
   else if (command[i] == '1')
    {
     digitalWrite(Sin_U, HIGH);
     digitalWrite(Sclk_U, LOW);
     delayMicroseconds(5); // Originally 5, I changed to 1
     digitalWrite(Sclk_U, HIGH);
     delayMicroseconds(5); // Originally 5, I changed to 1
    }
   else {Serial.print("E"); Serial.print(i);return;}
  }
  digitalWrite(Sin_U, LOW);
  digitalWrite(Sclk_U, LOW);
  delayMicroseconds(5);
//turn the phi pulses back on
  REG_PWM_ENA = PWM_ENA_CHID0;
  REG_PWM_ENA = PWM_ENA_CHID1;
  REG_PWM_CMR2 = 0;
  REG_PWM_CMR2 = PWM_CMR_CPRE_CLKA;
  REG_PWM_CMR2 &= ~PWM_CMR_CPOL;
  REG_PWM_ENA = PWM_ENA_CHID2;
  delayMicroseconds(20);
    delayMicroseconds(50);
    //if (!((digitalRead(Ca10_U)==LOW && command[11]=='0')||(digitalRead(Ca10_U)==HIGH &&
command[11]=='1'))) {Serial.print("EM");}
    //Serial.println(TERMINATION_CHAR);
    }

    else if (command[0]=='D') {digitalWrite(SEL_Nout,LOW); digitalWrite(Sel_Phi,LOW);
    if ( command[1]<'0' || command[1]>'9'){Serial.print("E1");return;}
    if ( command[2]<'0' || command[2]>'9'){Serial.print("E2");return;}
    if ( command[3]<'0' || command[3]>'9'){Serial.print("E3");return;}
    N_sample = 100*(command[1]-'0')+10*(command[2]-'0')+command[3]-'0'; // Convert N_sample_s to integer and
put in N_sample
    if (command[4]=='0') {digitalWrite(G_D,LOW);} else if (command[4]=='1') {digitalWrite(G_D,HIGH);} else
{Serial.print("E4");return;}
    if (command[5]=='0') {digitalWrite(F_D,LOW);} else if (command[5]=='1') {digitalWrite(F_D,HIGH);} else
{Serial.print("E5");return;}
    if (command[6]=='0') {digitalWrite(E_D,LOW);} else if (command[6]=='1') {digitalWrite(E_D,HIGH);} else
{Serial.print("E6");return;}
    if (command[7]=='0') {digitalWrite(D_D,LOW);} else if (command[7]=='1') {digitalWrite(D_D,HIGH);} else
{Serial.print("E7");return;}
    if (command[8]=='0') {digitalWrite(C_D,LOW);} else if (command[8]=='1') {digitalWrite(C_D,HIGH);} else
{Serial.print("E8");return;}
    if (command[9]=='0') {digitalWrite(B_D,LOW);} else if (command[9]=='1') {digitalWrite(B_D,HIGH);} else
{Serial.print("E9");return;}
    if (command[10]=='0') {digitalWrite(A_D,LOW);} else if (command[10]=='1') {digitalWrite(A_D,HIGH);} else
{Serial.print("E10");return;}
    // Configuring the CR for the Upper circuit
  //turn off the phi clock pulses
  REG_PWM_DIS = PWM_DIS_CHID0;
  REG_PWM_DIS = PWM_DIS_CHID1;
  REG_PWM_DIS = PWM_DIS_CHID2;
  delayMicroseconds(5);
  //fully reset the input registers
  for ( i = 13; i > 0 ; i--)
  {
   digitalWrite(Sin_D, LOW);
   digitalWrite(Sclk_D, LOW);
   delayMicroseconds(5); // Originally 5, I changed to 1
   digitalWrite(Sclk_D, HIGH);
   delayMicroseconds(5); // Originally 5, I changed to 1
  }
  //apply input sequences
  for ( i = 11; i <=21 ; i++)
  {
```

```
 if (command[i] == '0')
 {
  digitalWrite(Sin_D, LOW);
  digitalWrite(Sclk_D, LOW);
  delayMicroseconds(5); // Originally 5, I changed to 1
  digitalWrite(Sclk_D, HIGH);
  delayMicroseconds(5); // Originally 5, I changed to 1
 }
 else if (command[i] == '1')
 {
  digitalWrite(Sin_D, HIGH);
  digitalWrite(Sclk_D, LOW);
  delayMicroseconds(5); // Originally 5, I changed to 1
  digitalWrite(Sclk_D, HIGH);
  delayMicroseconds(5); // Originally 5, I changed to 1
 }
 else {Serial.print("E"); Serial.print(i);return;}
}
digitalWrite(Sin_D, LOW);
digitalWrite(Sclk_D, LOW);
delayMicroseconds(5);
//turn the phi pulses back on
REG_PWM_ENA = PWM_ENA_CHID0;
REG_PWM_ENA = PWM_ENA_CHID1;
REG_PWM_CMR2 = 0;
REG_PWM_CMR2 = PWM_CMR_CPRE_CLKA;
REG_PWM_CMR2 &= ~PWM_CMR_CPOL;
REG_PWM_ENA = PWM_ENA_CHID2;
delayMicroseconds(20);
   delayMicroseconds(50);
   //if (!((digitalRead(Ca10_D)==LOW && command[11]=='0')||(digitalRead(Ca10_D)==HIGH &&
command[11]=='1'))) {Serial.print("EM");}
   //Serial.println(TERMINATION_CHAR);
   }
   else {Serial.print("E0"); return;}
    /// Start Reading the data
    int buf[N_sample];
    REG_SPI0_CR = 1; //Enable SPI now to read the chip output
    for (int i = 0; i < N_sample;)
    {
     if (REG_SPI0_SR & 1)
     {
      Data = REG_SPI0_RDR | Mask;
      if (Data==0xFFF && Sat_Counter<=8*N_sample ) {Sat_Counter++; goto here_S;}  // Checks if the output is
4097(Saturated). If it happens it sets the signals again and captures new data
      //and increments Sat_Counter. If it happens again, again ignores and sets the signals and captures new data. If it
happens for the third time it means the output has a huge capacitance so it just records it in the buffer.
      else if (Data==0xFFF && Sat_Counter>8*N_sample){for (i=0; i<N_sample; i++) {buf[i]=0xFFF;}
Sat_Counter=0; goto Print_Data_S;}
      Data = Data - 2047;
      buf[i] = Data;
      //REG_SPI0_RDR=0;
      i++;
     }
    }
    Print_Data_S:
    char bufferr[4*N_sample+N_sample];
    int n = 0;
    for (int i = 0; i < N_sample; i++) {
      n += sprintf (&bufferr[n], "%04d,", buf[i]);
    }
    bufferr[4*N_sample+N_sample-1] = '\0';
    Serial.print(bufferr);
    Serial.print(',');
   }
   Serial.print(TERMINATION_CHAR);
  }
 else {Serial.print("E22");}
 }
}
```

Appendix D
Micro-hotplate

The major functionality of micro-hotplate is measuring and controlling temperature and detecting properties of deposited materials such as inorganic gases on sensitive layers at high temperatures under low power consumption. Figure D.1 illustrates the cross-sectional schematic of the micro-hotplate [1236]. It consists of a thermally isolated stage with a heater structure, a temperature sensor, and a set of contact electrodes for sensing thin films. The micro-hotplate membrane comprising CMOS dielectric layers, isolated the heated area of the membrane from the bulk chip. An additional temperature sensor is integrated close to the circuitry on the bulk chip to monitor the overall chip temperature. A micro-hotplate construction allows the heater element to heat up the sensing thin film up to a high temperature. A thermal sensor measures the temperature generated by materials exposure on the sensitive layers and reads out with the help of electrodes.

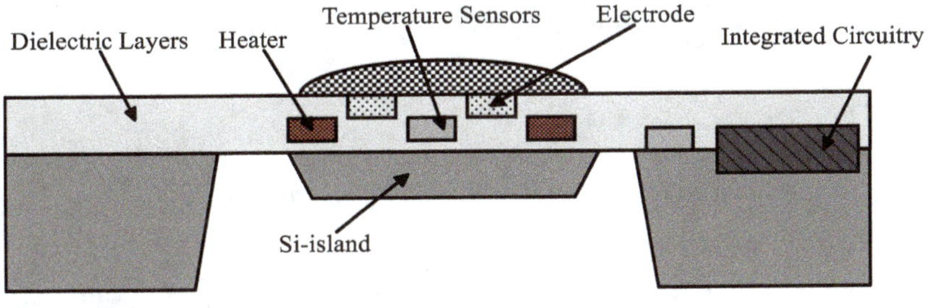

Figure D.1 Cross-sectional schematic of the micro-hotplate

The most important issues that should be considered in the design of micro-hotplate are the power consumption and temperature uniformity in the sensitive area. A non-uniform temperature profile causes an increment of ambiguous sensor operating temperature, which in turn degrades the sensor selectivity and sensitivity. A substantial amount of research has been suggested in an attempt to achieve the stated objectives [17,947,1236–1238]. In one effort, Semancik *et al.* [1237] investigated the use of such micro-hotplate arrays in statistically designed studies of conductometric sensing film process/performance correlations. Figure D.2 shows a single micro-hotplate. The four arrow-shaped top electrodes directly contacted the sensing films to measure the electrical characteristics of deposited sensing films and were isolated from the heater by a SiO_2 layer. As depicted in Figure D.2(b), a resistive heater made of a serpentine polysilicon element was embedded in SiO2. The metal plate situated above the heater ensures more uniform heat conduction throughout the device. The temperature measurement can be provided by the temperature coefficient of resistance of the metal plate or by measuring the polysilicon resistance. It is notable, that the metallization at a given foundry determines the maximum working temperature of the device. For instance, the maximum temperature is around 500 °C for aluminum metallization [1237]. In this work, the hotplate was held at 480 °C.

In another work, a gas sensor microsystem fabricated in industrial CMOS technology combined with post-CMOS micromachining with an operating temperature of up to 500 °C was proposed by Graf *et al.* [1238]. In this design, as shown in Figure D.3, a platinum temperature sensor was patterned on the micro-hotplate. Because platinum is a common material to establish contact with the sensitive layer, the Pt-metallization process step is employed to fabricate the

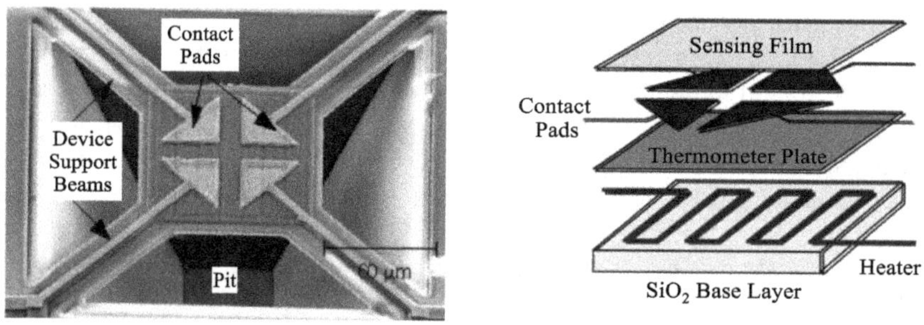

(a) (b)

Figure D.2 (a) Micrograph of a micro-hotplate, (b) schematic of functional layers constructed micro-hotplate by Semancik et al. [1237]

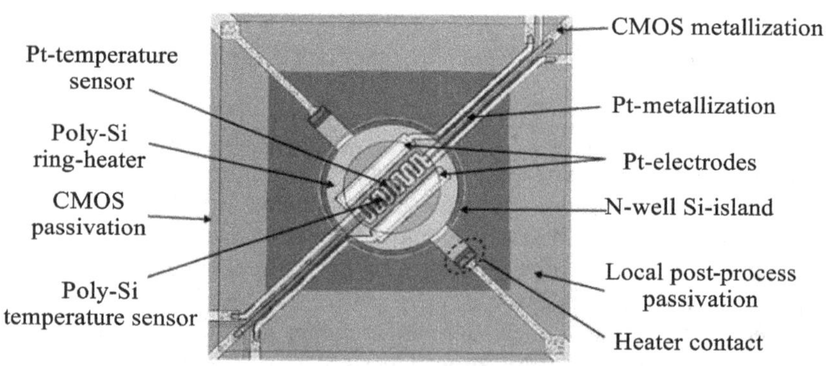

Figure D.3 Micrograph of the micro-hotplate with Pt-temperature sensor by Graf et al. [1238]

electrodes and the temperature sensor. The heater was realized as a ring heater with two semicircles in a parallel configuration and has been fabricated in polysilicon. The Pt-resistor of the temperature sensor was connected in a four-point configuration. An additional Poly-Si temperature sensor as a reference is integrated underneath the Pt-resistor.

A fully integrated gas sensor microsystem including a micro-hotplate and analog and digital circuitry on a single chip has been presented for the first time by Garf *et al.* [1236]. The micrograph of the micro-hotplate is shown in Figure D.4, which is designed with a ring heater and a square dielectric membrane on a CMOS substrate. A circular-shaped micro-hotplate provides comparatively long metal leads extending from the heated area to the corners of the membrane, so the heat dissipation is reduced. The width of the metal lines along each diagonal is approximately identical. So, a symmetric heat flow through the membrane is created.

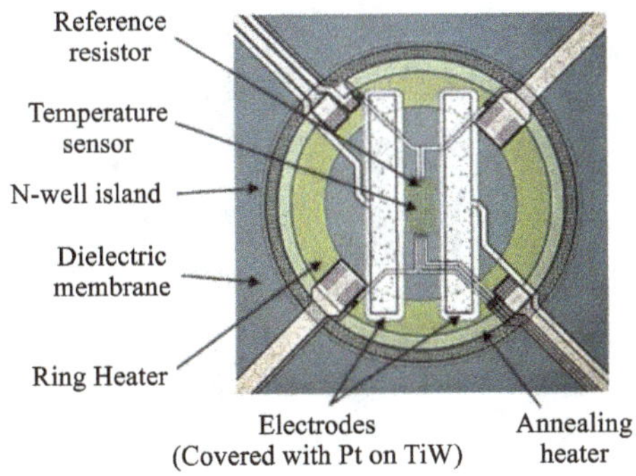

Figure D.4 Micrograph of a circular micro-hotplate by Graf et al. [1236]

References

[1] B. Jang and A. Hassibi, "Biosensor systems in standard CMOS processes: Fact or fiction?" *IEEE Transactions on Industrial Electronics*, vol. 56, no. 4, pp. 979–985, 2009.

[2] G. Nabovati, E. Ghafar-Zadeh, A. Letourneau, and M. Sawan, "Smart Cell Culture Monitoring and Drug Test Platform Using CMOS Capacitive Sensor Array," *IEEE Transactions on Biomedical Engineering*, vol. 66, no. 4, pp. 1094–1104, 2018.

[3] B. P. Senevirathna, S. Lu, M. P. Dandin, J. Basile, E. Smela, and P. A. Abshire, "Real-time measurements of cell proliferation using a lab-on-CMOS capacitance sensor array," *IEEE Transactions on Biomedical Circuits and Systems*, vol. 12, no. 3, pp. 510–520, 2018.

[4] N. Couniot, L. A. Francis, and D. Flandre, "A 16 × 16 CMOS Capacitive Biosensor Array Towards Detection of Single Bacterial Cell," *IEEE Transactions on Biomedical Circuits and Systems*, vol. 10, no. 2, pp. 364–374, 2016.

[5] C.-L. Hsu, A. Sun, Y. Zhao, E. Aronoff-Spencer, and D. A. Hall, "A 16 × 20 electrochemical CMOS biosensor array with in-pixel averaging using polar modulation," in *2018 IEEE Custom Integrated Circuits Conference (CICC)*, 2018: IEEE, pp. 1–4.

[6] J. Dragas, V. Viswam, A. Shadmani, *et al.*, "In Vitro Multi-Functional Microelectrode Array Featuring 59 760 Electrodes, 2048 Electrophysiology Channels, Stimulation, Impedance Measurement, and Neurotransmitter Detection Channels," *IEEE Journal of Solid-State Circuits*, vol. 52, no. 6, pp. 1576–1590, 2017.

[7] Y. Jiang, X. Liu, T. C. Dang, *et al.*, "A High-Sensitivity Potentiometric 65-nm CMOS ISFET Sensor for Rapid E. coli Screening," *IEEE Transactions on Biomedical Circuits and Systems*, vol. 12, no. 2, pp. 402–415, 2018.

[8] C. Giagkoulovits, M. A. Al-Rawhani, B. C. Cheah, *et al.*, "Hybrid amperometric and potentiometrie sensing based on a CMOS ISFET array," *in SENSORS*, 2017 *IEEE*, 2017: IEEE, pp. 1–3.

[9] S.-J. Wu, Y.-C. Wu, H.-H. Tsai, H.-H. Liao, Y.-Z. Juang, and C.-H. Lin, "ISFET-based pH sensor composed of a high transconductance CMOS chip and a disposable touch panel film as the sensing layer," in *2015 IEEE SENSORS*, 2015: IEEE, pp. 1–4.

[10] E. P. Papageorgiou, H. Zhang, S. Giverts, C. Park, B. E. Boser, and M. Anwar, "Real-time cancer detection with an integrated lensless

fluorescence contact imager," *Biomedical Optics Express*, vol. 9, no. 8, pp. 3607–3623, 2018.

[11] E. P. Papageorgiou, B. E. Boser, and M. Anwar, "An angle-selective CMOS imager with on-chip micro-collimators for blur reduction in near-field cell imaging," in *2016 IEEE 29th International Conference on Micro Electro Mechanical Systems (MEMS)*, 2016: IEEE, pp. 337–340.

[12] M. A. Al-Rawhani, B. C. Cheah, A. I. Macdonald, *et al.*, "A colorimetric CMOS-based platform for rapid total serum cholesterol quantification," *IEEE Sensors Journal*, vol. 17, no. 2, pp. 240–247, 2017.

[13] L. Hong, H. Li, H. Yang, and K. Sengupta, "Fully integrated fluorescence biosensors on-chip employing multi-functional nanoplasmonic optical structures in CMOS," *IEEE Journal of Solid-State Circuits*, vol. 52, no. 9, pp. 2388–2406, 2017.

[14] Y. Zheng, N. Shang, P. S. Haddad, and M. Sawan, "A microsystem for magnetic immunoassay based on planar microcoil array," *IEEE Transactions on Biomedical Circuits and Systems*, vol. 10, no. 2, pp. 477–486, 2016.

[15] P. Murali, A. M. Niknejad, and B. E. Boser, "CMOS Microflow Cytometer for Magnetic Label Detection and Classification," *IEEE Journal of Solid-State Circuits*, vol. 52, no. 2, pp. 543–555, 2017.

[16] H. Pourmodheji, E. Ghafar-Zadeh, and S. Magierowski, "A multi-disciplinary approach to high throughput nuclear magnetic resonance spectroscopy," *Sensors*, vol. 16, no. 6, p. 850, 2016.

[17] M. Graf, D. Barrettino, H. P. Baltes, and A. Hierlemann, *CMOS Hotplate Chemical Microsensors*. Berlin: Springer Science & Business Media, 2007.

[18] N. Manaresi, A. Romani, G. Medoro, *et al.*, "A CMOS chip for individual cell manipulation and detection," *IEEE Journal of Solid-State Circuits*, vol. 38, no. 12, pp. 2297–2305, 2003.

[19] Y. Liu, H. Lee, D. Ham, and R. M. Westervelt, "CMOS-based magnetic cell manipulation system," *CMOS Biotechnology*, pp. 103–144, 2007.

[20] A. Hierlemann and H. Baltes, "CMOS-based chemical microsensors," *Analyst*, vol. 128, no. 1, pp. 15–28, 2003.

[21] E. Ghafar-Zadeh, M. Sawan, and D. Therriault, "Novel direct-write CMOS-based laboratory-on-chip: Design, assembly and experimental results," *Sensors and Actuators A: Physical*, vol. 134, no. 1, pp. 27–36, 2007.

[22] E. Ghafar-Zadeh, M. Sawan, and D. Therriault, "A 0.18-µm CMOS capacitive sensor lab-on-chip," *Sensors and Actuators A: Physical*, vol. 141, no. 2, pp. 454–462, 2008.

[23] M. Zourob, S. Elwary, and A. Khademhosseini, *Recognition Receptors in Biosensors*. Springer, 2010.

[24] T. Shaffaf, S. Forouhi, and E. Ghafar-Zadeh, "Towards fully integrated portable sensing devices for COVID-19 and future global hazards: recent advances, challenges, and prospects," *Micromachines*, vol. 12, no. 8, p. 915, 2021.

[25] J. M. Rothberg, W. Hinz, T. M. Rearick, *et al.*, "An integrated semiconductor device enabling non-optical genome sequencing," *Nature*, vol. 475, no. 7356, pp. 348–352, 2011.

[26] A. Hassibi and T. H. Lee, "A programmable 0.18-μm CMOS electrochemical sensor microarray for biomolecular detection," *IEEE Sensors Journal*, vol. 6, no. 6, pp. 1380–1388, 2006.

[27] M. S.-C. Lu, Y.-C. Chen, and P.-C. Huang, "5 × 5 CMOS capacitive sensor array for detection of the neurotransmitter dopamine," *Biosensors and Bioelectronics*, vol. 26, no. 3, pp. 1093–1097, 2010.

[28] S. B. Prakash, P. Abshire, M. Urdaneta, and E. Smela, "A CMOS capacitance sensor for cell adhesion characterization," in *2005 IEEE International Symposium on Circuits and Systems*, 2005: IEEE, pp. 3495–3498.

[29] S. B. Prakash and P. Abshire, "Tracking cancer cell proliferation on a CMOS capacitance sensor chip," *Biosensors and Bioelectronics*, vol. 23, no. 10, pp. 1449–1457, 2008.

[30] S. B. Prakash and P. Abshire, "On-chip capacitance sensing for cell monitoring applications," *IEEE Sensors Journal*, vol. 7, no. 3, pp. 440–447, 2007.

[31] S. B. Prakash and P. Abshire, "A CMOS capacitance sensor that monitors cell viability," in *IEEE Sensors, 2005*, 2005: IEEE, p. 4.

[32] S.-W. Wang and M. S.-C. Lu, "CMOS capacitive sensors with sub-μm microelectrodes for biosensing applications," *IEEE Sensors Journal*, vol. 10, no. 5, pp. 991–996, 2010.

[33] N. Couniot, D. Bol, O. Poncelet, L. A. Francis, and D. Flandre, "A capacitance-to-frequency converter with on-chip passivated microelectrodes for bacteria detection in saline buffers up to 575 MHz," *IEEE Transactions on Circuits and Systems II: Express Briefs*, vol. 62, no. 2, pp. 159–163, 2014.

[34] N. Nikkhoo, C. Man, K. Maxwell, and P. G. Gulak, "A 0.18 μm CMOS Integrated Sensor for the Rapid Identification of Bacteria," in *Solid-State Circuits Conference, 2008. ISSCC 2008. Digest of Technical Papers. IEEE International*, 2008: IEEE, pp. 336–617.

[35] E. Ghafar-Zadeh, M. Sawan, V. P. Chodavarapu, and T. Hosseini-Nia, "Bacteria growth monitoring through a differential CMOS capacitive sensor," *IEEE Transactions on Biomedical Circuits and Systems*, vol. 4, no. 4, pp. 232–238, 2010.

[36] A. Tashtoush, *"Nano-amplification strategy using charge based capacitance measurement for pathogenic bacteria detection," Journal of Biotechnology and Strategic Health Research*, vol. 2, no. 2, pp. 87–100.

[37] G. Nabovati, E. Ghafar-Zadeh, A. Letourneau, and M. Sawan, "Towards high throughput cell growth screening: a new CMOS 8 × 8 biosensor array for life science applications," *IEEE Transactions on Biomedical Circuits and Systems*, 2016.

[38] C. Stagni, C. Guiducci, L. Benini, *et al.*, "CMOS DNA sensor array with integrated A/D conversion based on label-free capacitance measurement,"

IEEE Journal of Solid-State Circuits, vol. 41, no. 12, pp. 2956–2964, 2006.

[39] A. Manickam, A. Chevalier, M. McDermott, A. D. Ellington, and A. Hassibi, "A CMOS electrochemical impedance spectroscopy (EIS) biosensor array," *IEEE Transactions on Biomedical Circuits and Systems*, vol. 4, no. 6, pp. 379–390, 2010.

[40] F. Widdershoven, A. Cossettini, C. Laborde, *et al.*, "A CMOS pixelated nanocapacitor biosensor platform for high-frequency impedance spectroscopy and imaging," *IEEE Transactions on Biomedical Circuits and Systems*, 2018.

[41] J. Abbott, T. Ye, L. Qin, *et al.*, "CMOS nanoelectrode array for all-electrical intracellular electrophysiological imaging," *Nature Nanotechnology*, vol. 12, no. 5, pp. 460–467, 2017.

[42] U. Frey, J. Sedivy, F. Heer, *et al.*, "Switch-matrix-based high-density microelectrode array in CMOS technology," *IEEE Journal of Solid-State Circuits*, vol. 45, no. 2, pp. 467–482, 2010.

[43] M. Ballini, J. Müller, P. Livi, *et al.*, "A 1024-channel CMOS microelectrode array with 26,400 electrodes for recording and stimulation of electrogenic cells in vitro," *IEEE Journal of Solid-State Circuits*, vol. 49, no. 11, pp. 2705–2719, 2014.

[44] V. Viswam, R. Bounik, A. Shadmani, *et al.*, "Impedance spectroscopy and electrophysiological imaging of cells with a high-density CMOS microelectrode array system," *IEEE Transactions on Biomedical Circuits and Systems*, vol. 12, pp. 1356–1368, 2018.

[45] B. Miccoli, C. M. Lopez, E. Goikoetxea, *et al.*, "High-density electrical recording and impedance imaging with a multi-modal CMOS multi-electrode array chip," *Frontiers in Neuroscience*, vol. 13, p. 641, 2019.

[46] E. Ghafar-Zadeh, M. Sawan, and D. Therriault, "A microfluidic packaging technique for lab-on-chip applications," *IEEE Transactions on Advanced Packaging*, vol. 32, no. 2, pp. 410–416, 2009.

[47] E. Ghafar-Zadeh and M. Sawan, "Charge-based capacitive sensor array for CMOS-based laboratory-on-chip applications," *IEEE Sensors Journal*, vol. 8, no. 4, pp. 325–332, 2008.

[48] E. Ghafar-Zadeh, M. Sawan, and D. Therriault, "CMOS based capacitive sensor laboratory-on-chip: a multidisciplinary approach," *Analog Integrated Circuits and Signal Processing*, vol. 59, no. 1, pp. 1–12, 2009.

[49] A. M. Kummer, A. Hierlemann, and H. Baltes, "Tuning sensitivity and selectivity of complementary metal oxide semiconductor-based capacitive chemical microsensors," *Analytical Chemistry*, vol. 76, no. 9, pp. 2470–2477, 2004.

[50] C. Hagleitner, D. Lange, A. Hierlemann, O. Brand, and H. Baltes, "CMOS single-chip gas detection system comprising capacitive, calorimetric and mass-sensitive microsensors," *IEEE Journal of Solid-State Circuits*, vol. 37, no. 12, pp. 1867–1878, 2002.

[51] A. Hierlemann, "Integrated chemical microsensor systems in CMOS-technology," in *Solid-State Sensors, Actuators and Microsystems, 2005. Digest of Technical Papers. TRANSDUCERS'05. The 13th International Conference on*, 2005, vol. 2: IEEE, pp. 1134–1137.

[52] C. Cornila, A. Hierlemann, R. Lenggenhager, *et al.*, "Capacitive sensors in CMOS technology with polymer coating," *Sensors and Actuators B: Chemical*, vol. 25, no. 1–3, pp. 357–361, 1995.

[53] A. Hierlemann, U. Weimar, and H. Baltes, "Hand-held and palm-top chemical microsensor systems for gas analysis," *in Handbook of Machine Olfaction: Electronic Nose Technology. United states*, 2002, pp. 201–229.

[54] G. Nabovati, E. Ghafar-Zadeh, M. Mirzaei, G. Ayala-Charca, F. Awwad, and M. Sawan, "A new fully differential CMOS capacitance to digital converter for lab-on-chip applications," *IEEE Transactions on Biomedical Circuits and Systems*, vol. 9, no. 3, pp. 353–361, 2015.

[55] L. Moreno-Hagelsieb, B. Foultier, G. Laurent, *et al.*, "Electrical detection of DNA hybridization: three extraction techniques based on interdigitated Al/Al_2O_3 capacitors," *Biosensors and Bioelectronics*, vol. 22, no. 9–10, pp. 2199–2207, 2007.

[56] E. Ghafar-Zadeh and M. Sawan, *CMOS Capacitive Sensors for Lab-on-Chip Applications*. New York: Springer, 2010.

[57] X. Chen, X. Yan, K. A. Khor, and B. K. Tay, "Multilayer assembly of positively charged polyelectrolyte and negatively charged glucose oxidase on a 3D Nafion network for detecting glucose," *Biosensors and Bioelectronics*, vol. 22, no. 12, pp. 3256–3260, 2007.

[58] S. Zhang, W. Yang, Y. Niu, Y. Li, M. Zhang, and C. Sun, "Construction of glucose biosensor based on sorption of glucose oxidase onto multilayers of polyelectrolyte/nanoparticles," *Analytical and bioAnalytical Chemistry*, vol. 384, no. 3, pp. 736–741, 2006.

[59] B. Thierry, F. M. Winnik, Y. Merhi, J. Silver, and M. Tabrizian, "Bioactive coatings of endovascular stents based on polyelectrolyte multilayers," *Biomacromolecules*, vol. 4, no. 6, pp. 1564–1571, 2003.

[60] G. Nabovati, Y. W. Zhu, and M. Sawan, "Capacitive Sensor Arrays," *Wiley Encyclopedia of Electrical and Electronics Engineering*, pp. 1–12, 2015.

[61] A. Sun, E. Alvarez-Fontecilla, A. Venkatesh, E. Aronoff-Spencer, and D. A. Hall, "A 64 × 64 high-density redox amplified coulostatic discharge-based biosensor array in 180nm CMOS," *in ESSCIRC 2017–43rd IEEE European Solid State Circuits Conference*, 2017: IEEE, pp. 368–371.

[62] F. Hofmann, A. Frey, B. Holzapfl, *et al.*, "Passive DNA sensor with gold electrodes fabricated in a CMOS backend process," in *Solid-State Device Research Conference, 2002. Proceeding of the 32nd European*, 2002: IEEE, pp. 487–490.

[63] C. Guiducci, C. Stagni, A. Fischetti, U. Mastromatteo, L. Benini, and B. Riccoricco, "Microelectrodes on a silicon chip for label-free capacitive DNA sensing," *IEEE Sensors Journal*, vol. 6, no. 5, pp. 1084–1093, 2006.

[64] R. K. Shervedani, A. H. Mehrjardi, and N. Zamiri, “A novel method for glucose determination based on electrochemical impedance spectroscopy using glucose oxidase self-assembled biosensor,” *Bioelectrochemistry*, vol. 69, no. 2, pp. 201–208, 2006.

[65] K. Kerman, D. Ozkan, P. Kara, B. Meric, J. J. Gooding, and M. Ozsoz, “Voltammetric determination of DNA hybridization using methylene blue and self-assembled alkanethiol monolayer on gold electrodes,” *Analytica Chimica Acta*, vol. 462, no. 1, pp. 39–47, 2002.

[66] L. Berdondini, P. D. van der Wal, N. F. de Rooij, and M. Koudelka-Hep, “Development of an electroless post-processing technique for depositing gold as electrode material on CMOS devices,” *Sensors and Actuators B: Chemical*, vol. 99, no. 2–3, pp. 505–510, 2004.

[67] G. O. Mallory and J. B. Hajdu, *Electroless Plating: Fundamentals and Applications*. Orlando: William Andrew, 1990.

[68] J. W. Ko, H. C. Koo, D. W. Kim, *et al.*, “Electroless gold plating on aluminum patterned chips for CMOS-based sensor applications,” *Journal of The Electrochemical Society*, vol. 157, no. 1, pp. D46–D49, 2010.

[69] F. Widdershoven, D. Van Steenwinckel, J. Überfeld, *et al.*, “CMOS biosensor platform,” in *2010 IEEE International Electron Devices Meeting (IEDM)*, 2010: IEEE, pp. 36.1.1–36.1.4.

[70] S. G. Lemay, C. Laborde, C. Renault, A. Cossettini, L. Selmi, and F. P. Widdershoven, “High-frequency nanocapacitor arrays: Concept, recent developments, and outlook,” *Accounts of chemical research*, vol. 49, no. 10, pp. 2355–2362, 2016.

[71] F. Greve, J. Lichtenberg, K. Kirstein, U. Frey, J. Perriard, and A. Hierlemann, “A perforated CMOS microchip for immobilization and activity monitoring of electrogenic cells,” *Journal of Micromechanics and Microengineering*, vol. 17, no. 3, p. 462, 2007.

[72] F. Heer, S. Hafizovic, W. Franks, A. Blau, C. Ziegler, and A. Hierlemann, “CMOS microelectrode array for bidirectional interaction with neuronal networks,” *IEEE Journal of Solid-State Circuits*, vol. 41, no. 7, pp. 1620–1629, 2006.

[73] C. M. Lopez, H. S. Chun, S. Wang, *et al.*, “A multimodal CMOS MEA for high-throughput intracellular action potential measurements and impedance spectroscopy in drug-screening applications,” *IEEE Journal of Solid-State Circuits*, vol. 53, no. 11, pp. 3076–3086, 2018.

[74] C. Kokkinos, A. Economou, and M. I. Prodromidis, “Electrochemical immunosensors: Critical survey of different architectures and transduction strategies,” *TrAC Trends in Analytical Chemistry*, vol. 79, pp. 88–105, 2016.

[75] A. J. Bard and L. R. Faulkner, “Fundamentals and applications,” *Electrochemical Methods*, vol. 2, p. 482, 2001.

[76] Q. Liu and P. Wang, *Cell-based Biosensors: Principles and Applications*. Norwood, Massachusetts, United States: Artech House, 2009.

[77] M. Gouy, “Sur la constitution de la charge électrique à la surface d'un électrolyte,” *J. Phys. Theor. Appl*, vol. 9, no. 1, pp. 457–468, 1910.

[78] D. L. Chapman, "LI. A contribution to the theory of electrocapillarity," *The London, Edinburgh, and Dublin Philosophical Magazine and Journal of Science*, vol. 25, no. 148, pp. 475–481, 1913.

[79] V. Tsouti, C. Boutopoulos, I. Zergioti, and S. Chatzandroulis, "Capacitive microsystems for biological sensing," *Biosensors and Bioelectronics*, vol. 27, no. 1, pp. 1–11, 2011.

[80] W. M. Siu and R. S. Cobbold, "Basic properties of the electrolyte—SiO2—Si system: physical and theoretical aspects," *IEEE Transactions on Electron Devices*, vol. 26, no. 11, pp. 1805–1815, 1979.

[81] M. W. den Otter, "Approximate expressions for the capacitance and electrostatic potential of interdigitated electrodes," *Sensors and Actuators A: Physical*, vol. 96, no. 2–3, pp. 140–144, 2002.

[82] C. Berggren, B. Bjarnason, and G. Johansson, "Capacitive biosensors," *Electroanalysis: An International Journal Devoted to Fundamental and Practical Aspects of Electroanalysis*, vol. 13, no. 3, pp. 173–180, 2001.

[83] O. Laczka, E. Baldrich, F. X. Muñoz, and F. J. del Campo, "Detection of Escherichia coli and Salmonella typhimurium using interdigitated microelectrode capacitive immunosensors: the importance of transducer geometry," *Analytical Chemistry*, vol. 80, no. 19, pp. 7239–7247, 2008.

[84] P. Van Gerwen, W. Laureys, G. Huyberechts, *et al.*, "Nanoscaled interdigitated electrode arrays for biochemical sensors," in *Solid State Sensors and Actuators, 1997. TRANSDUCERS'97 Chicago, 1997 International Conference on*, 1997, vol. 2: IEEE, pp. 907–910.

[85] C. Jungreuthmayer, G. M. Birnbaumer, J. Zanghellini, and P. Ertl, "3D numerical simulation of a lab-on-a-chip—increasing measurement sensitivity of interdigitated capacitors by passivation optimization," *Lab on a Chip*, vol. 11, no. 7, pp. 1318–1325, 2011.

[86] S. O. Blume, R. Ben-Mrad, and P. E. Sullivan, "Modelling the capacitance of multi-layer conductor-facing interdigitated electrode structures," *Sensors and Actuators B: Chemical*, vol. 213, pp. 423–433, 2015.

[87] R. Igreja and C. Dias, "Extension to the analytical model of the interdigital electrodes capacitance for a multi-layered structure," *Sensors and Actuators A: Physical*, vol. 172, no. 2, pp. 392–399, 2011.

[88] R. Igreja and C. Dias, "Analytical evaluation of the interdigital electrodes capacitance for a multi-layered structure," *Sensors and Actuators A: Physical*, vol. 112, no. 2–3, pp. 291–301, 2004.

[89] S. O. Blume, R. Ben-Mrad, and P. E. Sullivan, "Characterization of coplanar electrode structures for microfluidic-based impedance spectroscopy," *Sensors and Actuators B: Chemical*, vol. 218, pp. 261–270, 2015.

[90] J. Wei, "Distributed capacitance of planar electrodes in optic and acoustic surface wave devices," *IEEE Journal of Quantum Electronics*, vol. 13, no. 4, pp. 152–158, 1977.

[91] R. Igreja and C. Dias, "Dielectric response of interdigital chemocapacitors: The role of the sensitive layer thickness," *Sensors and Actuators B: Chemical*, vol. 115, no. 1, pp. 69–78, 2006.

[92] D. Berdat, A. C. M. Rodríguez, F. Herrera, and M. A. Gijs, "Label-free detection of DNA with interdigitated micro-electrodes in a fluidic cell," *Lab on a Chip*, vol. 8, no. 2, pp. 302–308, 2008.

[93] A. Bratov, J. Ramón-Azcón, N. Abramova, *et al.*, "Three-dimensional interdigitated electrode array as a transducer for label-free biosensors," *Biosensors and Bioelectronics*, vol. 24, no. 4, pp. 729–735, 2008.

[94] H.-W. Jung, Y. W. Chang, G.-y. Lee, S. Cho, M.-J. Kang, and J.-C. Pyun, "A capacitive biosensor based on an interdigitated electrode with nanoislands," *Analytica Chimica Acta*, vol. 844, pp. 27–34, 2014.

[95] S. Rana, R. H. Page, and C. J. McNeil, "Comparison of planar and 3-D interdigitated electrodes as electrochemical impedance biosensors," *Electroanalysis*, vol. 23, no. 10, pp. 2485–2490, 2011.

[96] R. B. Schoch and P. Renaud, "Ion transport through nanoslits dominated by the effective surface charge," *Applied Physics Letters*, vol. 86, no. 25, p. 253111, 2005.

[97] D. Stein, M. Kruithof, and C. Dekker, "Surface-charge-governed ion transport in nanofluidic channels," *Physical Review Letters*, vol. 93, no. 3, p. 035901, 2004.

[98] R. B. Schoch, H. Van Lintel, and P. Renaud, "Effect of the surface charge on ion transport through nanoslits," *Physics of Fluids*, vol. 17, no. 10, p. 100604, 2005.

[99] M. Bäcker, F. Kramer, C. Huck, *et al.*, "Planar and 3D interdigitated electrodes for biosensing applications: The impact of a dielectric barrier on the sensor properties," *Physica Status Solidi (a),* vol. 211, no. 6, pp. 1357–1363, 2014.

[100] E. Ghafar-Zadeh, G. Ayala-Charca, B. Gholamzadeh, S. Ghasemi, and S. Magierowski, "Towards scalable capacitive cantilever arrays for emerging biomedical applications," *Sensors and Actuators A: Physical*, vol. 260, pp. 90–98, 2017.

[101] A. O. Manzanares and F. M. de Espinosa, "Air-coupled MUMPs capacitive micromachined ultrasonic transducers with resonant cavities," *Ultrasonics*, vol. 52, no. 4, pp. 482–489, 2012.

[102] E. T. Carlen, M. S. Weinberg, A. M. Zapata, and J. T. Borenstein, "A micromachined surface stress sensor with electronic readout," *Review of scientific instruments*, vol. 79, no. 1, pp. 015106 (1–7), 2008.

[103] E. T. Carlen, M. S. Weinberg, C. E. Dubé, A. M. Zapata, and J. T. Borenstein, "Micromachined silicon plates for sensing molecular interactions," *Applied Physics Letters*, vol. 89, no. 17, pp. 173123 (1–4), 2006.

[104] A.-Y. Chang and M. S.-C. Lu, "A CMOS magnetic microbead-based capacitive biosensor array with on-chip electromagnetic manipulation," *Biosensors and Bioelectronics*, vol. 45, pp. 6–12, 2013.

[105] A. Romani, N. Manaresi, L. Marzocchi, *et al.*, "Capacitive sensor array for localization of bioparticles in CMOS lab-on-a-chip," in *2004 IEEE International Solid-State Circuits Conference, 2004. Digest of Technical Papers. ISSCC*, 2004: IEEE, pp. 224–225.

[106] A. Alhoshany, S. Sivashankar, Y. Mashraei, H. Omran, and K. N. Salama, "A biosensor-CMOS platform and integrated readout circuit in 0.18-μm CMOS technology for cancer biomarker detection," *Sensors*, vol. 17, no. 9, pp. 1942 (1–12), 2017.

[107] J. S. Gaggatur and G. Banerjee, "High gain capacitance sensor interface for the monitoring of cell volume growth," in *2017 30th International Conference on VLSI Design and 2017 16th International Conference on Embedded Systems (VLSID)*, 2017: IEEE, pp. 201–206.

[108] K. Mohammad, D. A. Buchanan, and D. J. Thomson, "Integrated 0.35 pm CMOS capacitance sensor with atto-farad sensitivity for single cell analysis," *in Biomedical Circuits and Systems Conference (BioCAS)*, 2016 *IEEE*, 2016: IEEE, pp. 22–25.

[109] M. A. Miled, G. Massicotte, and M. Sawan, "Dielectrophoresis-based integrated lab-on-chip for nano and micro-particles manipulation and capacitive detection," *IEEE Transactions on Biomedical Circuits and Systems*, vol. 6, no. 2, pp. 120–132, 2012.

[110] S. B. Prakash and P. Abshire, "A fully differential rail-to-rail CMOS capacitance sensor with floating-gate trimming for mismatch compensation," *IEEE Transactions on Circuits and Systems I: Regular Papers*, vol. 56, no. 5, pp. 975–986, 2009.

[111] S. Forouhi, R. Dehghani, and E. Ghafar-Zadeh, "Toward high throughput core-CBCM CMOS capacitive sensors for life science applications: a novel current-mode for high dynamic range circuitry," *Sensors*, vol. 18, no. 10, pp. 3370 (1–29), 2018.

[112] K.-H. Lee, J.-O. Lee, M.-J. Sohn, *et al.*, "One-chip electronic detection of DNA hybridization using precision impedance-based CMOS array sensor," *Biosensors and Bioelectronics*, vol. 26, no. 4, pp. 1373–1379, 2010.

[113] P. Ciccarella, M. Carminati, M. Sampietro, and G. Ferrari, "Multichannel 65 zF rms resolution CMOS monolithic capacitive sensor for counting single micrometer-sized airborne particles on chip," *IEEE Journal of Solid-State Circuits*, vol. 51, no. 11, pp. 2545–2553, 2016.

[114] H. O. Tabrizi, S. Forouhi, and E. Ghafar-Zadeh, "A high dynamic range dual 8 × 16 capacitive sensor array for life science applications," *IEEE Transactions on Biomedical Circuits and Systems*, 2022.

[115] H. O. Tabrizi, O. Farhanieh, Q. Owen, S. Magierowski, and E. Ghafar-Zadeh, "Wide input dynamic range fully integrated capacitive sensor for life science applications," *IEEE Transactions on Biomedical Circuits and Systems*, vol. 15, no. 2, pp. 339–350, 2021.

[116] K. Mohammad and D. J. Thomson, "Differential ring oscillator based capacitance sensor for microfluidic applications," *IEEE Transactions on Biomedical Circuits and Systems*, vol. 11, no. 2, pp. 392–399, 2017.

[117] B. Senevirathna, S. Lu, and P. Abshire, "Characterization of a high dynamic range lab-on-CMOS capacitance sensor array," in *2017 IEEE International Symposium on Circuits and Systems (ISCAS)*, 2017: IEEE, pp. 1–4.

[118] B. Senevirathna, A. Castro, M. Dandin, E. Smela, and P. Abshire, "Lab-on-CMOS capacitance sensor array for real-time cell viability measurements with I2C readout," in *2016 IEEE International Symposium on Circuits and Systems (ISCAS)*, 2016: IEEE, pp. 2863–2866.

[119] E. Naviasky, T. Datta-Chaudhuri, and P. Abshire, "High resolution capacitance sensor array for real-time monitoring of cell viability," in *2014 IEEE International Symposium on Circuits and Systems (ISCAS)*, 2014: IEEE, pp. 634–637.

[120] A. A. Helmy, H.-J. Jeon, Y.-C. Lo, *et al.*, "A self-sustained CMOS microwave chemical sensor using a frequency synthesizer," *IEEE Journal of Solid-State Circuits*, vol. 47, no. 10, pp. 2467–2483, 2012.

[121] J. Nehring, M. Bartels, R. Weigel, and D. Kissinger, "A permittivity sensitive PLL based on a silicon-integrated capacitive sensor for microwave biosensing applications," in *2015 IEEE Topical Conference on Biomedical Wireless Technologies, Networks, and Sensing Systems (BioWireleSS)*, 2015: IEEE, pp. 1–3.

[122] S. Guha, K. Schmalz, C. Wenger, and F. Herzel, "Self-calibrating highly sensitive dynamic capacitance sensor: towards rapid sensing and counting of particles in laminar flow systems," *Analyst*, vol. 140, no. 9, pp. 3262–3272, 2015.

[123] O. Elhadidy, M. Elkholy, A. A. Helmy, S. Palermo, and K. Entesari, "A CMOS fractional-N PLL-based microwave chemical sensor with 1.5% permittivity accuracy," *IEEE Transactions on Microwave Theory and Techniques*, vol. 61, no. 9, pp. 3402–3416, 2013.

[124] T. Mitsunaka, D. Sato, N. Ashida, *et al.*, "CMOS biosensor IC focusing on dielectric relaxations of biological water with 120 and 60 GHz oscillator arrays," *IEEE Journal of Solid-State Circuits*, vol. 51, no. 11, pp. 2534–2544, 2016.

[125] J.-C. Chien, M. Anwar, E.-C. Yeh, L. P. Lee, and A. M. Niknejad, "A 6.5/17.5-GHz dual-channel interferometer-based capacitive sensor in 65-nm CMOS for high-speed flow cytometry," in *2014 IEEE MTT-S International Microwave Symposium (IMS)*, 2014: IEEE, pp. 1–4.

[126] J.-C. Chien, E.-C. Yeh, L. P. Lee, M. Anwar, and A. M. Niknejad, "A near-field modulation chopping stabilized injection-locked oscillator sensor for protein conformation detection at microwave frequency," in *2015 Symposium on VLSI Circuits (VLSI Circuits)*, 2015: IEEE, pp. C332–C333.

[127] J.-C. Chien and A. M. Niknejad, "Oscillator-based reactance sensors with injection locking for high-throughput flow cytometry using microwave dielectric spectroscopy," *J. Solid-State Circuits*, vol. 51, no. 2, pp. 457–472, 2016.

[128] O. Elhadidy, S. Shakib, K. Krenek, S. Palermo, and K. Entesari, "A wideband fully-integrated CMOS ring-oscillator PLL-based complex dielectric spectroscopy system," *IEEE Transactions on Circuits and Systems I: Regular Papers*, vol. 62, no. 8, pp. 1940–1949, 2015.

[129] J.-C. Chien, A. Ameri, E.-C. Yeh, A. N. Killilea, M. Anwar, and A. M. Niknejad, "A high-throughput flow cytometry-on-a-CMOS platform for single-cell dielectric spectroscopy at microwave frequencies," *Lab on a Chip*, vol. 18, no. 14, pp. 2065–2076, 2018.

[130] R. Nagulapalli, K. Hayatleh, S. Barker, S. Raparthy, and F. J. Lidgey, "A CMOS blood cancer detection sensor based on frequency deviation detection," *Analog Integrated Circuits and Signal Processing*, vol. 92, no. 3, pp. 437–442, 2017.

[131] X. Liu, L. Li, and A. J. Mason, *"High-throughput impedance spectroscopy biosensor array chip," Phil. Trans. R. Soc. A*, vol. 372, no. 2012, pp. 20130107 (1–14), 2014.

[132] A. Manickam, C. A. Johnson, S. Kavusi, and A. Hassibi, "Interface design for CMOS-integrated electrochemical impedance spectroscopy (EIS) biosensors," *Sensors*, vol. 12, no. 11, pp. 14467–14488, 2012.

[133] P. Ciccarella, M. Carminati, M. Sampietro, and G. Ferrari, "CMOS monolithic airborne-particulate-matter detector based on 32 capacitive sensors with a resolution of 65zF rms," in *2016 IEEE International Solid-State Circuits Conference (ISSCC)*, 2016: IEEE, pp. 486–488.

[134] S. Druart, D. Flandre, and L. A. Francis, "A self-oscillating system to measure the conductivity and the permittivity of liquids within a single triangular signal," *Journal of Sensors*, vol. 2014, 2014.

[135] D. De Venuto, S. Carrara, and B. Riccò, "Design of an integrated low-noise read-out system for DNA capacitive sensors," *Microelectronics Journal*, vol. 40, no. 9, pp. 1358–1365, 2009.

[136] N. Petrellis, C. Spathis, K. Georgakopoulou, and A. Birbas, "Capacitive sensor estimation based on self-configurable reference capacitance," *Recent Patents on Signal Processing*, vol. 3, no. 1, pp. 12–21, 2013.

[137] K. Georgakopoulou, C. Spathis, N. Petrellis, and A. Birbas, "A capacitive-to-digital converter with automatic range adaptation for readout instrumentation," *IEEE Transactions on Instrumentation and Measurement*, vol. 65, no. 2, pp. 336–345, 2016.

[138] H. Ha, D. Sylvester, D. Blaauw, and J.-Y. Sim, "12.6 A 160nW 63.9 fJ/conversion-step capacitance-to-digital converter for ultra-low-power wireless sensor nodes," in *Solid-State Circuits Conference Digest of Technical Papers (ISSCC), 2014 IEEE International*, 2014: IEEE, pp. 220–221.

[139] K. Tanaka, Y. Kuramochi, T. Kurashina, K. Okada, and A. Matsuzawa, "A 0.026 mm^2 capacitance-to-digital converter for biotelemetry applications using a charge redistribution technique," *in IEEE Asian Solid-State Circuits Conference*, 2007. *ASSCC'07*, 2007: IEEE, pp. 244–247.

[140] A. Alhoshany and K. N. Salama, "A precision, energy-efficient, oversampling, noise-shaping differential SAR capacitance-to-digital converter," *IEEE Transactions on Instrumentation and Measurement*, no. 99, 2018.

[141] J. C. Chen, B. W. McGaughy, D. Sylvester, and C. Hu, "An on-chip, attofarad interconnect charge-based capacitance measurement (CBCM)

technique," in *Electron Devices Meeting, 1996. IEDM'96, International*, 1996: IEEE, pp. 69–72.

[142] C. Guiducci, C. Stagni, G. Zuccheri, *et al.*, "DNA detection by integrable electronics," *Biosensors and Bioelectronics*, vol. 19, no. 8, pp. 781–787, 2004.

[143] R. Bach, B. Davis, and R. Laubhan, "Improvements to CBCM (charge-based capacitance measurement) for deep submicron CMOS technology," in *Proceedings of the 7th International Symposium on Quality Electronic Design*, 2006: IEEE Computer Society, pp. 324–329.

[144] Y.-W. Chang, H.-W. Chang, C.-H. Hsieh, *et al.*, "A novel simple CBCM method free from charge injection-induced errors," *IEEE Electron Device Letters*, vol. 25, no. 5, pp. 262–264, 2004.

[145] Y. Chang, H. Chang, T. Lu, *et al.*, "A novel CBCM method free from charge injection induced errors: Investigation into the impact of floating dummy-fills on interconnect capacitance," in *Microelectronic Test Structures, 2005. ICMTS 2005. Proceedings of the 2005 International Conference on*, 2005: IEEE, pp. 235–238.

[146] L. Vendrame, L. Bortesi, and A. Bogliolo, "Accuracy assessment and improvement of on-chip charge-based capacitance measurements," in *Proc. 7th IEEE SPI Workshop*, 2003, pp. 117–120.

[147] K. H. Dia, W. Tsao, C. H. Chien, and Z. Zeng, "A novel compact CBCM method for high resolution measurement in 28nm CMOS technology," in *Microelectronic Test Structures (ICMTS), 2012 IEEE International Conference on*, 2012: IEEE, pp. 119–121.

[148] H. O. Tabrizi, S. Forouhi, O. Farhanieh, A. Bozkurt, S. Magierowski, and E. Ghafar-Zadeh, "Calibration-free CMOS capacitive sensor for life science applications," *IEEE Transactions on Instrumentation and Measurement*, 2021.

[149] Z. Ning, H.-X. Delecourt, L. De Schepper, R. Gillon, and M. Tack, "Precise analogue characterization of MIM capacitors using an improved charge-based capacitance measurement (CBCM) technique," in Solid-State Device Research Conference, 2005. ESSDERC 2005. Proceedings of 35th European, 2005: IEEE, pp. 269–272.

[150] P. Zhang, Q. Wan, C. Feng, and H. Wang, "Gate capacitance measurement using a self-differential charge-based capacitance measurement method," *IEEE Electron Device Letters*, vol. 36, no. 12, pp. 1271–1273, 2015.

[151] X. P. Yu, R. Q. Tian, W. L. Xu, and Z. Shi, "A new on-chip signal generator for charge-based capacitance measurement circuit," *Journal of Electronic Testing*, vol. 31, no. 3, pp. 329–333, 2015.

[152] I. Evans and T. York, "Microelectronic capacitance transducer for particle detection," *IEEE Sensors Journal*, vol. 4, no. 3, pp. 364–372, 2004.

[153] S. Forouhi, R. Dehghani, and E. Ghafar-Zadeh, "High throughput core-CBCM CMOS capacitive sensor for life science applications," in *2018 IEEE Canadian Conference on Electrical & Computer Engineering (CCECE)*, 2018: IEEE, pp. 1–4.

[154] C. Guiducci, C. Stagni, G. Zuccheri, *et al.*, "A biosensor for direct detection of DNA sequences based on capacitance measurements," in *Proceeding of the 32nd European Solid-State Device Research Conference*, 2002, pp. 24–26.

[155] C. Stagni, C. Guiducci, M. Lanzoni, L. Benini, and B. Ricco, "Hardware-software design of a smart sensor for fully-electronic DNA hybridization detection," in *Proceedings of the conference on Design, Automation and Test in Europe-Volume 3*, 2005: *IEEE Computer Society*, pp. 198–203.

[156] C. Stagni, C. Guiducci, L. Benini, *et al.*, "A fully electronic label-free DNA sensor chip," *IEEE Sensors Journal*, vol. 7, no. 4, pp. 577–585, 2007.

[157] S. Carrara, V. Bhalla, C. Stagni, *et al.*, "Label-free cancer markers detection by capacitance biochip," *Sensors and Actuators B: Chemical*, vol. 136, no. 1, pp. 163–172, 2009.

[158] S. Carrara, L. Benini, V. Bhalla, *et al.*, "New insights for using self-assembly materials to improve the detection stability in label-free DNA-chip and immuno-sensors," *Biosensors and Bioelectronics*, vol. 24, no. 12, pp. 3425–3429, 2009.

[159] S. Carrara, A. Cavallini, Y. Leblebici, *et al.*, "Capacitance DNA bio-chips improved by new probe immobilization strategies," *Microelectronics Journal*, vol. 41, no. 11, pp. 711–717, 2010.

[160] A. M. Sampson, E. T. Peterson, and I. Papautsky, "Interdigitated array microelectrode capacitive sensor for detection of paraffinophilic myco-bacteria," in *Microfluidics, BioMEMS, and Medical Microsystems VI*, 2008, vol. 6886: *International Society for Optics and Photonics*, p. 68860X.

[161] T. York, T. Phua, L. Reichelt, A. Pawlowski, and R. Kneer, "A miniature electrical capacitance tomograph," *Measurement Science and Technology*, vol. 17, no. 8, p. 2119, 2006.

[162] C. Laborde, F. Pittino, H. Verhoeven, *et al.*, "Real-time imaging of microparticles and living cells with CMOS nanocapacitor arrays," *Nature Nanotechnology*, vol. 10, no. 9, pp. 791 (1–6), 2015.

[163] A. Tanskanen, B. Bahreyni, and M. Syrzycki, "Charge-based femto-farad capacitance measurement technique for MEMS applications," in *2016 IEEE Canadian Conference on Electrical and Computer Engineering (CCECE)*, 2016: IEEE, pp. 1–4.

[164] T. York, I. Evans, Z. Pokusevski, and T. Dyakowski, "Particle detection using an integrated capacitance sensor," *Sensors and Actuators A: Physical*, vol. 92, no. 1–3, pp. 74–79, 2001.

[165] Y. Yusof, K. Sugimoto, H. Ozawa, S. Uno, and K. Nakazato, "On-chip microelectrode capacitance measurement for biosensing applications," *Japanese Journal of Applied Physics*, vol. 49, no. 1S, pp. 01AG05 (1–6), 2010.

[166] B. Kim, S. Uno, and K. Nakazato, "Wireless charge based capacitance measurement circuits with on-chip spiral inductor for radio frequency identification biosensor," *Japanese Journal of Applied Physics*, vol. 51, no. 4S, p. 04DE08, 2012.

[167] K. Nakazato, "BioCMOS LSIs for portable gene-based diagnostic inspection system," in *Circuits and Systems (ISCAS), 2012 IEEE International Symposium on*, 2012: IEEE, pp. 2287–2290.

[168] K. Nakazato, "Potentiometric, amperometric, and impedimetric CMOS biosensor array," *in State of the Art in Biosensors-General Aspects*: InTech, 2013.

[169] Y. Yusof and N. Kazuo, "Ultra sensitive CMOS biosensor array for label free DNA detection: Circuit design consideration," in *Biomedical Engineering and Sciences (IECBES), 2014 IEEE Conference on*, 2014: IEEE, pp. 365–368.

[170] E. Ghafar-Zadeh and M. Sawan, "A high precision and linearity differential capacitive sensor circuit dedicated to bioparticles detection," in *The 3rd International IEEE-NEWCAS Conference, 2005*, 2005: IEEE, pp. 299–302.

[171] E. Ghafar-Zadeh and M. Sawan, "A Core-CBCM sigma delta capacitive sensor array dedicated to lab-on-chip applications," *Sensors and Actuators A: Physical*, vol. 144, no. 2, pp. 304–313, 2008.

[172] E. Ghafar-Zadeh, M. Sawan, and V. P. Chodavarapu, "Micro-organism-on-chip: Emerging direct-write CMOS-based platform for biological applications," *IEEE Transactions on Biomedical Circuits and Systems*, vol. 3, no. 4, pp. 212–219, 2009.

[173] L. Yao, M. Hajj-Hassan, E. Ghafar-Zadeh, A. Shabani, V. Chodavarapu, and M. Zourob, "CMOS capacitive sensor system for bacteria detection using phage organisms," *in Canadian Conference on Electrical and Computer Engineering. CCECE* 2008, 2008: IEEE, pp. 000877–000880.

[174] E. Ghafar-Zadeh and M. Sawan, "A hybrid microfluidic/CMOS capacitive sensor dedicated to lab-on-chip applications," *IEEE Transactions on Biomedical Circuits and Systems*, vol. 1, no. 4, pp. 270–277, 2007.

[175] M. A. Miled, M. Sawan, and E. Ghafar-Zadeh, "A dynamic decoder for first-order ΣΔ modulators dedicated to lab-on-chip applications," *IEEE Transactions on Signal Processing*, vol. 57, no. 10, pp. 4076–4084, 2009.

[176] M. A. Miled and M. Sawan, "Subthreshold transistor operation for a high sensitivity capacitive sensor," *in Canadian Conference on Electrical and Computer Engineering, 2008. CCECE 2008*, 2008: IEEE, pp. 001671–001674.

[177] M. A. Miled and M. Sawan, "A new fully integrated cmos interface for a dielectrophoretic lab-on-a-chip device," in *2011 IEEE International Symposium on Circuits and Systems (ISCAS), 2011*: IEEE, pp. 2349–2352.

[178] M. A. Miled, G. Massicotte, and M. Sawan, "Low-voltage lab-on-chip for micro and nanoparticles manipulation and detection: Experimental results," *Analog Integrated Circuits and Signal Processing*, vol. 73, no. 3, pp. 707–717, 2012.

[179] S. B. Prakash and P. Abshire, "A fully differential rail-to-rail capacitance measurement circuit for integrated cell sensing," in *2007 IEEE Sensors*, 2007: IEEE, pp. 1444–1447.

[180] T. Datta, E. Naviasky, and P. Abshire, "Floating-gate capacitance sensor array for cell viability monitoring," in *2013 IEEE Biomedical Circuits and Systems Conference (BioCAS)*, 2013: IEEE, pp. 65–68.

[181] Z. Jun-Rui, I. A. Nanolab, and M. Mazza, "Low-energy biomarker detection through charge-based impedance measurements," in *2016 IEEE Sensors*, 2016: IEEE, pp. 1–3.

[182] S. Forouhi, E. Ghafar-Zadeh, and R. Dehghani, "Multimodal CMOS based biosenors for Lab-on-Chip applications: A new multidisciplinary approach," in *Electrical and Computer Engineering (CCECE), 2017 IEEE 30th Canadian Conference on*, 2017: IEEE, pp. 1–4.

[183] G. Nabovati, E. Ghafar-Zadeh, and M. Sawan, "A novel multifunctional integrated biosensor array for simultaneous monitoring of cell growth and acidification rate," in *2016 IEEE International Symposium on Circuits and Systems (ISCAS)*, 2016: IEEE, pp. 2855–2858.

[184] D. Zhu, J. Mo, S. Xu, *et al.*, "A New Capacitance-to-Frequency Converter for On-Chip Capacitance Measurement and Calibration in CMOS Technology," *Journal of Electronic Testing*, pp. 1–5, 2016.

[185] R. Yamane, H. Iwasaki, Y. Dei, J. Cui, and T. Matsuoka, "A capacitance detection circuit for on-chip microparticle manipulation," in *2014 IEEE International Meeting for Future of Electron Devices, Kansai (IMFEDK)*, 2014: IEEE, pp. 1–2.

[186] A. M. Tashtoush, "Miniaturized IC biosensors on a single chip for biological/biomedical applications," *International Journal of Simulation–Systems, Science & Technology*, vol. 15, no. 6, pp. 8–18, 2014.

[187] S. Forouhi, O. Farhanieh, R. Dehghani, and E. Ghafar-Zadeh, "A current based capacitance-to-frequency converter for Lab-on-Chip applications," in *2017 IEEE 60th International Midwest Symposium on Circuits and Systems (MWSCAS)*, Boston, MA, USA, 2017: IEEE, pp. 92–95.

[188] X. Zhao, P. Gan, Q. Zhao, *et al.*, "A 124 fJ/bit cascode current mirror array based PUF with 1.50% native unstable bit ratio," *IEEE Transactions on Circuits and Systems I: Regular Papers*, vol. 66, no. 9, pp. 3494–3503, 2019.

[189] D. Harvey, *Analytical Chemistry 2.0. LibreTexts*, 2010.

[190] A. Hedayatipour, S. Aslanzadeh, and N. McFarlane, "CMOS based whole cell impedance sensing: Challenges and future outlook," *Biosensors and Bioelectronics*, vol. 143, p. 111600, 2019.

[191] H. Li, X. Liu, L. Li, X. Mu, R. Genov, and A. J. Mason, "CMOS electrochemical instrumentation for biosensor microsystems: A review," *Sensors*, vol. 17, no. 1, p. 74, 2016.

[192] H. M. Jafari and R. Genov, "Chopper-stabilized bidirectional current acquisition circuits for electrochemical amperometric biosensors," *IEEE Transactions on Circuits and Systems I: Regular Papers*, vol. 60, no. 5, pp. 1149–1157, 2013.

[193] B. N. Kim, A. D. Herbst, S. J. Kim, B. A. Minch, and M. Lindau, "Parallel recording of neurotransmitters release from chromaffin cells using a

10 × 10 CMOS IC potentiostat array with on-chip working electrodes," *Biosensors and Bioelectronics*, vol. 41, pp. 736–744, 2013.

[194] J.-C. Chien, S. W. Baker, H. T. Soh, and A. Arbabian, "Design and analysis of a sample-and-hold CMOS electrochemical sensor for aptamer-based therapeutic drug monitoring," *IEEE Journal of Solid-State Circuits*, vol. 55, no. 11, pp. 2914–2929, 2020.

[195] M. H. Nazari, H. Mazhab-Jafari, L. Leng, A. Guenther, and R. Genov, "CMOS neurotransmitter microarray: 96-channel integrated potentiostat with on-die microsensors," *IEEE Transactions on Biomedical Circuits and Systems*, vol. 7, no. 3, pp. 338–348, 2012.

[196] E. Ghodsevali, B. Gosselin, M. Boukadoum, and A. Miled, "High accuracy and sensitivity differential potentiostat with amplifier-based error cancellation feedback loop," *in Biomedical Circuits and Systems Conference (BioCAS)*, 2015 *IEEE*, 2015: IEEE, pp. 1–4.

[197] A. Manickam, K.-D. You, N. Wood, *et al.*, "A CMOS electrochemical biochip with 32 × 32 three-electrode voltammetry pixels," *IEEE Journal of Solid-State Circuits*, vol. 54, no. 11, pp. 2980–2990, 2019.

[198] S. S. Ghoreishizadeh, C. Baj-Rossi, A. Cavallini, S. Carrara, and G. De Micheli, "An integrated control and readout circuit for implantable multi-target electrochemical biosensing," *IEEE Transactions on Biomedical Circuits and Systems*, vol. 8, no. 6, pp. 891–898, 2014.

[199] L. Li, X. Liu, W. A. Qureshi, and A. J. Mason, "CMOS amperometric instrumentation and packaging for biosensor array applications," *IEEE Transactions on Biomedical Circuits and Systems*, vol. 5, no. 5, pp. 439–448, 2011.

[200] L. A. Buscaglia, O. N. Oliveira, and J. P. Carmo, "Roadmap for electrical impedance spectroscopy for sensing: a tutorial," *IEEE Sensors Journal*, vol. 21, no. 20, pp. 22246–22257, 2021.

[201] S. Sutula, J. P. Cuxart, J. Gonzalo-Ruiz, F. X. Muñoz-Pascual, L. Terés, and F. Serra-Graells, "A 25-μw all-MOS potentiostatic Delta-Sigma ADC for smart electrochemical sensors," *IEEE Transactions on circuits and systems I: Regular Papers*, vol. 61, no. 3, pp. 671–679, 2014.

[202] L. Busoni, M. Carla, and L. Lanzi, "A comparison between potentiostatic circuits with grounded work or auxiliary electrode," *Review of scientific instruments*, vol. 73, no. 4, pp. 1921–1923, 2002.

[203] W.-S. Wang, W.-T. Kuo, H.-Y. Huang, and C.-H. Luo, "Wide dynamic range CMOS potentiostat for amperometric chemical sensor," *Sensors*, vol. 10, no. 3, pp. 1782–1797, 2010.

[204] K. A. Al Mamun and N. McFarlane, "A CMOS potentiostatic glucose monitoring system for VACNF amperometric biosensors," in *Circuits and Systems (ISCAS), 2015 IEEE International Symposium on*, 2015: IEEE, pp. 477–480.

[205] S. M. Martin, F. H. Gebara, T. D. Strong, and R. B. Brown, "A low-voltage, chemical sensor interface for systems-on-chip: the fully-differential potentiostat," in *2004 IEEE International Symposium on*

Circuits and Systems (IEEE Cat. No. 04CH37512), 2004, vol. 4: IEEE, pp. IV-892.

[206] M. H. Nazari and R. Genov, "A fully differential CMOS potentiostat," in *Circuits and Systems, 2009. ISCAS 2009. IEEE International Symposium on*, 2009: IEEE, pp. 2177–2180.

[207] S. M. Martin, F. H. Gebara, T. D. Strong, and R. B. Brown, "A fully differential potentiostat," *IEEE Sensors Journal*, vol. 9, no. 2, pp. 135–142, 2009.

[208] S. M. Martin, F. H. Gebara, T. D. Strong, and R. B. Brown, "A low-voltage, chemical sensor interface for systems-on-chip: the fully-differential potentiostat," in *2004 IEEE International Symposium on Circuits and Systems (IEEE Cat. No. 04CH37512), 2004*, vol. 4: IEEE, pp. IV–892.

[209] E. Ghodsevali, B. Gosselin, M. Boukadoum, and A. Miled, "High accuracy and sensitivity differential potentiostat with amplifier-based error cancellation feedback loop," in *2015 IEEE Biomedical Circuits and Systems Conference (BioCAS)*, 2015: IEEE, pp. 1–4.

[210] S. M. Martin, F. H. Gebara, T. D. Strong, and R. B. Brown, "A fully differential potentiostat," *IEEE Sensors Journal*, vol. 9, no. 2, pp. 135–142, 2009.

[211] S. M. Martin, F. H. Gebara, B. J. Larivee, and R. B. Brown, "A CMOS-integrated microinstrument for trace detection of heavy metals," *IEEE Journal of Solid-State Circuits*, vol. 40, no. 12, pp. 2777–2786, 2005.

[212] X. Liu, "CMOS instrumentation for electrochemical biosensor array microsystems," Doctor of Philosophy *Electrical Engineering*, Michigan State University 2014.

[213] M. Augustyniak, C. Paulus, R. Brederlow, *et al.*, "A 24 × 16 CMOS-based chronocoulometric DNA microarray," in *2006 IEEE International Solid State Circuits Conference-Digest of Technical Papers, 2006*: IEEE, pp. 59–68.

[214] H. Li, S. Parsnejad, E. Ashoori, C. Thompson, E. K. Purcell, and A. J. Mason, "Ultracompact microwatt CMOS current readout with picoampere noise and kilohertz bandwidth for biosensor arrays," *IEEE Transactions on Biomedical Circuits and Systems*, vol. 12, no. 1, pp. 35–46, 2017.

[215] H. S. Narula and J. G. Harris, "A time-based VLSI potentiostat for ion current measurements," *IEEE Sensors Journal*, vol. 6, no. 2, pp. 239–247, 2006.

[216] G. Massicotte and M. Sawan, "An efficient time-based CMOS potentiostat for neurotransmitters sensing," in *2013 IEEE International Symposium on Medical Measurements and Applications (MeMeA)*, 2013: IEEE, pp. 274–277.

[217] J. Kim and W. Dunbar, "Nanopore-application CMOS potentiostat design with input parasitic compensation," *Electronics Letters*, vol. 50, no. 8, pp. 578–579, 2014.

[218] P.-A. Boutet and S. Manen, "Low power CMOS potentiostat for three electrodes amperometric chemical sensor," in *2011 Faible Tension Faible Consommation (FTFC)*, 2011: IEEE, pp. 15–18.

[219] S. Carrara, M. D. Torre, A. Cavallini, D. De Venuto, and G. De Micheli, "Multiplexing pH and temperature in a molecular biosensor," in *2010 Biomedical Circuits and Systems Conference (BioCAS)*, 2010: IEEE, pp. 146–149.

[220] W.-Y. Chung, A. A. Silverio, and V. F. Tsai, "An amperometric sensor readout circuit for multiple electrochemical sensor cells," in *2014 International Symposium on Integrated Circuits (ISIC)*, 2014: IEEE, pp. 528–531.

[221] D. De Venuto, M. D. Torre, C. Boero, S. Carrara, and G. De Micheli, "A novel multi-working electrode potentiostat for electrochemical detection of metabolites," in *Sensors*, 2010: IEEE, pp. 1572–1577.

[222] S. Ayers, K. D. Gillis, M. Lindau, and B. A. Minch, "Design of a CMOS potentiostat circuit for electrochemical detector arrays," *IEEE Transactions on Circuits and Systems I: Regular Papers*, vol. 54, no. 4, pp. 736–744, 2007.

[223] K. A. White, G. Mulberry, and B. N. Kim, "Parallel 1024-ch cyclic voltammetry on monolithic CMOS electrochemical detector array," *IEEE Sensors Journal*, vol. 20, no. 8, pp. 4395–4402, 2019.

[224] K. A. White, G. Mulberry, J. Smith, *et al.*, "Single-cell recording of vesicle release from human neuroblastoma cells using 1024-ch monolithic CMOS bioelectronics," *IEEE Transactions on Biomedical Circuits and Systems*, vol. 12, no. 6, pp. 1345–1355, 2018.

[225] M. Huang, C. I. Dorta-Quiñones, B. A. Minch, and M. Lindau, "On-chip cyclic voltammetry measurements using a compact 1024-electrode CMOS IC," *Analytical Chemistry*, vol. 93, no. 22, pp. 8027–8034, 2021.

[226] C.-W. Huang and M. S.-C. Lu, "Electrochemical detection of the neurotransmitter dopamine by nanoimprinted interdigitated electrodes and a CMOS circuit with enhanced collection efficiency," *IEEE Sensors Journal*, vol. 11, no. 9, pp. 1826–1831, 2011.

[227] M. Schienle, C. Paulus, A. Frey, *et al.*, "A fully electronic DNA sensor with 128 positions and in-pixel A/D conversion," *IEEE Journal of Solid-State Circuits*, vol. 39, no. 12, pp. 2438–2445, 2004.

[228] H. S. Narula and J. G. Harris, "A time-based VLSI potentiostat for ion current measurements," *Sensors Journal, IEEE*, vol. 6, no. 2, pp. 239–247, 2006.

[229] R. Thewes, F. Hofmann, A. Frey, *et al.*, "Sensor arrays for fully-electronic DNA detection on CMOS," in *Solid-State Circuits Conference, 2002. Digest of Technical Papers. ISSCC. 2002 IEEE International*, 2002, vol. 1: IEEE, pp. 350–473.

[230] W. Tedjo and T. Chen, "An integrated biosensor system with a high-density microelectrode array for real-time electrochemical imaging," *IEEE Transactions on Biomedical Circuits and Systems*, vol. 14, no. 1, pp. 20–35, 2019.

[231] P. M. Levine, P. Gong, R. Levicky, and K. L. Shepard, "Active CMOS sensor array for electrochemical biomolecular detection," *IEEE Journal of Solid-State Circuits*, vol. 43, no. 8, pp. 1859–1871, 2008.

[232] M. Schienle, C. Paulus, A. Frey, *et al.*, "A fully electronic DNA sensor with 128 positions and in-pixel A/D conversion," *IEEE Journal of Solid-State Circuits*, vol. 39, no. 12, pp. 2438–2445, 2004.

[233] P. M. Levine, P. Gong, R. Levicky, and K. L. Shepard, "Real-time, multiplexed electrochemical DNA detection using an active complementary metal-oxide-semiconductor biosensor array with integrated sensor electronics," *Biosensors and Bioelectronics*, vol. 24, no. 7, pp. 1995–2001, 2009.

[234] P. Kruppa, A. Frey, I. Kuehne, *et al.*, "A digital CMOS-based 24 × 16 sensor array platform for fully automatic electrochemical DNA detection," *Biosensors and Bioelectronics*, vol. 26, no. 4, pp. 1414–1419, 2010.

[235] C. Yang, Y. Huang, B. L. Hassler, R. M. Worden, and A. J. Mason, "Amperometric electrochemical microsystem for a miniaturized protein biosensor array," *IEEE Transactions on Biomedical Circuits and Systems*, vol. 3, no. 3, pp. 160–168, 2009.

[236] K. Y. Inoue, M. Matsudaira, M. Nakano, *et al.*, "Advanced LSI-based amperometric sensor array with light-shielding structure for effective removal of photocurrent and mode selectable function for individual operation of 400 electrodes," *Lab on a Chip*, vol. 15, no. 3, pp. 848–856, 2015.

[237] S. S. Ghoreishizadeh, I. Taurino, G. De Micheli, S. Carrara, and P. Georgiou, "A differential electrochemical readout ASIC with heterogeneous integration of bio-nano sensors for amperometric sensing," *IEEE Transactions on Biomedical Circuits and Systems*, vol. 11, no. 5, pp. 1148–1159, 2017.

[238] J. K. Rosenstein, M. Wanunu, C. A. Merchant, M. Drndic, and K. L. Shepard, "Integrated nanopore sensing platform with sub-microsecond temporal resolution," *Nature Methods*, vol. 9, no. 5, pp. 487–492, 2012.

[239] J. K. Rosenstein, S. Ramakrishnan, J. Roseman, and K. L. Shepard, "Single ion channel recordings with CMOS-anchored lipid membranes," *Nano letters*, vol. 13, no. 6, pp. 2682–2686, 2013.

[240] C. I. Dorta-Quiñones, M. Huang, J. C. Ruelas, *et al.*, "A bidirectional-current CMOS potentiostat for fast-scan cyclic voltammetry detector arrays," *IEEE Transactions on Biomedical Circuits and Systems*, vol. 12, no. 4, pp. 894–903, 2018.

[241] A. Laifi, A. Abaji, M. Adib, and R. Thewes, "A 96 dB SNDR current-mode continuous-time ΔΣ modulator for electrochemical sensor arrays," in *Mixed Design of Integrated Circuits & Systems (MIXDES), 2015 22nd International Conference*, 2015: IEEE, pp. 239–242.

[242] H. Jafari, L. Soleymani, and R. Genov, "16-channel CMOS impedance spectroscopy DNA analyzer with dual-slope multiplying ADCs," *IEEE Transactions on Biomedical Circuits and Systems*, vol. 6, no. 5, pp. 468–478, 2012.

[243] M. M. Bajestan, A. A. Helmy, H. Hedayati, and K. Entesari, "A 0.62–10GHz CMOS dielectric spectroscopy system for chemical/biological material characterization," in *2014 IEEE MTT-S International Microwave Symposium (IMS)*, 2014: IEEE, pp. 1–4.

[244] M. M. Bajestan, A. A. Helmy, H. Hedayati, and K. Entesari, "A 0.62–10 GHz complex dielectric spectroscopy system in CMOS," *IEEE Transactions on Microwave Theory and Techniques*, vol. 62, no. 12, pp. 3522–3537, 2014.

[245] G. Vlachogiannakis, M. A. Pertijs, M. Spirito, and L. C. de Vreede, "A 40-nm CMOS complex permittivity sensing pixel for material characterization at microwave frequencies," *IEEE Transactions on Microwave Theory and Techniques*, vol. 66, no. 3, pp. 1619–1634, 2018.

[246] P. Kassanos, I. F. Triantis, and A. Demosthenous, "A CMOS magnitude/phase measurement chip for impedance spectroscopy," *IEEE Sensors Journal*, vol. 13, no. 6, pp. 2229–2236, 2013.

[247] T. J. Hamilton, N. M. Nelson, D. Sander, and P. Abshire, "A cell impedance sensor based on a silicon cochlea," in *2009 IEEE Biomedical Circuits and Systems Conference*, 2009: IEEE, pp. 117–120.

[248] E. R. Kandel, J. H. Schwartz, T. M. Jessell, S. Siegelbaum, A. J. Hudspeth, and S. Mack, *Principles of Neural Science*. McGraw-hill New York, 2000.

[249] M. Mollazadeh, K. Murari, G. Cauwenberghs, and N. Thakor, "Wireless micropower instrumentation for multimodal acquisition of electrical and chemical neural activity," *IEEE Transactions on Biomedical Circuits and Systems*, vol. 3, no. 6, pp. 388–397, 2009.

[250] P. G. Patil and D. A. Turner, "The development of brain-machine interface neuroprosthetic devices," *Neurotherapeutics*, vol. 5, pp. 137–146, 2008.

[251] J. R. White, T. Levy, W. Bishop, and J. D. Beaty, "Real-time decision fusion for multimodal neural prosthetic devices," *PLOS One*, vol. 5, no. 3, p. e9493, 2010.

[252] M. Roham, D. P. Daberkow, E. S. Ramsson, *et al.*, "A wireless IC for wide-range neurochemical monitoring using amperometry and fast-scan cyclic voltammetry," *IEEE Transactions on Biomedical Circuits and Systems*, vol. 2, no. 1, pp. 3–9, 2008.

[253] K. Murari, M. Stanacevic, G. Cauwenberghs, and N. V. Thakor, "Integrated potentiostat for neurotransmitter sensing," *IEEE Engineering in Medicine and Biology Magazine*, vol. 24, no. 6, pp. 23–29, 2005.

[254] M. Johnson, R. Franklin, K. Scott, R. Brown, and D. Kipke, "Neural probes for concurrent detection of neurochemical and electrophysiological signals in vivo," in *2005 IEEE Engineering in Medicine and Biology 27th Annual Conference*, 2006: IEEE, pp. 7325–7328.

[255] B. Gosselin, "Recent advances in neural recording microsystems," *Sensors*, vol. 11, pp. 4572–4597, 2011.

[256] E. Neher and B. Sakmann, "Single-channel currents recorded from membrane of denervated frog muscle fibres," *Nature*, vol. 260, no. 5554, p. 799, 1976.

[257] L. Berdondini, P. Van Der Wal, O. Guenat, *et al.*, "High-density electrode array for imaging in vitro electrophysiological activity," *Biosensors and Bioelectronics*, vol. 21, no. 1, pp. 167–174, 2005.

[258] B. Eversmann, M. Jenkner, F. Hofmann, *et al.*, "A 128 × 128 CMOS biosensor array for extracellular recording of neural activity," *IEEE Journal of Solid-State Circuits*, vol. 38, no. 12, pp. 2306–2317, 2003.

[259] F. Heer, W. Franks, A. Blau, *et al.*, "CMOS microelectrode array for the monitoring of electrogenic cells," *Biosensors and Bioelectronics*, vol. 20, no. 2, pp. 358–366, 2004.

[260] J. Park, M. K. Aziz, S. Gonzalez, *et al.*, "A CMOS 22k-pixel single-cell resolution multi-modality real-time cellular sensing array," in *2017 IEEE Custom Integrated Circuits Conference (CICC)*, 2017: IEEE, pp. 1–4.

[261] G. Zeck and P. Fromherz, "Noninvasive neuroelectronic interfacing with synaptically connected snail neurons immobilized on a semiconductor chip," *Proceedings of the National Academy of Sciences*, vol. 98, no. 18, pp. 10457–10462, 2001.

[262] W. L. Rutten, "Selective electrical interfaces with the nervous system," *Annual Review of Biomedical Engineering*, vol. 4, no. 1, pp. 407–452, 2002.

[263] G. Gholmieh, W. Soussou, M. Han, *et al.*, "Custom-designed high-density conformal planar multielectrode arrays for brain slice electrophysiology," *Journal of Neuroscience Methods*, vol. 152, no. 1–2, pp. 116–129, 2006.

[264] R. Huys, D. Braeken, D. Jans, *et al.*, "Single-cell recording and stimulation with a 16k micro-nail electrode array integrated on a 0.18 μm CMOS chip," *Lab on a Chip*, vol. 12, no. 7, pp. 1274–1280, 2012.

[265] A. Lambacher, M. Jenkner, M. Merz, *et al.*, "Electrical imaging of neuronal activity by multi-transistor-array (MTA) recording at 7.8 μm resolution," *Applied Physics A*, vol. 79, no. 7, pp. 1607–1611, 2004.

[266] F. Heer, S. Hafizovic, T. Ugniwenko, *et al.*, "Single-chip microelectronic system to interface with living cells," *Biosensors and Bioelectronics*, vol. 22, no. 11, pp. 2546–2553, 2007.

[267] N. Manikandan, S. Muruganand, M. Divagar, and C. Viswanathan, "Design and fabrication of MEMS based intracranial pressure sensor for neurons study," *Vacuum*, vol. 163, pp. 204–209, 2019.

[268] G. Xiang, L. Pan, L. Huang, *et al.*, "Microelectrode array-based system for neuropharmacological applications with cortical neurons cultured in vitro," *Biosensors and Bioelectronics*, vol. 22, no. 11, pp. 2478–2484, 2007.

[269] A. F. Johnstone, G. W. Gross, D. G. Weiss, O. H.-U. Schroeder, A. Gramowski, and T. J. Shafer, "Microelectrode arrays: a physiologically based neurotoxicity testing platform for the 21st century," *Neurotoxicology*, vol. 31, no. 4, pp. 331–350, 2010.

[270] J. J. Pancrazio, S. A. Gray, Y. S. Shubin, *et al.*, "A portable microelectrode array recording system incorporating cultured neuronal networks for neurotoxin detection," *Biosensors and Bioelectronics*, vol. 18, no. 11, pp. 1339–1347, 2003.

[271] G. Piret, C. Hébert, J.-P. Mazellier, *et al.*, "3D-nanostructured boron-doped diamond for microelectrode array neural interfacing," *Biomaterials*, vol. 53, pp. 173–183, 2015.

[272] H. Charkhkar, C. Frewin, M. Nezafati, *et al.*, "Use of cortical neuronal networks for in vitro material biocompatibility testing," *Biosensors and Bioelectronics*, vol. 53, pp. 316–323, 2014.

[273] D. J. Bakkum, Z. C. Chao, and S. M. Potter, "Long-term activity-dependent plasticity of action potential propagation delay and amplitude in cortical networks," *PLOS One*, vol. 3, no. 5, p. e2088, 2008.

[274] L. B. Merabet, J. F. Rizzo, A. Amedi, D. C. Somers, and A. Pascual-Leone, "What blindness can tell us about seeing again: merging neuroplasticity and neuroprostheses," *Nature Reviews Neuroscience*, vol. 6, no. 1, pp. 71–77, 2005.

[275] Q. Liang, D. Zhang, G. Coppola, Y. Wang, S. Wei, and Y. Ge, "Multidimensional MEMS/micro sensor for force and moment sensing: A review," *IEEE Sensors Journal*, vol. 14, no. 8, pp. 2643–2657, 2014.

[276] J. Qu, H. Wu, P. Cheng, Q. Wang, and Q. Sun, "Recent advances in MEMS-based micro heat pipes," *International Journal of Heat and Mass Transfer*, vol. 110, pp. 294–313, 2017.

[277] P. J. Keller and M. B. Ahrens, "Visualizing whole-brain activity and development at the single-cell level using light-sheet microscopy," *Neuron*, vol. 85, no. 3, pp. 462–483, 2015.

[278] Q. Li, S. S. Wu, and K. C. Chou, "Subdiffraction-limit two-photon fluorescence microscopy for GFP-tagged cell imaging," *Biophysical Journal*, vol. 97, no. 12, pp. 3224–3228, 2009.

[279] R. Evans and K. Blackwell, "Calcium: amplitude, duration, or location?" *The Biological Bulletin*, vol. 228, no. 1, pp. 75–83, 2015.

[280] C. M. Díaz-García, R. Mongeon, C. Lahmann, D. Koveal, H. Zucker, and G. Yellen, "Neuronal stimulation triggers neuronal glycolysis and not lactate uptake," *Cell Metabolism*, vol. 26, no. 2, pp. 361–374. e4, 2017.

[281] M. J. Higley and B. L. Sabatini, "Calcium signaling in dendritic spines," *Cold Spring Harbor Perspectives in Biology*, vol. 4, no. 4, p. a005686, 2012.

[282] K. Svoboda and R. Yasuda, "Principles of two-photon excitation microscopy and its applications to neuroscience," *Neuron*, vol. 50, no. 6, pp. 823–839, 2006.

[283] A. C. Kreitzer, K. R. Gee, E. A. Archer, and W. G. Regehr, "Monitoring presynaptic calcium dynamics in projection fibers by in vivo loading of a novel calcium indicator," *Neuron*, vol. 27, no. 1, pp. 25–32, 2000.

[284] L. Filipis, K. Ait Ouares, P. Moreau, *et al.*, "A novel multisite confocal system for rapid Ca^{2+} imaging from submicron structures in brain slices," *Journal of Biophotonics*, vol. 11, no. 3, p. e201700197, 2018.

[285] T. Sasaki, N. Matsuki, and Y. Ikegaya, "Action-potential modulation during axonal conduction," *Science*, vol. 331, no. 6017, pp. 599–601, 2011.

[286] D. Engel, "Subcellular patch-clamp recordings from the somatodendritic domain of nigral dopamine neurons," *JoVE (Journal of Visualized Experiments)*, no. 117, p. e54601, 2016.

[287] K. J. Bender and L. O. Trussell, "Axon initial segment Ca^{2+} channels influence action potential generation and timing," *Neuron*, vol. 61, no. 2, pp. 259–271, 2009.

[288] F. O. Morin, Y. Takamura, and E. Tamiya, "Investigating neuronal activity with planar microelectrode arrays: achievements and new perspectives," *Journal of Bioscience and Bioengineering*, vol. 100, no. 2, pp. 131–143, 2005.

[289] B. D. DeBusschere and G. T. Kovacs, "Portable cell-based biosensor system using integrated CMOS cell-cartridges," *Biosensors and Bioelectronics*, vol. 16, no. 7–8, pp. 543–556, 2001.

[290] S. Martinoia, L. Bonzano, M. Chiappalone, M. Tedesco, M. Marcoli, and G. Maura, "In vitro cortical neuronal networks as a new high-sensitive system for biosensing applications," *Biosensors and Bioelectronics*, vol. 20, no. 10, pp. 2071–2078, 2005.

[291] S. Kintzios, E. Pistola, J. Konstas, *et al.*, "The application of the bioelectric recognition assay for the detection of human and plant viruses: Definition of operational parameters," *Biosensors and Bioelectronics*, vol. 16, no. 7–8, pp. 467–480, 2001.

[292] J. W. Parce, J. C. Owicki, K. M. Kercso, *et al.*, "Detection of cell-affecting agents with a silicon biosensor," *Science*, vol. 246, no. 4927, pp. 243–247, 1989.

[293] K. R. Rogers and C. L. Gerlach, "Peer reviewed: update on environmental biosensors," *Environmental Science & Technology*, vol. 33, no. 23, pp. 500A–506A, 1999.

[294] L. M. Valor, P. Charlesworth, L. Humphreys, C. N. Anderson, and S. G. Grant, "Network activity-independent coordinated gene expression program for synapse assembly," *Proceedings of the National Academy of Sciences*, vol. 104, no. 11, pp. 4658–4663, 2007.

[295] C. G. Dotti, C. A. Sullivan, and G. A. Banker, "The establishment of polarity by hippocampal neurons in culture," *Journal of Neuroscience*, vol. 8, no. 4, pp. 1454–1468, 1988.

[296] A. M. Taylor, M. Blurton-Jones, S. W. Rhee, D. H. Cribbs, C. W. Cotman, and N. L. Jeon, "A microfluidic culture platform for CNS axonal injury, regeneration and transport," *Nature Methods*, vol. 2, no. 8, pp. 599–605, 2005.

[297] G.-q. Bi and M.-m. Poo, "Synaptic modifications in cultured hippocampal neurons: dependence on spike timing, synaptic strength, and postsynaptic cell type," *Journal of Neuroscience*, vol. 18, no. 24, pp. 10464–10472, 1998.

[298] G.-q. Bi and M.-m. Poo, "Distributed synaptic modification in neural networks induced by patterned stimulation," *Nature*, vol. 401, no. 6755, pp. 792–796, 1999.

[299] R. B. Campenot, K. Lund, and S.-A. Mok, "Production of compartmented cultures of rat sympathetic neurons," *Nature Protocols*, vol. 4, no. 12, pp. 1869–1887, 2009.

[300] M. M. Segal and E. J. Furshpan, "Epileptiform activity in microcultures containing small numbers of hippocampal neurons," *Journal of Neurophysiology*, vol. 64, no. 5, pp. 1390–1399, 1990.

[301] D. Kleinfeld, K. Kahler, and P. Hockberger, "Controlled outgrowth of dissociated neurons on patterned substrates," *Journal of Neuroscience*, vol. 8, no. 11, pp. 4098–4120, 1988.

[302] B. Gähwiler, M. Capogna, D. Debanne, R. McKinney, and S. Thompson, "Organotypic slice cultures: a technique has come of age," *Trends in Neurosciences*, vol. 20, no. 10, pp. 471–477, 1997.

[303] Y. Nam and B. C. Wheeler, "In vitro microelectrode array technology and neural recordings," *Critical Reviews™ in Biomedical Engineering*, vol. 39, no.1, 2011.

[304] M. Hutzler, A. Lambacher, B. Eversmann, M. Jenkner, R. Thewes, and P. Fromherz, "High-resolution multitransistor array recording of electrical field potentials in cultured brain slices," *Journal of Neurophysiology*, vol. 96, no. 3, pp. 1638–1645, 2006.

[305] L. Berdondini, K. Imfeld, A. Maccione, *et al.*, "Active pixel sensor array for high spatio-temporal resolution electrophysiological recordings from single cell to large scale neuronal networks," *Lab on a Chip*, vol. 9, no. 18, pp. 2644–2651, 2009.

[306] H. P. Neves, T. Torfs, R. F. Yazicioglu, *et al.*, "The NeuroProbes project: a concept for electronic depth control," in *2008 30th Annual International Conference of the IEEE Engineering in Medicine and Biology Society*, 2008: IEEE, pp. 1857–1857.

[307] R. Fiáth, P. Beregszászi, D. Horváth, *et al.*, "Large-scale recording of thalamocortical circuits: in vivo electrophysiology with the two-dimensional electronic depth control silicon probe," *Journal of Neurophysiology*, vol. 116, no. 5, pp. 2312–2330, 2016.

[308] A. Hierlemann, U. Frey, S. Hafizovic, and F. Heer, "Growing cells atop microelectronic chips: interfacing electrogenic cells in vitro with CMOS-based microelectrode arrays," *Proceedings of the IEEE*, vol. 99, no. 2, pp. 252–284, 2010.

[309] X. Yuan, V. Emmenegger, M. E. J. Obien, A. Hierlemann, and U. Frey, "Dual-mode microelectrode array featuring 20k electrodes and high SNR for extracellular recording of neural networks," in *2018 IEEE Biomedical Circuits and Systems Conference (BioCAS)*, 2018: IEEE, pp. 1–4.

[310] H. Kang and Y. Nam, "In vitro neural recording by microelectrode arrays," *Stretchable Bioelectronics for Medical Devices and Systems*, pp. 275–291, 2016.

[311] A. Bhatti, K. H. Lee, H. Garmestani, and C. P. Lim, *Emerging Trends in Neuro Engineering and Neural Computation*. Springer, 2017.

[312] G. Bontorin, "Intelligent multielectrode arrays: improving spatiotemporal performances in hybrid (living-artificial), real-time, closed-loop systems," Université Sciences et Technologies-Bordeaux I, 2010.

[313] D. A. Robinson, "The electrical properties of metal microelectrodes," *Proceedings of the IEEE*, vol. 56, no. 6, pp. 1065–1071, 1968.

[314] C. Thomas Jr, P. Springer, G. Loeb, Y. Berwald-Netter, and L. Okun, "A miniature microelectrode array to monitor the bioelectric activity of cultured cells," *Experimental Cell Research*, vol. 74, no. 1, pp. 61–66, 1972.

[315] G. W. Gross, E. Rieske, G. Kreutzberg, and A. Meyer, "A new fixed-array multi-microelectrode system designed for long-term monitoring of extracellular single unit neuronal activity in vitro," *Neuroscience Letters*, vol. 6, no. 2–3, pp. 101–105, 1977.

[316] J. Pine, "Recording action potentials from cultured neurons with extracellular microcircuit electrodes," *Journal of Neuroscience Methods*, vol. 2, no. 1, pp. 19–31, 1980.

[317] L. Berdondini, A. Bosca, T. Nieus, and A. Maccione, "Active pixel sensor multielectrode array for high spatiotemporal resolution," *Nanotechnology and Neuroscience: Nano-electronic, Photonic and Mechanical Neuronal Interfacing*, pp. 207–238, 2014.

[318] W. G. Regehr, J. Pine, C. S. Cohan, M. D. Mischke, and D. W. Tank, "Sealing cultured invertebrate neurons to embedded dish electrodes facilitates long-term stimulation and recording," *Journal of Neuroscience Methods*, vol. 30, no. 2, pp. 91–106, 1989.

[319] J. N. Aziz, K. Abdelhalim, R. Shulyzki, *et al.*, "256-channel neural recording and delta compression microsystem with 3D electrodes," *IEEE Journal of Solid-State Circuits*, vol. 44, no. 3, pp. 995–1005, 2009.

[320] B. Johnson, S. T. Peace, T. A. Cleland, and A. Molnar, "A 50μm pitch, 1120-channel, 20kHz frame rate microelectrode array for slice recording," in *2013 IEEE Biomedical Circuits and Systems Conference (BioCAS)*, 2013: IEEE, pp. 109–112.

[321] B. Johnson, S. T. Peace, A. Wang, T. A. Cleland, and A. Molnar, "A 768-channel CMOS microelectrode array with angle sensitive pixels for neuronal recording," *IEEE Sensors Journal*, vol. 13, no. 9, pp. 3211–3218, 2013.

[322] A. Maccione, A. Simi, T. Nieus, *et al.*, "Sensing and actuating electrophysiological activity on brain tissue and neuronal cultures with a high-density CMOS-MEA," in *2013 Transducers & Eurosensors XXVII: The 17th International Conference on Solid-State Sensors, Actuators and Microsystems (TRANSDUCERS & EUROSENSORS XXVII)*, 2013: IEEE, pp. 752–755.

[323] G. Bertotti, D. Velychko, N. Dodel, *et al.*, "A CMOS-based sensor array for in-vitro neural tissue interfacing with 4225 recording sites and 1024 stimulation sites," in *2014 IEEE Biomedical Circuits and Systems Conference (BioCAS) Proceedings, 2014*: IEEE, pp. 304–307.

[324] B. Eversmann, A. Lambacher, T. Gerling, A. Kunze, P. Fromherz, and R. Thewes, "A neural tissue interfacing chip for in-vitro applications with 32k recording/stimulation channels on an active area of 2.6 mm^2," in *2011 Proceedings of the ESSCIRC (ESSCIRC)*, 2011: IEEE, pp. 211–214.

[325] E. Maeda, H. Robinson, and A. Kawana, "The mechanisms of generation and propagation of synchronized bursting in developing networks of cortical neurons," *Journal of Neuroscience*, vol. 15, no. 10, pp. 6834–6845, 1995.

[326] H. Oka, K. Shimono, R. Ogawa, H. Sugihara, and M. Taketani, "A new planar multielectrode array for extracellular recording: application to hippocampal acute slice," *Journal of Neuroscience Methods*, vol. 93, no. 1, pp. 61–67, 1999.

[327] J. Müller, M. Ballini, P. Livi, *et al.*, "High-resolution CMOS MEA platform to study neurons at subcellular, cellular, and network levels," *Lab on a Chip*, vol. 15, no. 13, pp. 2767–2780, 2015.

[328] M. Bigas, E. Cabruja, J. Forest, and J. Salvi, "Review of CMOS image sensors," *Microelectronics Journal*, vol. 37, no. 5, pp. 433–451, 2006.

[329] A. H. Marblestone, B. M. Zamft, Y. G. Maguire, *et al.*, "Physical principles for scalable neural recording," *Frontiers in Computational Neuroscience*, vol. 7, p. 137, 2013.

[330] M. Willemin, N. Blanc, G. Lang, *et al.*, "Optical characterization methods for solid-state image sensors," *Optics and Lasers in Engineering*, vol. 36, no. 2, pp. 185–194, 2001.

[331] F. Shahrokhi, K. Abdelhalim, D. Serletis, P. L. Carlen, and R. Genov, "The 128-channel fully differential digital integrated neural recording and stimulation interface," *IEEE Transactions on Biomedical Circuits and Systems*, vol. 4, no. 3, pp. 149–161, 2010.

[332] R. Muller, S. Gambini, and J. M. Rabaey, "A 0.013 mm, 5 W, DC-coupled neural signal acquisition IC with 0.5 V supply," *IEEE Journal of Solid-State Circuits*, vol. 47, no. 1, pp. 232–243, 2011.

[333] L. Berdondini, T. Overstolz, N. De Rooij, M. Koudelka-Hep, M. Wany, and P. Seitz, "High-density microelectrode arrays for electrophysiological activity imaging of neuronal networks," in *ICECS 2001. 8th IEEE International Conference on Electronics, Circuits and Systems (Cat. No. 01EX483)*, 2001, vol. 3: IEEE, pp. 1239–1242.

[334] J. Ogi, Y. Kato, Y. Matoba, *et al.*, "Twenty-four-micrometer-pitch microelectrode array with 6912-channel readout at 12 kHz via highly scalable implementation for high-spatial-resolution mapping of action potentials," *Biointerphases*, vol. 12, no. 5, p. 05F402, 2017.

[335] G. N. Angotzi, M. Malerba, A. Maccione, *et al.*, "A high temporal resolution multiscale recording system for in vivo neural studies," in *2017 IEEE International Symposium on Circuits and Systems (ISCAS)*, 2017: IEEE, pp. 1–4.

[336] E. W. Keefer, B. R. Botterman, M. I. Romero, A. F. Rossi, and G. W. Gross, "Carbon nanotube coating improves neuronal recordings," *Nature Nanotechnology*, vol. 3, no. 7, pp. 434–439, 2008.

[337] M. E. J. Obien, K. Deligkaris, T. Bullmann, D. J. Bakkum, and U. Frey, "Revealing neuronal function through microelectrode array recordings," *Frontiers in Neuroscience*, vol. 8, p. 423, 2015.

[338] P. Livi, F. Heer, U. Frey, D. J. Bakkum, and A. Hierlemann, "Compact voltage and current stimulation buffer for high-density microelectrode arrays," *IEEE Transactions on Biomedical Circuits and Systems*, vol. 4, no. 6, pp. 372–378, 2010.

[339] C. M. Lopez, A. Andrei, S. Mitra, *et al.*, "An implantable 455-active-electrode 52-channel CMOS neural probe," *IEEE Journal of Solid-State Circuits*, vol. 49, no. 1, pp. 248–261, 2013.

[340] C. M. Lopez, S. Mitra, J. Putzeys, *et al.*, "22.7 A 966-electrode neural probe with 384 configurable channels in 0.13 μm SOI CMOS," in *2016 IEEE International Solid-State Circuits Conference (ISSCC)*, 2016: IEEE, pp. 392–393.

[341] V. Viswam, J. Dragas, A. Shadmani, *et al.*, "22.8 Multi-functional microelectrode array system featuring 59,760 electrodes, 2048 electrophysiology channels, impedance and neurotransmitter measurement units," in *2016 IEEE International Solid-State Circuits Conference (ISSCC)*, 2016: IEEE, pp. 394–396.

[342] K. Seidl, M. Schwaerzle, I. Ulbert, H. P. Neves, O. Paul, and P. Ruther, "CMOS-based high-density silicon microprobe arrays for electronic depth control in intracortical neural recording–characterization and application," *Journal of Microelectromechanical Systems*, vol. 21, no. 6, pp. 1426–1435, 2012.

[343] E. Noorsal, K. Sooksood, H. Xu, R. Hornig, J. Becker, and M. Ortmanns, "A neural stimulator frontend with high-voltage compliance and programmable pulse shape for epiretinal implants," *IEEE Journal of Solid-State Circuits*, vol. 47, no. 1, pp. 244–256, 2011.

[344] N. Butz, A. Taschwer, S. Nessler, Y. Manoli, and M. Kuhl, "A 22 V compliant 56 μW twin-track active charge balancing enabling 100% charge compensation even in monophasic and 36% amplitude correction in biphasic neural stimulators," *IEEE Journal of Solid-State Circuits*, vol. 53, no. 8, pp. 2298–2310, 2018.

[345] H.-M. Lee, H. Park, and M. Ghovanloo, "A power-efficient wireless system with adaptive supply control for deep brain stimulation," *IEEE Journal of Solid-State Circuits*, vol. 48, no. 9, pp. 2203–2216, 2013.

[346] M. Ortmanns, A. Rocke, M. Gehrke, and H.-J. Tiedtke, "A 232-channel epiretinal stimulator ASIC," *IEEE Journal of Solid-State Circuits*, vol. 42, no. 12, pp. 2946–2959, 2007.

[347] J. Li, X. Liu, W. Mao, T. Chen, and H. Yu, "Advances in neural recording and stimulation integrated circuits," *Frontiers in Neuroscience*, vol. 15, p. 663204, 2021.

[348] W.-Y. Hsu and A. Schmid, "Compact, energy-efficient high-frequency switched capacitor neural stimulator with active charge balancing," *IEEE Transactions on Biomedical Circuits and Systems*, vol. 11, no. 4, pp. 878–888, 2017.

[349] M. N. Van Dongen, F. E. Hoebeek, S. Koekkoek, C. I. De Zeeuw, and W. A. Serdijn, "High frequency switched-mode stimulation can evoke post

synaptic responses in cerebellar principal neurons," *Frontiers in neuroengineering*, p. 2, 2015.

[350] H.-M. Lee, K. Y. Kwon, W. Li, and M. Ghovanloo, "A power-efficient switched-capacitor stimulating system for electrical/optical deep brain stimulation," *IEEE Journal of Solid-State Circuits*, vol. 50, no. 1, pp. 360–374, 2014.

[351] M. Ortmanns, "Charge balancing in functional electrical stimulators: A comparative study," in *2007 IEEE International Symposium on Circuits and Systems*, 2007: IEEE, pp. 573–576.

[352] A. Rothermel, L. Liu, N. P. Aryan, *et al.*, "A CMOS chip with active pixel array and specific test features for subretinal implantation," *IEEE Journal of Solid-State Circuits*, vol. 44, no. 1, pp. 290–300, 2008.

[353] L. Yao, P. Li, and M. Je, "A pulse-width-adaptive active charge balancing circuit with pulse-insertion based residual charge compensation and quantization for electrical stimulation applications," in *2015 IEEE Asian Solid-State Circuits Conference (A-SSCC)*, 2015: IEEE, pp. 1–4.

[354] Z. Chen, X. Liu, and Z. Wang, "A charge balancing technique for neurostimulators," *Analog Integrated Circuits and Signal Processing*, vol. 105, pp. 483–496, 2020.

[355] K. Sooksood, T. Stieglitz, and M. Ortmanns, *"An active approach for charge balancing in functional electrical stimulation," IEEE Transactions on Biomedical Circuits and Systems*, vol. 4, no. 3, pp. 162–170, 2010.

[356] S. Ha, C. Kim, J. Park, G. Cauwenberghs, and P. P. Mercier, "A fully integrated RF-powered energy-replenishing current-controlled stimulator," *IEEE Transactions on Biomedical Circuits and Systems*, vol. 13, no. 1, pp. 191–202, 2018.

[357] S.-Y. Park, J. Cho, K. Na, and E. Yoon, "Modular 128-channel Δ-$\Delta\sum$ analog front-end architecture using spectrum equalization scheme for 1024-channel 3-D neural recording microsystems," *IEEE Journal of Solid-State Circuits*, vol. 53, no. 2, pp. 501–514, 2017.

[358] R. R. Harrison and C. Charles, "A low-power low-noise CMOS amplifier for neural recording applications," *IEEE Journal of Solid-State Circuits*, vol. 38, no. 6, pp. 958–965, 2003.

[359] R. R. Harrison, P. T. Watkins, R. J. Kier, *et al.*, "A low-power integrated circuit for a wireless 100-electrode neural recording system," *IEEE Journal of Solid-State Circuits*, vol. 42, no. 1, pp. 123–133, 2006.

[360] J. Holleman and B. Otis, "A sub-microwatt low-noise amplifier for neural recording," in *2007 29th Annual International Conference of the IEEE Engineering in Medicine and Biology Society*, 2007: IEEE, pp. 3930–3933.

[361] R. Rieger, M. Schuettler, D. Pal, *et al.*, "Very low-noise ENG amplifier system using CMOS technology," *IEEE Transactions on Neural Systems and Rehabilitation Engineering*, vol. 14, no. 4, pp. 427–437, 2006.

[362] T. Jochum, T. Denison, and P. Wolf, "Integrated circuit amplifiers for multi-electrode intracortical recording," *Journal of Neural Engineering*, vol. 6, no. 1, p. 012001, 2009.

[363] P. Mohseni and K. Najafi, "A fully integrated neural recording amplifier with DC input stabilization," *IEEE Transactions on Biomedical Engineering*, vol. 51, no. 5, pp. 832–837, 2004.

[364] X. Zou, X. Xu, L. Yao, and Y. Lian, "A 1-V 450-nW fully integrated programmable biomedical sensor interface chip," *IEEE Journal of Solid-State Circuits*, vol. 44, no. 4, pp. 1067–1077, 2009.

[365] X. Liu, X. Huang, Y. Jiang, *et al.*, "A microfluidic cytometer for complete blood count with a 3.2-Megapixel, 1.1-μm-Pitch super-resolution image sensor in 65-nm BSI CMOS," *IEEE Transactions on Biomedical Circuits and Systems*, vol. 11, no. 4, pp. 794–803, 2017.

[366] H. Yu, M. Yan, and X. Huang, *CMOS integrated lab-on-a-chip system for personalized biomedical diagnosis*. John Wiley & Sons, 2018.

[367] S. B. Lee, H.-M. Lee, M. Kiani, U.-M. Jow, and M. Ghovanloo, *"An inductively powered scalable 32-channel wireless neural recording system-on-a-chip for neuroscience applications," IEEE Transactions on Biomedical Circuits and Systems*, vol. 4, no. 6, pp. 360–371, 2010.

[368] M. Mollazadeh, K. Murari, G. Cauwenberghs, and N. Thakor, "Micropower CMOS integrated low-noise amplification, filtering, and digitization of multimodal neuropotentials," *IEEE Transactions on Biomedical Circuits and Systems*, vol. 3, no. 1, pp. 1–10, 2009.

[369] S. Farshchi, A. Pesterev, P. Nuyujukian, E. Guenterberg, I. Mody, and J. W. Judy, "Embedded neural recording with TinyOS-based wireless-enabled processor modules," *IEEE Transactions on Neural Systems and Rehabilitation Engineering*, vol. 18, no. 2, pp. 134–141, 2010.

[370] Z. Xiao, C.-M. Tang, C. M. Dougherty, and R. Bashirullah, "A 20 μW neural recording tag with supply-current-modulated AFE in 0.13 μm CMOS," in *2010 IEEE International Solid-State Circuits Conference-(ISSCC),* 2010: IEEE, pp. 122–123.

[371] D. R. Kipke, W. Shain, G. Buzsáki, *et al.*, "Advanced neurotechnologies for chronic neural interfaces: new horizons and clinical opportunities," *Journal of Neuroscience*, vol. 28, no. 46, pp. 11830–11838, 2008.

[372] R. H. Olsson, D. L. Buhl, A. M. Sirota, G. Buzsáki, and K. D. Wise, "Band-tunable and multiplexed integrated circuits for simultaneous recording and stimulation with microelectrode arrays," *IEEE Transactions on Biomedical Engineering*, vol. 52, no. 7, pp. 1303–1311, 2005.

[373] W. Wattanapanitch, M. Fee, and R. Sarpeshkar, *"An energy-efficient micropower neural recording amplifier," IEEE Transactions on Biomedical Circuits and Systems*, vol. 1, no. 2, pp. 136–147, 2007.

[374] L. Shen, N. Lu, and N. Sun, "A 1-V 0.25-μW inverter stacking amplifier with 1.07 noise efficiency factor," *IEEE Journal of Solid-State Circuits*, vol. 53, no. 3, pp. 896–905, 2018.

[375] B. Gosselin, M. Sawan, and C. A. Chapman, "A low-power integrated bioamplifier with active low-frequency suppression," *IEEE Transactions on Biomedical Circuits and Systems*, vol. 1, no. 3, pp. 184–192, 2007.

[376] N. Verma, A. Shoeb, J. Bohorquez, J. Dawson, J. Guttag, and A. P. Chandrakasan, "A micro-power EEG acquisition SoC with integrated feature extraction processor for a chronic seizure detection system," *IEEE Journal of Solid-State Circuits*, vol. 45, no. 4, pp. 804–816, 2010.

[377] A.-T. Avestruz, W. Santa, D. Carlson, *et al.*, "A 5 µW/channel spectral analysis IC for chronic bidirectional brain–machine interfaces," *IEEE Journal of Solid-State Circuits*, vol. 43, no. 12, pp. 3006–3024, 2008.

[378] G. E. Perlin, A. M. Sodagar, and K. D. Wise, "A neural amplifier with high programmable gain and tunable bandwidth," in *2008 30th Annual International Conference of the IEEE Engineering in Medicine and Biology Society*, 2008: IEEE, pp. 3154–3157.

[379] M. Yin and M. Ghovanloo, "A low-noise preamplifier with adjustable gain and bandwidth for biopotential recording applications," in *2007 IEEE International Symposium on Circuits and Systems*, 2007: IEEE, pp. 321–324.

[380] T. Horiuchi, T. Swindell, D. Sander, and P. Abshire, "A low-power CMOS neural amplifier with amplitude measurements for spike sorting," in *2004 IEEE International Symposium on Circuits and Systems (ISCAS)*, 2004, vol. 4: IEEE, pp. IV–29.

[381] M. S. Chae, Z. Yang, M. R. Yuce, L. Hoang, and W. Liu, "A 128-channel 6 mW wireless neural recording IC with spike feature extraction and UWB transmitter," *IEEE Transactions on Neural Systems and Rehabilitation Engineering*, vol. 17, no. 4, pp. 312–321, 2009.

[382] X. Zou, W.-S. Liew, L. Yao, and Y. Lian, "A 1V 22µW 32-channel implantable EEG recording IC," in *2010 IEEE International Solid-State Circuits Conference-(ISSCC), 2010*: IEEE, pp. 126–127.

[383] B. Gosselin and M. Ghovanloo, "A high-performance analog front-end for an intraoral tongue-operated assistive technology," in *2011 IEEE International Symposium of Circuits and Systems (ISCAS)*, 2011: IEEE, pp. 2613–2616.

[384] J. Lee, M. Johnson, and D. Kipke, "A tunable biquad switched-capacitor amplifier-filter for neural recording," *IEEE Transactions on Biomedical Circuits and Systems*, vol. 4, no. 5, pp. 295–300, 2010.

[385] H. Jeon, J.-S. Bang, Y. Jung, I. Choi, and M. Je, "A high DR, DC-coupled, time-based neural-recording IC with degeneration R-DAC for bidirectional neural interface," *IEEE Journal of Solid-State Circuits*, vol. 54, no. 10, pp. 2658–2670, 2019.

[386] J. Xu, T. Wu, W. Liu, and Z. Yang, "A frequency shaping neural recorder with 3 pF input capacitance and 11 plus 4.5 bits dynamic range," *IEEE Transactions on Biomedical Circuits and Systems*, vol. 8, no. 4, pp. 510–527, 2014.

[387] W. Jiang, V. Hokhikyan, H. Chandrakumar, V. Karkare, and D. Marković, "A±50-mV linear-input-range VCO-based neural-recording front-end with digital nonlinearity correction," *IEEE Journal of Solid-State Circuits*, vol. 52, no. 1, pp. 173–184, 2016.

[388] R. Muller, H.-P. Le, W. Li, *et al.*, "A minimally invasive 64-channel wireless μECoG implant," *IEEE Journal of Solid-State Circuits*, vol. 50, no. 1, pp. 344–359, 2014.

[389] A. Nikas, S. Jambunathan, L. Klein, M. Voelker, and M. Ortmanns, "A continuous-time delta-sigma modulator using a modified instrumentation amplifier and current reuse DAC for neural recording," *IEEE Journal of Solid-State Circuits*, vol. 54, no. 10, pp. 2879–2891, 2019.

[390] S. Martinoia, M. Bove, G. Carlini, *et al.*, "A general-purpose system for long-term recording from a microelectrode array coupled to excitable cells," *Journal of Neuroscience Methods*, vol. 48, no. 1–2, pp. 115–121, 1993.

[391] P. Thiebaud, N. De Rooij, M. Koudelka-Hep, and L. Stoppini, "Microelectrode arrays for electrophysiological monitoring of hippocampal organotypic slice cultures," *IEEE Transactions on Biomedical Engineering*, vol. 44, no. 11, pp. 1159–1163, 1997.

[392] M. O. Heuschkel, M. Fejtl, M. Raggenbass, D. Bertrand, and P. Renaud, "A three-dimensional multi-electrode array for multi-site stimulation and recording in acute brain slices," *Journal of Neuroscience Methods*, vol. 114, no. 2, pp. 135–148, 2002.

[393] V. Lovat, D. Pantarotto, L. Lagostena, *et al.*, "Carbon nanotube substrates boost neuronal electrical signaling," *Nano Letters*, vol. 5, no. 6, pp. 1107–1110, 2005.

[394] N. A. Kotov, J. O. Winter, I. P. Clements, *et al.*, "Nanomaterials for neural interfaces," *Advanced Materials*, vol. 21, no. 40, pp. 3970–4004, 2009.

[395] L. Bareket-Keren and Y. Hanein, "Carbon nanotube-based multi electrode arrays for neuronal interfacing: progress and prospects," *Frontiers in Neural Circuits*, vol. 6, p. 122, 2013.

[396] T. Gabay, E. Jakobs, E. Ben-Jacob, and Y. Hanein, "Engineered self-organization of neural networks using carbon nanotube clusters," *Physica A: Statistical Mechanics and its Applications*, vol. 350, no. 2–4, pp. 611–621, 2005.

[397] S. Iijima, "Helical microtubules of graphitic carbon," *Nature*, vol. 354, no. 6348, pp. 56–58, 1991.

[398] G. H. Kim, K. Kim, E. Lee, *et al.*, "Recent progress on microelectrodes in neural interfaces," *Materials*, vol. 11, no. 10, p. 1995, 2018.

[399] W. S. Wong and A. Salleo, *Flexible Electronics: Materials and Applications*. Springer Science & Business Media, 2009.

[400] P. Ihalainen, A. Määttänen, and N. Sandler, "Printing technologies for biomolecule and cell-based applications," *International Journal of Pharmaceutics*, vol. 494, no. 2, pp. 585–592, 2015.

[401] U. Meyer, T. Meyer, J. Handschel, and H. P. Wiesmann, *Fundamentals of Tissue Engineering and Regenerative Medicine. Springer*, 2009.

[402] J. W. Lee, D. Kim, S. Yoo, H. Lee, G.-H. Lee, and Y. Nam, "Emerging neural stimulation technologies for bladder dysfunctions," *International Neurourology Journal*, vol. 19, no. 1, p. 3, 2015.

[403] D.-H. Kim, J. Viventi, J. J. Amsden, *et al.*, "Dissolvable films of silk fibroin for ultrathin conformal bio-integrated electronics," *Nature Materials*, vol. 9, no. 6, pp. 511–517, 2010.

[404] A. K. Soe, S. Nahavandi, and K. Khoshmanesh, "Neuroscience goes on a chip," *Biosensors and Bioelectronics*, vol. 35, no. 1, pp. 1–13, 2012.

[405] Y. Huang, J. C. Williams, and S. M. Johnson, "Brain slice on a chip: opportunities and challenges of applying microfluidic technology to intact tissues," *Lab on a Chip*, vol. 12, no. 12, pp. 2103–2117, 2012.

[406] L. Berdondini, M. Chiappalone, P. Van Der Wal, *et al.*, "A microelectrode array (MEA) integrated with clustering structures for investigating in vitro neurodynamics in confined interconnected sub-populations of neurons," *Sensors and Actuators B: Chemical*, vol. 114, no. 1, pp. 530–541, 2006.

[407] E. Claverol-Tinture, M. Ghirardi, F. Fiumara, X. Rosell, and J. Cabestany, "Multielectrode arrays with elastomeric microstructured overlays for extracellular recordings from patterned neurons," *Journal of Neural Engineering*, vol. 2, no. 2, p. L1, 2005.

[408] E. Marconi, T. Nieus, A. Maccione, *et al.*, "Emergent functional properties of neuronal networks with controlled topology," *PLOS One*, vol. 7, no. 4, p. e34648, 2012.

[409] P. Bergveld, "Development of an ion-sensitive solid-state device for neurophysiological measurements," *IEEE Transactions on Biomedical Engineering*, no. 1, pp. 70–71, 1970.

[410] M. Kaisti, "Detection principles of biological and chemical FET sensors," *Biosensors and Bioelectronics*, vol. 98, pp. 437–448, 2017.

[411] D. Xu, X. Huang, J. Guo, and X. Ma, "Automatic smartphone-based microfluidic biosensor system at the point of care," *Biosensors and Bioelectronics*, vol. 110, pp. 78–88, 2018.

[412] S. Su, Q. Sun, X. Gu, *et al.*, "Two-dimensional nanomaterials for biosensing applications," *TrAC Trends in Analytical Chemistry*, vol. 119, p. 115610, 2019.

[413] J. Sengupta and C. M. Hussain, *"Graphene-based field-effect transistor biosensors for the rapid detection and analysis of viruses: A perspective in view of COVID-19," Carbon Trends*, p. 100011, 2020.

[414] S. Pilehvar and K. De Wael, "Recent advances in electrochemical biosensors based on fullerene-C60 nano-structured platforms," *Biosensors*, vol. 5, no. 4, pp. 712–735, 2015.

[415] A. Panahi, D. Sadighbayan, S. Forouhi, and E. Ghafar-Zadeh, "Recent advances of field-effect transistor technology for infectious diseases," *Biosensors*, vol. 11, no. 4, p. 103, 2021.

[416] N. Nikkhoo, P. G. Gulak, and K. Maxwell, "Rapid detection of E. coli bacteria using potassium-sensitive FETs in CMOS," *IEEE Transactions on Biomedical Circuits and Systems*, vol. 7, no. 5, pp. 621–630, 2013.

[417] M. Milgrew, M. Riehle, and D. Cumming, "A large transistor-based sensor array chip for direct extracellular imaging," *Sensors and Actuators B: Chemical*, vol. 111, pp. 347–353, 2005.

[418] A. Errachid, N. Zine, J. Samitier, and J. Bausells, "FET-based chemical sensor systems fabricated with standard technologies," *Electroanalysis: An International Journal Devoted to Fundamental and Practical Aspects of Electroanalysis*, vol. 16, no. 22, pp. 1843–1851, 2004.

[419] P. Georgiou and C. Toumazou, "ISFET characteristics in CMOS and their application to weak inversion operation," *Sensors and Actuators B: Chemical*, vol. 143, no. 1, pp. 211–217, 2009.

[420] M. Sohbati and C. Toumazou, "Dimension and shape effects on the ISFET performance," *IEEE Sensors Journal*, vol. 15, no. 3, pp. 1670–1679, 2014.

[421] M.-H. Park, D. Han, R. Chand, D.-H. Lee, and Y.-S. Kim, *"Mechanism of label-free DNA detection using the floating electrode on pentacene thin film transistor," The Journal of Physical Chemistry C*, vol. 120, no. 9, pp. 4854–4859, 2016.

[422] M. Barbaro, A. Caboni, D. Loi, *et al.*, "Label-free, direct DNA detection by means of a standard CMOS electronic chip," *Sensors and Actuators B: Chemical*, vol. 171, pp. 148–154, 2012.

[423] J. Bausells, J. Carrabina, A. Errachid, and A. Merlos, "Ion-sensitive field-effect transistors fabricated in a commercial CMOS technology," *Sensors and Actuators B: Chemical*, vol. 57, no. 1–3, pp. 56–62, 1999.

[424] S. Martinoia, M. Grattarola, and G. Massobrio, "Modelling non-ideal behaviours in H^+-sensitive FETs with SPICE," *Sensors and Actuators B: Chemical*, vol. 7, no. 1–3, pp. 561–564, 1992.

[425] G. Massobrio, S. Martinoia, and M. Grattarola, "Use of SPICE for modeling silicon-based chemical sensors," *Sensors and Materials*, vol. 6, pp. 101–101, 1994.

[426] G. Massobrio and S. Martinoia, "Modelling the ISFET behaviour under temperature variations using BIOSPICE," *Electronics Letters*, vol. 32, no. 10, pp. 936–938, 1996.

[427] S. Martinoia, L. Lorenzelli, G. Massobrio, P. Conci, and A. Lui, "Temperature effects on the ISFET behaviour: simulations and measurements," *Sensors and Actuators B: Chemical*, vol. 50, no. 1, pp. 60–68, 1998.

[428] S. Martinoia and G. Massobrio, "A behavioral macromodel of the ISFET in SPICE," *Sensors and Actuators B: Chemical*, vol. 62, no. 3, pp. 182–189, 2000.

[429] P. G. Fernandes, H. J. Stiegler, M. Zhao, *et al.*, "SPICE macromodel of silicon-on-insulator-field-effect-transistor-based biological sensors," *Sensors and Actuators B: Chemical*, vol. 161, no. 1, pp. 163–170, 2012.

[430] D. E. Yates, S. Levine, and T. W. Healy, "Site-binding model of the electrical double layer at the oxide/water interface," *Journal of the Chemical Society, Faraday Transactions 1: Physical Chemistry in Condensed Phases*, vol. 70, pp. 1807–1818, 1974.

[431] S. Sinha, "Modeling and simulation of ISFET microsensor for different sensing films," in *ISSS International Conference on Smart Materials, Structures and Systems*, 2014.

[432] R. Van Hal, J. Eijkel, and P. Bergveld, "A novel description of ISFET sensitivity with the buffer capacity and double-layer capacitance as key parameters," *Sensors and Actuators B: Chemical*, vol. 24, no. 1–3, pp. 201–205, 1995.

[433] D. Ben-Yaakov, D. Andelman, R. Podgornik, and D. Harries, "Ion-specific hydration effects: Extending the Poisson-Boltzmann theory," *Current Opinion in Colloid & Interface Science*, vol. 16, no. 6, pp. 542–550, 2011.

[434] H. G. Ozcelik, Y. Sozen, H. Sahin, and M. Barisik, "Parametrizing non-bonded interactions between silica and water from first principles," *Applied Surface Science*, vol. 504, p. 144359, 2020.

[435] G. Jiang, C. Cheng, D. Li, and J. Z. Liu, "Molecular dynamics simulations of the electric double layer capacitance of graphene electrodes in mono-valent aqueous electrolytes," *Nano Research*, vol. 9, no. 1, pp. 174–186, 2016.

[436] L. Shepherd and C. Toumazou, "Weak inversion ISFETs for ultra-low power biochemical sensing and real-time analysis," *Sensors and Actuators B: Chemical*, vol. 107, no. 1, pp. 468–473, 2005.

[437] N. Miscourides and P. Georgiou, "Linear current-mode ISFET arrays," in *2016 IEEE International Symposium on Circuits and Systems (ISCAS)*, 2016: IEEE, pp. 2827–2830.

[438] K.-T. Park, D.-G. Cho, J.-W. Park, S. Hong, and J. Hwang, "Detection of airborne viruses using electro-aerodynamic deposition and a field-effect transistor," *Scientific Reports*, vol. 5, no. 1, 2015.

[439] J. Y. Kim, K. Choi, D. I. Moon, *et al.*, "Surface engineering for enhancement of sensitivity in an underlap-FET biosensor by control of wettability," (in Eng), *Biosens Bioelectron*, vol. 41, pp. 867–70, 2013.

[440] M. Shariati, "The field effect transistor DNA biosensor based on ITO nanowires in label-free hepatitis B virus detecting compatible with CMOS technology," *Biosens Bioelectron*, vol. 105, pp. 58–64, 2018.

[441] T. Dastagir, E. S. Forzani, R. Zhang, *et al.*, "Electrical detection of hepatitis C virus RNA on single wall carbon nanotube-field effect transistors," *Analyst*, vol. 132, no. 8, pp. 738–740, 2007.

[442] K. Malpartida-Cardenas, N. Miscourides, J. Rodriguez-Manzano, *et al.*, "Quantitative and rapid Plasmodium falciparum malaria diagnosis and artemisinin-resistance detection using a CMOS Lab-on-Chip platform," *Biosensors and Bioelectronics*, vol. 145, p. 111678, 2019.

[443] K. H. Lee, J. O. Lee, S. Choi, J. B. Yoon, and G. H. Cho, "A CMOS label-free DNA sensor using electrostatic induction of molecular charges," *Biosens Bioelectron*, vol. 31, no. 1, pp. 343–8, 2012.

[444] P. Bergveld, "The operation of an ISFET as an electronic device," *Sensors and Actuators*, vol. 1, pp. 17–29, 1981.

[445] M. J. Milgrew and D. R. Cumming, "Matching the transconductance characteristics of CMOS ISFET arrays by removing trapped charge," *IEEE Transactions on Electron Devices*, vol. 55, no. 4, pp. 1074–1079, 2008.

[446] P. Georgiou and C. Toumazou, "ISFET threshold voltage programming in CMOS using hot-electron injection," *Electronics Letters*, vol. 45, no. 22, pp. 1112–1113, 2009.

[447] A. Al-Ahdal and C. Toumazou, "ISFET threshold voltage programming in CMOS using electron tunnelling," *Electronics Letters*, vol. 47, no. 25, pp. 1398–1399, 2011.

[448] P. Georgiou and C. Toumazou, "CMOS-based programmable gate ISFET," *Electronics Letters*, vol. 44, no. 22, pp. 1289–1291, 2008.

[449] L. Yan, P. Georgiou, T. G. Constandinou, D. Garner, and C. Toumazou, "An auto-offset-removal circuit for chemical sensing based on the PG-ISFET," in *2009 IEEE International Symposium on Circuits and Systems*, 2009: IEEE, pp. 1165–1168.

[450] Y. Hu and P. Georgiou, "A robust ISFET pH-measuring front-end for chemical reaction monitoring," *IEEE Transactions on Biomedical Circuits and Systems*, vol. 8, no. 2, pp. 177–185, 2014.

[451] H.-S. Wong and M. H. White, "A CMOS-integrated 'ISFET-operational amplifier' chemical sensor employing differential sensing," *IEEE Transactions on Electron Devices*, vol. 36, no. 3, pp. 479–487, 1989.

[452] C.-Y. Aw and P. W. Cheung, *"A pH-ISFET sensor with on-chip temperature sensing," in Engineering in Medicine and Biology Society, 1988. Proceedings of the Annual International Conference of the IEEE*, 1988: IEEE, pp. 772–773.

[453] M. Sohbati and C. Toumazou, "A temperature insensitive continuous time ΔpH to digital converter," in *2014 IEEE International Symposium on Circuits and Systems (ISCAS)*, 2014: IEEE, pp. 37–40.

[454] Y.-L. Chin, J.-C. Chou, T.-P. Sun, W.-Y. Chung, and S.-K. Hsiung, "A novel pH sensitive ISFET with on chip temperature sensing using CMOS standard process," *Sensors and Actuators B: Chemical*, vol. 76, no. 1, pp. 582–593, 2001.

[455] W.-Y. Chung, Y.-T. Lin, D. G. Pijanowska, *et al.*, "New ISFET interface circuit design with temperature compensation," *Microelectronics Journal*, vol. 37, no. 10, pp. 1105–1114, 2006.

[456] C. Jakobson, M. Feinsod, and Y. Nemirovsky, "Low frequency noise and drift in ion sensitive field effect transistors," *Sensors and Actuators B: Chemical*, vol. 68, no. 1–3, pp. 134–139, 2000.

[457] C. Jakobson and Y. Nemirovsky, "1/f noise in ion sensitive field effect transistors from subthreshold to saturation," *IEEE Transactions on Electron Devices*, vol. 46, no. 1, pp. 259–261, 1999.

[458] A. Morgenshtein, L. Sudakov-Boreysha, U. Dinnar, C. G. Jakobson, and Y. Nemirovsky, "Wheatstone-Bridge readout interface for ISFET/REFET applications," *Sensors and Actuators B: Chemical*, vol. 98, no. 1, pp. 18–27, 2004.

[459] L. Ravezzi and P. Conci, "ISFET sensor coupled with CMOS read-out circuit microsystem," *Electronics Letters*, vol. 34, no. 23, pp. 2234–2235, 1998.

[460] B. Nemeth, M. S. Piechocinski, and D. R. Cumming, "High-resolution real-time ion-camera system using a CMOS-based chemical sensor array for proton imaging," *Sensors and Actuators B: Chemical*, vol. 171, pp. 747–752, 2012.

[461] T. Hizawa, K. Sawada, H. Takao, and M. Ishida, "Fabrication of a two-dimensional pH image sensor using a charge transfer technique," *Sensors and Actuators B: Chemical*, vol. 117, no. 2, pp. 509–515, 2006.

[462] Y.-N. Lee, T. Araki, Y. Kimura, *et al.*, "High-density 2-μm-pitch pH image sensor with high-speed operation up to 1933 fps," *IEEE Transactions on Biomedical Circuits and Systems*, vol. 13, no. 2, pp. 352–363, 2019.

[463] Y. Maruyama, S. Terao, and K. Sawada, "Label free CMOS DNA image sensor based on the charge transfer technique," *Biosensors and Bioelectronics*, vol. 24, no. 10, pp. 3108–3112, 2009.

[464] K. Sawada, T. Shimada, T. Ohshina, H. Takao, and M. Ishida, "Highly sensitive ion sensors using charge transfer technique," *Sensors and Actuators B: Chemical*, vol. 98, no. 1, pp. 69–72, 2004.

[465] M. Futagawa, R. Otake, F. Dasai, M. Ishida, and K. Sawada, "Realization of 1 million pixel charge transfer type ion image sensor with 12 μm pixel pitch," in *2015 Transducers-2015 18th International Conference on Solid-State Sensors, Actuators and Microsystems (TRANSDUCERS)*, 2015: IEEE, pp. 236–239.

[466] P. Georgiou and C. Toumazou, "An adaptive ISFET chemical imager chip," in *2008 IEEE International Symposium on Circuits and Systems*, 2008: IEEE, pp. 2078–2081.

[467] K. Seong-Jin and Y. Euisik, "Label-free CMOS bio sensor with on-chip noise reduction scheme for real-time quantitative monitoring of biomolecules," *IEEE Transactions on Biomedical Circuits and Systems*, vol. 6, no. 3, pp. 189–196, 2012.

[468] R. A. Croce, S. Vaddiraju, A. Legassey, *et al.*, "A highly miniaturized low-power CMOS-based pH monitoring platform," *IEEE Sensors Journal*, vol. 15, no. 2, pp. 895–901, 2014.

[469] N. Miscourides and P. Georgiou, "ISFET arrays in CMOS: A head-to-head comparison between voltage and current mode," *IEEE Sensors Journal*, vol. 19, no. 4, pp. 1224–1238, 2018.

[470] M. Duan, X. Zhong, J. Xu, Y.-K. Lee, and A. Bermak, "A high offset distribution tolerance high resolution ISFET array with auto-compensation for long-term bacterial metabolism monitoring," *IEEE Transactions on Biomedical Circuits and Systems*, vol. 14, no. 3, pp. 463–476, 2020.

[471] B. Premanode, N. Silawan, W. P. Chan, and C. Toumazou, "A composite ISFET readout circuit employing current feedback," *Sensors and Actuators B: Chemical*, vol. 127, no. 2, pp. 486–490, 2007.

[472] M. Douthwaite, E. Koutsos, D. C. Yates, P. D. Mitcheson, and P. Georgiou, "A thermally powered ISFET array for on-body pH measurement," *IEEE*

Transactions on Biomedical Circuits and Systems, vol. 11, no. 6, pp. 1324–1334, 2017.

[473] J. Zeng, L. Kuang, N. Miscourides, and P. Georgiou, "A 128 × 128 current-mode ultra-high frame rate ISFET array with in-pixel calibration for real-time ion imaging," *IEEE Transactions on Biomedical Circuits and Systems*, vol. 14, no. 2, pp. 359–372, 2020.

[474] Z. D. C. Goh, P. Georgiou, T. G. Constandinou, T. Prodromakis, and C. Toumazou, "Live demonstration: A CMOS-based lab-on-chip array for combined magnetic manipulation and opto-chemical sensing," in *2011 IEEE International Symposium of Circuits and Systems (ISCAS)*, 2011: IEEE, pp. 1997–2001.

[475] X. Huang, H. Yu, X. Liu, Y. Jiang, M. Yan, and D. Wu, "A dual-mode large-arrayed CMOS ISFET sensor for accurate and high-throughput pH sensing in biomedical diagnosis," *IEEE Transactions on Biomedical Engineering*, vol. 62, no. 9, pp. 2224–2233, 2015.

[476] N. Moser, T. S. Lande, and P. Georgiou, "A robust ISFET array with in-pixel quantisation and automatic offset calibration," in *2016 IEEE Biomedical Circuits and Systems Conference (BioCAS)*, 2016: IEEE, pp. 50–53.

[477] M. Cacho-Soblechero, K. Malpartida-Cardenas, N. Moser, and P. Georgiou, "Programmable ion-sensing using oscillator-based ISFET architectures," *IEEE Sensors Journal*, vol. 19, no. 19, pp. 8563–8575, 2019.

[478] K. Wang, Y. Liu, C. Toumazou, and P. Georgiou, "A TDC based ISFET readout for large-scale chemical sensing systems," in *2012 IEEE Biomedical Circuits and Systems Conference (BioCAS)*, 2012: IEEE, pp. 176–179.

[479] Y. Liu, T. G. Constandinou, and P. Georgiou, "Ultrafast large-scale chemical sensing with CMOS ISFETs: A level-crossing time-domain approach," *IEEE Transactions on Biomedical Circuits and Systems*, vol. 13, no. 6, pp. 1201–1213, 2019.

[480] C. Lee, Y.-W. Chen, and M. S.-C. Lu, "CMOS biosensors for the detection of dna hybridization in high ionic-strength solutions," *IEEE Sensors Journal*, vol. 21, no. 4, pp. 4135–4142, 2020.

[481] W. P. Chan, B. Premanode, and C. Toumazou, "An integrated ISFETs instrumentation system in standard CMOS technology," *IEEE Journal of Solid-State Circuits*, vol. 45, no. 9, pp. 1923–1934, 2010.

[482] C.-F. Chang and M. S.-C. Lu, "CMOS ion sensitive field effect transistors for highly sensitive detection of DNA hybridization," *IEEE Sensors Journal*, vol. 20, no. 16, pp. 8930–8937, 2020.

[483] P. Livi, A. Shadmani, M. Wipf, *et al.*, "Sensor system including silicon nanowire ion sensitive FET arrays and CMOS readout," *Sensors and Actuators B: Chemical*, vol. 204, pp. 568–577, 2014.

[484] A. Al-Ahdal and C. Toumazou, "High gain ISFET based νMOS chemical inverter," *Sensors and Actuators B: Chemical*, vol. 171, pp. 110–117, 2012.

[485] G. Nabovati, E. Ghafar-Zadeh, and M. Sawan, "A 64 pixel ISFET-based biosensor for extracellular pH gradient monitoring," in *2015 IEEE International Symposium on Circuits and Systems (ISCAS)*, 2015: IEEE, pp. 1762–1765.

[486] Y. Liu and C. Toumazou, "An ISFET based sensing array with sensor offset compensation and pH sensitivity enhancement," *in Proceedings of 2010 IEEE International Symposium on Circuits and Systems*, 2010: IEEE, pp. 2283–2286.

[487] M. Milgrew, P. Hammond, and D. Cumming, "The development of scalable sensor arrays using standard CMOS technology," *Sensors and Actuators B: Chemical*, vol. 103, no. 1–2, pp. 37–42, 2004.

[488] P. A. Hammond, D. Ali, and D. R. Cumming, "Design of a single-chip pH sensor using a conventional 0.6 μm CMOS process," *IEEE Sensors Journal*, vol. 4, no. 6, pp. 706–712, 2004.

[489] B. Palan, F. Santos, J. Karam, B. Courtois, and M. Husak, "New ISFET sensor interface circuit for biomedical applications," *Sensors and Actuators B: Chemical*, vol. 57, no. 1–3, pp. 63–68, 1999.

[490] G. Nabovati, E. Ghafarzadeh, F. Awwad, and M. Sawan, "A sigma delta ISFET readout circuit for Lab-on-Chip applications," in *New Circuits and Systems Conference (NEWCAS), 2013 IEEE 11th International*, 2013: IEEE, pp. 1–4.

[491] M. Kalofonou and C. Toumazou, "A low power sub-μW chemical Gilbert cell for ISFET differential reaction monitoring," *IEEE Transactions on Biomedical Circuits and Systems*, vol. 8, no. 4, pp. 565–574, 2013.

[492] P. Bergveld, "Thirty years of ISFETOLOGY: What happened in the past 30 years and what may happen in the next 30 years," *Sensors and Actuators B: Chemical*, vol. 88, no. 1, pp. 1–20, 2003.

[493] P. Georgiou and C. Toumazou, "An adaptive CMOS-based PG-ISFET for pH sensing," in *2009 IEEE International Symposium on Circuits and Systems*, 2009: IEEE, pp. 557–560.

[494] S. Jamasb, "An analytical technique for counteracting drift in ion-selective field effect transistors (ISFETs)," *IEEE Sensors Journal*, vol. 4, no. 6, pp. 795–801, 2004.

[495] N. Moser, T. S. Lande, C. Toumazou, and P. Georgiou, "ISFETs in CMOS and emergent trends in instrumentation: A review," *IEEE Sensors Journal*, vol. 16, no. 17, pp. 6496–6514, 2016.

[496] P. A. Hammond, D. Ali, and D. R. Cumming, "A system-on-chip digital pH meter for use in a wireless diagnostic capsule," *IEEE Transactions on Biomedical Engineering*, vol. 52, no. 4, pp. 687–694, 2005.

[497] J. M. Rothberg, W. Hinz, K. L. Johnson, and J. Bustillo, "Methods and apparatus for measuring analytes using large scale FET arrays," ed: Google Patents, 2011.

[498] M. M. Uzzal, P. Zarkesh-Ha, J. S. Edwards, E. Coelho, and P. Rawat, "A highly sensitive ISFET using pH-to-current conversion for real-time DNA sequencing," in *2014 27th IEEE International System-on-Chip Conference (SOCC)*, 2014: IEEE, pp. 410–414.

[499] P. Georgiou and C. Toumazou, "Chemical log-domain filter," *Electronics Letters*, vol. 45, no. 8, pp. 391–392, 2009.

[500] L. M. Shepherd and C. Toumazou, "A biochemical translinear principle with weak inversion ISFETs," *IEEE Transactions on Circuits and Systems I: Regular Papers*, vol. 52, no. 12, pp. 2614–2619, 2005.

[501] P. Georgiou and C. Toumazou, *"A silicon pancreatic beta cell for diabetes," IEEE Transactions on Biomedical Circuits and Systems*, vol. 1, no. 1, pp. 39–49, 2007.

[502] N. Miscourides and P. Georgiou, "A linear programmable-gate ISFET array operating in velocity saturation," in *2016 IEEE Biomedical Circuits and Systems Conference (BioCAS)*, 2016: IEEE, pp. 292–295.

[503] N. Miscourides, L.-S. Yu, J. Rodriguez-Manzano, and P. Georgiou, *"A 12.8 k current-mode velocity-saturation ISFET array for on-chip real-time DNA detection," IEEE Transactions on Biomedical Circuits and Systems*, vol. 12, no. 5, pp. 1202–1214, 2018.

[504] C. Toumazou, F. J. Lidgey, and D. Haigh, *Analogue IC Design: The Current-Mode Approach. Presbyterian Publishing Corp*, 1990.

[505] P. Pookaiyaudom, C. Toumazou, and F. Lidgey, "The chemical current-conveyor: a new microchip biosensor," in *2008 IEEE International Symposium on Circuits and Systems*, 2008: IEEE, pp. 3166–3169.

[506] K. Manetakis and C. Toumazou, "A new high-frequency very low output-impedance CMOS buffer," in *1996 IEEE International Symposium on Circuits and Systems. Circuits and Systems Connecting the World. ISCAS 96*, 1996, vol. 1: IEEE, pp. 485–488.

[507] B. Premanode, W. Chan, and C. Toumazou, "Ultra-low power precision ISFET readout using global current feedback," *Electronics Letters*, vol. 42, no. 22, pp. 1264–1265, 2006.

[508] W. Chan, B. Premanode, and C. Toumazou, "64 pH-ISFET averaging array employing global negative current feedback," *Electronics Letters*, vol. 45, no. 11, pp. 536–537, 2009.

[509] W. Juffali, P. Georgiou, and C. Toumazou, "ISFET based Urea: Creatinine translinear sensor," *Electronics Letters*, vol. 46, no. 11, pp. 746–748, 2010.

[510] M. Kalofonou, P. Georgiou, C.-P. Ou, and C. Toumazou, "An ISFET based translinear sensor for DNA methylation detection," *Sensors and Actuators B: Chemical*, vol. 161, no. 1, pp. 156–162, 2012.

[511] W. Wong, L. Shepherd, P. Georgiou, and C. Toumazou, "Towards ISFET based DNA logic for rapid nucleic acid detection,*" in Sensors*, 2009 *IEEE*, 2009: IEEE, pp. 1451–1454.

[512] A. AlAhdal and C. Toumazou, "ISFET-based chemical Schmitt trigger," *Electronics Letters*, vol. 48, no. 10, pp. 549–551, 2012.

[513] A. Al-Ahdal and C. Toumazou, "ISFET-based chemical switch," *IEEE Sensors Journal*, vol. 12, no. 5, pp. 1140–1146, 2011.

[514] Y. Liu, A. Al-Ahdal, P. Georgiou, and C. Toumazou, "Minimal readout scheme for ISFET sensing arrays based on pulse width modulation," *Electronics Letters*, vol. 48, no. 10, pp. 548–549, 2012.

[515] A. T. Do, J. Minkyu, and K. S. Yeo, "Improved inverter-based read-out scheme for low-power ISFET sensing array," *Electronics Letters*, vol. 49, no. 24, pp. 1517–1518, 2013.

[516] N. Moser, T. S. Lande, and P. Georgiou, "A novel pH-to-time ISFET pixel architecture with offset compensation," in *2015 IEEE International Symposium on Circuits and Systems (ISCAS)*, 2015: IEEE, pp. 481–484.

[517] N. Moser, J. Rodriguez-Manzano, T. S. Lande, and P. Georgiou, "A scalable ISFET sensing and memory array with sensor auto-calibration for on-chip real-time DNA detection," *IEEE Transactions on Biomedical Circuits and Systems*, vol. 12, no. 2, pp. 390–401, 2018.

[518] Y. Jiang, P. Coquet, and H. Yu, "Fast food safety screening with CMOS high-sensitivity large-arrayed ISFET sensor," in *2017 IEEE Biomedical Circuits and Systems Conference (BioCAS)*, 2017: IEEE, pp. 1–4.

[519] Y. Hu and P. Georgiou, "3-T ISFET front-end utilising parasitic device capacitance," *Electronics Letters*, vol. 50, no. 21, pp. 1507–1509, 2014.

[520] J. Chovelon, J. Fombon, P. Clechet, *et al.*, "Sensitization of dielectric surfaces by chemical grafting: application to pH ISFETs and REFETs," *Sensors and Actuators B: Chemical*, vol. 8, no. 3, pp. 221–225, 1992.

[521] H. Nakajima, M. Esashi, and T. Matsuo, "The cation concentration response of polymer gate ISFET," *Journal of The Electrochemical Society*, vol. 129, no. 1, p. 141, 1982.

[522] T. Matsuo and H. Nakajima, "Characteristics of reference electrodes using a polymer gate ISFET," *Sensors and Actuators*, vol. 5, no. 4, pp. 293–305, 1984.

[523] P. Bergveld, "Development, operation, and application of the ion-sensitive field-effect transistor as a tool for electrophysiology," *IEEE Transactions on Biomedical Engineering*, no. 5, pp. 342–351, 1972.

[524] A. Errachid, J. Bausells, and N. Jaffrezic-Renault, "A simple REFET for pH detection in differential mode," *Sensors and Actuators B: Chemical*, vol. 60, no. 1, pp. 43–48, 1999.

[525] P. Bergveld, A. Van Den Berg, P. Van Der Wal, M. Skowronska-Ptasinska, E. Sudhölter, and D. Reinhoudt, "How electrical and chemical requirements for REFETs may coincide," *Sensors and Actuators*, vol. 18, no. 3–4, pp. 309–327, 1989.

[526] M. Chudy, W. Wróblewski, and Z. Brzózka, "Towards REFET," *Sensors and Actuators B: Chemical*, vol. 57, no. 1–3, pp. 47–50, 1999.

[527] V. Chodavarapu, A. Titus, and A. Cartwright, "Differential read-out architecture for CMOS ISFET microsystems," *Electronics Letters*, vol. 41, no. 12, pp. 698–699, 2005.

[528] L. Shepherd, P. Georgiou, and C. Toumazou, "A novel voltage-clamped CMOS ISFET sensor interface," in *2007 IEEE International Symposium on Circuits and Systems*, 2007: IEEE, pp. 3331–3334.

[529] D. Petrovszki, S. Valkai, E. Gora, *et al.*, "An integrated electro-optical biosensor system for rapid, low-cost detection of bacteria," *Microelectronic Engineering*, vol. 239, p. 111523, 2021.

[530] A. T. T. Pham, A. Wallace, X. Zhang, *et al.*, "Optical-based biosensors and their portable healthcare devices for detecting and monitoring biomarkers in body fluids," *Diagnostics*, vol. 11, no. 7, p. 1285, 2021.

[531] G. Rajeev, E. Xifre-Perez, B. P. Simon, A. J. Cowin, L. F. Marsal, and N. H. Voelcker, "A label-free optical biosensor based on nanoporous anodic alumina for tumour necrosis factor-alpha detection in chronic wounds," *Sensors and Actuators B: Chemical*, vol. 257, pp. 116–123, 2018.

[532] K. Sadani, P. Nag, X. Y. Thian, and S. Mukherji, "Enzymatic optical biosensors for healthcare applications," *Biosensors and Bioelectronics: X*, p. 100278, 2022.

[533] I. M. Khan, S. Niazi, W. Akhtar, *et al.*, "Surface functionalized AuNCs optical biosensor as an emerging food safety indicator: Fundamental mechanism to future prospects," *Coordination Chemistry Reviews*, vol. 474, p. 214842, 2023.

[534] D. I. Meira, A. I. Barbosa, J. Borges, R. L. Reis, V. M. Correlo, and F. Vaz, "Recent advances in nanomaterial-based optical biosensors for food safety applications: Ochratoxin-A detection, as case study," *Critical Reviews in Food Science and Nutrition*, pp. 1–43, 2023.

[535] A. B. Pebdeni, A. Roshani, E. Mirsadoughi, S. Behzadifar, and M. Hosseini, "Recent advances in optical biosensors for specific detection of E. coli bacteria in food and water," *Food Control*, p. 108822, 2022.

[536] M. T. Giardi, V. Scognamiglio, G. Rea, *et al.*, "Optical biosensors for environmental monitoring based on computational and biotechnological tools for engineering the photosynthetic D1 protein of *Chlamydomonas reinhardtii*," *Biosensors and Bioelectronics*, vol. 25, no. 2, pp. 294–300, 2009.

[537] F. Long, A. Zhu, and H. Shi, "Recent advances in optical biosensors for environmental monitoring and early warning," *Sensors*, vol. 13, no. 10, pp. 13928–13948, 2013.

[538] F. Long, A. Zhu, and H. Shi, "Recent advances in optical biosensors for environmental monitoring and early warning," *Sensors*, vol. 13, no. 10, pp. 13928–13948, 2013.

[539] S. Rodriguez-Mozaz, M. J. L. de Alda, M.-P. Marco, and D. Barceló, "Biosensors for environmental monitoring: A global perspective," *Talanta*, vol. 65, no. 2, pp. 291–297, 2005.

[540] W.-J. Ho, J.-S. Chen, M.-D. Ker, *et al.*, "Fabrication of a miniature CMOS-based optical biosensor," *Biosensors and Bioelectronics*, vol. 22, no. 12, pp. 3008–3013, 2007.

[541] S. M. Borisov and O. S. Wolfbeis, "Optical biosensors," *Chemical Reviews*, vol. 108, no. 2, pp. 423–461, 2008.

[542] D. Dey and T. Goswami, "Optical biosensors: a revolution towards quantum nanoscale electronics device fabrication," *Journal of Biomedicine and Biotechnology*, vol. 2011, 2011.

[543] C. Chen and J. Wang, "Optical biosensors: An exhaustive and comprehensive review," *Analyst*, vol. 145, no. 5, pp. 1605–1628, 2020.

[544] Y.-T. Chen, Y.-C. Lee, Y.-H. Lai, *et al.*, "Review of integrated optical biosensors for point-of-care applications," *Biosensors*, vol. 10, no. 12, p. 209, 2020.

[545] B. Špačková, P. Wrobel, M. Bocková, and J. Homola, "Optical biosensors based on plasmonic nanostructures: a review," *Proceedings of the IEEE*, vol. 104, no. 12, pp. 2380–2408, 2016.

[546] Z. Xie, Z. Cao, Y. Liu, *et al.*, "Highly-sensitive optical biosensor based on equal FSR cascaded microring resonator with intensity interrogation for detection of progesterone molecules," *Optics Express*, vol. 25, no. 26, pp. 33193–33201, 2017.

[547] F. Aghimand, C. Hu, S. Sharma, K. K. Pochana, R. M. Murray, and A. Emami, "21.1 A 65nm CMOS living-cell dynamic fluorescence sensor with 1.05 fA sensitivity at 600/700nm wavelengths," in *2023 IEEE International Solid-State Circuits Conference (ISSCC), 2023*: IEEE, pp. 312–314.

[548] T. Arai, P. Kumar, C. Rockstuhl, K. Awazu, and J. Tominaga, "An optical biosensor based on localized surface plasmon resonance of silver nanostructured films," *Journal of Optics A: Pure and Applied Optics*, vol. 9, no. 7, p. 699, 2007.

[549] C. Adamopoulos, S. Buchbinder, P. Zarkos, *et al.*, "Fully integrated electronic-photonic sensor for label-free refractive index sensing in advanced zero-change CMOS-SOI process," in *2021 IEEE Custom Integrated Circuits Conference (CICC)*, 2021: IEEE, pp. 1–2.

[550] R. Al Mahmud, R. Sagor, and M. Khan, "Surface plasmon refractive index biosensors: A review of optical fiber, multilayer 2D material and gratings, and MIM configurations," *Optics & Laser Technology*, vol. 159, p. 108939, 2023.

[551] P. Steglich, S. Bondarenko, C. Mai, M. Paul, M. G. Weller, and A. Mai, "CMOS-compatible silicon photonic sensor for refractive index sensing using local back-side release," *IEEE Photonics Technology Letters*, vol. 32, no. 19, pp. 1241–1244, 2020.

[552] L. Hong, S. McManus, H. Yang, and K. Sengupta, "A fully integrated CMOS fluorescence biosensor with on-chip nanophotonic filter," in *2015 Symposium on VLSI Circuits (VLSI Circuits)*, 2015: IEEE, pp. C206–C207.

[553] B. Jang, P. Cao, A. Chevalier, A. Ellington, and A. Hassibi, "A CMOS fluorescent-based biosensor microarray," in *2009 IEEE International Solid-State Circuits Conference-Digest of Technical Papers*, 2009: *IEEE*, pp. 436–437,437 a.

[554] M. N. Khiarak, S. Martel, Y. De Koninck, and B. Gosselin, "High-DR CMOS fluorescence biosensor with extended counting ADC and noise cancellation," *IEEE Transactions on Circuits and Systems I: Regular Papers*, vol. 66, no. 6, pp. 2077–2087, 2019.

[555] A. Manickam, R. Singh, M. W. McDermott, *et al.*, "A fully integrated CMOS fluorescence biochip for DNA and RNA testing," *IEEE Journal of Solid-State Circuits*, vol. 52, no. 11, pp. 2857–2870, 2017.

[556] M. Strianese, M. Staiano, G. Ruggiero, T. Labella, C. Pellecchia, and S. D'Auria, "Fluorescence-based biosensors," *Methods in Molecular Biology*, vol. 875, pp. 193–216, 2012.

[557] C. Zhu, L. Hong, H. Yang, and K. Sengupta, "A packaged multiplexed fluorescent biomolecular sensor array and ultralow-power wireless interface in CMOS for ingestible electronic applications," *IEEE Sensors Journal*, vol. 22, no. 24, pp. 24060–24074, 2022.

[558] C. Zhu, Y. Wen, T. Liu, H. Yang, and K. Sengupta, "An ingestible pill with CMOS fluorescence sensor array, bi-directional wireless interface and packaged optics for in-vivo bio-molecular sensing," *IEEE Transactions on Biomedical Circuits and Systems*, 2023.

[559] M. L. Simpson, G. S. Sayler, G. Patterson, *et al.*, "An integrated CMOS microluminometer for low-level luminescence sensing in the bioluminescent bioreporter integrated circuit," *Sensors and Actuators B: Chemical*, vol. 72, no. 2, pp. 134–140, 2001.

[560] H. Eltoukhy, K. Salama, and A. E. Gamal, "A 0.18-/spl mu/m CMOS bioluminescence detection lab-on-chip," *IEEE Journal of Solid-State Circuits*, vol. 41, no. 3, pp. 651–662, 2006.

[561] Q. Liu, M. Jimenez, M. E. Inda, *et al.*, "A threshold-based bioluminescence detector with a CMOS-integrated photodiode array in 65 nm for a multi-diagnostic ingestible capsule," *IEEE Journal of Solid-State Circuits*, 2022.

[562] Y.-W. Chang, P.-C. Yu, Y.-T. Huang, and Y.-S. Yang, "A CMOS-compatible optical biosensing system based on visible absorption spectroscopy," in *2007 IEEE Conference on Electron Devices and Solid-State Circuits*, 2007: IEEE, pp. 1099–1102.

[563] Y.-W. Chang, P.-C. Yu, Y.-T. Huang, and Y.-S. Yang, "A high-sensitivity CMOS-compatible biosensing system based on absorption photometry," *IEEE Sensors Journal*, vol. 9, no. 2, pp. 120–127, 2009.

[564] A. Hofmann, M. Meister, A. Rolapp, P. Reich, F. Scholz, and E. Schäfer, "Light absorption measurement with a CMOS biochip for quantitative immunoassay based point-of-care applications," *IEEE Transactions on Biomedical Circuits and Systems*, vol. 15, no. 3, pp. 369–379, 2021.

[565] R. Yan, S. P. Mestas, G. Yuan, R. Safaisini, D. S. Dandy, and K. L. Lear, "Label-free silicon photonic biosensor system with integrated detector array," *Lab on a Chip*, vol. 9, no. 15, pp. 2163–2168, 2009.

[566] K. De Vos, "Label-free silicon photonics biosensor platform with microring resonators," *PhD dissertation*, 2010.

[567] E. Luan, H. Shoman, D. M. Ratner, K. C. Cheung, and L. Chrostowski, "Silicon photonic biosensors using label-free detection," *Sensors*, vol. 18, no. 10, p. 3519, 2018.

[568] C. Adamopoulos, P. Zarkos, S. Buchbinder, *et al.*, "Lab-on-chip for everyone: introducing an electronic-photonic platform for multiparametric biosensing using standard CMOS processes," *IEEE Open Journal of the Solid-State Circuits Society*, vol. 1, pp. 198–208, 2021.

[569] A. Mai, C. Mai, and P. Steglich, "From Lab-on-Chip to Lab-in-App: Challenges towards silicon photonic biosensors product developments," *Results in Optics*, vol. 9, p. 100317, 2022.

[570] S. Ning, H.-C. Chang, K.-C. Fan, *et al.*, "Silicon photonic chip-based biosensor for COVID-19 and flu detection with high sensitivity and specificity," in *Ultra-High-Definition Imaging Systems VI*, 2023, vol. 12444: SPIE, pp. 26–28.

[571] S. C. Sajan, A. Singh, P. K. Sharma, and S. Kumar, "Silicon photonics biosensors for cancer cells detection: a review," *IEEE Sensors Journal*, vol. 23, no. 4, pp. 3366–3377, 2023.

[572] S. Suarasan, J. Liu, M. Imanbekova, *et al.*, "Superhydrophobic bowl-like SERS substrates patterned from CMOS sensors for extracellular vesicle characterization," *Journal of Materials Chemistry B*, vol. 8, no. 38, pp. 8845–8852, 2020.

[573] K. Ehrlich, A. Kufcsák, S. McAughtrie, *et al.*, "pH sensing through a single optical fibre using SERS and CMOS SPAD line arrays," *Optics Express*, vol. 25, no. 25, pp. 30976–30986, 2017.

[574] L. Wang, G. Wang, K. Yang, and W. Liu, "Plasmonic biosensor with annular aperture array integrated on a resonant cavity LED," *Optics Communications*, vol. 535, p. 129336, 2023.

[575] A. K. Singh, M. Anwar, R. Pradhan, M. S. Ashar, N. Rai, and S. Dey, "Surface Plasmon Resonance based-optical biosensor: Emerging diagnostic tool for early detection of diseases," *Journal of Biophotonics*, p. e202200380, 2023.

[576] A. Shakoor, B. C. Cheah, D. Hao, *et al.*, "Plasmonic sensor monolithically integrated with a CMOS photodiode," *ACS Photonics*, vol. 3, pp. 1926–1933, 2016.

[577] A. Salazar, S. Camacho-Leon, S. O. Martínez-Chapa, and O. Rossetto, "Towards a SPR-based biosensing platform incorporating a CMOS active column sensor," *Analog Integrated Circuits and Signal Processing*, vol. 77, pp. 365–372, 2013.

[578] S. Koppa and Y. Joo, "Compact, low cost CMOS integrated SPR biosensor system,*" in Sensors*, 2013 *IEEE*, 2013: IEEE, pp. 1–4.

[579] L. Hong, H. Li, H. Yang, and K. Sengupta, "Integrated angle-insensitive nanoplasmonic filters for ultraminiaturized fluorescence microarray in a 65 nm digital CMOS process," *ACS Photonics*, vol. 5, pp. 4312–4322, 2018.

[580] L. Hong, H. Li, H. Yang, and K. Sengupta, "Fully integrated fluorescence biosensors on-chip employing multi-functional nanoplasmonic optical structures in CMOS," *IEEE Journal of Solid-State Circuits*, vol. 52, pp. 2388–2406, 2017.

[581] G. Patounakis, K. L. Shepard, and R. Levicky, "Active CMOS array sensor for time-resolved fluorescence detection," *IEEE Journal of Solid-State Circuits*, vol. 41, pp. 2521–2530, 2006.

[582] R. R. Singh, D. Ho, A. Nilchi, G. Gulak, P. Yau, and R. Genov, *"A CMOS/thin-film fluorescence contact imaging microsystem for DNA*

analysis," IEEE Transactions on Circuits and Systems I: Regular Papers, vol. 57, pp. 1029–1038, 2010.

[583] B. Jang, P. Cao, A. Chevalier, A. Ellington, and A. Hassibi, "A CMOS fluorescent-based biosensor microarray," in *2009 IEEE International Solid-State Circuits Conference-Digest of Technical Papers*, 2009, pp. 436–437.

[584] J. R. Lakowicz, *Principles of Fluorescence Spectroscopy*. Springer, 2006.

[585] B. T. Cunningham, "Label-free optical biosensors: An introduction," *in Label-Free Biosensors: Techniques and Applications*. Cambridge University Press, 2009, pp. 1–28.

[586] H. Eltoukhy, K. Salama, and A. E. Gamal, "A 0.18 μm CMOS bioluminescence detection lab-on-chip," *IEEE Journal of Solid-State Circuits*, vol. 41, no. 3, pp. 651–662, 2006.

[587] K. De Vos, *Label-Free Silicon Photonics Biosensor Platform with Microring Resonators. Ghent University*, 2010.

[588] C. Adamopoulos and V. Stojanovic, "A fully integrated electronic-photonic platform for label-free biosensing," *Thesis*, 2020.

[589] R.-B. Hwang, "A theoretical design of evanescent wave biosensors based on gate-controlled graphene surface plasmon resonance," *Scientific Reports*, vol. 11, pp. 1–10, 2021.

[590] J. Wang, M. M. Sanchez, Y. Yin, R. Herzer, L. Ma, and O. G. Schmidt, *"Silicon-based integrated label-free optofluidic biosensors: latest advances and roadmap," Advanced Materials Technologies*, vol. 5, p. 1901138, 2020.

[591] C. S. Huertas, O. Calvo-Lozano, A. Mitchell, and L. M. Lechuga, "Advanced evanescent-wave optical biosensors for the detection of nucleic acids: An analytic perspective," *Frontiers in Chemistry*, p. 724, 2019.

[592] Z. Xiao-Hong, L. Lan-Hua, X. Wei-Qi, *et al.*, "A reusable evanescent wave immunosensor for highly sensitive detection of bisphenol A in water samples," *Scientific Reports*, vol. 4, pp. 1–7, 2014.

[593] J. G. Wangüemert-Pérez, P. Cheben, A. Ortega-Moñux, *et al.*, "Evanescent field waveguide sensing with subwavelength grating structures in silicon-on-insulator," *Optics Letters*, vol. 39, pp. 4442–4445, 2014.

[594] C. R. Taitt, G. P. Anderson, and F. S. Ligler, "Evanescent wave fluorescence biosensors," *Biosensors and Bioelectronics*, vol. 20, pp. 2470–2487, 2005.

[595] A. M. Hutchinson, "Evanescent wave biosensors," *Molecular biotechnology*, vol. 3, pp. 47–54, 1995.

[596] H. Zhang, X. Lv, B. Huang, *et al.*, "In situ regeneration of silicon microring biosensors coated with parylene C," *Langmuir*, vol. 38, no. 1, pp. 504–513, 2022.

[597] P. Steglich, D. G. Rabus, C. Sada, *et al.*, "Silicon photonic micro-ring resonators for chemical and biological sensing: a tutorial," *IEEE Sensors Journal*, vol. 22, no. 11, pp. 10089–10105, 2022.

[598] P. Steglich, M. Hülsemann, B. Dietzel, and A. Mai, "Optical biosensors based on silicon-on-insulator ring resonators: A review," *Molecules*, vol. 24, p. 519, 2019.

[599] P. Steglich, C. Villringer, S. Pulwer, *et al.*, "Hybrid-waveguide ring resonator for biochemical sensing," *IEEE Sensors Journal*, vol. 17, pp. 4781–4790, 2017.

[600] M.-W. Seo, K. Kagawa, K. Yasutomi, *et al.*, "A 10 ps time-resolution CMOS image sensor with two-tap true-CDS lock-in pixels for fluorescence lifetime imaging," *IEEE Journal of Solid-State Circuits*, vol. 51, pp. 141–154, 2015.

[601] A. Gowri, N. A. Kumar, and B. S. Anand, "Recent advances in nanomaterials based biosensors for point of care (PoC) diagnosis of COVID-19–a minireview," *TrAC Trends in Analytical Chemistry*, vol. 137, p. 116205, 2021.

[602] C. Sun, M. T. Wade, Y. Lee, *et al.*, "Single-chip microprocessor that communicates directly using light," *Nature*, vol. 528, no. 7583, pp. 534–538, 2015.

[603] L. Zhang and A. M. Niknejad, "Design and analysis of a microwave-optical dual modality biomolecular sensing platform," *IEEE Journal of Solid-State Circuits*, vol. 55, pp. 639–649, 2019.

[604] M. N. Khiarak, K. Sasagawa, T. Tokuda, *et al.*, "A 17-bit 104-dB-DR high-precision low-power CMOS fluorescence biosensor with extended counting ADC and noise cancellation," in *2018 16th IEEE International New Circuits and Systems Conference (NEWCAS)*, 2018: IEEE, pp. 100–103.

[605] V. Gradišnik and D. Gumbarević, "a-Si: H pin photodiode as a biosensor," in *Advances in Photodetectors-Research and Applications. IntechOpen*, 2018.

[606] A. H. Titus, M. C. Cheung, and V. P. Chodavarapu, "CMOS photodetectors," in *Photodiodes-World Activities in 2011*. IntechOpen, 2011.

[607] G. Ghione, *Semiconductor Devices for High-Speed Optoelectronics*, vol. 116. Cambridge University Press, Cambridge, 2009.

[608] D. Mosconi, D. Stoppa, L. Pancheri, L. Gonzo, and A. Simoni, "CMOS single-photon avalanche diode array for time-resolved fluorescence detection," in *2006 Proceedings of the 32nd European Solid-State Circuits Conference*, 2006: IEEE, pp. 564–567.

[609] A. Hassibi, A. Manickam, R. Singh, *et al.*, "Multiplexed identification, quantification and genotyping of infectious agents using a semiconductor biochip," *Nature Biotechnology*, vol. 36, pp. 738–745, 2018.

[610] A. Manickam, R. Singh, M. W. McDermott, *et al.*, "A fully integrated CMOS fluorescence biochip for DNA and RNA testing," *IEEE Journal of Solid-State Circuits*, vol. 52, pp. 2857–2870, 2017.

[611] R. R. Singh, *"High dynamic range CMOS-integrated biosensors," PhD dissertation*, 2013.

[612] U. Lu, B. C.-P. Hu, Y.-C. Shih, C.-Y. Wu, and Y.-S. Yang, "The design of a novel complementary metal oxide semiconductor detection system for biochemical luminescence," *Biosensors and Bioelectronics*, vol. 19, pp. 1185–1191, 2004.

[613] K. K. Ng, *Complete Guide to Semiconductor Devices. Wiley Online Library*, 2002.

[614] N. Lovecchio, F. Costantini, E. Parisi, *et al.*, "Integrated optoelectronic device for detection of fluorescent molecules," *IEEE Transactions on Biomedical Circuits and Systems*, vol. 12, pp. 1337–1344, 2018.

[615] M. Gersbach, Y. Maruyama, R. Trimananda, *et al.*, "A time-resolved, low-noise single-photon image sensor fabricated in deep-submicron CMOS technology," *IEEE Journal of Solid-State Circuits*, vol. 47, pp. 1394–1407, 2012.

[616] C. Veerappan, J. Richardson, R. Walker, *et al.*, "A 160 × 128 single-photon image sensor with on-pixel 55ps 10b time-to-digital converter," in *2011 IEEE International Solid-State Circuits Conference*, 2011, pp. 312–314.

[617] D. Stoppa, D. Mosconi, L. Pancheri, and L. Gonzo, "Single-photon avalanche diode CMOS sensor for time-resolved fluorescence measurements," *IEEE Sensors Journal*, vol. 9, pp. 1084–1090, 2009.

[618] D. E. Schwartz, E. Charbon, and K. L. Shepard, "A single-photon avalanche diode array for fluorescence lifetime imaging microscopy," *IEEE Journal of Solid-State Circuits*, vol. 43, no. 11, pp. 2546–2557, 2008.

[619] W. J. Kindt and H. W. Van Zeijl, "Modelling and fabrication of Geiger mode avalanche photodiodes," *IEEE Transactions on Nuclear Science*, vol. 45, pp. 715–719, 1998.

[620] A. Rochas, M. Gani, B. Furrer, *et al.*, "Single photon detector fabricated in a complementary metal–oxide–semiconductor high-voltage technology," *Review of Scientific Instruments*, vol. 74, no. 7, pp. 3263–3270, 2003.

[621] R. R. Singh, L. Leng, A. Guenther, and R. Genov, "A CMOS-microfluidic chemiluminescence contact imaging microsystem," *IEEE Journal of Solid-State Circuits*, vol. 47, pp. 2822–2833, 2012.

[622] K. Salama, *CMOS Luminescence Detection Lab-on-Chip: Modeling, Design, and Characterization*. Stanford University, 2005.

[623] L. Gámiz-Gracia, A. M. García-Campaña, J. J. Soto-Chinchilla, J. F. Huertas-Pérez, and A. González-Casado, "Analysis of pesticides by chemiluminescence detection in the liquid phase," *TrAC Trends in Analytical Chemistry*, vol. 24, pp. 927–942, 2005.

[624] M. Strianese, M. Staiano, G. Ruggiero, T. Labella, C. Pellecchia, and S. D'Auria, "Fluorescence-based biosensors," in *Spectroscopic Methods of Analysis*. Springer, 2012, pp. 193–216.

[625] M. C. Morris, "Fluorescence-based biosensors from concepts to applications. Introduction," *Progress in Molecular Biology and Translational Science*, vol. 113, pp. xv–xviii, 2013.

[626] H. Mader, X. Li, S. Saleh, M. Link, P. Kele, and O. Wolfbeis, "Fluorescence methods and applications: Spectroscopy, imaging, and probes," *Annals of the New York Academy of Sciences*, vol. 1130, p. 218, 2008.

[627] L. Hong, S. McManus, H. Yang, and K. Sengupta, "A fully integrated CMOS fluorescence biosensor with on-chip nanophotonic filter," in *2015 Symposium on VLSI Circuits (VLSI Circuits)*, 2015, pp. C206–C207.

[628] W. L. Barnes, A. Dereux, and T. W. Ebbesen, "Surface plasmon subwavelength optics," *Nature*, vol. 424, no. 6950, pp. 824–830, 2003.

[629] L. Hong and K. Sengupta, "Fully integrated optical spectrometer in visible and near-IR in CMOS," *IEEE Transactions on Biomedical Circuits and Systems*, vol. 11, pp. 1176–1191, 2017.

[630] K. Sengupta, L. Hong, C. Zhu, and X. Lu, "Visible and near-IR nano-optical components and systems in CMOS," *IEEE Open Journal of the Solid-State Circuits Society*, vol. 1, pp. 247–262, 2021.

[631] L. Hong, H. Li, H. Yang, and K. Sengupta, "Nano-plasmonics and electronics co-integration in CMOS enabling a pill-sized multiplexed fluorescence microarray system," *Biomedical Optics Express*, vol. 9, pp. 5735–5758, 2018.

[632] Y. Jiang, H. Yu, X. Fu, *et al.*, "A nano-filter-integrated CMOS image sensor for fluorescent biomedical imaging," in *2018 IEEE Biomedical Circuits and Systems Conference (BioCAS)*, 2018: IEEE, pp. 1–4.

[633] S. Tang and D. Saeedkia, *Advances in Imaging and Sensing*. CRC Press, 2016.

[634] A. DeHennis, S. Getzlaff, D. Grice, and M. Mailand, *"An NFC-enabled CMOS IC for a wireless fully implantable glucose sensor," IEEE Journal of Biomedical and Health Informatics*, vol. 20, pp. 18–28, 2015.

[635] D. Tyndall, B. Rae, D. Li, J. Richardson, J. Arlt, and R. Henderson, "A 100Mphoton/s time-resolved mini-silicon photomultiplier with on-chip fluorescence lifetime estimation in 0.13 μm CMOS imaging technology," in *2012 IEEE International Solid-State Circuits Conference*, 2012, pp. 122–124.

[636] D. H. Ta-chien, S. Paul, P. Gong, *et al.*, "Gene expression analysis with an integrated CMOS microarray by time-resolved fluorescence detection," *Biosensors and Bioelectronics*, vol. 26, pp. 2660–2665, 2011.

[637] H.-J. Yoon, S. Itoh, and S. Kawahito, "A CMOS image sensor with in-pixel two-stage charge transfer for fluorescence lifetime imaging," *IEEE Transactions on Electron Devices*, vol. 56, pp. 214–221, 2009.

[638] D. H. Ta-chien, S. Sorgenfrei, P. Gong, R. Levicky, and K. L. Shepard, "A 0.18-μm CMOS array sensor for integrated time-resolved fluorescence detection," *IEEE Journal of Solid-State Circuits*, vol. 44, pp. 1644–1654, 2009.

[639] D. Ying and D. A. Hall, "Current sensing front-ends: A review and design guidance," *IEEE Sensors Journal*, vol. 21, no. 20, pp. 22329–22346, 2021.

[640] M. A. Awan, B. Wang, N. A. Quadir, and A. Bermak, "Review and analysis of CMOS current readout circuits for biosensing applications," in *2021 IEEE International Symposium on Circuits and Systems (ISCAS)*, 2021: IEEE, pp. 1–5.

[641] J. Musayev, C. Altiner, Y. Adiguzel, H. Kulah, S. Eminoglu, and T. Akin, "Capturing and detection of MCF-7 breast cancer cells with a CMOS

image sensor," *Sensors and Actuators A: Physical*, vol. 215, pp. 105–114, 2014.

[642] T. Tokuda, T. Noda, K. Sasagawa, and J. Ohta, "Optical and electric multifunctional CMOS image sensors for on-chip biosensing applications," *Materials*, vol. 4, no. 1, pp. 84–102, 2010.

[643] S. Parikh, G. Gulak, and P. Chow, "A CMOS image sensor for DNA microarrays," in *2007 IEEE Custom Integrated Circuits Conference*, 2007: IEEE, pp. 821–824.

[644] T. Tokuda, A. Yamamoto, K. Kagawa, M. Nunoshita, and J. Ohta, "A CMOS image sensor with optical and potential dual imaging function for on-chip bioscientific applications," *Sensors and Actuators A: Physical*, vol. 125, no. 2, pp. 273–280, 2006.

[645] T. Tokuda, D. C. Ng, A. Yamamoto, K. Kagawa, M. Nunoshita, and J. Ohta, "A CMOS optical/potential image sensor with 7.5 μm pixel size for on-chip neural and DNA spot sensing," in *2005 IEEE Engineering in Medicine and Biology 27th Annual Conference*, 2006: IEEE, pp. 7269–7272.

[646] Y. Maruyama, K. Sawada, H. Takao, and M. Ishida, "The fabrication of filter-less fluorescence detection sensor array using CMOS image sensor technique," *Sensors and Actuators A: Physical*, vol. 128, no. 1, pp. 66–70, 2006.

[647] A. El Gamal and H. Eltoukhy, "CMOS image sensors," *IEEE Circuits and Devices Magazine*, vol. 21, no. 3, pp. 6–20, 2005.

[648] A. Hofmann, M. Meister, A. Rolapp, P. Reich, F. Scholz, and E. Schäfer, "Light absorption measurement with a CMOS biochip for quantitative immunoassay based point-of-care applications," *IEEE Transactions on Biomedical Circuits and Systems*, vol. 15, pp. 369–379, 2021.

[649] B. Razavi, "The switched-capacitor integrator [a circuit for all seasons]," *IEEE Solid-State Circuits Magazine*, vol. 9, no. 1, pp. 9–11, 2017.

[650] A. Trabelsi and M. Boukadoum, "Comparison of two CMOS front-end transimpedance amplifiers for optical biosensors," *IEEE Sensors Journal*, vol. 13, no. 2, pp. 657–663, 2012.

[651] C. Zhu, L. Hong, H. Yang, and K. Sengupta, "Ingestible Bioelectronics: A packaged, bio-molecular, fluorescence-based sensor array with ultra-low-power wireless interface," in *2019 IEEE MTT-S International Microwave Symposium (IMS)*, 2019: IEEE, pp. 212–215.

[652] L. G. McIlrath, "A low-power low-noise ultrawide-dynamic-range CMOS imager with pixel-parallel A/D conversion," *IEEE Journal of Solid-State Circuits*, vol. 36, no. 5, pp. 846–853, 2001.

[653] A. Hassibi, "CMOS biochips for point-of-care molecular diagnostics," in *2014 IEEE Hot Chips 26 Symposium (HCS)*, 2014, pp. 1–32.

[654] K. Kriz, J. Gehrke, and D. Kriz, "Advancements toward magneto immunoassays," *Biosensors and Bioelectronics*, vol. 13, pp. 817–823, 1998.

[655] E. M. Materón, C. M. Miyazaki, O. Carr, *et al.*, "Magnetic nanoparticles in biomedical applications: A review," *Applied Surface Science Advances*, vol. 6, p. 100163, 2021.

[656] K. Wu, D. Su, R. Saha, J. Liu, V. K. Chugh, and J.-P. Wang, "Magnetic particle spectroscopy: a short review of applications using magnetic nanoparticles," *ACS Applied Nano Materials*, vol. 3, pp. 4972–4989, 2020.

[657] M. Pastucha, Z. Farka, K. Lacina, Z. Mikušová, and P. Skládal, "Magnetic nanoparticles for smart electrochemical immunoassays: a review on recent developments," *Microchimica Acta*, vol. 186, pp. 1–26, 2019.

[658] K. Skucha, S. Gambini, P. Liu, M. Megens, J. Kim, and B. E. Boser, "Design considerations for CMOS-integrated Hall-effect magnetic bead detectors for biosensor applications," *Journal of microelectromechanical systems*, vol. 22, pp. 1327–1338, 2013.

[659] P. P. Liu, K. Skucha, Y. Duan, *et al.*, "Magnetic relaxation detector for microbead labels," *IEEE Journal of Solid-State Circuits*, vol. 47, pp. 1056–1064, 2012.

[660] S. Gambini, K. Skucha, P. P. Liu, J. Kim, and R. Krigel, "A 10 kPixel CMOS Hall sensor array with baseline suppression and parallel readout for immunoassays," *IEEE Journal of Solid-State Circuits*, vol. 48, pp. 302–317, 2012.

[661] S. Gambini, K. Skucha, P. Liu, J. Kim, R. Krigel, R. Mathies, *et al.*, "A CMOS 10kpixel baseline-free magnetic bead detector with column-parallel readout for miniaturized immunoassays," in *2012 IEEE International Solid-State Circuits Conference*, 2012, pp. 126–128.

[662] P. Liu, K. Skucha, M. Megens, and B. Boser, "A CMOS Hall-effect sensor for the characterization and detection of magnetic nanoparticles for biomedical applications," *IEEE Transactions on Magnetics*, vol. 47, pp. 3449–3451, 2011.

[663] O. Florescu, M. Mattmann, and B. Boser, "Fully integrated detection of single magnetic beads in complementary metal-oxide-semiconductor." *Journal of Applied Physics*, vol. 103, no. 4, 046101, 2008.

[664] T. Aytur, J. Foley, M. Anwar, B. Boser, E. Harris, and P. R. Beatty, "A novel magnetic bead bioassay platform using a microchip-based sensor for infectious disease diagnosis," *Journal of Immunological Methods*, vol. 314, pp. 21–29, 2006.

[665] P.-A. Besse, G. Boero, M. Demierre, V. Pott, and R. Popovic, "Detection of a single magnetic microbead using a miniaturized silicon Hall sensor," *Applied Physics Letters*, vol. 80, pp. 4199–4201, 2002.

[666] C. Sideris, P. P. Khial, B. Ling, and A. Hajimiri, "A 0.3 ppm dual-resonance transformer-based drift-cancelling reference-free magnetic sensor for biosensing applications," in *2018 IEEE International Solid-State Circuits Conference-(ISSCC), 2018*, pp. 190–192.

[667] J. Geng, L. Zhang, C. Zhu, L. Dai, X. Shi, and H. Qian, "An oscillator-based CMOS magnetosensitive microarray biochip with on-chip inductor optimization methodology," *IEEE Transactions on Microwave Theory and Techniques*, vol. 66, pp. 2556–2569, 2018.

[668] P. Murali, A. M. Niknejad, and B. E. Boser, "CMOS microflow cytometer for magnetic label detection and classification," *IEEE Journal of Solid-State Circuits*, vol. 52, pp. 543–555, 2016.

[669] P. Murali, I. Izyumin, S. Prabhu, D. Cohen, and B. Boser. "A magnetic flow cytometer with integrated microfluidics," *in Solid-State Sensors, Actuators and Microsystems Workshop*, 2014: Transducer Research Foundation, pp. 159–162.

[670] P. Murali, I. Izyumin, D. Cohen, J.-C. Chien, A. M. Niknejad, and B. Boser. "24.6 A CMOS micro-flow cytometer for magnetic label detection and classification," in 2014 *IEEE International Solid-State Circuits Conference Digest of Technical Papers (ISSCC)*, 2014: IEEE, pp. 422–423.

[671] C. Sideris and A. Hajimiri, "Design and implementation of an integrated magnetic spectrometer for multiplexed biosensing," *IEEE Transactions on Biomedical Circuits and Systems*, vol. 7, pp. 773–784, 2013.

[672] C. Sideris and A. Hajimiri, "An integrated magnetic spectrometer for multiplexed biosensing," in *2013 IEEE International Solid-State Circuits Conference Digest of Technical Papers*, 2013, pp. 300–301.

[673] H. Wang, A. Mahdavi, D. A. Tirrell, and A. Hajimiri, "A magnetic cell-based sensor," *Lab on a Chip*, vol. 12, pp. 4465–4471, 2012.

[674] H. Wang, C. Sideris, and A. Hajimiri, "A frequency-shift based CMOS magnetic biosensor with spatially uniform sensor transducer gain," *in IEEE Custom Integrated Circuits Conference 2010*, 2010, pp. 1–4.

[675] H. Wang, S. Kosai, C. Sideris, and A. Hajimiri, "An ultrasensitive CMOS magnetic biosensor array with correlated double counting noise suppression," in *2010 IEEE MTT-S International Microwave Symposium*, 2010, pp. 616–619.

[676] H. Wang, Y. Chen, A. Hassibi, A. Scherer, and A. Hajimiri, "A frequency-shift CMOS magnetic biosensor array with single-bead sensitivity and no external magnet," in *2009 IEEE International Solid-State Circuits Conference-Digest of Technical Papers, 2009*, pp. 438–439.

[677] X. Zhou, E. Mai, M. Sveiven, *et al.*, "A 9.7-nTrms, 704-ms magnetic biosensor front-end for detecting magneto-relaxation," *IEEE Journal of Solid-State Circuits*, vol. 56, pp. 2171–2181, 2021.

[678] X. Zhou, *Magnetoresistive Biosensor Circuits and Systems for Ultrasensitive Point-of-Care Diagnostics*. University of California, San Diego, 2020.

[679] X. Zhou, M. Sveiven, and D. A. Hall, "A CMOS magnetoresistive sensor front-end with mismatch-tolerance and sub-ppm sensitivity for magnetic immunoassays," *IEEE Transactions on Biomedical Circuits and Systems*, vol. 13, pp. 1254–1263, 2019.

[680] X. Zhou, M. Sveiven, and D. A. Hall, "11.4 a fast-readout mismatch-insensitive magnetoresistive biosensor front-end achieving sub-ppm sensitivity," in *2019 IEEE International Solid-State Circuits Conference-(ISSCC)*, 2019, pp. 196–198.

[681] C.-C. Huang, X. Zhou, and D. A. Hall, "Giant magnetoresistive biosensors for time-domain magnetorelaxometry: A theoretical investigation and progress toward an immunoassay," *Scientific Reports*, vol. 7, pp. 1–10, 2017.

[682] T. Costa, F. A. Cardoso, J. Germano, P. P. Freitas, and M. S. Piedade, "A CMOS front-end with integrated magnetoresistive sensors for biomolecular recognition detection applications," *IEEE Transactions on Biomedical Circuits and Systems*, vol. 11, pp. 988–1000, 2017.

[683] M.-D. Cubells-Beltrán, C. Reig, A. De Marcellis, *et al.*, "Monolithic integration of Giant Magnetoresistance (GMR) devices onto standard processed CMOS dies," *Microelectronics Journal*, vol. 45, pp. 702–707, 2014.

[684] D. A. Hall, R. S. Gaster, K. A. A. Makinwa, S. X. Wang, and B. Murmann, "A 256 pixel magnetoresistive biosensor microarray in 0.18 μm CMOS," *IEEE Journal of Solid-State Circuits*, vol. 48, pp. 1290–1301, 2013.

[685] D. A. Hall, R. S. Gaster, S. J. Osterfeld, K. Makinwa, S. X. Wang, and B. Murmann, "A 256 channel magnetoresistive biosensor microarray for quantitative proteomics," in *2011 Symposium on VLSI Circuits-Digest of Technical Papers*, 2011, pp. 174–175.

[686] R. S. Gaster, L. Xu, S.-J. Han, *et al.*, "Quantification of protein interactions and solution transport using high-density GMR sensor arrays," *Nature Nanotechnology*, vol. 6, pp. 314–320, 2011.

[687] D. A. Hall, R. S. Gaster, T. Lin, *et al.*, "GMR biosensor arrays: A system perspective," *Biosensors and Bioelectronics*, vol. 25, pp. 2051–2057, 2010.

[688] S.-J. Han, H. Yu, B. Murmann, N. Pourmand, and S. X. Wang, "A high-density magnetoresistive biosensor array with drift-compensation mechanism," in *2007 IEEE International Solid-State Circuits Conference. Digest of Technical Papers*, 2007, pp. 168–594.

[689] B. M. De Boer, J. Kahlman, T. Jansen, H. Duric, and J. Veen, "An integrated and sensitive detection platform for magneto-resistive biosensors," *Biosensors and Bioelectronics*, vol. 22, pp. 2366–2370, 2007.

[690] S.-J. Han, L. Xu, H. Yu, R. J. Wilson, R. L. White, N. Pourmand, *et al.*, "CMOS integrated DNA microarray based on GMR sensors," in *2006 International Electron Devices Meeting*, 2006, pp. 1–4.

[691] B. Bao, Y. Hua, R. Wang, and D. Li, "Quantum-based magnetic field sensors for biosensing," *Advanced Quantum Technologies*, p. 2200146, 2023.

[692] H. Weinstock, *SQUID Sensors: Fundamentals, Fabrication, and Applications*. Springer Science & Business Media, 1996.

[693] J. Li, H. Liu, and T. Bi, "Tunnel magnetoresistance-based noncontact current sensing and measurement method," *IEEE Transactions on Instrumentation and Measurement*, vol. 71, pp. 1–9, 2022.

[694] A. Mohamed, M. Wagner, H. Heidari, and J. Anders, "A frontend for magnetoresistive sensors with a 2.2-pA/$\sqrt{}$ hz low-noise current source," *IEEE Solid-State Circuits Letters*, vol. 5, pp. 17–20, 2022.

[695] X. Wang, W. Li, L. Jin, *et al.*, "High-precision micro-displacement sensor based on tunnel magneto-resistance effect," *Scientific Reports*, vol. 12, pp. 1–10, 2022.

[696] W. Zhao, X. Tao, C. Ye, and Y. Tao, "Tunnel magnetoresistance sensor with AC modulation and impedance compensation for ultra-weak magnetic field measurement," *Sensors*, vol. 22, p. 1021, 2022.

[697] S. Andreev and P. Dimitrova, "Anisotropic-magnetoresistance integrated sensors," *Journal of Optoelectronics and Advanced Materials*, vol. 7, no. 1, pp. 199–206, 2005.

[698] I. Georgakopoulos, N. Hadjigeorgiou, and P. P. Sotiriadis, "A CMOS closed loop AMR sensor architecture," in *2017 Panhellenic Conference on Electronics and Telecommunications (PACET)*, 2017, pp. 1–4.

[699] C. Wang, W. Su, Z. Hu, *et al.*, "Highly sensitive magnetic sensor based on anisotropic magnetoresistance effect," *IEEE Transactions on Magnetics*, vol. 54, pp. 1–3, 2018.

[700] C. Han, M. Xu, J. Tang, Y. Liu, and Z. Zhou, "Giant magneto-impedance sensor with working point self adaptation for unshielded human bio-magnetic detection," *Virtual Reality & Intelligent Hardware*, vol. 4, no. 1, pp. 38–54, 2022.

[701] M. Hauser, L. Kraus, and P. Ripka, *"Giant magnetoimpedance sensors," IEEE Instrumentation & Measurement Magazine*, vol. 4, pp. 28–32, 2001.

[702] V. O. Jimenez, K. Y. Hwang, D. Nguyen, *et al.*, "Magnetoimpedance biosensors and real-time healthcare monitors: progress, opportunities, and challenges," *Biosensors*, vol. 12, p. 517, 2022.

[703] V. V. Kondalkar, X. Li, I. Park, S. S. Yang, and K. Lee, "Development of chipless, wireless current sensor system based on giant magnetoimpedance magnetic sensor and surface acoustic wave transponder," *Scientific Reports*, vol. 8, pp. 1–11, 2018.

[704] D. Su, K. Wu, K. Srinivasan, *et al.*, "Ultra-flexible giant magnetoresistance biosensors for lab-on-a-needle biosensing," *Advanced Materials Interfaces*, p. 2201417, 2023.

[705] J. Lenz and S. Edelstein, "Magnetic sensors and their applications," *IEEE Sensors Journal*, vol. 6, pp. 631–649, 2006.

[706] Morrish AH. *The Physical Principles of Magnetism*. New York: IEEE Press; 2001.

[707] N. Ida, *Engineering Electromagnetics*. Springer, 2015.

[708] E. P. Wohlfarth and K. H. J. Buschow, *Ferromagnetic Materials*. Volume 4, Elsevier, Amsterdam, 1989.

[709] W. P. Wolf, "Ferrimagnetism," *Reports on Progress in Physics*, vol. 24, p. 212, 1961.

[710] J. B. Haun, T. J. Yoon, H. Lee, and R. Weissleder, "Magnetic nanoparticle biosensors," *Wiley Interdisciplinary Reviews: Nanomedicine and Nanobiotechnology*, vol. 2, pp. 291–304, 2010.

[711] W. Marshall, "Antiferromagnetism," *Proceedings of the Royal Society of London. Series A. Mathematical and Physical Sciences*, vol. 232, no. 1188, pp. 48–68, 1955.

[712] A. Akbarzadeh, M. Samiei, and S. Davaran, "Magnetic nanoparticles: preparation, physical properties, and applications in biomedicine," *Nanoscale Research Letters*, vol. 7, pp. 1–13, 2012.

[713] K. Skucha, P. Liu, M. Megens, J. Kim, and B. Boser, "A compact Hall-effect sensor array for the detection and imaging of single magnetic beads in biomedical assays," in *2011 16th International Solid-State Sensors, Actuators and Microsystems Conference*, 2011, pp. 1833–1836.

[714] G. Li, V. Joshi, R. L. White, *et al.*, "Detection of single micron-sized magnetic bead and magnetic nanoparticles using spin valve sensors for biological applications," *Journal of Applied Physics*, vol. 93, pp. 7557–7559, 2003.

[715] D. Issadore, Y. I. Park, H. Shao, *et al.*, "Magnetic sensing technology for molecular analyses," *Lab on a Chip*, vol. 14, pp. 2385–2397, 2014.

[716] H. Lee, T.-J. Yoon, J.-L. Figueiredo, F. K. Swirski, and R. Weissleder, "Rapid detection and profiling of cancer cells in fine-needle aspirates," *Proceedings of the National Academy of Sciences*, vol. 106, no. 30, pp. 12459–12464, 2009.

[717] H. Lee, T. J. Yoon, and R. Weissleder, "Ultrasensitive detection of bacteria using Core–Shell nanoparticles and an NMR-Filter system," *Angewandte Chemie*, vol. 121, no. 31, pp. 5767–5770, 2009.

[718] M. Liong, C. Tassa, S. Y. Shaw, H. Lee, and R. Weissleder, "Multiplexed magnetic labeling amplification using oligonucleotide hybridization," *Advanced Materials*, vol. 23, pp. H254–H257, 2011.

[719] V. M. Peterson, C. M. Castro, H. Lee, and R. Weissleder, "Orthogonal amplification of nanoparticles for improved diagnostic sensing," *ACS Nano*, vol. 6, pp. 3506–3513, 2012.

[720] R. S. Gaster, D. A. Hall, and S. X. Wang, "Autoassembly protein arrays for analyzing antibody cross-reactivity," *Nano Letters*, vol. 11, pp. 2579–2583, 2011.

[721] P. C. Weber, D. H. Ohlendorf, J. J. Wendoloski, and F. R. Salemme, "Structural origins of high-affinity biotin binding to streptavidin," *Science*, vol. 243, pp. 85–88, 1989.

[722] D. A. Hall, "GMR spin-valve biochips and interface electronics for ultrasensitive in-vitro diagnostics," Stanford University, 2011.

[723] E. H. Hall, "On a new action of the magnet on electric currents," *American Journal of Mathematics*, vol. 2, pp. 287–292, 1879.

[724] R. Popović, *Hall Effect Devices: Magnetic Sensors and Characterization of Semiconductors*. Bristol: Adam Hilger, 1991.

[725] P. Liu, *A CMOS Magnetic Sensor Chip for Biomedical Applications*. University of California, Berkeley, 2012.

[726] H. Wang, "Magnetic sensors for diagnostic medicine: CMOS-based magnetic particle detectors for medical diagnosis applications," *IEEE Microwave Magazine*, vol. 14, pp. 110–130, 2013.

[727] R. Kötitz, W. Weitschies, L. Trahms, W. Brewer, and W. Semmler, "Determination of the binding reaction between avidin and biotin by relaxation measurements of magnetic nanoparticles," *Journal of Magnetism and Magnetic Materials*, vol. 194, pp. 62–68, 1999.

[728] S. Gambini, K. Skucha, P. P. Liu, J. Kim, and R. Krigel, "A 10 kPixel CMOS hall sensor array with baseline suppression and parallel readout for immunoassays," *IEEE Journal of Solid-State Circuits*, vol. 48, no. 1, pp. 302–317, 2012.

[729] A. Pai, A. Khachaturian, S. Chapman, A. Hu, H. Wang, and A. Hajimiri, "A handheld magnetic sensing platform for antigen and nucleic acid detection," *Analyst*, vol. 139, pp. 1403–1411, 2014.

[730] H. Wang, Y. Chen, A. Hassibi, A. Scherer, and A. Hajimiri, "A frequency-shift CMOS magnetic biosensor array with single-bead sensitivity and no external magnet," in *IEEE International Solid-State Circuits Conference-2009 (ISSCC 2009), Digest of Technical Papers*, 2009: IEEE, pp. 438–439.

[731] H. Wang and A. Hajimiri, "Design of a frequency-shift-based CMOS magnetic sensor array for point-of-care (PoC) biomolecular diagnosis applications," in *Integrated Microsystems: Electronics, Photonics, and Biotechnology (Devices, Circuits, and Systems)*. CRC Press, 2011, pp. 109–131.

[732] J. El-Ali, P. K. Sorger, and K. F. Jensen, "Cells on chips," *Nature*, vol. 442, pp. 403–411, 2006.

[733] G. T. A. Kovacs, "Electronic sensors with living cellular components," *Proceedings of the IEEE*, vol. 91, pp. 915–929, 2003.

[734] P. C. Fannin, L. Cohen-Tannoudji, E. Bertrand, A. T. Giannitsis, C. Mac Oireachtaigh, and J. Bibette, "Investigation of the complex susceptibility of magnetic beads containing maghemite nanoparticles," *Journal of Magnetism and Magnetic Materials*, vol. 303, pp. 147–152, 2006.

[735] A. Pai, A. Khachaturian, S. Chapman, A. Hu, H. Wang, and A. Hajimiri, "A handheld magnetic sensing platform for antigen and nucleic acid detection," *Analyst*, vol. 139, no. 6, pp. 1403–1411, 2014.

[736] M. Brown and C. Wittwer, "Flow cytometry: principles and clinical applications in hematology," *Clinical Chemistry*, vol. 46, pp. 1221–1229, 2000.

[737] P. C. Fannin, T. Relihan, and S. W. Charles, "Experimental and theoretical profiles of the frequency-dependent complex susceptibility of systems containing nanometer-sized magnetic particles," *Physical Review B*, vol. 55, p. 14423, 1997.

[738] C. Sideris, P. P. Khial, and A. Hajimiri, "Design and implementation of reference-free drift-cancelling CMOS magnetic sensors for biosensing applications," *IEEE Journal of Solid-State Circuits*, vol. 53, no. 11, pp. 3065–3075, 2018.

[739] M. Gal-Katziri and A. Hajimiri, "A coupled inductive bridge for magnetic sensing applications," *in ESSCIRC 2018-IEEE 44th European Solid State Circuits Conference (ESSCIRC)*, 2018, pp. 342–345.

[740] M. Gal-Katziri and A. Hajimiri, "Analysis and design of coupled inductive bridges for magnetic sensing applications," *IEEE Journal of Solid-State Circuits*, vol. 54, pp. 1883–1894, 2019.

[741] M. N. Baibich, J. M. Broto, A. Fert, *et al.*, "Giant magnetoresistance of (001) Fe/(001) Cr magnetic superlattices," *Physical Review Letters*, vol. 61, p. 2472, 1988.

[742] G. Binasch, P. Grünberg, F. Saurenbach, and W. Zinn, "Enhanced magnetoresistance in layered magnetic structures with antiferromagnetic interlayer exchange," *Physical Review B*, vol. 39, p. 4828, 1989.

[743] E. E. Fullerton and I. K. Schuller, "The 2007 Nobel Prize in physics: Magnetism and transport at the nanoscale," *ACS Nano*, vol. 1, pp. 384–389, 2007.

[744] S. Adem, S. Jain, M. Sveiven, X. Zhou, A. J. o'Donoghue, and D. A. Hall, "Giant magnetoresistive biosensors for real-time quantitative detection of protease activity," *Scientific Reports*, vol. 10, pp. 1–10, 2020.

[745] D. A. Hall, R. S. Gaster, S. J. Osterfeld, B. Murmann, and S. X. Wang, "GMR biosensor arrays: Correction techniques for reproducibility and enhanced sensitivity," *Biosensors and Bioelectronics*, vol. 25, pp. 2177–2181, 2010.

[746] X. Zhou, C.-C. Huang, and D. A. Hall, "Giant magnetoresistive biosensor array for detecting magnetorelaxation," *IEEE Transactions on Biomedical Circuits and Systems*, vol. 11, pp. 755–764, 2017.

[747] F. A. Cardoso, T. Costa, J. Germano, *et al.*, "Integration of magnetoresistive biochips on a CMOS circuit," *IEEE Transactions on Magnetics*, vol. 48, pp. 3784–3787, 2012.

[748] T. Costa, J. Germano, M. S. Piedade, F. A. Cardoso, and P. P. Freitas, "Design and optimization of a CMOS front-end for magnetoresistive sensor based biomolecular recognition detection," in *2016 IEEE International Symposium on Circuits and Systems (ISCAS)*, 2016, pp. 2859–2862.

[749] K. Kim, D. A. Hall, C. Yao, *et al.*, "Magnetoresistive biosensors with on-chip pulsed excitation and magnetic correlated double sampling," *Scientific Reports*, vol. 8, pp. 1–10, 2018.

[750] S.-J. Yun, S.-B. Shin, H.-C. Choi, and S.-G. Lee, "A 1mW current-reuse CMOS differential LC-VCO with low phase noise," in *ISSCC. 2005 IEEE International Digest of Technical Papers. Solid-State Circuits Conference, 2005*, 2005: IEEE, pp. 540–616.

[751] T. Azadmousavi and E. N. Aghdam, "A low power current-reuse LC-VCO with an adaptive body-biasing technique," *AEU-International Journal of Electronics and Communications*, vol. 89, pp. 56–61, 2018.

[752] T. Azadmousavi and E. N. Aghdam, "Adaptive body biasing circuit for reliability and variability compensation of a low power RF amplifier,"

IEEE Transactions on Device and Materials Reliability, vol. 19, no. 1, pp. 226–232, 2019.

[753] I. I. Rabi, J. R. Zacharias, S. Millman, and P. Kusch, "A new method of measuring nuclear magnetic moment," *Physical Review*, vol. 53, no. 4, p. 318, 1938.

[754] H. Lee, E. Sun, D. Ham, and R. Weissleder, "Chip–NMR biosensor for detection and molecular analysis of cells," *Nature Medicine*, vol. 14, pp. 869–874, 2008.

[755] K.-M. Lei, P.-I. Mak, M.-K. Law, and R. P. Martins, "A μNMR CMOS transceiver using a butterfly-coil input for integration with a digital microfluidic device inside a portable magnet," *IEEE Journal of Solid-State Circuits*, vol. 51, pp. 2274–2286, 2016.

[756] A. M. Hartel and A. C. Moore, "Extraction and 1H NMR analysis of fats from convenience foods: A laboratory experiment for organic chemistry," *Journal of Chemical Education*, vol. 91, no. 10, pp. 1702–1705, 2014.

[757] J. D. Roberts, *Nuclear Magnetic Resonance: Applications to Organic Chemistry*. McGraw-Hill Book Company, Inc., 1959.

[758] I. Speciale, A. Notaro, P. Garcia-Vello, *et al.*, "Liquid-state NMR spectroscopy for complex carbohydrate structural analysis: A hitchhiker's guide," *Carbohydrate Polymers*, vol. 277, p. 118885, 2022.

[759] N. Sun, Y. Liu, H. Lee, R. Weissleder, and D. Ham, "CMOS RF biosensor utilizing nuclear magnetic resonance," *IEEE Journal of Solid-State Circuits*, vol. 44, no. 5, pp. 1629–1643, 2009.

[760] N. Sun, T.-J. Yoon, H. Lee, W. Andress, R. Weissleder, and D. Ham, "Palm NMR and 1-chip NMR," *IEEE Journal of Solid-State Circuits*, vol. 46, no. 1, pp. 342–352, 2010.

[761] C. Göbl, T. Madl, B. Simon, and M. Sattler, "NMR approaches for structural analysis of multidomain proteins and complexes in solution," *Progress in Nuclear Magnetic Resonance Spectroscopy*, vol. 80, pp. 26–63, 2014.

[762] K.-M. Lei, P.-I. Mak, M.-K. Law, and R. P. Martins, "A palm-size μNMR relaxometer using a digital microfluidic device and a semiconductor transceiver for chemical/biological diagnosis," *Analyst*, vol. 140, no. 15, pp. 5129–5137, 2015.

[763] K.-M. Lei, P.-I. Mak, M.-K. Law, and R. P. Martins, "A uNMR CMOS transceiver using a butterfly-coil input for integration with a digital microfluidic device inside a portable magnet," *IEEE Journal of Solid-State Circuits*, vol. 51, no. 10, pp. 2274–2286, 2016.

[764] P. Mansfield and A. A. Maudsley, "Medical imaging by NMR," *The British Journal of Radiology*, vol. 50, no. 591, pp. 188–194, 1977.

[765] P. Mansfield and I. Pykett, "Biological and medical imaging by NMR," *Journal of Magnetic Resonance* (1969), vol. 29, no. 2, pp. 355–373, 1978.

[766] D. Funck, M. Sinn, J. R. Fleming, *et al.*, "Discovery of a Ni^{2+}-dependent guanidine hydrolase in bacteria," *Nature*, vol. 603, no. 7901, pp. 515–521, 2022.

[767] H. Pourmodheji, E. Ghafar-Zadeh, and S. Magierowski, "A CMOS differential receiver dedicated to nuclear magnetic resonance applications," *Analog Integrated Circuits and Signal Processing*, vol. 91, pp. 97–109, 2017.

[768] A. Dupré, K.-M. Lei, P.-I. Mak, R. P. Martins, and W. K. Peng, "Micro- and nanofabrication NMR technologies for point-of-care medical applications–a review," *Microelectronic Engineering*, vol. 209, pp. 66–74, 2019.

[769] J. Handwerker, M. Pérez-Rodas, M. Beyerlein, *et al.*, "A CMOS NMR needle for probing brain physiology with high spatial and temporal resolution," *Nature Methods*, vol. 17, no. 1, pp. 64–67, 2020.

[770] F. Dreyer, Q. Yang, D. Krüger, and J. Anders, "A chip-based NMR relaxometry system for point-of-care analysis," in *2022 IEEE Biomedical Circuits and Systems Conference (BioCAS)*, 2022: IEEE, pp. 183–187.

[771] D. Krüger, A. Zhang, B. Aghelnejad, *et al.*, "A portable CMOS-based spin resonance system for high-resolution spectroscopy and imaging," *IEEE Journal of Solid-State Circuits*, vol. 58, no. 7, pp. 1838–1849, 2023.

[772] Y. Liu, N. Sun, H. Lee, R. Weissleder, and D. Ham, "CMOS mini nuclear magnetic resonance system and its application for biomolecular sensing," in *2008 IEEE International Solid-State Circuits Conference-Digest of Technical Papers, 2008*: IEEE, pp. 140–602.

[773] J. M. Perez, L. Josephson, T. O'Loughlin, D. Högemann, and R. Weissleder, "Magnetic relaxation switches capable of sensing molecular interactions," *Nature Biotechnology*, vol. 20, pp. 816–820, 2002.

[774] K.-M. Lei, H. Heidari, P.-I. Mak, M.-K. Law, F. Maloberti, and R. P. Martins, "A handheld high-sensitivity micro-NMR CMOS platform with B-field stabilization for multi-type biological/chemical assays," *IEEE Journal of Solid-State Circuits*, vol. 52, no. 1, pp. 284–297, 2016.

[775] B. Sporrer, L. Wu, L. Bettini, *et al.*, "A fully integrated dual-channel on-coil CMOS receiver for array coils in 1.5–10.5 T MRI," *IEEE Transactions on Biomedical Circuits and Systems*, vol. 11, no. 6, pp. 1245–1255, 2017.

[776] K.-M. Lei, D. Ha, Y.-Q. Song, *et al.*, "Portable NMR with parallelism," *Analytical Chemistry*, vol. 92, no. 2, pp. 2112–2120, 2020.

[777] N. Sahin Solmaz, M. Grisi, A. V. Matheoud, G. Gualco, and G. Boero, "Single-chip dynamic nuclear polarization microsystem," *Analytical Chemistry*, vol. 92, no. 14, pp. 9782–9789, 2020.

[778] J. Borggräfe, J. Victor, H. Rosenbach, *et al.*, "Time-resolved structural analysis of an RNA-cleaving DNA catalyst," *Nature*, vol. 601, no. 7891, pp. 144–149, 2022.

[779] M. V. Gomez, S. Baas, and A. H. Velders, "Multinuclear 1D and 2D NMR with 19F-Photo-CIDNP hyperpolarization in a microfluidic chip with untuned microcoil," *Nature Communications*, vol. 14, p. 3885, 2023.

[780] D. Kruger, A. Zhang, H. Hinton, *et al.*, "A portable CMOS-based MRI system with 67 × 67 × 83 μm 3 image resolution," *in ESSCIRC*

2022-IEEE 48th European Solid State Circuits Conference (ESSCIRC), 2022: IEEE, pp. 225–228.

[781] H. Lee, E. Sun, D. Ham, and R. Weissleder, "Chip–NMR biosensor for detection and molecular analysis of cells," *Nature Medicine*, vol. 14, no. 8, pp. 869–874, 2008.

[782] S. Z. Kiss, N. MacKinnon, and J. G. Korvink, "Microfluidic Overhauser DNP chip for signal-enhanced compact NMR," *Scientific Reports*, vol. 11, no. 1, p. 4671, 2021.

[783] "Minispec", http://www.theminispec.com/ (accessed November 19, 2007).

[784] N. Sun, T.-J. Yoon, H. Lee, *et al.*, "Palm NMR and one-chip NMR," in *2010 IEEE International Solid-State Circuits Conference-(ISSCC)*, 2010: *IEEE*, pp. 488–489.

[785] N. Sun, T. J. Yoon, H. Lee, W. Andress, R. Weissleder, and D. Ham, "Palm NMR and 1-chip NMR," *IEEE Journal of Solid-State Circuits*, vol. 46, pp. 342–352, 2010.

[786] K.-M. Lei, N. Sun, P.-I. Mak, R. P. Martins, and D. Ham, "Micro-NMR on CMOS for biomolecular sensing," *in CMOS Circuits for Biological Sensing and Processing*. Springer, 2018, pp. 101–132.

[787] V. Srinivasan, V. K. Pamula, and R. B. Fair, "An integrated digital microfluidic lab-on-a-chip for clinical diagnostics on human physiological fluids," *Lab on a Chip*, vol. 4, pp. 310–315, 2004.

[788] I. Barbulovic-Nad, H. Yang, P. S. Park, and A. R. Wheeler, "Digital microfluidics for cell-based assays," *Lab on a Chip*, vol. 8, pp. 519–526, 2008.

[789] J. Gao, X. Liu, T. Chen, *et al.*, "An intelligent digital microfluidic system with fuzzy-enhanced feedback for multi-droplet manipulation," *Lab on a Chip*, vol. 13, pp. 443–451, 2013.

[790] F. Lapierre, M. Harnois, Y. Coffinier, R. Boukherroub, and V. Thomy, "Split and flow: Reconfigurable capillary connection for digital microfluidic devices," *Lab on a Chip*, vol. 14, pp. 3589–3593, 2014.

[791] K.-M. Lei, P.-I. Mak, M.-K. Law, and R. P. Martins, "A μNMR CMOS Transceiver using a butterfly-coil input for integration with a digital microfluidic device inside a portable magnet," *IEEE Journal of Solid-State Circuits*, vol. 51, pp. 2274–2286, 2016.

[792] K. R. Metz, M. M. Lam, and A. G. Webb, "Reference deconvolution: a simple and effective method for resolution enhancement in nuclear magnetic resonance spectroscopy," *Concepts in Magnetic Resonance*, vol. 12, pp. 21–42, 2000.

[793] Ē. Kupče and R. Freeman, "Molecular structure from a single NMR sequence (fast-PANACEA)," *Journal of Magnetic Resonance*, vol. 206, pp. 147–153, 2010.

[794] K.-M. Lei, H. Heidari, P.-I. Mak, M.-K. Law, F. Maloberti, and R. P. Martins, "A handheld high-sensitivity micro-NMR CMOS platform with B-field stabilization for multi-type biological/chemical assays," *IEEE Journal of Solid-State Circuits*, vol. 52, pp. 284–297, 2016.

[795] K.-M. Lei, H. Heidari, P.-I. Mak, M.-K. Law, F. Maloberti, and R. P. Martins, "28.1 A handheld 50pM-sensitivity micro-NMR CMOS platform with B-field stabilization for multi-type biological/chemical assays," in *2016 IEEE International Solid-State Circuits Conference (ISSCC)*, 2016: IEEE, pp. 474–475.

[796] S. Z. Kiss, N. MacKinnon, and J. Korvink, "Microfluidic Overhauser DNP chip for signal-enhanced compact NMR," *Scientific Reports*, vol. 11, no. 1, p. 4671, 2021.

[797] M. G. Saunders, "Enhancing the sensitivity of NMR by Dynamic Nuclear Polarisation," University of Birmingham, *PhD dissertation*, 2011.

[798] A. W. Overhauser, "Polarization of nuclei in metals," *Physical Review*, vol. 92, no. 2, p. 411, 1953.

[799] T. R. Carver and C. P. Slichter, "Polarization of nuclear spins in metals," *Physical Review*, vol. 92, no. 1, p. 212, 1953.

[800] J. H. Ardenkjaer-Larsen, "On the present and future of dissolution-DNP," *Journal of Magnetic Resonance*, vol. 264, pp. 3–12, 2016.

[801] D. P. Cistola and M. D. Robinson, "Compact NMR relaxometry of human blood and blood components," *TrAC Trends in Analytical Chemistry*, vol. 83, pp. 53–64, 2016.

[802] S. N. Fricke, J. Pourtabib, J. Madsen, *et al.*, "Estimates of blood plasma water content using portable NMR relaxometry," *Measurement Science and Technology*, vol. 31, no. 3, p. 035701, 2019.

[803] D. Y. Lee, S. Kang, Y. Lee, *et al.*, "PEGylated bilirubin-coated iron oxide nanoparticles as a biosensor for magnetic relaxation switching-based ROS detection in whole blood," *Theranostics*, vol. 10, no. 5, p. 1997, 2020.

[804] D. Ha, J. Paulsen, N. Sun, Y.-Q. Song, and D. Ham, "Scalable NMR spectroscopy with semiconductor chips," *Proceedings of the National Academy of Sciences*, vol. 111, no. 33, pp. 11955–11960, 2014.

[805] J. Anders, G. Chiaramonte, P. SanGiorgio, and G. Boero, "A single-chip array of NMR receivers," *Journal of Magnetic Resonance*, vol. 201, no. 2, pp. 239–249, 2009.

[806] M. Grisi, G. Gualco, and G. Boero, "A broadband single-chip transceiver for multi-nuclear NMR probes," *Review of Scientific Instruments*, vol. 86, no. 4, 2015.

[807] T. Cherifi, N. Abouchi, G.-N. Lu, *et al.*, "A CMOS microcoil-associated preamplifier for NMR spectroscopy," *IEEE Transactions on Circuits and Systems I: Regular Papers*, vol. 52, no. 12, pp. 2576–2583, 2005.

[808] S. Fan, Q. Zhou, K.-M. Lei, P.-I. Mak, and R. P. Martins, "A miniaturized 3-D-MRI scanner featuring an HV-SOI ASIC and achieving a $10 \times 8 \times 8$ mm^3 field of view," *IEEE Journal of Solid-State Circuits*, vol. 58, no. 7, pp. 2028–2039, 2023.

[809] J.-J. Ho, "The design and fabrication of a micro-thermal/pressure-sensor for medical electro-skin application," *Solid-State Electronics*, vol. 46, no. 8, pp. 1205–1209, 2002.

[810] A. Vaz, A. Ubarretxena, I. Zalbide, *et al.*, "Full passive UHF tag with a temperature sensor suitable for human body temperature monitoring," *IEEE Transactions on Circuits and Systems II: Express Briefs*, vol. 57, no. 2, pp. 95–99, 2010.

[811] B. Mimoun, A. Van Der Horst, R. Dekker, D. Van Der Voort, M. Rutten, and F. Van de Vosse, "Thermal flow sensors on flexible substrates for minimally invasive medical instruments," *in Sensors*, 2012 *IEEE*, 2012: IEEE, pp. 1–4.

[812] W. P. Chan, M. Narducci, Y. Gao, *et al.*, "A monolithically integrated pressure/oxygen/temperature sensing SoC for multimodality intracranial neuromonitoring," *IEEE Journal of Solid-State Circuits*, vol. 49, no. 11, pp. 2449–2461, 2014.

[813] N. Van Helleputte, M. Konijnenburg, H. Kim, *et al.*, "18.3 A multi-parameter signal-acquisition SoC for connected personal health applications," in *2014 IEEE International Solid-State Circuits Conference Digest of Technical Papers (ISSCC)*, 2014: IEEE, pp. 314–315.

[814] Y. Luo, K.-H. Teng, Y. Li, W. Mao, C.-H. Heng, and Y. Lian, "A 93μW 11Mbps wireless vital signs monitoring SoC with 3-lead ECG, bio-impedance, and body temperature," in *2017 IEEE Asian Solid-State Circuits Conference (A-SSCC)*, 2017: IEEE, pp. 29–32.

[815] P. Schönle, F. Glaser, T. Burger, G. Rovere, L. Benini, and Q. Huang, "A multi-sensor and parallel processing SoC for miniaturized medical instrumentation," *IEEE Journal of Solid-State Circuits*, vol. 53, no. 7, pp. 2076–2087, 2018.

[816] S. Pan and K. A. Makinwa, "A 6.6-μW Wheatstone-bridge temperature sensor for biomedical applications," *IEEE Solid-State Circuits Letters*, vol. 3, pp. 334–337, 2020.

[817] E. Moisello, P. Malcovati, and E. Bonizzoni, "Thermal sensors for contactless temperature measurements, occupancy detection, and automatic operation of appliances during the Covid-19 pandemic: A review," *Micromachines*, vol. 12, no. 2, p. 148, 2021.

[818] Y.-J. An, D.-H. Jung, K. Ryu, S.-H. Woo, and S.-O. Jung, "An energy-efficient all-digital time-domain-based CMOS temperature sensor for SoC thermal management," *IEEE Transactions on Very Large Scale Integration (VLSI) Systems*, vol. 23, no. 8, pp. 1508–1517, 2014.

[819] Y. Bao and W. Li, "A high-speed temperature sensor with low supply sensitivity for SoC thermal monitoring," *Journal of Circuits, Systems and Computers*, vol. 27, no. 07, p. 1850116, 2018.

[820] C.-Y. Ku and T.-T. Liu, "A voltage-scalable low-power all-digital temperature sensor for on-chip thermal monitoring," *IEEE Transactions on Circuits and Systems II: Express Briefs*, vol. 66, no. 10, pp. 1658–1662, 2019.

[821] O. Bass and J. Shor, "A miniaturized 0.003 *mm 2 PNP-based thermal sensor for dense CPU thermal monitoring,"* IEEE Transactions on *Circuits and Systems I: Regular Papers*, vol. 67, no. 9, pp. 2984–2992, 2020.

[822] J. Shor, "Compact thermal sensors for dense cpu thermal monitoring and regulation: A review," *IEEE Sensors Journal*, vol. 21, no. 11, pp. 12774–12788, 2020.

[823] N. El-Zarif, M. Ali, M. Amer, *et al.*, "Investigation of different integrated temperature monitoring sensors for high-voltage SoC DC-DC converters," in *2022 20th IEEE Interregional NEWCAS Conference (NEWCAS)*, 2022: IEEE, pp. 178–182.

[824] M. K. Law, A. Bermak, and H. C. Luong, "A sub-μ W embedded CMOS temperature sensor for RFID food monitoring application," *IEEE Journal of Solid-State Circuits*, vol. 45, no. 6, pp. 1246–1255, 2010.

[825] J. Virtanen, L. Ukkonen, T. Björninen, L. Sydänheimo, and A. Z. Elsherbeni, "Temperature sensor tag for passive UHF RFID systems," in *2011 IEEE Sensors Applications Symposium*, 2011: IEEE, pp. 312–317.

[826] B. Wang, M.-K. Law, A. Bermak, and H. C. Luong, "A passive RFID tag embedded temperature sensor with improved process spreads immunity for a −30°C to 60°C sensing range," *IEEE Transactions on Circuits and Systems I: Regular Papers*, vol. 61, no. 2, pp. 337–346, 2013.

[827] F. Deng, Y. He, B. Li, *et al.*, "Design of an embedded CMOS temperature sensor for passive RFID tag chips," *Sensors*, vol. 15, no. 5, pp. 11442–11453, 2015.

[828] M. A. Pertijs and J. Huijsing, *Precision Temperature Sensors in CMOS Technology*. Springer Science & Business Media, 2006.

[829] G. C. Meijer, "Thermal sensors based on transistors," *Sensors and Actuators*, vol. 10, no. 1–2, pp. 103–125, 1986.

[830] T. Verster, "pn junction as an ultralinear calculable thermometer," *Electronics Letters*, vol. 4, p. 175, 1968.

[831] R. Dobkin, "Monolithic temperature transducer," in *1974 IEEE International Solid-State Circuits Conference. Digest of Technical Papers*, 1974, vol. 17: IEEE, pp. 126–127.

[832] A. Boomkamp and G. Meijer, "An accurate biomedical temperature transducer with on-chip microcomputer interfacing," in *ESSCIRC'85: 11th European Solid-State Circuits Conference*, 1985: IEEE, pp. 420–423.

[833] G. C. Meijer, G. Wang, and F. Fruett, "Temperature sensors and voltage references implemented in CMOS technology," *IEEE Sensors Journal*, vol. 1, no. 3, pp. 225–234, 2001.

[834] M. Pertijs, A. Niederkorn, X. Ma, B. McKillop, A. Bakker, and J. Huijsing, "A CMOS temperature sensor with a 3σ inaccuracy of ±0.5°C from –50°C to 120°C," *ISSCC Dig. Tech. Papers*, pp. 200–201, 2003.

[835] M. A. Pertijs, G. C. Meijer, and J. H. Huijsing, "Precision temperature measurement using CMOS substrate PNP transistors," *IEEE Sensors Journal*, vol. 4, no. 3, pp. 294–300, 2004.

[836] M. A. Pertijs, K. A. Makinwa, and J. H. Huijsing, "A CMOS smart temperature sensor with a 3σ inaccuracy of ±0.1°C from -55°C to 125°C," *IEEE Journal of Solid-State Circuits*, vol. 40, no. 12, pp. 2805–2815, 2005.

[837] M. A. Pertijs, A. Niederkorn, X. Ma, B. McKillop, A. Bakker, and J. H. Huijsing, "A CMOS smart temperature sensor with a 3σ inaccuracy of ±0.5°C from 50°C to 120°C," *IEEE Journal of Solid-State Circuits*, vol. 40, no. 2, pp. 454–461, 2005.

[838] M. Pertijs, A. Aita, K. Makinwa, and J. Huijsing, "Voltage calibration of smart temperature sensors," *in Sensors*, 2008 *IEEE*, 2008: IEEE, pp. 756–759.

[839] F. Sebastiano, L. J. Breems, K. A. Makinwa, S. Drago, D. M. Leenaerts, and B. Nauta, "A 1.2V 10μW NPN-Based Temperature Sensor in 65nm CMOS With an Inaccuracy of 0.2°C (3σ) From -70°C to 125°C," *IEEE Journal of Solid-State Circuits*, vol. 45, no. 12, pp. 2591–2601, 2010.

[840] K. Souri, Y. Chae, and K. A. Makinwa, "A CMOS temperature sensor with a voltage-calibrated inaccuracy of ± 0.15°C (3σ) from − 55°C to 125°C," *IEEE Journal of Solid-State Circuits*, vol. 48, no. 1, pp. 292–301, 2012.

[841] A. L. Aita, M. A. Pertijs, K. A. Makinwa, J. H. Huijsing, and G. C. Meijer, "Low-power CMOS smart temperature sensor with a batch-calibrated inaccuracy of ±0.25°C (±3σ) from −70°C to 130°C," *IEEE Sensors Journal*, vol. 13, no. 5, pp. 1840–1848, 2013.

[842] A. Heidary, G. Wang, K. Makinwa, and G. Meijer, "12.8 A BJT-based CMOS temperature sensor with a 3.6 pJ· K 2-resolution FoM," in *2014 IEEE International Solid-State Circuits Conference Digest of Technical Papers (ISSCC)*, 2014: IEEE, pp. 224–225.

[843] G. Wang, A. Heidari, K. A. Makinwa, and G. C. Meijer, "An accurate BJT-based CMOS temperature sensor with duty-cycle-modulated output," *IEEE Transactions on Industrial Electronics*, vol. 64, no. 2, pp. 1572–1580, 2016.

[844] B. Yousefzadeh, S. Shalmany, and K. Makinwa, "A BJT-based temperature-to-digital converter with ±60mK (30) inaccuracy from−70°C to 125°C in 160nm CMOS," in *IEEE Symp. VLSI Circuits*, 2016.

[845] B. Yousefzadeh, S. H. Shalmany, and K. A. Makinwa, "A BJT-based temperature-to-digital converter with ±60 mK (3σ) inaccuracy from −55°C to +125°C in 0.16-μm CMOS," *IEEE Journal of Solid-State Circuits*, vol. 52, no. 4, pp. 1044–1052, 2017.

[846] Y. Xu, M.-K. Law, P.-I. Mak, and R. P. Martins, "A curvature compensated BJT-based time-domain temperature sensor with an inaccuracy of ±0.7°C from -40°C to 125°C," in *2019 IEEE International Conference on Electron Devices and Solid-State Circuits (EDSSC)*, 2019: IEEE, pp. 1–3.

[847] Z. Huang, Z. Tang, X.-P. Yu, Z. Shi, L. Lin, and N. N. Tan, "A BJT-based CMOS temperature sensor with duty-cycle-modulated output and ±0.5 C (3σ) inaccuracy from −40°C to 125°C," *IEEE Transactions on Circuits and Systems II: Express Briefs,* vol. 68, no. 8, pp. 2780–2784, 2021.

[848] J.-H. Park, J.-H. Hwang, C. Shin, and S.-J. Kim, "A 3.1-μW BJT-based CMOS temperature-to-frequency converter with untrimmed inaccuracy of ±1°C (3σ) from −40°C to 140°C," in *2021 IEEE Asian Solid-State Circuits Conference (A-SSCC)*, 2021: IEEE, pp. 1–3.

[849] J.-H. Park, J.-H. Hwang, C. Shin, and S.-J. Kim, "A BJT-based temperature-to-frequency converter with ±1°C (3\Σ) inaccuracy from –40°C to 140°C for on-chip thermal monitoring," *IEEE Journal of Solid-State Circuits*, vol. 57, no. 10, pp. 2909–2918, 2022.

[850] B. Wang and M.-K. Law, "Subranging BJT-based CMOS temperature sensor with a ±0.45°C inaccuracy (3σ) from −50°C to 180°C and a resolution-FoM of 7.2 pJ·K^2 at 150°C," *IEEE Journal of Solid-State Circuits*, vol. 57, no. 12, pp. 3693–3703, 2022.

[851] R. Wei, H. Lin, Q. Chen, G. Huang, and W. Hu, "A CMOS temperature sensor with a smart calibrated inaccuracy of ±0.11 (3σ)," *Sensors*, vol. 23, no. 11, p. 5132, 2023.

[852] E. A. Vittoz, "MOS transistors operated in the lateral bipolar mode and their application in CMOS technology," *IEEE Journal of Solid-State Circuits*, vol. 18, no. 3, pp. 273–279, 1983.

[853] K. S. Szajda, C. G. Sodini, and H. F. Bowman, "A low noise, high resolution silicon temperature sensor," *IEEE Journal of Solid-State Circuits*, vol. 31, no. 9, pp. 1308–1313, 1996.

[854] L. Xu, J. H. Huijsing, and K. A. Makinwa, "A 12μW NPN-based temperature sensor with a 18.4 pJ·K^2 FOM in 0.18 μm BCD CMOS," in *2017 7th IEEE International Workshop on Advances in Sensors and Interfaces (IWASI)*, 2017: IEEE, pp. 180–182.

[855] S. Pan and K. A. Makinwa, "Energy-efficient high-resolution resistor-based temperature sensors," *Hybrid ADCs, Smart Sensors for the IoT, and Sub-1V & Advanced Node Analog Circuit Design: Advances in Analog Circuit Design 2017*, pp. 183–200, 2018.

[856] T. Someya, V. Van Hoek, J. Angevare, S. Pan, and K. Makinwa, "A 210 nW NPN-based temperature sensor with an inaccuracy of ±0.15°C (3σ) from −15°C to 85°C utilizing dual-mode frontend," *IEEE Solid-State Circuits Letters*, vol. 5, pp. 272–275, 2022.

[857] P. Krummenacher and H. Oguey, "Smart temperature sensor in CMOS technology," *Sensors and Actuators A: Physical*, vol. 22, no. 1–3, pp. 636–638, 1990.

[858] R. Bianchi, F. V. Dos Santos, J. Karam, B. Courtois, F. Pressecq, and S. Sifflet, "CMOS compatible temperature sensor based on the lateral bipolar transistor for very wide temperature range applications," *Sensors and Actuators A: Physical*, vol. 71, no. 1–2, pp. 3–9, 1998.

[859] R. Bianchi, J. Karam, B. Courtois, R. Nadal, F. Pressecq, and S. Sifflet, "CMOS-compatible temperature sensor with digital output for wide temperature range applications," *Microelectronics Journal*, vol. 31, no. 9–10, pp. 803–810, 2000.

[860] A. Bakker and J. H. Huijsing, "Micropower CMOS temperature sensor with digital output," *IEEE Journal of Solid-State Circuits*, vol. 31, no. 7, pp. 933–937, 1996.

[861] G. C. Meijer, "Concepts for bandgap references and voltage measurement systems," *Analog Circuit Design: Low-Noise, Low-Power, Low-Voltage;*

Mixed-Mode Design with CAD Tools; Voltage, Current and Time References, pp. 243–268, 1996.

[862] F. Fruett, G. Wang, and G. C. Meijer, "The piezojunction effect in NPN and PNP vertical transistors and its influence on silicon temperature sensors," *Sensors and Actuators A: Physical*, vol. 85, no. 1–3, pp. 70–74, 2000.

[863] A. L. Aita and K. A. Makinwa, "Low-power operation of a precision CMOS temperature sensor based on substrate PNPs," *in Sensors*, 2007 *IEEE*, 2007: IEEE, pp. 856–859.

[864] A. L. Aita, M. A. Pertijs, K. A. Makinwa, and J. H. Huijsing, "A CMOS smart temperature sensor with a batch-calibrated inaccuracy of $\pm 0.25°C$ (3σ) from $-70°C$ to 130°C," in *2009 IEEE International Solid-State Circuits Conference-Digest of Technical Papers*, 2009: *IEEE*, pp. 342–343,343 a.

[865] J. Pathrose, C. Liu, K. T. Chai, and Y. P. Xu, "A time-domain band-gap temperature sensor in SOI CMOS for high-temperature applications," *IEEE Transactions on Circuits and Systems II: Express Briefs*, vol. 62, no. 5, pp. 436–440, 2014.

[866] Y.-C. Hsu, C.-L. Tai, M.-C. Chuang, A. Roth, and E. Soenen, "5.9 An 18.75 μW dynamic-distributing-bias temperature sensor with 0.87°C (3σ) untrimmed inaccuracy and 0.00946 mm^2 area," in *2017 IEEE International Solid-State Circuits Conference (ISSCC)*, 2017: IEEE, pp. 102–103.

[867] B. Wang, M.-K. Law, C.-Y. Tsui, and A. Bermak, "A 10.6 pJ· K^2 resolution fom temperature sensor using a stable multivibrator," *IEEE Transactions on Circuits and Systems II: Express Briefs*, vol. 65, no. 7, pp. 869–873, 2017.

[868] Z. Tang, Y. Fang, X. Yu, Z. Shi, and N. Tan, "Capacitor-reused CMOS temperature sensor with duty-cycle-modulated output and 0.38 °C (3σ) inaccuracy," *Electronics Letters*, vol. 54, no. 9, pp. 568–570, 2018.

[869] R. K. Kumar, H. Jiang, and K. A. Makinwa, "An energy-efficient BJT-based temperature-to-digital converter with $\pm 0.13°C$ (3σ) inaccuracy from -40 to 125°C," in *2019 IEEE Asian Solid-State Circuits Conference (A-SSCC)*, 2019: IEEE, pp. 107–108.

[870] S. H. Shalmany, K. Souri, U. Sonmez, *et al.*, "3.7 A 620μW BJT-based temperature-to-digital converter with 0.65 mK resolution and FoM of 190 fJ·K^2," in *2020 IEEE International Solid-State Circuits Conference-(ISSCC), 2020*: IEEE, pp. 70–72.

[871] K. Souri, Y. Chae, Y. Ponomarev, and K. A. Makinwa, "A precision DTMOST-based temperature sensor," in *2011 Proceedings of the ESSCIRC (ESSCIRC)*, 2011: IEEE, pp. 279–282.

[872] Z. Tang, S. Pan, and K. A. Makinwa, "23.5 A sub-1V 810nW capacitively-biased BJT-based temperature sensor with an inaccuracy of $\pm 0.15°C$ (3σ) from $-55°C$ to 125°C," in *2023 IEEE International Solid-State Circuits Conference (ISSCC), 2023*: IEEE, pp. 22–24.

[873] M. Eberlein and H. Pretl, "A no-trim, scaling-friendly thermal sensor in 16nm FinFET using bulk diodes as sensing elements," *IEEE Solid-State Circuits Letters*, vol. 2, no. 9, pp. 63–66, 2019.

[874] Z. Tang, Y. Fang, X.-P. Yu, N. N. Tan, Z. Shi, and P. Harpe, "An energy-efficient capacitively biased diode-based temperature sensor in 55-nm CMOS," *IEEE Solid-State Circuits Letters*, vol. 4, pp. 210–213, 2021.

[875] Q. Chen, M. Meterelliyoz, and K. Roy, "A CMOS thermal sensor and its applications in temperature adaptive design," in *7th International Symposium on Quality Electronic Design (ISQED'06),* 2006: IEEE, pp. 6 pp.-248.

[876] M. Sasaki, M. Ikeda, and K. Asada, "A temperature sensor with an inaccuracy of −1/+0.8°C Using 90-nm 1-V CMOS for online thermal monitoring of VLSI circuits," *IEEE Transactions on Semiconductor Manufacturing*, vol. 21, no. 2, pp. 201–208, 2008.

[877] C. Zhao, "CMOS on-chip temperature sensors for power management," *PhD dissertation*, 2014.

[878] Q. Huang, H. Joo, J. Kim, C. Zhan, and J. Burm, "An energy-efficient frequency-domain CMOS temperature sensor with switched vernier time-to-digital conversion," *IEEE Sensors Journal*, vol. 17, no. 10, pp. 3001–3011, 2017.

[879] S. Jung, T. Ji, and V. K. Varadan, "Temperature sensor using thermal transport properties in the subthreshold regime of an organic thin film transistor," *Applied Physics Letters*, vol. 90, no. 6, p. 062105, 2007.

[880] K. Ueno, T. Asai, and Y. Amemiya, "Temperature-to-frequency converter consisting of subthreshold MOSFET circuits for smart temperature-sensor LSIs," *in TRANSDUCERS 2009–2009 International Solid-State Sensors, Actuators and Microsystems Conference*, 2009: IEEE, pp. 2433–2436.

[881] K. Ueno, T. Asai, and Y. Amemiya, "Low-power temperature-to-frequency converter consisting of subthreshold CMOS circuits for integrated smart temperature sensors," *Sensors and Actuators A: Physical*, vol. 165, no. 1, pp. 132–137, 2011.

[882] L. Lu, B. Vosooghi, J. Chen, and C. Li, "A subthreshold-MOSFETs-based scattered relative temperature sensor front-end with a non-calibrated ±2.5°C (3σ) relative inaccuracy from −40°C to 100°C," *IEEE Transactions on Circuits and Systems I: Regular Papers*, vol. 60, no. 5, pp. 1104–1112, 2013.

[883] J.-S. Na, W. Shin, B.-C. Kwak, S.-K. Hong, and O.-K. Kwon, "A CMOS-based temperature sensor with subthreshold operation for low-voltage and low-power on-chip thermal monitoring," *Journal of Semiconductor Technology and Science*, vol. 17, *no.* 1, pp. 29–34, 2017.

[884] T. Someya, A. M. Islam, T. Sakurai, and M. Takamiya, "A 13nW temperature-to-digital converter utilizing sub-threshold MOSFET operation at sub-thermal drain voltage," in *2018 IEEE Custom Integrated Circuits Conference (CICC)*, 2018: IEEE, pp. 1–4.

[885] A. Wang, B. H. Calhoun, and A. P. Chandrakasan, *Sub-Threshold Design for Ultra Low-Power Systems. Springer*, 2006.

[886] W. Zhao, R. Pan, Y. Ha, and Z. Yang, "A 0.4 V 280-nW frequency reference-less nearly all-digital hybrid domain temperature sensor," in *2014 IEEE Asian Solid-State Circuits Conference (A-SSCC)*, 2014: IEEE, pp. 301–304.

[887] C. Azcona, B. Calvo, N. Medrano, and S. Celma, "1.2 V–0.18μm CMOS temperature sensors with quasi-digital output for portable systems," *IEEE Transactions on Instrumentation and Measurement*, vol. 64, no. 9, pp. 2565–2573, 2015.

[888] M. Bashir and P. Sreehari Rao, "A low power, miniature temperature sensor with one-point calibrated accuracy of ±0.25°C from −55° to 125°C in 65 nm CMOS process," *Analog Integrated Circuits and Signal Processing*, vol. 99, no. 2, pp. 311–323, 2019.

[889] T. Someya, A. M. Islam, T. Sakurai, and M. Takamiya, "An 11-nW CMOS temperature-to-digital converter utilizing sub-threshold current at sub-thermal drain voltage," *IEEE Journal of Solid-State Circuits*, vol. 54, no. 3, pp. 613–622, 2019.

[890] R. Krishna, A. K. Mal, and R. Mahapatra, "Time-domain smart temperature sensor using current starved inverters and switched ring oscillator-based time-to-digital converter," *Circuits, Systems, and Signal Processing*, vol. 39, no. 4, pp. 1751–1769, 2020.

[891] E. Moisello, C. M. Ippolito, G. Bruno, P. Malcovati, and E. Bonizzoni, "A MOS-based temperature sensor with inherent inaccuracy reduction enabled by time-domain operation," *IEEE Transactions on Instrumentation and Measurement*, 2023.

[892] K. Souri, Y. Chae, F. Thus, and K. Makinwa, "12.7 A 0.85 V 600nW all-CMOS temperature sensor with an inaccuracy of ±0.4 °C (3σ) from −40° to 125°C," in *2014 IEEE International Solid-State Circuits Conference Digest of Technical Papers (ISSCC)*, 2014: IEEE, pp. 222–223.

[893] H. Homulle, L. Song, E. Charbon, and F. Sebastiano, "The cryogenic temperature behavior of bipolar, MOS, and DTMOS transistors in standard CMOS," *IEEE Journal of the Electron Devices Society*, vol. 6, pp. 263–270, 2018.

[894] T. Huizinga, M. Babaie, A. Vladimirescu, and F. Sebastiano, "Integrated cryo-CMOS temperature sensors for quantum control ICs," in *2022 IEEE 15th Workshop on Low Temperature Electronics (WOLTE)*, 2022: IEEE, pp. 1–4.

[895] R. Khanpeth, M. Kongpoon, and A. Thanachayanont, "A DTMOS-based temperature sensor with an inaccuracy of ±0.25°C (3σ) from −20°C to 85°C," in *2023 20th International Conference on Electrical Engineering/Electronics, Computer, Telecommunications and Information Technology (ECTI-CON)*, 2023: IEEE, pp. 1–4.

[896] F. Assaderaghi, D. Sinitsky, S. A. Parke, J. Bokor, P. K. Ko, and C. Hu, "Dynamic threshold-voltage MOSFET (DTMOS) for ultra-low voltage

VLSI," *IEEE Transactions on Electron Devices*, vol. 44, no. 3, pp. 414–422, 1997.

[897] M. Terauchi, "Temperature dependence of the subthreshold characteristics of dynamic threshold metal–oxide–semiconductor field-effect transistors and its application to an absolute-temperature sensing scheme for low-voltage operation," *Japanese Journal of Applied Physics*, vol. 46, no. 7R, p. 4102, 2007.

[898] M. Terauchi, A. Hamasaki, and A. Suketa, "Selectable logarithmic/linear response active pixel sensor cell with reduced fixed-pattern-noise based on dynamic threshold MOS operation," *Japanese Journal of Applied Physics*, vol. 44, no. 4S, p. 2347, 2005.

[899] A.-J. Annema, "Low-power bandgap references featuring DTMOSTs," *IEEE Journal of Solid-State Circuits*, vol. 34, no. 7, pp. 949–955, 1999.

[900] Y. Touloukian, R. Powell, C. Ho, and M. Nicolaou, "Thermophysical properties of matter-The TPRC data series. Volume 10. Thermal Diffusivity," *DTIC Document*, 1974.

[901] K. A. Makinwa and M. F. Snoeij, "A CMOS temperature-to-frequency converter with an inaccuracy of less than ±0.5 °C (3σ) from −40 °C to 105 °C," *IEEE Journal of Solid-State Circuits*, vol. 41, no. 12, pp. 2992–2997, 2006.

[902] S. Kashmiri and K. Makinwa, "Measuring the thermal diffusivity of CMOS chips," in *Sensors*, 2009 IEEE, 2009: IEEE, pp. 45–48.

[903] K. A. Makinwa and M. F. Snoeij, "A CMOS temperature-to-frequency converter with an inaccuracy of less than ±0.5°C (3σ) from −40°C to 105°C," *IEEE Journal of Solid-State Circuits*, vol. 41, no. 12, pp. 2992–2997, 2006.

[904] C. P. van Vroonhoven and K. A. Makinwa, "A CMOS temperature-to-digital converter with an inaccuracy of ±0.5°C (3σ) from −55°C to 125°C," in *2008 IEEE International Solid-State Circuits Conference-Digest of Technical Papers, 2008*: IEEE, pp. 576–637.

[905] K. Souri and K. A. Makinwa, *Energy-Efficient Smart Temperature Sensors in CMOS Technology*. Springer, 2018.

[906] C. P. van Vroonhoven, D. d'Aquino, and K. A. Makinwa, "A thermal-diffusivity-based temperature sensor with an untrimmed inaccuracy of ±0.2°C (3σ) from −55°C to 125°C," in *2010 IEEE International Solid-State Circuits Conference-(ISSCC)*, 2010: IEEE, pp. 314–315.

[907] U. Sönmez, F. Sebastiano, and K. A. Makinwa, "11.4 1650μm^2 thermal-diffusivity sensors with inaccuracies down to ±0.75°C in 40nm CMOS," in *2016 IEEE International Solid-State Circuits Conference (ISSCC)*, 2016: IEEE, pp. 206–207.

[908] J. Angevare, L. Pedalà, U. Sönmez, F. Sebastiano, and K. A. Makinwa, "A 2800-μm^2 thermal-diffusivity temperature sensor with VCO-based readout in 160-nm CMOS," in *2015 IEEE Asian Solid-State Circuits Conference (A-SSCC)*, 2015: IEEE, pp. 1–4.

[909] C. P. van Vroonhoven and K. A. Makinwa, "Thermal diffusivity sensing: A new temperature sensing paradigm," in *2011 IEEE Custom Integrated Circuits Conference (CICC)*, 2011: IEEE, pp. 1–6.

[910] R. Quan, U. Sonmez, F. Sebastiano, and K. A. Makinwa, "27.8 A 4600μm^2 1.5°C (3σ) 0.9 kS/s thermal-diffusivity temperature sensor with VCO-based readout," in *2015 IEEE International Solid-State Circuits Conference-(ISSCC) Digest of Technical Papers*, 2015: IEEE, pp. 1–3.

[911] S. Pan, J. A. Angevare, and K. A. Makinwa, "5.4 A hybrid thermal-diffusivity/resistor-based temperature sensor with a self-calibrated inaccuracy of ±0.25°C (3σ) from −55°C to 125°C," in *2021 IEEE International Solid-State Circuits Conference (ISSCC), 2021, vol. 64*: IEEE, pp. 78–80.

[912] S. Pan, Y. Luo, S. H. Shalmany, and K. A. Makinwa, "A resistor-based temperature sensor with a 0.13 pJK^2 resolution FoM," *IEEE Journal of Solid-State Circuits*, vol. 53, no. 1, pp. 164–173, 2017.

[913] S. Pan, J. A. Angevare, and K. A. Makinwa, "A self-calibrated hybrid thermal-diffusivity/resistor-based temperature sensor," *IEEE Journal of Solid-State Circuits*, vol. 56, no. 12, pp. 3551–3559, 2021.

[914] S. Pan, H. Jiang, and K. A. Makinwa, "A CMOS temperature sensor with a 49 fJK^2 resolution FoM," in *2017 Symposium on VLSI Circuits*, 2017: IEEE, pp. C82–C83.

[915] S. Pan and K. A. Makinwa, "A 0.25 mm^2-resistor-based temperature sensor with an inaccuracy of 0.12°C (3σ) from −55°C to 125°C," *IEEE Journal of Solid-State Circuits*, vol. 53, no. 12, pp. 3347–3355, 2018.

[916] H. Park and J. Kim, "A 0.8-V resistor-based temperature sensor in 65-nm CMOS with supply sensitivity of 0.28 C/V," *IEEE Journal of Solid-State Circuits*, vol. 53, no. 3, pp. 906–912, 2018.

[917] H. Xin, M. Andraud, P. Baltus, E. Cantatore, and P. Harpe, "A 174 pW–488.3 nW 1 S/s–100 kS/s all-dynamic resistive temperature sensor with speed/resolution/resistance adaptability," *IEEE Solid-State Circuits Letters*, vol. 1, no. 3, pp. 70–73, 2018.

[918] S. Pan and K. A. Makinwa, "10.4 A Wheatstone bridge temperature sensor with a resolution FoM of 20 fJK^2," in *2019 IEEE International Solid-State Circuits Conference-(ISSCC), 2019*: IEEE, pp. 186–188.

[919] H. Jiang, C.-C. Huang, M. R. Chan, and D. A. Hall, "A 2-in-1 temperature and humidity sensor with a single FLL wheatstone-bridge front-end," *IEEE Journal of Solid-State Circuits*, vol. 55, no. 8, pp. 2174–2185, 2020.

[920] S. Pan and K. A. Makinwa, "A 10 fJ·K^2 Wheatstone bridge temperature sensor with a tail-resistor-linearized OTA," *IEEE Journal of Solid-State Circuits*, vol. 56, no. 2, pp. 501–510, 2020.

[921] M. Shahmohammadi, K. Souri, and K. A. Makinwa, "A resistor-based temperature sensor for MEMS frequency references," in *2013 Proceedings of the ESSCIRC (ESSCIRC)*, 2013: IEEE, pp. 225–228.

[922] P. Park, D. Ruffieux, and K. A. Makinwa, "A thermistor-based temperature sensor for a real-time clock with ± 2 ppm frequency stability," *IEEE Journal of Solid-State Circuits*, vol. 50, no. 7, pp. 1571–1580, 2015.

[923] C.-H. Weng, C.-K. Wu, and T.-H. Lin, "A CMOS thermistor-embedded continuous-time delta-sigma temperature sensor with a resolution FoM of 0.65 pJ°C^2," *IEEE Journal of Solid-State Circuits*, vol. 50, no. 11, pp. 2491–2500, 2015.

[924] J. Angevare and K. A. Makinwa, "A 6800-μm^2 resistor-based temperature sensor in 180-nm CMOS," in *2018 IEEE Asian Solid-State Circuits Conference (A-SSCC)*, 2018: IEEE, pp. 43–46.

[925] S. Pan, Ç. Gürleyük, M. F. Pimenta, and K. A. Makinwa, "10.3 A 0.12 mm^2 Wien-Bridge temperature sensor with 0.1°C (3σ) inaccuracy from −40°C to 180°C," in *2019 IEEE International Solid-State Circuits Conference-(ISSCC), 2019*: IEEE, pp. 184–186.

[926] W. Choi, Y. Lee, S. Kim, *et al.*, "A compact resistor-based CMOS temperature sensor with an inaccuracy of 0.12°C (3σ) and a resolution FoM of 0.43 pJK^2 in 65-nm CMOS," *IEEE Journal of Solid-State Circuits*, vol. 53, no. 12, pp. 3356–3367, 2018.

[927] Y. Lee, W. Choi, T. Kim, S. Song, K. A. Makinwa, and Y. Chae, "A 5800-μm^2 resistor-based temperature sensor with a one-point trimmed inaccuracy of ±1.2°C (3σ) from −50°C to 105°C in 65-nm CMOS," in *ESSCIRC 2019-IEEE 45th European Solid State Circuits Conference (ESSCIRC)*, 2019: IEEE, pp. 68–71.

[928] A. Khashaba, J. Zhu, A. Elmallah, M. Ahmed, and P. K. Hanumolu, "3.2 A 0.0088 mm^2 resistor-based temperature sensor achieving 92fJ·K^2 FoM in 65nm CMOS," in *2020 IEEE International Solid-State Circuits Conference-(ISSCC)*, 2020: IEEE, pp. 60–62.

[929] J. A. Angevare, Y. Chae, and K. A. Makinwa, "5.3 A highly digital 2210μm^2 resistor-based temperature sensor with a 1-point trimmed inaccuracy of ±1.3°C (3σ) from −55°C to 125°C in 65nm CMOS," in *2021 IEEE International Solid-State Circuits Conference (ISSCC), 2021, vol. 64*: IEEE, pp. 76–78.

[930] Y. Lee, T. Kim, and Y. Chae, "A 0.9 V 6,400-μm^2 resistor-based temperature sensor with a one-point trimmed 3σ inaccuracy of 10.64°C from −50°C to 125°C," *IEEE Transactions on Circuits and Systems II: Express Briefs*, 2023.

[931] A. Jain, H. Jiang, C. Pochet, and D. A. Hall, "A 310 nW temperature sensor achieving 9.8 mK resolution using a DFLL-based readout circuit," *IEEE Transactions on Circuits and Systems II: Express Briefs*, vol. 69, no. 3, pp. 704–708, 2021.

[932] T. M. Tritt, *Thermoelectric Materials: Principles, Structure, Properties, and Applications*. Amsterdam, The Netherlands: Elsevier, 2002.

[933] A. Van *Herwaarden, "The Seebeck effect in silicon ICs," Sensors and Actuators*, vol. 6, no. 4, pp. 245–254, 1984.

[934] N. Kerness, A. Koll, A. Schaufelbuhl, *et al.*, "N-well based CMOS calorimetric chemical sensors," in *Proceedings IEEE Thirteenth Annual International Conference on Micro Electro Mechanical Systems (Cat. No. 00CH36308)*, 2000: IEEE, pp. 96–101.

[935] H. Muro, T. Mitamura, and C. J. Schaeffer, "Diaphragm-type IR sensors using a polysilicon thermopile," *IEEE Transactions on Fundamentals and Materials*, vol. 112, no. 12, pp. 968–973, 1992.

[936] D. Moser and H. Baltes, "A high sensitivity CMOS gas flow sensor on a thin dielectric membrane," *Sensors and Actuators A: Physical*, vol. 37, pp. 33–37, 1993.

[937] E. Aldrete-Vidrio, D. Mateo, and J. Altet, "Differential temperature sensors fully compatible with a 0.35-um CMOS process," *IEEE Transactions on Components and Packaging Technologies*, vol. 30, no. 4, pp. 618–626, 2007.

[938] S. M. Sze, "Semicondutor sensors," in *Semicondutor Sensors*, 1994, pp. 550–550.

[939] S. Middelhoek, A. Bellekom, U. Dauderstadt, *et al.*, "Silicon sensors," *Measurement Science and Technology*, vol. 6, no. 12, p. 1641, 1995.

[940] T. Akin, "CMOS-based thermal sensors," *CMOS—MEMS*, pp. 479–512, 2005.

[941] A. Van Herwaarden and P. Sarro, "Thermal sensors based on the Seebeck effect," *Sensors and Actuators*, vol. 10, no. 3–4, pp. 321–346, 1986.

[942] E. Moisello, M. Vaiana, M. E. Castagna, G. Bruno, E. Bonizzoni, and P. Malcovati, "A chopper interface circuit for thermopile-based thermal sensors," in *2019 IEEE International Symposium on Circuits and Systems (ISCAS)*, 2019: IEEE, pp. 1–5.

[943] A. Van Herwaarden, D. Van Duyn, B. Van Oudheusden, and P. Sarro, "Integrated thermopile sensors," *Sensors and Actuators A: Physical*, vol. 22, no. 1–3, pp. 621–630, 1990.

[944] P. Bataillard, E. Steffgen, S. Haemmerli, A. Manz, and H. Widmer, "An integrated silicon thermopile as biosensor for the thermal monitoring of glucose, urea and penicillin," *Biosensors and Bioelectronics*, vol. 8, no. 2, pp. 89–98, 1993.

[945] P. Sarro, A. Van Herwaarden, and W. Van der Vlist, "A silicon-silicon nitride membrane fabrication process for smart thermal sensors," *Sensors and Actuators A: Physical*, vol. 42, no. 1–3, pp. 666–671, 1994.

[946] D. Jaeggi, J. Funk, A. Halberli, and H. Baltes, "Overall system analysis of a CMOS thermal converter," in *Proceedings of the International Solid-State Sensors and Actuators Conference-TRANSDUCERS'95*, 1995, vol. 2: IEEE, pp. 112–115.

[947] A. Koll, A. Schaufelbuhl, N. Scheeberger, *et al.*, "Micromachined CMOS calorimetric chemical sensor with on-chip low noise amplifier," in *Technical Digest. IEEE International MEMS 99 Conference. Twelfth IEEE International Conference on Micro Electro Mechanical Systems (Cat. no. 99CH36291)*, 1999: IEEE, pp. 547–551.

[948] I. Voiculescu, M. Zaghloul, and R. A. McGill, "Design and fabrication of a temperature sensor based on thermopile in CMOS technology," *in ASME International Mechanical Engineering Congress and Exposition*, 2003, vol. 37211, pp. 597–601.

[949] M. v. Arx and H. Baltes, *Thermal Properties of CMOS Thin Films. ETH Zurich*, 1998.

[950] T. Müller, M. Brandl, O. Brand, and H. Baltes, "An industrial CMOS process family adapted for the fabrication of smart silicon sensors," *Sensors and Actuators A: Physical*, vol. 84, no. 1–2, pp. 126–133, 2000.

[951] N. G. Green, H. Morgan, and J. J. Milner, "Manipulation and trapping of sub-micron bioparticles using dielectrophoresis," *Journal of Biochemical and Biophysical Methods*, vol. 35, no. 2, pp. 89–102, 1997.

[952] L. V. Panina, A. Gurevich, A. Beklemisheva, A. Omelyanchik, K. Levada, and V. Rodionova, "Spatial manipulation of particles and cells at micro- and nanoscale via magnetic forces," *Cells*, vol. 11, no. 6, p. 950, 2022.

[953] X.-B. Wang, Y. Huang, P. R. Gascoyne, and F. F. Becker, "Dielectrophoretic manipulation of particles," *IEEE Transactions on Industry Applications*, vol. 33, no. 3, pp. 660–669, 1997.

[954] E. O. Adekanmbi and S. K. Srivastava, "Dielectrophoretic applications for disease diagnostics using lab-on-a-chip platforms," *Lab on a Chip*, vol. 16, no. 12, pp. 2148–2167, 2016.

[955] H. Li, Y. Zheng, D. Akin, and R. Bashir, "Characterization and modeling of a microfluidic dielectrophoresis filter for biological species," *Journal of Microelectromechanical Systems*, vol. 14, no. 1, pp. 103–112, 2005.

[956] G. Medoro, N. Manaresi, A. Leonardi, L. Altomare, M. Tartagni, and R. Guerrieri, "A lab-on-a-chip for cell detection and manipulation," *IEEE Sensors Journal*, vol. 3, no. 3, pp. 317–325, 2003.

[957] E. A. Henslee, M. B. Sano, A. D. Rojas, E. M. Schmelz, and R. V. Davalos, "Selective concentration of human cancer cells using contactless dielectrophoresis," *Electrophoresis*, vol. 32, no. 18, pp. 2523–2529, 2011.

[958] K. Park, S. Kabiri, and S. Sonkusale, "Dielectrophoretic lab-on-CMOS platform for trapping and manipulation of cells," *Biomedical Microdevices*, vol. 18, pp. 1–11, 2016.

[959] J. Suehiro, G. Zhou, M. Imamura, and M. Hara, "Dielectrophoretic filter for separation and recovery of biological cells in water," *IEEE Transactions on Industry Applications*, vol. 39, no. 5, pp. 1514–1521, 2003.

[960] G. J. Fiechtner, E. B. Cummings, and A. K. Singh, "Apparatus and method for concentrating and filtering particles suspended in a fluid," Google Patents, 2009.

[961] G. R. Pesch, M. Lorenz, S. Sachdev, *et al.*, "Bridging the scales in high-throughput dielectrophoretic (bio-) particle separation in porous media," *Scientific Reports*, vol. 8, no. 1, p. 10480, 2018.

[962] M. Kandemir, R. Wagterveld, D. Yntema, and K. Keesman, "Selective particle filtering in a large acoustophoretic serpentine channel," *Scientific Reports*, vol. 9, no. 1, p. 7156, 2019.

[963] H. Matbaechi Ettehad, "Dielectrophoretic manipulation of yeast cells using CMOS integrated microfluidic," *BTU Cottbus-Senftenberg*, 2021.

[964] B. Landenberger, H. Höfemann, S. Wadle, and A. Rohrbach, "Microfluidic sorting of arbitrary cells with dynamic optical tweezers," *Lab on a Chip*, vol. 12, no. 17, pp. 3177–3183, 2012.

[965] K. Heon Lee, K. Soo Lee, J. Ho Jung, C. Bong Chang, and H. Jin Sung, "Optical mobility of blood cells for label-free cell separation applications," *Applied Physics Letters*, vol. 102, no. 14, p. 141911, 2013.

[966] X. Zhang, L. Ma, and Y. Zhang, "High-resolution optical tweezers for single-molecule manipulation," *The Yale Journal of Biology and Medicine*, vol. 86, no. 3, p. 367, 2013.

[967] J. Gieseler, J. R. Gomez-Solano, A. Magazzù, *et al.*, "Optical tweezers—from calibration to applications: a tutorial," *Advances in Optics and Photonics*, vol. 13, no. 1, pp. 74–241, 2021.

[968] H. Jin, J. Ge, J. Yang, and Y. Zhang, "Single-molecule manipulation study of chaperoned SNARE folding and assembly with optical tweezers," in *Optical Tweezers: Methods and Protocols*. Springer, 2022, pp. 461–481.

[969] H. Lee, Y. Liu, E. Alsberg, D. E. Ingber, R. M. Westervelt, and D. Ham, "An IC/microfluidic hybrid microsystem for 2D magnetic manipulation of individual biological cells," in *ISSCC. 2005 IEEE International Digest of Technical Papers. Solid-State Circuits Conference, 2005*, 2005: IEEE, pp. 80–586.

[970] H. Lee, Y. Liu, R. M. Westervelt, and D. Ham, "IC/microfluidic hybrid system for magnetic manipulation of biological cells," *IEEE Journal of Solid-State Circuits*, vol. 41, no. 6, pp. 1471–1480, 2006.

[971] H. Lee, Y. Liu, D. Ham, and R. M. Westervelt, "Integrated cell manipulation system—CMOS/microfluidic hybrid," *Lab on a Chip*, vol. 7, no. 3, pp. 331–337, 2007.

[972] F. T. Abu-Nimeh and F. M. Salem, "An integrated open-cavity system for magnetic bead manipulation," *IEEE Transactions on Biomedical Circuits and Systems*, vol. 7, no. 1, pp. 31–42, 2012.

[973] Q. Cao, X. Han, and L. Li, "Configurations and control of magnetic fields for manipulating magnetic particles in microfluidic applications: magnet systems and manipulation mechanisms," *Lab on a Chip*, vol. 14, no. 15, pp. 2762–2777, 2014.

[974] Y. Viero, Q. He, and A. Bancaud, "Hydrodynamic manipulation of DNA in nanopost arrays: Unhooking dynamics and size separation," *Small*, vol. 7, no. 24, pp. 3508–3518, 2011.

[975] D. B. Phillips, L. Debono, S. H. Simpson, and M. J. Padgett, "Optically controlled hydrodynamic micro-manipulation," in *Optical Trapping and Optical Micromanipulation XII*, 2015, vol. 9548: SPIE, pp. 234–241.

[976] A. Ahmad Khalili, M. R. Ahmad, M. Takeuchi, M. Nakajima, Y. Hasegawa, and R. Mohamed Zulkifli, "A microfluidic device for hydrodynamic trapping and manipulation platform of a single biological cell," *Applied Sciences*, vol. 6, no. 2, p. 40, 2016.

[977] U. G. Būtaitė, G. M. Gibson, Y.-L. D. Ho, M. Taverne, J. M. Taylor, and D. B. Phillips, "Indirect optical trapping using light driven micro-rotors

for reconfigurable hydrodynamic manipulation," *Nature Communications*, vol. 10, no. 1, p. 1215, 2019.

[978] Y.-S. Chan, T.-T. Hsu, Y.-P. Chen, H.-M. Wu, Y.-M. Wang, and C.-Y. Lee, "A traveling-wave dielectrophoresis bio-chip for cell manipulation in standard CMOS *Process," IEEE Transactions on Circuits and Systems II: Express Briefs*, vol. 69, no. 3, pp. 1582–1586, 2021.

[979] Y. Li, Y. Wang, K. Wan, *et al.*, "On the design, functions, and biomedical applications of high-throughput dielectrophoretic micro-/nanoplatforms: a review," *Nanoscale*, vol. 13, no. 8, pp. 4330–4358, 2021.

[980] Q. A. Pankhurst, J. Connolly, S. K. Jones, and J. Dobson, "Applications of magnetic nanoparticles in biomedicine," *Journal of Physics D: Applied Physics*, vol. 36, no. 13, p. R167, 2003.

[981] M. Marcus, M. Karni, K. Baranes, *et al.*, "Iron oxide nanoparticles for neuronal cell applications: uptake study and magnetic manipulations," *Journal of Nanobiotechnology*, vol. 14, no. 1, pp. 1–12, 2016.

[982] L. Agiotis, I. Theodorakos, S. Samothrakitis, S. Papazoglou, I. Zergioti, and Y. Raptis, "Magnetic manipulation of superparamagnetic nanoparticles in a microfluidic system for drug delivery applications," *Journal of Magnetism and Magnetic Materials*, vol. 401, pp. 956–964, 2016.

[983] Q. Pankhurst, N. Thanh, S. Jones, and J. Dobson, *"Progress in applications of magnetic nanoparticles in biomedicine," Journal of Physics D: Applied Physics*, vol. 42, no. 22, p. 224001, 2009.

[984] Q. Cao, X. Han, and L. Li, "Enhancement of the efficiency of magnetic targeting for drug delivery: development and evaluation of magnet system," *Journal of Magnetism and Magnetic Materials*, vol. 323, no. 15, pp. 1919–1924, 2011.

[985] O. Veiseh, J. W. Gunn, and M. Zhang, "Design and fabrication of magnetic nanoparticles for targeted drug delivery and imaging," *Advanced Drug Delivery Reviews*, vol. 62, no. 3, pp. 284–304, 2010.

[986] S. Jiang, K. Y. Win, S. Liu, C. P. Teng, Y. Zheng, and M.-Y. Han, "Surface-functionalized nanoparticles for biosensing and imaging-guided therapeutics," *Nanoscale*, vol. 5, no. 8, pp. 3127–3148, 2013.

[987] K. Smistrup, B. Kjeldsen, J. Reimers, M. Dufva, J. Petersen, and M. F. Hansen, "On-chip magnetic bead microarray using hydrodynamic focusing in a passive magnetic separator," *Lab on a Chip*, vol. 5, no. 11, pp. 1315–1319, 2005.

[988] S.-Y. Yang, K.-Y. Lien, K.-J. Huang, H.-Y. Lei, and G.-B. Lee, "Micro flow cytometry utilizing a magnetic bead-based immunoassay for rapid virus detection," *Biosensors and Bioelectronics*, vol. 24, no. 4, pp. 855–862, 2008.

[989] T. G. Constandinou, P. Georgiou, T. Prodromakis, and C. Toumazou, "A CMOS-based lab-on-chip array for the combined magnetic stimulation and opto-chemical sensing of neural tissue," in *2010 12th International Workshop on Cellular Nanoscale Networks and their Applications (CNNA 2010)*, 2010: *IEEE*, pp. 1–6.

[990] A.-Y. Chang and M. S.-C. Lu, "A CMOS micromanipulation and capacitive immunosensor array towards single magnetic microbead detection," in *2013 Transducers & Eurosensors XXVII: The 17th International Conference on Solid-State Sensors, Actuators and Microsystems (TRANSDUCERS & EUROSENSORS XXVII)*, 2013: IEEE, pp. 345–348.

[991] X.-B. Wang, Y. Huang, F. Becker, and P. Gascoyne, *"A unified theory of dielectrophoresis and travelling wave dielectrophoresis," Journal of Physics D: Applied Physics*, vol. 27, no. 7, p. 1571, 1994.

[992] N. G. Green and H. Morgan, "Dielectrophoretic separation of nanoparticles," *Journal of Physics D: Applied Physics*, vol. 30, no. 11, p. L41, 1997.

[993] R. Hölzel, N. Calander, Z. Chiragwandi, M. Willander, and F. F. Bier, "Trapping single molecules by dielectrophoresis," *Physical Review Letters*, vol. 95, no. 12, p. 128102, 2005.

[994] P. Mayer, G. W. Slater, and G. Drouin, "Theory of DNA sequencing using free-solution electrophoresis of protein-DNA complexes," *Analytical Chemistry*, vol. 66, no. 10, pp. 1777–1780, 1994.

[995] M. P. Hughes, R. Pethig, and X.-B. Wang, "Dielectrophoretic forces on particles in travelling electric fields," *Journal of Physics D: Applied Physics*, vol. 29, no. 2, p. 474, 1996.

[996] R. Pethig, M. S. Talary, and R. S. Lee, "Enhancing traveling-wave dielectrophoresis with signal superposition," *IEEE Engineering in Medicine and Biology Magazine*, vol. 22, no. 6, pp. 43–50, 2003.

[997] Y. Zhao, U.-C. Yi, and S. K. Cho, "Microparticle concentration and separation by traveling-wave dielectrophoresis (twDEP) for digital microfluidics," *Journal of Microelectromechanical Systems*, vol. 16, no. 6, pp. 1472–1481, 2007.

[998] E. Choi, B. Kim, and J. Park, "High-throughput microparticle separation using gradient traveling wave dielectrophoresis," *Journal of Micromechanics and Microengineering*, vol. 19, no. 12, p. 125014, 2009.

[999] Y. Li and K. V. Kaler, "DEP based cell separation utilizing planar microelectrode array," in *Annual Report Conference on Electrical Insulation and Dielectric Phenomena*, 2002: IEEE, pp. 680–684.

[1000] C. H. Kua, Y. C. Lam, C. Yang, K. Youcef-Toumi, and I. Rodriguez, "Modeling of dielectrophoretic force for moving dielectrophoresis electrodes," *Journal of Electrostatics*, vol. 66, no. 9–10, pp. 514–525, 2008.

[1001] T. P. Hunt, D. Issadore, and R. M. Westervelt, "Integrated circuit/microfluidic chip to programmably trap and move cells and droplets with dielectrophoresis," *Lab on a Chip*, vol. 8, no. 1, pp. 81–87, 2008.

[1002] M. A. Miled and M. Sawan, "A new fully integrated CMOS interface for a dielectrophoretic lab-on-a-chip device," in *2011 IEEE International Symposium of Circuits and Systems (ISCAS)*, 2011: IEEE, pp. 2349–2352.

[1003] R. Siddique and N. Thakor, "Investigation of nerve injury through microfluidic devices," *Journal of The Royal Society Interface*, vol. 11, no. 90, p. 20130676, 2014.

[1004] K. Lee, J. He, R. Clement, S. Massia, and B. Kim, "Biocompatible benzocyclobutene (BCB)-based neural implants with micro-fluidic channel," *Biosensors and Bioelectronics*, vol. 20, no. 2, pp. 404–407, 2004.

[1005] E. P. Kartalov and S. R. Quake, "Microfluidic device reads up to four consecutive base pairs in DNA sequencing-by-synthesis," *Nucleic Acids Research*, vol. 32, no. 9, pp. 2873–2879, 2004.

[1006] E. P. Kartalov, J. F. Zhong, A. Scherer, S. R. Quake, C. R. Taylor, and W. French Anderson, "High-throughput multi-antigen microfluidic fluorescence immunoassays," *BioTechniques*, vol. 40, no. 1, pp. 85–90, 2006.

[1007] A. Y. Fu, H.-P. Chou, C. Spence, F. H. Arnold, and S. R. Quake, "An integrated microfabricated cell sorter," *Analytical Chemistry*, vol. 74, no. 11, pp. 2451–2457, 2002.

[1008] S. K. Dertinger, X. Jiang, Z. Li, V. N. Murthy, and G. M. Whitesides, "Gradients of substrate-bound laminin orient axonal specification of neurons," *Proceedings of the National Academy of Sciences*, vol. 99, no. 20, pp. 12542–12547, 2002.

[1009] W. Tan and T. A. Desai, "Layer-by-layer microfluidics for biomimetic three-dimensional structures," *Biomaterials*, vol. 25, no. 7–8, pp. 1355–1364, 2004.

[1010] W. Tan and T. A. Desai, "Microscale multilayer cocultures for biomimetic blood vessels," *Journal of Biomedical Materials Research Part A: An Official Journal of The Society for Biomaterials, The Japanese Society for Biomaterials, and The Australian Society for Biomaterials and the Korean Society for Biomaterials, vol.* 72, no. 2, pp. 146–160, 2005.

[1011] E. V. Romanova, K. A. Fosser, S. S. Rubakhin, R. G. Nuzzo, and J. V. Sweedler, "Engineering the morphology and electrophysiological parameters of cultured neurons by microfluidic surface patterning," *The FASEB Journal*, vol. 18, no. 11, pp. 1267–1269, 2004.

[1012] N. Ma, K. W. Koelling, and J. J. Chalmers, "Fabrication and use of a transient contractional flow device to quantify the sensitivity of mammalian and insect cells to hydrodynamic forces," *Biotechnology and Bioengineering*, vol. 80, no. 4, pp. 428–437, 2002.

[1013] B. L. Gray, D. K. Lieu, S. D. Collins, R. L. Smith, and A. I. Barakat, "Microchannel platform for the study of endothelial cell shape and function," *Biomedical Microdevices*, vol. 4, pp. 9–16, 2002.

[1014] A. Sin, K. C. Chin, M. F. Jamil, Y. Kostov, G. Rao, and M. L. Shuler, "The design and fabrication of three-chamber microscale cell culture analog devices with integrated dissolved oxygen sensors," *Biotechnology Progress*, vol. 20, no. 1, pp. 338–345, 2004.

[1015] A. Rothert, S. K. Deo, L. Millner, L. G. Puckett, M. J. Madou, and S. Daunert, "Whole-cell-reporter-gene-based biosensing systems on a compact disk microfluidics platform," *Analytical Biochemistry*, vol. 342, no. 1, pp. 11–19, 2005.

[1016] A. R. Wheeler, W. R. Throndset, R. J. Whelan, *et al.*, "Microfluidic device for single-cell analysis," *Analytical Chemistry*, vol. 75, no. 14, pp. 3581–3586, 2003.

[1017] J. V. Rocheleau, G. M. Walker, W. S. Head, O. P. McGuinness, and D. W. Piston, "Microfluidic glucose stimulation reveals limited coordination of intracellular Ca^{2+} activity oscillations in pancreatic islets," *Proceedings of the National Academy of Sciences*, vol. 101, no. 35, pp. 12899–12903, 2004.

[1018] M. Rothbauer, H. Zirath, and P. Ertl, "Recent advances in microfluidic technologies for cell-to-cell interaction studies," *Lab on a Chip*, vol. 18, no. 2, pp. 249–270, 2018.

[1019] S. M. Khan, A. Gumus, J. M. Nassar, and M. M. Hussain, "CMOS enabled microfluidic systems for healthcare based applications," *Advanced Materials*, pp. 1705759 (1–26), 2018.

[1020] L. Li and A. J. Mason, "Post-CMOS parylene packaging for on-chip biosensor arrays," in *2010 IEEE Sensors*, 2010: IEEE, pp. 1613–1616.

[1021] J. S. Park, M. K. Aziz, S. Li, *et al.*, "1024-pixel CMOS multimodality joint cellular sensor/stimulator array for real-time holistic cellular characterization and cell-based drug screening," *IEEE Transactions on Biomedical Circuits and Systems*, vol. 12, no. 1, pp. 80–94, 2017.

[1022] T. B. Datta-Chaudhuri, "Integration of CMOS technology into lab-on-chip systems applied to the development of a bioelectronic nose," *Electrical and Computer Engineering, University of Maryland, College Park*, 2015.

[1023] T. Prodromakis, K. Michelakis, T. Zoumpoulidis, R. Dekker, and C. Toumazou, "Biocompatible encapsulation of CMOS based chemical sensors," *in Sensors*, 2009 *IEEE*, 2009: IEEE, pp. 791–794.

[1024] T. Datta-Chaudhuri, P. Abshire, and E. Smela, "Packaging commercial CMOS chips for lab on a chip integration," *Lab on a Chip*, vol. 14, no. 10, pp. 1753–1766, 2014.

[1025] J. Musayev, Y. Adlguzel, H. Kulah, S. Eminoglu, and T. Akln, "Label-free DNA detection using a charge sensitive CMOS microarray sensor chip," *IEEE Sensors Journal*, vol. 14, no. 5, pp. 1608–1616, 2014.

[1026] N. Halonen, J. Kilpijärvi, M. Sobocinski, *et al.*, "Low temperature co-fired ceramic packaging of CMOS capacitive sensor chip towards cell viability monitoring," *Beilstein Journal of Nanotechnology*, vol. 7, p. 1871, 2016.

[1027] A. H. Graham, C. R. Bowen, S. M. Surguy, J. Robbins, and J. Taylor, "New prototype assembly methods for biosensor integrated circuits," *Medical Engineering and Physics*, vol. 33, no. 8, pp. 973–979, 2011.

[1028] H. Wang, A. Mahdavi, D. A. Tirrell, and A. Hajimiri, "A magnetic cell-based sensor," *Lab on a Chip*, vol. 12, no. 21, pp. 4465–4471, 2012.

[1029] P. K. Challa, T. Kartanas, J. Charmet, and T. P. Knowles, "Microfluidic devices fabricated using fast wafer-scale LED-lithography patterning," *Biomicrofluidics*, vol. 11, no. 1, p. 014113, 2017.

[1030] H. Yin, L. Li, and A. J. Mason, "Screen-printed planar metallization for lab-on-CMOS with epoxy carrier," in *2016 IEEE International Symposium on Circuits and Systems (ISCAS)*, 2016: IEEE, pp. 2887–2890.

[1031] D. Welch and J. B. Christen, "Seamless integration of CMOS and microfluidics using flip chip bonding," *Journal of Micromechanics and Microengineering*, vol. 23, no. 3, p. 035009, 2013.

[1032] L. Li, H. Yin, and A. J. Mason, "Epoxy chip-in-carrier integration and screen-printed metalization for multichannel microfluidic lab-on-CMOS microsystems," *IEEE Transactions on Biomedical Circuits and Systems*, vol. 12, no. 2, pp. 416–425, 2018.

[1033] B. Zhang, Q. Dong, C. E. Korman, Z. Li, and M. E. Zaghloul, "Flexible packaging of solid-state integrated circuit chips with elastomeric microfluidics," *Scientific Reports*, vol. 3, p. 1098, 2013.

[1034] A. Wu, L. Wang, E. Jensen, R. Mathies, and B. Boser, "Modular integration of electronics and microfluidic systems using flexible printed circuit boards," *Lab on a Chip*, vol. 10, no. 4, pp. 519–521, 2010.

[1035] I. Burdallo, C. Jimenez-Jorquera, C. Fernández-Sánchez, and A. Baldi, "Integration of microelectronic chips in microfluidic systems on printed circuit board," *Journal of Micromechanics and Microengineering*, vol. 22, no. 10, pp. 105022 (1–7), 2012.

[1036] Y. Huang and A. J. Mason, "Lab-on-CMOS integration of microfluidics and electrochemical sensors," *Lab on a Chip*, vol. 13, no. 19, pp. 3929–3934, 2013.

[1037] H. Norian, R. M. Field, I. Kymissis, and K. L. Shepard, "An integrated CMOS quantitative-polymerase-chain-reaction lab-on-chip for point-of-care diagnostics," *Lab on a Chip*, vol. 14, no. 20, pp. 4076–4084, 2014.

[1038] M. Huang, J. B. Delacruz, J. C. Ruelas, S. S. Rathore, and M. Lindau, "Surface-modified CMOS IC electrochemical sensor array targeting single chromaffin cells for highly parallel amperometry measurements," *Pflügers Archiv-European Journal of Physiology*, vol. 470, no. 1, pp. 113–123, 2018.

[1039] Y. Huang and A. J. Mason, "Lab-on-CMOS: Integrating microfluidics and electrochemical sensor on CMOS," in *2011 6th IEEE International Conference on Nano/Micro Engineered and Molecular Systems*, Kaohsiung, Taiwan, 2011: IEEE, pp. 690–693.

[1040] P. Man, D. Jones, and C. Mastrangelo, "Microfluidic plastic capillaries on silicon substrates: a new inexpensive technology for bioanalysis chips," in *Proceedings IEEE The Tenth Annual International Workshop on Micro Electro Mechanical Systems, MEMS'97*, 1997: IEEE, pp. 311–316.

[1041] N. Bhattacharjee, A. Urrios, S. Kang, and A. Folch, "The upcoming 3D-printing revolution in microfluidics," *Lab on a Chip*, vol. 16, no. 10, pp. 1720–1742, 2016.

[1042] R. Amin, S. Knowlton, A. Hart, *et al.*, "3D-printed microfluidic devices," *Biofabrication*, vol. 8, no. 2, p. 022001, 2016.

[1043] S. Waheed, J. M. Cabot, N. P. Macdonald, *et al.*, "3D printed microfluidic devices: enablers and barriers," *Lab on a Chip*, vol. 16, no. 11, pp. 1993–2013, 2016.

[1044] D. Therriault, S. R. White, and J. A. Lewis, "Chaotic mixing in three-dimensional microvascular networks fabricated by direct-write assembly," *Nature Materials*, vol. 2, no. 4, p. 265, 2003.

[1045] M. Lindsay, K. Bishop, S. Sengupta, *et al.*, "Heterogeneous integration of CMOS sensors and fluidic networks using wafer-level molding," *IEEE Transactions on Biomedical Circuits and Systems*, vol. 12, no. 5, pp. 1046–1055, 2018.

[1046] W. J. Karl, M. Schikowski, J.-E. Thon, and R. Knechtel, "Adhesive wafer bonding for CMOS based lab-on-a-chip devices," *Japanese Journal of Applied Physics*, vol. 59, no. SB, p. SBBD04, 2020.

[1047] M. Mujika, S. Arana, E. Castano, *et al.*, "Magnetoresistive immunosensor for the detection of Escherichia coli O157: H7 including a microfluidic network," *Biosensors and Bioelectronics*, vol. 24, no. 5, pp. 1253–1258, 2009.

[1048] M. Mujika, S. Arana, E. Castano, *et al.*, "Microsystem for the immunomagnetic detection of Escherichia coli O157: H7," *Physica Status Solidi (a)*, vol. 205, no. 6, pp. 1478–1483, 2008.

[1049] M. Agirregabiria, F. Blanco, J. Berganzo, *et al.*, "Fabrication of SU-8 multilayer microstructures based on successive CMOS compatible adhesive bonding and releasing steps," *Lab on a Chip*, vol. 5, no. 5, pp. 545–552, 2005.

[1050] M. Inac, M. Wietstruck, A. Göritz, *et al.*, "Bicmos integrated microfluidic packaging by wafer bonding for lab-on-chip applications," in *2017 IEEE 67th Electronic Components and Technology Conference (ECTC)*, 2017: IEEE, pp. 786–791.

[1051] H. Matbaechi Ettehad, P. Soltani Zarrin, R. Hölzel, and C. Wenger, "Dielectrophoretic immobilization of yeast cells using CMOS integrated microfluidics," *Micromachines*, vol. 11, no. 5, p. 501, 2020.

[1052] A. Rasmussen, M. Gaitan, L. E. Locascio, and M. E. Zaghloul, "Fabrication techniques to realize CMOS-compatible microfluidic microchannels," *Journal of Microelectromechanical Systems*, vol. 10, no. 2, pp. 286–297, 2001.

[1053] J. A. Chediak, Z. Luo, J. Seo, N. Cheung, L. P. Lee, and T. D. Sands, "Heterogeneous integration of CdS filters with GaN LEDs for fluorescence detection microsystems," *Sensors and Actuators A: Physical*, vol. 111, no. 1, pp. 1–7, 2004.

[1054] I. Chartier, C. Bory, A. Fuchs, *et al.*, "Fabrication of hybrid plastic-silicon microfluidic devices for individual cell manipulation by dielectrophoresis," in *Microfluidics, BioMEMS, and Medical Microsystems II,* 2003, vol. 5345: International Society for Optics and Photonics, pp. 7–16.

[1055] A. Uddin, K. Milaninia, C.-H. Chen, and L. Theogarajan, "Wafer scale integration of CMOS chips for biomedical applications via self-aligned masking," *IEEE Transactions on Components, Packaging and Manufacturing Technology*, vol. 1, no. 12, pp. 1996–2004, 2011.

[1056] L. Hartley, K. V. Kaler, and O. Yadid-Pecht, "Hybrid integration of an active pixel sensor and microfluidics for cytometry on a chip," *IEEE Transactions on Circuits and Systems I: Regular Papers*, vol. 54, no. 1, pp. 99–110, 2007.

[1057] M. A. Burns, B. N. Johnson, S. N. Brahmasandra, *et al.*, "An integrated nanoliter DNA analysis device," *Science*, vol. 282, no. 5388, pp. 484–487, 1998.

[1058] N. Ho, J. Kratochvil, G. Blackburn, and J. Janata, "Encapsulation of polymeric membrane-based ion-selective field effect transistors," *Sensors and Actuators*, vol. 4, pp. 413–421, 1983.

[1059] H. Li, X. Liu, L. Li, X. Mu, R. Genov, and A. J. Mason, "CMOS electrochemical instrumentation for biosensor microsystems: a review," *Sensors*, vol. 17, no. 1, p. 74, 2016.

[1060] A. Miled and M. Sawan, "High throughput microfluidic rapid and low cost prototyping packaging methods," *JoVE (Journal of Visualized Experiments)*, no. 82, p. e50735, 2013.

[1061] M. Dandin, I. D. Jung, M. Piyasena, *et al.*, "Post-CMOS packaging methods for integrated biosensors," *in Sensors*, 2009 *IEEE*, 2009: IEEE, pp. 795–798.

[1062] T. Chi, J. S. Park, J. C. Butts, *et al.*, "A multi-modality CMOS sensor array for cell-based assay and drug screening," *IEEE Transactions on Biomedical Circuits and Systems*, vol. 9, no. 6, pp. 801–814, 2015.

[1063] A. Balasubramanian, B. Bhuva, R. Mernaugh, and F. R. Haselton, "Si-based sensor for virus detection," *IEEE Sensors Journal*, vol. 5, no. 3, pp. 340–344, 2005.

[1064] A. Mahadhy, "Development of an ultrasensitive capacitive DNA-sensor: a promising tool towards microbial diagnostics," *Lund University (Media-Tryck)*, 2015.

[1065] L. Wang, M. Veselinovic, L. Yang, B. J. Geiss, D. S. Dandy, and T. Chen, "A sensitive DNA capacitive biosensor using interdigitated electrodes," *Biosensors and Bioelectronics*, vol. 87, pp. 646–653, 2017.

[1066] A. Qureshi, Y. Gurbuz, and J. H. Niazi, "Capacitive aptamer–antibody based sandwich assay for the detection of VEGF cancer biomarker in serum," *Sensors and Actuators B: Chemical*, vol. 209, pp. 645–651, 2015.

[1067] J. Tian, Y. Bai, X. Tang, S. Sang, and F. Wang, "A capacitive surface stress biosensor for CSFV detection," *Microelectronic Engineering*, vol. 159, pp. 55–59, 2016.

[1068] P. Zhurauski, S. K. Arya, P. Jolly, *et al.*, "Sensitive and selective Affimer-functionalised interdigitated electrode-based capacitive biosensor for Her4 protein tumour biomarker detection," *Biosensors and Bioelectronics*, vol. 108, pp. 1–8, 2018.

[1069] H. A. Zeinabad, H. Ghourchian, M. Falahati, M. Fathipour, M. Azizi, and S. M. Boutorabi, "Ultrasensitive interdigitated capacitance immunosensor using gold nanoparticles," *Nanotechnology*, vol. 29, no. 26, p. 265102, 2018.

[1070] G. Ertürk, M. Hedström, M. A. Tümer, A. Denizli, and B. Mattiasson, "Real-time prostate-specific antigen detection with prostate-specific antigen imprinted capacitive biosensors," *Analytica Chimica Acta*, vol. 891, pp. 120–129, 2015.

[1071] G. Ertürk, M. Hedström, and B. Mattiasson, "A sensitive and real-time assay of trypsin by using molecular imprinting-based capacitive biosensor," *Biosensors and Bioelectronics*, vol. 86, pp. 557–565, 2016.

[1072] F. Canfarotta, J. Czulak, A. Guerreiro, *et al.*, "A novel capacitive sensor based on molecularly imprinted nanoparticles as recognition elements," *Biosensors and Bioelectronics*, vol. 120, pp. 108–114, 2018.

[1073] A. Foubert, N. V. Beloglazova, M. Hedström, and S. De Saeger, "Antibody immobilization strategy for the development of a capacitive immunosensor detecting zearalenone," *Talanta*, vol. 191, pp. 202–208, 2019.

[1074] P. Lenain, S. De Saeger, B. Mattiasson, and M. Hedström, "Affinity sensor based on immobilized molecular imprinted synthetic recognition elements," *Biosensors and Bioelectronics*, vol. 69, pp. 34–39, 2015.

[1075] M. Hedström and B. Mattiasson, "Bioimprinting as a tool for the detection of aflatoxin B1 using a capacitive biosensor," *Biotechnology Reports*, vol. 11, pp. 12–17, 2016.

[1076] A. Hartono, E. Sanjaya, and R. Ramli, "Glucose sensing using capacitive biosensor based on polyvinylidene fluoride thin film," *Biosensors*, vol. 8, no. 1, p. 12, 2018.

[1077] C. I. Justino, A. C. Freitas, R. Pereira, A. C. Duarte, and T. A. R. Santos, "Recent developments in recognition elements for chemical sensors and biosensors," *TrAC Trends in Analytical Chemistry*, vol. 68, pp. 2–17, 2015.

[1078] M. C. Rodriguez, A.-N. Kawde, and J. Wang, "Aptamer biosensor for label-free impedance spectroscopy detection of proteins based on recognition-induced switching of the surface charge," *Chemical Communications*, no. 34, pp. 4267–4269, 2005.

[1079] C. I. Justino, A. C. Duarte, and T. A. Rocha-Santos, "Analytical applications of affibodies," *TrAC Trends in Analytical Chemistry*, vol. 65, pp. 73–82, 2015.

[1080] B. Renberg, J. Nordin, A. Merca, *et al.*, "Affibody molecules in protein capture microarrays: evaluation of multidomain ligands and different detection formats," *Journal of Proteome Research*, vol. 6, no. 1, pp. 171–179, 2007.

[1081] A. Johnson, Q. Song, P. Ko Ferrigno, P. R. Bueno, and J. J. Davis, "Sensitive Affimer and antibody based impedimetric label-free assays for C-reactive protein," *Analytical Chemistry*, vol. 84, no. 15, pp. 6553–6560, 2012.

[1082] P. K. Ferrigno, "Non-antibody protein-based biosensors," *Essays in Biochemistry*, vol. 60, no. 1, pp. 19–25, 2016.

[1083] Z. Cheng, E. Wang, and X. Yang, "Capacitive detection of glucose using molecularly imprinted polymers," *Biosensors and Bioelectronics*, vol. 16, no. 3, pp. 179–185, 2001.

[1084] A. Aghaei, M. R. M. Hosseini, and M. Najafi, "A novel capacitive biosensor for cholesterol assay that uses an electropolymerized molecularly imprinted polymer," *Electrochimica Acta*, vol. 55, no. 5, pp. 1503–1508, 2010.

[1085] T. Cohen, J. Starosvetsky, U. Cheruti, and R. Armon, "Whole cell imprinting in sol-gel thin films for bacterial recognition in liquids: Macromolecular fingerprinting," *International Journal of Molecular Sciences*, vol. 11, no. 4, pp. 1236–1252, 2010.

[1086] R. Ouyang, J. Lei, and H. Ju, "Surface molecularly imprinted nanowire for protein specific recognition," *Chemical Communications*, no. 44, pp. 5761–5763, 2008.

[1087] E. Yildirim, E. Turan, and T. Caykara, "Construction of myoglobin imprinted polymer films by grafting from silicon surface," *Journal of Materials Chemistry*, vol. 22, no. 2, pp. 636–642, 2012.

[1088] L. J. Guo, "Recent progress in nanoimprint technology and its applications," *Journal of Physics D: Applied Physics*, vol. 37, no. 11, p. R123, 2004.

[1089] A. Mujahid, N. Iqbal, and A. Afzal, "Bioimprinting strategies: From soft lithography to biomimetic sensors and beyond," *Biotechnology Advances*, vol. 31, no. 8, pp. 1435–1447, 2013.

[1090] E. Alipour, H. Ghourchian, and S. M. Boutorabi, "Gold nanoparticle based capacitive immunosensor for detection of hepatitis B surface antigen," *Analytical Methods*, vol. 5, no. 17, pp. 4448–4453, 2013.

[1091] K. Besteman, J.-O. Lee, F. G. Wiertz, H. A. Heering, and C. Dekker, "Enzyme-coated carbon nanotubes as single-molecule biosensors," *Nano Letters*, vol. 3, no. 6, pp. 727–730, 2003.

[1092] P. Sanguino, T. Monteiro, S. Bhattacharyya, C. Dias, R. Igreja, and R. Franco, "ZnO nanorods as immobilization layers for interdigitated capacitive immunosensors," *Sensors and Actuators B: Chemical*, vol. 204, pp. 211–217, 2014.

[1093] L. Yang, W. Wei, X. Gao, J. Xia, and H. Tao, "A new antibody immobilization strategy based on electrodeposition of nanometer-sized hydroxyapatite for label-free capacitive immunosensor," *Talanta*, vol. 68, no. 1, pp. 40–46, 2005.

[1094] L. C. Shriver-Lake, W. B. Gammeter, S. S. Bang, and M. Pazirandeh, "Covalent binding of genetically engineered microorganisms to porous glass beads," *Analytica Chimica Acta*, vol. 470, no. 1, pp. 71–78, 2002.

[1095] G. Tsekenis, M. Chatzipetrou, J. Tanner, *et al.*, "Surface functionalization studies and direct laser printing of oligonucleotides toward the fabrication of a micromembrane DNA capacitive biosensor," *Sensors and Actuators B: Chemical*, vol. 175, pp. 123–131, 2012.

[1096] F. Malvano, L. Maritato, G. Carapella, *et al.*, "Fabrication of $SrTiO_3$ layer on Pt electrode for label-free capacitive biosensors," *Biosensors*, vol. 8, no. 1, p. 26, 2018.

[1097] G.-Y. Lee, Y.-H. Choi, H.-W. Chung, H. Ko, S. Cho, and J.-C. Pyun, "Capacitive immunoaffinity biosensor based on vertically paired ring-electrodes," *Biosensors and Bioelectronics*, vol. 40, no. 1, pp. 227–232, 2013.

[1098] N.-P. Huang, J. Vörös, S. M. De Paul, M. Textor, and N. D. Spencer, "Biotin-derivatized poly (L-lysine)-g-poly (ethylene glycol): A novel polymeric interface for bioaffinity sensing," *Langmuir*, vol. 18, no. 1, pp. 220–230, 2002.

[1099] P. Sanguino, T. Monteiro, F. Marques, C. J. Dias, R. Igreja, and R. Franco, "Interdigitated capacitive immunosensors with PVDF immobilization layers," *IEEE Sensors Journal*, vol. 14, no. 4, pp. 1260–1265, 2014.

[1100] D. A. Butterfield, J. Colvin, J. Liu, J. Wang, L. Bachas, and D. Bhattacharrya, "Electron paramagnetic resonance spin label titration: a novel method to investigate random and site-specific immobilization of enzymes onto polymeric membranes with different properties," *Analytica Chimica Acta*, vol. 470, no. 1, pp. 29–36, 2002.

[1101] S. Cosnier, "Biomolecule immobilization on electrode surfaces by entrapment or attachment to electrochemically polymerized films. A review," *Biosensors and Bioelectronics*, vol. 14, no. 5, pp. 443–456, 1999.

[1102] F. Gardies, C. Martelet, B. Colin, and B. Mandrand, "Feasibility of an immunosensor based upon capacitance measurements," *Sensors and Actuators*, vol. 17, no. 3–4, pp. 461–464, 1989.

[1103] P. Bataillard, F. Gardies, N. Jaffrezic-Renault, C. Martelet, B. Colin, and B. Mandrand, "Direct detection of immunospecies by capacitance measurements," *Analytical Chemistry*, vol. 60, no. 21, pp. 2374–2379, 1988.

[1104] E. Souteyrand, J. Martin, and C. Martelet, "Direct detection of biomolecules by electrochemical impedance measurements," *Sensors and Actuators B: Chemical*, vol. 20, no. 1, pp. 63–69, 1994.

[1105] C. Saby, N. Jaffrezic-Renault, C. Martelet, *et al.*, "Immobilization of antibodies onto a capacitance silicon-based transducer," *Sensors and Actuators B: Chemical*, vol. 16, no. 1–3, pp. 458–462, 1993.

[1106] H. Maupas, C. Saby, C. Martelet, *et al.*, "Impedance analysis of Si/SiO_2 heterostructures grafted with antibodies: an approach for immunosensor development," *Journal of ElectroAnalytical Chemistry*, vol. 406, no. 1–2, pp. 53–58, 1996.

[1107] A. Barraud, H. Perrot, V. Billard, C. Martelet, and J. Therasse, "Study of immunoglobulin G thin layers obtained by the Langmuir-Blodgett method: application to immunosensors," *Biosensors and Bioelectronics*, vol. 8, no. 1, pp. 39–48, 1993.

[1108] V. Billard, C. Martelet, P. Binder, and J. Therasse, "Toxin detection using capacitance measurements on immunospecies grafted onto a semiconductor substrate," *Analytica Chimica Acta*, vol. 249, no. 2, pp. 367–372, 1991.

[1109] A. Sibai, K. Elamri, D. Barbier, N. Jaffrezic-Renault, and E. Souteyrand, "Analysis of the polymer-antibody-antigen interaction in a capacitive immunosensor by FTIR difference spectroscopy," *Sensors and Actuators B: Chemical*, vol. 31, no. 1–2, pp. 125–130, 1996.

[1110] H. Berney, J. Alderman, W. Lane, and J. K. Collins, *"A differential capacitive biosensor using polyethylene glycol to overlay the biolayer,"* *Sensors and Actuators B: Chemical*, vol. 44, no. 1–3, pp. 578–584, 1997.

[1111] C. Berggren and G. Johansson, "Capacitance measurements of antibody–antigen interactions in a flow system," *Analytical Chemistry*, vol. 69, no. 18, pp. 3651–3657, 1997.

[1112] C. Berggren, B. Bjarnason, and G. Johansson, "An immunological Interleukine-6 capacitive biosensor using perturbation with a potentiostatic step," *Biosensors and Bioelectronics*, vol. 13, no. 10, pp. 1061–1068, 1998.

[1113] C. E. Nwankire, A. Venkatanarayanan, T. Glennon, T. E. Keyes, R. J. Forster, and J. Ducrée, "Label-free impedance detection of cancer cells from whole blood on an integrated centrifugal microfluidic platform," *Biosensors and Bioelectronics*, vol. 68, pp. 382–389, 2015.

[1114] M. Knichel, P. Heiduschka, W. Beck, G. Jung, and W. Göpel, "Utilization of a self-assembled peptide monolayer for an impedimetric immunosensor," *Sensors and Actuators B: Chemical*, vol. 28, no. 2, pp. 85–94, 1995.

[1115] A. Gebbert, M. Alvarez-Icaza, W. Stoecklein, and R. D. Schmid, "Real-time monitoring of immunochemical interactions with a tantalum capacitance flow-through cell," *Analytical Chemistry*, vol. 64, no. 9, pp. 997–1003, 1992.

[1116] H. Maupas, A. Soldatkin, C. Martelet, N. Jaffrezic-Renault, and B. Mandrand, "Direct immunosensing using differential electrochemical measurements of impedimetric variations," *Journal of ElectroAnalytical Chemistry*, vol. 421, no. 1–2, pp. 165–171, 1997.

[1117] S. Ameur, H. Maupas, C. Martelet, *et al.*, "Impedimetric measurements on polarized functionalized platinum electrodes: application to direct immunosensing," *Materials Science and Engineering: C*, vol. 5, no. 2, pp. 111–119, 1997.

[1118] K. K. Ng, *Complete Guide to Semiconductor Devices. Wiley Online Library*, 2002.

[1119] C.-H. Chiang, M.-C. Chou, P.-H. Hsieh, and M. S.-C. Lu, "Design and characterization of a CMOS MEMS capacitive oscillator for resonant sensing in liquids," *IEEE Sensors Journal*, vol. 16, no. 5, pp. 1136–1142, 2015.

[1120] N. Moser, C. L. Leong, Y. Hu, *et al.*, "Complementary metal–oxide–semiconductor potentiometric field-effect transistor array platform using sensor learning for multi-ion imaging," *Analytical Chemistry*, vol. 92, no. 7, pp. 5276–5285, 2020.

[1121] R. R. Singh, L. Leng, A. Guenther, and R. Genov, "A CMOS-microfluidic chemiluminescence contact imaging microsystem," *IEEE Journal of Solid-State Circuits*, vol. 47, no. 11, pp. 2822–2833, 2012.

[1122] A. Sibbald, P. Whalley, and A. Covington, "A miniature flow-through cell with a four-function chemfet integrated circuit for simultaneous measurements of potassium, hydrogen, calcium and sodium ions," *Analytica Chimica Acta*, vol. 159, pp. 47–62, 1984.

[1123] G. Högg, O. Lutze, and K. Cammann, "Novel membrane material for ion-selective field-effect transistors with extended lifetime and improved selectivity," *Analytica Chimica Acta*, vol. 335, no. 1–2, pp. 103–109, 1996.

[1124] A. Van den Berg, A. Grisel, and E. Verney-Norberg, "An ISFET-based calcium sensor using a photopolymerized polysiloxane membrane," *Sensors and Actuators B: Chemical*, vol. 4, no. 3–4, pp. 235–238, 1991.

[1125] A. Bratov, N. Abramova, and C. Domínguez, "Investigation of chloride sensitive ISFETs with different membrane compositions suitable for medical applications," *Analytica Chimica Acta*, vol. 514, no. 1, pp. 99–106, 2004.

[1126] N. Moser, C. L. Leong, Y. Hu, M. Boutelle, and P. Georgiou, "An ion imaging ISFET array for potassium and sodium detection," in *2016 IEEE International Symposium on Circuits and Systems (ISCAS)*, 2016: IEEE, pp. 2847–2850.

[1127] G.-Y. Lee, J.-H. Park, Y. W. Chang, M.-J. Kang, S. Cho, and J.-C. Pyun, "Capacitive biosensor based on vertically paired electrode with controlled parasitic capacitance," *Sensors and Actuators B: Chemical*, 2018.

[1128] K. Bunnfors, N. Abrikossova, J. Kilpijärvi, *et al.*, "Nanoparticle activated Neutrophils-on-a chip: A label-free capacitive sensor to monitor cells at work," *Sensors and Actuators B: Chemical*, p. 128020, 2020.

[1129] H. Osouli Tabrizi, A. Panahi, S. Forouhi, *et al.*, "Oral cells-on-chip: design, modeling and experimental results," *Bioengineering*, vol. 9, no. 5, p. 218, 2022.

[1130] B. Senevirathna, S. Lu, M. Dandin, J. Basile, E. Smela, and P. Abshire, "High resolution monitoring of chemotherapeutic agent potency in cancer cells using a CMOS capacitance biosensor," *Biosensors and Bioelectronics*, vol. 142, p. 111501, 2019.

[1131] K. Hu, C. E. Arcadia, and J. K. Rosenstein, "Super-resolution electrochemical impedance imaging with a 100 × 100 CMOS sensor array," in *2021 IEEE Biomedical Circuits and Systems Conference (BioCAS)*, 2021: IEEE, pp. 1–4.

[1132] N. Nelson, D. Sander, M. Dandin, S. B. Prakash, A. Sarje, and P. Abshire, "Handheld fluorometers for lab-on-a-chip applications," *IEEE Transactions on Biomedical Circuits and Systems*, vol. 3, no. 2, pp. 97–107, 2009.

[1133] P.-H. Kuo, J.-C. Kuo, H.-T. Hsueh, *et al.*, "A smart CMOS assay SoC for rapid blood screening test of risk prediction," *IEEE Transactions on Biomedical Circuits and Systems*, vol. 9, no. 6, pp. 790–800, 2015.

[1134] H. Ma, R. W. Wallbank, R. Chaji, *et al.*, "An impedance-based integrated biosensor for suspended DNA characterization," *Scientific Reports*, vol. 3, no. 1, pp. 1–7, 2013.

[1135] H. O. Tabrizi, S. Forouhi, M. Ghafar-Zadeh, *et al.*, "CMOS capacitive dry DNA storage monitoring: design, implementation and experimental results," *IEEE Sensors Journal*, vol. 22, no. 6, pp. 5521–5530, 2022.

[1136] E. Ghafar-Zadeh, S. F. Chowdhury, A. Aliakbar, *et al.*, "Handheld impedance biosensor system using engineered proteinaceous receptors," *Biomedical Microdevices*, vol. 12, no. 6, pp. 967–975, 2010.

[1137] M. Schena, DNA Microarrays: *A Practical Approach (no. 205)*. New York: Oxford University Press, 1999.

[1138] M. Schena, *Microarray Analysis*. Wiley-Liss, 2003.

[1139] L. Gonzalez-Macia, A. Morrin, M. R. Smyth, and A. J. Killard, "Advanced printing and deposition methodologies for the fabrication of biosensors and biodevices," *Analyst*, vol. 135, no. 5, pp. 845–867, 2010.

[1140] D. Bowtell, J. Sambrook, *and A. D. C. Manual, DNA Microarrays. Cold Spring* Habor Laboratory Press, New York, vol. 79, 2003.

[1141] B. Huang, H. Wu, S. Kim, B. K. Kobilka, and R. N. Zare, "Phospholipid biotinylation of polydimethylsiloxane (PDMS) for protein immobilization," *Lab on a Chip*, vol. 6, no. 3, pp. 369–373, 2006.

[1142] G. T. Hermanson, *Bioconjugate Techniques*. Academic Press, 2013.

[1143] D. A. Wagenaar, R. Madhavan, J. Pine, and S. M. Potter, "Controlling bursting in cortical cultures with closed-loop multi-electrode stimulation," *Journal of Neuroscience*, vol. 25, no. 3, pp. 680–688, 2005.

[1144] S. B. Jun, M. R. Hynd, N. Dowell-Mesfin, *et al.*, "Low-density neuronal networks cultured using patterned poly-l-lysine on microelectrode arrays," *Journal of Neuroscience Methods*, vol. 160, no. 2, pp. 317–326, 2007.

[1145] J. C. Chang, G. J. Brewer, and B. C. Wheeler, "Modulation of neural network activity by patterning," *Biosensors and Bioelectronics*, vol. 16, no. 7–8, pp. 527–533, 2001.

[1146] I. Suzuki, Y. Sugio, Y. Jimbo, and K. Yasuda, "Stepwise pattern modification of neuronal network in photo-thermally-etched agarose architecture on multi-electrode array chip for individual-cell-based electrophysiological measurement," *Lab on a Chip*, vol. 5, no. 3, pp. 241–247, 2005.

[1147] H. Golan, K. Mikenberg, V. Greenberger, and M. Segal, "GABA withdrawal modifies network activity in cultured hippocampal neurons," *Neural Plasticity*, vol. 7, no. 1–2, pp. 31–42, 1999.

[1148] J. V. Selinger, J. J. Pancrazio, and G. W. Gross, "Measuring synchronization in neuronal networks for biosensor applications," *Biosensors and Bioelectronics*, vol. 19, no. 7, pp. 675–683, 2004.

[1149] Y. Jimbo, A. Kawana, P. Parodi, and V. Torre, "The dynamics of a neuronal culture of dissociated cortical neurons of neonatal rats," *Biological Cybernetics*, vol. 83, no. 1, pp. 1–20, 2000.

[1150] S. Morefield, E. Keefer, K. Chapman, and G. Gross, "Drug evaluations using neuronal networks cultured on microelectrode arrays," *Biosensors and Bioelectronics*, vol. 15, no. 7–8, pp. 383–396, 2000.

[1151] D. A. Wagenaar, J. Pine, and S. M. Potter, "Searching for plasticity in dissociated cortical cultures on multi-electrode arrays," *Journal of Negative Results in Biomedicine*, vol. 5, pp. 1–19, 2006.

[1152] D. Eytan and S. Marom, "Dynamics and effective topology underlying synchronization in networks of cortical neurons," *Journal of Neuroscience*, vol. 26, no. 33, pp. 8465–8476, 2006.

[1153] R. Segev, Y. Shapira, M. Benveniste, and E. Ben-Jacob, "Observations and modeling of synchronized bursting in two-dimensional neural networks," *Physical Review E*, vol. 64, no. 1, p. 011920, 2001.

[1154] M. E. Ruaro, P. Bonifazi, and V. Torre, "Toward the neurocomputer: image processing and pattern recognition with neuronal cultures," *IEEE Transactions on Biomedical Engineering*, vol. 52, no. 3, pp. 371–383, 2005.

[1155] S. Illes, S. Theiss, H.-P. Hartung, M. Siebler, and M. Dihné, "Niche-dependent development of functional neuronal networks from embryonic stem cell-derived neural populations," *BMC Neuroscience*, vol. 10, no. 1, pp. 1–16, 2009.

[1156] S. Illes, W. Fleischer, M. Siebler, H.-P. Hartung, and M. Dihné, "Development and pharmacological modulation of embryonic stem cell-derived neuronal network activity," *Experimental Neurology*, vol. 207, no. 1, pp. 171–176, 2007.

[1157] J. Van Pelt, P. S. Wolters, M. A. Corner, W. L. Rutten, and G. J. Ramakers, "Long-term characterization of firing dynamics of spontaneous bursts in cultured neural networks," *IEEE Transactions on Biomedical Engineering*, vol. 51, no. 11, pp. 2051–2062, 2004.

[1158] C. D. James, A. J. Spence, N. M. Dowell-Mesfin, *et al.*, "Extracellular recordings from patterned neuronal networks using planar microelectrode arrays," *IEEE Transactions on Biomedical Engineering*, vol. 51, no. 9, pp. 1640–1648, 2004.

[1159] Y. Nam, D. W. Branch, and B. C. Wheeler, "Epoxy-silane linking of biomolecules is simple and effective for patterning neuronal cultures," *Biosensors and Bioelectronics*, vol. 22, no. 5, pp. 589–597, 2006.

[1160] B. K. Sinha, *Principles of Animal Cell Culture. International Book*, 2008.

[1161] S. Przyborski, *Technology Platforms for 3D Cell Culture: A User's Guide*. John Wiley & Sons, 2017.

[1162] O. Uysal, T. Sevimli, M. Sevimli, S. Gunes, and A. E. Sariboyaci, "Cell and tissue culture: the base of biotechnology," in *Omics Technologies and Bio-Engineering*, 2018, pp. 391–429.

[1163] L. Voloboueva, X. Sun, Y.-B. Ouyang, and R. Giffard, "Cell culture: primary neural cells," *in Reference Module in Neuroscience and Biobehavioral Psychology*, Elsevier, 2017.

[1164] K. W. Ng and J.-T. Schantz, *A Manual for Primary Human Cell Culture*. World Scientific, 2010.

[1165] D. O. Fauza and M. Bani, *Fetal Stem Cells in Regenerative Medicine: Principles and Translational Strategies*. Springer, 2016.

[1166] H. McIlwain, L. Buchel, and J. Cheshire, "The inorganic phosphate and phosphocreatine of brain especially during metabolism in vitro," *Biochemical Journal*, vol. 48, no. 1, p. 12, 1951.

[1167] G. L. Collingridge, "The brain slice preparation: a tribute to the pioneer Henry McIlwain," *Journal of Neuroscience Methods*, vol. 59, no. 1, pp. 5–9, 1995.

[1168] M. Vinci, S. Gowan, F. Boxall, *et al.*, "Advances in establishment and analysis of three-dimensional tumor spheroid-based functional assays for target validation and drug evaluation," *BMC Biology*, vol. 10, no. 1, pp. 1–21, 2012.

[1169] M. Ravi, V. Paramesh, S. Kaviya, E. Anuradha, and F. P. Solomon, "3D cell culture systems: advantages and applications," *Journal of Cellular Physiology*, vol. 230, no. 1, pp. 16–26, 2015.

[1170] E. Ferrea, A. Maccione, L. Medrihan, *et al.*, "Large-scale, high-resolution electrophysiological imaging of field potentials in brain slices with microelectronic multielectrode arrays," *Frontiers in Neural Circuits*, vol. 6, p. 80, 2012.

[1171] L. Stoppini, P.-A. Buchs, and D. Muller, "A simple method for organotypic cultures of nervous tissue," *Journal of Neuroscience Methods*, vol. 37, no. 2, pp. 173–182, 1991.

[1172] Y. Buskila, P. P. Breen, J. Tapson, A. van Schaik, M. Barton, and J. W. Morley, "Extending the viability of acute brain slices," *Scientific Reports*, vol. 4, no. 1, p. 5309, 2014.

[1173] B. H. Gähwiler, "Organotypic slice cultures: a model for interdisciplinary studies," *Progress in Clinical and Biological Research*, vol. 253, pp. 13–18, 1987.

[1174] B. Heimrich and M. Frotscher, "Slice cultures as a model to study entorhinal-hippocampal interaction," *Hippocampus*, vol. 3, no. S1, pp. 11–17, 1993.

[1175] M. Finley, D. Fairman, D. Liu, P. Li, A. Wood, and S. Cho, "Functional validation of adult hippocampal organotypic cultures as an in vitro model of brain injury," *Brain Research*, vol. 1001, no. 1–2, pp. 125–132, 2004.

[1176] N. Toni, P.-A. Buchs, I. Nikonenko, P. Povilaitite, L. Parisi, and D. Muller, "Remodeling of synaptic membranes after induction of long-term potentiation," *Journal of Neuroscience*, vol. 21, no. 16, pp. 6245–6251, 2001.

[1177] D. H. Brager, M. Capogna, and S. M. Thompson, "Short-term synaptic plasticity, simulation of nerve terminal dynamics, and the effects of protein kinase C activation in rat hippocampus," *The Journal of Physiology*, vol. 541, no. 2, pp. 545–559, 2002.

[1178] S. F. Chen, C. C. Huang, H. M. Wu, S. H. Chen, Y. C. Liang, and K. S. Hsu, "Seizure, neuron loss, and mossy fiber sprouting in herpes simplex virus type 1–infected organotypic hippocampal cultures," *Epilepsia*, vol. 45, no. 4, pp. 322–332, 2004.

[1179] C. O. Aptowicz, P. E. Kunkler, and R. P. Kraig, "Homeostatic plasticity in hippocampal slice cultures involves changes in voltage-gated Na^+ channel expression," *Brain Research*, vol. 998, no. 2, pp. 155–163, 2004.

[1180] A. P. Silva, P. S. Pinheiro, A. P. Carvalho, *et al.*, "Activation of neuropeptide Y receptors is neuroprotective against excitotoxicity in organotypic hippocampal slice cultures," *The FASEB Journal*, vol. 17, no. 9, pp. 1118–1120, 2003.

[1181] R. Kovács, S. Schuchmann, S. Gabriel, O. Kann, J. Kardos, and U. Heinemann, "Free radical–mediated cell damage after experimental status epilepticus in hippocampal slice cultures," *Journal of Neurophysiology*, vol. 88, no. 6, pp. 2909–2918, 2002.

[1182] H. L. Cater, A. Chandratheva, C. D. Benham, B. Morrison III, and L. E. Sundstrom, "Lactate and glucose as energy substrates during, and after, oxygen deprivation in rat hippocampal acute and cultured slices," *Journal of Neurochemistry*, vol. 87, no. 6, pp. 1381–1390, 2003.

[1183] T. Cronberg, A. Rytter, F. Asztély, A. Söder, and T. Wieloch, "Glucose but not lactate in combination with acidosis aggravates ischemic neuronal death in vitro," *Stroke*, vol. 35, no. 3, pp. 753–757, 2004.

[1184] J. L. Perez Velazquez, L. Kokarovtseva, M. Weisspapir, and M. V. Frantseva, "Anti-porin antibodies prevent excitotoxic and ischemic damage to brain tissue," *Journal of Neurotrauma*, vol. 20, no. 7, pp. 633–647, 2003.

[1185] A. Barth, L. Barth, R. S. Morrison, and D. W. Newell, "bFGF enhances the protective effects of MK-801 against ischemic neuronal injury in vitro," *Neuroreport*, vol. 7, no. 9, pp. 1461–1464, 1996.

[1186] Y.-S. Lee, J. Baratta, J. Yu, V. W. Lin, and R. T. Robertson, "AFGF promotes axonal growth in rat spinal cord organotypic slice co-cultures," *Journal of Neurotrauma*, vol. 19, no. 3, pp. 357–367, 2002.

[1187] B. Teter, P. T. Xu, J. Gilbert, A. Roses, D. Galasko, and G. Cole, "Human apolipoprotein E isoform-specific differences in neuronal sprouting in organotypic hippocampal culture," *Journal of Neurochemistry*, vol. 73, no. 6, pp. 2613–2616, 1999.

[1188] R. A. McKinney, A. Lüthi, C. E. Bandtlow, B. H. Gähwiler, and S. M. Thompson, "Selective glutamate receptor antagonists can induce or prevent axonal sprouting in rat hippocampal slice cultures," *Proceedings of the National Academy of Sciences*, vol. 96, no. 20, pp. 11631–11636, 1999.

[1189] L. Stoppini, P.-A. Buchs, and D. Muller, "Lesion-induced neurite sprouting and synapse formation in hippocampal organotypic cultures," *Neuroscience*, vol. 57, no. 4, pp. 985–994, 1993.

[1190] G. A. Roth, V. Spada, K. Hamill, and M. B. Bornstein, "Insulin-like growth factor I increases myelination and inhibits demyelination in cultured organotypic nerve tissue," *Developmental Brain Research*, vol. 88, no. 1, pp. 102–108, 1995.

[1191] L. Notterpek, P. Bullock, S. Malek-Hedayat, R. Fisher, and L. Rome, "Myelination in cerebellar slice cultures: development of a system amenable to biochemical analysis," *Journal of Neuroscience Research*, vol. 36, no. 6, pp. 621–634, 1993.

[1192] A. M. Ghoumari, C. Ibanez, M. El-Etr, *et al.*, "Progesterone and its metabolites increase myelin basic protein expression in organotypic slice cultures of rat cerebellum," *Journal of Neurochemistry*, vol. 86, no. 4, pp. 848–859, 2003.

[1193] C. Demerens, B. Stankoff, M. Logak, *et al.*, "Induction of myelination in the central nervous system by electrical activity," *Proceedings of the National Academy of Sciences*, vol. 93, no. 18, pp. 9887–9892, 1996.

[1194] R. G. Keynes, S. Duport, and J. Garthwaite, "Hippocampal neurons in organotypic slice culture are highly resistant to damage by endogenous and exogenous nitric oxide," *European Journal of Neuroscience*, vol. 19, no. 5, pp. 1163–1173, 2004.

[1195] J. C. Vis, R. T. de Boer-van Huizen, M. M. Verbeek, R. M. de Waal, H. J. ten Donkelaar, and B. Kremer, "3-Nitropropionic acid induces cell death and mitochondrial dysfunction in rat corticostriatal slice cultures," *Neuroscience Letters*, vol. 329, no. 1, pp. 86–90, 2002.

[1196] R. P. Brendza, K. Simmons, K. R. Bales, S. M. Paul, M. P. Goldberg, and D. M. Holtzman, "Use of YFP to study amyloid-β associated neurite alterations in live brain slices," *Neurobiology of Aging*, vol. 24, no. 8, pp. 1071–1077, 2003.

[1197] J. Bendiske and B. A. Bahr, "Lysosomal activation is a compensatory response against protein accumulation and associated synaptic pathogenesis—an approach for slowing Alzheimer disease?" *Journal of Neuropathology & Experimental Neurology*, vol. 62, no. 5, pp. 451–463, 2003.

[1198] K. Duff, W. Noble, K. Gaynor, and Y. Matsuoka, "Organotypic slice cultures from transgenic mice as disease model systems," *Journal of Molecular Neuroscience*, vol. 19, pp. 317–320, 2002.

[1199] D. G. Hay, K. Sathasivam, S. Tobaben, *et al.*, "Progressive decrease in chaperone protein levels in a mouse model of Huntington's disease and induction of stress proteins as a therapeutic approach," *Human Molecular Genetics*, vol. 13, no. 13, pp. 1389–1405, 2004.

[1200] R. C. Murphy and A. Messer, "A single-chain Fv intrabody provides functional protection against the effects of mutant protein in an organotypic slice culture model of Huntington's disease," *Molecular Brain Research*, vol. 121, no. 1–2, pp. 141–145, 2004.

[1201] T. B. Sherer, R. Betarbet, C. M. Testa, *et al.*, "Mechanism of toxicity in rotenone models of Parkinson's disease," *Journal of Neuroscience*, vol. 23, no. 34, pp. 10756–10764, 2003.

[1202] J. K. Leutgeb, J. U. Frey, and T. Behnisch, "LTP in cultured hippocampal–entorhinal cortex slices from young adult (P25–30) rats," *Journal of Neuroscience Methods*, vol. 130, no. 1, pp. 19–32, 2003.

[1203] T. P. Yu, H. A. Lester, and N. Davidson, "Requirement of a critical period of GABAergic receptor blockade for induction of a cAMP-mediated long-term depression at CA3-CA1 synapses," *Synapse*, vol. 49, no. 1, pp. 12–19, 2003.

[1204] P. Alix, J. Winterer, and W. Müller, "New illumination technique for IR-video guided patch-clamp recording from neurons in slice cultures on biomembrane," *Journal of Neuroscience Methods*, vol. 128, no. 1–2, pp. 79–84, 2003.

[1205] M. P. Thomas, M. I. Davis, D. T. Monaghan, and R. A. Morrisett, "Organotypic brain slice cultures for functional analysis of alcohol-related disorders: novel versus conventional preparations," *Alcoholism: Clinical and Experimental Research*, vol. 22, no. 1, pp. 51–59, 1998.

[1206] S. Kasparov, A. Teschemacher, and J. Paton, "Dynamic confocal imaging in acute brain slices and organotypic slice cultures using a spectral confocal microscope with single photon excitation," *Experimental Physiology*, vol. 87, no. 6, pp. 715–724, 2002.

[1207] M. E. Dailey and M. Waite, "Confocal imaging of microglial cell dynamics in hippocampal slice cultures," *Methods*, vol. 18, no. 2, pp. 222–230, 1999.

[1208] A. K. McAllister, "Biolistic transfection of cultured organotypic brain slices," *Gene Delivery to Mammalian Cells: Volume 1: Nonviral Gene Transfer Techniques*, pp. 197–205, 2004.

[1209] M. Ibrahim, A. Si-Ammour, M. R. Celio, F. Mauch, and P.-A. Menoud, "Construction and application of a microprojectile system for the transfection of organotypic brain slices," *Journal of Neuroscience Methods*, vol. 101, no. 2, pp. 171–179, 2000.

[1210] A. Thomas, D. Kim, R. Fields, H. Chin, and H. Gainer, "Quantitative analysis of gene expression in organotypic slice-explant cultures by particle-mediated gene transfer," *Journal of Neuroscience Methods*, vol. 84, no. 1–2, pp. 181–191, 1998.

[1211] P. J. Bergold, P. Casaccia-Bonnefil, X. Zeng, and H. J. Federoff, "Transsynaptic neuronal loss induced in hippocampal slice cultures by a herpes simplex virus vector expressing the GluR6 subunit of the kainate receptor," *Proceedings of the National Academy of Sciences*, vol. 90, no. 13, pp. 6165–6169, 1993.

[1212] M. U. Ehrengruber, S. Hennou, H. Büeler, H. Y. Naim, N. Déglon, and K. Lundstrom, "Gene transfer into neurons from hippocampal slices: comparison of recombinant Semliki Forest Virus, adenovirus, adeno-associated virus, lentivirus, and measles virus," *Molecular and Cellular Neuroscience*, vol. 17, no. 5, pp. 855–871, 2001.

[1213] B. Morrison III, J. Eberwine, D. Meaney, and T. McIntosh, "Traumatic injury induces differential expression of cell death genes in organotypic brain slice cultures determined by complementary DNA array hybridization," *Neuroscience*, vol. 96, no. 1, pp. 131–139, 2000.

[1214] B. Morrison III, D. F. Meaney, S. S. Margulies, and T. K. McIntosh, "Dynamic mechanical stretch of organotypic brain slice cultures induces differential genomic expression: relationship to mechanical parameters," *Journal of Biomechanical Engineering*, vol. 122, no. 3, pp. 224–230, 2000.

[1215] R. Fan and A. J. Tenner, "Complement C1q expression induced by Aβ in rat hippocampal organotypic slice cultures," *Experimental Neurology*, vol. 185, no. 2, pp. 241–253, 2004.

[1216] F. Mörl, M. Gröschel, J. Leemhuis, and D. K. Meyer, "Intrinsic GABA neurons inhibit proenkephalin gene expression in slice cultures of rat neostriatum," *European Journal of Neuroscience*, vol. 15, no. 7, pp. 1115–1124, 2002.

[1217] C. M. Colbert, "Preparation of cortical brain slices for electrophysiological recording," *Ion Channels: Methods and Protocols*, pp. 117–125, 2006.

[1218] H. J. Luhmann and W. Kilb, "Intact in vitro preparations of the neonatal rodent cortex: Analysis of cellular properties and network activity," *Isolated Central Nervous System Circuits*, pp. 301–314, 2012.

[1219] A. Agmon and B. Connors, "Thalamocortical responses of mouse somatosensory (barrel) cortexin vitro," *Neuroscience*, vol. 41, no. 2–3, pp. 365–379, 1991.

[1220] A. Agmon and B. W. Connors, "Correlation between intrinsic firing patterns and thalamocortical synaptic responses of neurons in mouse barrel cortex," *Journal of Neuroscience*, vol. 12, no. 1, pp. 319–329, 1992.

[1221] G. Nabovati, E. Ghafar-Zadeh, A. Letourneau, and M. Sawan, "Towards high throughput cell growth screening: a new CMOS 8 × 8 biosensor array for life science applications," *IEEE Transactions on Biomedical Circuits and Systems*, vol. 11, no. 2, pp. 380–391, 2017.

[1222] P. Eriksson, A. A. Tal, A. Skallberg, *et al.*, "Cerium oxide nanoparticles with antioxidant capabilities and gadolinium integration for MRI contrast enhancement," *Scientific Reports*, vol. 8, no. 1, pp. 1–12, 2018.

[1223] D. Oda and E. Watson, "Human oral epithelial cell culture I. Improved conditions for reproducible culture in serum-free medium," *In Vitro Cellular & Developmental Biology*, vol. 26, no. 6, pp. 589–595, 1990.

[1224] R. Sheth, M. H. Neale, A. J. Shortt, I. Massie, A. J. Vernon, and J. T. Daniels, "Culture and characterization of oral mucosal epithelial cells on a fibrin gel for ocular surface reconstruction," *Current Eye Research*, vol. 40, no. 11, pp. 1077–1087, 2015.

[1225] C. E. Arcadia, K. Hu, S. Epstein, M. Wanunu, A. Adler, and J. K. Rosenstein, "CMOS electrochemical imaging arrays for the detection and classification of microorganisms," in *2021 IEEE International Symposium on Circuits and Systems (ISCAS)*, 2021: IEEE, pp. 1–5.

[1226] L. C. Clark Jr and C. Lyons, "Electrode systems for continuous monitoring in cardiovascular surgery," *Annals of the New York Academy of Sciences*, vol. 102, no. 1, pp. 29–45, 1962.

[1227] J. Liu and J. Wang, "A novel improved design for the first-generation glucose biosensor," *Food Technology and Biotechnology*, vol. 39, no. 1, pp. 55–58, 2001.

[1228] G. F. Khan, M. Ohwa, and W. Wernet, "Design of a stable charge transfer complex electrode for a third-generation amperometric glucose sensor," *Analytical Chemistry*, vol. 68, no. 17, pp. 2939–2945, 1996.

[1229] C. Jang, H.-J. Lee, and J.-G. Yook, "Radio-frequency biosensors for real-time and continuous glucose detection," *Sensors*, vol. 21, no. 5, p. 1843, 2021.

[1230] "Ion Torrent | Thermo Fisher Scientific - CA." https://www.thermofisher.com/ca/en/home/brands/ion-torrent.html (accessed August 7, 2023).

[1231] "CMOS-MEA5000-System | www.multichannelsystems.com." https://www.multichannelsystems.com/products/cmos-mea5000-system (accessed August 7, 2023).

[1232] "Bioimager | Bioimager Advanced Custom Microscopes." https://www.bioimager.com/ (accessed August 7, 2023).

[1233] M. Helou, M. Reisbeck, S. F. Tedde, *et al.*, "Time-of-flight magnetic flow cytometry in whole blood with integrated sample preparation," *Lab on a Chip*, vol. 13, no. 6, pp. 1035–1038, 2013.

[1234] K. Niitsu, S. Ota, K. Gamo, H. Kondo, M. Hori, and K. Nakazato, "Development of microelectrode arrays using electroless plating for CMOS-based direct counting of bacterial and HeLa cells," *IEEE Transactions on Biomedical Circuits and Systems*, vol. 9, no. 5, pp. 607–619, 2015.

[1235] K. Raghunandan, T. L. Viswanathan, and T. R. Viswanathan, "Linear current-controlled oscillator for analog to digital conversion," *in Proceedings of the IEEE 2014 Custom Integrated Circuits Conference*, 2014: IEEE, pp. 1–4.

[1236] M. Graf, D. Barrettino, S. Taschini, C. Hagleitner, A. Hierlemann, and H. Baltes, "Metal oxide-based monolithic complementary metal oxide semiconductor gas sensor microsystem," *Analytical Chemistry*, vol. 76, no. 15, pp. 4437–4445, 2004.

[1237] S. Semancik, R. E. Cavicchi, M. Wheeler, *et al.*, "Microhotplate platforms for chemical sensor research," *Sensors and Actuators B: Chemical*, vol. 77, no. 1–2, pp. 579–591, 2001.

[1238] M. Graf, D. Barrettino, K.-U. Kirstein, and A. Hierlemann, "CMOS microhotplate sensor system for operating temperatures up to 500 C," *Sensors and Actuators B: Chemical*, vol. 117, no. 2, pp. 346–352, 2006.

Index

www.ingramcontent.com/pod-product-compliance
Lightning Source LLC
LaVergne TN
LVHW011537200926
848484LV00001B/279
9781839536731